The 1989 Neuro-Computing Bibliography

The 1989 Neuro-Computing Bibliography

edited by Casimir C. Klimasauskas

second edition

A Bradford Book

The MIT Press
Cambridge, Massachusetts
London, England

First edition published 1987, Neuro-Connection

This book was printed and bound in the United States of America

Library of Congress Cataloging-in-Publication Data

Klimasauskas, Casimir C.
The 1989 neuro-computing bibliography / edited by Casimir C. "Casey" Klimasauskas. — 2nd ed.
p. cm.
Rev. ed. of: The 1987 annotated neuro-computing bibliography. 1st ed. 1987.
ISBN 0-262-11134-9
1. Neural computers—Bibliography. 2. Neural circuitry--Bibliography. I. Klimasauskas, Casimir C. 1987 annotated neuro-computing bibliography. II. Title
Z5642.2.K57 1989 QA76.5
016.0063—dc19 89-12145
CIP

Contents

Preface

I began this project in 1986 with the intent of learning about neuro-computing. As it progressed, I began to realize how difficult it was to get an overview of the field. Friends and associates kept asking about various subjects in neuro-computing, and I would refer them to the bibliography. It became apparent that the market was ready and eager for a resource book on neuro-computing, a book about everything you ever wanted to know about neuro-computing, but didn't know whom to ask. That is how this book came about.

The first edition of the bibliography was published in late 1987. That edition contained 1,500 references. Over the past year, we have identified an additional 2,000 references for inclusion. The pace has picked up. In particular, papers describing applications of neuro-computing are appearing in ever more diverse publications.

The Bibliography. In preparing this second edition, we have continued to review the literature as well as to write to each of the individuals we have identified as researchers in the field. In most instances, we have acquired copies of the works referenced.

An exciting aspect of neuro-computing is the diversity of fields it brings together. In including or excluding particular works, we have had to make a judgment as to their applicability to neuro-computing. Some works have been included because they closely parallel ideas and techniques in neuro-computing. Others have been included for their background information on the brain and its functions. It has been difficult to decide upon the inclusion or exclusion of works in two borderline areas: Neuro-Physiology and Modeling, and Connectionism. (Comprehensive treatment of both of these areas would more than double the size of the source book.) Our general criteria for including these works was that they were referenced in two or more papers already included in the bibliography.

In an attempt to make this as up-to-date as possible, we identified 2,000 individuals who have published works on neuro-computing and written to them requesting copies of their most recent papers. Six hundred responded. Their individual contributions are a major contribution to this current work.

A Final Note. If we have failed to include your work or any works which you feel should be a part of this source book, please send us a copy of the works. We will be happy to consider them for future inclusion. All errors and omissions are mine.

A Special Note of Recognition. There are several people who have contributed to this book in very important ways. Nemour Landaiche has been instrumental in researching current papers as well as tracking down hard to find references. Pat Fitzgerald, assistant director for public services, Carnegie Mellon University Libraries, has checked every reference for completeness and consistency. Darde Khan has done a yeoman's job of handling correspondences, entering references, and making corrections. Finally, my wife Jane has been a constant motivation to keep the process moving through many of the long hours and sometimes overwhelming magnitude of the task. This work is as much that of Nemour Landaiche, Pat Fitzgerald, Darde Khan and Jane's as my own.

Casimir C. Klimasauskas
Sewickley, Pennsylvania
April 1989

Introduction

This source book has been designed as a tool for beginners as well as experts in the field of neuro-computing. Its intent is to provide references to a broad overview of the practical techniques and applications of neuro-computing as well as background material on the underpinnings of the field.

More than a compendium, it is designed to be used as a handy desk-top reference by engineers and researchers in the field. Special features include multiple listings of articles and papers when multiply published, cross-referencing of all secondary authors, and a comprehensive keyword index of all titles. Every feature of the source book has been carefully chosen to make it as easy to use as possible.

The source book covers a wide variety of fields including Forecasting, Mathematics, Probability Theory, Neural Physiology, Neural Modeling, Expert Systems, Adaptive (Process) Control, Robotics, Computer Vision, Speech Recognition, Character Recognition, Signal Processing, Neuro-Computing, Connectionism, and Psychology. Though providing a comprehensive bibliography of each of these topics is beyond the scope of this work, references have been included when appropriate.

Every care has been taken to insure that the references are accurate; however no guarantee can be made as to the accuracy, applicability, or availability of any article or reference.

How to Use This Bibliography

The source book has been divided into two sections. The first section contains a listing of authors, titles, and related material. The second section is a comprehensive keyword cross reference of the material.

Bibliographic Listing. The bibliographic listing contains authors and titles. Secondary authors are cross-referenced intermixed with primary authors and titles. The format of each is described below.

Primary References. Primary References are uniquely identified by the author's last name, initials, the year of publication, and a lower case letter to disambiguate publications in the same year. A modified APA format has been used to enhance the clarity and consistency of references. A sample reference follows.

> [Carpenter, GA 87a] Carpenter, Gail A.; Grossberg, Stephen. (1987) *A Massively Parallel Architecture for a Self-Organizing Neural Pattern Recognition Machine.* Computer Vision, Graphics, and Image Processing. Volume 37. pp 54–115.

Authors are listed in order, last name first, followed by the first name and/or initials. This is followed by the publication data. The title of the article (if applicable) follows in italics. The book or journal is underlined. For articles, the volume and number of the journal is spelled out to prevent any confusion. Page numbers likewise are spelled out. The city where the publisher is located and publisher follow the page number.

Secondary References use a similar format. The secondary author's name, initials, and year of publication are followed by an asterisk. The exact reference of the article follows, including the title. (The titles are included to make it easier to determine if you want to look at the article.) A sample reference follows.

> [Grossberg, S 87*] SEE: Carpenter, GA 87a. *A Massively Parallel Architecture for a Self-Organizing Neural Pattern Recognition Machine.*

Keyword Cross-Reference. This is a Keyword In Context (KWIC) index of all articles and titles. The particular keyword is aligned in the center of the page, highlighted by a double space. As much of the title as will fit to the right and left are printed. The exact reference to the article is in the right margin of the index. Periods are used to fill between the article or book title and the author reference to the right.

Hints on Use. The source book has been designed for use. In particular, it was designed to be used in the following ways.

An Introduction to Neuro-Computing. The Keyword Cross-Reference provides an easy way for a beginner in neuro-computing to find articles of interest.

Bibliographic Reference. The bibliography is ideal for research papers, research projects. Through the Keyword Cross-Reference, it is possible to locate all of the papers in a particular area of interest. The author cross-references in the main bibliography then assists in finding other papers by the author which may be of interest. Those titles in turn may suggest other topics for research in the Keyword Cross-Reference.

Author Research. The author cross-references provide a mechanism for finding everything that has been written on neuro-computing by a particular author. This is helpful for forming a perspective or view of a particular researcher's work.

Specific Problem Solving. When working in a particular problem area, the Keyword Cross-Reference provides an easy mechanism to find out how other researchers have applied Neuro-Computing to that problem.

The source book is a powerful and flexible tool for learning about neuro-computing. I trust it will more than pay for itself.

Bibliographic Listing

[Abbiss, JB 88a] Abbiss, J. B.; Brames, B. J.; Bayley, J. S.; Fiddy, M. A. (1988). *Super-Resolution and Neural Computing.* <u>High Speed Computing, Los Angeles, CA, January 11-12, 1988</u>. SPIE Volume 880. pp 100-106. Bellingham, WA: Society of Photo-Optical Instrumentation Engineers.

[Abbruzzese, F 88a] Abbruzzese, Francesco. (1988). *A Transputer Implementation of a McCulloch & Pitts Network.* <u>International Conference on Parallel Processing and Applications, September 1987, L'Aquila, Italy</u>. pp 135-140. New York: Elsevier.

[Abend, K 65a] Abend, K.; Harley, T. J.; Kanal, Laveen N. (1965). *Classification of Binary Random Patterns.* <u>IEEE Transactions on Information Theory</u>. Volume IT-11. pp 538-544.

[Abrahams, E 84*] SEE: Palmer, RG 84a. *Models of Hierarchically Constrained Dynamics for Glassy Relaxation.*

[Abrahams, E 86*] SEE: Anderson, CH 86a. *Bayesian Probability Network.*

[Abrahams, E 87*] SEE: Anderson, CH 87b. *The Bayes Connection.*

[Abrams, IT 86*] SEE: Cohen, NJ 86a. *Skill Learning and Repetition Priming in Symmetry Detection: Parallel Studies of Human Subjects and Connectionist Models.*

[Abramson, N 62a] Abramson, N.; Braverman, David. (1962). *Learning to Recognize Patterns in a Random Environment.* <u>IRE Transactions on Information Theory</u>. Volume IT-8. pp 58-63.

[Abu-Mostafa, YS 85a] Abu-Mostafa, Yaser S.; Psaltis, Demetri. (1985). *Computation Power of Parallelism in Optical Architectures.* <u>Workshop on Computer Architecture for Pattern Analysis and Image Database Management (CAPAIDM), Miami, FL, 1985</u>. pp 42-47. Washington: IEEE Computer Society Press.

[Abu-Mostafa, YS 85b] Abu-Mostafa, Yaser S.; St Jacques, Jeannine-Marie. (1985). *Information Capacity of the Hopfield Model.* <u>IEEE Transactions on Information Theory</u>. Volume IT-31. pp 461-464.

[Abu-Mostafa, YS 86a] Abu-Mostafa, Yaser S. (1986). *Complexity of Information Extraction.* <u>IEEE Transactions on Information Theory</u>. Volume IT-32. pp 513-525.

[Abu-Mostafa, YS 86b] Abu-Mostafa, Yaser S. (1986). *Neural Networks for Computing?.* In Denker, John S. (ed.). <u>Neural Networks for Computing, Snowbird, Utah, 1986</u>. AIP Proceedings 151. pp 1-6. New York: American Institute of Physics.

[Abu-Mostafa, YS 87a] Abu-Mostafa, Yaser S.; Psaltis, Demetri. (1987). *Optical Neural Computers.* <u>Scientific American</u>. Volume 255. pp 88-95.

[Abu-Mostafa, YS 88*] SEE: Olafsson, S 88a. *The Capacity of Multilevel Threshold Functions.*

[Abu-Mostafa, YS 88a] Abu-Mostafa, Yaser S. (1988). *Connectivity Versus Entropy.* In Anderson, Dana Z. (ed.). <u>Neural Information Processing Systems</u>. pp 1-8. New York: American Institute of Physics.

[Abu-Mostafa, YS 88b] Abu-Mostafa, Yaser S. (September, 1988). *Lower Bound for Connectivity in Local-Learning Neural Networks.* <u>Journal of Complexity</u>. Volume 4. pp 246-255.

[Abu-Mostafa, YS 88c] Abu-Mostafa, Yaser S. (December, 1988). *Random Problems.* <u>Journal of Complexity</u>. Volume 4. pp 277-286.

[Aburdene, MF 88a] Aburdene, Maurice F. (1988). *Simulation of a Distributed Algorithm Using the Connection Machine.* <u>Modeling and Simulation, 19th, Pittsburgh, PA, 1988: Proceedings</u>. Volume 19. pp 1039-1048. Research Triangle Park, NC: Instrument Society of America.

[Ackley, DH 82*] SEE: Berliner, HJ 82a. *The QBKG System: Generating Explanations from a Non-Discrete Knowledge*

Representation.

[Ackley, DH 84*] SEE: Hinton, GE 84a. Boltzmann Machines: Constraint Satisfaction Networks that Learn.

[Ackley, DH 85a] Ackley, David H.; Hinton, Geoffrey E.; Sejnowski, Terrence J. (1985). *A Learning Algorithm for Boltzmann Machines.* Cognitive Science. Volume 9. pp 147-169.

[Ackley, DH 85b] Ackley, David H.; Hinton, Geoffrey E.; Sejnowski, Terrence J. (1988). *A Learning Algorithm for Boltzmann Machines.* In Anderson, James A. (ed.); Rosenfeld, Edward (ed.). Neurocomputing: Foundations of Research. pp 638-650. Cambridge, MA: MIT Press.

[Ackley, DH 88a] Ackley, David H.; Hinton, Geoffrey E.; Sejnowski, Terrence J. (1988). *A Learning Algorithm for Boltzmann Machines.* In Waltz, David L. (ed.); Feldman, Jerome A. (ed.). Connectionist Models and Their Implications: Readings from Cognitive Science. pp 285-308. Norwood, NJ: Ablex.

[Adachi, N 78*] SEE: Takeuchi, Y 78a. *The Stability of Generalized Volterra Equations.*

[Adams, MJ 79a] Adams, Marilyn Jager. (1979). *Models of Word Recognition.* Cognitive Psychology. Volume 11. pp 133-176.

[Addanki, S 83a] Addanki, S. (1983). A Connectionist Approach to Motor Control. PhD Dissertation. Rochester, NY: University of Rochester.

[Adelman, G 87a] Adelman, George (ed.). (1987). Encyclopedia of Neuroscience. Boston: Birkhauser.

[Adelson, B 86*] SEE: McClelland, JL 86c. *Connectionist Models and Cognitive Science: Goals, Directions and Implications.*

[Adelson, EH 85*] SEE: Movshon, JA 85a. *The Analysis of Moving Visual Patterns.*

[Aderman, D 71a] Aderman, David; Smith, Edward E. (1971). *Expectancy as a Determinant of Functional Units in Perceptual Recognition.* Cognitive Psychology. Volume 2. pp 117-129.

[Adler, JE 87*] SEE: Black, IB 87a. *Biochemistry of Information Storage in the Nervous System.*

[Afshartous, N 87*] SEE: Chun, HW 87c. *Network Regions: Alternatives to the Winner-Take-All Structure.*

[Agha, GA 86a] Agha, Gul A. (1986). ACTORS: A Model of Concurrent Computation in Distributed Systems. Cambridge, MA: MIT Press.

[Agranat, A 87a] Agranat, A.; Yariv, A. (1987). *A New Architecture for a Microelectronic Implementation of Neural Network Models.* IEEE International Conference on Neural Networks (ICNN-87), San Diego, CA, June 21-24, 1987: Proceedings. Volume 3. pp 403-410. New York: Institute of Electrical and Electronics Engineers.

[Agrawala, AK 68*] SEE: Ho, Y 68a. *On Pattern Classification Algorithms: Introduction and Survey.*

[Ahmed, MS 86a] Ahmed, Mohammed S. (1986). Enhancement of Noisy Speech by Adaptive MMSE Filtering. CMU-CS-86-182. Pittsburgh, PA: Carnegie Mellon University, Computer Science Department.

[Ahn, SM 74a] Ahn, S. M.; Freeman, Walter J. (1974). *Steady-State and Limit Cycle Activity of Mass of Neurons Forming Simple Feedback Loops (I): Lumped Circuit Model.* Kybernetik. Volume 16. pp 87-91.

[Ahn, SM 74b] Ahn, S. M.; Freeman, Walter J. (1974). *Steady-State and Limit Cycle Activity of Mass of Neurons Forming Simple Feedback Loops (II): Lumped Circuit Model.* Kybernetik. Volume 16. pp 127-132.

[Ahumada, AJ 87a] Ahumada, Albert J. (1987). *Simulated Annealing in Networks for Computing Possible Arrangements for Red and*

Green Cones. IEEE International Conference on Neural Networks (ICNN-87), San Diego, CA, June 21-24, 1987: Proceedings. Volume 4. pp 107-114. New York: Institute of Electrical and Electronics Engineers.

[Aibara, T 81*] SEE: Murakami, K 81a. *Construction of a Distributed Associative Memory on the Basis of Bayes Discriminant Rule*.

[Aibara, T 82*] SEE: Murakami, K 82a. *Optimal Association with Partly Missing Key Vectors*.

[Aibara, T 88*] SEE: Murakami, K 88a. *Construction of Associative Memory from Incomplete Memorized Information Using Space Reduction Method*.

[Aicardi, F 87a] Aicardi, F.; Borsellino, A. I. (1987). *Statistical Properties of Flip-Flop Processes Associated to the Chaotic Behavior of Systems with Strange Attractors*. Biological Cybernetics. Volume 55. pp 377-385.

[Aiello, A 70a] Aiello, A.; Burattini, E.; Caianiello, E. R. (1970). *Synthesis of Reverberatory Neural Networks*. Kybernetik. Volume 7. pp 191-195.

[Aihara, K 80a] Aihara, Kazuyuki; Utsunomiya, Toshio. (1980). *A Macroscopic Analysis on Large Scale Networks on Neuronlike Elements with Refractory*. Conference on Medical Informatics (MEDINFO 80), 3rd, Tokyo, September 29-October 4, 1980: Proceedings. pp 364-368. Amsterdam: North-Holland.

[Aipperspach, AG 84a] Aipperspach, A. G.; Wu, P. T. (1984). *Content-Addressable memory Cell Having Only Six Transitions*. IBM Technical Disclosure Bulletin. Volume 26. pp 4208-4209.

[Aitkenhead, AM 85a] Aitkenhead, A. M. (ed.); Slack, J. M. (ed.). (1985). Issues in Cognitive Modeling: A Reader. Hillsdale, NJ: Lawrence Erlbaum.

[Ajmera, RC 80a] Ajmera, R. C.; Kohli, C. K.; Newcomb, R. W. (1980). *Retinal-Type Neuristor Sections*. IEEE Region 3 Conference and Exhibit.

[Akers, LA 88a] Akers, L. A.; Walker, M. R.; Ferry, D. K.; Grondin, R. O. (1988). *Limited Interconnectivity in Synthetic Neural Systems*. In Eckmiller, Rolf (ed.); von der Malsburg, Christoph (ed.). Neural Computers. NATO ASI Series, Volume F41. pp 407-416. Berlin: Springer-Verlag.

[Albert, A 72a] Albert, A. (1972). Regression and The Moore-Penrose Pseudoinverse. New York: Academic Press.

[Albus, JS 71a] Albus, James S. (1971). *A Theory of Cerebellar Function*. Mathematical Biosciences. Volume 10. pp 25-61.

[Albus, JS 72a] Albus, James S. (1972). *The Cerebellum: A Substrate for List-Processing in the Brain*. Cybernetics, Artificial Intelligence, and Ecology. pp 67-94. New York: Spartan.

[Albus, JS 72b] Albus, James S. (1972). Theoretical and Experimental Aspects of a Cerebellar Model. Ph.D. Dissertation. College Park, MD: University of Maryland.

[Albus, JS 75a] Albus, James S. (September, 1975). *A New Approach to Manipulator Control: The Cerebellar Model Articulation Controller (CMAC)*. Journal of Dynamic Systems, Measurement, and Control. Volume 97. pp 220-227.

[Albus, JS 75b] Albus, James S. (September, 1975). *Data Storage in the Cerebellar Model Articulation Controller (CMAC)*. Journal of Dynamic Systems, Measurement, and Control. Volume 97. pp 228-233.

[Albus, JS 76a] Albus, James S.; Evans, Jr, John M. (April, 1974). *Robot Systems*. Scientific American. Volume 234. pp 77-86b.

[Albus, JS 79*] SEE: Barbera, AJ 79a. *Hierarchical Control of Robots Using Microcomputers*.

[Albus, JS 79a] Albus, James S. (June, 1979). *A Model of the Brain for Robot Control, Part 1: Defining Notation*. BYTE.

Volume 4. pp 10-34.

[Albus, JS 79b] Albus, James S. (July, 1979). *A Model of the Brain for Robot Control, Part 2: A Neurological Model.* BYTE. Volume 4. pp 54-95.

[Albus, JS 79c] Albus, James S. (August, 1979). *A Model of the Brain for Robot Control, Part 3: A Comparison of the Brain and our Model.* BYTE. Volume 4. pp 66-80.

[Albus, JS 79d] Albus, James S. (September, 1979). *A Model of the Brain for Robot Control, Part 4: Mechanisms of Choice.* BYTE. Volume 4. pp 130-148.

[Albus, JS 79e] Albus, James S. (1979). *Mechanisms of Planning and Problem Solving in the Brain.* Mathematical Biosciences. Volume 45. pp 247-293.

[Albus, JS 80a] Albus, James S.; Barbera, Anthony J.; Fitzgerald, M. L.; Nagel, Roger N.; VanderBrug, G. J.; Wheatley, T. E. (1980). *A Measurement and Control Model for Adaptive Robots.* International Symposium on Industrial Robots, 10th, Milan, March 5-7, 1980: Proceedings. pp 303-309. Kempston: IFS (Publications) Ltd..

[Albus, JS 81a] Albus, James S. (1981). Brains, Behavior and Robotics.

[Albus, JS 81b] Albus, James S.; Barbera, Anthony J.; Nagel, Roger N. (1981). *Theory and Practice of Hierarchical Control.* IEEE Computer Society International Conference, 23rd, 1981: Proceedings. pp 18-39. New York: Institute of Electrical and Electronics Engineers.

[Albus, JS 85a] Albus, James S.; McLean, C. R.; Barbera, Anthony J.; Fitzgerald, M. L. (1985). *Hierarchical Control for Robots, and Teleoperators.* In Saridis, A. (ed.); Meystel, A. (ed.). IEEE Workshop on Intelligent Control, Troy, NY, August 26, 1985: Proceedings. pp 39-49. New York: Institute of Electrical and Electronics Engineers.

[Albus, JS 88a] Albus, James S. (1988). *The Central Nervous System as a Low and High Level Control System.* Sensors and Sensory Systems for Advanced Robots. NATO ASI Series, Volume F43. pp 1-20. New York: Springer-Verlag.

[Alder, B 86*] SEE: Ceperley, D 86a. *Quantum Monte Carlo.*

[Aldridge, NB 88*] SEE: White, HJ 88a. *Digital and Analogue Holographic Associative Memories.*

[Aleksander, I 87a] Aleksander, Igor. (1987). *Adaptive Pattern Recognition Systems and Boltzmann Machines: A Rapprochement.* Pattern Recognition Letters. Volume 6. pp 113-120.

[Aleksander, I 88*] SEE: Dobson, VG 88b. *Differences in Performance of Binary Hetero-Associative Networks Using Link-Enabling vs. Link-Disabling Local Modification Rules.*

[Aleksander, I 88*] SEE: Gera, M 88a. *A Logic Neuron Model of Language Acquisition.*

[Aleksander, I 88a] Aleksander, Igor. (1988). *Are Special Chips Necessary for Neural Computers?.* Conference on VLSI for AI, Oxford, 1988.

[Aleksander, I 88b] Aleksander, Igor. (1988). *Learning and Distributed Memory: Getting Close to Neurons.* In Forsyth, Richard (ed.). Machine Learning. London: Addison-Wesley.

[Aleksander, I 88c] Aleksander, Igor. (1988). *Logical Connectionist Systems.* In Eckmiller, Rolf (ed.); von der Malsburg, Christoph (ed.). Neural Computers. NATO ASI Series, Volume F41. pp 189-197. Berlin: Springer-Verlag.

[Aleksander, I 88d] Aleksander, Igor. (1988). *Neural Computing: Hype or Reality?.* European Management Journal. Volume 6. pp 114-117.

[Aleksander, I 88e] Aleksander, Igor. (March, 1988). *Neural Computing: Ideas from the Brain.* Computer Bulletin.

[Aleksander, I 88f] Aleksander, Igor. (1988). *The Logic of Connectionist Systems.* In Aleksander, Igor (ed.). Neural Computing Architectures. London: Kogan Page.

[Alexander, ST 86a] Alexander, S. Thomas. (1986). Adaptive Signal Processing: Theory and Applications. New York: Springer-Verlag.

[Ali, M 87*] SEE: Dietz, WE 87a. *Qualitative and Temporal Reasoning in Engine Behavior Analysis and Fault Diagnosis.*

[Ali, M 88*] SEE: Maren, AJ 88b. *Hierarchical Scene Structure Representations to Facilitate Image Understanding.*

[Ali, M 88*] SEE: Maren, AJ 88a. *A Multilayer Cooperative/Competitive Method for Creating Hierarchical Structures by Clustering Maximally-Related Nodes.*

[Ali, M 88*] SEE: Dietz, WE 88a. *Pattern-Based Fault Diagnosis Using Neural Networks.*

[Ali, M 89*] SEE: Dietz, WE 89a. *Neural Network Models Applied to Real-Time Fault Diagnosis.*

[Alkon, DL 79a] Alkon, Daniel L. (1979). *Voltage-Dependent Calcium and Potassium Ion Conductances: A Contingency Mechanism for an Associative Learning Model.* Science. Volume 205. pp 810-816.

[Alkon, DL 87*] SEE: Lederhendler, II 87a. *Associatively Reduced Withdrawal from Shadows in Hermissenda: A Direct Behavioral Analog of Photoreceptor Responses to Brief Light Steps.*

[Alkon, DL 87*] SEE: Coulter, DA 87a. *Persistent Conditioning-Specific Changes of Post-Synaptic Membrane Currents in Heermissenda and Hippocampus.*

[Alkon, DL 87a] Alkon, Daniel L.; Disterhoft, John F.; Coulter, Douglas A. (1987). *Conditioning-Specific Modification of Postsynaptic Membrane Currents in Mollusc and Mammal.* The Neural and Molecular Bases of Learning. pp 205-237. New York: Wiley.

[Alkon, DL 88*] SEE: Vogl, TP 88a. *Accelerating the Convergence of the Back-Propagation Method.*

[Alkon, DL 88*] SEE: Disterhoft, JF 88a. *Conditioning-Specific Biophysical Alterations in Rabbit Hippocampus.*

[Alkon, DL 88*] SEE: LoTurco, JJ 88a. *Enhancement of Synaptic Potentials in Rabbit CA1 Pyramidal Neurons Following Classical Conditioning.*

[Alkon, DL 88*] SEE: Naito, S 88a. *Transient and Persistent Depolarization-Induced Changes of Protein Phosphorylation in a Molluscan Nervous System.*

[Alkon, DL 88a] Alkon, Daniel L. (1988). *Conservation of Cellular Mechanisms for Models of Learning and Memory.* Cellular Mechanisms of Conditioning and Behavioral Plasticity. pp 37-44. New York: Plenum.

[Alkon, DL 88b] Alkon, Daniel L.; Naito, Shigetaka; Kubota, M.; Chen, C.; Bank, Barry; Smallwood, J. (1988). *Regulation of Hermissenda K+ Channels by Cytoplasmic and Membrane-Associated C-Kinase.* Journal of Neurochemistry. Volume 51. pp 903-917.

[Allanson, JT 56a] Allanson, J. T. (1956). *Some Properties of a Randomly Connected Neural Network.* Information Theory, London Symposium, 3rd, Held at the Royal Institution, London, September 12-16, 1955. pp 303-313. New York: Academic Press.

[Allemang, D 89*] SEE: Goel, A 89a. *Learning in Connectionist Networks: Has the Credit Assignment Problem Been Solved?.*

[Allen, RB 87*] SEE: Alspector, J 87a. *A Neuromorphic VLSI Learning System.*

[Allen, RB 87a] Allen, Robert B. (1987). *Several Studies on Natural Language and Back-Propagation.* IEEE International

Conference on Neural Networks (ICNN-87), San Diego, CA, June 21-24, 1987: Proceedings. Volume 2. pp 335-341. New York: Institute of Electrical and Electronics Engineers.

[Allen, RB 88*] SEE: Alspector, J 88b. *Performance of a Stochastic Learning Microchip.*

[Allen, RB 88*] SEE: Alspector, J 88c. *Stochastic Learning Networks and their Electronic Implementation.*

[Allen, RB 88a] Allen, Robert B.; Riecken, Mark E. (1988). *Interacting and Communicating Connectionist Agents.* Neural Networks. Volume 1.

[Allen, RB 88b] Allen, Robert B. (1988). *Sequential Connectionist Networks for Answering Simple Questions about a Microworld.* Cognitive Science Society: Proceedings. pp 489-495.

[Allman, WF 86a] Allman, William F. (May 1986). *Mindworks.* Science. pp 22-31.

[Allman, WF 87a] Allman, William F. (May / June 1987). *Designing Computers That Think the Way We Do.* Technology Review.

[Allmeyer, DH 76*] SEE: Estes, WK 76a. *Serial Position Functions for Letter Identification at Brief and Extended Exposure Durations.*

[Allouche, JP 86a] Allouche, J. P. (1986). *The Ising Model and The Rudin Shapiro Sequence.* In Bienenstock, Elie L. (ed.); Fogelman Soulie, F. (ed.); Weisbuch, Gerard (ed.). Disordered Systems and Biological Organization. NATO ASI Series, Volume F20. pp 59-62. New York: Springer-Verlag.

[Almeida, LB 87a] Almeida, Luis B. (1987). *A Learning Rule for Asynchronous Perceptrons with Feedback in a Combinatorial Environment.* IEEE International Conference on Neural Networks (ICNN-87), San Diego, CA, June 21-24, 1987: Proceedings. Volume 2. pp 609-618. New York: Institute of Electrical and Electronics Engineers.

[Almeida, LB 88a] Almeida, Luis B. (1988). *Backpropagation in Nonfeedforward Networks.* In Aleksander, Igor (ed.). Neural Computing. London: Kogan Page.

[Almeida, LB 88b] Almeida, Luis B.; Redol, R. Alves. (1988). *Backpropagation in Perceptrons with Feedback.* In Eckmiller, Rolf (ed.); von der Malsburg, Christoph (ed.). Neural Computers. NATO ASI Series, Volume F41. pp 199-208. Berlin: Springer-Verlag.

[Alpert, BK 87*] SEE: Mjolsness, E 87b. *Recursively Generated Neural Networks.*

[Alpert, BK 88*] SEE: Mjolsness, E 88b. Scaling, Machine Learning, and Genetic Neural Nets.

[Alspector, J 87a] Alspector, Joshua; Allen, Robert B. (1987). *A Neuromorphic VLSI Learning System.* Advanced Research in VLSI: Proceedings of the 1987 Stanford Conference. pp 313-349. Cambridge, MA: MIT Press.

[Alspector, J 88a] Alspector, Joshua. (1988). *A VLSI Approach to Neural-Style Information Processing.* IEEE Workshop on VLSI Signal Processing, Monterey, CA, November 2-4, 1988: Proceedings. New York: Institute of Electrical and Electronics Engineers.

[Alspector, J 88b] Alspector, Joshua; Gupta, Bhusan; Allen, Robert B. (1988). *Performance of a Stochastic Learning Microchip.* Neural Information Processing Systems. New York: American Institute of Physics.

[Alspector, J 88c] Alspector, Joshua; Allen, Robert B.; Hu, Victor; Satyanarayana, Srinagesh. (1988). *Stochastic Learning Networks and their Electronic Implementation.* In Anderson, Dana Z. (ed.). Neural Information Processing Systems. pp 9-21. New York: American Institute of Physics.

[Altes, RA 75a] Altes, Richard A.; Reese, W. D. (1975). *Doppler Tolerant Classification of Distributed Targets - a Bionic Sonar.* IEEE Transactions on Aerospace and Electronic Systems. Volume AES-11. pp 708-724.

[Altes, RA 76a] Altes, Richard A. (1976). *Sonar for Generalized Target Description and its Similarity to Animal Echolocation Systems*. Journal of the Acoustical Society of America. Volume 59. pp 97-105.

[Altes, RA 78a] Altes, Richard A. (1978). *Fourier-Mellin Transform and Mammalian Hearing*. Journal of the Acoustic Society of America. Volume 63. pp 174-183.

[Altes, RA 78b] Altes, Richard A. (1978). Spectrogram Correlation. Technical Report TR-404. Hawaii: Naval Ocean Systems Center.

[Altes, RA 80a] Altes, Richard A. (1980). *Detection, Estimation, and Classification with Spectrograms*. Journal of the Acoustic Society of America. Volume 67. pp 1232-1246.

[Altes, RA 88a] Altes, Richard A. (1988). *Unconstrained Minimum Mean-Square Error Parameter Estimation with Hopfield Networks*. IEEE International Conference on Neural Networks (ICNN-88), San Diego, CA July 24-27, 1988: Proceedings. Volume 2. pp 541-548. New York: Institute of Electrical and Electronics Engineers.

[Altes, RA 89a] Altes, Richard A. (1989). *An Interpretation of Cortical Maps in Echolocating Bats*. Journal of the Acoustical Society of America. Volume 85.

[Altman, JL 77a] Altman, J. L. (1977). Pattern Recognition Using Associative Memories. PhD Dissertation. Cleveland, OH: Case Western Reserve University.

[Amari, S 67a] Amari, Shun-ichi. (1967). *A Theory of Adaptive Pattern Classifiers*. IEEE Transactions on Electronic Computers. Volume EC-16. pp 299-307.

[Amari, S 71a] Amari, Shun-ichi. (1971). *Characteristics of Randomly Connected Threshold-Element Networks and Network Systems*. Proceedings of IEEE. Volume 59. pp 35-47.

[Amari, S 72a] Amari, Shun-ichi. (1972). *Characteristics of Random Nets of Analog Neuron-Like Elements*. IEEE Transactions on Systems, Man and Cybernetics. Volume SMC-2. pp 643-657.

[Amari, S 72b] Amari, Shun-ichi. (1972). *Learning Patterns and Pattern Sequences by Self-Organizing Nets of Threshold Elements*. IEEE Transactions on Computers. Volume C-21. pp 1197-1206.

[Amari, S 74a] Amari, Shun-ichi. (1974). *A Mathematical Theory of Nerve Nets*. In Kotani, M. (ed.). Advances in Biophysics. pp 75-120. Tokyo: University of Tokyo Press.

[Amari, S 74b] Amari, Shun-ichi. (1974). *A Method of Statistical Neurodynamics*. Biological Cybernetics. Number 14. pp 201-215.

[Amari, S 75a] Amari, Shun-ichi. (1975). *Homogeneous Nets of Neuron-Like Elements*. Biological Cybernetics. Volume 17. pp 211-220.

[Amari, S 77a] Amari, Shun-ichi. (1977). *A Mathematical Approach to Neural Systems*. Systems Neuroscience. pp 67-117. New York: Academic Press.

[Amari, S 77b] Amari, Shun-ichi. (1977). *Dynamics of Pattern Formation in Lateral-Inhibition Type Neural Fields*. Biological Cybernetics. Volume 27. pp 77-87.

[Amari, S 77c] Amari, Shun-ichi; Yoshida, Kiyonori; Kanatani, Ken-Ichi. (1977). *Mathematical Foundation for Statistical Neurodynamics*. SIAM Journal of Applied Mathematics. Volume 33.

[Amari, S 77d] Amari, Shun-ichi. (1977). *Neural Theory of Association and Concept Formation*. Biological Cybernetics. Volume 26. pp 175-185.

[Amari, S 77e] Amari, Shun-ichi. (1988). *Neural Theory of Association and Concept Formation*. In Anderson, James A. (ed.);

Rosenfeld, Edward (ed.). Neurocomputing: Foundations of Research. pp 271-282. Cambridge, MA: MIT Press.

[Amari, S 78*] SEE: Lieblich, I 78a. *An Extended First Approximation Model for the Amygdaloid Kindling Phenomenon.*

[Amari, S 78a] Amari, Shun-ichi; Takeuchi, Akikazu. (1978). *Mathematical Theory on Formation of Category Detecting Nerve Cells.* Biological Cybernetics. Volume 29. pp 127-136.

[Amari, S 79*] SEE: Kishimoto, K 79a. *Existence and Stability of Local Excitations in Neural Fields.*

[Amari, S 79*] SEE: Takeuchi, A 79a. *Formation of Topographic Maps and Columnar Microstructures in Nerve Fields.*

[Amari, S 79a] Amari, Shun-ichi; Takeuchi, Akikazu. (1979). *Theory of Self-Organizing Nerve Nets.* Transactions of the Institute of Electronics and Communication Engineers of Japan. Volume E62. pp 896-897.

[Amari, S 79b] Amari, Shun-ichi. (1979). *Theory of Self-Organizing Nerve Nets with Special Reference to Association and Concept Formation.* International Joint Conference on Artificial Intelligence (IJCAI-79), 6th, Tokyo, August 20-23, 1979: Proceedings. Volume 1. pp 13-15. International Joint Conference on Artificial Intelligence.

[Amari, S 80a] Amari, Shun-ichi. (1980). *Topographic Organization of Nerve Fields.* Bulletin of Mathematical Biology. Volume 42. pp 339-364.

[Amari, S 82a] Amari, Shun-ichi. (1982). *A Mathematical Theory of Self-Organizing Nerve Nets.* Progress in Cybernetics and Systems Research. Volume 9. pp 195-204.

[Amari, S 82b] Amari, Shun-ichi. (1982). *A Mathematical Theory of Self-Organizing Nerve Systems.* Biomathematics: Current Status and Perspectives. pp 159-197. Amsterdam: North-Holland.

[Amari, S 82c] Amari, Shun-ichi; Arbib, Michael A. (1982). Competition and Cooperation in Neural Nets, Kyoto, 1982: Proceedings. New York: Springer-Verlag.

[Amari, S 83a] Amari, Shun-ichi. (1983). *Field Theory of Self-Organizing Neural Networks.* IEEE Transactions on Systems, Man and Cybernetics. Volume SMC-13. pp 741-748.

[Amari, S 85a] Amari, Shun-ichi. (1985). *Formation of Retinotopy and Columnar Microstructures by Self-Organization: A Mathematical Model.* Models of the Visual Cortex. pp 157-163. New York: Wiley.

[Amari, S 87a] Amari, Shun-ichi. (1987). *On Mathematical Methods in the Theory of Neural Networks.* IEEE International Conference on Neural Networks (ICNN-87), San Diego, CA, June 21-24, 1987: Proceedings. Volume 3. pp 3-10. New York: Institute of Electrical and Electronics Engineers.

[Amari, S 87b] Amari, Shun-ichi; Maruyama, Minoru. (1987). *On the Topological Representation of Signals in Self-Organizing Nerve Fields.* Mathematical Topics in Population Biology, Morphogenesis and Neurosciences, International Symposium, Kyoto, November 10-15, 1985. Volume 71. pp 282-291. New York: Springer-Verlag.

[Amari, S 88a] Amari, Shun-ichi. (1988). *Mathematical Theory of Self-Organization in Neural Nets.* Organization of Neural Networks, Structures and Models. pp 52-66.

[Amari, S 88b] Amari, Shun-ichi; Maginu, Kenjiro. (1988). *Statistical Neurodynamics of Associative Memory.* Neural Networks. Volume 1. pp 63-73.

[Amari, S 88c] Amari, Shun-ichi. (1988). *Statistical Neurodynamics of Various Versions of Correlation Associative Memory.* IEEE International Conference on Neural Networks (ICNN-88), San Diego, CA, July 24-27, 1988: Proceedings. Volume 1. pp 633-640. New York: Institute of Electrical and Electronics Engineers.

[Amatniek, E 58a] Amatniek, Ernest. (1958). *Measurement of Bioelectric Potentials with Microelectrodes and Neutralized*

Input Capacity Amplifiers. IRE Transactions on Medical Electronics. Volume PGME-10. pp 3-14.

[Amato, I 87a] Amato, Ivan. (January 24, 1987). *Growing 'Brains' in a Computer*. Science News. Volume 131. pp 60-61.

[Ambros-Ingerson, J 88*] SEE: Granger, R 88a. *Derivation of Encoding Characteristics of Layer II Cerebral Cortex*.

[Ambros-Ingerson, J 88*] SEE: Granger, R 88b. *Partitioning of Sensory Data by a Cortical Network*.

[Ambroziak, A 64a] Ambroziak, A. (1964). *Semi-Distributed Neuristor Line Using Unijunction Transistors*. Solid State Electronics. Volume 7. pp 259-265.

[Amit, DJ 85a] Amit, Daniel J.; Gutfreund, Hanoch; Sompolinsky, H. (1985). *Spin-Glass Models of Neural Networks*. Physical Review A. Volume 32. pp 1007-1018.

[Amit, DJ 85b] Amit, Daniel J.; Gutfreund, Hanoch. (1985). *Storing Infinite Numbers of Patterns in a Spin-Glass Model of Neural Networks*. Physical Review Letters. Volume 55. pp 1530-1533.

[Amit, DJ 87*] SEE: Meidan, M 87a. Electronic Analog Neural Network with Noise and Time Delays.

[Ammon, K 86a] Ammon, Kurt. (1986). *An Architecture for Mathematical Cognition*. Cognitive Science Society, 8th, Amherst, MA, August 15-17, 1986: Proceedings. pp 585-594. Hillsdale, NJ: Lawrence Erlbaum.

[an der Heiden, U 80a] an der Heiden, Uwe. (1980). Analysis of Neural Networks. Biomathematics Number 35. Berlin: Springer-Verlag.

[Anandan, P 84*] SEE: Barto, AG 84a. Pattern Recognizing Stochastic Learning Automata.

[Anandan, P 85*] SEE: Barto, AG 85e. *Pattern Recognizing Stochastic Learning Automata*.

[Anandan, P 86*] SEE: Barto, AG 86a. *Cooperativity in Networks of Pattern Recognizing Stochastic Learning Automata*.

[Anandan, P 88*] SEE: Mjolsness, E 88a. Optimization in Model Matching and Perceptual Organization: A First Look.

[Anastassiou, D 87*] SEE: Tsividis, Y 87c. *Switched-Capacitor Neural Networks*.

[Anastassiou, D 88*] SEE: Kollias, S 88a. *Adaptive Training of Multilayer Neural Networks Using a Least Squares Estimation Technique*.

[Anastassiou, D 88a] Anastassiou, Dimitris; Kollias, Stefanos. (1988). *Digital Image Halftoning Using Neural Networks*. Visual Communications and Image Processing '88. SPIE Volume 1001. pp 1062-1069. Bellingham, WA: Society of Photo-Optical Instrumentation Engineers.

[Anastassiou, D 89*] SEE: Kollias, S 89a. *Image Halftoning and Reconstruction Using a Neural Network*.

[Andersen, K 88*] SEE: Karsai, G 88a. Gas Tungsten Arc Weld Modeling Using a Mapping Network.

[Andersen, P 66a] Andersen, P.; Gillow, M.; Rudjord, T. (1966). *Rhythmic Activity in a Simulated Neuronal Network*. Journal of Physiology. Volume 185. pp 418-428.

[Anderson, CH 86a] Anderson, Charles H.; Abrahams, Elihu. (1986). *Bayesian Probability Network*. In Denker, John S. (ed.). Neural Networks for Computing, Snowbird, Utah, 1986. AIP Proceedings 151. pp 7-11. New York: American Institute of Physics.

[Anderson, CH 87a] Anderson, Charles H.; Van Essen, D. C. (1987). *Shifter Circuits: A Computational Strategy for Dyanmic Aspects of Visual Processing*. Proceedings of the National Academy of Sciences. Volume 84. pp 6297-6301.

[Anderson, CH 87b] Anderson, Charles H.; Abrahams, Elihu. (1987). *The Bayes Connection.* IEEE International Conference on Neural Networks (ICNN-87), San Diego, CA, June 21-24, 1987: Proceedings. Volume 3. pp 105-112. New York: Institute of Electrical and Electronics Engineers.

[Anderson, CW 82*] SEE: Barto, AG 82a. Adaptive Neurons-Like Elements that Can Solve Difficult Learning Control Problems.

[Anderson, CW 82*] SEE: Barto, AG 82c. *Synthesis of Non-Linear Control Surfaces by a Layered Associative Search Network.*

[Anderson, CW 82a] Anderson, Charles W. (1982). Feature Generation and Selection by a Layered Network of Reinforcement Learning Elements: Some Initial Experiments. COINS Technical Report 82-12. Amherst, MA: University of Massachusetts.

[Anderson, CW 83*] SEE: Barto, AG 83b. *Neuron-like Adaptive Elements that Can Solve Difficult Learning Control Problems.*

[Anderson, CW 83*] SEE: Barto, AG 83c. *Neuron-like Elements That Can Solve Difficult Learning and Control Problems.*

[Anderson, CW 85*] SEE: Barto, AG 85f. *Structural Learning in Connectionist Systems.*

[Anderson, CW 86*] SEE: Barto, AG 86a. *Cooperativity in Networks of Pattern Recognizing Stochastic Learning Automata.*

[Anderson, CW 86a] Anderson, Charles W. (1986). Learning and Problem Solving with Multilayer Connectionist Systems. COINS Technical Report 86-50. Amherst, MA: University of Massachusetts.

[Anderson, DZ 86a] Anderson, Dana Z. (1986). *Coherent Optical Eigenstate Memory.* Optics Letters. Volume 11. p 56 f.

[Anderson, DZ 86b] Anderson, Dana Z. (1986). *Optical Resonators and Neural Networks.* In Denker, John S. (ed.). Neural Networks for Computing, Snowbird, Utah, 1986. AIP Proceedings 151. pp 12-16. New York: American Institute of Physics.

[Anderson, DZ 87a] Anderson, Dana Z. (1987). *Adaptable Interconnects for Optical Neuromorphs: Demonstration of a Photorefractive Projection Operator.* IEEE International Conference on Neural Networks (ICNN-87), San Diego, CA, June 21-24, 1987: Proceedings. Volume 3. pp 577-584. New York: Institute of Electrical and Electronics Engineers.

[Anderson, DZ 87b] Anderson, Dana Z.; Lininger, Diana M. (1987). *Dynamic Optical Interconnects: Volume Holograms as Optical Two-Port Operators.* Applied Optics. Volume 26. pp 5031-5038.

[Anderson, DZ 88*] SEE: Loos, H 88a. *Reflexive Associative Memories.*

[Anderson, DZ 88a] Anderson, Dana Z. (ed.). (1988). Neural Information Processing Systems. New York: American Institute of Physics.

[Anderson, DZ 88b] Anderson, Dana Z. (1988). *Nonlinear Optical Neural Networks: Dynamic Ring Oscillations.* In Eckmiller, Rolf (ed.); von der Malsburg, Christoph (ed.). Neural Computers. NATO ASI Series, Volume F41. pp 417-424. Berlin: Springer-Verlag.

[Anderson, JA 68a] Anderson, James A. (1968). *A Memory Storage Model Utilizing Spatial Correlation Functions.* Kybernetik. Volume 5. pp 113-119.

[Anderson, JA 70a] Anderson, James A. (1970). *Two Models for Memory Organization Using Interacting Traces.* Mathematical Biosciences. Volume 8. pp 137-160.

[Anderson, JA 72a] Anderson, James A. (1988). *A Simple Neural Network Generating an Interactive Memory.* In Anderson, James A. (ed.); Rosenfeld, Edward (ed.). Neurocomputing: Foundations of Research. pp 181-192. Cambridge, MA: MIT Press.

[Anderson, JA 72b] Anderson, James A. (1972). *A Simple Neural Network Generating an Interactive Memory.* Mathematical Biosciences. Volume 14. pp 197-220.

[Anderson, JA 72c] Anderson, James A. (1972). *Models of Interactive Memory.* Asilomar Conference on Circuits and Systems, 5th, November 8-10, 1971: Conference Record. pp 242-247. New York: Institute of Electrical and Electronics Engineers.

[Anderson, JA 74a] Anderson, James A. (1974). What Is a Distinctive Feature?. Technical Report 74-1. Providence, RI: Brown University, Center for Neural Studies.

[Anderson, JA 77a] Anderson, James A.; Silverstein, Jack W.; Ritz, Stephen A.; Jones, Randall S. (1988). *Distinctive Features, Categorical Perception, and Probability Learning: Some Applications of a Neural Model.* In Anderson, James A. (ed.); Rosenfeld, Edward (ed.). Neurocomputing: Foundations of Research. pp 287-326. Cambridge, MA: MIT Press.

[Anderson, JA 77b] Anderson, James A.; Silverstein, Jack W.; Ritz, Stephen A.; Jones, Randall S. (1977). *Distinctive Features, Categorical Perception, and Probability Learning: Some Applications of a Neural Model.* Psychological Review. Volume 84. pp 413-451.

[Anderson, JA 77c] Anderson, James A. (1977). *Neural Models with Cognitive Implications.* Basic Processes in Reading: Perception and Comprehension. Hillsdale, NJ: Lawrence Erlbaum.

[Anderson, JA 77d] Anderson, James A.; Silverstein, Jack W.; Ritz, Stephen A. (1977). *Vowel Pre-Processing with a Neurally Based Model.* IEEE International Conference on Acoustics, Speech, and Signal Processing (ICASSP-77), Hartford, CT, May 9-11, 1977: Proceedings. pp 265-269. New York: Institute of Electrical and Electronics Engineers.

[Anderson, JA 78a] Anderson, James A.; Cooper, Leon N. (1978). *Les Modeles Mathematiques de l'Organization Biologique de la Memoire.* Pluriscience, Encyclopaedia Univeralis. pp 168-175.

[Anderson, JA 79a] Anderson, James A. (1979). *Parallel Computation with Simple Neural Networks.* Cognition and Brain Theory. Volume 3. pp 45-53.

[Anderson, JA 79b] Anderson, James A. (1979). *States: Computer & Brain.* Contemporary Psychology. Volume 24. pp 172-173.

[Anderson, JA 80a] Anderson, James A. (1980). *Review.* Contemporary Psychology. Volume 25. pp 644-646.

[Anderson, JA 81*] SEE: Hinton, GE 81c. Parallel Models of Associative Memory.

[Anderson, JA 81a] Anderson, James A.; Mozer, Michael C. (1981). *Categorization and Selective Neurons.* In Hinton, Geoffrey E. (ed.). Parallel Models of Associative Memory. pp 213-236. Hillsdale, NJ: Lawrence Erlbaum.

[Anderson, JA 81b] Anderson, James A.; Hinton, Geoffrey E. (1981). *Models of Information Processing in the Brain.* In Hinton, Geoffrey E. (ed.). Parallel Models of Associative Memory. pp 9-48. Hillsdale, NJ: Lawrence Erlbaum.

[Anderson, JA 83a] Anderson, James A. (1983). *Cognitive and Psychological Computation with Neural Models.* IEEE Transactions on Systems, Man and Cybernetics. Volume SMC-13. pp 799-815.

[Anderson, JA 84*] SEE: Kawamoto, AH 84a. *Lexical Access Using a Neural Network.*

[Anderson, JA 84*] SEE: Knapp, AG 84a. *A Signal Averaging Model for Concept Formation.*

[Anderson, JA 84*] SEE: Levy, WB 84a. Synaptic Change in the Nervous System.

[Anderson, JA 84*] SEE: Knapp, AG 84b. *Theory of Categorization Based on Distributed Memory Storage.*

[Anderson, JA 84*] SEE: Knapp, AG 84c. *Theory of Categorization Based on Distributed Memory Storage.*

[Anderson, JA 85*] SEE: Kawamoto, AH 85a. *A Neural Network Model of Multistable Perception.*

[Anderson, JA 85a] Anderson, James A. (1985). *What Hebb Synapses Build.* In Levy, William B. (ed.); Anderson, James A. (ed.); Lehmkuhle, Stephen (ed.). Synaptic Modifications, Neuron Selectivity, and Nervous System Organization. pp 153-174. Hillsdale, NJ: Lawrence Erlbaum.

[Anderson, JA 86a] Anderson, James A. (1986). *Cognitive Capabilities of a Parallel System.* In Bienenstock, Elie L. (ed.). Disordered Systems and Biological Organization. pp 209-226. New York: Springer-Verlag.

[Anderson, JA 86b] Anderson, James A.; Murphy, Gregory L. (1986). *Concepts in Connectionist Models.* In Denker, John S. (ed.). Neural Networks for Computing, Snowbird, Utah, 1986. AIP Proceedings 151. pp 17-22. New York: American Institute of Physics.

[Anderson, JA 86c] Anderson, James A.; Golden, Richard M.; Murphy, Gregory L. (1986). *Concepts in Distributed Systems.* In Szu, Harold H. (ed.). Optical and Hybrid Computing. SPIE 634. pp 260-276. Bellingham, WA: Society of Photo-Optical Instrumentation Engineers.

[Anderson, JA 86d] Anderson, James A. (1986). *Neural Models: Networks for Fun and Profit.* Nature. Volume 322. pp 406-407.

[Anderson, JA 86e] Anderson, James A.; Murphy, Gregory L. (1986). *Psychological Concepts in Parallel Systems.* Physica. Volume 22D. pp 318-336.

[Anderson, JA 87*] SEE: Kersten, D 87a. *Associative Learning of Scene Parameters from Images.*

[Anderson, JA 88*] SEE: Munro, PW 88a. *Tools for Connectionist Modeling: The Dynamical Systems Methodology.*

[Anderson, JA 88a] Anderson, James A. (1988). *Concept Formation of Neural Networks: Implications for Evolution of Cognitive Functions.* Human Evolution. Volume 3. pp 81-97.

[Anderson, JR 73a] Anderson, John R.; Bower, Gordon H. (1979). Human Associative Memory. Hillsdale, NJ: Lawrence Erlbaum.

[Anderson, JR 82a] Anderson, John R. (1982). *Acquisition of Cognitive Skill.* Psychological Review. Volume 89. pp 369-406.

[Anderson, JR 83a] Anderson, John R. (1983). *Retrieval of Information from Long-Term Memory.* Science. Number 220. pp 25-30.

[Anderson, JR 83b] Anderson, John R. (1983). The Architecture of Cognition. Cognitive Science Series, 5. Cambridge, MA: Harvard University Press.

[Anderson, PW 84*] SEE: Palmer, RG 84a. *Models of Hierarchically Constrained Dynamics for Glassy Relaxation.*

[Andrews, HC 72a] Andrews, H. C. (1972). Introduction to Mathematical Techniques in Pattern Recognition. New York: Wiley.

[Angell, JB 61*] SEE: Widrow, B 61a. *Birth, Life, and Death in Microelectronic Systems.*

[Angell, JB 62*] SEE: Widrow, B 62c. *Reliable, Trainable Networks for Computing and Control.*

[Angeniol, B 88a] Angeniol, Bernard; De La Croix Vaubois, Gael; Le Texier, Jean-Yves. (1988). *Self-Organizing Feature Maps and the Travelling Salesman Problem.* Neural Networks. Volume 1. pp 289-293.

[Angyan, AJ 60*] SEE: Zemanek, H 60a. *A Model for Neurophysiological Functions.*

[Annaratone, M 86a] Annaratone, Marco; Arnould, Emmanuel; Gross, Thomas; King, H. T.; Lam, Monica S.; Menzilcioglu,

Onat. (1986). *Warp Architecture and Implementation.* Computer Architecture News. Volume 14. pp 346-356.

[Anne, A 67*] SEE: Molnar, CE 67a. *The CHASM: A Macromodular Computer for Analyzing Neuron Models.*

[Anninos, PA 70a] Anninos, Photios A.; Beek, B.; Csermely, T. J.; Harth, Erich M.; Pertile, G. (1970). *Dynamics of Neural Structures.* Journal of Theoretical Biology. Volume 26. pp 121-148.

[Anninos, PA 72a] Anninos, Photios A. (1972). *Cyclic Modes in Artificial Neural Nets.* Kybernetik. Volume 11. pp 5-14.

[Anninos, PA 72b] Anninos, Photios A. (1972). *Mathematical Models of Memory Trace and Forgetfulness.* Kybernetik. Volume 10. pp 165-167.

[Anninos, PA 73a] Anninos, Photios A. (1973). *Evoked Potentials in Artificial Neural Nets.* Kybernetik. Volume 13. pp 24-29.

[Anninos, PA 75a] Anninos, Photios A.; Raman, S. (1975). *Derivation of a Mathematical Equation for the EEG and the General Solution within the Brain and in Space.* International Journal of Theoretical Physics. Volume 12. pp 1-9.

[Anninos, PA 76*] SEE: Leake, B 76a. *Effect of Connectivity on the Activity of Neural Net Models.*

[Anninos, PA 79*] SEE: Papadopoullos, MS 79a. *Memory Topology of Neural Nets.*

[Anninos, PA 80a] Anninos, Photios A.; Zenone, Silvio. (1980). *A Neural Net Model for the a-Rhythm.* Biological Cybernetics. Volume 36. pp 187-191.

[Anninos, PA 82a] Anninos, Photios A.; Zenone, Silvio; Cyrulnik, R. A.; Elul, R. (1982). *Encoding of Visual Information: Correlation of EEG Response to Intracellular Response.* Brain Research. Volume 244. pp 182-185.

[Anninos, PA 83a] Anninos, Photios A.; Argyrakis, Panos. (1983). *A Mathematical Model for the Decay of Short-Term Memory with Age.* Journal of Theoretical Biology. Volume 102. pp 191-197.

[Anninos, PA 83b] Anninos, Photios A.; Zenone, Silvio; Elul, R. (1983). *Artificial Neural Nets: Dependence of the EEG Amplitude's Probability Distribution on Statistical Parameters.* Journal of Theoretical Biology. Volume 103. pp 339-348.

[Anninos, PA 84a] Anninos, Photios A.; Argyrakis, Panos; Skouras, Andreas. (1984). *A Computer Model for Learning Processes and the Role of the Cerebral Commissures.* Biological Cybernetics. Volume 50. pp 329-336.

[Anninos, PA 84b] Anninos, Photios A.; Kokkinidis, Michael. (1984). *A Neural Net Model for Multiple Memory Domains.* Journal of Theoretical Biology. Volume 109. pp 95-110.

[Anninos, PA 84c] Anninos, Photios A.; Kokkinidis, Michael; Skouras, Andreas. (1984). *Noisy Neural Nets Exhibiting Memory Domains.* Journal of Theoretical Biology. Volume 109. pp 581-594.

[Anninos, PA 85*] SEE: Kokkinidis, M 85a. *Noisy Neural Nets Exhibiting Epileptic Features.*

[Anninos, PA 87a] Anninos, Photios A.; Kokkinidis, Michael. (1987). *A Mathematical Formalism for the Finite Difference Equation in Neural Nets with Chemical Markers.* Journal of Theoretical Neurobiology. Volume 5. pp 107-126.

[Anninos, PA 87b] Anninos, Photios A. (1987). *A Neural Model Theory Leading to Kindling Effect of Epilepsy.* International Journal of Neuroscience. Volume 36. pp 177-198.

[Anninos, PA 87c] Anninos, Photios A.; Anogianakis, G.; Apostolakis, M.; Thomas, G. (1987). *A Neural Net Model for the Release of Prolactin During Pseudopregnancy in the Rat.* Journal of Theoretical Biology. Volume 129. pp 359-368.

[Anninos, PA 88a] Anninos, Photios A. (1988). *A Neural Model with Multiple Memory Domains.* In Eckmiller, Rolf (ed.); von der Malsburg, Christoph (ed.). Neural Computers. NATO ASI Series, Volume F41. pp 209-217. Berlin: Springer-Verlag.

[Anninos, PA 88b] Anninos, Photios A.; Cook, Norman D. (1988). *Neural Net Simulation of the Corpus Callosum.* International Journal of Neuroscience. Volume 38. pp 381-391.

[Anogianakis, G 87*] SEE: Anninos, PA 87c. *A Neural Net Model for the Release of Prolactin During Pseudopregnancy in the Rat.*

[Ansari, A 87*] SEE: David, GW 87a. *Sensitivity Analysis of Hopfield Net.*

[Ansari, A 88*] SEE: Kowalski, JM 88a. *Coupled Rotators as a Model for Spontaneous Activity in Neural Networks.*

[Anthes, GH 88a] Anthes, Gary H. (May 9, 1988). *NeuralWare Expert System Classified Stock Patterns.* Federal Computer Week. pp 21.

[Antosiewicz, HA 55*] SEE: Cole, KS 55a. *Automatic Computation of Nerve Excitation.*

[Antosiewicz, HA 59*] SEE: FitzHugh, R 59a. *Automatic Computation of Nerve Excitation - Detailed Corrections and Additions.*

[Apostolakis, M 87*] SEE: Anninos, PA 87c. *A Neural Net Model for the Release of Prolactin During Pseudopregnancy in the Rat.*

[Appenzeller, T 87*] SEE: Mishkin, M 87a. *The Anatomy of Memory.*

[Arbib, MA 72a] Arbib, Michael A. (1972). The Metaphorical Brain. New York: Wiley.

[Arbib, MA 74*] SEE: Szentagothai, J 74a. *Conceptual Models of Neural Organization.*

[Arbib, MA 76a] Arbib, Michael A.; Kilmer, W. L.; Spinelli, D. N. (1976). *Neural Models and Memory.* Neural Mechanisms and Memory. Cambridge, MA: MIT Press.

[Arbib, MA 79a] Arbib, Michael A.; Caplan, David. (1979). *Neuro-Linguistics Must be Computational.* The Brain and Behavioral Sciences. Volume 2. pp 449-483.

[Arbib, MA 79b] Arbib, Michael A. (1979). Perceptual Structures and Distributed Motor Control. COINS Report 79-11. Amherst, MA: University of Massachusetts.

[Arbib, MA 82*] SEE: Amari, S 82c. Competition and Cooperation in Neural Nets, Kyoto, 1982: Proceedings.

[Arbib, MA 82a] Arbib, Michael A. (1982). *Modeling Neural Mechanisms of Visual-Motor Coordination in the Frog and Toad.* In Amari, Shun-ichi (ed.); Arbib, Michael A. (ed.). Competition and Cooperation in Neural Nets. pp 342-370. New York: Springer-Verlag.

[Arbib, MA 82b] Arbib, Michael A. (ed.); Caplan, David (ed.); Marshall, John C. (ed.). (1982). Neural Models of Language Processes. New York: Academic Press.

[Arbib, MA 84*] SEE: Selfridge, OG 84a. Adaptive Control of Ill-Defined Systems.

[Arbib, MA 85*] SEE: Lyons, D 85a. A Task-Level Model of Distributed Computation for Sensory-Based Control of Complex Robot Systems.

[Arbib, MA 85a] Arbib, Michael A. (1985). *Brain Theory and Cooperative Computation.* Human Neurobiology. Volume 4. pp 201-218.

[Arbib, MA 85b] Arbib, Michael A.; Iberall, Thea; Lyons, Damian. (1985). *Coordinated Control Programs for Movements of the Hand*. Experimental Brain Research. Supplement 10. pp 113-129. New York: Springer-Verlag.

[Arbib, MA 85c] Arbib, Michael A.; House, Donald H. (1985). Depth and Detours: An Essay on Visually Guided Behavior. COINS Technical Report 85-28. Amherst, MA: University of Massachusetts.

[Arbib, MA 85d] Arbib, Michael A. (1985). Levels of Modeling of Neural Interactions Underlying Visuomoter Coordination. COINS Technical Report 85-36. Amherst, MA: University of Massachusetts.

[Arbib, MA 85e] Arbib, Michael A. (1985). Modularity and Interaction of Brain Regions Underlying Visuomoter Coordination. COINS Technical Report 85-29. Amherst, MA: University of Massachusetts.

[Arbib, MA 86*] SEE: Dakin, GA 86a. *The Hierarchical Control of Skilled Hand Movements*.

[Arbib, MA 86*] SEE: Iberall, T 86a. *Opposition Space as a Structuring Concept for the Analysis of Skilled Hand Movements*.

[Arbib, MA 86*] SEE: Vijaykumar, R 86a. A Task-Level Robot Planner for Assembly Operations.

[Arbib, MA 87a] Arbib, Michael A. (1987). Brains, Machines, and Mathematics. 2nd ed. New York: Springer-Verlag.

[Arbib, MA 87b] Arbib, Michael A.; House, Donald H. (1987). *Depth and Detours: An Essay on Visually Guided Behavior*. In Arbib, Michael A. (ed.); Hanson, Allen R. (ed.). Vision, Brain, and Cooperative Computation. pp 129-164. Cambridge, MA: MIT Press.

[Arbib, MA 87c] Arbib, Michael A.; Iberall, Thea; Lyons, Damian. (1987). *Schemas That Integrate Vision and Touch for Hand Control*. In Arbib, Michael A. (ed.); Hanson, Allen R. (ed.). Vision, Brain, and Cooperative Computation. pp 419-488. Cambridge, MA: MIT Press.

[Arbib, MA 87d] Arbib, Michael A. (ed.); Hanson, Allen R. (ed.). (1987). Vision, Brain, and Cooperative Computation. Cambridge, MA: MIT Press.

[Arbib, MA 88a] Arbib, Michael A. (1988). *Schemas and Neurons: Two Levels of Neural Computing*. In Eckmiller, Rolf (ed.); von der Malsburg, Christoph (ed.). Neural Computers. NATO ASI Series, Volume F41. pp 311-320. Berlin: Springer-Verlag.

[Argyrakis, P 83*] SEE: Anninos, PA 83a. *A Mathematical Model for the Decay of Short-Term Memory with Age*.

[Argyrakis, P 84*] SEE: Anninos, PA 84a. *A Computer Model for Learning Processes and the Role of the Cerebral Commissures*.

[Arimoto, S 62*] SEE: Nagumo, J 62a. *An Active Pulse Transmission Line Simulating Nerve Axon*.

[Arimoto, S 87*] SEE: Kawamura, S 87a. *Intelligent Control of Robot Motion Based on Learning Method*.

[Arimoto, S 87*] SEE: Kawamura, S 87b. *Technical Issues in Learning Control for Robot Motions*.

[Arimoto, S 88*] SEE: Kawamura, S 88a. *Realization of Robot Motion Based on a Learning Method*.

[Arimoto, S 88*] SEE: Suzuji, H 88a. *Self-Organization of Associative Database and Its Applications*.

[Arimoto, S 88*] SEE: Suzuki, H 88a. *Self-Organizing Model for Pattern Learning and Its Application to Robot Eyesight*.

[Arimoto, S 88*] SEE: Suzuki, H 88b. *Self-Organizing Model of Learning and Its Application to Eyesight System of Robot*.

[Arimoto, S 88*] SEE: Suzuki, H 88c. *Visual Control of Autonomous Mobile Robot Based on Self-Organizing Model for Pattern*

Learning.

[Arimoto, S 88a] Arimoto, Suguru. (1988). *Motion Control of Robotic Manipulator Based on Motor Program Learning.* IFAC Symposium on Robot Control, 2nd, Karlsruhe, October 5-7, 1988: Proceedings.

[Arnould, E 86*] SEE: Annaratone, M 86a. *Warp Architecture and Implementation.*

[Arsenault, HH 87*] SEE: Macukow, B 87a. *Modification of the Threshold Condition for a Content-Addressable Memory Based on the Hopfield Model.*

[Arsenault, HH 87*] SEE: Macukow, B 87b. *Optical Associative Memory Model Based on Neural Networks Having Variable Interconnection Weights.*

[Asakawa, K 88*] SEE: Hosogi, S 88a. *Cerebellar Neural Network Model for Manipulator Control.*

[Asakawa, K 88*] SEE: Watanabe, N 88a. *Mobile Robot Control by Neural Networks and Their Associated Learning Algorithm.*

[Asanuma, C 86*] SEE: Crick, FHC 86a. *Certain Aspects of the Anatomy and Physiology of the Cerebral Cortex.*

[Ashby, WR 62a] Ashby, W. R.; von Foerster, Heinz; Walker, C. C. (1962). *Instability of Pulse Activity in a Net with Threshold.* Nature. Volume 196. pp 561-562.

[Assefi, T 72*] SEE: Nahi, NE 72a. *Bayesian Recursive Image Estimation.*

[Atal, B 79*] SEE: Schroeder, MR 79a. *Optimizing Digital Speech Coders by Exploiting Masking Properties of the Human Ear.*

[Athale, RA 82a] Athale, Ravindra A.; Collins, W. C. (1982). *Optical Matrix-Matrix Multiplication Based Outer Product Decomposition.* Applied Optics. Volume 21. p 2089 f.

[Athale, RA 84*] SEE: Goodman, JW 84a. *Optical Interconnections for VLSI Systems.*

[Athale, RA 84a] Athale, Ravindra A.; Lee, John N. (1984). *Optical Processing Using Outer-Product Concepts.* Proceedings of the IEEE. Volume 72. pp 931-941.

[Athale, RA 86a] Athale, Ravindra A.; Friedlander, C. B.; Kushner, B. G. (1986). *Attentive Associative Architectures and Their Implications to Optical Computing.* Optical Computing. SPIE Volume 625. pp 179-188. Bellingham, WA: Society of Photo-Optical Instrumentation Engineers.

[Athale, RA 86b] Athale, Ravindra A.; Szu, Harold H.; Friedlander, Charles B. (July 1986). *Attentive Associative Memory and Its Optical Implementation.* Optics Letters.

[Athale, RA 86c] Athale, Ravindra A.; Szu, Harold H.; Friedlander, Carl B. (1986). *Optical Implementation of Associative Memory with Controlled Nonlinearity in the Correlation Domain.* Optics Letters. Volume 11. pp 482-484.

[Athale, RA 87*] SEE: Stirk, CW 87a. *Effects of System Noise on an Optical Implementation of an Additive Lateral Inhibitory Network.*

[Athale, RA 88*] SEE: Rovnyak, SM 88a. *Optical Implementation of Winner-Take-All Networks: Noise Considerations.*

[Athale, RA 88a] Athale, Ravindra A.; Stirk, Charles W. ((submitted)). *Compact Architectures for Adaptive Neural Nets.* Optical Engineering.

[Atiya, AF 87a] Atiya, Amir F. (1987). *Learning on a General Network.* In Anderson, Dana Z. (ed.). Neural Information Processing Systems. pp 22-30. New York: American Institute of Physics.

[Atiya, AF 87b] Atiya, Amir F. (1987). *Optimal Neural Spike Classification.* In Anderson, Dana Z. (ed.). Neural

Information Processing Systems. pp 22-30. New York: American Institute of Physics.

[Atiya, AF 88*] SEE: Bower, JM 88c. *Optimal Neural Spike Classification.*

[Atiya, AF 88a] Atiya, Amir F. (1988). *A Neural Unsupervised Learning Technique.* Neural Networks. Volume 1.

[Atkinson, RC 65a] Atkinson, Reilly C.; Bower, Gordon H.; Crothers, E. J. (1965). An Introduction to Mathematical Learning Theory. New York: Wiley.

[Atkinson, RC 68a] Atkinson, Reilly C.; Shiffron, R. M. (1968). *Human Memory: A Proposed System and its Control Processes.* The Psychology of Learning and Motivation: Advances in Research and Theory. Volume 2. pp 89-195. New York: Academic Press.

[Atkinson, RC 71*] SEE: Juola, JF 71a. *Recognition Time for Information Stored in Long-Term Memory.*

[Atkinson, RC 88a] Atkinson, Reilly C. (1988). *A Simple Model for Spike-Train-Output Neurons.* Neural Networks. Volume 1.

[Atkinson, RC 88b] Atkinson, Reilly C. (1988). Exact Solutions for Feedforward Network Parameters. Waltham, MA: Arcon Corporation.

[Atkinson, RC 88c] Atkinson, Reilly C. (1988). *Experience vs. Architecture in Feedforward Networks: A Case Study.* IEEE International Conference on Neural Networks (ICNN-88), San Diego, CA, July 24-27, 1988: Proceedings.

[Atlan, H 88*] SEE: Weisbuch, G 88a. *Control of the Immune Response.*

[Atlas, LE 87*] SEE: Marks II, RJ 87a. *A Class of Continuous Level Neural Nets.*

[Atlas, LE 87*] SEE: Marks II, RJ 87c. *Composite Matched Filter Output Partitioning.*

[Atlas, LE 87*] SEE: Marks II, RJ 87d. *Composite Matched Filtering with Error Correction.*

[Atlas, LE 87*] SEE: Cheung, KF 87a. *Conventional and Composite Matched Filters with Error Correction: A Comparison.*

[Atlas, LE 87*] SEE: Cheung, KF 87b. *Neural Net Associative Memories Based on Convex Set Projections.*

[Atlas, LE 87*] SEE: Cheung, KF 87c. *Synchronous vs Asynchronous Behavior of Hopfield's CAM Neural Net.*

[Atlas, LE 88*] SEE: Marks II, RJ 88a. *Architectures for a Continuous Level Neural Network Based on Alternating Orthogonal Projections.*

[Atlas, LE 88*] SEE: Homma, T 88a. *An Artificial Neural Network for Spatio-Temporal Bipolar Patterns: Application to Phoneme Classification.*

[Atlas, LE 88*] SEE: Oh, S 88a. *Effect of Clock Skew in Iterative Neural Network and Optical Feedback Processors.*

[Atlas, LE 88*] SEE: Marks II, RJ 88f. *The Effect of Stochastic Interconnects in Artificial Neural Network Classification.*

[Atlas, LE 88*] SEE: Marks II, RJ 88b. *Generalization in Layered Classification Neural Networks.*

[Atlas, LE 88*] SEE: Marks II, RJ 88c. *Nonlinearity Requirements for Correlation Based Associative Memories.*

[Atlas, LE 88*] SEE: Marks II, RJ 88d. *Optical-Processor Architectures for Alternating-Projection Neural Networks.*

[Atlas, LE 88*] SEE: Marks II, RJ 88e. *Performance Analysis of Associative Memories with Nonlinearities in the Correlation*

Domain.

[Atlas, LE 88*] SEE: Marks II, RJ 88g. *The Performance of Convex Set Projection Based Neural Networks.*

[Atlas, LE 88a] Atlas, Les E.; Homma, Toshiteru; Marks II, Robert Jackson. (1988). *An Artificial Neural Network for Spatio-Temporal Bipolar Patterns: Application to Phoneme Classification.* In Anderson, Dana Z. (ed.). Neural Information Processing Systems. pp 31-40. New York: American Institute of Physics.

[Austin, J 87a] Austin, James. (1987). *ADAM: A Distributed Associative Memory for Scene Analysis.* IEEE International Conference on Neural Networks (ICNN-87), San Diego, CA, June 21-24, 1987: Proceedings. Volume 4. pp 285-292. New York: Institute of Electrical and Electronics Engineers.

[Austin, J 87b] Austin, James; Stonham, T. J. (1987). *Distributed Associative Memory for Use in Scene Analysis.* Image and Vision Computing. Volume 5. pp 251-260.

[Austin, J 87c] Austin, James; Stonham, T. J. (1987). The ADAM Associative Memory. YCS 94. Heslington, York: University of York, Department of Computer Science.

[Austin, J 88a] Austin, James. (1988). Grey Scale N Tuple Processing. Computer Science 301. pp 110-120. New York: Springer-Verlag.

[Austin, J 88b] Austin, James. (1988). Image Pre-Processing with a Generalised Grey Scale N-Tuple Operator. YCS 102. Heslington, York: University of York, Department of Computer Science.

[Averbukh, DYA 69a] Averbukh, D. Y. A. (1969). *Random Nets of Analog Neurons.* Automatic Remote Control. pp 116-123.

[Awad, TA 88*] SEE: Bode, PM 88a. *Develpment of the Two-Part Pattern During Regeneration of the Head in Hydra.*

[Axelrad, H 86a] Axelrad, H.; Bernard, C.; Giraud, B.; Marc, M. E. (1986). *Recurrent Collateral Inhibition Simulated in a Simple Neuronal Automata Assembly.* In Bienenstock, Elie L. (ed.). Disordered Systems and Biological Organization. pp 389-398. New York: Springer-Verlag.

[Axelrad, H 87a] Axelrad, H.; Bernard, C.; Cottrell, M.; Giraud, B. (1987). *The Use of an Artificial Neural Network Model to Analyze the Information Transfer Properties of a Simplified Model of the Cerebellar Cortex.* IEEE International Conference on Neural Networks (ICNN-87), San Diego, CA, June 21-24, 1987: Proceedings. Volume 4. pp 59-66. New York: Institute of Electrical and Electronics Engineers.

[Babcock, KL 86a] Babcock, K. L.; Westervelt, G. L. (1986). *Complex Dynamics in Simple Neural Circuits.* In Denker, John S. (ed.). Neural Networks for Computing, Snowbird, Utah, 1986. AIP Proceedings 151. pp 23-28. New York: American Institute of Physics.

[Babloyantz, A 85a] Babloyantz, A.; Salazar, J. M. (1985). *Evidence of Chaotic Dynamics of Brain Activity During the Sleep Cycle.* Physics Letters. Volume 111A. pp 152-156.

[Babloyantz, A 86a] Babloyantz, A.; Destexhe, A. (1986). *Low-Dimensional Chaos in an Instance of Epilepsy.* Proceedings of the National Academy of Sciences. Volume 83. pp 3513-3517.

[Babloyantz, A 87a] Babloyantz, A.; Destexhe, A. (1987). *Chaos in Neural Networks.* IEEE International Conference on Neural Networks (ICNN-87), San Diego, CA, June 21-24, 1987: Proceedings. Volume 4. pp 31-40. New York: Institute of Electrical and Electronics Engineers.

[Babloyantz, A 87b] Babloyantz, A.; Destexhe, A. (1987). *Strange Attractors in the Human Cortex.* Temporal Disorder in Human Oscillatory Systems. Synergetics Volume 36. pp 48-56. New York: Springer-Verlag.

[Bachmann, CM 87a] Bachmann, Charles M.; Cooper, Leon N.; Dembo, Amir; Zeitouni, Ofer. (1987). *A Relaxation Model for*

Memory with High Storage Density. Proceedings of the National Academy of Sciences. Volume 84. pp 7529-7531.

[Bacon, J 73*] SEE: Kaufman, L 73a. *Stereopsis Without Image Segregation*.

[Bagherzadeh, N 87a] Bagherzadeh, Nader; Kerola, Teemu; Leddy, Bill; Brice, Richard. (1987). *On Parallel Execution of the Traveling Salesman Problem on a Neural Network Model*. IEEE International Conference on Neural Networks (ICNN-87), San Diego, CA, June 21-24, 1987: Proceedings. Volume 3. pp 317-324. New York: Institute of Electrical and Electronics Engineers.

[Bahl, LR 83a] Bahl, Lalit R.; Jelinek, Frederick; Mercer, Robert L. (1983). *A Maximum Likelihood Approach to Continuous Speech Recognition*. IEEE Transactions on Pattern Matching and Machine Intelligence. Volume PAMI-5. pp 179-190.

[Bahr, C 86*] SEE: Hammerstrom, D 86e. The OGC Cognitive Architecture Project: Silicon Implementation of Connectionist/Neural Networks.

[Bahr, C 88a] Bahr, Casey. (1988). ANNE: Another Neural Network Emulator. CS/E-88-028. Beaverton, OR: Oregon Graduate Center, Computer Science/Engineering Department.

[Bahr, C 88b] Bahr, Casey. (1988). Anne User Manual. CS/E-88-029. Beaverton, OR: Oregon Graduate Center, Computer Science/Engineering Department.

[Bahr, C 88c] Bahr, Casey; Hammerstrom, Daniel; Jagla, Kevin. ((to be presented)). *Concurrent Neural Network Simulation: Two Examples within a Single, Integrated Neural Network Hardware Development Environment*. IASTED Applied Simulation and Modelling Conference, Galveston, TX, May 18, 1988.

[Bailey, J 86*] SEE: Hammerstrom, D 86e. The OGC Cognitive Architecture Project: Silicon Implementation of Connectionist/Neural Networks.

[Bailey, J 86a] Bailey, Jim; Hammerstrom, Daniel. (1986). How to Make a Billion Connections. CS/E-86-007. Beaverton, OR: Oregon Graduate Center, Computer Science/Engineering Department.

[Bailey, J 88a] Bailey, Jim. (1988). A VLSI Interconnect Structure for Neural Networks. CS/E-88027. Beaverton, OR: Oregon Graduate Center, Computer Science/Engineering Department.

[Baird, B 86a] Baird, Bill. (1986). *Bifurcation Analysis of Oscillating Network Model of Pattern Recognition in the Rabbit Olefactory Bulb*. In Denker, John S. (ed.). Neural Networks for Computing, Snowbird, Utah, 1986. AIP Proceedings 151. pp 29-34. New York: American Institute of Physics.

[Baird, B 86b] Baird, Bill. (1986). *Nonlinear Dynamics of Pattern Formation and Pattern Recognition in the Rabbit Olfactory Bulb*. Physica. Volume 22D. pp 150-175.

[Baird, B 87*] SEE: Freeman, WJ 87c. *Relation of Olfactory EEG to Behavior: Spatial Analysis*.

[Baird, B 87a] Baird, Bill. (1987). *Bifurcation Analysis of a Network Model of Rabit Olfactory Bulb with Periodic Attractors Stored by a Sequence Learning Algorithm*. IEEE International Conference on Neural Networks (ICNN-87), San Diego, CA, June 21-24, 1987: Proceedings. Volume 2. pp 147-152. New York: Institute of Electrical and Electronics Engineers.

[Baird, B 88a] Baird, Bill. (1988). *Bifurcation Theory Methods for Programming Static or Periodic Attractors and Their Bifurcations in Dynamic Neural Networks*. IEEE International Conference on Neural Networks (ICNN-88), San Diego, CA, July 24-27, 1988: Proceedings. New York: Institute of Electrical and Electronics Engineers.

[Bak, CS 86*] SEE: Little, MJ 86a. *Enhanced Memory Capacity of a Hopfield Neural Network*.

[Baker, F 59*] SEE: Stark, L 59a. *Stability and Oscillations in a Neurological Servomechanism*.

[Baker, JK 75a] Baker, James K. (1975). Stochastic Modeling as a Means of Automatic Speech Recognition. PhD Dissertation. Pittsburgh, PA: Carnegie Mellon University.

[Baker, RG 76*] SEE: Smith, PT 76a. *The Influence of English Spelling Patterns on Pronunciation.*

[Baker, T 86*] SEE: Hammerstrom, D 86e. The OGC Cognitive Architecture Project: Silicon Implementation of Connectionist/Neural Networks.

[Balachandra, R 86*] SEE: Gallant, SI 86d. *Using Automated Techniques to Generate an Expert System for R&D Project Monitoring.*

[Baldi, P 86a] Baldi, Pierre; Baum, Eric B. (1986). *Bounds on the Size of Ultrametric Structures.* Physical Review Letters. Volume 56. pp 1598-1600.

[Baldi, P 86b] Baldi, Pierre; Baum, Eric B. (1986). *Caging and Exhibiting Ultrametric Structures.* In Denker, John S. (ed.). Neural Networks for Computing, Snowbird, Utah, 1986. AIP Proceedings 151. pp 35-40. New York: American Institute of Physics.

[Baldi, P 87a] Baldi, Pierre; Venkatesh, Santosh S. (1987). *Number of Stable Points for Spin-Glasses and Neural Networks of Higher Orders.* Physical Review Letters. Volume 58. pp 913-916.

[Baldi, P 87b] Baldi, Pierre. (1987). *Symmetries and Learning in Neural Network Models.* Physical Review Letters. Volume 59. pp 1976-1978.

[Baldi, P 88a] Baldi, Pierre; Rinott, Yosef. (in press). *Asymptotic Normality of Some Graph Related Statistics.* Journal of Applied Probability.

[Baldi, P 88b] Baldi, Pierre. (1988). *Group Actions and Learning for a Family of Automata.* Journal of Computer and System Science. Volume 36.

[Baldi, P 88c] Baldi, Pierre; Heiligenberg, Walter. (1988). *How Sensory Maps Could Enhance Resolution Through Ordered Arrangements of Broadly Tuned Receivers.* Biological Cybernetics. Volume 59.

[Baldi, P 88d] Baldi, Pierre. (1988). *Neural Networks, Acyclic Orientations of the Hypercube and Sets of Orthogonal Vectors.* SIAM Journal of Discrete Mathematics. Volume 1.

[Baldi, P 88e] Baldi, Pierre; Hornik, Kurt. (in press). *Neural Networks and Principal Component Analysis: Learning from Examples Without Local Minima.* Neural Networks.

[Baldi, P 88f] Baldi, Pierre. (1988). *Neural Networks, Orientations of the Hypercube and Algebraic Threshold Functions.* IEEE Transactions on Information Theory. Volume 34.

[Baldi, P 88g] Baldi, Pierre; Rinott, Yosef. (in press). *On Normal Approximations of Distributions in Terms of Depenency Graphs.* Annals of Probability.

[Baldi, P 88h] Baldi, Pierre; Venkatesh, Santosh S. (1988). *On Properties of Networks of Neuron-Like Elements.* In Anderson, Dana Z. (ed.). Neural Information Processing Systems. pp 41-51. New York: American Institute of Physics.

[Ball, GH 62a] Ball, G. H. (1962). An Invariant Input for a Pattern Recognition Machine. Technical Report 2003-4. Stanford, CA: Stanford University.

[Ball, M 75*] SEE: Widrow, B 75b. *The Complex LMS Algorithm.*

[Ballard, DH 75*] SEE: Kimme, C 75a. *Finding Circles by an Array of Accumulators.*

[Ballard, DH 81*] SEE: Feldman, JA 81a. Computing with Connections.

[Ballard, DH 81a] Ballard, Dana H.; Sabbah, Daniel. (1981). *On Shapes*. International Joint Conference on Artificial Intelligence (IJCAI-81), 7th, Vancouver, August 24-28, 1981: Proceedings. Volume 2. pp 607-612. International Joint Conference on Artificial Intelligence.

[Ballard, DH 81b] Ballard, Dana H. (1981). Parameter Networks. Report TR-75. Rochester, NY: University of Rochester.

[Ballard, DH 81c] Ballard, Dana H. (1981). *Parameter Networks: Towards a Theory of Low-Level Vision*. International Joint Conference on Artificial Intelligence (IJCAI-81), 7th, Vancouver, August 24-28, 1981: Proceedings. Volume 2. pp 1068-1078. International Joint Conference on Artificial Intelligence.

[Ballard, DH 82*] SEE: Hrechanyk, LM 82a. *A Connectionist Model of Form Perception*.

[Ballard, DH 82*] SEE: Feldman, JA 82b. *Connectionist Models and their Properties*.

[Ballard, DH 82*] SEE: Feldman, JA 82a. *Connectionist Models and Their Properties*.

[Ballard, DH 82a] Ballard, Dana H.; Brown, C. M. (1982). Computer Vision. Englewood Cliffs, NJ: Prentice-Hall.

[Ballard, DH 83a] Ballard, Dana H.; Hinton, Geoffrey E.; Sejnowski, Terrence J. (1983). *Parallel Visual Computation*. Nature. Volume 306. pp 21-26.

[Ballard, DH 84a] Ballard, Dana H.; Hayes, P. J. (1984). *Parallel Logical Inference*. Cognitive Science Society, 6th, Boulder, CO, June 28-30, 1984: Proceedings. Hillsdale, NJ: Lawrence Erlbaum.

[Ballard, DH 84b] Ballard, Dana H. (1984). *Parameter Nets*. Artificial Intelligence. Volume 22. pp 235-267.

[Ballard, DH 84c] Ballard, Dana H. (1983). *Task Frames in Robot Manipulation*. National Conference on Artificial Intelligence (AAAI-84), 3rd, Austin, TX, August 6-10, 1984: Proceedings. pp 16-22. Washington, DC: American Association for Artificial Intelligence.

[Ballard, DH 85a] Ballard, Dana H.; Brown, Christopher M. (1985). *Vision*. BYTE. Volume 10. pp 245-261.

[Ballard, DH 86a] Ballard, Dana H. (1986). *Cortical Connections and Parallel Processing: Structure and Function*. The Behavioral and Brain Sciences. Volume 9. pp 67-120.

[Ballard, DH 86b] Ballard, Dana H. (1986). Parallel Logical Inference and Energy Minimization. Technical Report TR-142. Rochester, NY: University of Rochester.

[Ballard, DH 86c] Ballard, Dana H. (1986). Value Unit Interpolation: A Receptive Field Model. Technical Report TR-175. Rochester, NY: University of Rochester.

[Ballard, DH 87a] Ballard, Dana H. (1987). *Cortical Connections and Parallel Processing: Structure and Function*. In Arbib, Michael A. (ed.); Hanson, Allen R. (ed.). Vision, Brain and Cooperative Computation. pp 563-622. Cambridge, MA: MIT Press.

[Ballard, DH 87b] Ballard, Dana H. (1987). *Interpolation Coding: A Representation for Numbers in Neural Models*. Biological Cybernetics. Volume 57. pp 389-402.

[Ballard, DH 88*] SEE: Feldman, JA 88b. *Connectionist Models and Their Properties*.

[Baloh, RW 88a] Baloh, Robert W.; Beykirch, Karl; Honrubia, Vicente; Yee, Robert D. (1988). *Eye Movements Induced by Linear Acceleration on a Parallel Swing*. Journal of Neurophysiology. Volume 60.

[Bandara, KMSV 87a] Bandara, K. M. S. V.; Coon, D. D.; Karunasiri, R. P. G. (1987). *Interfacing Multispectral Sensors to*

Real Time Processors Based on Neural Network Models. Applied Physics Letters. Volume 51. pp 961-963.

[Bandara, KMSV 88*] SEE: Coon, DD 88a. *Infrared Transient Sensing.*

[Bandler, W 80a] Bandler, W.; Kohout, L. (1980). *Fuzzy Power Sets and Fuzzy Implication Operators*. Fuzzy Sets and Systems. Volume 4. pp 13-30.

[Banik, J 87*] SEE: Wong, Y 87a. *Neural Networks for Template Matching: Application to Real-Time Classification of the Action Potentials of Real Neurons.*

[Banik, J 88*] SEE: Bower, JM 88b. *Neural Networks for Template Matching: Application to Real-Time Classification of the Action Potentials of Real Neurons.*

[Bank, B 87*] SEE: Coulter, DA 87a. *Persistent Conditioning-Specific Changes of Post-Synaptic Membrane Currents in Heermissenda and Hippocampus.*

[Bank, B 88*] SEE: Alkon, DL 88b. *Regulation of Hermissenda K+ Channels by Cytoplasmic and Membrane-Associated C-Kinase.*

[Bank, B 88*] SEE: Naito, S 88a. *Transient and Persistent Depolarization-Induced Changes of Protein Phosphorylation in a Molluscan Nervous System.*

[Banquet, J 86a] Banquet, Jean-Paul; Grossberg, Stephen. (1986). *Structure of Event-Related Potentials During Learning: An Experimental and Theoretical Analysis*. Boston, MA: Boston University, Center for Adaptive Systems.

[Banquet, J 87a] Banquet, Jean-Paul; Grossberg, Stephen. (1987). *Probing Cognitive Processes Through the Structure of Event-Related Potentials During Learning: An Experimental and Theoretical Analysis*. Applied Optics. Volume 26. pp 4931-4946.

[Banzhaf, W 87a] Banzhaf, Wolfgang. (1987). *Spontaneous Pattern Classification by a Dynamical System*. Journal de Physique. Volume 48. pp 2027-2035.

[Banzhaf, W 87b] Banzhaf, Wolfgang. (1987). *Towards Continuous Models of Memory*. IEEE International Conference on Neural Networks (ICNN-87), San Diego, CA, June 21-24, 1987: Proceedings. Volume 2. pp 223-230. New York: Institute of Electrical and Electronics Engineers.

[Banzhaf, W 88a] Banzhaf, Wolfgang; Haken, Hermann. (1988). *A Network for Recognition and Classification of Continuous Patterns*. Neural Networks. Volume 1.

[Banzhaf, W 89a] Banzhaf, Wolfgang. (in press). *A Network of Multistate Units Capable of Associative Memory and Pattern Classification*. Physica D.

[Barbera, AJ 79a] Barbera, Anthony J.; Albus, James S.; Fitzgerald, M. L. (1979). *Hierarchical Control of Robots Using Microcomputers*. International Symposium on Industrial Robots, 9th, Washington, DC, March 13-15, 1979: Proceedings. pp 405-422. Dearborn, MI: Society of Manufacturing Engineers.

[Barbera, AJ 80*] SEE: Albus, JS 80a. *A Measurement and Control Model for Adaptive Robots.*

[Barbera, AJ 81*] SEE: Albus, JS 81b. *Theory and Practice of Hierarchical Control.*

[Barbera, AJ 85*] SEE: Albus, JS 85a. *Hierarchical Control for Robots, and Teleoperators.*

[Barhen, J 87*] SEE: Gulati, S 87a. *Nonlinear Neural Networks for Deterministic Scheduling.*

[Barhen, J 87a] Barhen, Jacob; Toomarian, N.; Protopopescu, Vladimir; Clinard, M. (1987). *Concurrent Neuromorphic Algorithms for Optimization of the Computational Load of a Hypercube Supercomputer*. IEEE International Conference on Neural Networks (ICNN-87), San Diego, CA, June 21-24, 1987: Proceedings. Volume 4. pp 687-696. New York: Institute of Electrical and Electronics Engineers.

[Barhen, J 87b] Barhen, Jacob; Toomarian, N.; Protopopescu, Vladimir. (1987). *Optimization of the Computational Load of a Hypercube Supercomputer Onboard a Mobile Robot.* Applied Optics. Volume 26. pp 5007-5014.

[Barhen, J 88*] SEE: Guez, A 88b. *On the Stability, Storage Capacity, and Design of Nonlinear Continuous Neural Networks.*

[Barhen, J 88b] Barhen, Jacob; Dress, William B.; Jorgensen, C. C. (1988). *Concurrent Neuromorphic Algorithms for Autonomous Robots.* Neural Computers. NATO ASI Series Volume F41.

[Barlow, HB 72a] Barlow, H. B. (1972). *Single Units and Sensation: A Neuron Doctrine for Perceptual Psychology?.* Perception. Volume 1. pp 371-394.

[Barlow, HB 82*] SEE: Sakitt, B 82a. *A Model for the Economical Encoding of the Visual Image in Cerebral Cortex.*

[Barnard, P 84*] SEE: Hammond, N 84a. *Evaluating the Interface of a Document Processor: A Comparison of Expert Judgement and User Observation.*

[Barnden, JA 84a] Barnden, John A. (1984). *On Short-Term Information-Processing in Connectionist Theories.* Cognition and Brain Theory. Volume 7. pp 25-59.

[Barnden, JA 86a] Barnden, John A. (1986). Complex Cognitive Information-Processing: A Computational Architecture with a Connectionist Implementation. Technical Report No. 211. Bloomington, IN: Indiana University, Computer Science Department.

[Barnden, JA 86b] Barnden, John A. (1986). *Connectionist Value Units: Some Concerns.* Behavioral and Brain Sciences. Volume 9.

[Barnden, JA 87a] Barnden, John A. (1987). *Chaos, Symbols, and Connectionism.* Behavioral and Brain Sciences. Volume 10. pp 174-175.

[Barnden, JA 87b] Barnden, John A. (1987). *Simulation of an Array-Based Neural Net Model.* Frontiers of Massively Parallel Scientific Computation: Proceedings. NASA Publication 2478. pp 59-68.

[Barnden, JA 87c] Barnden, John A. (1987). *The Centrality of Instantiations.* Behavioral and Brain Sciences. Volume 10. pp 437-438.

[Barnden, JA 88a] Barnden, John A. (1988). *Simulations of Composit, A Supra-Connectionist Architecture for Commonsense Reasoning.* IEEE Symposium on the Frontiers of Massively Parallel Computation, Fairfax, VA, October 10-12, 1988: Proceedings. New York: Institute of Electrical and Electronics Engineers.

[Barnden, JA 88b] Barnden, John A. (1988). *The Right of Free Association: Relative-Position Encoding for Connectionist Data Structures.* Conference of the Cognitive Science Society, 10th, August, 1988: Proceedings. Hillsdale, NJ: Lawrence Erlbaum.

[Barnes, DM 87a] Barnes, Deborah M. (1987). *Neural Models Yield Data on Learning.* Science. Volume 236. pp 1628-1629.

[Baron, RJ 87a] Baron, Robert J. (1987). The Cerebral Computer: An Introduction to the Computational Structure of the Human Brain. Hillsdale, NJ: Lawrence Erlbaum.

[Barr, AH 88*] SEE: Platt, JC 88b. Constrained Differential Optimization for Neural Networks.

[Barr, AH 88*] SEE: Platt, JC 88a. *Constrained Differential Optimization.*

[Barrett, HH 85*] SEE: Smith, WE 85a. *Image Reconstruction from Coded Data: I. Reconstruction Algorithms and Experimental*

Results.

[Barrioneuvo, G 88*] SEE: Krieger, D 88a. *A Neurobiologically Based Neural Net Model of the Hippocampus.*

[Barron, AR 84*] SEE: Barron, RL 84a. *Adaptive Learning Networks: Development and Application in the United States of Algorithms Related to GMDH.*

[Barron, AR 84a] Barron, Andrew R. (1984). *Predicted Squared Error: A Criterion for Automatic Model Selection.* In Farlow, Stanley J. (ed.). Self-Organizing Methods in Modeling: GMDH Type Algorithms. pp 87-104. New York: Marcel Dekker.

[Barron, RL 84a] Barron, Roger L.; Mucciardi, Anthony N.; Cook, Francis J.; Craig, Joseph Nelson; Barron, Andrew R. (1984). *Adaptive Learning Networks: Development and Application in the United States of Algorithms Related to GMDH.* In Farlow, Stanley J. (ed.). Self-Organizing Methods in Modeling, GMDH type Algorithms. pp 25-65. New York: Marcel Dekker.

[Barron, RL 87*] SEE: Shrier, S 87a. *Polynomial and Neural Networks: Analogies and Engineering Applications.*

[Barroso, F 73*] SEE: Kaufman, L 73a. *Stereopsis Without Image Segregation.*

[Barrow, HG 87a] Barrow, Harry G. (1987). *Learning Receptive Fields.* IEEE International Conference on Neural Networks (ICNN-87), San Diego, CA, June 21-24, 1987: Proceedings. Volume 4. pp 115-122. New York: Institute of Electrical and Electronics Engineers.

[Bartholdi, JJ 83a] Bartholdi, John J.; Platzman, Loren K.; Collins, R. Lee; Warden, William H. (1983). *A Minimal Technology Routing System for Meals on Wheels.* Interfaces. Volume 13. pp 1-8.

[Bartholdi, JJ 84a] Bartholdi, John J.; Platzman, Loren K. (1984). Heuristics Based on Space-Filling Curves for Combinatorial Problems in the Plane. PDRC Report Series 84-08. Atlanta, GA: Georgia Institute of Technology.

[Bartholdi, JJ 86a] Bartholdi, John J.; Platzman, Loren K. (June 1986). *Retrieval Strategies for a Carousel Conveyor.* IEEE Transactions. pp 166-173.

[Bartholdi, JJ 86b] Bartholdi, John J.; Trick, Michael A. (1986). *Stable Matching with Preferences Derived From a Psychological Model.* Operations Research Letters. Volume 5. pp 165-169.

[Barto, AG 76a] Barto, Andrew G. (1976). *A Neural Network Simulation Method Using the Fast Fourier Transform.* IEEE Transactions on Systems, Man and Cybernetics. Volume SMC-6. pp 863-867.

[Barto, AG 81*] SEE: Sutton, RS 81a. *An Adaptive Network that Constructs and Uses an Internal Model of its World.*

[Barto, AG 81*] SEE: Sutton, RS 81b. *Toward A Modern Theory of Adaptive Networks: Expectation and Prediction.*

[Barto, AG 81a] Barto, Andrew G.; Sutton, Richard S.; Brouwer, Peter S. (1981). *Associative Search Network: A Reinforcement Learning Associative Memory.* Biological Cybernetics. Volume 40. pp 201-211.

[Barto, AG 81b] Barto, Andrew G.; Sutton, Richard S. (1984). Goal Seeking Components for Adaptive Intelligence: An Initial Assessment. AFWAL-TR-81-1070, ADA-101476. Wright Patterson Air Force Base, OH: Wright Aeronautical Laboratories, Avionics Laboratory.

[Barto, AG 81c] Barto, Andrew G.; Sutton, Richard S. (1981). *Landmark Learning: An Illustration of Associative Search.* Biological Cybernetics. Volume 42. pp 1-8.

[Barto, AG 82a] Barto, Andrew G.; Sutton, Richard S.; Anderson, Charles W. (1982). Adaptive Neurons-Like Elements that Can Solve Difficult Learning Control Problems. COINS Technical Report 82-20. Amherst, MA: University of

Massachusetts.

[Barto, AG 82b] Barto, Andrew G.; Sutton, Richard S. (1982). *Simulation of Anticipatory Responses in Classical Conditioning by a Neuron-Like Adaptive Element*. Behavioural Brain Research. Volume 4. pp 221-235.

[Barto, AG 82c] Barto, Andrew G.; Anderson, Charles W.; Sutton, Richard S. (1982). *Synthesis of Non-Linear Control Surfaces by a Layered Associative Search Network*. Biological Cybernetics. Volume 43. pp 175-185.

[Barto, AG 83a] Barto, Andrew G.; Sutton, Richard S. (1983). Neural Problem Solving. COINS Technical Report 83-03. Amherst, MA: University of Massachusetts.

[Barto, AG 83b] Barto, Andrew G.; Sutton, Richard S.; Anderson, Charles W. (1988). *Neuron-like Adaptive Elements that Can Solve Difficult Learning Control Problems*. In Anderson, James A. (ed.); Rosenfeld, Edward (ed.). Neurocomputing: Foundations of Research. pp 537-550. Cambridge, MA: MIT Press.

[Barto, AG 83c] Barto, Andrew G.; Sutton, Richard S.; Anderson, Charles W. (1983). *Neuron-like Elements That Can Solve Difficult Learning and Control Problems*. IEEE Transactions on Systems, Man and Cybernetics. Volume SMC-13. pp 834-846.

[Barto, AG 84a] Barto, Andrew G.; Anandan, P. (1984). Pattern Recognizing Stochastic Learning Automata. Coins Technical Report 84-30. Amherst, MA: University of Massachusetts.

[Barto, AG 84b] Barto, Andrew G.; Sutton, Richard S. (1984). Simulation Experiments with Goal-Seeking Adaptive Elements. AFWAL-TR-84-1022, ADA-140295. Wright Patterson Air Force Base, OH: Wright Aeronautical Laboratories, Avionics Laboratory.

[Barto, AG 85*] SEE: Moore, JW 85a. *Connectionistic Learning in Real Time: Sutton-Barto Adaptive Element and Classical Conditioning of the Rabbit Nictitating Membrane Response*.

[Barto, AG 85*] SEE: Utgoff, PE 85a. Learning Efficient Recognizers for Analytically Derived Concepts.

[Barto, AG 85a] Barto, Andrew G. (1985). *Adaptive Neural Networks for Learning Control: Some Computational Experiments*. IEEE Workshop on Intelligent Control, New Orleans, LA, August 25, 1985: Proceedings. pp 170-175. Washington, DC: Institute of Electrical and Electronics Engineers.

[Barto, AG 85b] Barto, Andrew G. (1985). *Learning by Statistical Cooperation of Self-Interested Neuron-Like Computing Elements*. Human Neurobiology. Volume 4. pp 229-256.

[Barto, AG 85c] Barto, Andrew G. (1985). Learning by Statistical Cooperation of Self-Interested Neuron-Like Computing Elements. COINS Technical Report 85-11. Amherst, MA: University of Massachusetts.

[Barto, AG 85d] Barto, Andrew G.; Sutton, Richard S. (1985). *Neural Problem Solving*. In Levy, William B. (ed.); Anderson, James A. (ed.); Lehmkuhle, Stephen (ed.). Synaptic Modifications, Neuron Selectivity, and Nervous System Organization. pp 123-152. Hillsdale, NJ: Lawrence Erlbaum.

[Barto, AG 85e] Barto, Andrew G.; Anandan, P. (1985). *Pattern Recognizing Stochastic Learning Automata*. IEEE Transactions on Systems, Man and Cybernetics. Volume SMC-15. pp 360-375.

[Barto, AG 85f] Barto, Andrew G.; Anderson, Charles W. (1985). *Structural Learning in Connectionist Systems*. Cognitive Science Society, 7th, Irvine, CA, August 15-17, 1985: Proceedings. pp 43-53.

[Barto, AG 86a] Barto, Andrew G.; Anandan, P.; Anderson, Charles W.; Narendra, Kumpati S. (1986). *Cooperativity in Networks of Pattern Recognizing Stochastic Learning Automata*. Adpative and Learning Systems. pp 235-246. New York: Plenum.

[Barto, AG 86b] Barto, Andrew G. (1986). *Game-Theoretic Cooperativity in Networks of Self-Interested Units*. In Denker, John

S. (ed.). Neural Networks for Computing, Snowbird, Utah, 1986. AIP Proceedings 151. pp 41-46. New York: American Institute of Physics.

[Barto, AG 87*] SEE: Sutton, RS 87a. *A Temporal-Difference Model of Classical Conditioning.*

[Barto, AG 87a] Barto, Andrew G. (1987). *An Approach to Learning Control Surfaces by Connectionist Systems.* In Arbib, Michael A. (ed.); Hanson, Allen R. (ed.). Vision, Brain and Cooperative Computation. pp 665-702. Cambridge, MA: MIT Press.

[Barto, AG 87b] Barto, Andrew G.; Jordan, Michael I. (1987). *Gradient Following Without Back-Propagation in Layered Networks.* IEEE International Conference on Neural Networks (ICNN-87), San Diego, CA, June 21-24, 1987: Proceedings. Volume 2. pp 629-636. New York: Institute of Electrical and Electronics Engineers.

[Barto, AG 88a] Barto, Andrew G. (1988). From Chemotaxis to Cooperativity: Abstract Exercises in Neuronal Learning Strategies. COINS Technical Report 88-65. Amherst, MA: University of Massachusetts, Computer and Information Science.

[Baskett, F 75*] SEE: Friedman, JH 75a. *An Algorithm for Finding Nearest Neighbors.*

[Basti, G 86a] Basti, Gianfranco. (1986). *L'Istanza Analogica Nelle Scienze Del Cervello Considerazioni Filosofiche.* Epistemologia. Volume IX. pp 233-264.

[Basti, G 86b] Basti, Gianfranco. (1986). *L'Uso di Modelli Computazionali in Psicofisiologia Ed Il Principio Gnoseologico di Rappresentazione.* Epistemologia. Volume IX. pp 123-150.

[Basti, G 87a] Basti, Gianfranco. (1987). *Cognitive Neuroscience, Intentionality and Realism: The Philosophical Importance of the Connectionist Approach.* IEEE International Conference on Neural Networks (ICNN-87), San Diego, CA, June 21-24, 1987: Proceedings. Volume 4. pp 795-818. New York: Institute of Electrical and Electronics Engineers.

[Batali, J 81a] Batali, John. (1981). A Vision Chip. AI Memos 869. Cambridge, MA: Massachusetts Institute of Technology.

[Batalina, MA 80a] Batalina, M. A.; Burlakov, I. D.; Vlasov, A. N.; Stafeev, V. I. (1980). *Neuristor Line Made of S-Type Germanium Diodes.* Soviet Physics Semiconductors. Volume 14. pp 1251-1254.

[Baudrenghien, P 87*] SEE: Widrow, B 87b. *Fundamental Relations Between the LMS Algorithm and the DFT.*

[Baudry, M 84*] SEE: Lynch, G 84a. *The Biochemistry of Memory: A New and Specific Hypothesis.*

[Baudry, M 89*] SEE: Lynch, G 89a. *Cortical Encoding of Memory: Hypotheses Derived from Analysis and Simulation of Physiological Learning Rules in Anatomical Structures.*

[Baum, EB 86*] SEE: Baldi, P 86a. *Bounds on the Size of Ultrametric Structures.*

[Baum, EB 86*] SEE: Baldi, P 86b. *Caging and Exhibiting Ultrametric Structures.*

[Baum, EB 86a] Baum, Eric B. (1986). *Generalizing Back Propagation to Computation.* In Denker, John S. (ed.). Neural Networks for Computing, Snowbird, Utah, 1986. AIP Proceedings 151. pp 47-52. New York: American Institute of Physics.

[Baum, EB 86b] Baum, Eric B. (1986). *Towards Practical 'Neural' Computation for Combinatorial Optimization Problems.* In Denker, John S. (ed.). Neural Networks for Computing, Snowbird, Utah, 1986. AIP Proceedings 151. pp 53-58. New York: American Institute of Physics.

[Baum, EB 87a] Baum, Eric B.; Moody, John; Wilczek, Frank. (1987). *New "Neural" Algorithms for Associative Memory.*

IEEE International Conference on Neural Networks (ICNN-87), San Diego, CA, June 21-24, 1987: Proceedings. Volume 2. pp 505-506. New York: Institute of Electrical and Electronics Engineers.

[Baum, EB 88a] Baum, Eric B.; Moody, John; Wilczek, Frank. (1988). *Internal Representations for Associative Memory.* Biological Cybernetics. Volume 59. pp 217-228.

[Baum, EB 88b] Baum, Eric B. (1988). *Neural Nets for Economists.* The Economy as an Evolving Complex System, SFI Studies in the Sciences of Complexity. pp 33-48. Reading, MA: Addison-Wesley.

[Baum, EB 88c] Baum, Eric B. (1988). *On the Capabilities of Multilayer Perceptrons.* Journal of Complexity. Volume 4. pp 193-215.

[Baum, EB 88d] Baum, Eric B.; Wilczek, Frank. (1988). *Supervised Learning of Probability Distributions by Neural Networks.* In Anderson, Dana Z. (ed.). Neural Information Processing Systems. pp 52-61. New York: American Institute of Physics.

[Baum, LE 70a] Baum, Leonard E.; Petrie, Ted; Soules, George; Weiss, Norman. (1970). *A Maximization Technique Occurring in the Statistical Analysis of Probabilistic Functions of Markov Chains.* Annals of Mathematical Statistics. Volume 41. pp 164-171.

[Baumal, R 64*] SEE: Tulving, E 64a. *Interaction of Two Sources of Information in Tachistoscopic Word Recognition.*

[Bavan, S 88*] SEE: Pacheco, M 88a. *A Simple VLSI Architecture for Neurocomputing.*

[Baxter, RA 87*] SEE: Widrow, B 87c. *Learning Phenomena in Layered Neural Networks.*

[Baxter, RA 88*] SEE: Widrow, B 88a. *Layered Neural Nets for Pattern Recognition.*

[Baxter, W 88a] Baxter, William; Dow, Bruce. (1988). *Centric Models of the Orientation Map in Primary Visual Cortex.* In Anderson, Dana Z. (ed.). Neural Information Processing Systems. pp 62-71. New York: American Institute of Physics.

[Bayley, JS 87a] Bayley, J. S.; Fiddy, M. A. (1987). *On the Use of the Hopfield Model for Optical Pattern Recognition.* Optics Communications. Volume 64. pp 105-110.

[Bayley, JS 88*] SEE: Abbiss, JB 88a. *Super-Resolution and Neural Computing.*

[Bear, MF 87a] Bear, Mark F.; Cooper, Leon N.; Ebner, Ford F. (1987). *A Physiological Basis for a Theory of Synapse Modification.* Science. Volume 237. pp 42-48.

[Beaver, G 86*] SEE: Hammerstrom, D 86e. The OGC Cognitive Architecture Project: Silicon Implementation of Connectionist/Neural Networks.

[Beck, J 66a] Beck, Jacob. (1966). *Effect of Orientation and of Shape Similarity on Perceptual Grouping.* Perception and Psychophysics. Volume 1. pp 300-302.

[Beck, J 66b] Beck, Jacob. (1966). *Perceptual Grouping Produced by Changes in Orientation and Shape.* Science. Volume 154. pp 538-540.

[Beck, J 71a] Beck, Jacob. (1971). *Surface Lightness and Cues for the Illumination.* American Journal of Psychology. Volume 84. pp 1-11.

[Beck, J 72a] Beck, Jacob. (1972). *Similarity Grouping and Peripheral Discriminability Under Uncertainty.* American Journal of Psychology. Volume 85. pp 1-19.

[Beck, J 81a] Beck, Jacob; Prazdny, Slava. (1981). *Highlights and the Perception of Glossiness.* Perception and

Psychophysics. Volume 30. pp 407-410.

[Beck, J 83a] Beck, Jacob; Prazdny, K.; Rosenfeld, Azriel. (1983). *A Theory of Textural Segmentation.* In Beck, Jacob (ed.); Hope, Barbara (ed.); Rosenfeld, Azriel (ed.). Human and Machine Vision. pp 1-38. New York: Academic Press.

[Beck, J 83b] Beck, Jacob. (1983). *Textural Segmentation, Second-Order Statistics, and Textural Elements.* Biological Cybernetics. Volume 48. pp 125-130.

[Beck, J 84a] Beck, Jacob; Prazdny, K.; Ivry, Richard. (1984). *The Perception of Transparency with Achromatic Colors.* Perception & Psychophysics. Volume 35. pp 407-422.

[Becker, CA 77a] Becker, Curtis A.; Killion, Thomas H. (1977). *Interaction of Visual and Cognitive Effects in Word Recognition.* Journal of Experimental Psychology: Human Perception and Performance. Volume 3. pp 389-401.

[Becker, G 88*] SEE: Bock, P 88a. *A Design Tool for Modeling Large Networks of Simple Parallel Processors.*

[Becker, LA 87a] Becker, Lee A.; Peng, Jing. (1987). *Network Processing of Hierarchical Knowledge for Classification and Diagnosis.* IEEE International Conference on Neural Networks (ICNN-87), San Diego, CA, June 21-24, 1987: Proceedings. Volume 2. pp 309-318. New York: Institute of Electrical and Electronics Engineers.

[Becker, LA 87b] Becker, Lee A.; Peng, Jing. (1987). *Using Activation Networks for Analogical Ordering of Consideration: One Method for Integrating Connectionist and Symbolic Processing.* IEEE International Conference on Neural Networks (ICNN-87), San Diego, CA, June 21-24, 1987: Proceedings. Volume 2. 367-372. New York: Institute of Electrical and Electronics Engineers.

[Becker, LA 88a] Becker, Lee A.; Immes, Luke. (1988). *Committing Facts to Long Term Memory.* Neural Networks. Volume 1.

[Beek, B 70*] SEE: Harth, EM 70a. *Brain Functions and Neural Dynamics.*

[Beek, B 70*] SEE: Anninos, PA 70a. *Dynamics of Neural Structures.*

[Beis, JS 88*] SEE: Sutton, JP 88a. *Hierarchical Model of Memory and Memory Loss.*

[Belew, RK 87a] Belew, Richard K. (1987). *Designing Appropriate Learning Rules for Connectionist Systems.* IEEE International Conference on Neural Networks (ICNN-87), San Diego, CA, June 21-24, 1987: Proceedings. Volume 2. pp 479-486. New York: Institute of Electrical and Electronics Engineers.

[Bell, T 88a] Bell, Tony. (1988). *Sequential Processing Using Attractor Transitions.* Proceedings of Connectionist Models Summer School.

[Bellman, RE 87a] Bellman, R. E.; Zadeh, Lofti A. (1987). *Decision-Making in a Fuzzy Environment.* In Yager, Ronald R. (ed.); Ovchinnikov, S. (ed.); Tong, R. M. (ed.); Nguyen, H. T. (ed.). Fuzzy Sets and Applications: Selected papers by L A Zadeh. pp 53-79. New York: Wiley.

[Benito, S 80a] Benito, S.; Ferrate, G. A.; Galvan, J. (1980). *Simulation and Evaluation of Ill Defined Neuronal Networks with Random Structure and Parameter Setting.* In Dekker, L. (ed.); Savastano, G. (ed.); Vansteenkiste, G. C. (ed.). IMACS Congress on Simulation of Systems, 9th, Sorrento, September 24-28, 1979: Proceedings. pp 1077-1086. Amsterdam: North-Holland.

[Bennett, A 88a] Bennett, Adam. (in press). *An Exactly Solvable Model for Competitive Networks.* Journal of Physics.

[Bennett, EL 76*] SEE: Rosenzweig, MR 76a. Neural Mechanisms of Learning and Memory.

[Benson, MW 86*] SEE: Hoffman, GW 86a. *Neurons with Hysteresis Form a Network that Can Learn without Any Changes in*

Synaptic Connection Strengths.

[Bentley, JL 77*] SEE: Friedman, JH 77a. *An Algorithm for Finding Best Matches in Logarithmic Expected Time.*

[Berg, G 87a] Berg, George. (1987). *A Parallel Natural Language Processing with Distributed Control.* Cognitive Science Society, 9th, Seattle, WA, July 16-18, 1987: Proceedings. pp 487-495. Hillsdale, NJ: Lawrence Erlbaum.

[Bergen, JR 79*] SEE: Wilson, HR 79a. *A Four Mechanism Model for Threshold Spatial Vision.*

[Bergen, JR 83a] Bergen, James R.; Julesz, Bela. (1983). *Parallel Versus Serial Processing in Rapid Pattern Discrimination.* Nature. Volume 303. pp 696-698.

[Berger, B 88*] SEE: Mann, J 88a. *A Self-Organizing Neural Net Chip.*

[Berger, R 87*] SEE: Raffel, JI 87a. *A Generic Architecture for Wafer-Scale Neuromorphic Systems.*

[Berger, T 88*] SEE: Krieger, D 88a. *A Neurobiologically Based Neural Net Model of the Hippocampus.*

[Berger, TW 88*] SEE: Sclabassi, RJ 88b. *Nonlinear Systems Analysis of the Hippocampal Perforant Path-Dentate Projection. I. Theoretical and Interpretational Considerations.*

[Berger, TW 88*] SEE: Sclabassi, RJ 88a. *A Systems Theoretic Approach to the Study of CNS Function.*

[Berger, TW 88a] Berger, Theodore W.; Eriksson, Jan L.; Ciarolla, David A.; Sclabassi, Robert J. (1988). *Nonlinear Systems Analysis of the Hippocampal Perforant Path-Dentate Projection. II. Effects of Random Impulse Train Stimulation.* Journal of Neurophysiology. Volume 60. pp 1077-1094.

[Berger, TW 88b] Berger, Theodore W.; Eriksson, Jan L.; Ciarolla, David A.; Sclabassi, Robert J. (1988). *Nonlinear Systems Analysis of the Hippocampal Perforant Path-Dentate Projection. III. Comparison of Random Train and Paired Impulse Stimulation.* Journal of Neurophysiology. Volume 60. pp 1095-1109.

[Bergeron, DW 87*] SEE: Brill, MH 87a. *Retinal Model with Adaptive Contrast Sensitivity and Resolution.*

[Bergholm, F 88*] SEE: Waxman, AM 88a. *Convected Activation Profiles and the Measurement of Visual Motion.*

[Bergman, A 86a] Bergman, Aviv; Kerszberg, Michel. (1986). *The Evolution of Data Processing Abilities in Competing Automata.* Computer Simulation in Brain Science, Copenhagen, August, 1986.

[Bergman, A 87a] Bergman, Aviv; Kerszberg, Michel. (1987). *Breeding Intelligent Automata.* IEEE International Conference on Neural Networks (ICNN-87), San Diego, CA, June 21-24, 1987: Proceedings. Volume 2. pp 63-70. New York: Institute of Electrical and Electronics Engineers.

[Bergstrom, SS 73a] Bergstrom, Sten Sture. (1973). *A Note on the Neural Unit Model for Contrast Phenomena.* Vision Research. Volume 13. pp 2087-2092.

[Berke, P 87a] Berke, Peter. (1987). *That Does Not Compute.* IEEE International Conference on Neural Networks (ICNN-87), San Diego, CA, June 21-24, 1987: Proceedings. Volume 4. pp 819-826. New York: Institute of Electrical and Electronics Engineers.

[Berliner, HJ 82a] Berliner, Hans J.; Ackley, David H. (1982). *The QBKG System: Generating Explanations from a Non-Discrete Knowledge Representation.* National Conference on Artificial Intelligence (AAAI-82), 1st, Pittsburgh, PA, August 18-20, 1982: Proceedings. pp 213-216. Washington, DC: American Association for Artificial Intelligence.

[Berlizot, BB 82*] SEE: McGill, KC 82a. *On the Nature and Elimination of Stimulus Artifact in Nerve Signals Evoked and*

Recorded Using Surface Electrodes.

[Berman, M 56a] Berman, Mones; Schoenfeld, Robert. (1956). *Invariants in Experimental Data on Linear Kinetics and the Formulation of Models.* Journal of Applied Physics. Volume 27. pp 1361-1370.

[Bernard, C 86*] SEE: Axelrad, H 86a. *Recurrent Collateral Inhibition Simulated in a Simple Neuronal Automata Assembly.*

[Bernard, C 87*] SEE: Axelrad, H 87a. *The Use of an Artificial Neural Network Model to Analyze the Information Transfer Properties of a Simplified Model of the Cerebellar Cortex.*

[Bernasconi, J 88a] Bernasconi, J. (1988). *Analysis and Comparison of Different Learning Algorithms for Pattern Association Problems.* In Anderson, Dana Z. (ed.). Neural Information Processing Systems. pp 72-81. New York: American Institute of Physics.

[Bernstein, J 81a] Bernstein, J. (1981). *Profiles: Marvin Minsky.* The New Yorker. pp 50-126.

[Beroule, D 87a] Beroule, Dominique. (1987). *Guided Propagation Inside a Topographic Memory.* IEEE International Conference on Neural Networks (ICNN-87), San Diego, CA, June 21-24, 1987: Proceedings. Volume 4. pp 469-476. New York: Institute of Electrical and Electronics Engineers.

[Beroule, D 87b] Beroule, Dominique. (1987). *The Adaptive, Dynamic, and Associative Memory Model: A Possible Future Tool for Vocal Human-Computer Communication.* The Structure of Multimodal Dialogue. pp 13.1-13.15. Amsterdam: North-Holland.

[Beroule, D 88a] Beroule, Dominique. (1988). *The Never-Ending Learning.* In Eckmiller, Rolf (ed.); von der Malsburg, Christoph (ed.). Neural Computers. NATO ASI Series, Volume F41. pp 219-230. Berlin: Springer-Verlag.

[Berthier, NE 85*] SEE: Moore, JW 85a. *Connectionistic Learning in Real Time: Sutton-Barto Adaptive Element and Classical Conditioning of the Rabbit Nictitating Membrane Response.*

[Berthier, NE 86*] SEE: Blazis, DEJ 86a. *Simulation of the Classically Conditioned Nictitating Membrane Response by a Neuron-Like Adaptive Element.*

[Berthoz, A 88*] SEE: Droulez, J 88a. *Spatial and Temporal Transformations in Visuo-Motor Coordination.*

[Besag, J 74a] Besag, Julian. (1974). *Spatial Interaction and the Statistical Analysis of Lattice Systems.* Journal of the Royal Statistical Society, Series B. Volume 36. pp 192-236.

[Besag, JE 72a] Besag, J. E. (1972). *Nearest-Neighbour Systems and the Auto-Logistic Model for Binary Data.* Journal of the Royal Statistical Society, Series B. Volume 34. pp 75-83.

[Betts, B 88a] Betts, Bill. (1988). *The T5 Base Modulator Hypothesis: A Dynamic Model of T5 Neuron Function in Toads.* In Ewert, J-P (ed.); Arbib, Michael A. (ed.). Visuomotor Coordination: Amphibians, Comparisons, Models, and Robots. New York: Plenum.

[Betts, B 88b] Betts, Bill. (1988). *The Toad Optic Tectum as a Recurrent On-Center Off-Surround Neural Net with Quenching Threshold.* IEEE International Conference on Neural Networks (ICNN-88), San Diego, CA, July 24-27, 1988: Proceedings. Volume 2. pp 47-54. New York: Institute of Electrical and Electronics Engineers.

[Beurle, RL 62a] Beurle, R. L. (1962). *Functional Organization in Random Networks.* Principles of Self-Organization. pp 291-314. New York: Pergamon.

[Beurle, RL 62b] Beurle, R. L. (1962). *Storage and Manipulation of Information in Random Networks.* Aspects of the Theory of Artificial Intelligence. pp 19-43. New York: Plenum.

[Bever, TG 88*] SEE: Lachter, J 88a. *The Relation Between Linguistic Structure and Associative Theories of Language*

Learning: A Constructive Critique of Some Connectionist Learning Models.

[Beykirch, K 88*] SEE: Baloh, RW 88a. *Eye Movements Induced by Linear Acceleration on a Parallel Swing.*

[Bezdek, JC 87*] SEE: Zhang, W 87a. *NPN Relations and Cognitive Map Development.*

[Bezdek, JC 88*] SEE: Lee, JSJ 88a. *A Computer Vision Architecture for Intelligence Control of Neural Recognition in Dynamic Environments.*

[Bezdek, JC 88*] SEE: Lee, JSJ 88b. *A Feature Projection Based Adaptive Pattern Recognition Network.*

[Bezdek, JC 88*] SEE: Zhang, W 88a. *On NPN Logic.*

[Bharucha, JJ 87a] Bharucha, Jamshed J. (1987). *MUSACT: A Connectionist Model of Musical Harmony.* Cognitive Science Society, 9th, Seattle, WA, July 16-18, 1987: Proceedings. pp 508-517. Hillsdale, NJ: Lawrence Erlbaum.

[Bialko, M 75*] SEE: Wilamowski, BM 75a. *Novel Inductorless Neuristor Line.*

[Bienenstock, EL 82a] Bienenstock, Elie L.; Cooper, Leon N.; Munro, Paul W. (1982). *A Theory for the Development of Neuron Selectivity: Orientation Selectivity and Binocular Interactions in the Visual Cortex.* Journal of Neuroscience. Volume 2. pp 32-48.

[Bienenstock, EL 82b] Bienenstock, Elie L.; Cooper, Leon N.; Munro, Paul W. (1988). *Theory for the Development of Neuron Selectivity: Orientation Specificity and Binocular Interaction in Visual Cortex.* In Anderson, James A. (ed.); Rosenfeld, Edward (ed.). Neurocomputing: Foundations of Research. pp 439-456. Cambridge, MA: MIT Press.

[Bienenstock, EL 86*] SEE: von der Malsburg, C 86b. *Statistical Coding and Short-Term Synaptic Plasticity: A Scheme for Knowledge Representation in the Brain.*

[Bienenstock, EL 86a] Bienenstock, Elie L. (ed.); Soulie, F. Fogelman (ed.); Weisbuch, Gerard (ed.). (1986). Disordered Systems and Biological Organization. New York: Springer-Verlag.

[Bienenstock, EL 87*] SEE: von der Malsburg, C 87a. *A Neural Network for the Retrieval of Superimposed Connection Patterns.*

[Bienenstock, EL 87a] Bienenstock, Elie L.; von der Malsburg, Christoph. (1987). *A Neural Network for Invariant Pattern Recognition.* Europhysics Letters. Volume 4. pp 121-126.

[Bienenstock, EL 88a] Bienenstock, Elie L. (1988). *Relational Models in Natural and Artificial Vision.* In Eckmiller, Rolf (ed.); von der Malsburg, Christoph (ed.). Neural Computers. NATO ASI Series, Volume F41. pp 61-70. Berlin: Springer-Verlag.

[Bilbro, GL 88a] Bilbro, G. L.; White, Mark; Snyder, Wesley. (1988). *Image Segmentation with Neurocomputers.* In Eckmiller, Rolf (ed.); von der Malsburg, Christoph (ed.). Neural Computers. NATO ASI Series, Volume F41. pp 71-79. Berlin: Springer-Verlag.

[Billings, SA 81*] SEE: Harris, CJ 81a. Self-Tuning and Adaptive Control: Theory and Applications.

[Bingham, G 86*] SEE: Iberall, T 86a. *Opposition Space as a Structuring Concept for the Analysis of Skilled Hand Movements.*

[Bischof, WF 88a] Bischof, Walter F.; Caelli, Terry M. (1988). *Parsing Scale-Space and Spatial Stability Analysis.* Computer Vision, Graphics, and Image Processing. Volume 42. pp 192-205.

[Bizzi, E 88*] SEE: Mussa-Ivaldi, FA 88c. *Theoretical and Experimental Perspectives on Arm Trajectory Formation: A*

Distributed Model of Motor Redundancy.

[Black, IB 87a] Black, Ira B.; Adler, Joshua E.; Dreyfus, Cheryl F.; Friedman, Wilma F.; LaGamma, Edmund F.; Roach, Arthur H. (1987). *Biochemistry of Information Storage in the Nervous System.* Science. Volume 236. pp 1263-1268.

[Blackburn, MR 87a] Blackburn, Michael R. (1987). *Entrainment in a Neural Network Simulation of Spontaneously Active Units with Accommodation.* IEEE International Conference on Neural Networks (ICNN-87), San Diego, CA, June 21-24, 1987: Proceedings. Volume 2. pp 153-160. New York: Institute of Electrical and Electronics Engineers.

[Blackburn, MR 88a] Blackburn, Michael R.; Nguyen, H. G. (1988). An Artificial Neural System for Autonomous Undersea Vehicles. Technical Document 1318. San Deigo, CA: Naval Ocean Systems Center.

[Blackman, D 87*] SEE: Mueller, P 87a. *Neural Computation of Visual Images.*

[Blackmore, WR 81a] Blackmore, W. R. (1981). *Human Software.* Behavior Research Methods & Instrumentation. Volume 13. pp 553-570.

[Blank, MA 80*] SEE: Foss, DJ 80a. *Identifying the Speech Codes.*

[Blaum, M 88*] SEE: Bruck, J 88d. *Neural Networks, Error-Correcting Codes and Polynomials Over the Binary n-Cube.*

[Blaydon, CC 70*] SEE: Kashyap, RL 70a. *Stochastic Approximation.*

[Blazis, DEJ 85*] SEE: Moore, JW 85a. *Connectionistic Learning in Real Time: Sutton-Barto Adaptive Element and Classical Conditioning of the Rabbit Nictitating Membrane Response.*

[Blazis, DEJ 86a] Blazis, Diane E. J.; Desmond, John E.; Moore, John W.; Berthier, Neil E. (1986). *Simulation of the Classically Conditioned Nictitating Membrane Response by a Neuron-Like Adaptive Element.* Cognitive Science Society, 8th, Amherst, MA, August 15-17, 1986:Proceedings. pp 176-186. Hillsdale, NJ: Lawrence Erlbaum.

[Bledsoe, WW 59a] Bledsoe, W. W.; Browning, Iben. (1959). *Pattern Recognition and Reading by Machine.* Eastern Joint Computer Conference, Boston, MA, December 1-3, 1959: Convention Record. Number 16. pp 225-232.

[Blelloch, G 88*] SEE: Rosenberg, CR 88a. *An Implementation of Network Learning on the Connection Machine.*

[Block, HD 61a] Block, H. D. (1961). *Analysis of Perceptrons.* Western Joint Computer Conference, Los Angeles, CA, May 9-11, 1961: Proceedings. pp 281-289.

[Block, HD 62a] Block, H. D.; Knight, Bruce W.; Rosenblatt, Frank. (1962). *Analysis of a Four-Layer Series-Coupled Perceptron, II.* Reviews of Modern Physics. Volume 34. pp 135-142.

[Block, HD 62b] Block, H. D. (1988). *The Perceptron: A Model for Brain Functioning. I.* In Anderson, James A. (ed.); Rosenfeld, Edward (ed.). Neurocomputing: Foundations of Research. pp 138-150. Cambridge, MA: MIT Press.

[Block, HD 64a] Block, H. D.; Nilsson, Nils J.; Duda, R. O. (1964). *Determination and Detection of Features in Patterns.* Computer and Information Sciences: Collected Papers on Learning, Adaptation, and Control in Information Systems. pp 75-110. Washington, DC: Spartan.

[Blom, J 87*] SEE: Toet, A 87a. *The Construction of a Simultaneous Functional Order in Nervous Systems I. Relevance of Signal Covariances and Signal Coincidences.*

[Blom, J 87*] SEE: Toet, A 87b. *The Construction of a Simultaneous Functional Order in Nervous Systems II. Computing Geometrical Structures.*

[Blom, J 87*] SEE: Toet, A 87c. *The Construction of a Simultaneous Functional Order in Nervous Systems III. The Influence* *of Environmental Constraints on the Resulting Functional Order.*

[Blomfield, S 74a] Blomfield, Stephen. (1974). *Arithmetical Operations Performed by Nerve Cells.* Brain Research. Volume 69. pp 115-124.

[Bloom, BH 70a] Bloom, B. H. (1970). *Space / Time Trade-Offs in Hash Coding with Allowable Errors.* Communications of the ACM. Volume 13. pp 422-426.

[Blum, J 69*] SEE: Kilmer, WL 69a. *A Model of the Vertebrate Central Command System.*

[Bobrow, DG 79*] SEE: Norman, DA 79a. *Descriptions: An Intermediate Stage in Memory Retrieval.*

[Bobrowski, L 78a] Bobrowski, Leon. (1978). *Learning Processes in Multi-Layer Threshold Nets.* Biological Cybernetics. Volume 31. pp 1-6.

[Bobrowski, L 80a] Bobrowski, Leon; Caianiello, E. R. (1980). *A Comparison of Two Unsupervised Learning Algorithms.* Biological Cybernetics. Volume 37. pp 1-7.

[Bobrowski, L 82a] Bobrowski, Leon. (1982). *Rules for Forming Receptive Fields of Formal Neurons During Unsupervised Learning Processes.* Biological Cybernetics. Volume 43. pp 23-28.

[Bock, P 88a] Bock, Peter; Becker, Glenn. (1988). *A Design Tool for Modeling Large Networks of Simple Parallel Processors.* Modeling and Simulation Annual Conference, 19th, Pittsburgh, PA, May 5-6, 1988: Proceedings. Volume 19. pp 1365-1370. Research Triangle Park, NC: Instrument Society of America.

[Bode, HR 86*] SEE: Koizumi, O 86a. *Plasticity in the Nervous System of Adult Hydra I. The Position-Dependent Expression of FMRFamide-Like Immunoreactivity.*

[Bode, HR 88*] SEE: Koizumi, O 88a. *Plasticity in the Nervous System of Adult Hydra II. Conversion of Ganglion Cells of the Body Column into Epidermal Sensory Cells of the Hypostome.*

[Bode, PM 88a] Bode, P. M.; Awad, T. A.; Koizumi, Osamu; Nakashima, Y.; Grimmelikhuijzen, C. J. P. (1988). *Develpment of the Two-Part Pattern During Regeneration of the Head in Hydra.* Development. Volume 102. pp 223-235.

[Boltyanskil, VG 62*] SEE: Pontryagin, LS 62a. The Mathematical Theory of Optimal Processes.

[Bonhoeffer, T 84*] SEE: Palm, G 84b. *Parallel Processing for Associative and Neuronal Networks.*

[Bonomi, E 86a] Bonomi, E.; Lutton, J. L. (1986). *Statistical Mechanics: A General Approach to Combinatorial Optimization.* In Bienenstock, Elie L. (ed.). Disordered Systems and Biological Organization. pp 295-300. New York: Springer-Verlag.

[Bookman, L 88a] Bookman, L.; Zhang, X. (1988). *A Data Flow Analysis of Connectionist Networks.* Neural Networks.

[Bookman, LA 87*] SEE: Chun, HW 87c. *Network Regions: Alternatives to the Winner-Take-All Structure.*

[Bookman, LA 88a] Bookman, Lawrence A. (1988). *A Connectionist Scheme for Modelling Context.* In Touretzky, David S. (ed.); Hinton, Geoffrey E. (ed.); Sejnowski, Terrence J. (ed.). Proceedings of the 1988 Connectionist Models Summer School. Los Altos, CA: Morgan Kaufman.

[Bookstein, A 80a] Bookstein, A. (1980). *Fuzzy Requests: An Approach to Weighted Boolean Searches.* Journal of the American Society for Information Science. Volume 31. pp 240-247.

[Borg-Graham, LJ 87a] Borg-Graham, Lyle J. (1987). *Modeling the Electrical Behavior of Cortical Neurons-Simulation of Hippocampal Pyramidal Cells.* Computer Simulations in Brain Science, Copenhagen, 1986: Proceedings. Cambridge:

Cambridge University Press.

[Borg-Graham, LJ 88a] Borg-Graham, Lyle J. (1988). *Simulations Suggest Information Processing Roles for the Diverse Currents in Hippocampal Neurons.* In Anderson, Dana Z. (ed.). Neural Information Processing Systems. pp 82-94. New York: American Institute of Physics.

[Borsellino, AI 68a] Borsellino, A. I.; Fuortes, M. G. F. (1968). *Interpretation of Responses of Visual Cells of Limulus.* Proceedings of the IEEE. Volume 56. pp 1024-1032.

[Borsellino, AI 87*] SEE: Aicardi, F 87a. *Statistical Properties of Flip-Flop Processes Associated to the Chaotic Behavior of Systems with Strange Attractors.*

[Bortz, AB 75a] Bortz, A. B.; Kalos, M. H.; Lebowitz, J. L. (1975). *A New Algorithm for Monte Carlo Simulation of Ising Spin Systems.* Journal of Computational Physics. Volume 17. pp 10-18.

[Bottini, S 87a] Bottini, S. (1987). *Noise-Like Coding in Associative Memories.* IEEE International Conference on Neural Networks (ICNN-87), San Diego, CA, June 21-24, 1987: Proceedings. Volume 3. pp 243-244. New York: Institute of Electrical and Electronics Engineers.

[Bou-Ghazale, S 88a] Bou-Ghazale, Silvio; Razavi, Hassan; Makki, Rafic Z. (1988). *Design and Implementation of an Associative Memory for Intensive Pattern Matching.* Modeling and Simulation Annual Conference, 19th, Pittsburgh, PA, May 5-6, 1988: Proceedings. Volume 19. pp 1119-1124. Research Triangle Park, NC: Instrument Society of America.

[Boules, AN 88*] SEE: Deshmukh, VD 88b. *Computer Simulation of Event Related Potentials During Voluntary Cognition.*

[Bouligand, Y 86a] Bouligand, Y. (1986). *Fibroblasts, Morphogenesis and Cellular Automata.* In Bienenstock, Elie L. (ed.). Disordered Systems and Biological Organization. pp 367-380. New York: Springer-Verlag.

[Bounds, DG 86a] Bounds, D. G. (1986). *Numerical Simulations of Boltzmann Machines.* In Denker, John S. (ed.). Neural Networks for Computing, Snowbird, Utah, 1986. AIP Proceedings 151. pp 59-64. New York: American Institute of Physics.

[Bourland, H 87a] Bourland, H.; Wellekens, C. J. (1987). *Multilayer Perceptrons and Automatic Speech Recognition.* IEEE International Conference on Neural Networks (ICNN-87), San Diego, CA, June 21-24, 1987: Proceedings. Volume 4. pp 407-416. New York: Institute of Electrical and Electronics Engineers.

[Bourland, H 88a] Bourland, H.; Wellekens, C. J. (1988). Links Between Markov Models and Multilayer Perceptrons. Manuscript M 263. Brussels: Philips Research Laboratory.

[Bourlard, H 88*] SEE: Wellekens, CJ 88b. Les Applications des Modeles Connexionistes a la Reconnaissance de la Parole.

[Bourne, JR 86*] SEE: Shiavi, RG 86a. *Methods of Biological Signal Processing.*

[Bower, GH 65*] SEE: Atkinson, RC 65a. An Introduction to Mathematical Learning Theory.

[Bower, GH 73*] SEE: Anderson, JR 73a. Human Associative Memory.

[Bower, GH 86*] SEE: McClelland, JL 86c. *Connectionist Models and Cognitive Science: Goals, Directions and Implications.*

[Bower, JM 87*] SEE: Wilson, MA 87a. *A Computer Simulation of Olfactory Cortex with Functional Implications for Storage and Retrieval of Olfactory Information.*

[Bower, JM 87*] SEE: Wong, Y 87a. *Neural Networks for Template Matching: Application to Real-Time Classification of the Action Potentials of Real Neurons.*

[Bower, JM 87*] SEE: Carman, GJ 87a. *Optimal Mapping of the Configuration Space of a Network of Neurons.*

[Bower, JM 87a] Bower, James M.; Nelson, M. E.; Wilson, Matthew A. (1987). Piriform (Olfactory) Cortex Model on the Hypercube. Report C(3)P-404. Pasadena, CA: California Institute of Technology.

[Bower, JM 88a] Bower, James M.; Wilson, Matthew A. (1988). *A Computer Simulation of Olfactory Cortex with Functional Implications for Storage and Retrieval of Olfactory Information.* In Anderson, Dana Z. (ed.). Neural Information Processing Systems. pp 114-126. New York: American Institute of Physics.

[Bower, JM 88b] Bower, James M.; Wong, Yiu-fai; Banik, Jashojiban. (1988). *Neural Networks for Template Matching: Application to Real-Time Classification of the Action Potentials of Real Neurons.* In Anderson, Dana Z. (ed.). Neural Information Processing Systems. pp 103-113. New York: American Institute of Physics.

[Bower, JM 88c] Bower, James M.; Atiya, Amir F. (1988). *Optimal Neural Spike Classification.* In Anderson, Dana Z. (ed.). Neural Information Processing Systems. pp 95-102. New York: American Institute of Physics.

[Box, G 76a] Box, G.; Jenkins, G. (1976). Time Series Analysis: Forecasting and Control. San Francisco, CA: Holden Day.

[Boyce, WE 81*] SEE: Brannan, JR 81a. *Spatially Localized Interactive Neural Populations-I, A Mathematical Model.*

[Boyd, GD 87a] Boyd, Gary D. (1987). *Optically Excited Synapse for Neural Networks.* Applied Optics. Volume 26. pp 2712-2719.

[Braae, M 79a] Braae, M.; Rutherford, D. A. (1979). *Theoretical and Linguistic Aspects of the Fuzzy Logic Controller.* Automatica. Volume 15. pp 553-577.

[Bracewell, R 78a] Bracewell, R. (1978). The Fourier Transform and its Applications. New York: McGraw-Hill.

[Brachman, RJ 85a] Brachman, Ronald J. (1985). *"I Lied About the Trees" Or, Defaults and Definitions in Knowledge Representation.* AI Magazine. Volume 6. pp 80-93.

[Brady, D 87*] SEE: Psaltis, D 87a. *Learning in Optical Neural Computers.*

[Brady, D 88*] SEE: Psaltis, D 88a. *Adaptive Optical Networks Using Photorefractive Crystals.*

[Brady, D 88*] SEE: Hsu, K 88a. *Experimental Demonstrations of Optical Neural Computers.*

[Brady, M 87a] Brady, Michael; Yuille, Alan L. (1987). *An Extremum Principle for Shape from Contour.* In Arbib, Michael A. (ed.); Hanson, Allen R. (ed.). Vision, Brain and Cooperative Computation. pp 329-360. Cambridge, MA: MIT Press.

[Brain, AE 61a] Brain, A. E. (1961). *The Simulation of Neural Elements by Electrical Networks Based on Multi-Aperture Magnetic Cores.* Proceedings of the IRE. Volume 49. pp 49-52.

[Braitenberg, V 89a] Braitenberg, Valentino. (1989). *Some Arguments for a Theory of Cell Assemblies in the Cerebral Cortex.* In Nadel, Lynn (ed.); Cooper, Lynn A. (ed.); Culicover, Peter (ed.); Harnish, R. Michael (ed.). Neural Connections, Mental Computation. pp 104-134. Cambridge, MA: MIT Press.

[Brames, BJ 88*] SEE: Abbiss, JB 88a. *Super-Resolution and Neural Computing.*

[Brandeis, D 87*] SEE: Lehmann, D 87b. *Human Brain EEG Fields: Micro-States and Their Functional Significance.*

[Brandeis, D 88*] SEE: Lehmann, D 88a. *Spontaneous and Information-Triggered Segments of Series of Human Brain Electric Field Maps.*

[Brannan, JR 81a] Brannan, James R.; Boyce, William E. (1981). *Spatially Localized Interactive Neural Populations-I, A*

Mathematical Model. Bulletin of Mathematical Biology. Volume 43. pp 427-446.

[Brannan, JR 82a] Brannan, James R. (1982). *Spatially Localized Interactive Neural Populations-II: Stability and Dynamics of Excitatory Sets.* Bulletin of Mathematical Biology. Volume 44. pp 43-56.

[Bras, H 87a] Bras, Helene; Destombes, J.; Gogan, Paul; Tyc-Dumont, Suzanne. (1987). *The Dendrites of Single Brain-Stem Motoneurons Intracellularly Labelled with Horseradish Peroxidase in the Cat: An Ultrastructural Analysis.* Neuroscience. Volume 22. pp 971-981.

[Bras, H 87b] Bras, Helene; Gogan, Paul; Tyc-Dumont, Suzanne. (1987). *The Dendrites of Single Brain-Stem Motoneurons Intracellularly Labelled with Horseradish Peroxidase in the Cat: Morphological and Electrical.* Neuroscience. Volume 22. pp 947-970.

[Bras, H 87c] Bras, Helene; Gogan, Paul; Tyc-Dumont, Suzanne. (1987). *The Mammalian Central Neurone is a Complex Computing Device.* IEEE International Conference on Neural Networks (ICNN-87), San Diego, CA, June 21-24, 1987: Proceedings. Volume 4. pp 123-126. New York: Institute of Electrical and Electronics Engineers.

[Braverman, D 62*] SEE: Abramson, N 62a. *Learning to Recognize Patterns in a Random Environment.*

[Braverman, D 62a] Braverman, David. (1962). *Learning Filters for Optimum Pattern Recognition.* IRE Transactions on Information Theory. Volume IT-8. pp 280-285.

[Breiman, L 86*] SEE: Grajski, KA 86a. *Classification of EEG Spatial Patterns with a Tree-Structured Methodology: CART.*

[Brice, R 87*] SEE: Bagherzadeh, N 87a. *On Parallel Execution of the Traveling Salesman Problem on a Neural Network Model.*

[Brill, DR 85*] SEE: Stork, DG 85a. Seeing the Light: Optics in Nature, Photography, Color, Vision, and Holography.

[Brill, MH 87a] Brill, Michael H.; Bergeron, Doreen W.; Stoner, William W. (1987). *Retinal Model with Adaptive Contrast Sensitivity and Resolution.* Applied Optics. Volume 26. pp 4993-4998.

[Brillinger, DR 79a] Brillinger, David R.; Segundo, Jose P. (1979). *Empirical Examination of the Threshold Model of Neuron Firing.* Biological Cybernetics. Volume 35. pp 213-220.

[Brindley, GS 69a] Brindley, G. S. (1969). *Nerve Net Models of Plausible Size that Perform Many Simple Learning Tasks.* Proceedings of the Royal Society of London, Series B. Volume 174. pp 173-191.

[Broadbent, D 85a] Broadbent, Donald. (1985). *A Question of Levels: Comment on McClelland and Rumelhart.* Journal of Experimental Psychology: General. Volume 114. pp 189-192.

[Broadwell, W 86*] SEE: Oyster, JM 86a. Associative Network Applications to Robot Vision.

[Brodie, SE 78a] Brodie, Scott E.; Knight, Bruce W.; Ratliff, Floyd. (1988). *The Response of the Limulus Retina to Moving Stimuli: A Prediction by Fourier Synthesis.* In Anderson, James A. (ed.); Rosenfeld, Edward (ed.). Neurocomputing: Foundations of Research. pp 330-346. Cambridge, MA: MIT Press.

[Brooks, HA 84a] Brooks, Hugh A.; Probert, Thomas H. (1984). *Let's Ask GMDH What Effect the Environment Has on Fisheries.* In Farlow, Stanley J. (ed.). Self-Organizing Methods in Modeling: GMDH Type Algorithms. pp 169-178. New York: Marcel Dekker.

[Brousil, JK 67a] Brousil, James K.; Smith, David R. (1967). *A Threshold Logic Network for Shape Invariance.* IEEE Transactions on Electronic Computers. Volume EC-16. pp 818-828.

[Brouwer, PS 81*] SEE: Barto, AG 81a. *Associative Search Network: A Reinforcement Learning Associative Memory.*

[Brown, C 86a] Brown, Chappell. (October 6, 1986). *B.U. Center Researching Neural Nets.* Electronic Engineering Times. p 53 f.

[Brown, C 86b] Brown, Chappell. (September 8, 1986). *CCDs, NMOS Team to Build Neural Net Models On Chip.* Electronic Engineering Times. p 33 f.

[Brown, C 86c] Brown, Chappell. (April 21, 1986). *First Commercial Nerve Nets Challenge Traditional AI.* Electronic Engineering Times. p 1 f.

[Brown, C 86d] Brown, Chappell. (April 7, 1986). *Hopfield's Nerve Nets Realize Biocomputing.* Electronic Engineering Times. pp 53 f.

[Brown, C 86e] Brown, Chappell. (April 7, 1986). *Neural Chips Emulate Brain Functions (ATT VLSI).* Electronic Engineering Times. pp 53 f.

[Brown, C 86f] Brown, Chappell. (April 28, 1986). *Neural Circuits Will Improve Vision, Speech Systems.* Electronic Engineering Times. p 71 f.

[Brown, C 86g] Brown, Chappell. (August 4, 1986). *Neural Networks Perfect For Wafer-Scale Integration?.* Electronic Engineering Times. p 51.

[Brown, C 86h] Brown, Chappell. (December 1, 1986). *Parallel Optical Computer Solves Multivariable Problems in Real-Time.* Electronic Engineering Times. pp 31-32.

[Brown, C 86i] Brown, Chappell. (November 24, 1986). *Parallel Optics: Solution to Von Neumann Bottleneck?.* Electronic Engineering Times. p 35.

[Brown, C 86j] Brown, Chappell. (April 21, 1986). *Parallel-Processors Bypass Von Neumann Bottleneck.* Electronic Engineering Times. p 58 f.

[Brown, C 86k] Brown, Chappell. (September 1, 1986). *Startup Nestor Readies Neural Network-Based Products.* Electronic Engineering Times. p 1.

[Brown, C 87a] Brown, Chappell. (January 12, 1987). *Brain Drain Siphons Neural Net Pioneers into Industry.* Electronic Engineering Times.

[Brown, C 87b] Brown, Chappell. (April 13, 1987). *Cell Automata: Key to Science Spreadsheet?.* Electronic Engineering Times.

[Brown, C 87c] Brown, Chappell; Johnson, R. Colin. (May 4, 1987). *Engineer A Black-Box Solution in the First "Real Cognizer" Contest.* Electronic Engineering Times. p 51f.

[Brown, C 87d] Brown, Chappell. (May 25, 1987). *Fastest Multipliers Zip Along at 210ps.* Electronic Engineering Times. p 37f.

[Brown, C 87e] Brown, Chappell; Johnson, R. Colin. (July 13, 1987). *In Back-to-Basics Movement, VLSI Expert Calls for More Simplicity in Computation.* Electronic Engineering Times. p 41f.

[Brown, C 87f] Brown, Chappell. (May 11, 1987). *Machine Vision Lurches Ahead.* Electronic Engineering Times. p 43f.

[Brown, C 87g] Brown, Chappell; Johnson, R. Colin. (July 13, 1987). *Neural Net Pioneer Grossberg Extends ART-2 Architecture into Analog Domain.* Electronic Engineering Times. p 41f.

[Brown, C 87h] Brown, Chappell. (February 23, 1987). *Neural Net Startup Takes Teaching Role.* Electronic Engineering Times.

[Brown, C 87i] Brown, Chappell. (January 5, 1987). *Neural Network Startups Backed by Venture Capital.* Electronic Engineering Times. 23.

[Brown, C 87j] Brown, Chappell; Johnson, R. Colin. (June 29, 1987). *Neural Networks Get a Conference to Call Their Own.* Electronic Engineering Times. pp 39f.

[Brown, C 87k] Brown, Chappell. (May 11, 1987). *Scientists Focus on "Early Vision" for Smart Sensors.* Electronic Engineering Times. p 43f.

[Brown, C 87l] Brown, Chappell; Johnson, R. Colin. (June 22, 1987). *Theories of Neural Nets Expose the Limits of AI.* Electronic Engineering Times. p 41.

[Brown, C 88*] SEE: Johnson, RC 88a. Cognizers: Neural Networks and Machines That Think.

[Brown, CM 82*] SEE: Ballard, DH 82a. Computer Vision.

[Brown, CM 85*] SEE: Ballard, DH 85a. *Vision.*

[Brown, NH 88a] Brown, Nathan H. (1988). *Neural Network Implementation Approaches for the Connection Machine.* In Anderson, Dana Z. (ed.). Neural Information Processing Systems. pp 127-136. New York: American Institute of Physics.

[Brown, PB 87a] Brown, Paul B.; Millecchia, Ronald; Stiney, Michael. (1987). *Analog Memory for Continuous-Voltage, Discrete-Time Implementations of Neural Networks.* IEEE International Conference on Neural Networks (ICNN-87), San Diego, CA, June 21-24, 1987: Proceedings. Volume 3. pp 523-530. New York: Institute of Electrical and Electronics Engineers.

[Brown, PF 83a] Brown, Peter F.; Lee, Chin-Hui; Spohrer, James C. (1983). *Bayesian Adaptation in Speech Recognition.* IEEE International Conference on Acoustics, Speech, and Signal Processing (ICASSP-83), April, 1983: Proceedings. pp 761-764.

[Brown, PJ 63*] SEE: Talbert, LR 63a. A Real-Time Adaptive Speech-Recognition System.

[Brown, RJ 87a] Brown, Robert Jay. (1987). *An Artificial Neural Network Experiment.* Dr. Dobb's Journal. Volume 12. pp 16-29.

[Browning, I 59*] SEE: Bledsoe, WW 59a. *Pattern Recognition and Reading by Machine.*

[Browning, I 64a] Browning, Iben. (1964). A Self-Organizing System Called 'Mickey Mouse'. Palo Alto, CA: Panoramic Research.

[Brownstein, M 87a] Brownstein, Mark. (June 29, 1987). *Developers Show Neural Networks: "Intelligent" Systems Learn, Reason Like Humans.* Info World. p 8.

[Bruce, AD 86a] Bruce, A. D.; Canning, Andrew; Forrest, Bruce M.; Gardner, E.; Wallace, D. J. (1986). *Learning and Memory Properties in Fully Connected Networks.* In Denker, John S. (ed.). Neural Networks for Computing, Snowbird, Utah, 1986. AIP Proceedings 151. pp 65-70. New York: American Institute of Physics.

[Bruce, AD 87a] Bruce, A. D.; Gardner, E. J.; Wallace, D. J. (1987). *Dynamics and Statistical Mechanics of the Hopfield Model.* Journal of Physics A, Mathematical and General. Volume 20. pp 2909-2934.

[Bruck, J 87a] Bruck, Jehoshua; Goodman, Joseph W. (1987). *A Generalized Convergence Theorem for Neural Networks and Its Applications in Combinatorial Optimization.* IEEE International Conference on Neural Networks (ICNN-87), San Diego, CA, June 21-24, 1987: Proceedings. Volume 3. pp 649-656. New York: Institute of Electrical and Electronics Engineers.

[Bruck, J 88a] Bruck, Jehoshua; Goodman, Joseph W. (1988). *A Generalized Convergence Theorem for Neural Networks.* IEEE

Transactions on Information Theory. Volume IT-34.

[Bruck, J 88b] Bruck, Jehoshua; Sanz, Jorge. (1988). *A Study on Neural Networks.* International Journal of Intelligent Systems. Volume 3. pp 59-75.

[Bruck, J 88c] Bruck, Jehoshua. (in press). *Harmonic Analysis of Polynomial Threshold Functions.* SIAM Journal on Discrete Mathematics.

[Bruck, J 88d] Bruck, Jehoshua; Blaum, Mario. ((to appear)). *Neural Networks, Error-Correcting Codes and Polynomials Over the Binary n-Cube.* IEEE Transactions on Information Theory.

[Bruck, J 88e] Bruck, Jehoshua; Goodman, Joseph W. (1988). *On the Power of Neural Networks for Solving Hard Problems.* In Anderson, Dana Z. (ed.). Neural Information Processing Systems. pp 137-143. New York: American Institute of Physics.

[Brucks, M 82*] SEE: Small, SL 82b. ISCON: An Interactive Simulator for Connectionist Networks.

[Bruckstein, AM 79a] Bruckstein, Alfred M.; Zeevi, Yehoshua Y. (1979). *Analysis of "Integrate-to-Threshold" Neural Coding Schemes.* Biological Cybernetics. Volume 34. pp 63-79.

[Bruckstein, AM 80*] SEE: Zeevi, YY 80a. *Modeling of Two-Phase Adaptation in Neuronal Encoders.*

[Bruckstein, AM 81*] SEE: Zeevi, YY 81a. *Adaptive Neural Encoder Model with Selfinhibition and Threshold Control.*

[Bruckstein, AM 81a] Bruckstein, Alfred M.; Zeevi, Yehoshua Y. (1981). *Stochastic Pulse Frequency Modulation Models of Neural Coding Processes.* International Conference on Cybernetics and Society, Atlanta, GA, October 26-28, 1981: Proceedings. pp 102-105. New York: Institute of Electrical and Electronics Engineers.

[Bruckstein, AM 85a] Bruckstein, Alfred M.; Zeevi, Yehoshua Y. (1985). *An Adaptive Stochastic Model for the Neural Coding Process.* IEEE Transactions on Systems, Man and Cybernetics. Volume SMC-15. pp 343-351.

[Bruner, JS 51a] Bruner, Jerome S.; Postman, Leo; Rodrigues, John. (1951). *Expectation and the Perception of Color.* American Journal of Psychology. Volume 64. pp 216-227.

[Bruner, JS 54*] SEE: Miller, GA 54a. *Familiarity of Letter Sequences and Tachistoscopic Identification.*

[Bry, K 78*] SEE: Haltsonen, S 78a. *Application of Novelty Filter to Segmentation of Speech.*

[Bry, K 78*] SEE: Jalanko, M 78a. *Application of Orthogonal Projection Principles to Simultaneous Phonemic Segmentation and Labeling of Continuous Speech.*

[Bry, K 79*] SEE: Kohonen, T 79a. *Spectral Classification of Phonemes by Learning Subspaces.*

[Buchanan, BG 75*] SEE: Shortliffe, EH 75a. *A Model of Inexact Reasoning in Medicine.*

[Buchanan, BG 81*] SEE: Dietterich, TD 81a. The Role of the Critic in Learning Systems.

[Buchanan, JT 88*] SEE: Grillner, S 88a. *Simulation of the Segmental Burst Generating Network for Locomotion in Lamprey.*

[Buell, DA 81a] Buell, D. A.; Kraft, D. H. (1981). *A Model for a Weighted Retrieval System.* Journal of the American Society for Information Science. Volume 32. pp 211-216.

[Buhmann, J 86a] Buhmann, Joachim; Schulten, Klaus. (1986). *A Physiological Neural Network as an Autoassociative Memory.* In Bienenstock, Elie L. (ed.). Disordered Systems and Biological Organization. pp 273-280. New York: Springer-Verlag.

[Buhmann, J 86b] Buhmann, Joachim; Schulten, Klaus. (1986). *Influence of Noise on the Behavior of an Autoassociative Neural Network*. In Denker, John S. (ed.). Neural Networks for Computing, Snowbird, Utah, 1986. AIP Proceedings 151. pp 71-76. New York: American Institute of Physics.

[Buhmann, J 86c] Buhmann, Joachim; Divko, R.; Ritter, Helge; Schulten, Klaus. (1986). *Physicists Explore Human and Artificial Intelligence*. In Clementi, E. (ed.); Chin, S. (ed.). Structure and Dynamics of Nucleic Acids, Proteins, and Membranes. pp 301-327. New York: Plenum.

[Buhmann, J 87a] Buhmann, Joachim; Schulten, Klaus. (1987). *Influence of Noise on the Function of a "Physiological" Neural Network*. Biological Cybernetics. Volume 56. pp 313-327.

[Buhmann, J 88a] Buhmann, Joachim; Schulten, Klaus. (1988). *Storing Sequences of Biased Patterns in Neural Networks with Stochastic Dynamics*. In Eckmiller, Rolf (ed.); von der Malsburg, Christoph (ed.). Neural Computers. NATO ASI Series, Volume F41. pp 231-242. Berlin: Springer-Verlag.

[Bullock, D 87a] Bullock, Daniel; Grossberg, Stephen. (1987). *A Neural Network Architecture for Automatic Trajectory Formation and Coordination of Multiple Effectors During Variable-Speed Arm Movements*. IEEE International Conference on Neural Networks (ICNN-87), San Diego, CA, June 21-24, 1987: Proceedings. Volume 4. pp 559-566. New York: Institute of Electrical and Electronics Engineers.

[Bullock, D 87b] Bullock, Daniel; Liederman, Jacqueline; Todorovic, Dejan. (1987). *Reconciling Stable Asymmetry with Recovery of Function: An Adaptive Systems Perspective on Functional Plasticity*. Child Development. Volume 58. pp 689-697.

[Bullock, D 88a] Bullock, Daniel. (1988). *How Neural Networks Factor Problems of Sensory-Motor Control*. American Control Conference,7th, Atlanta, GA, June 15-17, 1988: Proceedings. Volume 3. pp 2271-2275.

[Bullock, D 88b] Bullock, Daniel; Grossberg, Stephen. (1988). *Neural Dynamics of Planned Arm Movements: Emergent Invariants and Speed-Accuracy Properties During Trajectory Formation*. Psychological Review. Volume 95. pp 49-90.

[Bullock, D 88c] Bullock, Daniel; Grossberg, Stephen. (1988). *Neural Dynamics of Planned Arm Movements: Emergent Invariants and Speed-Accuracy Properties During Trajectory Formation*. Neural Networks and Natural Intelligence. pp 553-622. Cambridge, MA: MIT Press.

[Bullock, D 88d] Bullock, Daniel; Grossberg, Stephen. (1988). *The Vite Model: A Neural Command Circuit for Generating Arm and Articulator Trajectories*. Dynamic Patterns in Complex Systems. pp 305-326. Singapore: World Scientific.

[Bullock, TH 59a] Bullock, Theodore Holmes. (1959). *Neuron Doctrine and Electrophysiology*. Science. Volume 129. pp 997-1002.

[Buneman, OP 69*] SEE: Willshaw, DJ 69b. *Non-Holographic Associative Memory*.

[Buneman, OP 69*] SEE: Willshaw, DJ 69a. *Non-Holographic Associative Memory*.

[Buneman, OP 70*] SEE: Longuet-Higgins, HC 70a. *Theories of Associative Recall*.

[Buneman, OP 72*] SEE: Willshaw, DJ 72a. *Parallel and Serial Methods of Pattern Matching*.

[Bunge, E 77a] Bunge, Ernst. (1977). *Speaker Recognition by Computer*. Philips Technical Review. Volume 37. pp 207-219.

[Bunge, E 77b] Bunge, Ernst; Hofker, Ulrich; Dohne, Hans D.; Jesorsky, Peter; Kriener, Bernhard; Wesseling, Dieter. (1977). *The AUROS Project: Automatic Recognition of Speakers by Computers*. Frequenz. Volume 31. pp 345-351.

[Burattini, E 70*] SEE: Aiello, A 70a. *Synthesis of Reverberatory Neural Networks*.

[Burattini, E 72a] Burattini, E.; Liesis, V. (1972). *A Method of Analysis of the Models of Neural Nets.* Kybernetik. Volume 10. pp 38-44.

[Burger, B 87*] SEE: Levy, WB 87a. *Electrophysiological Observations Which Help Describe an Associative Synaptic Modification Rule.*

[Burke, B 87*] SEE: Freeman, WJ 87a. *Hardware Simulation of Brain Dynamics in Learning: The SPOCK.*

[Burke, B 88*] SEE: Freeman, WJ 88a. *Central Pattern Generating and Recognizing in Olfactory Bulb: A Correlation Learning Rule.*

[Burlakov, ID 80*] SEE: Batalina, MA 80a. *Neuristor Line Made of S-Type Germanium Diodes.*

[Burnod, Y 87*] SEE: Korn, H 87a. *Spontaneous Quantal Currents in a Central Neuron Match Predictions from Binomial Analysis of Evoked Responses.*

[Burnod, Y 88a] Burnod, Yves. (1988). An Adaptive Neural Network, The Cerebral Cortex. Biologie Theorique 3. Paris: Masson.

[Burr, DJ 83a] Burr, David J. (1983). *Designing a Handwriting Reader.* IEEE Transactions on Pattern Analysis and Machine Intelligence. Volume PAMI-5. pp 554-559.

[Burr, DJ 86a] Burr, David J. (1986). *A Neural Network Digit Recognizer.* IEEE International Conference on Systems, Man, and Cybernetics, Atlanta, GA, October 14-17, 1986: Proceedings. pp 1621-1625. New York: Institute of Electrical and Electronics Engineers.

[Burr, DJ 87a] Burr, David J. (1987). *Experiments with a Connectionist Text Reader.* IEEE International Conference on Neural Networks (ICNN-87), San Diego, CA, June 21-24, 1987: Proceedings. Volume 4. pp 717-725. New York: Institute of Electrical and Electronics Engineers.

[Burr, DJ 88*] SEE: Hanson, SJ 88a. *Minkowski-r Back-Propagation: Learning in Connectionist Models with Non-Euclidian Error Signals.*

[Burr, DJ 88a] Burr, David J. (1988). *An Improved Elastic Net Method for the Traveling Salesman Problem.* IEEE International Conference on Neural Networks (ICNN-88), San Diego, CA, July 24-27, 1988: Proceedings. New York: Institute of Electrical and Electronics Engineers.

[Burr, DJ 88b] Burr, David J. (1988). *Experiments on Neural Net Recognition of Spoken and Written Text.* IEEE Transactions on Acoustics, Speech, and Signal Processing. Volume 36. pp 1162-1168.

[Burr, DJ 88c] Burr, David J. (1988). *Speech Recognition Experiments with Perceptrons.* In Anderson, Dana Z. (ed.). Neural Information Processing Systems. pp 144-153. New York: American Institute of Physics.

[Burton, MJ 88a] Burton, Michael J. (1988). *A Comparison of Optical Character Recognition Algorithms.* Modeling and Simulation Conference, 19th, Pittsburgh, PA, May 5-6, 1988: Proceedings. Volume 19. pp 2245-2250. Research Triangle Park, NC: Instrument Society of America.

[Bush, RR 51a] Bush, Robert R.; Mosteller, Frederick. (1951). *A Mathematical Model for Simple Learning.* Psychological Review. Volume 58. pp 313-323.

[Bush, RR 51b] Bush, Robert R.; Mosteller, Frederick. (1963). *A Mathematical Model for Simple Learning.* In Luce, R. Duncan (ed.); Bush, Robert R. (ed.); Galanter, Eugene (ed.). Readings in Mathematical Psychology. Volume 1. pp 278-288. New York: Wiley.

[Bush, RR 51c] Bush, Robert R.; Mosteller, Frederick. (1951). *A Model for Stimulus Generalization and Discrimination.*

Psychological Review. Volume 58. pp 413-423.

[Bush, RR 51d] Bush, Robert R.; Mosteller, Frederick. (1963). *A Model for Stimulus Generalization and Discrimination.* In Luce, R. Duncan (ed.); Bush, Robert R. (ed.); Galanter, Eugene (ed.). Readings in Mathematical Psychology. Volume 1. pp 289-299. New York: Wiley.

[Bush, RR 55a] Bush, Robert R.; Mosteller, Frederick. (1955). Stochastic Models for Learning. New York: Wiley.

[Bush, RR 59a] Bush, Robert R. (ed.); Estes, W. K. (ed.). (1959). Studies in Mathematical Learning Theory. Stanford, CA: Stanford University Press.

[Butler, C 88*] SEE: Caudill, M 88a. *Comparison of Internal Processing of Different Network Paradigms.*

[Butler, ZF 88*] SEE: Murray, AF 88c. *Bit-Serial Neural Networks.*

[Butler, ZF 88*] SEE: Murray, AF 88e. *VLSI Neural Networks.*

[Butler, ZF 88a] Butler, Zoe F.; Murray, Alan F.; Smith, V. W. (1988). *VLSI Bit-Serial Neural Networks.* Workshop on AI, Oxford, July, 88.

[Butsch, LM 62*] SEE: Ostgaard, MA 62a. *Adaptive and Self-Organizing Flight Control Systems.*

[Caelli, TM 86*] SEE: Oguztoreli, NN 86a. *An Inverse Problem in Neural Processing.*

[Caelli, TM 87a] Caelli, Terry M.; Nagendran, Shyamala. (1987). *Fast Edge-Only Matching Techniques for Robot Pattern Recognition.* Computer Vision, Graphics, and Image Processing. Volume 39. pp 131-143.

[Caelli, TM 87b] Caelli, Terry M.; Rentschler, I.; Scheidler, W. (1987). *Visual Pattern Recognition in Humans.* Biological Cybernetics. Volume 57. pp 233-240.

[Caelli, TM 88*] SEE: Bischof, WF 88a. *Parsing Scale-Space and Spatial Stability Analysis.*

[Caelli, TM 88a] Caelli, Terry M. (1988). *An Adaptive Computational Model for Texture Segmentation.* IEEE Transactions on Systems, Man and Cybernetics. Volume 18. pp 9-17.

[Caianiello, ER 61a] Caianiello, E. R. (1961). *Outline of a Theory of Thought Processes and Thinking Machines.* Journal of Theoretical Biology. Volume 1. pp 204-235.

[Caianiello, ER 66a] Caianiello, E. R. (1966). *Decision Equation and Reverberations.* Kybernetik. Volume 5. pp 98-100.

[Caianiello, ER 67a] Caianiello, E. R.; De Luca, A.; Ricciardi, Luigi M. (1967). *Reverberations and Control of Neural Networks.* Kybernetik. Volume 4. pp 10-18.

[Caianiello, ER 70*] SEE: Aiello, A 70a. *Synthesis of Reverberatory Neural Networks.*

[Caianiello, ER 75a] Caianiello, E. R.; Grimson, W. E. L. (1975). *Synthesis of Boolean Nets and Time Behavior of a General Mathematical Neuron.* Biological Cybernetics. Volume 18. pp 111-117.

[Caianiello, ER 80*] SEE: Bobrowski, L 80a. *A Comparison of Two Unsupervised Learning Algorithms.*

[Caianiello, ER 88a] Caianiello, E. R.; Marinaro, M.; Tagliaferri, R. (1988). *The Inverse Problem for Linear Boolean Nets.* In Eckmiller, Rolf (ed.); von der Malsburg, Christoph (ed.). Neural Computers. NATO ASI Series, Volume F41. pp 243-250. Berlin: Springer-Verlag.

[Calvert, TW 74*] SEE: Young, TY 74a. Classification, Estimation, and Pattern Recognition.

[Calvin, WH 72a] Calvin, William H. (1972). *Synaptic Potential Summation and Repetitive Firing Mechanisms: Input-Output Theory for the Recruitment of Neurons into Epileptic Bursting Firing Patterns*. Brain Research. Volume 39. pp 71-94.

[Cameron, SH 60a] Cameron, S. H. (1960). An Estimate of the Complexity Requisite in a Universal Decision Network. Report 60-600. pp 197-212. Wright Patterson Air Force Base, OH: Wright Patterson Air Force Base.

[Campbell, FW 68a] Campbell, Fergus W. (1968). *The Human Eye as an Optical Filter*. Proceedings of the IEEE. Volume 56. pp 1009-1014.

[Canning, A 86*] SEE: Bruce, AD 86a. *Learning and Memory Properties in Fully Connected Networks*.

[Canning, A 88b] Canning, Andrew; Gardner, E. (in press). *Partially Connected Models of Neural Networks*. Journal of Physics A.

[Cannon, SC 83a] Cannon, Stephen C.; Robinson, David A.; Shamma, Shihab. (1983). *A Proposed Neural Network for the Integrator of the Oculomotor System*. Biological Cybernetics. Volume 49. pp 127-136.

[Cannon, SC 85a] Cannon, Stephen C.; Robinson, David A. (1985). *An Improved Neural-Network Model for the Neural Integrator of the Oculomotor System: More Realistic Neuron Behavior*. Biological Cybernetics. Volume 53. pp 93-108.

[Caplan, D 79*] SEE: Arbib, MA 79a. *Neuro-Linguistics Must be Computational*.

[Caplan, D 82*] SEE: Arbib, MA 82b. Neural Models of Language Processes.

[Capranica, RR 68*] SEE: Frishkopf, LS 68a. *Neural Coding in the Bullfrog's Auditory System: A Teleological Approach*.

[Caprio, J 88*] SEE: Kanwal, JS 88b. *Overlapping Taste and Tactile Maps of the Oropharynx in the Vagal Lobe of the Channel Catfish, Ictalurus punctatus*.

[Carbonell, JG 85a] Carbonell, Jaime G.; Tomita, Masaru. (1985). New Approaches to Machine Translation. CMU-CS-85-143. Pittsburgh, PA: Carnegie Mellon University, Computer Science Department.

[Carley, LR 88a] Carley, L. Richard. (1988). *Presynaptic Neural Information Processing*. In Anderson, Dana Z. (ed.). Neural Information Processing Systems. pp 154-163. New York: American Institute of Physics.

[Carman, GJ 87a] Carman, George J.; Rasnow, Brian; Bower, James M. (1987). *Optimal Mapping of the Configuration Space of a Network of Neurons*. IEEE Conference on Neural Information Processing, Natural and Synthetic, Denver, CO, November 8-12, 1987.

[Carman, GJ 87b] Carman, George J. (1987). *Reduction of Visual Processes to a Common Algorithm*. Society for Neuroscience Abstracts. Volume 13.

[Carman, GJ 88a] Carman, George J. (1988). *Reduction of Visual Processes to a Common Algorithm*. Conference on Neural Networks for Computation, Snowbird, Utah.

[Carnevali, P 87*] SEE: Patarnello, S 87a. *Learning Networks of Neurons with Boolean Logic*.

[Carnevali, P 87a] Carnevali, Paolo; Patarnello, S. (1987). *Exhaustive Thermodynamical Analysis of Boolean Learning Networks*. Europhysics Letters. Volume 4. pp 1199-1204.

[Carnevali, P 88*] SEE: Patarnello, S 88a. *Learning Networks of Neurons with Boolean Logic*.

[Carpenter, GA 76a] Carpenter, Gail A. (1976). *Nerve Impulse Equations*. Structural Stability, the Theory of Catastrophes, and Applications in the Sciences. Mathematics Volume 525. pp 58-76. New York: Springer-Verlag.

[Carpenter, GA 77a] Carpenter, Gail A. (1977). *A Geometric Approach to Singular Perturbation Problems with Applications to Nerve Impulse Equations*. Journal of Differential Equations. Volume 23. pp 335-367.

[Carpenter, GA 77b] Carpenter, Gail A. (1977). *Periodic Solutions of Nerve Impulse Equations*. Journal of Mathematical Analysis and Applications. Volume 58. pp 152-173.

[Carpenter, GA 78a] Carpenter, Gail A. (1978). *A Mathematical Analysis of Excitable Membrane Phenomena*. Progress in Cybernetics and Systems Research. Volume 3. pp 505-514.

[Carpenter, GA 81a] Carpenter, Gail A. (1981). *Normal and Abnormal Signal Patterns in Nerve Cells*. Mathematical Psychology and Psychophysiology. SIAM-AMS Proceedings Vol 13. pp 49-90. Providence, RI: American Mathematical Society.

[Carpenter, GA 83a] Carpenter, Gail A.; Grossberg, Stephen. (1983). *A Neural Theory of Circadian Rhythms: The Gated Pacemaker*. Biological Cybernetics. Volume 48. pp 35-59.

[Carpenter, GA 84a] Carpenter, Gail A.; Grossberg, Stephen. (1984). *A Neural Theory of Circadian Rhythms: Aschoff's Rule in Diurnal and Nocturnal Mammals*. American Journal of Physiology. Number 247. pp R1067-R1082.

[Carpenter, GA 84b] Carpenter, Gail A. (1984). *Gated Dipoles: Photoreceptors and Circadian Rhythms*. Behavioral Processes. Volume 9. pp 90-91.

[Carpenter, GA 85a] Carpenter, Gail A.; Grossberg, Stephen. (1985). *Category Learning and Adaptive Pattern Recognition: A Neural Network Model*. Proceedings of the Third Army Conference on Applied Mathematics and Computing. ARO Report 86-1. pp 37-56.

[Carpenter, GA 85b] Carpenter, Gail A.; Grossberg, Stephen. (1985). *Neural Dynamics of Adaptive Pattern Recognition: Priming, Search, Attention, and Category Formation*. Society for Neuroscience Abstracts. Volume 11. pp 1110 f.

[Carpenter, GA 85c] Carpenter, Gail A.; Grossberg, Stephen. (1985). *Neural Dynamics of Circadian Rhythms: The Mammalian Hypothalamic Pacemaker*. Mathematics and Computers in Biomedical Applications, A Publication of the IMACS Technical Committee TC-5 on "Modelling of BiomedicalSystems". pp 79-102. Amsterdam: Elsevier.

[Carpenter, GA 85d] Carpenter, Gail A. (1985). *The Non-Parametric Influence of Light on Mammalian Circadian Rhythms*. Ordinary and Partial Differential Equations Proceedings of the Eighth Conference held at Dundee, Scotland, June 25-29, 1984. Mathematics Volume 1151. pp 90-108. New York: Springer-Verlag.

[Carpenter, GA 86b] Carpenter, Gail A.; Grossberg, Stephen. (1986). *Absolutely Stable Learning of Recognition Codes by a Self-Organizing Neural Network*. In Denker, John S. (ed.). Neural Networks for Computing, Snowbird, Utah, 1986. AIP Proceedings 151. pp 77-85. New York: American Institute of Physics.

[Carpenter, GA 86c] Carpenter, Gail A.; Grossberg, Stephen. (1986). *Adaptation and Transmitter Gating in Vertebrate Photoreceptors*. In Grossberg, Stephen (ed.). Adaptive Brain. Volume II. pp 271-310. New York: Elsevier.

[Carpenter, GA 86d] Carpenter, Gail A.; Grossberg, Stephen. (1986). *Adaptive Resonance Theory: Stable Self-organization of Neural Recognition Codes in Response to Arbitrary Lists of Input Patterns*. Cognitive Science Society, 8th, Amherst, MA, August 15-17, 1986: Proceedings. pp 45-62. Hillsdale, NJ: Lawrence Erlbaum.

[Carpenter, GA 86e] Carpenter, Gail A.; Grossberg, Stephen. (1986). *Associative Learning, Adaptive Pattern Recognition, and Cooperative-Competitive Decision Making by Neural Networks*. Hybrid and Optical Computing. SPIE Volume 634. pp 218-247. Bellingham, WA: Society of Photo-Optical Instrumentation Engineers.

[Carpenter, GA 86f] Carpenter, Gail A. (1986). *Book Review*. Mathematical Biosciences. Volume 79. pp 231-233.

[Carpenter, GA 86g] Carpenter, Gail A.; Cohen, Michael A.; Grossberg, Stephen. (1986). Computing with Neural Networks: The Role of Symmetry. Boston, MA: Boston University, Center for Adaptive Systems.

[Carpenter, GA 86h] Carpenter, Gail A.; Grossberg, Stephen. (1986). *Discovering Order in Chaos: Stable Self-Organization of Neural Recognition Codes.* Perspectives in Biological Dynamics and Theoretical Medicine. Volume 441.

[Carpenter, GA 86i] Carpenter, Gail A.; Grossberg, Stephen. (1986). *Neural Dynamics of Category Learning and Recognition: Attention, Memory Consolidation, and Amnesia.* In Grossberg, Stephen (ed.). Adaptive Brain. Volume 1. pp 238-286. New York: Elsevier.

[Carpenter, GA 86j] Carpenter, Gail A.; Grossberg, Stephen. (1986). *Neural Dynamics of Category Learning and Recognition: Attention, Memory Consolidation, and Amnesia.* In Davis, J. (ed.); Newburgh, R. (ed.); Wegman, E. (ed.). Brain Structure, Learning, and Memory. AAAS Symposium Series.

[Carpenter, GA 86k] Carpenter, Gail A.; Grossberg, Stephen. (1986). *Neural Dynamics of Category Learning and Recognition: Structural Invariants, Evoked Potentials, and Reinforcement.* In Commons, Michael L. (ed.); Kosslyn, Stephen M. (ed.); Herrnstein, Richard J. (ed.). Pattern Recognition and Concepts in Animals, People and Machines: Qualitative Analyses of Behavior Series. Volume III. Hillsdale, NJ: Lawrence Erlbaum.

[Carpenter, GA 86l] Carpenter, Gail A.; Grossberg, Stephen. (1986). *Neural Theory of Circadian Rhythms: Aschoff's Rule in Diurnal and Nocturnal Mammals.* In Grossberg, Stephen (ed.). Adaptive Brain. Volume I. pp 352-385. New York: Elsevier.

[Carpenter, GA 86m] Carpenter, Gail A.; Grossberg, Stephen. (1986). *Neural Theory of Circadian Rhythms: Split Rhythms, After-Effects, and Motivational Interactions.* In Grossberg, Stephen (ed.). Adaptive Brain. Volume I. pp 386-447. New York: Elsevier.

[Carpenter, GA 86n] Carpenter, Gail A.; Grossberg, Stephen. (1986). *Neural Theory of Circadian Rhythms: The Gated Pacemaker.* In Grossberg, Stephen (ed.). Adaptive Brain. Volume I. pp 309-351. New York: Elsevier.

[Carpenter, GA 87a] Carpenter, Gail A.; Grossberg, Stephen. (1987). *A Massively Parallel Architecture for a Self-Organizing Neural Pattern Recognition Machine.* Computer Vision, Graphics, and Image Processing. Volume 37. pp 54-115.

[Carpenter, GA 87b] Carpenter, Gail A. (1987). *Adpative Resonance Networks.* IEEE International Conference on Neural Networks (ICNN-87), San Diego, CA, June 21-24, 1987: Proceedings. Tutorial Number 3. San Diego: Institute of Electrical and Electronics Engineers.

[Carpenter, GA 87c] Carpenter, Gail A.; Grossberg, Stephen. (1987). *ART-2: Self-Organization of Stable Category Recognition Codes for Analog Input Patterns.* IEEE International Conference on Neural Networks (ICNN-87), San Diego, CA, June 21-24, 1987: Proceedings. Volume 2. pp 727-736. New York: Institute of Electrical and Electronics Engineers.

[Carpenter, GA 87d] Carpenter, Gail A.; Grossberg, Stephen. (1987). *Invariant Pattern Recognition and Recall by an Attentive Self-Organizing ART Architecture in a Nonstationary World.* IEEE International Conference on Neural Networks (ICNN-87), San Diego, CA, June 21-24, 1987: Proceedings. Volume 2. pp 737-746. New York: Institute of Electrical and Electronics Engineers.

[Carpenter, GA 87e] Carpenter, Gail A.; Grossberg, Stephen. (1987). *Neural Networks: Introduction to the 1 December 1987 Issue of Applied Optics.* Applied Optics. Volume 26. pp 4909.

[Carpenter, GA 87f] Carpenter, Gail A.; Cohen, Michael A.; Grossberg, Stephen. (1987). *Technical Comments: Computing with Neural Networks.* Science. Volume 235. pp 1226-1227.

[Carpenter, GA 88a] Carpenter, Gail A.; Grossberg, Stephen. (1988). *The ART of Adaptive Pattern Recognition by a Self-Organizing Neural Network.* Computer. Volume 21. pp 77-88.

[Carr, TH 78a] Carr, Thomas H.; Davidson, Brian J.; Hawkins, Harold L. (1978). *Perceptual Flexibility in Word*

Recognition: Strategies Affect Orthographic Computation But Not Lexical Access. Journal of Experimental Psychology: Human Perception and Performance. Volume 4. pp 674-690.

[Casalino, G 88*] SEE: De Rossi, D 88a. *Skin & Muscles: The Basic Wet-Ware for Sensory-Motor Control.*

[Casasent, DP 76a] Casasent, David P.; Psaltis, Demetri. (1976). *Position, Rotation, and Scale Invariant Optical Correlation.* Applied Optics. Volume 15. pp 1795-1799.

[Casasent, DP 84a] Casasent, David P.; Fetterly, Donald. (1984). *Recent Advances in Optical Pattern Recognition.* Optical Computing. SPIE Volume 456. pp 105-115. Bellingham, WA: Society of Photo-Optical Instrumentation Engineers.

[Casasent, DP 85a] Casasent, David P. (1985). *Coherent Optical Pattern Recognition: A Review.* Optical Engineering. Volume 24. pp 26-32.

[Casasent, DP 86a] Casasent, David P. (1986). *A Connectionist Architecture for Matching 3-D Models to Moving Edge Features.* Intelligent Robots and Computer Vision. SPIE Volume 726. pp 240-245. Bellingham, WA: Society of Photo-Optical Instrumentation Engineers.

[Casasent, DP 86b] Casasent, David P. (1986). *Scene Analysis Research: Optical Pattern Recognition and Artificial Intelligence.* Hybrid and Optical Computing. SPIE Volume 634. Bellingham, WA: Society of Photo-Optical Instrumentation Engineers.

[Cater, JP 87a] Cater, John P. (1987). *Successfully Using Peak Learning Rates of 10 (and Greater) in Back-Propagation Networks with the Heuristic Learning Algorithm.* IEEE International Conference on Neural Networks (ICNN-87), San Diego, CA, June 21-24, 1987: Proceedings. Volume 2. pp 645-652. New York: Institute of Electrical and Electronics Engineers.

[Caudill, M 87a] Caudill, Maureen. (December, 1987). *Neural Networks Primer, Part I.* AI Expert. Volume 2. pp 46-52.

[Caudill, M 88a] Caudill, Maureen; Butler, Charles. (1988). *Comparison of Internal Processing of Different Network Paradigms.* Neural Networks. Volume 1.

[Caudill, M 88b] Caudill, Maureen. (February, 1988). *Neural Networks Primer, Part II.* AI Expert. Volume 3. pp 55-61.

[Caudill, M 88c] Caudill, Maureen. (June, 1988). *Neural Networks Primer, Part III.* AI Expert. Volume 3. pp 53-59.

[Caulfield, HJ 85*] SEE: Eichmann, G 85a. *Optical Learning (Inference) Machines.*

[Caulfield, HJ 87*] SEE: Szu, HH 87g. *Optical Expert Systems.*

[Caulfield, HJ 87a] Caulfield, H. John. (1987). *Parallel N**4 Weighted Optical Interconnections.* IEEE International Conference on Neural Networks (ICNN-87), San Diego, CA, June 21-24, 1987: Proceedings. Volume 3. pp 595-598. New York: Institute of Electrical and Electronics Engineers.

[Caulfield, HJ 88*] SEE: Kinser, JM 88a. *Design for a Massive All-Optical Bidirectional Associative Memory: the Big BAM.*

[Caulfield, HJ 88*] SEE: Fiesler, E 88a. Weight Discretization in Backward Error Propagation Neural Networks.

[Cavanagh, JP 72a] Cavanagh, J. P. (1972). Holographic Processes Realizable in the Neural Realm: Prediction of Short Term Memory Performance. PhD Dissertation. Pittsburgh, PA: Carnegie Mellon University.

[Cavedon, A 85*] SEE: Metelli, F 85a. *Balanced and Unbalanced, Complete and Partial Transparency.*

[Cederquist, J 80a] Cederquist, Jack; Lee, Sing H. (1980). *Coherent Optical Feedback for the Analog Solution of Partial*

Differential Equations. Journal of the Optical Society of America. Volume 70. pp 944-953.

[Ceperley, D 86a] Ceperley, David; Alder, Berni. (1986). *Quantum Monte Carlo*. Science. Volume 231. pp 555-560.

[Cervantes, JH 87a] Cervantes, James H.; Hildebrant, Richard R. (1987). *Comparison of Three Neuron-Based Computation Schemes*. IEEE International Conference on Neural Networks (ICNN-87), San Diego, CA, June 21-24, 1987: Proceedings. Volume 3. pp 657-672. New York: Institute of Electrical and Electronics Engineers.

[Cervantes-Perez, F 85a] Cervantes-Perez, Francisco. (1985). Modeling and Analysis of Neural Networks in the Visuomotor System of the Anuran Amphibian. COINS Technical Report 85-27. Amherst, MA: University of Massachusetts.

[Chacc, EG 86*] SEE: Cosnard, M 86a. *Dynamical Properties of an Automaton With Memory*.

[Chacc, EG 86a] Chacc, E. Goles. (1986). *Positive Automata Networks*. In Bienenstock, Elie L. (ed.). Disordered Systems and Biological Organization. pp 101-112. New York: Springer-Verlag.

[Chader, GJ 86*] SEE: Osborne, NN 86a. Progress in Retinal Research.

[Chamberlin, RV 84a] Chamberlin, R. V.; Mozurkewich, George; Orbach, R. (1984). *Time Decay of the Remanent Magnetization in Spin-Glasses*. Physical Review Letters. Volume 52. pp 867-870.

[Chambers, RA 68*] SEE: Michie, D 68a. *BOXES: An Experiment in Adaptive Control*.

[Chambers, RA 68*] SEE: Michie, D 68b. *Boxes as a Model of Pattern-Formation*.

[Chan, L 87a] Chan, Lai-Wan; Fallside, Frank. (1987). *An Adaptive Training Algorithm for Back Propagation Networks*. Computer Speech and Language. Volume 2. pp 205-218.

[Chan, L 88*] SEE: Fallside, F 88a. Connectionist Models and Geometric Reasoning.

[Chande, A 83a] Chande, A.; DeClaris, N.; Newcomb, R. W. (1983). *Neural-Type Robotics: An Overview*. Midwest Symposium on Circuits and Systems, 26th, Pueblo, CO, August, 1983: Proceedings. pp 68-69.

[Chande, A 85*] SEE: Clema, JK 85a. *Expert Systems and Machine Learning*.

[Chande, A 86*] SEE: Clema, JK 86a. *Knowledge Bases and Machine Learning*.

[Chandrasekaran, B 81a] Chandrasekaran, B. (1981). *Natural and Social System Metaphors for Distributed Problem Solving: Introduction to the Issue*. IEEE Transactions on Systems, Man and Cybernetics. Volume SMC-11. pp 1-5.

[Chang, C 74a] Chang, Chin-Liang. (1974). *Finding Prototypes for Nearest Neighbor Classifiers*. IEEE Transactions on Computers. Volume C-23. pp 1179-1184.

[Chang, E 87*] SEE: Gullichsen, E 87a. *Pattern Classification by Neural Network: An Experimental System for Icon Recognition*.

[Chang, H 80a] Chang, H.; Chen, T. C. (1980). *Fast and Compact Associative Memory Cell*. IBM Technical Disclosure Bulletin. Volume 23. pp 3021-3023.

[Changeux, J 85a] Changeux, Jean-Pierre. (1985). *Remarks About the "Singularity" of Nerve Cells and its Ontogenesis*. In Levy, William B. (ed.); Anderson, James A. (ed.); Lehmkuhle, Stephen (ed.). Synaptic Modifications, Neuron Selectivity, and Nervous System Organization. pp 89-104. Hillsdale, NJ: Lawrence Erlbaum.

[Char, JM 88a] Char, Jois Malathi; Cherkassky, Vladimir; Wechsler, Harry; Zimmerman, George Lee. (1988). *Distributed and Fault-Tolerant Computation for Retrieval Tasks Using Distributed Associative Memories*. IEEE Transactions on

Computers. Volume 37. pp 484-490.

[Charnes, A 63a] Charnes, Abraham. (1963). *On Some Fundamental Theorems of Perceptron Theory and Their Geometry.* American Mathematical Society Notices. Volume 10. pp 60.

[Charniak, E 83a] Charniak, Eugene. (1983). *Passing Markers: A Theory of Contextual Influence in Language Comprehension.* Cognitive Science. Volume 7. pp 171-190.

[Charniak, E 86a] Charniak, Eugene. (1986). *A Neat Theory of Marker Passing.* National Conference on Artificial Intelligence (AAAI-86), 5th, Philadelphia, PA, August 11-15, 1986: Proceedings. Volume 1. pp 584-588.

[Charniak, E 86b] Charniak, Eugene. (1987). *Connectionism and Explanation.* Theoretical Issues in Natural Language Processing-3 (TINLAP-3), New Mexico State University, January 7-9, 1987: Proceedings. New York: Association for Computing Machinery.

[Charniak, E 87a] Charniak, Eugene; Sanots, Eugene. (1987). *A Connectionist Context-Free Parser Which is not Context-Free, But Then it is not Really Connectionist Either.* Cognitive Science Society, 9th, Seattle, WA, July 16-18, 1987: Proceedings. pp 70-77. Hillsdale, NJ: Lawrence Erlbaum.

[Chauvet, G 86a] Chauvet, G. (1986). *Cerebellum Models: An Interpretation of Some Features.* In Bienenstock, Elie L. (ed.). Disordered Systems and Biological Organization. pp 399-402. New York: Springer-Verlag.

[Chay, TR 87a] Chay, Teresa Ree. (1987). *Efficient Method for Estimating Kinetic Constants from Single Channel Patch Clamp Recordings.* IEEE International Conference on Neural Networks (ICNN-87), San Diego, CA, June 21-24, 1987: Proceedings. Volume 4. pp 127-136. New York: Institute of Electrical and Electronics Engineers.

[Chay, TR 87b] Chay, Teresa Ree. (1987). *Efficient Method for Estimating Kinetic Constants from Single Channel Patch Clamp Recordings.* IEEE International Conference on Neural Networks (ICNN-87), San Diego, CA, June 21-24, 1987: Proceedings. Volume 3. pp 205-214. New York: Institute of Electrical and Electronics Engineers.

[Cheeseman, P 85a] Cheeseman, Peter. (1985). *In Defense of Probability.* International Joint Conference on Artificial Intelligence (IJCAI-85), 9th, Los Angeles, CA, August 18-23, 1985: Proceedings. Volume 2. pp 1002-1009. Los Altos, CA: Morgan Kaufmann.

[Chellappa, R 82a] Chellappa, Ramalingam; Kashyap, Rangasami L. (1982). *Digital Image Restoration Using Spatial Interaction Models.* IEEE Transactions on Acoustics, Speech, and Signal Processing. Volume ASSP-30. pp 461-472.

[Chellappa, R 83*] SEE: Kashyap, RL 83a. *Estimation and Choice of Neighbors in Spatial-Interaction Models of Images.*

[Chellappa, R 87*] SEE: Zhou, Y 87a. *A Novel Approach to Image Restoration Based on a Neural Network.*

[Chellappa, R 88*] SEE: Zhou, Y 88a. *Computation of Optical Flow Using a Neural Network.*

[Chellappa, R 88*] SEE: Zhou, Y 88b. *Image Restoration Using a Neural Network.*

[Chellappa, R 88*] SEE: Zhou, Y 88c. *Stereo Matching Using a Neural Network.*

[Chen, C 88*] SEE: Alkon, DL 88b. *Regulation of Hermissenda K+ Channels by Cytoplasmic and Membrane-Associated C-Kinase.*

[Chen, CH 86a] Chen, C. H. (1986). *Seismic and Underwater Acoustic Waveform Analysis.* In Young, Tzay Y. (ed.); Fu, King-Sun (ed.). Handbook of Pattern Recognition and Image Processing. pp 527-545. New York: Academic Press.

[Chen, CH 87*] SEE: Ersoy, OK 87a. *Information Processing and Storage Based on Transforms and Neural Networks.*

[Chen, HH 86*] SEE: Maxwell, T 86a. *Nonlinear Dynamics of Artificial Neural Systems.*

[Chen, HH 86a] Chen, H. H.; Lee, Y. C.; Sun, G. Z.; Lee, H. Y.; Maxwell, Tom. (1986). *High Order Correlation Model for Associative Memory*. In Denker, John S. (ed.). Neural Networks for Computing, Snowbird, Utah, 1986. AIP Proceedings 151. pp 86-99. New York: American Institute of Physics.

[Chen, HH 87*] SEE: Sun, GZ 87a. *Learning Stereopsis with Neural Networks.*

[Chen, HH 88*] SEE: Sun, GZ 88a. *A Novel Net that Learns Sequential Decision Process.*

[Chen, S 87a] Chen, Su-Shing. (1987). *Automated Reasoning on Neural Networks: A Probabilistic Approach.* IEEE International Conference on Neural Networks (ICNN-87), San Diego, CA, June 21-24, 1987: Proceedings. Volume 2. pp 373-378. New York: Institute of Electrical and Electronics Engineers.

[Chen, S 87b] Chen, Su-Shing. (1987). *Neural Networks and Computational Vision.* IEEE International Conference on Neural Networks (ICNN-87), San Diego, CA, June 21-24, 1987: Proceedings. Volume 4. pp 659-666. New York: Institute of Electrical and Electronics Engineers.

[Chen, S 88*] SEE: Zhang, W 88a. *On NPN Logic.*

[Chen, S 88a] Chen, S.; Xu, X.; Tsai, W. T.; Huang, Nai-Kuan. (1988). A Case Study of Solving Optimization Problems Using Neural Network. Minneapolis, MN: University of Minnesota, Computer Science and Electrical Engineering.

[Chen, T 88*] SEE: Takefuji, Y 88a. *Multinomial Conjunctoid Statistical Learning Machines.*

[Chen, TC 80*] SEE: Chang, H 80a. *Fast and Compact Associative Memory Cell.*

[Chen, V 88*] SEE: Klassen, M 88a. *Characteristics of the Functional Link Net: A Higher Order Delta Rule Net.*

[Cheng, R 87*] SEE: Kam, M 87a. *On the Design of a Content-Addressable Memory via Binary Neural Networks.*

[Cherkassky, V 88*] SEE: Char, JM 88a. *Distributed and Fault-Tolerant Computation for Retrieval Tasks Using Distributed Associative Memories.*

[Cherniak, C 88a] Cherniak, Christopher. (1988). *Undebuggability and Cognitive Science.* Communications of the ACM. Volume 31. pp 402-412.

[Chester, M 86a] Chester, M. (June 1986). *Fuzzy Logic's Soft Start Belies a Cutting Edge.* Electronic Products. pp 19-20.

[Cheung, JY 88a] Cheung, John Y.; Omidvar, Massoud. (1988). *Mathematical Analysis of Learning Behavior of Neuronal Models.* In Anderson, Dana Z. (ed.). Neural Information Processing Systems. pp 164-173. New York: American Institute of Physics.

[Cheung, KF 87*] SEE: Marks II, RJ 87a. *A Class of Continuous Level Neural Nets.*

[Cheung, KF 87*] SEE: Marks II, RJ 87c. *Composite Matched Filter Output Partitioning.*

[Cheung, KF 87a] Cheung, Kwan F.; Atlas, Les E.; Ritcey, James A.; Green, Charles A.; Marks II, Robert Jackson. (1987). *Conventional and Composite Matched Filters with Error Correction: A Comparison.* Applied Optics. Volume 26. pp 4235-4239.

[Cheung, KF 87b] Cheung, Kwan F.; Oh, Seho; Marks II, Robert Jackson; Atlas, Les E. (1987). *Neural Net Associative Memories Based on Convex Set Projections.* IEEE International Conference on Neural Networks (ICNN-87), San Diego, CA, June 21-24, 1987: Proceedings. Volume 3. pp 245-252. New York: Institute of Electrical and Electronics Engineers.

[Cheung, KF 87c] Cheung, Kwan F.; Atlas, Les E.; Marks II, Robert Jackson. (1987). *Synchronous vs Asynchronous Behavior of Hopfield's CAM Neural Net*. Applied Optics. Volume 26. pp 4808-4813.

[Cheung, KF 88*] SEE: Marks II, RJ 88a. *Architectures for a Continuous Level Neural Network Based on Alternating Orthogonal Projections*.

[Cheung, KF 88*] SEE: Marks II, RJ 88c. *Nonlinearity Requirements for Correlation Based Associative Memories*.

[Cheung, KF 88*] SEE: Marks II, RJ 88d. *Optical-Processor Architectures for Alternating-Projection Neural Networks*.

[Cheung, KF 88*] SEE: Marks II, RJ 88e. *Performance Analysis of Associative Memories with Nonlinearities in the Correlation Domain*.

[Chin, RT 86a] Chin, Roland T. (1986). *Algorithms and Techniques for Automated Visual Inspection*. In Young, Tzay Y. (ed.); Fu, King-Sun (ed.). Handbook of Pattern Recognition and Image Processing. pp 588-612. New York: Academic Press.

[Chiueh, T 88a] Chiueh, Tzi-Dar; Goodman, Rodney. (1988). *A Neural Network Classifier Based on Coding Theory*. In Anderson, Dana Z. (ed.). Neural Information Processing Systems. pp 174-183. New York: American Institute of Physics.

[Chiueh, T 88b] Chiueh, Tzi-Dar; Goodman, Rodney M. (1988). *High-Capacity Exponential Associative Memories*. IEEE International Conference on Neural Networks (ICNN-88), San Diego, CA, July 24-27, 1988: Proceedings. New York: Institute of Electrical and Electronics Engineers.

[Cho, YB 87*] SEE: Takefuji, Y 87a. *Error Correcting System Based on Neural Circuits*.

[Cho, YB 88*] SEE: Takefuji, Y 88a. *Multinomial Conjunctoid Statistical Learning Machines*.

[Choi, JJ 88*] SEE: Marks II, RJ 88c. *Nonlinearity Requirements for Correlation Based Associative Memories*.

[Choi, JJ 88*] SEE: Marks II, RJ 88e. *Performance Analysis of Associative Memories with Nonlinearities in the Correlation Domain*.

[Chong, MWH 88a] Chong, Michael W. H.; Fallside, Frank. (1988). Implementation of Neural Networks for Speech Recognition on a Transputer Array. CUED/F-INFENG/TR.8. Cambridge: Cambridge University, Engineering Department.

[Chopping, PT 68a] Chopping, P. T. (1968). *Holographic Model of Temporal Recall*. Nature. Volume 217. pp 781-782.

[Chou, PA 88a] Chou, Philip A. ((to appear)). *The Capacity of the Kanerva Associative Memory*. IEEE Transactions on Information Theory.

[Chou, PA 88b] Chou, Philip A. (1988). *The Capacity of the Kanerva Associative Memory is Exponential*. In Anderson, Dana Z. (ed.). Neural Information Processing Systems. pp 184-191. New York: American Institute of Physics.

[Choudry, A 88*] SEE: Fiesler, E 88a. Weight Discretization in Backward Error Propagation Neural Networks.

[Chover, J 88a] Chover, Joshua. (1988). *Phase Transitions in Neural Networks*. In Anderson, Dana Z. (ed.). Neural Information Processing Systems. pp 192-200. New York: American Institute of Physics.

[Chow, CK 66a] Chow, C. K. (1966). *A Class of Non-Linear Recognition Procedures*. IEEE Transactions on Systems, Sciences and Cybernetics. Volume SSC-2. pp 101-109.

[Chow, CK 66b] Chow, C. K.; Liu, C. N. (1966). *An Approach to Structure Adaptation in Pattern Recognition*. IEEE Transactions on Systems Sciences and Cybernetics. Volume SSC-2. pp 73-80.

[Chow, YL 84*] SEE: Schwartz, RM 84a. *Improved Hidden Markov Modeling of Phonemes for Continuous Speech Recognition*.

[Chow, YL 85*] SEE: Schwartz, RM 85a. *Context-Dependent Modeling for Acoustic-Phonetic Recognition of Continuous Speech.*

[Chow, YL 87*] SEE: Schwartz, RM 87a. *Rapid Speaker Adaptation Using a Probabilistic Spectral Mapping.*

[Chua, LO 87*] SEE: Kennedy, MP 87a. *Circuit Theoretic Solutions for Neural Networks - An Old Approach to a New Problem.*

[Chua, LO 87*] SEE: Kennedy, MP 87b. *Unifying the Tank and Hopfield Linear Programming Circuit and the Canonical Nonlinear Programming Circuit of Chua and Lin.*

[Chua, LO 88*] SEE: Kennedy, MP 88a. *Neural Networks for Nonlinear Programming.*

[Chua, LO 88a] Chua, L. O.; Lin, T. (1988). *A Neural Network Approach to Transform Image Coding.* International Journal of Circuit Theory and Applications. Volume 16. pp 317-324.

[Chua, LO 88b] Chua, Leon O.; Yang, Lin. (1988). *Cellular Neural Networks: Theory.* IEEE Transactions on Circuits and Systems. Volume CAS 35. pp 1257-1290.

[Chun, HW 85a] Chun, Hon Wai. (1985). Schemata in a Massively Parallel Model of Computation: A Non-Committing Approach to Schema Selection. PhD Dissertation Proposal. Urbana-Champaign, IL: University of Illinois.

[Chun, HW 86*] SEE: Wong, MK 86a. *Toward a Massively Parallel System for Word Recognition.*

[Chun, HW 86a] Chun, Hon Wai. (1986). *A Representation for Temporal Sequence and Duration in Massively Parallel Networks: Exploiting Link Interactions.* National Conference on Artificial Intelligence (AAAI-86), 5th, Philadelphia, PA, August 11-15, 1986: Proceedings. Volume 1. pp 372-376.

[Chun, HW 87*] SEE: Li, T 87a. *Massively Parallel Network-Based Natural Language Parsing System.*

[Chun, HW 87a] Chun, Hon Wai; Mimo, Alejandro. (1987). *A Massively Parallel Model of Schema Selection.* IEEE International Conference on Neural Networks (ICNN-87), San Diego, CA, June 21-24, 1987: Proceedings. Volume 2. pp 379-386. New York: Institute of Electrical and Electronics Engineers.

[Chun, HW 87b] Chun, Hon Wai; Mimo, Alejandro. (1987). *A Model of Schema Selection Using Marker Passing and Connectionist Spreading Activation.* Cognitive Science Society, 9th, Seattle, WA, July 16-18, 1987: Proceedings. pp 887-896. Hillsdale, NJ: Lawrence Erlbaum.

[Chun, HW 87c] Chun, Hon Wai; Bookman, Lawrence A.; Afshartous, Niki. (1987). *Network Regions: Alternatives to the Winner-Take-All Structure.* International Joint Conference on Artificial Intelligence (IJCAI-87), 10th, Milan, August 22-28, 1987: Proceedings.

[Churchland, PS 88*] SEE: Sejnowski, TJ 88a. *Computational Neuroscience.*

[Churchland, PS 89a] Churchland, Patricia Smith; Sejnowski, Terrence J. (1989). *Neural Representation and Neural Computation.* In Nadel, Lynn (ed.); Cooper, Lynn A. (ed.); Culicover, Peter (ed.); Harnish, R. Michael (ed.). Neural Connections, Mental Computation. pp 15-48. Cambridge, MA: MIT Press.

[Ciarolla, DA 88*] SEE: Berger, TW 88a. *Nonlinear Systems Analysis of the Hippocampal Perforant Path-Dentate Projection. II. Effects of Random Impulse Train Stimulation.*

[Ciarolla, DA 88*] SEE: Berger, TW 88b. *Nonlinear Systems Analysis of the Hippocampal Perforant Path-Dentate Projection. III. Comparison of Random Train and Paired Impulse Stimulation.*

[Clark, DW 88*] SEE: Hassoun, MH 88a. *An Adaptive Attentive Learning Algorithm for Single-Layer Neural Networks.*

[Clark, JW 84a] Clark, John W.; Winston, Jeffrey V.; Rafelski, Johann. (1984). *Self-Organization of Neural Networks.*

Physics Letters. Volume 102A. pp 207-211.

[Clark, LT 87a] Clark, L. T.; Grondin, R. O. (1987). *Comparison of a Pipelined "Best Match" Content Addressable Memory with Neural Networks*. IEEE International Conference on Neural Networks (ICNN-87), San Diego, CA, June 21-24, 1987: Proceedings. Volume 3. pp 411-418. New York: Institute of Electrical and Electronics Engineers.

[Clark, WA 54*] SEE: Farley, BG 54a. *Simulation of Self-Organizing Systems by Digital Computer*.

[Clark, WA 55a] Clark, W. A.; Farley, B. G. (1955). *Generalization of Pattern Recogntion in a Self-Organizing System*. Western Joint Computer Conference, 1955: Proceedings. pp 86-91. New York: Institute of Radio Engineers.

[Clark, WA 60*] SEE: Farley, BG 60a. *Activity in Networks of Neuron-Like Elements*.

[Cleary, JG 87a] Cleary, John G. (1987). *A Simple VLSI Connectionist Architecture*. IEEE International Conference on Neural Networks (ICNN-87), San Diego, CA, June 21-24, 1987: Proceedings. Volume 3. pp 419-426. New York: Institute of Electrical and Electronics Engineers.

[Cleeremans, A 88*] SEE: Servan-Schreiber, D 88a. Encoding Sequential Structure in Simple Recurrent Networks.

[Clema, JK 71a] Clema, Joe K.; Kirkham, John. (1971). *CONSIM (Conflict Simulator): Risk, Cost and Benefit in Political Simulations*. Conference of the Association for Computing Machinery, Chicago, IL, August 3-5, 1971: Proceedings. pp 226-235. New York: Association for Computing Machinery.

[Clema, JK 72a] Clema, Joe K.; Fynewever, Mark. (1972). *The Dynamic Re-Evaluation of Alternatives and the Emulation of Human Decision Making*. Conference of the Association for Computing Machinery, Boston, MA, August, 1972: Proceedings. pp 852-859. New York: Association for Computing Machinery.

[Clema, JK 85a] Clema, Joe K. (ed.); Werling, Richard; Chande, Alhad. (1985). *Expert Systems and Machine Learning*. Summer Computer Simulation Conference, Chicago, IL, July 22-24, 1985: Proceedings. pp 683-686. Amsterdam: North-Holland.

[Clema, JK 86a] Clema, Joe K.; Werling, Richard; Chande, Alhad. (1986). *Knowledge Bases and Machine Learning*. Applications of Artificial Intelligence in Engineering Problems: Proceedings of the 1st International Conference, Southampton University, U.K., April, 1986. Volume 1. pp 67-75. New York: Springer-Verlag.

[Clifford, SP 88a] Clifford, Sandra P.; Nasrabadi, Nasser M. (1988). *Integration of Stereo Vision and Optical Flow Using Markov Random Fields*. IEEE International Conference on Neural Networks (ICNN-88), San Diego, CA, July 24-27, 1988: Proceedings. Volume 1. pp 577-584. New York: Institute of Electrical and Electronics Engineers.

[Clinard, M 87*] SEE: Barhen, J 87a. *Concurrent Neuromorphic Algorithms for Optimization of the Computational Load of a Hypercube Supercomputer*.

[Coates, CL 67*] SEE: Lewis, PM 67a. Threshold Logic.

[Cochet, Y 88a] Cochet, Yves; Paget, Gerard. (1988). *Reseaux Neuronzux Pour Bases D'Images*. Neuro-Nimes'88, Nimes, November 15-17, 1988: Proceedings.

[Cochet, Y 88b] Cochet, Yves; Paget, Gerard. (1988). *ZZENN (Zig Zag Epigenetic Neural Networks), A New Approach to Connectionist Machine Learning*. International Computer Science Conference '88, Hong Kong, December 19-21, 1988: Proceedings.

[Cochet, Y 88c] Cochet, Yves; Paget, Gerard. (1988). *ZZENN: Zig Zag Epigenetic Neural Networks and Their Use in Information Systems*. ESPCI European Conference on Neural Networks (nEURO'88), 1st, Paris, June 6-9, 1988: Proceedings.

[Cohen, A 86a] Cohen, Arnon; Froind, Igal. (1986?). Software Package for Interactive Text Independent, Speaker

Verification. Negev: Ben-Gurion University.

[Cohen, B 85a] Cohen, Benjamin; Murphy, Gregory L. (1984). *Models of Concepts*. Cognitive Science. Volume 8. pp 27-58.

[Cohen, E 87*] SEE: Sterling, P 87a. *Microcircuitry of the On-Beta Ganglion Cell in Daylight, Twilight, and Starlight*.

[Cohen, JC 88a] Cohen, J. C.; Dunbar, K.; McClelland, James L. (1988). On the Control of Automatic Processes: A Parallel Distributed Processing Model of the Stroop Effect. Technical Report AIP-40. Pittsburgh, PA: Carnegie Mellon University, Computer Science Department.

[Cohen, L 88*] SEE: Spitzer, AR 88a. A Method for Determining Optimal Interelectrode Spacing for Cerebral Topographic Mapping.

[Cohen, MA 83a] Cohen, Michael A.; Grossberg, Stephen. (1986). *Absolute Stability of Global Pattern Formation and Parallel Memory Storage by Competitive Neural Networks*. In Grossberg, Stephen (ed.). Adaptive Brain, Volume I. pp 287-308. New York: Elsevier.

[Cohen, MA 83b] Cohen, Michael A.; Grossberg, Stephen. (1983). *Absolute Stability of Global Pattern Formation and Parallel Memory Storage by Competitive Neural Networks*. IEEE Transactions on Systems, Man and Cybernetics. Volume SMC-13. pp 815-826.

[Cohen, MA 84*] SEE: Grossberg, S 84b. *Some Global Properties of Binocular Resonances: Disparity Matching, Filling-In, and Figure-Ground Synthesis*.

[Cohen, MA 84a] Cohen, Michael A.; Grossberg, Stephen. (1984). *Neural Dynamics of Brightness Perception: Features, Boundaries, Diffusion, and Resonance*. Perception and Psychophysics. Volume 36. pp 428-456.

[Cohen, MA 84b] Cohen, Michael A.; Grossberg, Stephen. (1984). *Some Global Properties of Binocular Resonances: Disparity Matching, Filling-in, and Figure-Ground Synthesis*. Figural Synthesis. Hillsdale, NJ: Lawrence Erlbaum.

[Cohen, MA 85a] Cohen, Michael A.; Levine, Daniel S. (1985). *Unbounded Oscillatory Solutions for a System of Interactive Populations*. SIAM Journal of Applied Mathematics. Volume 45. pp 268-279.

[Cohen, MA 86*] SEE: Carpenter, GA 86g. Computing with Neural Networks: The Role of Symmetry.

[Cohen, MA 86a] Cohen, Michael A. (1986). *Coupled Mode Theory for Neural Networks*. In Denker, John S. (ed.). Neural Networks for Computing, Snowbird, Utah, 1986. AIP Proceedings 151. pp 100-109. New York: American Institute of Physics.

[Cohen, MA 86b] Cohen, Michael A.; Grossberg, Stephen. (1986). *Neural Dynamics of Brightness Perception: Features, Boundaries, Diffusion and Resonance*. In Grossberg, Stephen (ed.). Adaptive Brain, Volume II. pp 211-270. New York: Elsevier.

[Cohen, MA 86c] Cohen, Michael A.; Grossberg, Stephen. (1986). *Neural Dynamics of Speech and Language Coding: Developmental Programs, Perceptual Grouping, and Competition for Short-Term Memory*. In Grossberg, Stephen (ed.). Adaptive Brain, Volume II. pp 456-498. New York: Elsevier.

[Cohen, MA 86d] Cohen, Marcus A. (1986). *Self Organization, Association, and Categorization in a Phase Conjugating Resonator*. Optical Computing. SPIE Volume 625. pp 214-219. Bellingham, WA: Society of Photo-Optical Instrumentation Engineers.

[Cohen, MA 86e] Cohen, Michael A.; Grossberg, Stephen. (1986). *Unitized Recognition Codes for Parts and Wholes: The Unique Cue in Configural Discriminations*. In Commons, Michael L. (ed.); Kosslyn, Stephen M. (ed.); Herrnstein, Richard J. (ed.). Pattern Recognition and Concepts in Animals, People, and Machines. Hillsdale, NJ: Lawrence Erlbaum.

[Cohen, MA 87*] SEE: Carpenter, GA 87f. *Technical Comments: Computing with Neural Networks.*

[Cohen, MA 87a] Cohen, Michael A.; Grossberg, Stephen. (1987). *Masking Fields: A Massively Parallel Architecture for Learning, Recognizing, and Predicting Multiple Groupings of Patterned Data.* IEEE International Conference on Neural Networks (ICNN-87), San Diego, CA, June 21-24, 1987: Proceedings. Volume 2. pp 787-794. New York: Institute of Electrical and Electronics Engineers.

[Cohen, MA 87b] Cohen, Michael A.; Grossberg, Stephen. (1987). *Masking Fields: A Massively Parallel Neural Architecture for Learning, Recognizing, and Predicting Multiple Groupings of Patterned Data.* Applied Optics. Volume 26. pp 1866-1891.

[Cohen, MA 87c] Cohen, Michael A.; Grossberg, Stephen; Stork, David G. (1987). *Recent Developments in a Neural Model of Real-Time Speech Analysis and Synthesis.* IEEE International Conference on Neural Networks (ICNN-87), San Diego, CA, June 21-24, 1987: Proceedings. Volume 4. pp 443-454. New York: Institute of Electrical and Electronics Engineers.

[Cohen, MA 87d] Cohen, Michael A. (1987). *Sustained Oscillation in Symmetric Cooperative-Competitive Neural Networks.* IEEE International Conference on Neural Networks (ICNN-87), San Diego, CA, June 21-24, 1987: Proceedings. Volume 2. pp 127-130. New York: Institute of Electrical and Electronics Engineers.

[Cohen, MM 87*] SEE: Massaro, DW 87a. *Process and Connectionist Models of Pattern Recognition.*

[Cohen, MS 86a] Cohen, Marcus S. (1986). Volume Holographic Information Processing in a Phase-Conjugating Resonator. CRL Pre-print. Las Cruces, NM: New Mexico State University, Computing Research Laboratory.

[Cohen, MS 87a] Cohen, Marcus S.; Julian, William H. (1987). *Neural Networks with a Hopf Bifurcation: Slowly Modulated Waves.* IEEE International Conference on Neural Networks (ICNN-87), San Diego, CA, June 21-24, 1987: Proceedings. Volume 2. pp 161-168. New York: Institute of Electrical and Electronics Engineers.

[Cohen, NJ 86a] Cohen, Neal J.; Abrams, Irene T.; Harley, Walter S.; Tabor, Lisa; Sejnowski, Terrence J. (1986). *Skill Learning and Repetition Priming in Symmetry Detection: Parallel Studies of Human Subjects and Connectionist Models.* Cognitive Science Society, 8th, Amherst, MA, August 15-17, 1986: Proceedings. pp 23-44. Hillsdale, NJ: Lawrence Erlbaum.

[Cohen, PR 85*] SEE: Sullivan, M 85a. *An Endorsement-Based Plan Recognition Program.*

[Cole, KS 55a] Cole, K. S.; Antosiewicz, H. A.; Rabinowitz, P. (1955). *Automatic Computation of Nerve Excitation.* Journal of the Society for Industrial and Applied Mathematics. Volume 3. pp 153-172.

[Cole, M 65a] Cole, Michael. (1965). *Search Behavior: A Correction Procedure for Three-Choice Probability Learning.* Journal of Mathematical Psychology. Volume 2. pp 145-170.

[Cole, RA 73a] Cole, Richard A. (1973). *Listening for Mispronunciations: A Measure of What We Hear During Speech.* Perception and Psychophysics. Volume 13. pp 153-156.

[Coleman, BD 71a] Coleman, Bernard D. (1971). *A Mathematical Theory of Lateral Sensory Inhibition.* Archive for Rational Mechanics and Analysis. Volume 43. pp 79-100. New York: Springer-Verlag.

[Collier, RJ 66a] Collier, Robert J. (1966). *Some Current Views on Holography.* IEEE Spectrum. Volume 3. pp 67-74.

[Collins, AM 69a] Collins, Allan M.; Quillian, M. Ross. *Retrieval Time from Semantic Memory.* Journal of Verbal Learning and Verbal Behavior. Volume 8. pp 240-247.

[Collins, AM 75a] Collins, Allan M.; Loftus, Elizabeth F. (1975). *A Spreading-Activation Theory of Semantic Processing.* Psychological Review. Volume 82. pp 407-425.

[Collins, DH 88a] Collins, David H. (1988). *Two-Dimensional Pattern Sequence Prediction Using High-Order Neural Networks.* Neural Networks. Volume 1, Supplement 1. pp 19.

[Collins, DR 87a] Collins, Dean R.; Sampsell, Jeffrey B.; Penz, P. Andrew; Gately, Michael T. (1987). *Optical Neurocomputers Implementation Using Deformable Mirror Arrays.* IEEE International Conference on Neural Networks (ICNN-87), San Diego, CA, June 21-24, 1987: Proceedings. Volume 3. pp 631-638. New York: Institute of Electrical and Electronics Engineers.

[Collins, JW 87a] Collins, John W. (1987). *A Neural Network Based on Co-Occurrence Probabilities.* IEEE International Conference on Neural Networks (ICNN-87), San Diego, CA, June 21-24, 1987: Proceedings. Volume 2. pp 487-494. New York: Institute of Electrical and Electronics Engineers.

[Collins, RL 83*] SEE: Bartholdi, JJ 83a. *A Minimal Technology Routing System for Meals on Wheels.*

[Collins, WC 82*] SEE: Athale, RA 82a. *Optical Matrix-Matrix Multiplication Based Outer Product Decomposition.*

[Colvin, ME 86a] Colvin, M. E.; Eeckman, Frank H.; Tromp, J. W. (1986). *Model for Cortical Function.* In Denker, John S. (ed.). Neural Networks for Computing, Snowbird, Utah, 1986. AIP Proceedings 151. pp 110-114. New York: American Institute of Physics.

[Combs, N 87*] SEE: Tillery, SI 87a. *The Problem of Internal Representations in Neural Nets: Concepts, Implications and a Proposed Metric.*

[Conrad, M 72a] Conrad, Michael. (1972). *Information Processing in Molecular Systems.* Currents In Modern Biology. Volume 5. pp 1-14.

[Conrad, M 74a] Conrad, Michael. (1974). *Molecular Automata.* In Conrad, Michael (ed.); Guttinger, W. (ed.); Dal Cin, M. (ed.). Physics and Mathematics of the Nervous System. pp 419-430. New York: Springer-Verlag.

[Conrad, M 74b] Conrad, Michael. (1974). *Molecular Information Processing in the Central Nervous System, Part I.* In Conrad, Michael (ed.); Guttinger, W. (ed.); Dal Cin, M. (ed.). Physics and Mathematics of the Nervous System. pp 82-107. New York: Springer-Verlag.

[Conrad, M 74c] Conrad, Michael. (1974). *The Limits of Biological Simulation.* Journal of Theoretical Biology. Volume 45. pp 585-590.

[Conrad, M 76a] Conrad, Michael. (1976). *Molecular Information Structures in the Brain.* Journal of Neuroscience Research. Volume 2. pp 233-254.

[Conrad, M 78a] Conrad, Michael. (1978). *Evolution of the Adaptive Landscape.* In Heim, R. (ed.); Palm, Gunther (ed.). Theoretical Approaches to Complex Systems. New York: Springer-Verlag.

[Conrad, M 83*] SEE: Kampfner, RR 83a. *Computational Modeling of Evolutionary Learning Processes in the Brain.*

[Conrad, M 83*] SEE: Kampfner, RR 83b. *Sequential Behavior and Stability Properties of Enzymatic Neuron Networks.*

[Conrad, M 84a] Conrad, Michael. (1984). *Microscopic-Macroscopic Interface in Biological Information Processing.* BioSystems. Volume 16. pp 345-363.

[Conrad, M 85a] Conrad, Michael. (1985). *On Design Principles for a Molecular Computer.* Communications of the ACM. Volume 28. pp 464-480.

[Conrad, M 86*] SEE: Kirby, KG 86a. *Intraneuronal Dynamics as a Substrate for Evolutionary Learning.*

[Conrad, M 87*] SEE: Trenary, R 87a. *A Neuron Model of a Memory System for Autonomous Exploration of an Environment.*

[Conrad, M 87a] Conrad, Michael. (October, 1987). *Biomolecular Information Processing.* IEEE Potentials. pp 12-15.

[Conrad, M 87b] Conrad, Michael. (1987). *Rapprochement of Artificial Intelligence and Dynamics.* European Journal of Operational Research. Volume 30. pp 280-290.

[Conrad, M 88*] SEE: Kirby, KG 88a. *Bit-Vector Optimization Algorithms for Control of Learning in Neurons with Second-Messenger Dynamics.*

[Conrad, M 88a] Conrad, Michael. (1988). *Neuromolecular Computing: Mechanisms and Architectures.* IEEE International Conference on Engineering in Medicine and Biology, 10th, New Orleans, November 4-7, 1988: Proceedings. pp 1015-1016. New York: Institute of Electrical and Electronics Engineers.

[Conrad, M 88b] Conrad, Michael; Kampfner, Roberto R.; Kirby, Kevin G. (1988). *Neuronal Dynamics and Evolutionary Learning.* Advances in Cognitive Science, Steps Toward Convergence. AAAS Selected Symposium 104. pp 169-189. Boulder, CO: American Association for the Advancement of Science.

[Conrad, M 88c] Conrad, Michael. (1988). *Prolegomena to Evolutionary Programming.* Advances in Cognitive Science, Steps Toward Convergence. AAAS Selected Symposium 104. pp 150-168. Boulder, CO: American Association for the Advancement of Science.

[Conrad, M 88d] Conrad, Michael. (1988). *The Price of Progammability.* The Universal Turing Machine a Half-Century Survey. pp 285-307. New York: Oxford University Press.

[Conwell, PR 87a] Conwell, Peter R. (1987). *Effects of Connection Delays in Two State Model Neural Circuits.* IEEE International Conference on Neural Networks (ICNN-87), San Diego, CA, June 21-24, 1987: Proceedings. Volume 3. pp 95-104. New York: Institute of Electrical and Electronics Engineers.

[Cook, FJ 84*] SEE: Barron, RL 84a. *Adaptive Learning Networks: Development and Application in the United States of Algorithms Related to GMDH.*

[Cook, G 67a] Cook, Gerald; Stark, Lawrence. (1967). *Derivation of a Model for the Human Eye-Positioning Mechanism.* Bulletin of Mathematical Biophysics. Volume 29. pp 153-174.

[Cook, GE 88*] SEE: Karsai, G 88a. Gas Tungsten Arc Weld Modeling Using a Mapping Network.

[Cook, ND 88*] SEE: Anninos, PA 88b. *Neural Net Simulation of the Corpus Callosum.*

[Coon, DD 86a] Coon, D. D.; Perera, A. G. U. (1986). *Spectral Information Coding by Infrared Photoreceptors.* International Journal of Infrared and Millimeter Waves. Volume 7. pp 1571-1583.

[Coon, DD 87*] SEE: Bandara, KMSV 87a. *Interfacing Multispectral Sensors to Real Time Processors Based on Neural Network Models.*

[Coon, DD 87a] Coon, D. D.; Ma, S. N.; Perera, A. G. U. (1987). *Farey-Fraction Frequency Modulation in the Neuronlike Output of Silicon p-i-n Diodes at 4.2 K.* Physical Review Letters. Volume 58. pp 1139-1142.

[Coon, DD 87b] Coon, D. D.; Perera, A. G. U. (1987). *Parallel Asynchronous Focal Plane Array Processing.* International Journal of Infrared and Millimeter Waves. Volume 8. pp 1037-1052.

[Coon, DD 87c] Coon, D. D.; Perera, A. G. U. (1987). *Semiconductor Electronic Concepts for Neural Network Emulation.* International Journal of Electronics. Volume 63. pp 61-69.

[Coon, DD 88a] Coon, D. D.; Karunasiri, R. P. G.; Bandara, K. M. S. V. (1988). *Infrared Transient Sensing.* Optical Engineering. Volume 27. pp 471-474.

[Coon, DD 88b] Coon, D. D.; Perera, A. G. U. (1988). *Integrate-and-Fire Coding in Hodgkin-Huxley Circuits Employing Silicon*

Diodes. Neural Networks. Volume 1.

[Coon, DD 88c] Coon, D. D.; Perera, A. G. U. (1988). *Neuron-Like Transient Phenomena in Silicon p-i-n Structures*. Solid State Electronics. Volume 31. pp 851-862.

[Coon, DD 88d] Coon, D. D.; Perera, A. G. U. (1988). *New Hardware for Massive Neural Networks*. In Anderson, Dana Z. (ed.). Neural Information Processing Systems. pp 201-210. New York: American Institute of Physics.

[Cooper, DB 70a] Cooper, David B.; Freeman, John H. (1970). *On the Asymptotic Improvement in the Outcome of Supervised Learning Provided by Additional Non-Supervised Learning*. IEEE Transactions on Computers. Volume C-19. pp 1055-1063.

[Cooper, DB 83a] Cooper, David B.; Sung, Francis P. (1983). *Multiple-Window Parallel Adaptive Boundary Finding in Computer Vision*. IEEE Transactions on Pattern Analysis and Machine Intelligence. Volume PAMI-5. pp 299-316.

[Cooper, LA 89*] SEE: Nadel, L 89a. Neural Connections, Mental Computation.

[Cooper, LN 73a] Cooper, Leon N. (1988). *A Possible Organization of Animal Memory and Learning*. In Anderson, James A. (ed.); Rosenfeld, Edward (ed.). Neurocomputing: Foundations of Research. pp 195-208. Cambridge, MA: MIT Press.

[Cooper, LN 73b] Cooper, Leon N. (1973). *A Possible Organization of Animal Memory and Learning*. Nobel Symposium on Collective Properties of Physical Systems, 24th, Sweden, June 12-16, 1973: Proceedings. Volume 24. pp 252-264. New York: Academic Press.

[Cooper, LN 75*] SEE: Nass, MM 75a. *A Theory for the Development of Feature Detecting Cells in the Visual Cortex*.

[Cooper, LN 76a] Cooper, Leon N.; Elbaum, Charles. (April 13, 1976). *Information Processing System*. United States Patent 3,950,733.

[Cooper, LN 78*] SEE: Anderson, JA 78a. *Les Modeles Mathematiques de l'Organization Biologique de la Memoire*.

[Cooper, LN 79a] Cooper, Leon N.; Liberman, Fishel; Oja, Erkki. (1979). *A Theory for the Acquisition and Loss of Neuron Specificity in Visual Cortex*. Biological Cybernetics. Volume 33. pp 9-28.

[Cooper, LN 81a] Cooper, Leon N. (1981). *Distributed Memory in the Central Nervous Systems: Possible Test of Assumptions in Visual Cortex*. In Schmitt, Francis O. (ed.); Wordern, F. G. (ed.); Adelman, George (ed.); Dennis, S. G. (ed.). The Organization of the Cerebral Cortex. Cambridge, MA: MIT Press.

[Cooper, LN 81b] Cooper, Leon N.; Elbaum, Charles. (March 3, 1981). Information Processing System Using Threshold Passive Modification. United States Patent 4,254,474.

[Cooper, LN 82*] SEE: Reilly, DL 82a. *A Neural Model for Category Learning*.

[Cooper, LN 82*] SEE: Bienenstock, EL 82a. *A Theory for the Development of Neuron Selectivity: Orientation Selectivity and Binocular Interactions in the Visual Cortex*.

[Cooper, LN 82*] SEE: Bienenstock, EL 82b. *Theory for the Development of Neuron Selectivity: Orientation Specificity and Binocular Interaction in Visual Cortex*.

[Cooper, LN 82a] Cooper, Leon N.; Elbaum, Charles; Reilly, Douglas L. (April 20, 1982). Self Organizing General Pattern Class Separator and Identifier. United States Patent 4,326,259.

[Cooper, LN 84a] Cooper, Leon N. (1984). *Neuron Learning to Network Organization*. J C Maxwell, The Sesquicentennial Symposium: New Vistas in Mathematics, Science and Technology. pp 41-90. Amsterdam: Elsevier.

[Cooper, LN 85a] Cooper, Leon N.; Munro, Paul W.; Scofield, Christopher L. (1985). *Neuron Selectivity: Single Neuron and Neuron Networks.* In Levy, William B. (ed.); Anderson, James A. (ed.); Lehmkuhle, Stephen (ed.). Synaptic Modifications, Neuron Selectivity, and Nervous System Organization. pp 175-192. Hillsdale, NJ: Lawrence Erlbaum.

[Cooper, LN 87*] SEE: Reilly, DL 87a. *Learning System Architectures Composed of Multiple Learning Modules.*

[Cooper, LN 87*] SEE: Bear, MF 87a. *A Physiological Basis for a Theory of Synapse Modification.*

[Cooper, LN 87*] SEE: Bachmann, CM 87a. *A Relaxation Model for Memory with High Storage Density.*

[Cooper, LN 87a] Cooper, Leon N. (1987). *Optimization of Neural Network Connectivity.* IEEE International Conference on Neural Networks (ICNN-87), San Diego, CA, June 21-24, 1987: Proceedings. Volume 3. pp 767-768. New York: Institute of Electrical and Electronics Engineers.

[Cooper, LN 88*] SEE: Scofield, CL 88b. *Pattern Class Degeneracy in an Unrestricted Storage Density Memory.*

[Corbit, JD 73*] SEE: Eimas, PD 73a. *Selective Adaptation of Linguistic Feature Detectors.*

[Corkill, DD 81*] SEE: Lesser, VR 81a. *Functionally-Accurate, Cooperative Distributed Systems.*

[Corkill, DD 85*] SEE: Durfee, EH 85a. Coherent Cooperation among Communicating Problem Solvers.

[Corkill, DD 85*] SEE: Durfee, EH 85b. Increasing Coherence in a Distributed Problem Solving Network.

[Cornell, D 73*] SEE: Homa, D 73a. *Prototype Abstraction and Classification of New Instances as a Function of the Number of Instances Defining the Prototype.*

[Corsi, P 82a] Corsi, Patrick. (1982). *Speaker Recognition: A Survey.* Automatic Speech Analysis and Recognition. pp 277-308. Boston: D Reidel.

[Cortes, C 87a] Cortes, C.; Krogh, A.; Hertz, J. A. (1987). *Hierarchical Associative Networks.* Journal of Physics A: Mathematical and General. Volume 20. pp 4449-4455.

[Cosic, C 88a] Cosic, Cynthia; Munro, Paul W. (1988). *Learning to Represent and Understand Locative Prepositional Phrases.* Research Report LIS002/IS88002. Pittsburgh, PA: University of Pittsburgh.

[Cosnard, M 86a] Cosnard, M.; Chacc, E. Goles. (1986). *Dynamical Properties of an Automaton With Memory.* In Bienenstock, Elie L. (ed.). Disordered Systems and Biological Organization. pp 63-66. New York: Springer-Verlag.

[Coste, J 86a] Coste, J.; Henon, M. (1986). *Invariant Cycles in the Random Mapping of n Integers into Themselves. Comparison with Kauffman Binary Network..* In Bienenstock, Elie L. (ed.). Disordered Systems and Biological Organization. pp 361-366. New York: Springer-Verlag.

[Cotter, CA 88*] SEE: Ryan, TW 88b. Vector Quantization Training by a Self-Organizing Neural Network.

[Cotterill, RMJ 88a] Cotterill, Rodney M. J. (1988). *Dynamical Properties of a New Type of Neural Network.* In Eckmiller, Rolf (ed.); von der Malsburg, Christoph (ed.). Neural Computers. NATO ASI Series, Volume F41. pp 425-433. Berlin: Springer-Verlag.

[Cottrell, GW 82*] SEE: Small, SL 82c. *Toward Connectionist Parsing.*

[Cottrell, GW 83a] Cottrell, Garrison W.; Small, Steven L. (1983). *A Connectionist Scheme for Modeling Word Sense Disambiguation.* Cognition and Brain Theory. Volume 6. pp 89-120.

[Cottrell, GW 84a] Cottrell, Garrison W.; Small, Steven L. (1984). *Viewing Parsing as Word Sense Discrimination: A*

Connectionist Approach. In Guida, G. (ed.). Computational Models of Natural Language Processing. pp 91-119. New York: Elsevier.

[Cottrell, GW 85a] Cottrell, Garrison W. (1985). *Connectionist Parsing*. Cognitive Science Society, 7th, Irvine, CA, August 15-17, 1985: Proceedings. pp 201-211. Hillsdale, NJ: Lawrence Erlbaum.

[Cottrell, GW 87a] Cottrell, Garrison W.; Munro, Paul W.; Zipser, David. (1987). Image Compression by Back Propagation: An Example of Extensional Programming. ICS-8702. San Diego, CA: University of California.

[Cottrell, GW 87b] Cottrell, Garrison W.; Munro, Paul W.; Zipser, David. (1987). *Learning Internal Representations from Gray-Scale Images: An Example of Extensional Programming*. Cognitive Science Society, 9th, Seattle, WA, July 16-18, 1987: Proceedings. pp 461-473. Hillsdale, NJ: Lawrence Erlbaum.

[Cottrell, M 87*] SEE: Axelrad, H 87a. *The Use of an Artificial Neural Network Model to Analyze the Information Transfer Properties of a Simplified Model of the Cerebellar Cortex*.

[Cottrell, M 88a] Cottrell, M. (1988). *Stability and Attractivity in Associative Memory Networks*. Biological Cybernetics. Volume 58. pp 129-139.

[Coulter, DA 87*] SEE: Alkon, DL 87a. *Conditioning-Specific Modification of Postsynaptic Membrane Currents in Mollusc and Mammal*.

[Coulter, DA 87a] Coulter, Douglas A.; Disterhoft, John F.; Bank, Barry; Alkon, Daniel L. (1987). *Persistent Conditioning-Specific Changes of Post-Synaptic Membrane Currents in Heermissenda and Hippocampus*. Neuroplasticity, Learning, and Memory. pp 17-43. Alan R Liss.

[Coulter, DA 88*] SEE: Disterhoft, JF 88a. *Conditioning-Specific Biophysical Alterations in Rabbit Hippocampus*.

[Coulter, DA 88*] SEE: LoTurco, JJ 88a. *Enhancement of Synaptic Potentials in Rabbit CA1 Pyramidal Neurons Following Classical Conditioning*.

[Cover, TM 65a] Cover, Thomas M. (1965). *Geometrical and Statistical Properties of Systems of Linear Inequalities with Applications in Pattern Recognition*. IEEE Transactions on Electronic Computers. Volume EC-14. pp 326-334.

[Cover, TM 67a] Cover, Thomas M. (1967). *Nearest Neighbor Pattern Classification*. IEEE Transactions on Information Theory. Volume IT-13. pp 21-27.

[Cover, TM 81*] SEE: van Campenhout, JM 81a. *Maximum Entropy and Conditional Probability*.

[Cowan, JD 70a] Cowan, Jack D. (1970). *A Statistical Mechanics of Nervous Activity*. Some Mathematical Questions in Biology. Lectures on Mathematics Vol 2. Providence, RI: American Mathematical Society.

[Cowan, JD 73*] SEE: Wilson, HR 73a. *A Mathematical Theory of the Functional Dynamics of Cortical and Thalamic Nervous Tissue*.

[Cowan, JD 75*] SEE: Feldman, JL 75a. *Large-Scale Activity in Neural-Nets II: A Model for the Brainstem Respiratory Oscillator*.

[Cowan, JD 79*] SEE: Ermentrout, GB 79a. *A Mathematical Theory of Visual Hallucination Patterns*.

[Cowan, JD 80*] SEE: Ermentrout, GB 80a. *Large Scale Spatially Organized Activity in Neural Nets*.

[Cowan, JD 80*] SEE: Ermentrout, GB 80b. *Secondary Bifurcation in Neuronal Nets*.

[Cowan, JD 82*] SEE: von der Malsburg, C 82a. *Outline of a Theory for the Ontogenesis of Iso-Orientation Domains in Visual*

Cortex.

[Cowan, JD 85a] Cowan, Jack D.; von der Malsburg, Christoph. (1985). *A Proposed Mechanism for the Origin and Development of Iso-Orientation Columns.* Models of the Visual Cortex. pp 462-472. New York: Wiley.

[Cowan, JD 85b] Cowan, Jack D. (1985). *What Do Drug-Induced Visual Hallucinations Tell Us About the Brain?.* In Levy, William B. (ed.); Anderson, James A. (ed.); Lehmkuhle, Stephen (ed.). Synaptic Modifications, Neuron Selectivity, and Nervous System Organization. pp 223-242. Hillsdale, NJ: Lawrence Erlbaum.

[Cowan, JD 87a] Cowan, Jack D.; Sharp, David H. (1987). Neural Nets. LA-UR-87-4098. Los Alamos, NM: Los Alamos National Laboratory, Theoretical Division.

[Cowan, JD 88a] Cowan, Jack D.; Sharp, David H. (1988). *Neural Nets and Artificial Intelligence.* Daedalus. Volume 117. pp 85-121.

[Cox, RT 46a] Cox, R. T. (1946). *Probability, Frequency, and Reasonable Expectations.* American Journal of Physics. Volume 14. pp 1-13.

[Crafts, HS 62a] Crafts, Harold S. (1962). A Magnetic Variable-Gain Component for Adaptive Networks. Technical Report 1851-2. Stanford, CA: Stanford University.

[Crafts, HS 63a] Crafts, Harold S. (1963). *Components that Learn and How to Use Them.* Electronics Magazine. Volume 36. pp 49-53.

[Craig, JN 84*] SEE: Barron, RL 84a. *Adaptive Learning Networks: Development and Application in the United States of Algorithms Related to GMDH.*

[Crane, HD 60a] Crane, H. D. (1960). Neuristor Studies. Report 1506-2. Stanford, CA: Stanford University.

[Crane, HD 62a] Crane, H. D. (1962). *Neuristor: A Novel Device and System Concept.* Proceedings of the IRE. Volume 50. pp 2048-2060.

[Crane, HD 64a] Crane, H. D. (1964). *Possibilities for Signal Processing in Axon Systems.* Neural Theory and Modeling, Ojai Symposium, 1962: Proceedings. pp 138-153. Stanford, CA: Stanford University Press.

[Crawford, WJ 87*] SEE: Myers, MH 87a. Application of New Artificial Neural System Information Processing Principles to Pattern Classification: Final Report, US Army Research Office.

[Crawford, WJ 87*] SEE: Kuczewski, RM 87a. Exploration of Backward Error Propagation as a Self-Organizational Structure.

[Crawford, WJ 87*] SEE: Kuczewski, RM 87b. *Exploration of Backward Error Propagation as a Self-Organized Structure.*

[Crawford, WJ 87*] SEE: Kuczewski, RM 87e. *Neurocomputer Workstations and Processors: Approaches and Applications.*

[Crawford, WJ 87*] SEE: Kuczewski, RM 87f. Neurocomputer Workstations and Processors: Approaches and Applications.

[Crick, F 83a] Crick, Francis; Mitchison, Graeme. (1983). *The Function of Dream Sleep.* Nature. Volume 304. pp 111-114.

[Crick, F 84a] Crick, Francis. (1988). *Function of the Thalamic Reticular Complex: The Searchlight Hypothesis.* In Anderson, James A. (ed.); Rosenfeld, Edward (ed.). Neurocomputing: Foundations of Research. pp 571-576. Cambridge, MA: MIT Press.

[Crick, FHC 79a] Crick, F. H. C. (1979). *Thinking About the Brain.* Scientific American. Volume 241. pp 219-232.

[Crick, FHC 86a] Crick, F. H. C.; Asanuma, C. (1986). *Certain Aspects of the Anatomy and Physiology of the Cerebral*

Cortex. In Rumelhart, David E. (ed.). Parallel Distributed Processing: Psychological and Biological Models. Volume 2. pp 333-371. Cambridge, MA: MIT Press.

[Crisman, J 86*] SEE: Wallace, R 86a. *Progress in Robot Road-Following*.

[Cromie, WJ 88a] Cromie, William J. (December/January 1988-89). *Neural Networks: What Can They Do?*. The MIT Report. pp 4-15.

[Cronin, J 77a] Cronin, Jane. (1977). *Mathematical Aspects of Periodic Catatonic Schizophrenia*. Bulletin of Mathematical Biology. Volume 39. pp 187-199.

[Cross, GR 83a] Cross, George R.; Jain, Anil K. (1983). *Markov Random Field Texture Models*. IEEE Transactions on Pattern Analysis and Machine Intelligence. Volume PAMI-5. pp 25-39.

[Cross, J 73*] SEE: Homa, D 73a. *Prototype Abstraction and Classification of New Instances as a Function of the Number of Instances Defining the Prototype*.

[Crothers, EJ 65*] SEE: Atkinson, RC 65a. An Introduction to Mathematical Learning Theory.

[Crutchfield, JP 86a] Crutchfield, James P.; Farmer, J. Doyne; Packard, Norman H.; Shaw, Robert S. (1986). *Chaos*. Scientific American. Volume 255. pp 46-57.

[Cruz, CA 82a] Cruz, Claude A.; Myers, H. Joseph. (1982). Associative Networks. Report No G320-3474. Palo Alto, CA: IBM Palo Alto Scientific Center.

[Cruz, CA 83a] Cruz, Claude A.; Myers, H. Joseph. (1983). Associative Networks, II. Report No G320-3446. Palo Alto, CA: IBM Palo Alto Scientific Center.

[Cruz, CA 85a] Cruz, Claude A.; Tam, Jason Y. (1985). NEP: An Emulation-Assist Processor for Parallel Associative Networks. Report No G320-3475. Palo Alto, CA: IBM Palo Alto Scientific Center.

[Cruz, CA 86a] Cruz, Claude A.; Hanson, William A.; Tam, Jason Y. (1986). *Flow-of-Activation Processing: Parallel Associative Networks (PAN)*. In Denker, John S. (ed.). Neural Networks for Computing, Snowbird, Utah, 1986. AIP Proceedings 151. pp 115-120. New York: American Institute of Physics.

[Cruz, CA 87*] SEE: Hanson, WA 87b. *CONE - Computational Network Environment*.

[Cruz, CA 87b] Cruz, Claude A.; Hanson, William A.; Tam, Jason Y. (1987). *Knowledge Processing Through Flow-of-Activation*. IEEE International Conference on Neural Networks (ICNN-87), San Diego, CA, June 21-24, 1987: Proceedings. Volume 2. pp 343-350. New York: Institute of Electrical and Electronics Engineers.

[Cruz, CA 87c] Cruz, Claude A.; Hanson, William A.; Tam, Jason Y. (1987). *Neural Network Emulation Hardware Design Considerations*. IEEE International Conference on Neural Networks (ICNN-87), San Diego, CA, June 21-24, 1987: Proceedings. Volume 3. pp 427-434. New York: Institute of Electrical and Electronics Engineers.

[Cruz, CA 88a] Cruz, Claude A. (1988). Understanding Neural Networks, a Primer. Amherst, NH: Graeme Publishing Corporation.

[Csermely, TJ 70*] SEE: Harth, EM 70a. *Brain Functions and Neural Dynamics*.

[Csermely, TJ 70*] SEE: Anninos, PA 70a. *Dynamics of Neural Structures*.

[Csermely, TJ 78*] SEE: Finette, S 78a. *Anisotropic Connectivity and Cooperative Phenomena as a Basis for Orientation Sensitivity in the Visual Cortex*.

[Culbertson, JT 48a] Culbertson, James T. (1948). *A Mechanism for Optic Nerve Conduction and Form Perception: I*.

Bulletin of Mathematical Biophysics. Volume 10. pp 31-40.

[Culbertson, JT 63a] Culbertson, James T. (1963). The Minds of Robots, Sense Data, Memory Images, and Behavior in Conscious Automata. Urbana, IL: University of Illinois Press.

[Culhane, AD 88a] Culhane, A. D.; Peckerar, M. C.; Marrian, C. R. K. (1988). A Neural Net Approach to Discrete Fourier Transforms. College Park, MD: University of Maryland.

[Culicover, P 89*] SEE: Nadel, L 89a. Neural Connections, Mental Computation.

[Cummins, KL 82*] SEE: McGill, KC 82a. *On the Nature and Elimination of Stimulus Artifact in Nerve Signals Evoked and Recorded Using Surface Electrodes.*

[Cunningham, R 88*] SEE: Waxman, AM 88b. *The Neural Analog Diffusion-Enhancement Layer (NADEL) and Early Visual Processing.*

[Curlander, JC 83a] Curlander, John C.; Marmarelis, Vasilis Z. (1983). *Processing of Visual Information in the Distal Neurons of the Vertebrate Retina.* IEEE Transactions on Systems, Man and Cybernetics. Volume SMC-13. pp 934-943.

[Curtis, SR 85a] Curtis, Susan R.; Oppenheim, Alan V.; Lim, Jae S. (1985). *Signal Reconstruction from Fourier Transform Sign Information.* IEEE Transactions on Acoustics, Speech, and Signal Processing. Volume ASSP-33. pp 643-657.

[Cutler, L 88*] SEE: Ross, MD 88a. *Macular Bioaccelerometers on Earth and in Space.*

[Cutler, L 88*] SEE: Ross, MD 88c. *The Neuroanatomical Substrate for Information Processing in Macular Endorgans.*

[Cuykendall, R 88a] Cuykendall, R.; Reese, R. (1987). *Scaling Techniques for Neural Algorithms.* Neural Networks. Volume 1, Supplement 1. pp 87.

[Cuykendall, R 89a] Cuykendall, R.; Reese, R. ((in press)). *Scaling the Neural TSP Algorithm.* Biological Cybernetics. Volume 60.

[Cyrulnik, RA 82*] SEE: Anninos, PA 82a. *Encoding of Visual Information: Correlation of EEG Response to Intracellular Response.*

[Czarnul, Z 75*] SEE: Wilamowski, BM 75a. *Novel Inductorless Neuristor Line.*

[D'Autrechy, CL 88a] D'Autrechy, C. Lynne; Reggia, James A.; Sutton III, Granger G.; Goodall, Sharon M. (1988). *A General-Purpose Simulation Environment for Developing Connectionist Models.* Simulation. Volume 51. pp 5-19.

[d'Humieres, D 86*] SEE: Weisbuch, G 86a. *Determining the Dynamic Landscape of Hopfield Networks.*

[d'Humieres, D 86a] d'Humieres, D. (1986). *Layered Networks for Unsupervised Learning.* In Bienenstock, Elie L. (ed.). Disordered Systems and Biological Organization. pp 241-246. New York: Springer-Verlag.

[Da Pos, O 85*] SEE: Metelli, F 85a. *Balanced and Unbalanced, Complete and Partial Transparency.*

[Da Silva, KMC 80*] SEE: Fernandes, PML 80a. *Nerve Cell Soma Model with High Reliability and Low Power Consumption.*

[Dahl, ED 87a] Dahl, Edward Denning. (1987). *Accelerated Learning Using the Generalized Delta Rule.* IEEE International Conference on Neural Networks (ICNN-87), San Diego, CA, June 21-24, 1987: Proceedings. Volume 2. pp 523-530. New York: Institute of Electrical and Electronics Engineers.

[Dahl, ED 87b] Dahl, Edward Denning. (1987). *Neural Network Algorithm for an NP-Complete Problem: Map and Graph Coloring.* IEEE International Conference on Neural Networks (ICNN-87), San Diego, CA, June 21-24, 1987:

Proceedings. Volume 3. pp 113-120. New York: Institute of Electrical and Electronics Engineers.

[Dakin, GA 86a] Dakin, Gordon A.; Arbib, Michael A. (1986). *The Hierarchical Control of Skilled Hand Movements.* COINS Technical Report 86-35. Amherst, MA: University of Massachusetts.

[Dammann, JE 65*] SEE: Mattson, RL 65a. *A Technique for Determining and Coding Subclasses in Pattern Recognition Problems.*

[Dammasch, IE 84a] Dammasch, Ingolf E.; Wagner, G. P. (1984). *On the Properties of Randomly Connected McCulloch-Pitts Networks: Differences Between Input-Constant and Input-Variant Networks.* Cybernetics and Systems. Volume 15. pp 91-117.

[Dammasch, IE 86a] Dammasch, Ingolf E. (1986). *Morphogenesis and Properties of Neuronal Model Networks.* Cybernetics and Systems '86. pp 327-334. Boston: D Reidel.

[Dammasch, IE 86b] Dammasch, Ingolf E.; Wagner, G. P.; Wolff, J. R. (1986). *Self-Stabilization of Neuronal Networks, I. The Compensation Algorithm for Synaptogenesis.* Biological Cybernetics. Volume 54. pp 211-222.

[Dammasch, IE 87a] Dammasch, Ingolf E. (1987). *A Mechanism for Stabilizing Intelligent Properties of Neural Networks.* International Congress of Cybernetics and Systems, Cybernetics and Systems, Present and Future, 7th, London, September 7-11, 1987: Proceedings. Volume 2. pp 864-871. Annes: Thales Publications.

[Dammasch, IE 87b] Dammasch, Ingolf E. (1987). *Reactive Synaptogenesis in Small Neuronal Circuits: Simulations Based on the Compensation Algorithm.* IEEE International Conference on Neural Networks (ICNN-87), San Diego, CA, June 21-24, 1987: Proceedings. Volume 4. pp 23-29. New York: Institute of Electrical and Electronics Engineers.

[Dammasch, IE 88a] Dammasch, Ingolf E.; Fischer, S.; Wolff, J. R. (1988). *A Model of Synaptogenetic Consequences of Afferent Input in Neuronal Nets.* Sense Organs, Interfaces between Environment and Behavior, Proceedings of the 16th Gottingen Neurobiology Conference. Georg Thieme Verlag.

[Dammasch, IE 88b] Dammasch, Ingolf E. (1988). *Computer-Aided Design of Neurobiological Experiments.* Computer Simulation in Brain Science. pp 495-503. Cambridge: Cambridge University Press.

[Dammasch, IE 88c] Dammasch, Ingolf E.; Wagner, G. P.; Wolff, J. R. (1988). *Self-Stabilization of Neuronal Networks, II. Stability Conditions for Synaptogenesis.* Biological Cybernetics. Volume 58. pp 149-158.

[Damper, RI 88*] SEE: Pont, MJ 88a. *A Neural Model of Infant Speech Perception.*

[Daniloff, RG 71*] SEE: Moll, KL 71a. *Investigation of the Timing of Velar Movements during Speech.*

[Danker, AJ 81a] Danker, Alan J.; Rosenfeld, Azriel. (1981). *Blob Detection by Relaxation.* IEEE Transactions on Pattern Analysis and Machine Intelligence. Volume PAMI-3. pp 79-92.

[Darianian, M 88*] SEE: Schulz, M 88a. *An Associative Co-Processor Architecture with the UNIX Operating System for the Acceleration of CAD Tools.*

[Darken, C 88*] SEE: Moody, J 88a. Fast Learning in Networks of Locally-Tuned Processing Units.

[Darken, C 88*] SEE: Moody, J 88b. Learning with Localized Receptive Fields.

[DARPA 88a] DARPA. (1988). DARPA Neural Network Study. Fairfax, VA: AFCEA International Press.

[Das, R 88a] Das, R.; Goldberg, D. E. (1988). *Discrete-Time Parameter Estimation with Genetic Algorithms.* Modeling and Simulation, 19th, Pittsburgh, PA, May 5-6, 1988: Proceedings. Volume 19. pp 2391-2396. Research Triangle Park, NC: Instrument Society of America.

[Dasgupta, S 87*] SEE: Shrivastava, Y 87a. *Convergence Issues in Perceptron Based Adaptive Neural Network Models.*

[Dattatreya, GR 86*] SEE: Kanal, LN 86a. *Problem Solving Methods for Pattern Recognition.*

[Daud, T 87a] Daud, T.; Moopenn, Alexander; Lamb, J. L.; Ramesham, R.; Thakoor, Anilkumar P. (1987). *Neural Network Based Feed-Forward High Density Associative Memory.* International Electron Devices Meeting, Washington, DC, December 6-9, 1987: Proceedings. Volume 5.6. pp 107-110. New York: Institute of Electrical and Electronics Engineers.

[Daugman, JG 85a] Daugman, John G. (1985). *Uncertainty Relation for Resolution in Space, Spatial Frequency, and Orientation Optimized by Two-Dimensional Visual Cortical Filters.* Journal of the Optical Society of America A. Volume 2. pp 1160-1169.

[Daugman, JG 87a] Daugman, John G.; Kammen, Daniel M. (1987). *Image Statistics, Gases, and Visual Neural Primitives.* IEEE International Conference on Neural Networks (ICNN-87), San Diego, CA, June 21-24, 1987: Proceedings. Volume 4. pp 163-176. New York: Institute of Electrical and Electronics Engineers.

[Daugman, JG 88a] Daugman, John G. (1988). *Complete Discrete 2-D Gabor Transforms by Neural Networks for Image Analysis and Compression.* IEEE Transactions on Acoustics, Speech, and Signal Processing. Volume 36. pp 1169-1179.

[Daunicht, WJ 87a] Daunicht, Wolfgang J.; Pellionisz, A. J. (1987). *Spatial Arrangement of the Vestibular and the Oculomotor System in the Rat.* Brain Research. Volume 435. pp 48-56.

[Daunicht, WJ 88a] Daunicht, Wolfgang J. (1988). *A Biophysical Approach to the Spatial Function of Eye Movements, Extraocular Proprioception and the Vestibulo-Ocular Reflex.* Biological Cybernetics. Volume 58. pp 225-233.

[Daunicht, WJ 88b] Daunicht, Wolfgang J. (1988). *Neural Networks Mediating Linearizable Dynamic Redundant Sensori-Motor Reflexes Characterized by Minimum of Hermitian Norm.* IEEE International Conference on Neural Networks (ICNN-88), San Diego, CA, July 24-27, 1988: Proceedings. Volume 2. pp 611-616. New York: Institute of Electrical and Electronics Engineers.

[Daunicht, WJ 88c] Daunicht, Wolfgang J. (1988). *The Relevance of Mechanical Sensors for Neural Generation of Movements in Space.* In Eckmiller, Rolf (ed.); von der Malsburg, Christoph (ed.). Neural Computers. NATO ASI Series, Volume F41. pp 335-344. Berlin: Springer-Verlag.

[Daunicht, WJ 88d] Daunicht, Wolfgang J. (1988). *The Relevance of Mechanical Sensors for Neural Generation of Movements in Space.* Neural Computers. pp 335-344. Heidelberg: Springer-Verlag.

[David, GW 87a] David, Gerald W.; Ansari, Assad. (1987). *Sensitivity Analysis of Hopfield Net.* IEEE International Conference on Neural Networks (ICNN-87), San Diego, CA, June 21-24, 1987: Proceedings. Volume 3. pp 325-328. New York: Institute of Electrical and Electronics Engineers.

[Davidson, BJ 78*] SEE: Carr, TH 78a. *Perceptual Flexibility in Word Recognition: Strategies Affect Orthographic Computation But Not Lexical Access.*

[Davidson, CH 62*] SEE: Smith, DR 62a. *Maintained Activity in Neural Nets.*

[Davies, TR 88a] Davies, Todd R. (1988). *Some Notes on the Probabilistic Semantics of Logistic Function Parameters in Neural Networks.* Neural Networks. Volume 1, Supplement 1.

[Davis, BR 82a] Davis, Barry R. (1982). A Neurobiological Approach to Machine Intelligence. Ph.D. Dissertation. Providence, RI: Brown University.

[Davis, BR 84a] Davis, Barry R. (1984). *Non-Parametric Prediction Analysis for Binary Data.* Communications in Statistics. Volume 13. pp 1161-1175.

[Davis, BR 85a] Davis, Barry R. (1985). *An Associative Hierarchical Self-Organizing System.* IEEE Transactions on Systems, Man and Cybernetics. Volume SMC-15. pp 570-579.

[Davis, LS 81a] Davis, Larry S.; Rosenfeld, Azriel. (1981). *Cooperating Processes for Low-Level Vision: A Survey.* Artificial Intelligence. Volume 17. pp 245-263.

[Davis, TL 84*] SEE: Solnick, B 84a. *Numbers of Specific Types of Neuron in Layer IVab of Cat Striate Cortex.*

[Dawes, R 88*] SEE: Kowalski, JM 88a. *Coupled Rotators as a Model for Spontaneous Activity in Neural Networks.*

[Dawes, RL 87a] Dawes, Robert L. (1987). *Biomasscomp: A Procedure for Mapping the Architecture of a Living Neural Network into a Machine.* IEEE International Conference on Neural Networks (ICNN-87), San Diego, CA, June 21-24, 1987: Proceedings. Volume 3. pp 215-226. New York: Institute of Electrical and Electronics Engineers.

[Dawes, RL 88a] Dawes, Robert L. (1988). *DMORPH: An Entropy-Based Measure of the Structural Similarity Between Two Random Systems.* Midcon/88, Dallas, TX, August 30-September 1, 1988: Conference Record.

[Dawes, RL 88b] Dawes, Robert L. (1988). *Theory and Feasibility of the Parametric Avalanche for Advanced Electronic Warfare Applications.* SBIR Phase I Final Report for the period 25 September 1987 to 18 April 1988. MRC-CECOM-88-004.

[Day, DS 87a] Day, David S. (1987). *Towards Integrating Automatic and Controlled Problem Solving.* IEEE International Conference on Neural Networks (ICNN-87), San Diego, CA, June 21-24, 1987: Proceedings. Volume 2. pp 661-670. New York: Institute of Electrical and Electronics Engineers.

[Dayhoff, JE 87a] Dayhoff, Judith E. (1987). *Detection of Favored Patterns in the Temporal Structure of Nerve Cell Connections.* IEEE International Conference on Neural Networks (ICNN-87), San Diego, CA, June 21-24, 1987: Proceedings. Volume 3. pp 63-78. New York: Institute of Electrical and Electronics Engineers.

[Dayhoff, JE 87b] Dayhoff, Judith E. (1987). *Graphical Folding and Distributed Failures in Large Network Simulations.* IEEE International Conference on Neural Networks (ICNN-87), San Diego, CA, June 21-24, 1987: Proceedings. Volume 2. pp 231-242. New York: Institute of Electrical and Electronics Engineers.

[Dayhoff, JE 88*] SEE: Dayhoff, RE 88a. *Neural Networks for Medical Image Processing: A Study of Feature Identification.*

[Dayhoff, RE 88a] Dayhoff, Ruth E.; Dayhoff, Judith E. (1988). *Neural Networks for Medical Image Processing: A Study of Feature Identification.* Proceedings of the Symposium on Computer Applications in Medical Care. pp 271-275. New York: Institute of Electrical and Electronics Engineers.

[De Kwaadsteniet, JW 82a] De Kwaadsteniet, J. W. (1982). *Statistical Analysis and Stochastic Modeling of Neuronal Spike-Train Activity.* Mathematical Biosciences. Volume 60. pp 17-71.

[De La Croix Vaubois, G 88*] SEE: Angeniol, B 88a. *Self-Organizing Feature Maps and the Travelling Salesman Problem.*

[De Luca, A 67*] SEE: Caianiello, ER 67a. *Reverberations and Control of Neural Networks.*

[De Luca, A 72a] De Luca, A.; Termini, S. (1972). *A Definition of a Nonprobabilistic Entropy in the Setting of Fuzzy Sets Theory.* Information and Control. Volume 20. pp 301-312.

[de Mori, R 86a] de Mori, Renato; Probst, David. (1986). *Computer Recognition of Speech.* In Young, Tzay Y. (ed.); Fu, King-Sun (ed.). Handbook of Pattern Recognition and Image Processing. pp 499-526. New York: Academic Press.

[De Rossi, D 88a] De Rossi, D.; Casalino, G.; Morasso, P.; Mussa Ivaldi, F. A. (1988). *Skin & Muscles: The Basic Wet-Ware for Sensory-Motor Control.* Neural Networks. Volume 1. pp 332.

[de Ruyter van Steveninck, R 83*] SEE: Zaagman, WH 83a. *Adaptive Strategies in Fly Vision: On Their Image-Processing*

Qualities.

[de Vegvar, PGN 87*] SEE: Graf, HP 87b. A CMOS Associative Memory Chip Based on Neural Networks.

[de Vegvar, PGN 87*] SEE: Graf, HP 87a. *A CMOS Associative Memory Chip.*

[de Vegvar, PGN 87*] SEE: Graf, HP 87c. *A CMOS Implementation of a Neural Network Model.*

[DeClaris, N 76a] DeClaris, N. (1976). *Neural-Type Junctions: A New Circuit Concept.* IEEE Midwestern Symposium on Circuits and Systems: Proceedings. pp 268-271.

[DeClaris, N 83*] SEE: Chande, A 83a. *Neural-Type Robotics: An Overview.*

[Deffner, G 87a] Deffner, Gerhard. (1987). *Simulation of Task Performance as a Guide to the Identification of Solution Strategy from Eye-Movement Recordings.* Cognitive Science Society, 9th, Seattle, WA, July 16-18, 1987: Proceedings. pp 427-441. Hillsdale, NJ: Lawrence Erlbaum.

[Degoulet, P 88*] SEE: Pham, KM 88a. *MOSAIC: A Macro-Connectionist Organization System for Artificial Intelligence Computation.*

[Deich, RO 88*] SEE: Taber, WR 88a. *A Comparison of Feedforward Neural Networks: Fuzzy Operators and Acoustic Ship Signatures.*

[Deich, RO 88*] SEE: Taber, WR 88b. *Fuzzy Sets and Neural Networks.*

[Deich, RO 88*] SEE: Taber, WR 88d. *The Recognition of Orca Calls with a Neural Network.*

[Deich, RO 88*] SEE: Taber, WR 88c. *Risk Weighting in Knowledge Bases Using Fuzzy Cognitive Maps.*

[Deich, RO 88a] Deich, R. O.; Taber, W. R. (1988). *Applications of Neural Networks to Underwater Acoustic Signatures: Fuzzy Decision Making and the Neuron Ring.* Intelligence Community AI Symposium, Washington, DC, October 12-14, 1988: Proceedings.

[Delcomyn, F 80a] Delcomyn, Fred. (1980). *Neural Basis of Rhythmic Behavior in Animals.* Science. Volume 210. pp 492-498.

[Delgutte, B 84a] Delgutte, Bertrand. (1984). *Speech Coding in the Auditory Nerve: II. Processing Schemes for Vowel-Like Sounds.* Journal of the Acoustical Society of America. Volume 75. pp 879-886.

[Dell, GS 88a] Dell, Gary S. (1988). *Positive Feedback in Hierarchical Connectionist Models: Applications in Language Production.* In Waltz, David L. (ed.); Feldman, Jerome A. (ed.). Connectionist Models and Their Implications: Readings from Cognitive Science. pp 97-118. Norwood, NJ: Ablex.

[Delman, S 83*] SEE: Gilchrist, AL 83a. *The Classification and Integration of Edges as Critical to the Perception of Reflectance and Illumination.*

[Delsarte, P 87a] Delsarte, P.; Kamp, Y. (to appear). *Low Rank Matrices with a Given Sign Pattern.* SIAM Journal on Discrete Mathematics.

[Dembo, A 87*] SEE: Bachmann, CM 87a. *A Relaxation Model for Memory with High Storage Density.*

[Dembo, A 88a] Dembo, Amir; Zeitouni, Ofer. (1988). *General Potential Surfaces and Neural Networks.* Physical Review A. Volume 37. pp 2134-2143.

[Dembo, A 88b] Dembo, Amir; Zeitouni, Ofer. (1988). *High Density Associative Memories.* In Anderson, Dana Z. (ed.).

Neural Information Processing Systems. pp 211-218. New York: American Institute of Physics.

[Demongeot, J 86a] Demongeot, J.; Fricot, J. (1986). *Random Fields and Spatial Renewal Potentials.* In Bienenstock, Elie L. (ed.). Disordered Systems and Biological Organization. pp 71-84. New York: Springer-Verlag.

[Dempster, AP 67a] Dempster, A. P. (1967). *Upper and Lower Probabilities Induced by a Multi-Valued Mapping.* Annals of Mathematical Statistics. Volume 38. pp 325-339.

[Dempster, AP 68a] Dempster, A. P. (1968). *A Generalization of Bayesian Inference.* Journal of the Royal Statistical Society, Series B. Volume 30. pp 205-247.

[den Uyl, MJ 86a] den Uyl, Marten J. (1986). *Representing Magnitude by Memory Resonance.* Cognitive Science Society, 8th, Amherst, MA, August 15-17, 1986: Proceedings. pp 63-71. Hillsdale, NJ: Lawrence Erlbaum.

[Denes, P 55a] Denes, P. (1955). *Effect of Duration on the Perception of Voicing.* Journal of the Acoustical Society of America. Volume 27. pp 761-764.

[Denes, P 57*] SEE: Fry, DB 57a. *On Presenting the Output of a Mechanical Speech Recognizer.*

[Denker, JS 86*] SEE: Jackel, LD 86a. *Artificial Neural Networks for Computing.*

[Denker, JS 86*] SEE: Hubbard, W 86a. *Electronic Neural Networks.*

[Denker, JS 86*] SEE: Jackel, LD 86b. *High Resolution Micro-Fabrication and Neural Networks.*

[Denker, JS 86*] SEE: Graf, HP 86a. *VLSI Implementation of a Neural Network Memory with Several Hundreds of Neurons.*

[Denker, JS 86a] Denker, John S. (1986). *Neural Network Refinements and Extensions.* In Denker, John S. (ed.). Neural Networks for Computing, Snowbird, Utah, 1986. AIP Proceedings 151. pp 121-128. New York: American Institute of Physics.

[Denker, JS 86b] Denker, John S. (ed.). (1986). Neural Networks for Computing, Snowbird, Utah, 1986. AIP Proceedings 151. New York: American Institute of Physics.

[Denker, JS 87*] SEE: Howard, RE 87a. *An Associative Memory Based on an Electronic Neural Network Architecture.*

[Denker, JS 87*] SEE: Jackel, LD 87a. *Building a Hierarchy with Neural Networks: An Example-Image Vector Quantization.*

[Denker, JS 87a] Denker, John S.; Schwartz, Daniel B.; Wittner, Ben S.; Solla, Sara; Hopfield, John J.; Howard, Richard E. (1987). *Automatic Learning, Rule Extraction, and Generalization.* Complex Systems. Volume 1. pp 877-922.

[Denker, JS 87b] Denker, John S. (1987). Neural Network Models of Learning and Adaptation. Holmdel, NJ: AT&T Bell Laboratories.

[Denker, JS 88*] SEE: Mackie, S 88a. *A Digital Implementation of a Best Match Classifier.*

[Denker, JS 88*] SEE: Mackie, S 88c. *Microelectronic Implementations of Connectionist Neural Networks.*

[Denker, JS 88*] SEE: Wittner, BS 88a. *Strategies for Teaching Layered Networks Classification Tasks.*

[Denker, JS 88a] Denker, John S.; Wittner, Ben S. (1988). *Network Generality, Traning Required, and Precision Required.* In Anderson, Dana Z. (ed.). Neural Information Processing Systems. pp 219-222. New York: American Institute of Physics.

[Denq, D 85*] SEE: Stiles, GS 85a. *On the Effect of Noise on the Moore-Penrose Generalized Inverse Associative Memory.*

[Dentino, M 78a] Dentino, Mauro; McCool, John M.; Widrow, Bernard. (1978). *Adaptive Filtering in the Frequency Domain.* Proceedings of the IEEE. Volume 66. pp 1658-1659.

[Derman, C 70a] Derman, C. (1970). Finite State Markovian Decision Processes. New York: Academic Press.

[Derr, MA 83*] SEE: Rosenbaum, DA 83a. *Hierarchical Control of Rapid Movement Sequences.*

[Derrida, B 87*] SEE: Gardner, E 87a. *Zero Temperature Parallel Dynamics for Infinite Range Spin Glasses and Neural Networks.*

[Derrida, B 87a] Derrida, B.; Gardner, E.; Zippelius, A. (1987). *An Exactly Solvable Asymmetric Neural Network Model.* Europhysics Letters. Volume 4. pp 167-173.

[Derthick, MA 84a] Derthick, Mark A. (1984). Variations on the Boltzmann Machine Learning Algorithm. CMU-CS-84-120. Pittsburgh, PA: Carnegie Mellon University.

[Derthick, MA 86a] Derthick, Mark A.; Plaut, David C. (1986). *Is Distributed Connectionism Compatible with the Physical Symbol System Hypothesis.* Cognitive Science Society, 8th, Amherst, MA, August 15-17, 1986: Proceedings. pp 639-644. Hillsdale, NJ: Lawrence Erlbaum.

[Derthick, MA 87*] SEE: Touretzky, DS 87c. *Symbol Structures in Connectionist Networks: Five Properties and Two Architectures.*

[Derthick, MA 87a] Derthick, Mark A. (1987). *A Connectionist Architecture for Representing and Reasoning About Structured Knowledge.* Cognitive Science Society, 9th, Seattle, WA, July 16-18, 1987: Proceedings. pp 131-142. Hillsdale, NJ: Lawrence Erlbaum.

[Derthick, MA 87b] Derthick, Mark A. (1987). *Counterfactual Reasoning with Direct Models.* National Conference on Artificial Intelligence (AAAI-87), 6th, July 13-17, 1987: Proceedings.

[Derthick, MA 87c] Derthick, Mark A. (1987). *Factual and Counterfactual Reasoning by Constructing Plausible Models.* National Conference on Artificial Intelligence (AAAI-87), 6th, July 13-17, 1987: Proceedings. Volume 1. pp 346-351.

[Derthick, MA 88a] Derthick, Mark A.; Tebelski, Joe. (1988). *`Ensemble' Boltzmann Units have Collective Computational Properties like those of Hopfield and Tank Neurons.* In Anderson, Dana Z. (ed.). Neural Information Processing Systems. pp 223-232. New York: American Institute of Physics.

[Deshmukh, VD 87a] Deshmukh, Vinod D. (1987). *Oscillatory Neural Ectoderm (One): The "One" Hypothesis for Human Holoencephalon and Behavior.* Medical Hypotheses. Volume 23. pp 107-108.

[Deshmukh, VD 88a] Deshmukh, Vinod D. (1988). *A Chain of Suprasegmental Neuroscillatory Circuits: A Human Brain Theory.* Clinical Electroencephalography. Volume 19. pp 7-13.

[Deshmukh, VD 88b] Deshmukh, Vinod D.; Russo, Jr, Louis S.; Boules, Adel N. (1988). *Computer Simulation of Event Related Potentials During Voluntary Cognition.* Neural Networks. Volume 1, Supplement 1. pp 170.

[Desmond, JE 85*] SEE: Moore, JW 85a. *Connectionistic Learning in Real Time: Sutton-Barto Adaptive Element and Classical Conditioning of the Rabbit Nictitating Membrane Response.*

[Desmond, JE 86*] SEE: Blazis, DEJ 86a. *Simulation of the Classically Conditioned Nictitating Membrane Response by a Neuron-Like Adaptive Element.*

[Desmond, JE 88a] Desmond, John E.; Moore, John W. (1988). *Adaptive Timing in Neural Networks: The Conditioned Response.* Biological Cybernetics. Volume 58. pp 405-415.

[Desmond, NL 85*] SEE: Levy, WB 85b. *The Rules of Elemental Synaptic Plasticity.*

[Destexhe, A 86*] SEE: Babloyantz, A 86a. *Low-Dimensional Chaos in an Instance of Epilepsy.*

[Destexhe, A 87*] SEE: Babloyantz, A 87a. *Chaos in Neural Networks.*

[Destexhe, A 87*] SEE: Babloyantz, A 87b. *Strange Attractors in the Human Cortex.*

[Destombes, J 87*] SEE: Bras, H 87a. *The Dendrites of Single Brain-Stem Motoneurons Intracellularly Labelled with Horseradish Peroxidase in the Cat: An Ultrastructural Analysis.*

[Deutsch, D 63*] SEE: Deutsch, JA 63a. *Attention: Some Theoretical Considerations.*

[Deutsch, JA 63a] Deutsch, J. A.; Deutsch, D. (1963). *Attention: Some Theoretical Considerations.* Psychological Review. Volume 70. pp 80-90.

[Deutsch, S 81a] Deutsch, Sid. (1981). *A Simplified Version of Kunihiko Fukushima's Neocognitron.* Biological Cybernetics. Volume 42. pp 17-21.

[Dev, P 75a] Dev, Parvati. (1975). *Perception of Depth Surfaces in Random-Dot Stereograms: A Neural Model.* International Journal of Man-Machine Studies. Volume 7. pp 511-528.

[Dev, P 76*] SEE: Schmitt, FO 76a. *Electrotonic Processing of Information by Brain Cells.*

[Devaney, RL 87a] Devaney, Robert L. (1987). An Introduction to Chaotic Dynamical Systems. Redwood City, CA: Addison-Wesley.

[Devijver, PA 79a] Devijver, Pierre A. (1979). *New Error Bounds with the Nearest Neighbor Rule.* IEEE Transactions on Information Theory. Volume IT-25. pp 749-753.

[Devijver, PA 80a] Devijver, Pierre A.; Kittler, Josef. (1980). On the Nearest Neighbor Rule. Oxford, England: University of Oxford.

[Devijver, PA 82a] Devijver, Pierre A.; Kittler, Josef. (1982). Pattern Recognition: A Statistical Approach. London: Prentice-Hall.

[Devos, MR 88a] Devos, M. R.; Orban, G. A. (1988). *Self-Adapting Back-Propagation.* Neuro-Nimes '88, Nimes, November 15-17, 1988: Proceedings.

[Devroye, L 81a] Devroye, Luc. (1981). *On the Inequality of Cover and Hart in Nearest Neighbor Discrimination.* IEEE Transactions on Pattern Analysis and Machine Intelligence. Volume PAMI-3. pp 75-78.

[DeYoe, EA 88a] DeYoe, E. A.; Van Essen, D. C. (1988). *Concurrent Processing Streams in Monkey Visual Cortex.* Trends in NeuroSciences. Volume 11. pp 219-226.

[Dickinson, BW 87*] SEE: Petsche, T 87a. *A Locally Interconnected Neural Network for Deconvolution.*

[Dickinson, BW 88*] SEE: Kuh, A 88a. *Information Capacity of Associative Memories.*

[Dickinson, BW 88*] SEE: Petsche, T 88a. *A Trellis-Structured Neural Network.*

[Dietterich, TD 81a] Dietterich, T. D.; Buchanan, Bruce G. (1981). The Role of the Critic in Learning Systems. Report STAN-CS-81-891. Stanford, CA: Stanford University.

[Dietz, WE 87a] Dietz, W. E.; Ali, M. (1987). *Qualitative and Temporal Reasoning in Engine Behavior Analysis and Fault*

Diagnosis. Conference on Artificial Intelligence for Space Applications, 3rd, Huntsville, AL, November 2-3, 1987: Proceedings.

[Dietz, WE 88a] Dietz, W. E.; Kiech, E. L.; Ali, M. (1988). *Pattern-Based Fault Diagnosis Using Neural Networks.* International Conference on Industrial and Engineering Applications of Artificial Intelligence and Expert Systems, 1st, Tullahoma, TN, June 1-3, 1988. Tullahoma, TN: University of Tennessee, Space Institute.

[Dietz, WE 89a] Dietz, W. E.; Kiech, E. L.; Ali, M. (1989). *Neural Network Models Applied to Real-Time Fault Diagnosis.* Journal of Neural Network Computing. Volume 1.

[Dikaiakos, M 88*] SEE: Stafylopatis, A 88a. *Spatial Organization of Neural Networks: A Probabilistic Modeling Approach.*

[Dill, JC 68a] Dill, J. C.; Randall, D. L.; Richer, I. (1968). *PLEXUS: An On-Line System for Modeling Neural Networks.* Communications of the ACM. Volume 11. pp 622-629.

[Dimopoulos, N 79a] Dimopoulos, N.; Newcomb, R. W. (1979). *Modeling Networks of Morishita Neurons With Application to the Cerebellum.* International Conference on Cybernetics and Society, Denver, CO, October 8-10, 1979: Proceedings. pp 597-602.

[Dimopoulos, N 80a] Dimopoulos, N.; Newcomb, R. W. (1980). *Stability Properties of a Class of Large Scale Neural Networks.* IEEE International Symposium on Circuits and Systems, 1980: Proceedings.

[Dimopoulos, N 82a] Dimopoulos, N. (1982). *A Study of the Asymptotic Behavior of Neural Networks.* IEEE International Symposium on Circuits and Systems, 1982: Proceedings. Volume 2. pp 280-283.

[Dinneen, GP 55a] Dinneen, G. P. (1955). *Programming Pattern Recognition.* Western Joint Computer Conference, 1955: Proceedings. pp 94-100. New York: Institute of Radio Engineers.

[Disterhoft, JF 87*] SEE: Alkon, DL 87a. *Conditioning-Specific Modification of Postsynaptic Membrane Currents in Mollusc and Mammal.*

[Disterhoft, JF 87*] SEE: Coulter, DA 87a. *Persistent Conditioning-Specific Changes of Post-Synaptic Membrane Currents in Heermissenda and Hippocampus.*

[Disterhoft, JF 88a] Disterhoft, John F.; Coulter, Douglas A.; Alkon, Daniel L. (1988). *Conditioning-Specific Biophysical Alterations in Rabbit Hippocampus.* Cellular Mechanisms of Conditioning and Behavioral Plasticity. pp 89-104. New York: Plenum.

[Divko, R 86*] SEE: Buhmann, J 86c. *Physicists Explore Human and Artificial Intelligence.*

[Divko, R 86a] Divko, R.; Schulten, Klaus. (1986). *Stochastic Spin Models for Pattern Recognition.* In Denker, John S. (ed.). Neural Networks for Computing, Snowbird, Utah, 1986. AIP Proceedings 151. pp 129-134. New York: American Institute of Physics.

[Dobruschin, PL 68a] Dobruschin, P. L. (1968). *The Description of a Random Field by Means of Conditional Probabilities and Conditions of Its Regularity.* Theory of Probability and Its Applications. Volume 13. pp 197-224.

[Dobson, VG 87a] Dobson, Vernon G. (1987). *Superior Accuracy of Decrementing over Incrementing Associative Networks when Operating in Initially Random Connectivities.* Journal of Intelligent Systems. Volume 1. pp 43-77.

[Dobson, VG 88a] Dobson, Vernon G. ((in press)). *Decrementing Associative Networks.* European Conference on Neural Networks (nEuro'88), 1st, Paris, June, 1988: Proceedings.

[Dobson, VG 88b] Dobson, Vernon G.; Johnston, Alan; Wright, Michael J.; Aleksander, Igor. ((to appear)). *Differences in Performance of Binary Hetero-Associative Networks Using Link-Enabling vs. Link-Disabling Local Modification*

Rules. Neural Computing Architecture. Cambridge, MA: MIT Press.

[Dobson, VG 88c] Dobson, Vernon G. ((in press)). *Visual Search with Decrementing Network Competitive Windows*. In Brogan, D. (ed.). First International Conference on Visual Search. London: Taylor and Francis.

[Dodd, N 87a] Dodd, Nigel. (1987). *Multi-Layer Perceptron Inversion Applied to Image Neighbourhood Operations*. IEEE International Conference on Neural Networks (ICNN-87), San Diego, CA, June 21-24, 1987: Proceedings. Volume 4. pp 293-300. New York: Institute of Electrical and Electronics Engineers.

[Doddington, GR 81a] Doddington, George R.; Schalk, Thomas B. (1981). *Speech Recognition: Turning Theory into Practice*. IEEE Spectrum. Volume 18. pp 26-32.

[Doddington, GR 86*] SEE: Gass, W 86a. *A Parallel Signal Processor System*.

[Dohne, HD 77*] SEE: Bunge, E 77b. *The AUROS Project: Automatic Recognition of Speakers by Computers*.

[Dolan, CP 87a] Dolan, Charles P.; Dyer, Michael G. (1987). *Symbolic Schemata, Role Binding, and the Evolution of Structure in Connectionist Memories*. IEEE International Conference on Neural Networks (ICNN-87), San Diego, CA, June 21-24, 1987: Proceedings. Volume 2. pp 287-298. New York: Institute of Electrical and Electronics Engineers.

[Donchin, E 77*] SEE: Squires, NK 77a. *Bisensory Stimulation: Inferring Decision-Related Processes from the P300 Component*.

[Donoghue, JP 87*] SEE: Singer, A 87a. *A Computer Simulation of Cerebral Neocortex: Computational Capabilities of Nonlinear Neural Networks*.

[Donoghue, JP 88*] SEE: Singer, A 88a. *A Computer Simulation of Cerebral Neocortex: Computational Capabilities of Nonlinear Neural Networks*.

[Donoghue, JP 88*] SEE: Sanes, JN 88a. *Rapid Reorganization of Adult Rat Motor Cortex Somatic Representation Patterns after Motor Nerve Injury*.

[Dorfman, LJ 82*] SEE: McGill, KC 82a. *On the Nature and Elimination of Stimulus Artifact in Nerve Signals Evoked and Recorded Using Surface Electrodes*.

[Dorofeyuk, AM 71a] Dorofeyuk, A. M. (1971). *Automatic Classification Algorithms*. Automatic Remote Control. Volume 32. pp 1928-1958.

[Dorsett, DA 73*] SEE: Willows, AOD 73a. *The Neuronal Basis of Behavior in Tritonia, III Neuronal Mechanism of a Fixed Action Pattern*.

[Douglas, P 85a] Douglas, Paul. (1985). Training of HMM Recognizers by Simulated Annealing. pp 13-16. New York: Institute of Electrical and Electronics Engineers.

[Douglas, SA 87a] Douglas, Sarah A.; Novick, David G.; Tomlin, Russel S. (1987). *Consistency and Variation in Spatial Reference*. Cognitive Science Society, 9th, Seattle, WA, July 16-18, 1987: Proceedings. pp 417-426. Hillsdale, NJ: Lawrence Erlbaum.

[Dow, B 88*] SEE: Baxter, W 88a. *Centric Models of the Orientation Map in Primary Visual Cortex*.

[Dow, RJF 88*] SEE: Sietsma, J 88a. *Neural Net Pruning - Why and How*.

[Draper, JS 87a] Draper, James Stark; Frankel, Donald S.; Hancock, Hunter; Mize, Anthony Scott. (1987). *A Microcomputer Neural Net Benchmarked Against Standard Classification Techniques*. IEEE International Conference on Neural Networks (ICNN-87), San Diego, CA, June 21-24, 1987: Proceedings. Volume 4. pp 651-658. New York: Institute of

Electrical and Electronics Engineers.

[Dress, WB 87a] Dress, William B.; Knisley, J. R. (1987). *A Darwinian Approach to Artificial Neural Systems.* IEEE International Conference on Systems, Man and Cybernetics, 1987: Proceedings. pp 572-577.

[Dress, WB 87b] Dress, William B. (1987). Alternative Knowledge Acquisition: Developing a Pulse-Coded Neural Network. Oak Ridge, TN: Oak Ridge National Laboratory, Instrumentation and Controls Division.

[Dress, WB 87c] Dress, William B. (1987). *Darwinian Optimization of Synthetic Neural Systems.* IEEE International Conference on Neural Networks (ICNN-87), San Diego, CA, June 21-24, 1987: Proceedings. Volume 3. pp 769-775. New York: Institute of Electrical and Electronics Engineers.

[Dress, WB 87d] Dress, William B. (1987). *Frequency-Coded Artificial Neural Networks: An Approach to Self-Organizing Systems.* IEEE International Conference on Neural Networks (ICNN-87), San Diego, CA, June 21-24, 1987: Proceedings. Volume 2. pp 47-54. New York: Institute of Electrical and Electronics Engineers.

[Dress, WB 87e] Dress, William B. (1987). High-Performance Neural Networks. Oak Ridge, TN: Oak Ridge National Laboratory, Instrumentation and Controls Division.

[Dress, WB 88*] SEE: Barhen, J 88b. *Concurrent Neuromorphic Algorithms for Autonomous Robots.*

[Dress, WB 88a] Dress, William B. (1988). Electronic Life and Synthetic Intelligent Systems. Oak Ridge, TN: Oak Ridge National Laboratory, Instrumentation and Controls Division.

[Dreyfus, CF 87*] SEE: Black, IB 87a. *Biochemistry of Information Storage in the Nervous System.*

[Dreyfus, G 85*] SEE: Personnaz, L 85a. *Information Storage and Retrieval in Spin-Glass Like Neural Networks.*

[Dreyfus, G 86*] SEE: Personnaz, L 86a. *A Biologically Constrained Learning Mechanism in Networks of Formal Neurons.*

[Dreyfus, G 86*] SEE: Personnaz, L 86b. *Collective Computational Properties of Neural Networks: New Learning Mechanisms.*

[Dreyfus, G 86*] SEE: Personnaz, L 86c. *Designing a Neural Network Satisfying a Given Set of Constraints.*

[Dreyfus, G 86*] SEE: Guyon, I 86a. *Engineering Applications of Spin Glass Concepts.*

[Dreyfus, G 86*] SEE: Personnaz, L 86e. *Neural Network Design for Efficient Information Retrieval.*

[Dreyfus, G 86*] SEE: Personnaz, L 86d. *Neural Network Design for Efficient Information Retrieval.*

[Dreyfus, G 86*] SEE: Personnaz, L 86f. *Simple Selectionist Learning Rule for Neural Networks.*

[Dreyfus, G 87*] SEE: Personnaz, L 87a. *High-Order Neural Networks: Information Storage without Errors.*

[Dreyfus, G 88*] SEE: Guyon, I 88a. *Of Points and Loops.*

[Dreyfus, G 88a] Dreyfus, G.; Guyon, I.; Nadal, J. P.; Personnaz, L. (1988). *High Order Neural Networks for Efficient Associative Memory Design.* In Anderson, Dana Z. (ed.). Neural Information Processing Systems. pp 233-241. New York: American Institute of Physics.

[Dreyfus, H 86a] Dreyfus, Hubert; Dreyfus, Stuart. (1986). *Why Computers May Never Think Like People.* Technology Review. Volume 89. pp 42-61.

[Dreyfus, S 86*] SEE: Dreyfus, H 86a. *Why Computers May Never Think Like People.*

[Droulez, J 88a] Droulez, J.; Berthoz, A. (1988). *Spatial and Temporal Transformations in Visuo-Motor Coordination.* In

Eckmiller, Rolf (ed.); von der Malsburg, Christoph (ed.). Neural Computers. NATO ASI Series, Volume F41. pp 345-357. Berlin: Springer-Verlag.

[Duda, RO 64*] SEE: Block, HD 64a. *Determination and Detection of Features in Patterns.*

[Duda, RO 72a] Duda, R. O.; Hart, Peter E. (1972). *Use of the Hough Transform to Detect Lines and Curves in Pictures.* Communications of the ACM. Volume 15. pp 11-15.

[Duda, RO 73a] Duda, R. O.; Hart, Peter E. (1973). Pattern Classification and Scene Analysis. New York: Wiley.

[Duda, WL 56*] SEE: Rochester, N 56b. *Tests on a Cell Assembly Theory of the Action of the Brain, Using a Large Digital Computer.*

[Duda, WL 56*] SEE: Rochester, N 56a. *Tests on a Cell Assembly Theory of the Action of the Brain, Using a Large Digital Computer.*

[Dunbar, K 88*] SEE: Cohen, JC 88a. On the Control of Automatic Processes: A Parallel Distributed Processing Model of the Stroop Effect.

[Dunning, GJ 85a] Dunning, G. J.; Marom, E.; Owechko, Yuri; Soffer, Bernard H. (1985). *All-Optical Associative Holographic Memory with Feedback Using Phase Conjugate Mirrors.* Optical Society of America, 1985: Proceedings. pp 58 f.

[Dunning, GJ 86*] SEE: Soffer, BH 86a. *Associative Holographic Memory with Feedback Using Phase-Conjugate Mirrors.*

[Dunning, GJ 86*] SEE: Owechko, Y 86b. *Associative Memory in a Phase Conjugate Resonator Cavity Utilizing a Hologram.*

[Dunning, GJ 86*] SEE: Owechko, Y 86a. *Associative Memory in a Phase Conjugate Resonator Cavity Utilizing a Hologram.*

[Dunning, GJ 86a] Dunning, G. J.; Marom, E.; Owechko, Yuri; Soffer, Bernard H. (1986). *Optical Holographic Associative Memory Using a Phase Conjugate Resonator.* Optical Computing. SPIE Volume 625. pp 205-213. Bellingham, WA: Society of Photo-Optical Instrumentation Engineers.

[Dunning, GJ 87*] SEE: Owechko, Y 87a. *Holographic Associative Memory with Nonlinearities in the Correlation Domain.*

[Dunning, GJ 87*] SEE: Owechko, Y 87c. *Video Rate Optical Associative Memory with Shift Invariance.*

[Dunning, GJ 87a] Dunning, G. J.; Marom, E.; Owechko, Yuri; Soffer, Bernard H. (1987). *All-Optical Associative Memory with Shift Invariance and Multiple-Image Recall.* Optics Letters. Volume 12. pp 346-348.

[Dunning, GJ 88*] SEE: Owechko, Y 88a. *Optoelectronic Neural Networks Based on Holographically Interconnected Image Processors.*

[Duong, T 87*] SEE: Moopenn, A 87a. *A Neurocomputer Based on an Analog-Digital Hybrid Architecture.*

[Durbin, R 87a] Durbin, Richard; Willshaw, David J. (1987). *An Analogue Approach to the Traveling Salesman Problem Using an Elastic Net Method.* Nature. Volume 326. pp 689-691.

[Durfee, EH 84a] Durfee, Edmund H. (1984). A Parallel Simulation of a Distributed Problem Solving Network. COINS Technical Report 84-19. Amherst, MA: University of Massachusetts, Computer Science Department.

[Durfee, EH 85a] Durfee, Edmund H.; Lesser, Victor R.; Corkill, Daniel D. (1985). Coherent Cooperation among Communicating Problem Solvers. COINS Technical Report 85-15. Amherst, MA: University of Massachusetts, Computer Science Department.

[Durfee, EH 85b] Durfee, Edmund H.; Lesser, Victor R.; Corkill, Daniel D. (1985). Increasing Coherence in a Distributed

Problem Solving Network. COINS Technical Report 85-14. Amherst, MA: University of Massachusetts, Computer Science Department.

[Durfee, EH 86a] Durfee, Edmund H. (1986). An Approach to Cooperation: Planning and Communication in a Distributed Problem Solving Network. COINS Technical Report 86-09. Amherst, MA: University of Massachusetts, Computer Science Department.

[Dutta, S 88a] Dutta, Soumitra; Shekhar, Shashi. (1988). *Bond Rating: A Non-Conservative Application of Neural Networks.* IEEE International Conference on Neural Networks (ICNN-88), San Diego, CA, July 24-27, 1988: Proceedings. Volume 2. pp 443-450. New York: Institute of Electrical and Electronics Engineers.

[Dutta, S 88b] Dutta, Soumitra; Shekhar, Shashi. (1988). *Using Neural Networks for Generalization Problems.* Neural Networks. Volume 1.

[Dutta, S 89a] Dutta, Soumitra; Shekhar, Shashi. (1989). *An Artificial Intelligence Approach to Predicting Bond Ratings.* Workshop on Artificial Intelligence in Economics and Management, Singapore, January, 1989: Proceedings.

[Dyer, CR 87*] SEE: Stewart, CV 87a. *A Connectionist Model for Stereo Vision.*

[Dyer, CR 87*] SEE: Stewart, CV 87b. Local Constraint Integration in a Connectionist Model of Stereo Vision.

[Dyer, CR 88*] SEE: Stewart, CV 88a. Simulation of a Connectionist Stereo Algorithm on a Shared-Memory Multiprocessor.

[Dyer, CR 88*] SEE: Stewart, CV 88b. The Trinocular General Support Algorithm: A Three Camera Stereo Algorithm for Overcoming Binocular Matching Errors.

[Dyer, MG 87*] SEE: Dolan, CP 87a. *Symbolic Schemata, Role Binding, and the Evolution of Structure in Connectionist Memories.*

[Dyer, MG 88*] SEE: Miikkulainen, R 88a. *Encoding Input/Output Representations in Connectionist Cognitive Systems.*

[Dyer, MG 88*] SEE: Miikkulainen, R 88b. Forming Global Representations with Extended Backpropagation.

[Dyer, MG 88a] Dyer, Michael G. (1988). Symbolic NeuroEngineering for Natural Language Processing: A Multilevel Research Approach. Technical Report UCLA-AI-88-14. Los Angeles, CA: University of California, Computer Science Department.

[Ebner, FF 87*] SEE: Bear, MF 87a. *A Physiological Basis for a Theory of Synapse Modification.*

[Eccles, JC 67a] Eccles, J. C.; Ito, Masao; Szentagothai, John. (1967). The Cerebellum as a Neuronal Machine. New York: Springer-Verlag.

[Eccles, JC 72a] Eccles, J. C. (1972). *Possible Synaptic Mechanism Subserving Learning.* Brain and Human Behavior. pp 39-61. New York: Springer-Verlag.

[Eckmann, J 81a] Eckmann, J-P. (1981). *Roads to Turbulence in Dissipative Dynamical Systems.* Reviews of Modern Physics. Volume 53. pp 643-654.

[Eckmiller, R 75a] Eckmiller, Rolf. (1975). *Electronic Simulation of the Vertebrate Retina.* IEEE Transactions on Biomedical Engineering. Volume BME-22. pp 305-311.

[Eckmiller, R 83a] Eckmiller, Rolf. (1983). *Neural Control of Foveal Pursuit Versus Saccadic Eye Movements in Primates - Single-Unit Data and Models.* IEEE Transactions on Systems, Man and Cybernetics. Volume SMC-13. pp 980-989.

[Eckmiller, R 86a] Eckmiller, Rolf. (1986). *Computational Properties of a Neural Net with a Triangular Lattice Structure and a Traveling Activity Peak.* IEEE International Conference on Systems, Man, and Cybernetics, Atlanta, GA, October 14-17, 1986: Proceedings. pp 633-637. New York: Institute of Electrical and Electronics Engineers.

[Eckmiller, R 86b] Eckmiller, Rolf. (1986). *Topological and Dynamical Aspects of a Neural Network Model for Generation of Pursuit Motor Programs.* Cognitive Science Society, 8th, Amherst, MA, August 15-17, 1986: Proceedings. pp 645-651. Hillsdale, NJ: Lawrence Erlbaum.

[Eckmiller, R 87a] Eckmiller, Rolf. (1987). *Computational Model of the Motor Program Generator for Pursuit.* Journal of Neuroscience Methods. Volume 21. pp 127-138.

[Eckmiller, R 87b] Eckmiller, Rolf. (1987). *Neural Network Mechanisms for Generation and Learning of Motor Programs.* IEEE International Conference on Neural Networks (ICNN-87), San Diego, CA, June 21-24, 1987: Proceedings. Volume 4. pp 545-550. New York: Institute of Electrical and Electronics Engineers.

[Eckmiller, R 88a] Eckmiller, Rolf. (1988). *Neural Networks for Motor Program Generation.* In Eckmiller, Rolf (ed.); von der Malsburg, Christoph (ed.). Neural Computers. NATO ASI Series, Volume F41. pp 359-370. Berlin: Springer-Verlag.

[Edelman, GM 82a] Edelman, Gerald M.; Reeke Jr, George N. (1982). *Selective Networks Capable of Representative Transformations, Limited Generalizations, and Associative Memory.* Proceedings of the National Academy of Sciences. Volume 79. pp 2091-2095.

[Edelman, GM 84*] SEE: Reeke Jr, GN 84a. *Selective Networks and Recognition Automata.*

[Edelman, GM 85*] SEE: Finkel, LH 85a. *Interaction of Synaptic Modification Rules within Populations of Neurons.*

[Edelman, GM 86*] SEE: Reeke Jr, GN 86a. *Recognition Automata Based on Selective Neural Networks.*

[Edelman, GM 87*] SEE: Pearson, JC 87a. *Plasticity in the Organization of Adult Cerebral Cortical Maps: A Computer Simulation Based on Neuronal Group Selection.*

[Edelman, GM 87*] SEE: Finkel, LH 87a. *Population Rules for Synapses in Networks.*

[Edelman, GM 87*] SEE: Reeke Jr, GN 87a. *Selective Neural Networks and Their Implications for Recognition Automata.*

[Edelman, GM 87a] Edelman, Gerald M.; Finkel, Leif H.; Pearson, John C. (1987). *Computer Simulation of Somatotopic Map Plasticity.* Journal of Physiology. Volume 385. pp 27-28.

[Edelman, GM 88*] SEE: Reeke Jr, GN 88a. *Real Brains and Artificial Intelligence.*

[Edelman, GM 89*] SEE: Finkel, LH 89a. *A Population Approach to the Neural Basis of Perceptual Categorization.*

[Eeckman, FH 86*] SEE: Colvin, ME 86a. *Model for Cortical Function.*

[Eeckman, FH 86a] Eeckman, Frank H.; Freeman, Walter J. (1986). *Sigmoid Nonlinearity in Neural Computation: An Experimental Approach.* In Denker, John S. (ed.). Neural Networks for Computing, Snowbird, Utah, 1986. AIP Proceedings 151. pp 135-139. New York: American Institute of Physics.

[Eeckman, FH 88a] Eeckman, Frank H. (1988). *The Sigmoid Nonlinearity in Prepyriform Cortex.* In Anderson, Dana Z. (ed.). Neural Information Processing Systems. pp 242-248. New York: American Institute of Physics.

[Eich, JM 82a] Eich, J. M. (1982). *A Composite Holographic Associative Recall Model.* Psychological Review. Volume 89. pp 627-661.

[Eichel, JE 87*] SEE: Honrubia, V 87a. *Morphological Aspects of the Vestibular Nerve of the Squirrel Monkey.*

[Eichmann, G 85a] Eichmann, George; Caulfield, H. John. (1985). *Optical Learning (Inference) Machines.* Applied Optics. Volume 24. pp 2051-2054.

[Eilbert, JL 87*] SEE: Guez, A 87a. *Neuromorphic Architecture for Adaptive Robot Control: A Preliminary Analysis.*

[Eilbert, JL 87a] Eilbert, James L.; Guez, Allon. (1987). *The Control of Behavior in a Hierarchical Neural Network: Modelling Attention Channels and Behavior Modes.* IEEE International Conference on Neural Networks (ICNN-87), San Diego, CA, June 21-24, 1987: Proceedings. Volume 3. pp 121-130. New York: Institute of Electrical and Electronics Engineers.

[Eilbert, JL 88*] SEE: Guez, A 88a. *Neural Network Architecture for Control.*

[Eimas, PD 73a] Eimas, Peter D.; Corbit, John D. (1973). *Selective Adaptation of Linguistic Feature Detectors.* Cognitive Psychology. Volume 4. pp 99-109.

[Eisenberg, J 87*] SEE: Freeman, WJ 87a. *Hardware Simulation of Brain Dynamics in Learning: The SPOCK.*

[Ekeberg, O 83*] SEE: Lansner, A 83a. A Program Package Implementing a System of Interacting Adaptive Associative Networks.

[Ekeberg, O 85*] SEE: Lansner, A 85b. *Reliability and Speed of Recall in an Associative Network.*

[Ekeberg, O 87*] SEE: Lansner, A 87a. *An Associative Network Solving the "4-Bit ADDER Problem".*

[Elbaum, C 76*] SEE: Cooper, LN 76a. *Information Processing System.*

[Elbaum, C 81*] SEE: Cooper, LN 81b. Information Processing System Using Threshold Passive Modification.

[Elbaum, C 82*] SEE: Reilly, DL 82a. *A Neural Model for Category Learning.*

[Elbaum, C 82*] SEE: Cooper, LN 82a. Self Organizing General Pattern Class Separator and Identifier.

[Elbaum, C 87*] SEE: Reilly, DL 87a. *Learning System Architectures Composed of Multiple Learning Modules.*

[Elbaum, C 88*] SEE: Scofield, CL 88b. *Pattern Class Degeneracy in an Unrestricted Storage Density Memory.*

[Elidan, J 87a] Elidan, Josef; Langhofer, Laurn; Honrubia, Vicente. (1987). *The Neural Generators of the Vestibular Evoked Response.* Brain Research. Volume 423. pp 385-390.

[Elidan, J 89a] Elidan, Josef; Langhofer, Laurn; Honrubia, Vicente. (1989). *The Firing Properties of Second-Order Vestibular Neurons in Correlation with the Far-Field Recorded Vestibular-Evoked Response.* Laryngoscope. Volume 99.

[Elinson, MI 81a] Elinson, M. I.; Sharov, A. M. (1981). *Creation of Efficient Recognition Systems and the Problems of Microelectronics.* Soviet Microelectronics. Volume 10. pp 13-19.

[El-Leithy, N 85a] El-Leithy, N.; Newcomb, R. W. (1985). *Overview of Neural-Type Electronics.* Midwest Symposium on Circuits and Systems, 28th, Louisville, KY, August 19-20, 1985: Proceedings.

[El-Leithy, N 87*] SEE: Newcomb, RW 87a. *Semistate Description of an MOS Neural-Type Cell.*

[El-Leithy, N 87a] El-Leithy, N.; Newcomb, R. W.; Zaghoul, M. (1987). *A Basic MOS Neural-Type Junction: A Perspective on Neural-Type Microsystems.* IEEE International Conference on Neural Networks (ICNN-87), San Diego, CA, June 21-24, 1987: Proceedings. Volume 3. pp 469-478. New York: Institute of Electrical and Electronics Engineers.

[El-Leithy, N 87b] El-Leithy, N.; Newcomb, R. W. (1987). *A Semistate Model for Neural-Type Junction Circuits.*

Mathematical Theory Networks and Systems (MTNS-87), Phoenix, AZ, June 15-19, 1987: Proceedings.

[Ellias, SA 75a] Ellias, S. A.; Grossberg, Stephen. (1975). *Pattern Formation, Contrast Control, and Oscillations in the Short Term Memory of Shunting On-Center Off-Surround Networks.* Biological Cybernetics. Volume 20. pp 69-98.

[Elliot, DF 87a] Elliot, Douglas F. (ed.). (1987). Handbook of Digital Signal Processing. Orlando, FL: Academic Press.

[Elliott, H 82*] SEE: Hansen, FR 82a. *Image Segmentation Using Simple Markov Field Models.*

[Elman, JL 83a] Elman, Jeffrey L.; McClelland, James L. (1983). Speech Perception as a Cognitive Process: The Interactive Activation Model. ICS-8302. San Diego, CA: University of California, Institute for Cognitive Science.

[Elman, JL 86*] SEE: McClelland, JL 86e. *Interactive Processes in Speech Perception: The TRACE Model.*

[Elman, JL 86*] SEE: McClelland, JL 86k. *The TRACE Model of Speech Perception.*

[Elman, JL 86a] Elman, Jeffrey L.; McClelland, James L. (1986). *Exploiting Lawful Variability in the Speech Wave.* Invariance and Variability in Speech Processes. pp 360-380. Hillsdale, NJ: Lawrence Erlbaum.

[Elman, JL 87a] Elman, Jeffrey L.; Zipser, David. (1987). Learning the Hidden Structure of Speech. ICS-8701. San Diego, CA: University of California, Institute for Cognitive Science.

[Elul, R 82*] SEE: Anninos, PA 82a. *Encoding of Visual Information: Correlation of EEG Response to Intracellular Response.*

[Elul, R 83*] SEE: Anninos, PA 83b. *Artificial Neural Nets: Dependence of the EEG Amplitude's Probability Distribution on Statistical Parameters.*

[Emerling, MR 85*] SEE: Sivilotti, MA 85a. *A Novel Associative Memory Implemented Using Collective Computation.*

[Emerling, MR 86*] SEE: Sivilotti, MA 86a. *VLSI Architectures for Implementation of Neural Networks.*

[Enger, PS 69a] Enger, P. S.; Jansen, J. K. S.; Walloe, Lars. (1969). *A Biological Model of the Excitation of a Second Order Sensory Neurone.* Kybernetik. Volume 6. pp 141-149.

[Epworth, RW 87*] SEE: Howard, RE 87a. *An Associative Memory Based on an Electronic Neural Network Architecture.*

[Eriksen, CW 70a] Eriksen, Charles W.; Rohrbaugh, John. (1970). *Visual Masking in Multielement Displays.* Journal of Experimental Psychology. Volume 83. pp 147-154.

[Eriksson, JL 88*] SEE: Sclabassi, RJ 88b. *Nonlinear Systems Analysis of the Hippocampal Perforant Path-Dentate Projection. I. Theoretical and Interpretational Considerations.*

[Eriksson, JL 88*] SEE: Berger, TW 88a. *Nonlinear Systems Analysis of the Hippocampal Perforant Path-Dentate Projection. II. Effects of Random Impulse Train Stimulation.*

[Eriksson, JL 88*] SEE: Berger, TW 88b. *Nonlinear Systems Analysis of the Hippocampal Perforant Path-Dentate Projection. III. Comparison of Random Train and Paired Impulse Stimulation.*

[Ermentrout, GB 79a] Ermentrout, G. B.; Cowan, Jack D. (1979). *A Mathematical Theory of Visual Hallucination Patterns.* Biological Cybernetics. Volume 34. pp 137-150.

[Ermentrout, GB 80a] Ermentrout, G. B.; Cowan, Jack D. (1980). *Large Scale Spatially Organized Activity in Neural Nets.* SIAM Journal on Applied Mathematics. Volume 38. pp 1-21.

[Ermentrout, GB 80b] Ermentrout, G. B.; Cowan, Jack D. (1980). *Secondary Bifurcation in Neuronal Nets.* SIAM Journal on

Applied Mathematics. Volume 39. pp 323-340.

[Ersoy, OK 87a] Ersoy, O. K.; Chen, C. H.; Yu, P-T; Yegani, P. (1987). *Information Processing and Storage Based on Transforms and Neural Networks.* IEEE International Conference on Neural Networks (ICNN-87), San Diego, CA, June 21-24, 1987: Proceedings. Volume 3. pp 777-784. New York: Institute of Electrical and Electronics Engineers.

[Ersu, E 81a] Ersu, Enis. (1981). *A Learning Mechanism for an Associative Storage System.* International Conference on Cybernetics and Society, Atlanta, GA October 26-28, 1981: Proceedings. pp 511-515. New York: Institute of Electrical and Electronics Engineers.

[Ersu, E 86a] Ersu, Enis; Tolle, Henning. (1986). *Acceleration of Learning by Variable Generalization for on-Line Self-Organizing Control.* Polish-English Seminar on Real-Time Process Control, 5th, Warsaw, September, 1986.

[Ersu, E 87a] Ersu, Enis; Tolle, Henning. (1988). *Learning Control Structures with Neuron-Like Associative Memory Systems.* In von Seelen, Werner (ed.); Shaw, G. (ed.); Leinhos, U. M. (ed.). Organization of Neural Networks. Weinheim: VCH Verlagsges.

[Ersu, E 88a] Ersu, Enis; Tolle, Henning. (1988). *Hierarchical Learning Control: An Approach with Neuron-Like Associative Memories.* In Anderson, Dana Z. (ed.). Neural Information Processing Systems. pp 249-261. New York: American Institute of Physics.

[Eshelman, L 88*] SEE: Tan, M 88a. The Impact of Noise on Learning.

[ESNC 88a] ESNC. (1988). European Seminar on Neural Computing. London: IBC Technical Services.

[ESNC 89a] ESNC. (1989). The Second European Seminar on Neural Computing: Commercial Prospects. London: IBC Technical Services.

[Estes, WK 50a] Estes, W. K. (1950). *Toward a Statistical Theory of Learning.* Psychological Review. Volume 57. pp 94-107.

[Estes, WK 54a] Estes, W. K.; Straughan, J. H. (1954). *Analysis of a Verbal Conditioning Situation in Terms of Statistical Learning Theory.* Journal of Experimental Psychology. Volume 47. pp 225-234.

[Estes, WK 57a] Estes, W. K. (1957). *Theory of Learning with Constant, Variable, or Contingent Probabilities of Reinforcement.* Psychometrika. Volume 22. pp 113-132.

[Estes, WK 59*] SEE: Bush, RR 59a. Studies in Mathematical Learning Theory.

[Estes, WK 59a] Estes, W. K. (1959). *The Statistical Approach to Learning Theory.* In Koch, Sigmund (ed.). Psychology: A Study of a Science. Volume 2. pp 380-491. New York: McGraw-Hill.

[Estes, WK 72a] Estes, W. K. (1972). *Research and Theory on the Learning of Probabilities.* Journal of the American Statistical Association. Volume 67. pp 81-102.

[Estes, WK 76a] Estes, W. K.; Allmeyer, D. H.; Reder, S. M. (1976). *Serial Position Functions for Letter Identification at Brief and Extended Exposure Durations.* Perception and Psychophysics. Volume 19. pp 1-15.

[Evans, EF 74a] Evans, E. F. (1974). *Neural Processes for the Detection of Acoustic Patterns and for Sound Localization.* The Neurosciences Third Study Program. pp 131-145. Cambridge, MA: MIT Press.

[Evans, Jr, JM 76*] SEE: Albus, JS 76a. *Robot Systems.*

[Everitt, B 74a] Everitt, B. (1974). Cluster Analysis. Reviews of Current Res Vol 11. London: Heinemann Educational.

[Faber, DS 87*] SEE: Korn, H 87a. *Spontaneous Quantal Currents in a Central Neuron Match Predictions from Binomial Analysis* *of Evoked Responses*.

[Fabrikant, J 88*] SEE: Spitzer, AR 88a. A Method for Determining Optimal Interelectrode Spacing for Cerebral Topographic Mapping.

[Fagg, AH 88*] SEE: Taber, WR 88d. *The Recognition of Orca Calls with a Neural Network*.

[Fahlman, SE 75a] Fahlman, Scott E. (1975). The Intersection Problem. Working Paper 115. Cambridge, MA: Massachusetts Institute of Technology, AI Laboratory.

[Fahlman, SE 79a] Fahlman, Scott E. (1979). NETL: A System for Representing and Using Real-World Knowledge. Cambridge, MA: MIT Press.

[Fahlman, SE 80a] Fahlman, Scott E. (1980). Design Sketch for a Million-Element NETL Machine. Pittsburgh, PA: Carnegie Mellon University, Computer Science Department.

[Fahlman, SE 80b] Fahlman, Scott E. (1980). Hashnet Interconnection Scheme. CMU-CS-80-125. Pittsburgh, PA: Carnegie Mellon University, Computer Science Department.

[Fahlman, SE 81a] Fahlman, Scott E.; Touretzky, David S.; van Roggen, Walter. (1981). *Cancellation in a Parallel Semantic Network*. International Joint Conference on Artificial Intelligence (IJCAI-81), 7th, Vancouver, August 24-28, 1981: Proceedings.

[Fahlman, SE 81b] Fahlman, Scott E. (1981). *Representing Implicit Knowledge*. In Hinton, Geoffrey E. (ed.). Parallel Models of Associative Memory. pp 145-160. Hillsdale, NJ: Lawrence Erlbaum.

[Fahlman, SE 82a] Fahlman, Scott E. (1982). *Three Flavors of Parallelism*. Canadian Society for Computational Studies of Intelligence, 4th, Saskatoon, Saskatchewan, May 17-19, 1982: Proceedings. pp 230-235. Canadian Society for Computational Studies of Intelligence.

[Fahlman, SE 83a] Fahlman, Scott E.; Hinton, Geoffrey E.; Sejnowski, Terrence J. (1983). *Massively Parallel Architectures for AI: Netl, Thistle, and Boltzmann Machines*. National Conference on Artificial Intelligence (AAAI-83), 2nd, Washington, DC, August 22-26, 1983: Proceedings. pp 109-113. Washington, DC: American Association for Artificial Intelligence.

[Fahlman, SE 87a] Fahlman, Scott E.; Hinton, Geoffrey E. (1987). *Connectionist Architectures for Artificial Intelligence*. IEEE Computer. Volume 20. pp 100-109.

[Fahlman, SE 88a] Fahlman, Scott E. (1988). An Empirical Study of Learning Speed in Back-Propagation Networks. CMU-CS-88-162. Pittsburgh, PA: Carnegie Mellon University, Computer Science Department.

[Fairhurst, MC 76a] Fairhurst, M. C.; Stonham, T. J. (1976). *A Classification System for Alpha-Numeric Characters Based on Learning Network Techniques*. Digital Processes. Volume 2. pp 321-339.

[Fairhurst, MC 80a] Fairhurst, M. C.; Goutos, G. P. (1980). *An Evaluation of Cell Function in Relation to Reverberatory Modes in Cellular Networks*. IEEE Transactions on Systems, Man and Cybernetics. Volume SMC-10. pp 655-659.

[Falk, DS 85*] SEE: Stork, DG 85a. Seeing the Light: Optics in Nature, Photography, Color, Vision, and Holography.

[Fallside, F 86*] SEE: Prager, RW 86a. *Boltzmann Machines for Speech Recognition*.

[Fallside, F 87*] SEE: Chan, L 87a. *An Adaptive Training Algorithm for Back Propagation Networks*.

[Fallside, F 87*] SEE: Prager, RW 87a. *A Comparison of the Boltzmann Machine and the Back Propagation Network as Recognizers of Static Speech patterns*.

[Fallside, F 87*] SEE: Robinson, AJR 87a. The Utility Driven Dynamic Error Propagation Network.

[Fallside, F 88*] SEE: Harrison, TD 88a. A Connectionist Structure for Phoneme Recognition.

[Fallside, F 88*] SEE: Robinson, AJR 88a. A Dynamic Connectionist Model for Phoneme Recognition: Preliminary Results.

[Fallside, F 88*] SEE: Chong, MWH 88a. Implementation of Neural Networks for Speech Recognition on a Transputer Array.

[Fallside, F 88*] SEE: Plumbley, MD 88a. An Information-Theoretic Approach to Unsupervised Connectionist Models.

[Fallside, F 88*] SEE: Prager, RW 88a. The Modified Kanerva Model for Automatic Speech Recognition.

[Fallside, F 88*] SEE: Niranjan, M 88a. Neural Networks and Radial Basis Functions in Classifying Static Speech Patterns.

[Fallside, F 88*] SEE: Robinson, AJR 88b. *Static and Dynamic Error Propagation Networks with Application to Speech Coding.*

[Fallside, F 88a] Fallside, Frank; Chan, Lai-Wan. (1986). Connectionist Models and Geometric Reasoning. CUED/F-CAMS/TR.266. Cambridge: Cambridge University, Engineering Department.

[Fallside, F 88b] Fallside, Frank. (1988). Connectionist Models in Speech Recognition: A Brief Summary and Bibliography. CUED/F-INFENG/TR.18. Cambridge: Cambridge University, Engineering Department.

[Fanardjian, VV 82*] SEE: Melkonian, DS 82a. *Simulation of Learning Processes in Neuronal Networks of the Cerebellum.*

[Fanty, MA 85a] Fanty, Mark A. (1985). Context-Free Parsing in Connectionist Networks. TR-174. Rochester, NY: University of Rochester, Computer Science Department.

[Fanty, MA 86a] Fanty, Mark A. (1986). A Connectionist Simulator for the BBN Butterfly Multi-Processor. TR-164. Rochester, NY: University of Rochester, Computer Science Department.

[Fanty, MA 86b] Fanty, Mark A. (1986). *Context-Free Parsing with Connectionist Networks.* In Denker, John S. (ed.). Neural Networks for Computing, Snowbird, Utah, 1986. AIP Proceedings 151. pp 140-145. New York: American Institute of Physics.

[Fanty, MA 88*] SEE: Feldman, JA 88a. *Computing with Structured Neural Networks.*

[Farber, R 86*] SEE: Lapedes, A 86a. *Programming a Massively Parallel, Computation Universal System: Static Behavior.*

[Farber, R 87*] SEE: Lapedes, A 87a. Nonlinear Signal Processing Using Neural Networks: Prediction and System Modelling.

[Farber, R 88*] SEE: Lapedes, A 88a. *How Neural Networks Work.*

[Farhat, NH 84*] SEE: Psaltis, D 84a. *A New Approach to Optical Information Processing Based on the Hopfield Model.*

[Farhat, NH 85*] SEE: Psaltis, D 85a. *Optical Information Processing Based on an Associative Memory Model of Neural Nets with Thresholding and Feedback.*

[Farhat, NH 85a] Farhat, Nabil H.; Psaltis, Demetri; Prata, Aluizio; Paek, Eung Gi. (1988). *Optical Implementation of the Hopfield Model.* In Anderson, James A. (ed.); Rosenfeld, Edward (ed.). Neurocomputing: Foundations of Research. pp 653-660. Cambridge, MA: MIT Press.

[Farhat, NH 85b] Farhat, Nabil H.; Psaltis, Demetri; Prata, Aluizio; Paek, Eung Gi. (1985). *Optical Implementation of the Hopfield Model.* Applied Optics. Volume 24. pp 1469-1475.

[Farhat, NH 86a] Farhat, Nabil H. (1986). *Neural Net Models and Optical Computing: A Brief Overview.* In Szu, Harold H.

(ed.). Optical and Hybrid Computing. SPIE Volume 634. pp 307-311. Bellingham, WA: Society of Photo-Optical Instrumentation Engineers.

[Farhat, NH 86b] Farhat, Nabil H.; Miyahara, S.; Lee, K. S. (1986). *Optical Analog of Two-Dimensional Neural Networks and Their Application in Recognition of Radar Targets.* In Denker, John S. (ed.). Neural Networks for Computing, Snowbird, Utah, 1986. AIP Proceedings 151. pp 146-152. New York: American Institute of Physics.

[Farhat, NH 87a] Farhat, Nabil H.; Shae, Z. Y. (1987). *Architectures and Methodologies for Self-Organization and Stochastic Learning in Opto-Electronic Analogs of Neural Nets.* IEEE International Conference on Neural Networks (ICNN-87), San Diego, CA, June 21-24, 1987: Proceedings. Volume 3. pp 565-576. New York: Institute of Electrical and Electronics Engineers.

[Farhat, NH 87b] Farhat, Nabil H. (1987). *Optoelectronic Analogs of Self-Programming Neural Nets: Architecture and Methodologies for Implementing Fast Stochastic Learning by Simulated Annealing.* Applied Optics. Volume 26. pp 5093-5103.

[Farley, BG 54a] Farley, B. G.; Clark, W. A. (September, 1954). *Simulation of Self-Organizing Systems by Digital Computer.* IRE Transactions on Information Theory. PGIT-4. pp 76-84.

[Farley, BG 55*] SEE: Clark, WA 55a. *Generalization of Pattern Recogntion in a Self-Organizing System.*

[Farley, BG 60a] Farley, B. G.; Clark, W. A. (1961). *Activity in Networks of Neuron-Like Elements.* Information Theory, London Symposium, 4th, Held at the Royal Institution, London, August 29th-September 2nd, 1960. pp 242-251. Belfast: Universities Press.

[Farley, E 87*] SEE: Hennessey, AK 87b. *The Gamma Network as an Interconnection Architecture for Neural Networks.*

[Farley, ET 87a] Farley, E. T.; Hennessey, A. Kathleen. (1987). A Prototype Crossbar Switch for the Gamma Network. Lubbock, TX: Texas Tech University.

[Farlow, SJ 84a] Farlow, Stanley J. (1984). *A FORTRAN Program for the GMDH Algorithm.* In Farlow, Stanley J. (ed.). Self-Organizing Methods in Modeling: GMDH Type Algorithms. pp 277-290. New York: Marcel Dekker.

[Farlow, SJ 84b] Farlow, Stanley J. (ed.). (1984). Self-Organizing Methods in Modeling: GMDH Type Algorithms. New York: Marcel Dekker.

[Farlow, SJ 84c] Farlow, Stanley J. (1984). *The GMDH Algorithm.* In Farlow, Stanley J. (ed.). Self-Organizing Methods in Modeling: GMDH Type Algorithms. pp 1-24. New York: Marcel Dekker.

[Farmer, JD 86*] SEE: Crutchfield, JP 86a. *Chaos.*

[Farr, T 86*] SEE: Stromberg, W 86a. *A Fourier-Based Textual Feature Extraction Procedure.*

[Farrell, JA 88*] SEE: Michel, AN 88a. *Qualitative Analysis of Neural Networks.*

[Farrell, JA 88*] SEE: Michel, AN 88b. *Stability Results for Neural Networks.*

[Fattu, JM 86*] SEE: Patrick, EA 86a. Artificial Intelligence with Statistical Pattern Recognition.

[Feigenbaum, MJ 78a] Feigenbaum, Mitchell J. (1978). *Quantitative Universality for a Class of Nonlinear Transformations.* Journal of Statistical Physics. Volume 19. pp 25-52.

[Feigenbaum, MJ 79a] Feigenbaum, Mitchell J. (1979). *The Universal Metric Properties of Nonlinear Transformations.* Journal of Statistical Physics. Volume 21. pp 669-706.

[Feild Jr, WB 88a] Feild Jr, William B.; Navlakha, Jainendra K. (in press). On Choosing Non-Random Input Exemplars to Form

Stable States in a Hopfield Neural Net.

[Feild Jr, WB 88b] Feild Jr, William B.; Navlakha, Jainendra K. (in press). Transputer Implementation of Hopfield Neural Network.

[Feinstein, DI 83*] SEE: Hopfield, JJ 83a. *Unlearning Has a Stabilizing Effect in Collective Memories.*

[Feldman, JA 74a] Feldman, Jerome A.; Yakimovsky, Yoram. (1974). *Decision Theory and Artificial Intelligence: I. A Semantics-Based Region Analyzer.* Artificial Intelligence. Volume 5. pp 349-371.

[Feldman, JA 80a] Feldman, Jerome A.; Nigam, A. (1980). *A Model and Proof Technique for Message-Based Systems.* SIAM Journal on Computing. Volume 9. pp 768-784.

[Feldman, JA 81a] Feldman, Jerome A.; Ballard, Dana H. (1981). Computing with Connections. TR-72. Rochester, NY: University of Rochester, Computer Science Department.

[Feldman, JA 81b] Feldman, Jerome A. (1981). *Connectionist Model of Visual Memory.* In Hinton, Geoffrey E. (ed.). Parallel Models of Associative Memory. pp 49-82. Hillsdale, NJ: Lawrence Erlbaum.

[Feldman, JA 81c] Feldman, Jerome A. (1981). Memory and Change in Connection Networks. Report TR-96. Rochester, NY: University of Rochester, Computer Science Department.

[Feldman, JA 82a] Feldman, Jerome A.; Ballard, Dana H. (1982). *Connectionist Models and Their Properties.* Cognitive Science. Volume 6. pp 205-254.

[Feldman, JA 82b] Feldman, Jerome A.; Ballard, Dana H. (1988). *Connectionist Models and their Properties.* In Anderson, James A. (ed.); Rosenfeld, Edward (ed.). Neurocomputing: Foundations of Research. pp 484-508. Cambridge, MA: MIT Press.

[Feldman, JA 82c] Feldman, Jerome A. (1982). *Dynamic Connections in Neural Networks.* Biological Cybernetics. Volume 46. pp 27-39.

[Feldman, JA 82d] Feldman, Jerome A. (1982). Four Frames Suffice. TR-99. Rochester, NY: University of Rochester, Computer Science Department.

[Feldman, JA 84*] SEE: Shastri, L 84a. Semantic Networks and Neural Nets.

[Feldman, JA 85*] SEE: Shastri, L 85a. *Evidential Reasoning in Semantic Networks: A Formal Theory.*

[Feldman, JA 85a] Feldman, Jerome A. (1985). *Connections.* BYTE. Volume 10. pp 277-284.

[Feldman, JA 85b] Feldman, Jerome A. (1985). Energy and the Behavior of Connectionist Models. TR-155. Rochester, NY: University of Rochester, Department of Computer Science.

[Feldman, JA 85c] Feldman, Jerome A. (1985). *Four Frames Suffice: A Provisional Model of Vision and Space.* Behavioral and Brain Sciences. Volume 8. pp 265-289.

[Feldman, JA 85d] Feldman, Jerome A. (ed.). (1985). *Special Issue on Connectionist Models and Their Application.* Cognitive Science. Volume 9.

[Feldman, JA 86*] SEE: McClelland, JL 86c. *Connectionist Models and Cognitive Science: Goals, Directions and Implications.*

[Feldman, JA 86a] Feldman, Jerome A. (1986). Neural Representation of Conceptual Knowledge. TR-189. Rochester, NY: University of Rochester, Department of Computer Science.

[Feldman, JA 87a] Feldman, Jerome A. (1987). *A Functional Model of Vision and Space.* In Arbib, Michael A. (ed.); Hanson, Allen R. (ed.). Vision, Brain and Cooperative Computation. pp 531-562. Cambridge, MA: MIT Press.

[Feldman, JA 88*] SEE: Waltz, DL 88b. Connectionist Models and Their Implications: Readings from Cognitive Science.

[Feldman, JA 88*] SEE: Waltz, DL 88a. *Connectionist Models and Their Implications.*

[Feldman, JA 88a] Feldman, Jerome A.; Fanty, Mark A.; Goddard, Nigel H. (1988). *Computing with Structured Neural Networks.* Computer. Volume 21. pp 91-103.

[Feldman, JA 88b] Feldman, Jerome A.; Ballard, Dana H. (1988). *Connectionist Models and Their Properties.* In Waltz, David L. (ed.); Feldman, Jerome A. (ed.). Connectionist Models and Their Implications: Readings from Cognitive Science. pp 13-62. Norwood, NJ: Ablex.

[Feldman, JA 88c] Feldman, Jerome A. (1988). *Connectionist Representation of Concepts.* In Waltz, David L. (ed.); Feldman, Jerome A. (ed.). Connectionist Models and Their Implications: Readings from Cognitive Science. pp 341-364. Norwood, NJ: Ablex.

[Feldman, JA 88d] Feldman, Jerome A. (1988). *Structured Neural Networks in Nature and in Computer Science.* In Eckmiller, Rolf (ed.); von der Malsburg, Christoph (ed.). Neural Computers. NATO ASI Series, Volume F41. pp 17-21. Berlin: Springer-Verlag.

[Feldman, JA 89a] Feldman, Jerome A. (1989). *Neural Representation of Conceptual Knowledge.* In Nadel, Lynn (ed.); Cooper, Lynn A. (ed.); Culicover, Peter (ed.); Harnish, R. Michael (ed.). Neural Connections, Mental Computation. pp 68-103. Cambridge, MA: MIT Press.

[Feldman, JL 75a] Feldman, J. L.; Cowan, Jack D. (1975). *Large-Scale Activity in Neural-Nets II: A Model for the Brainstem Respiratory Oscillator.* Biological Cybernetics. Volume 17. pp 39-51.

[Fenanzo Jr, AJ 86a] Fenanzo Jr, A. J. (1986). *Darwinian Evolution as a Paradigm for AI Research.* SIGART Newsletter. Number 97. pp 22-23.

[Fender, DH 62a] Fender, Derek H. (1964). *The Eye-Movement Control System: Evolution of a Model.* Neural Theory and Modeling, Proceedings of the 1962 Ojai Symposium. pp 306-324. Stanford, CA: Stanford University Press.

[Feng, Y 88*] SEE: Nasrabadi, NM 88a. *Vector Quantization of Images Based Upon the Kohonen Self-Organizing Feature Maps.*

[Fernandes, PML 80a] Fernandes, P. M. L.; Da Silva, K. M. C. (1980). *Nerve Cell Soma Model with High Reliability and Low Power Consumption.* Medical & Biological Engineering & Computing. Volume 18. pp 261-264.

[Fernandez, MF 88a] Fernandez, Manuel F. (1988). *On Tropistic Processing and Its Applications.* In Anderson, Dana Z. (ed.). Neural Information Processing Systems. pp 262-269. New York: American Institute of Physics.

[Ferrara, Jr, ER 81a] Ferrara, Jr, Earl R.; Widrow, Bernard. (1981). *Multichannel Adaptive Filtering for Signal Enhancement.* IEEE Transactions on Circuits and Systems. Volume CAS-28. pp 606-610.

[Ferrara, Jr, ER 81b] Ferrara, Jr, Earl R.; Widrow, Bernard. (1981). *The Time-Sequenced Adaptive Filter.* IEEE Transactions on Circuits and Systems. Volume CAS-28. pp 519-523.

[Ferrate, GA 80*] SEE: Benito, S 80a. *Simulation and Evaluation of Ill Defined Neuronal Networks with Random Structure and Parameter Setting.*

[Ferry, DK 88*] SEE: Akers, LA 88a. *Limited Interconnectivity in Synthetic Neural Systems.*

[Ferster, D 87a] Ferster, David; Koch, Christof. (1987). *Neuronal Connections Underlying Orientation Selectivity in Cat*

Visual Cortex. Trends in Neurosciences. Volume 10. pp 487-492.

[Fetterly, D 84*] SEE: Casasent, DP 84a. *Recent Advances in Optical Pattern Recognition.*

[Fetz, EE 88a] Fetz, Eberhard E. (1988). *Correlational Strength and Computational Algebra of Synaptic Connections Between Neurons*. In Anderson, Dana Z. (ed.). Neural Information Processing Systems. pp 270-277. New York: American Institute of Physics.

[Feustel, TC 85*] SEE: Salasoo, A 85a. *Building Permanent Memory Codes: Codification and Repetition Effects in Word Identification.*

[Fiddy, MA 87*] SEE: Bayley, JS 87a. *On the Use of the Hopfield Model for Optical Pattern Recognition.*

[Fiddy, MA 88*] SEE: Abbiss, JB 88a. *Super-Resolution and Neural Computing.*

[Fienberg, SE 74a] Fienberg, Stephen E. (1974). *Stochastic Models for Single Neuron Firing Trains: A Survey*. Biometrics, Journal of the Biometric Society. Volume 30. pp 399-427.

[Fiesler, E 88a] Fiesler, E.; Choudry, A.; Caulfield, H. John. (1988). Weight Discretization in Backward Error Propagation Neural Networks. Huntsville, AL: University of Alabama, Computer Science Department.

[Fikes, RE 71a] Fikes, Richard E.; Nilsson, Nils J. (1971). *STRIPS: A New Approach to the Application of Theorem Proving to Problem Solving*. Artificial Intelligence. Volume 2. pp 189-208.

[Findler, NV 79a] Findler, Nicholas V. (1979). Associative Networks. New York: Academic Press.

[Findler, NV 79b] Findler, Nicholas V. (ed.). (1979). Associative Networks: Representation and Use of Knowledge by Computers. New York: Academic Press.

[Finette, S 78a] Finette, S.; Harth, Erich M.; Csermely, T. J. (1978). *Anisotropic Connectivity and Cooperative Phenomena as a Basis for Orientation Sensitivity in the Visual Cortex*. Biological Cybernetics. Volume 30. pp 231-240.

[Finkel, LH 85a] Finkel, Leif H.; Edelman, Gerald M. (1985). *Interaction of Synaptic Modification Rules within Populations of Neurons*. Proceedings of the National Academy of Sciences. Volume 82. pp 1291-1295.

[Finkel, LH 87*] SEE: Edelman, GM 87a. *Computer Simulation of Somatotopic Map Plasticity.*

[Finkel, LH 87*] SEE: Pearson, JC 87a. *Plasticity in the Organization of Adult Cerebral Cortical Maps: A Computer Simulation Based on Neuronal Group Selection.*

[Finkel, LH 87a] Finkel, Leif H.; Edelman, Gerald M. (1987). *Population Rules for Synapses in Networks*. Synaptic Function. pp 711-757. New York: Wiley.

[Finkel, LH 88a] Finkel, Leif H. (1988). *Creativity, Selection, and Categorization*. Journal of Social and Biological Structures. Volume 11. pp 86-89.

[Finkel, LH 88b] Finkel, Leif H. (1988). *Neuronal Group Selection: A Basis for Categorization by the Nervous System*. The Alphabet and the Brain. pp 51-69. New York: Springer-Verlag.

[Finkel, LH 89a] Finkel, Leif H.; Reeke Jr, George N.; Edelman, Gerald M. (1989). *A Population Approach to the Neural Basis of Perceptual Categorization*. Neural Connections, Mental Computation. pp 146-179. Cambridge, MA: MIT Press.

[Finkel, RA 77*] SEE: Friedman, JH 77a. *An Algorithm for Finding Best Matches in Logarithmic Expected Time.*

[Firschein, O 62*] SEE: Fischler, M 62a. *An Approach to General Pattern Recognition.*

[Firschein, O 63*] SEE: Mattson, RL 63a. *An Experimental Investigation of a Class of Pattern Recognition Synthesis Algorithms.*

[Firschein, O 63a] Firschein, O.; Fischler, M. (1963). *Automatic Subclass Determination for Pattern-Recognition Applications.* IEEE Transactions on Electronic Computers. Volume EC-12. pp 137-141.

[Fischer, FP 70a] Fischer, F. P.; Patrick, Edward A. (1970). *A Preprocessing Algorithm for Nearest Neighbor Decision Rules.* National Electronics Conference, 26th: Proceedings. pp 481-485.

[Fischer, S 88*] SEE: Dammasch, IE 88a. *A Model of Synaptogenetic Consequences of Afferent Input in Neuronal Nets.*

[Fischler, I 71*] SEE: Juola, JF 71a. *Recognition Time for Information Stored in Long-Term Memory.*

[Fischler, M 62a] Fischler, M.; Mattson, R. L.; Firschein, O.; Healy, L. D. (1962). *An Approach to General Pattern Recognition.* IRE Transactions on Information Theory. Volume IT-8. pp 64-73.

[Fischler, M 63*] SEE: Firschein, O 63a. *Automatic Subclass Determination for Pattern-Recognition Applications.*

[Fischler, M 63*] SEE: Mattson, RL 63a. *An Experimental Investigation of a Class of Pattern Recognition Synthesis Algorithms.*

[Fischler, MA 81*] SEE: Garvey, TD 81a. *An Inference Technique for Integrating Knowledge from Disparate Sources.*

[Fisher, AD 77a] Fisher, Arthur D. (December 1977). Plain Talk About Neurodevelopmental Epistemology and Brain Theory. (Unpublished Report). Cambridge, MA: Massachusetts Institute of Technology.

[Fisher, AD 81*] SEE: Warde, C 81a. *Optical Information Processing Characteristics of the Microchannel Spatial Light Modulator.*

[Fisher, AD 84a] Fisher, Arthur D.; Giles, C. Lee; Lee, John N. (1984). *Associative Processor Architectures for Optical Computing.* Journal of the Optical Society of America, A. Volume 1. pp 1337-1338.

[Fisher, AD 85a] Fisher, Arthur D.; Giles, C. Lee; Lee, John N. (March 1985). *An Adaptive Optical Computing Element.* Proceedings of the OSA Topical Meeting.

[Fisher, AD 85b] Fisher, Arthur D.; Giles, C. Lee. (1985). *Optical Adaptive Associative Computer Architectures.* Technological Leverage: A Competitive Necessity, Spring CompCon IEEE Computer Society International Conference, 30th, San Francisco, CA, February 25-28, 1985. pp 342-344. Institute of Electrical and Electronics Engineers.

[Fisher, AD 86a] Fisher, Arthur D.; Fukuda, Robert C.; Lee, John N. (1986). *Implementations of Adaptive Associative Optical Computing Elements.* Optical Computing. SPIE Volume 625. pp 196-204. Bellingham, WA: Society of Photo-Optical Instrumentation Engineers.

[Fisher, AD 87a] Fisher, Arthur D. (1987). *On Applying Associative Networks: An Approach for Representation and Goal-Directed Learning.* IEEE International Conference on Neural Networks (ICNN-87), San Diego, CA, June 21-24, 1987: Proceedings. Volume 4. pp 677-686. New York: Institute of Electrical and Electronics Engineers.

[Fisher, AD 87b] Fisher, Arthur D.; Lippincott, Wendy L.; Lee, John N. (1987). *Optical Implementations of Associative Networks with Versatile Adaptive Learning Capabilities.* Applied Optics. Volume 26. pp 5039-5054.

[Fisher, ME 80a] Fisher, Michael E.; Selke, Walter. (1980). *Infinitely Many Commensurate Phases in a Simple Ising Model.* Physical Review Letters. Volume 44. pp 1502-1505.

[Fisher, ME 81a] Fisher, Michael E.; Selke, Walter. (1981). *Low Temperature Analysis of the Axial Next-Nearest Neighbor Ising Model Near its Multiphase Point.* Philosophical Transactions of the Royal Society of London, Series-A.

Volume 302. pp 1-44.

[Fitzgerald, ML 79*] SEE: Barbera, AJ 79a. *Hierarchical Control of Robots Using Microcomputers.*

[Fitzgerald, ML 80*] SEE: Albus, JS 80a. *A Measurement and Control Model for Adaptive Robots.*

[Fitzgerald, ML 85*] SEE: Albus, JS 85a. *Hierarchical Control for Robots, and Teleoperators.*

[FitzHugh, R 59a] FitzHugh, Richard; Antosiewicz, H. A. (1959). *Automatic Computation of Nerve Excitation - Detailed Corrections and Additions.* Journal of the Society for Industrial and Applied Mathematics. Volume 7. pp 447-458.

[FitzHugh, R 69a] FitzHugh, Richard. (1969). *Mathematical Models of Excitation and Propagation in Nerve.* Biological Engineering. Electrical Series Volume 9. pp 1-85. New York: McGraw-Hill.

[Fleisher, M 88a] Fleisher, Michael. (1988). *The Hopfield Model with Multi-Level Neurons.* In Anderson, Dana Z. (ed.). Neural Information Processing Systems. pp 278-289. New York: American Institute of Physics.

[Fletcher, R 80a] Fletcher, Roger. (1980). Practical Methods of Optimization, Volume 1: Unconstrained Optimization. New York: Wiley.

[Flinn, PA 74a] Flinn, P. A. (1974). *Monte Carlo Calculation of Phase Separation in a Two-Dimensional Ising System.* Journal of Statistical Physics. Volume 10. pp 89-97.

[Flood, MM 87a] Flood, Merrill M. (1987). *A Screening Algorithm for Levenglick's Consensus Ranking Model.* IEEE International Conference on Neural Networks (ICNN-87), San Diego, CA, June 21-24, 1987: Proceedings. Volume 4. pp 741-744. New York: Institute of Electrical and Electronics Engineers.

[Flores, I 60a] Flores, I.; Grey, L. (1960). *Optimization of Reference Signals for Character Recognition Systems.* IRE Transactions on Electronic Computers. Volume EC-9. pp 54-61.

[Fodor, JA 88a] Fodor, Jerry A.; Pylyshyn, Zenon W. (1988). *Connectionism and Cognitive Architecture: A Critical Analysis.* In Pinker, Steven (ed.); Mehler, Jacques (ed.). Connections and Symbols. pp 3-72. Cambridge, MA: MIT Press.

[Fohlmeister, JF 73a] Fohlmeister, Jurgen F. (1973). *A Model for Phasic and Tonic Repetitively Firing Neuronal Encoders.* Kybernetik. Volume 13. pp 104-112.

[Fong, Y 86*] SEE: Johnson, K 86a. *A Simplified Model of Retinal Function with Parallel Architecture and Integrated Space-Time Filtering for Low Level Visual Processing.*

[Fong, Y 87*] SEE: Johnson, K 87a. *Two New Theories of Retinal Function.*

[Fong, Y 88*] SEE: Hayduk, JW 88a. Architecture for Fast Learning of Decision Regions in Neural Networks.

[Foo, YS 87*] SEE: Takefuji, Y 87a. *Error Correcting System Based on Neural Circuits.*

[Foo, YS 87a] Foo, Yoon-Pin Simon; Takefuji, Yoshiyasu. (1987). *A Neural Network Scheme for 3-Dimensional Pattern Recognition and Construction.* IEEE International Conference on Neural Networks (ICNN-87), San Diego, CA, June 21-24, 1987: Proceedings. Volume 4. pp 301-308. New York: Institute of Electrical and Electronics Engineers.

[Foo, YS 88a] Foo, Yoon-Pin Simon; Takefuji, Yoshiyasu. (1988). *Integer Linear Programming Neural Networks for Job-Shop Scheduling.* IEEE International Conference on Neural Networks (ICNN-88), San Diego, CA, July 24-27, 1988: Proceedings. Volume 2. pp 341-348. New York: Institute of Electrical and Electronics Engineers.

[Foo, YS 88b] Foo, Yoon-Pin Simon; Takefuji, Yoshiyasu. (1988). *Stochastic Neural Networks for Solving Job-Shop*

Scheduling: Part 1. Problem Representation. IEEE International Conference on Neural Networks (ICNN-88), San Diego, CA, July 24-27, 1988: Proceedings. Volume 2. pp 275-282. New York: Institute of Electrical and Electronics Engineers.

[Foo, YS 88c] Foo, Yoon-Pin Simon; Takefuji, Yoshiyasu. (1988). *Stochastic Neural Networks for Solving Job-Shop Scheduling: Part 2. Architecture and Simulations.* IEEE International Conference on Neural Networks (ICNN-88), San Diego, CA, July 24-27, 1988: Proceedings. Volume 2. pp 283-290. New York: Institute of Electrical and Electronics Engineers.

[Forrest, BM 86*] SEE: Bruce, AD 86a. *Learning and Memory Properties in Fully Connected Networks.*

[Forrest, BM 87*] SEE: Rohwer, R 87a. *Training Time-Dependence in Neural Networks.*

[Forrest, BM 87a] Forrest, Bruce M.; Roweth, D.; Stroud, N.; Wallace, D. J.; Wilson, G. V. (1987). *Implementing Neural Network Models on Parallel Computers.* Computer Journal. Volume 30. pp 413-419.

[Foss, DJ 80a] Foss, Donald J.; Blank, Michelle A. (1980). *Identifying the Speech Codes.* Cognitive Psychology. Volume 12. pp 1-31.

[Fox, EA 83*] SEE: Salton, G 83a. *Extended Boolean Information Retrieval.*

[Fralick, SC 67a] Fralick, Stanley C. (1967). *Learning to Recognize Patterns Without a Teacher.* IEEE Transactions on Information Theory. Volume IT-13. pp 57-64.

[Frankel, DS 84a] Frankel, Donald S. (1984). *Pattern Recognition of Fourier Transform Infared Spectra of Organic Compounds.* Analytical Chemistry. Volume 56. pp 1011-1014.

[Frankel, DS 87*] SEE: Draper, JS 87a. *A Microcomputer Neural Net Benchmarked Against Standard Classification Techniques.*

[Franzini, MA 86a] Franzini, Michael A. (1986). Consciousness and the Goals of Artificial Intelligence. Pittsburgh, PA: Carnegie Mellon University, Psychology Department.

[Franzini, MA 87a] Franzini, Michael A. (1987). A Connectionist Model for Unsupervised Feature Discovery and Pattern Recognition. Pittsburgh, PA: Carnegie Mellon University, Psychology Department.

[Fraser, MD 88a] Fraser, Martin D.; Gagliano, Ross A. (1988). *The Simulation of Energy and Memory in Neural Networks.* Modeling and Simulation Conference, 19th, Pittsburgh, PA, May 5-6, 1988: Proceedings. Volume 19. pp 637-642. Research Triangle Park, NC: Instrument Society of America.

[Freed, MA 86*] SEE: Smith, RG 86a. *Microcircuitry of the Dark-Adapted Cat Retina: Functional Architecture of the Rod-Cone Network.*

[Freed, MA 87*] SEE: Sterling, P 87a. *Microcircuitry of the On-Beta Ganglion Cell in Daylight, Twilight, and Starlight.*

[Freed, MA 87a] Freed, Michael A.; Smith, Robert G.; Sterling, Peter. (1987). *Rod Bipolar Array in the Cat Retina: Pattern of Input from Rods and GABA-Accumulating Amacrine Cells.* Journal of Comparative Neurology. Volume 266. pp 445-455.

[Freed, MA 88*] SEE: Sterling, P 88a. *Architecture of Rod and Cone Circuits to the On-Beta Ganglion Cell.*

[Freed, MA 88a] Freed, Michael A.; Sterling, Peter. (1988). *The On-Alpha Ganglion Cell of the Cat Retina and Its Presynaptic Cell Types.* Journal of Neuroscience. Volume 8. pp 2303-2320.

[Freeman, JH 70*] SEE: Cooper, DB 70a. *On the Asymptotic Improvement in the Outcome of Supervised Learning Provided by Additional Non-Supervised Learning.*

[Freeman, WJ 74*] SEE: Ahn, SM 74a. *Steady-State and Limit Cycle Activity of Mass of Neurons Forming Simple Feedback Loops (I): Lumped Circuit Model.*

[Freeman, WJ 74*] SEE: Ahn, SM 74b. *Steady-State and Limit Cycle Activity of Mass of Neurons Forming Simple Feedback Loops (II): Lumped Circuit Model.*

[Freeman, WJ 75b] Freeman, Walter J. (1975). *Parallel Processing of Signals in Neural Nets as Manifested in the EEG.* International Journal of Man-Machine Studies. Volume 7. pp 347-369.

[Freeman, WJ 79a] Freeman, Walter J. (1979). *EEG Analysis Gives Models of Neuronal Template-Matching Mechanism for Sensory Search with Olfactory Bulb.* Biological Cybernetics. Volume 35. pp 221-234.

[Freeman, WJ 81a] Freeman, Walter J. (1981). *A Neural Mechanism for Generalization over Equivalent Stimuli in the Olfactory System.* Mathematical Psychology and Psychophysiology. SIAM-AMS Proceedings Volume 13. pp 25-38. Providence, RI: American Mathematical Society.

[Freeman, WJ 83a] Freeman, Walter J. (1983). *Dynamics of Image Formation by Nerve Cell Assemblies.* Synergetics of the Brain, International Symposium on Synergetics, Schloss Elmau, Bavaria, May 2-7, 1983: Proceedings. pp 102-121. New York: Springer-Verlag.

[Freeman, WJ 86*] SEE: Grajski, KA 86a. *Classification of EEG Spatial Patterns with a Tree-Structured Methodology: CART.*

[Freeman, WJ 86*] SEE: Grajski, KA 86b. *Dynamic Model of Olfactory Discrimination.*

[Freeman, WJ 86*] SEE: Eeckman, FH 86a. *Sigmoid Nonlinearity in Neural Computation: An Experimental Approach.*

[Freeman, WJ 86a] Freeman, Walter J.; Viana Di Prisco, Gonzalo. (1986). *Relation of Olfactory EEG to Behavior: Time Series Analysis.* Behavioral Neuroscience. Volume 100. pp 753-763.

[Freeman, WJ 87*] SEE: Skarda, CA 87a. *How Brains Make Chaos in Order to Make Sense of the World.*

[Freeman, WJ 87a] Freeman, Walter J.; Eisenberg, Joe; Burke, Brian. (1987). *Hardware Simulation of Brain Dynamics in Learning: The SPOCK.* IEEE International Conference on Neural Networks (ICNN-87), San Diego, CA, June 21-24, 1987: Proceedings. Volume 3. pp 435-442. New York: Institute of Electrical and Electronics Engineers.

[Freeman, WJ 87b] Freeman, Walter J.; Grajski, Kamil A. (1987). *Relation of Olfactory EEG to Behavior: Factor Analysis.* Behavioral Neuroscience. Volume 101. pp 766-777.

[Freeman, WJ 87c] Freeman, Walter J.; Baird, Bill. (1987). *Relation of Olfactory EEG to Behavior: Spatial Analysis.* Behavioral Neuroscience. Volume 101. pp 393-408.

[Freeman, WJ 87d] Freeman, Walter J. (1987). *Simulation of Chaotic EEG Patterns with a Dynamic Model of the Olfactory System.* Biological Cybernetics. Volume 56. pp 139-150.

[Freeman, WJ 87e] Freeman, Walter J.; van Dijk, Bob W. (1987). *Spatial Patterns of Visual Cortical Fast EEG During Conditioned Reflex in a Rhesus Monkey.* Brain Research. Volume 422. pp 267-276.

[Freeman, WJ 88*] SEE: Grajski, KA 88b. *Spatial EEG Correlates of Non-Associative and Associative Olfactory Learning in Rabbits.*

[Freeman, WJ 88a] Freeman, Walter J.; Yao, Yong; Burke, Brian. ((in press)). *Central Pattern Generating and Recognizing in Olfactory Bulb: A Correlation Learning Rule.* Neural Networks.

[Freeman, WJ 88b] Freeman, Walter J. (1988). *Why Neural Networks Don't Fly: Inquiry into the Neurodynamics of Biological Intelligence.* IEEE International Conference on Neural Networks (ICNN-88), San Diego, CA, July 24-27, 1988:

Proceedings. Volume 2. pp 1-8. New York: Institute of Electrical and Electronics Engineers.

[French, T 88*] SEE: Gonzales, MEQ 88a. A Naturalistic Approach to Mental Representation.

[Fricot, J 86*] SEE: Demongeot, J 86a. *Random Fields and Spatial Renewal Potentials.*

[Friedberg, RM 58a] Friedberg, R. M. (1958). *A Learning Machine: Part 1.* IBM Journal of Research and Development. Volume 2. pp 2-13.

[Frieden, BR 72a] Frieden, B. Roy. (1972). *Restoring with Maximum Likelihood and Maximum Entropy.* Journal of the Optical Society of America. Volume 62. pp 511-518.

[Friedlander, CB 86*] SEE: Athale, RA 86a. *Attentive Associative Architectures and Their Implications to Optical Computing.*

[Friedlander, CB 86*] SEE: Athale, RA 86b. *Attentive Associative Memory and Its Optical Implementation.*

[Friedlander, CB 86*] SEE: Athale, RA 86c. *Optical Implementation of Associative Memory with Controlled Nonlinearity in the Correlation Domain.*

[Friedman, D 88*] SEE: Pati, YC 88a. *Neural Networks for Tactile Perception.*

[Friedman, JH 75a] Friedman, Jerome H.; Baskett, Forest; Shustek, Leonard J. (1975). *An Algorithm for Finding Nearest Neighbors.* IEEE Transactions on Computers. Volume C-24. pp 1000-1006.

[Friedman, JH 77a] Friedman, Jerome H.; Bentley, Jon Louis; Finkel, Raphael Ari. (1977). *An Algorithm for Finding Best Matches in Logarithmic Expected Time.* ACM Transactions on Mathematical Software. Volume 3. pp 209-226.

[Friedman, WF 87*] SEE: Black, IB 87a. *Biochemistry of Information Storage in the Nervous System.*

[Friedrich, R 87*] SEE: Fuchs, A 87a. *Spatio-Temporal Analysis of Multi-channel Alpha EEG Map Series.*

[Friesen, WO 78a] Friesen, W. Otto; Stent, Gunther S. (1978). *Neural Circuits for Generating Rhythmic Movements.* Annual Review of Biophysics and Bioengineering. Volume 7. pp 37-61.

[Friesen, WO 81*] SEE: Mitchell, CE 81a. *A Neuromime System for Neural Circuit Analysis.*

[Frishkopf, LS 68a] Frishkopf, L. S.; Capranica, Robert R.; Goldstein, Jr., Moise H. (1968). *Neural Coding in the Bullfrog's Auditory System: A Teleological Approach.* Proceedings of the IEEE. Volume 56. pp 969-980.

[Frohn, H 87a] Frohn, H.; Geiger, H.; Singer, W. (1987). *Addendum: A Self-Organizing Neural Network Sharing Features of the Mammalian Visual System.* Biological Cybernetics. Volume 57. pp 145.

[Froind, I 86*] SEE: Cohen, A 86a. Software Package for Interactive Text Independent, Speaker Verification.

[Fry, DB 57a] Fry, D. B.; Denes, P. (1957). *On Presenting the Output of a Mechanical Speech Recognizer.* Journal of the Acoustical Society of America. Volume 29. pp 364-367.

[Fu, K 66*] SEE: Nicolic, ZJ 66a. *An Algorithm for Learning without External Supervision and its Application to Learning Control Systems.*

[Fu, K 70*] SEE: Mendel, JM 70a. Adaptation, Learning, and Pattern Recognition Systems: Theory and Applications.

[Fu, K 70*] SEE: Kashyap, RL 70a. *Stochastic Approximation.*

[Fu, K 70a] Fu, King-Sun. (1970). *Learning Control Systems: A Review and Outlook.* IEEE Transactions on Automatic

Control. Volume AC-15. pp 210-221.

[Fu, K 71*] SEE: Gilstad, DW 71a. *Two-Dimensional Adaptive Model of a Human Controller Using Pattern Recognition Techniques.*

[Fu, K 86*] SEE: Young, TY 86a. Handbook of Pattern Recognition and Image Processing.

[Fu, K 86a] Fu, King-Sun. (1986). *Syntactic Pattern Recognition.* In Young, Tzay Y. (ed.); Fu, King-Sun (ed.). Handbook of Pattern Recognition and Image Processing. pp 85-118. New York: Academic Press.

[Fuchs, A 87a] Fuchs, A.; Friedrich, R.; Haken, Hermann; Lehmann, Dietrich. (1987). *Spatio-Temporal Analysis of Multi-channel Alpha EEG Map Series.* In Haken, Hermann (ed.). Computational Systems - Natural and Artificial. pp 74-83. Heidelberg: Springer.

[Fuchs, A 88a] Fuchs, A.; Haken, Hermann. (1988). *Computer Simulations of Pattern Recognition as a Dynamical Process of a Synergetic System.* Neural and Synergetic Computers. Series in Synergetics Vol 42. pp 16-28. New York: Springer-Verlag.

[Fuhrmann, G 81a] Fuhrmann, Gy. (1981). *Modelling the Visual Cortex with "Modulo System" Concept.* Biological Cybernetics. Volume 40. pp 39-48.

[Fujita, M 87a] Fujita, Masahiko. ((in press)). *A Model for Oblique Saccade Generation and Adaptation.* In Arbib, Michael A. (ed.); Amari, Shun-ichi (ed.). Dynamic Interaction in Neural Networks: Models and Data. New York: Springer-Verlag.

[Fujita, S 84a] Fujita, Shin-ichi; Koi, Hiroshi. (1984). *Application of GMDH to Environmental System Modeling and Management.* In Farlow, Stanley J. (ed.). Self-Organizing Methods in Modeling: GMDH Type Algorithms. pp 257-276. New York: Marcel Dekker.

[Fujiwara, S 87*] SEE: Okajima, K 87a. *A Heteroassociative Memory Network with Feedback Connection.*

[Fukaya, M 87a] Fukaya, M.; Kitagawa, M.; Okabe, Y. (1987). *Two-Level Neural Networks: Learning by Interaction with Environment.* IEEE International Conference on Neural Networks (ICNN-87), San Diego, CA, June 21-24, 1987: Proceedings. Volume 2. pp 531-539. New York: Institute of Electrical and Electronics Engineers.

[Fukuda, RC 86*] SEE: Fisher, AD 86a. *Implementations of Adaptive Associative Optical Computing Elements.*

[Fukunaga, K 73a] Fukunaga, Keinosuke; Hostetler, Larry D. (1973). *Optimization of k-Nearest-Neighbor Density Estimates.* IEEE Transactions on Information Theory. Volume IT-19. pp 320-326.

[Fukunaga, K 75a] Fukunaga, Keinosuke; Narendra, Patrenahalli M. (1975). *A Branch and Bound Algorithm for Computing k-Nearest Neighbors.* IEEE Transactions on Computers. Volume C-24. pp 750-753.

[Fukunaga, K 86a] Fukunaga, Keinosuke. (1986). *Statistical Pattern Classification.* In Young, Tzay Y. (ed.); Fu, King-Sun (ed.). Handbook of Pattern Recognition and Image Processing. pp 3-32. New York: Academic Press.

[Fukushima, K 69a] Fukushima, Kunihiko. (1969). *Visual Feature Extraction by a Multi-Layered Network of Analog Threshold Elements.* IEEE Transactions on Systems, Science and Cybernetics. Volume SSC-5. pp 322-333.

[Fukushima, K 70a] Fukushima, Kunihiko. (1970). *A Feature Extractor for Curvilinear Patterns: A Design Suggested by the Mammalian Visual System.* Kybernetik. Volume 7. pp 153-160.

[Fukushima, K 71a] Fukushima, Kunihiko. (1971). *Feature-Extracting Network for Geometrical Patterns: A Design with Reference to Visual Neural Systems.* NHK Technical Reports. Volume 23. p 351 f.

[Fukushima, K 73a] Fukushima, Kunihiko. (1973). *A Model of Associative Memory in the Brain.* Kybernetik. Volume 12. pp

58-63.

[Fukushima, K 75*] SEE: Hirai, Y 75a. *A Model of Neural Network Extracting Binocular Parallax.*

[Fukushima, K 75a] Fukushima, Kunihiko; Miyake, Sei. (1975). *A Self-Organizing Multilayer Neural Network.* Systems, Computers, Controls. Volume 6. pp 15-22.

[Fukushima, K 75b] Fukushima, Kunihiko. (1975). *Cognitron: A Self-Organizing Multilayered Neural Network.* Biological Cybernetics. Volume 20. pp 121-136.

[Fukushima, K 77a] Fukushima, Kunihiko; Miyake, Sei. (1977). *Self-Organizing Neural Network with a Function of Associative Memory - Feedback Type Cognitron -.* Transactions of the Institute of Electronics and Communication Engineers of Japan. Volume E60. pp 143-150.

[Fukushima, K 78*] SEE: Hirai, Y 78a. *An Inference upon the Neural Network Finding Binocular Correspondence.*

[Fukushima, K 78a] Fukushima, Kunihiko; Miyake, Sei. (1978). *Self-Organizing Neural Network with a Function of Associative Memory: Feedback Type Cognitron.* Biological Cybernetics. Volume 28. pp 201-208.

[Fukushima, K 79a] Fukushima, Kunihiko. (1979). *Improvement in Pattern-Selectivity of a Cognitron.* Transactions of the Institute of Electronics and Communication Engineers of Japan. Volume E62. pp 650-657.

[Fukushima, K 79b] Fukushima, Kunihiko. (1979). *Neural Network Model for a Mechanism of Pattern Recogition Unaffected by Shift in Position - Neocognitron -.* Transactions of the Institute of Electronics and Communication Engineers of Japan. Volume E62. pp 658-665.

[Fukushima, K 79c] Fukushima, Kunihiko. (1979). *Self-Organization of a Neural Network Which Gives Position-Invariant Response.* International Joint Conference on Artificial Intelligence (IJCAI-79), 6th, Tokyo, August 20-23, 1979: Proceedings. pp 291-293.

[Fukushima, K 80a] Fukushima, Kunihiko. (1980). *Neocognitron: A Self-Organizing Neural Network Model for a Mechanism of Pattern Recognition Unaffected by Shift in Position.* Biological Cybernetics. Volume 36. pp 193-202.

[Fukushima, K 80b] Fukushima, Kunihiko; Miyake, Sei. (1980). *Neocognitron: Self-Organizing Network Capable of Position-Invariant Recognition of Patterns.* International Conference on Pattern Recognition, 5th, Miami Beach, FL, December 1-4, 1980: Proceedings. Volume 1. pp 459-461.

[Fukushima, K 80c] Fukushima, Kunihiko; Miyake, Sei. (1980). *Neocognitron: Self-Organizing Network Capable of Position-Invariant Recognition of Patterns.* International Conference on Pattern Recognition, 5th, Miami Beach, FL, December 1-4, 1980: Proceedings. Volume 1. pp 459-461. New York: Institute of Electrical and Electronics Engineers.

[Fukushima, K 81a] Fukushima, Kunihiko. (1981). Cognitron: A Self-Organizing Multi-Layered Neural Network Model. Technical Monograph Number 30. Tokyo: NHK Broadcasting Science Research Laboratories.

[Fukushima, K 82a] Fukushima, Kunihiko; Miyake, Sei. (1982). *Neocognitron: A New Algorithm for Pattern Recognition Tolerant of Deformation and Shifts in Position.* Pattern Recognition. Volume 15. pp 455-469.

[Fukushima, K 82b] Fukushima, Kunihiko; Miyake, Sei. (1982). *Neocognitron: A Self-Organizing Neural Network Model for a Mechanism of Visual Pattern Recognition.* In Amari, Shun-ichi (ed.); Arbib, Michael A. (ed.). Competition and Cooperation in Neural Nets. pp 248-266. New York: Springer-Verlag.

[Fukushima, K 83*] SEE: Miyake, S 83a. *A Neural Network Model for the Mechanism of Feature Extraction - A Self-Organizing Network with Feedback Inhibition.*

[Fukushima, K 83a] Fukushima, Kunihiko. (1983). *A Hierarchical Neural Network Model for Associative Memory.* Electronics and Communications in Japan. Volume 66-A. pp 43-52.

[Fukushima, K 83b] Fukushima, Kunihiko; Miyake, Sei; Ito, Takayuki. (1983). *Neocognitron: A Neural Network Model for a Mechanism of Visual Pattern Recognition.* IEEE Transactions on Systems, Man and Cybernetics. Volume SMC-13. pp 826-834.

[Fukushima, K 83c] Fukushima, Kunihiko; Miyake, Sei; Ito, Takayuki. (1988). *Neocognitron: A Neural Network Model for a Mechanism of Visual Pattern Recognition.* In Anderson, James A. (ed.); Rosenfeld, Edward (ed.). Neurocomputing: Foundations of Research. pp 526-534. Cambridge, MA: MIT Press.

[Fukushima, K 84a] Fukushima, Kunihiko. (1984). *A Hierarchical Neural Network Model for Associative Memory.* Biological Cybernetics. Volume 50. pp 105-113.

[Fukushima, K 84b] Fukushima, Kunihiko; Miyake, Sei. (1984). *A Neural Network Model for the Mechanism of Feature Extraction.* Biological Cybernetics. Volume 50. pp 377-384.

[Fukushima, K 86*] SEE: Miyake, S 86a. *Neural Network Model for the Mechanism of Pattern Information Processing.*

[Fukushima, K 86a] Fukushima, Kunihiko. (1986). *A Neural Network Model for Selective Attention in Visual Pattern Recognition.* Biological Cybernetics. Volume 55. pp 5-15.

[Fukushima, K 87a] Fukushima, Kunihiko. (1987). *A Neural Network Model for Selective Attention.* IEEE International Conference on Neural Networks (ICNN-87), San Diego, CA, June 21-24, 1987: Proceedings. Volume 2. pp 11-18. New York: Institute of Electrical and Electronics Engineers.

[Fukushima, K 87b] Fukushima, Kunihiko. (1987). *A Neural Network Model for Selective Attention.* IEEE International Conference on Neural Networks (ICNN-87), San Diego, CA, June 21-24, 1987: Proceedings. Volume 2. pp 11-18. New York: Institute of Electrical and Electronics Engineers.

[Fukushima, K 87c] Fukushima, Kunihiko. (1987). *Neural Network Model for Selective Attention in Visual Pattern Recognition and Associative Recall.* Applied Optics. Volume 26. pp 4985-4992.

[Fukushima, K 87d] Fukushima, Kunihiko. (1987). *Neural Network Model for Selective Attention in Visual Pattern Recognition Associative Recall.* Applied Optics. Volume 26. pp 4985-4992.

[Fukushima, K 87e] Fukushima, Kunihiko. (1987). *Self-Organizing Neural Network Models for Visual Pattern Recognition.* Acta Neurochirurgica. Supplement 41. pp 51-67. New York: Springer-Verlag.

[Fukushima, K 88*] SEE: Miyake, S 88c. *Self-Organizing Neural Network with the Mechanism of Feedback Information Processing.*

[Fukushima, K 88a] Fukushima, Kunihiko. (1988). *A Hierarchical Neural Network Model for Selective Attention.* In Eckmiller, Rolf (ed.); von der Malsburg, Christoph (ed.). Neural Computers. NATO ASI Series, Volume F41. pp 81-90. Berlin: Springer-Verlag.

[Fukushima, K 88b] Fukushima, Kunihiko. (March, 1988). *A Neural Network for Visual Pattern Recognition.* Computer. Volume 21. pp 65-74.

[Fukushima, K 88c] Fukushima, Kunihiko. (1988). *Neocognitron: A Hierarchical Neural Network Capable of Visual Pattern Recognition.* Neural Networks. Volume 1. pp 119-130.

[Fukushima, Y 81a] Fukushima, Yasuo. (1981). *Single Channel Potassium Currents of the Anomalous Rectifier.* Nature. Volume 294. pp 368-371.

[Fuller, MS 83a] Fuller, Marc S.; Williams, William J. (1983). *A Continuous Information Theoretic Approach to the Analysis of Cutaneous Receptor Neurons*. Biological Cybernetics. Volume 47. pp 13-16. New York: Springer-Verlag.

[Fuortes, MGF 68*] SEE: Borsellino, AI 68a. *Interpretation of Responses of Visual Cells of Limulus*.

[Furman, RE 87*] SEE: Mueller, P 87a. *Neural Computation of Visual Images*.

[Furukawa, K 87*] SEE: Kawato, M 87b. *A Hierarchical Neural-Network Model for Control and Learning of Voluntary Movement*.

[Futami, R 82*] SEE: Hoshimiya, N 82a. *A Preliminary Study of the Phoneme Recognition with Hardware Neural Network*.

[Fynewever, M 72*] SEE: Clema, JK 72a. *The Dynamic Re-Evaluation of Alternatives and the Emulation of Human Decision Making*.

[Gabor, D 46a] Gabor, Denis. (1946). *Theory of Communications*. Journal of the IEE. Volume 93. pp 429-457.

[Gabor, D 48a] Gabor, Denis. (1948). *A New Microscopic Principle*. Nature. Volume 161. pp 777-778.

[Gabor, D 51a] Gabor, Denis. (1951). *Microscopy by Reconstructed Wave Fronts: II*. Proceedings of the Physical Society. Volume 64. pp 449-469.

[Gabor, D 60a] Gabor, Denis; Wilby, W. P. L.; Woodcock, R. (1960). *A Universal Non-Linear Filter, Predictor, and Simulator which Optimizes Itself by a Learning Process*. Proceedings of the IEE. Volume 108. pp 422-435.

[Gabor, D 65a] Gabor, Denis. (1965). *Character Recognition by Holography*. Nature. Volume 208. pp 422-423.

[Gabor, D 68a] Gabor, Denis. (1968). *Improved Holographic Model of Temporal Recall*. Nature. Volume 217. pp 1288-1289.

[Gabor, D 69a] Gabor, Denis. (1969). *Associative Holographic Memories*. IBM Journal of Research and Development. Volume 156. pp 156-159.

[Gagliano, RA 88*] SEE: Fraser, MD 88a. *The Simulation of Energy and Memory in Neural Networks*.

[Gagne, S 87*] SEE: Ogmen, H 87a. *Dynamic Pattern Matching with Filling-In*.

[Gagne, S 87*] SEE: Ogmen, H 87b. *Early Spatio-Temporal Integration by Gated Competitive Networks*.

[Gagne, S 87*] SEE: Ogmen, H 87c. *Low Level Encoding of Motion by Asynchronous Shunting Networks*.

[Gagne, S 87*] SEE: Ogmen, H 87d. *Phototransduction in Invertebrates*.

[Gaines, BR 83a] Gaines, Brian R. (1983). *Precise Past, Fuzzy Future*. International Journal of Man-Machine Studies. Volume 19. pp 117-134.

[Gaines, BR 87a] Gaines, Brian R. (1987). *Positive Feedback Processes Underlying Functional Differentiation*. IEEE International Conference on Neural Networks (ICNN-87), San Diego, CA, June 21-24, 1987: Proceedings. Volume 2. pp 387-394. New York: Institute of Electrical and Electronics Engineers.

[Gaines, BR 87b] Gaines, Brian R. (1987). *Uncertainty as a Foundation of Computational Power in Neural Networks*. IEEE International Conference on Neural Networks (ICNN-87), San Diego, CA, June 21-24, 1987: Proceedings. Volume 3. pp 51-58. New York: Institute of Electrical and Electronics Engineers.

[Gallant, AR 88a] Gallant, A. Ronald; White, Halbert. (1988). There Exists a Neural Network That Does Not Make Avoidable Mistakes. #88-22. San Diego, CA: University of California, Department of Economics.

[Gallant, SI 85a] Gallant, Stephen I. (1985). *Automatic Generation of Expert Systems from Examples*. International

Conference on Artificial Intelligence Applications, 2nd, 1985: Proceedings.

[Gallant, SI 86a] Gallant, Stephen I. (1986). *Brittleness and Machine Learning.* International Meeting on Advances in Learning, Les Arcs, July, 1986.

[Gallant, SI 86b] Gallant, Stephen I. (1986). *Optimal Linear Discriminants.* International Conference on Pattern Recognition, 8th, Paris, October 27-31, 1986: Proceedings. pp 849-852. Washington, DC: Institute of Electrical and Electronics Engineers.

[Gallant, SI 86c] Gallant, Stephen I. (1986). *Three Constructive Algorithms for Network Learning.* Cognitive Science Society, 8th, Amherst, MA, August 15-17, 1986: Proceedings. pp 652-660. Hillsdale, NJ: Lawrence Erlbaum.

[Gallant, SI 86d] Gallant, Stephen I.; Balachandra, Ramaiya. (1986). *Using Automated Techniques to Generate an Expert System for R&D Project Monitoring.* IFAC International Symposium on Economics and Artificial Intelligence, 1st, Aix-En-Provence, September 2-4, 1986.

[Gallant, SI 87a] Gallant, Stephen I. (1987). *Automated Generation of Connectionist Expert Systems for Problems Involving Noise and Redundancy.* AAAI Workshop on Uncertainty, Seattle, WA, 1987.

[Gallant, SI 87b] Gallant, Stephen I.; Smith, Donald. (1987). *Random Cells: An Idea Whose Time Has Come and Gone... And Come Again?.* IEEE International Conference on Neural Networks (ICNN-87), San Diego, CA, June 21-24, 1987: Proceedings. Volume 2. pp 671-678. New York: Institute of Electrical and Electronics Engineers.

[Gallant, SI 87c] Gallant, Stephen I. (1987). Sequential Associative Memories. NU-CCS-87-20. Boston: Northeastern University, College of Computer Science.

[Gallant, SI 88a] Gallant, Stephen I. (in print). *Bayesian Assessment of a Connectionist Model for Fault Detection.* AAAI Workshop on Uncertainty, St Paul, MN, August 19-21, 1988.

[Gallant, SI 88b] Gallant, Stephen I. (1988). *Connectionist Expert Systems.* Communications of the ACM. Volume 31. pp 152-169.

[Gallant, SI 88c] Gallant, Stephen I. (in print). *Example-Based Knowledge Engineering with Connectionist Expert Systems.* IEEE Midcon, Dallas, TX, August 30-September 1, 1988.

[Gallant, SI 88d] Gallant, Stephen I.; King, Donna J. (in print). *Experiments with Sequential Associative Memories.* Cognitive Science Society, 10th, Montreal, August, 1988: Proceedings.

[Gallinari, P 87*] SEE: Soulie, FF 87a. *Evaluation of Network Architectures on Test Learning Tasks.*

[Galvan, J 80*] SEE: Benito, S 80a. *Simulation and Evaluation of Ill Defined Neuronal Networks with Random Structure and Parameter Setting.*

[Gamba, A 61a] Gamba, A. (1961). *Optimum Performance of Learning Machines.* Proceedings of the IRE. Volume 49. pp 349-350.

[Gamkrelidze, RV 62*] SEE: Pontryagin, LS 62a. The Mathematical Theory of Optimal Processes.

[Gantz, J 88a] Gantz, John. (April 4, 1988). *IBM Neural Network Chips Make Everything Obsolete.* Info World. pp 31.

[Garcia, GA 88*] SEE: Shimabukuro, RL 88a. *Dual-Polarity Nonvolatile MOS Analogue Memory (MAM) Cell for Neural-Type Circuitry.*

[Gardner, D 81*] SEE: White, RL 81a. *Self-Inhibition Alters Firing Patterns of Neurons in Aplysia Buccal Ganglia.*

[Gardner, D 86a] Gardner, Daniel. (1986). *Variations in Amplitude and Time Course of Inhibitory Postsynaptic Currents.*

Journal of Neurophysiology. Volume 56. pp 1424-1438.

[Gardner, D 87a] Gardner, Daniel. (1987). *Synaptic Diversity Characterizes Biological Neural Networks.* IEEE International Conference on Neural Networks (ICNN-87), San Diego, CA, June 21-24, 1987: Proceedings. Volume 4. pp 17-22. New York: Institute of Electrical and Electronics Engineers.

[Gardner, D 88a] Gardner, Daniel. (1988). *A Test for Modulation of Synaptic Efficacy by Induced Membrane Potential Variance.* Brain Research Bulletin. Volume 21. pp 521-528.

[Gardner, D 89a] Gardner, Daniel. (1989). *Noise Modulation of Synaptic Weights in a Biological Neural Network.* Neural Networks. Volume 2. pp 69-76.

[Gardner, E 86*] SEE: Bruce, AD 86a. *Learning and Memory Properties in Fully Connected Networks.*

[Gardner, E 86a] Gardner, E. (1986). *Structure of Metastable States in the Hopfield Model.* Journal of Physics A: Mathematical and General. Volume 19. pp L1047-L1052.

[Gardner, E 87*] SEE: Derrida, B 87a. *An Exactly Solvable Asymmetric Neural Network Model.*

[Gardner, E 87a] Gardner, E.; Derrida, B.; Mottishaw, P. (1987). *Zero Temperature Parallel Dynamics for Infinite Range Spin Glasses and Neural Networks.* Journal de Physique. Volume 48. pp 741-755.

[Gardner, E 88*] SEE: Canning, A 88b. *Partially Connected Models of Neural Networks.*

[Gardner, E 88a] Gardner, E.; Stroud, N.; Wallace, D. J. (1988). *Training with Noise: Application to Word and Text Storage.* In Eckmiller, Rolf (ed.); von der Malsburg, Christoph (ed.). Neural Computers. NATO ASI Series, Volume F41. pp 251-260. Berlin: Springer-Verlag.

[Gardner, EJ 87*] SEE: Bruce, AD 87a. *Dynamics and Statistical Mechanics of the Hopfield Model.*

[Gardner Lepp, M 88*] SEE: Vichniac, GY 88a. *A Neural Network for the Optimization of Communications Network Design.*

[Gardner, S 86a] Gardner, Sheldon. (1986). *Application of Neural Network Algorithms and Architectures to Correlation / Tracking and Identification.* In Denker, John S. (ed.). Neural Networks for Computing, Snowbird, Utah, 1986. AIP Proceedings 151. pp 153-157. New York: American Institute of Physics.

[Gardner, S 87a] Gardner, Sheldon. (1987). *Statistical Dynamics of Ultradiffusion in Hierarchical Systems.* IEEE International Conference on Neural Networks (ICNN-87), San Diego, CA, June 21-24, 1987: Proceedings. Volume 3. pp 759-766. New York: Institute of Electrical and Electronics Engineers.

[Gardner, S 88b] Gardner, Sheldon. (1988). *Morphology of Neural Networks in the Mammalian Visual System.* Neural Networks. Volume 1.

[Gardner, S 88c] Gardner, Sheldon. (1988). *Ultradiffusion, Scale Space Transformation, and the Morphology of Neural Networks.* IEEE International Conference on Neural Networks (ICNN-88), San Diego, CA, July 24-27, 1988: Proceedings. Volume 1. pp 617-623. New York: Institute of Electrical and Electronics Engineers.

[Gardner, WA 73a] Gardner, W. A. (1973). *The Structure of Least Mean Square Linear Estimators for Synchronous M-ary Signals.* IEEE Transactions on Information Theory. Volume IT-19. pp 240-243.

[Gardner, WA 81a] Gardner, W. A. (1981). *Design of Nearest Prototype Signal Classifiers.* IEEE Transactions on Information Theory. Volume IT-27. pp 521-522.

[Garey, MR 79a] Garey, Michael R.; Johnson, David S. (1979). Computers and Intractability: A Guide to the Theory of NP-Completeness. San Francisco: Freeman.

[Garth, S 88a] Garth, Simon. (1988). *A Dedicated Computer for Simulation of Large Systems of Neural Nets.* In Eckmiller, Rolf (ed.); von der Malsburg, Christoph (ed.). Neural Computers. NATO ASI Series, Volume F41. pp 435-444. Berlin: Springer-Verlag.

[Garth, SCJ 87a] Garth, Simon C. J. (1987). *A Chipset for High Speed Simulation of Neural Network Systems.* IEEE International Conference on Neural Networks (ICNN-87), San Diego, CA, June 21-24, 1987: Proceedings. Volume 3. pp 443-452. New York: Institute of Electrical and Electronics Engineers.

[Garvey, TD 81a] Garvey, Thomas D.; Lowrance, John D.; Fischler, Martin A. (1981). *An Inference Technique for Integrating Knowledge from Disparate Sources.* International Joint Conference on Artificial Intelligence (IJCAI-81), 7th, Vancouver, August 24-28, 1981: Proceedings. Volume 1. pp 319-325. International Joint Conference on Artificial Intelligence.

[Gass, W 86a] Gass, Wanda; Tarrant, Richard; Doddington, George R. (1986). *A Parallel Signal Processor System.* IEEE International Conference on Acoustics, Speech, and Signal Processing (ICASSP-86), Tokyo, April 7-11, 1986: Proceedings. Volume 4. pp 2887-2890. New York: Institute of Electrical and Electronics Engineers.

[Gat, E 87*] SEE: Norrod, FE 87a. *Feedback-Induced Sequentiality in Neural Networks.*

[Gately, MT 87*] SEE: Collins, DR 87a. *Optical Neurocomputers Implementation Using Deformable Mirror Arrays.*

[Gatley, MT 88a] Gatley, Michael T. (1988). *Cycles: A Simulation Tool for Studying Cyclic Neural Networks.* In Anderson, Dana Z. (ed.). Neural Information Processing Systems. pp 290-296. New York: American Institute of Physics.

[Gaudiano, P 87a] Gaudiano, Paolo; MacGregor, Ronald J. (1987). *Dynamic Activity and Memory Traces in Computer-Simulated Recurrently-Connected Neural Networks.* IEEE International Conference on Neural Networks (ICNN-87), San Diego, CA, June 21-24, 1987: Proceedings. Volume 2. pp 177-186. New York: Institute of Electrical and Electronics Engineers.

[Gaudiano, P 88a] Gaudiano, Paolo. (1988). *Temporal Patterns of Activity in Neural Networks.* In Anderson, Dana Z. (ed.). Neural Information Processing Systems. pp 297-300. New York: American Institute of Physics.

[Gawronski, RP 87a] Gawronski, Richard P.; Love, William R. (1987). *Artina: A Flexible Computer Model of Retinal Processes for Moving Images.* Southeast Regional Conference, 25th, Birmingham, AL, April 1-3, 1987: Proceedings. pp 21-28. New York: Association for Computing Machinery.

[Gawronski, RR 87a] Gawronski, Richard R.; Love, William R. (1987). *Modelling of an Adaptive Neuronlike Network for the Construction of a Learning Control System.* Southeast Regional Conference, 25th, Birmingham, AL, April 1-3, 1987: Proceedings. pp 151-156. New York: Association for Computing Machinery.

[Gawronski, RR 88a] Gawronski, Richard R. (1988). *A Learning Algorithm for Multilayer Neuronlike Network.* Florida Artificial Intelligence Research Symposium, 1st, Orlando, FL, May 4-6, 1988: Proceedings. pp 263-268.

[Gawronski, RR 88b] Gawronski, Richard R. (1988). *Heuristics.* International Encyclopedia of Robotics Applications and Automation. pp 631-637. New York: Wiley.

[Gaylord, TK 84*] SEE: Guest, CC 84a. *EXCLUSIVE OR Processing (Binary Image Subtraction) Using Thick Fourier Holograms.*

[Geffner, H 87a] Geffner, Hector; Pearl, Judea. (1987). *On the Probabilistic Semantics of Connectionist Networks.* IEEE International Conference on Neural Networks (ICNN-87), San Diego, CA, June 21-24, 1987: Proceedings. Volume 2. pp 187-196. New York: Institute of Electrical and Electronics Engineers.

[Geiger, H 87*] SEE: Frohn, H 87a. *Addendum: A Self-Organizing Neural Network Sharing Features of the Mammalian Visual System.*

[Gelade, G 80*] SEE: Treisman, AM 80a. *A Feature-Integration Theory of Attention.*

[Gelatt, CD 83*] SEE: Kirkpatrick, S 83a. *Optimization by Simulated Annealing.*

[Gelatt, CD 83*] SEE: Kirkpatrick, S 83b. *Optimization by Simulated Annealing.*

[Gelenbe, E 87a] Gelenbe, Erol; Stafylopatis, Andreas. (1987). *Temporal Behavior of Neural Networks.* IEEE International Conference on Neural Networks (ICNN-87), San Diego, CA, June 21-24, 1987: Proceedings. Volume 3. pp 267-276. New York: Institute of Electrical and Electronics Engineers.

[Gelfand, IM 71a] Gelfand, I. M.; Tsetlin, M. L. (1971). *Mathematical Modeling of Mechanisms of the Central Nervous System.* In Gelfand, I. M. (ed.); Gurfinkel, V. S. (ed.); Fomin, S. V. (ed.); Tsetlin, M. L. (ed.). Models of the Structural-Functional Organization of Certain Biological Systems. Cambridge, MA: MIT Press.

[Gelfand, JJ 88*] SEE: Handelman, DA 88a. *Integration of Knowledge-Based System and Neural Network Techniques for Robotic Control.*

[Gelfand, JJ 88*] SEE: Pearson, JC 88a. *Neural Network Approach to Sensory Fusion.*

[Gelfand, JJ 88*] SEE: Lane, SH 88a. *A Neural Network Computational Map Approach to Reflexive Motor Control.*

[Gelfand, JJ 88a] Gelfand, Jack J.; Pearson, John C.; Spence, Clay D.; Sullivan, W. E. (1988). *Multisensor Integration in Biological Systems.* IEEE Symposium on Intelligent Control, 3rd, Arlington, VA, August, 1988: Proceedings. New York: Institute of Electrical and Electronics Engineers.

[Gelperim, AE 85a] Gelperim, A. E.; Hopfield, John J.; Tank, David W. (1985). *The Logic of LIMAX Learning.* Model Neural Networks and Behavior. New York: Plenum.

[Geman, D 84*] SEE: Geman, S 84b. *Stochastic Relaxation, Gibbs Distributions, and the Bayesian Restoration of Images.*

[Geman, D 84*] SEE: Geman, S 84a. *Stochastic Relaxation, Gibbs Distributions, and the Bayesian Restoration of Images.*

[Geman, D 86a] Geman, Donald; Geman, Stuart. (1986). *Bayesian Image Analysis.* In Bienenstock, Elie L. (ed.). Disordered Systems and Biological Organization. pp 301-320. New York: Springer-Verlag.

[Geman, S 78a] Geman, Stuart. (1978). *Application of Stochastic Averaging to Learning Systems.* Brain Theory Newsletter. Volume 3. pp 69-71.

[Geman, S 79a] Geman, Stuart. (1979). *Some Averaging and Stability Results for Random Differential Equations.* SIAM Journal of Applied Mathematics. Volume 36. pp 86-105.

[Geman, S 81a] Geman, Stuart. (1981). *Notes on a Self-Organizing Machine.* In Hinton, Geoffrey E. (ed.). Parallel Models of Associative Memory. pp 237-264. Hillsdale, NJ: Lawrence Erlbaum.

[Geman, S 81b] Geman, Stuart. (1981). *The Law of Large Numbers in Neural Modeling.* Mathematical Psychology and Psychophysiology. Providence, RI: American Mathematical Society.

[Geman, S 84a] Geman, Stuart; Geman, Donald. (1984). *Stochastic Relaxation, Gibbs Distributions, and the Bayesian Restoration of Images.* IEEE Transactions on Pattern Analysis and Machine Intelligence. Volume PAMI-6. pp 721-741.

[Geman, S 84b] Geman, Stuart; Geman, Donald. (1988). *Stochastic Relaxation, Gibbs Distributions, and the Bayesian Restoration of Images.* In Anderson, James A. (ed.); Rosenfeld, Edward (ed.). Neurocomputing: Foundations of Research. pp 614-634. Cambridge, MA: MIT Press.

[Geman, S 86*] SEE: Geman, D 86a. *Bayesian Image Analysis.*

[Geoffrion, AM 61a] Geoffrion, Arthur M. (1961). *The Origin and Significance of Perceptron Theory.* Journal of Industrial

Engineering. Volume 12. pp 399-405.

[George, DA 73*] SEE: Kaye, AR 73a. *Relationships Between LMS Compensators for Intersymbol Interference in Linear and Non-Linear Modulation Schemes.*

[Gera, M 88a] Gera, Michael; Aleksander, Igor. (1988). *A Logic Neuron Model of Language Acquisition.* International Neural Nets Conference, Nimes, 1988.

[Gerstein, GL 62a] Gerstein, George L. (1962). *Mathematical Models for the All-Or-None Activity of Some Neurons.* IRE Transactions on Information Theory. Volume IT-8. pp S137-S143.

[Geva, S 87*] SEE: Touretzky, DS 87a. *A Distributed Connectionist Representation for Concept Structures.*

[Gevins, AS 87a] Gevins, Alan S.; Morgan, Nelson H. (1987). *"Ignorance-Based" Neural-Network Signal Processing in Brain Research.* IEEE International Conference on Neural Networks (ICNN-87), San Diego, CA, June 21-24, 1987: Proceedings. Volume 4. pp 705-716. New York: Institute of Electrical and Electronics Engineers.

[Gevins, AS 87b] Gevins, Alan S. (1987). *How the Human Brain Thinks and Acts: Results of Latest Research Using Neural Networks to Analyze Brain Signals.* IEEE International Conference on Neural Networks (ICNN-87), San Diego, CA, June 21-24, 1987: Proceedings. Volume 4. pp 41-48. New York: Institute of Electrical and Electronics Engineers.

[Gevins, AS 88a] Gevins, Alan S.; Morgan, Nelson H. (1988). *Applications of Neural-Network (NN) Signal Processing in Brain Research.* IEEE Transactions on Acoustics, Speech, and Signal Processing. Volume 36. pp 1152-1161.

[Gharavi, H 88*] SEE: Paek, EG 88a. *A Derivation of Neural Network Models for Associative Memory.*

[Giannakopoulos, E 87*] SEE: von Seelen, W 87b. *Properties of Cortical Systems - A Basis for Computer Vision.*

[Giannakopoulos, F 87*] SEE: von Seelen, W 87a. *Characteristics of Neuronal Systems in the Visual Cortex.*

[Gibbs, Jr, RW 81*] SEE: Goodman, GO 81a. *The Role of Syntactic Context in Word Recognition.*

[Gibson, JJ 33a] Gibson, James J. (1933). *Adaptation, After-Effect and Contrast in the Perception of Curved Lines.* Journal of Experimental Psychology. Volume 16. pp 1-31.

[Gibson, JJ 50a] Gibson, James J. (1950). Perception of the Visual World. Boston: Houghton-Mifflin.

[Gibson, JJ 66a] Gibson, James J. (1966). The Senses Considered as Perceptual Systems. Boston: Houghton-Mifflin.

[Gibson, JJ 79a] Gibson, James J. (1979). The Ecological Approach to Visual Perception. Boston: Houghton-Mifflin.

[Gidas, B 86a] Gidas, B. (1986). *Langevin Equation as a Global Minimization Algorithm.* In Bienenstock, Elie L. (ed.). Disordered Systems and Biological Organization. pp 321-326. New York: Springer-Verlag.

[Giebel, H 70*] SEE: Marko, H 70a. *Recognition of Handwritten Characters with a System of Homogeneous Layers.*

[Giebel, H 71a] Giebel, H. (1971). *Feature Extraction and Recognition of Handwritten Characters by Homogeneous Layers.* Pattern Recognition in Biological and Technical Systems. pp 162-169. New York: Springer-Verlag.

[Gielen, CCAM 86a] Gielen, C. C. A. M.; Houk, James C. *Simple Changes in Reflex Threshold Cannot Explain All Aspects of Rapid Voluntary Movements.* Behavioral and Brain Sciences. Volume 9. pp 605-607.

[Gielen, CCAM 87a] Gielen, C. C. A. M.; Houk, James C. (1987). *A Model of the Motor Servo: Incorporating Nonlinear Spindle Receptor and Muscle Mechanical Properties.* Biological Cybernetics. Volume 57. pp 217-231.

[Gilbert, CD 83a] Gilbert, C. D.; Wiesel, T. N. (1983). *Clustered Intrinsic Connections in Cat Visual Cortex.* Journal of

Neuroscience. Volume 3. pp 1116-1133.

[Gilbert, PFC 74a] Gilbert, Peter F. C. (1974). *A Theory of Memory that Explains the Function and Structure of the Cerebellum*. Brain Research. Volume 70. pp 1-18.

[Gilbert, S 87*] SEE: Raffel, JI 87a. *A Generic Architecture for Wafer-Scale Neuromorphic Systems*.

[Gilchrist, AL 77a] Gilchrist, Alan L. (1977). *Perceived Lightness Depends on Perceived Spatial Arrangement*. Science. Volume 195. pp 185-187.

[Gilchrist, AL 80a] Gilchrist, Alan L. (1980). *When Does Perceived Lightness Depend on Perceived Spatial Arrangement?*. Perception and Psychophysics. Volume 28. pp 527-538.

[Gilchrist, AL 83a] Gilchrist, Alan L.; Delman, Stanley; Jacobsen, Alan. (1983). *The Classification and Integration of Edges as Critical to the Perception of Reflectance and Illumination*. Perception and Psychophysics. Volume 33. pp 425-436.

[Giles, CL 84*] SEE: Fisher, AD 84a. *Associative Processor Architectures for Optical Computing*.

[Giles, CL 85*] SEE: Fisher, AD 85a. *An Adaptive Optical Computing Element*.

[Giles, CL 85*] SEE: Fisher, AD 85b. *Optical Adaptive Associative Computer Architectures*.

[Giles, CL 86*] SEE: Maxwell, T 86a. *Nonlinear Dynamics of Artificial Neural Systems*.

[Giles, CL 87*] SEE: Maxwell, T 87a. *Generalization in Neural Networks: The Contiguity Problem*.

[Giles, CL 87a] Giles, C. Lee; Maxwell, Tom. (1987). *Learning Invariance, and Generalization in High-Order Neural Networks*. Applied Optics. Volume 26. pp 4972-4978.

[Giles, CL 88a] Giles, C. Lee; Griffin, R. D.; Maxwell, Tom. (1988). *Encoding Geometric Invariances in Higher-Order Neural Networks*. In Anderson, Dana Z. (ed.). Neural Information Processing Systems. pp 301-309. New York: American Institute of Physics.

[Gill, PE 81a] Gill, P. E.; Murray, W.; Wright, M. H. (1981). Practical Optimization. New York: Academic Press.

[Gillow, M 66*] SEE: Andersen, P 66a. *Rhythmic Activity in a Simulated Neuronal Network*.

[Gilstad, DW 71a] Gilstad, Dennis W.; Fu, King-Sun. (1971). *Two-Dimensional Adaptive Model of a Human Controller Using Pattern Recognition Techniques*. IEEE Transactions on Systems, Man and Cybernetics. Volume SMC-1. pp 261-266.

[Gilstrap, LO 87*] SEE: Shrier, S 87a. *Polynomial and Neural Networks: Analogies and Engineering Applications*.

[Gindi, GR 87*] SEE: Gmitro, AF 87a. *Optical Neurocomputer for Implementation of the Marr-Poggio Stereo Algorithm*.

[Gindi, GR 87a] Gindi, Gene R.; Gmitro, Arthur F.; Parthasarathy, Kannan. (1987). *Winner-Take-All Networks and Associative Memory: Analysis and Optical Realization*. IEEE International Conference on Neural Networks (ICNN-87), San Diego, CA, June 21-24, 1987: Proceedings. Volume 3. pp 607-614. New York: Institute of Electrical and Electronics Engineers.

[Gindi, GR 88*] SEE: Mjolsness, E 88a. Optimization in Model Matching and Perceptual Organization: A First Look.

[Ginosar, R 88*] SEE: Zeevi, YY 88a. *Neural Computers in Vision: Processing of High Dimensional Data*.

[Giraud, B 86*] SEE: Axelrad, H 86a. *Recurrent Collateral Inhibition Simulated in a Simple Neuronal Automata Assembly*.

[Giraud, B 87*] SEE: Axelrad, H 87a. *The Use of an Artificial Neural Network Model to Analyze the Information Transfer Properties of a Simplified Model of the Cerebellar Cortex.*

[Gish, H 85*] SEE: Makhoul, J 85a. *Vector Quantization in Speech Coding.*

[Gizzi, MS 85*] SEE: Movshon, JA 85a. *The Analysis of Moving Visual Patterns.*

[Glanz, FH 87*] SEE: Miller, WT 87b. *Application of a General Learning Algorithm to the Control of Robotic Manipulators.*

[Glanz, FH 87a] Glanz, F. H.; Miller, W. Thomas. (1987). *Shape Recognition Using a CMAC Based Learning.* Intelligent Robots and Computer Vision. SPIE Volume 848. pp 294-298. Bellingham, WA: Society of Photo-Optical Instrumentation Engineers.

[Glanz, FH 88a] Glanz, F. H.; Miller, W. Thomas. (1988). Deconvolution Using a CMAC Neural Network. Durham, NH: University of New Hampshire, Electrical and Computer Engineering.

[Glass, L 73a] Glass, Leon; Perez, Rafael. (1973). *Perception of Random Dot Interference Patterns.* Nature. Volume 246. pp 360-362.

[Glass, L 83*] SEE: Guevara, MR 83a. *Chaos in Neurobiology.*

[Glickstein, M 88a] Glickstein, Mitchell. (1988). *The Discovery of the Visual Cortex.* Scientific American. Volume 259. pp 118-127.

[Glover, JR 75*] SEE: Widrow, B 75a. *Adaptive Noise Canceling: Principles and Applications.*

[Glushko, RJ 79a] Glushko, Robert J. (1979). *The Organization and Activation of Orthographic Knowledge in Reading Aloud.* Journal of Experimental Psychology: Human Perception and Performance. Volume 5. pp 674-691.

[Gmitro, AF 87*] SEE: Gindi, GR 87a. *Winner-Take-All Networks and Associative Memory: Analysis and Optical Realization.*

[Gmitro, AF 87a] Gmitro, Arthur F.; Gindi, Gene R. (1987). *Optical Neurocomputer for Implementation of the Marr-Poggio Stereo Algorithm.* IEEE International Conference on Neural Networks (ICNN-87), San Diego, CA, June 21-24, 1987: Proceedings. Volume 3. pp 599-606. New York: Institute of Electrical and Electronics Engineers.

[Goddard, NH 88*] SEE: Feldman, JA 88a. *Computing with Structured Neural Networks.*

[Goel, A 89a] Goel, Ashok; Kolen, John F.; Allemang, Dean. ((to appear)). *Learning in Connectionist Networks: Has the Credit Assignment Problem Been Solved?.* Conference on Aerospace Applications of Artificial Intelligence, 4th, Dayton, OH, January, 1989: Proceedings.

[Gogan, P 87*] SEE: Bras, H 87a. *The Dendrites of Single Brain-Stem Motoneurons Intracellularly Labelled with Horseradish Peroxidase in the Cat: An Ultrastructural Analysis.*

[Gogan, P 87*] SEE: Bras, H 87b. *The Dendrites of Single Brain-Stem Motoneurons Intracellularly Labelled with Horseradish Peroxidase in the Cat: Morphological and Electrical.*

[Gogan, P 87*] SEE: Bras, H 87c. *The Mammalian Central Neurone is a Complex Computing Device.*

[Gogel, WC 56a] Gogel, Walter C. (1956). *The Tendency to See Objects as Equidistant and Its Inverse Relation to Lateral Separation.* Psychological Monographs: General and Applied. Volume 70. pp 1-17.

[Gogel, WC 65a] Gogel, Walter C. (1965). *Equidistance Tendency and Its Consequences.* Psychological Bulletin. Volume 64. pp 153-163.

[Gol'tsev, AD 82a] Gol'tsev, A. D.; Kussul, E. M. (1982). *Controlling the Motion of a Moving Robot Using a Neuroid*

Network. Automation and Remote Control. Volume 43. pp 606-611.

[Gold, B 86a] Gold, Ben. (1986). *Hopfield Model Applied to Vowel and Consonant Discrimination.* In Denker, John S. (ed.). Neural Networks for Computing, Snowbird, Utah, 1986. AIP Proceedings 151. pp 165-181. New York: American Institute of Physics.

[Gold, B 86b] Gold, Ben. (1986). Hopfield Model Applied to Vowel and Consonant Discrimination. TR-747. Lexington, MA: Massachusetts Institute of Technology, Lincoln Laboratory.

[Gold, B 87*] SEE: Lippman, RP 87a. A Comparison of Hamming and Hopfield Neural Nets for Pattern Classification.

[Gold, B 87*] SEE: Lippmann, RP 87d. *Neural-Net Classifiers Useful for Speech Recognition.*

[Gold, B 87a] Gold, Ben; Lippmann, Richard P.; Malpass, M. L. (1987). *Some Neural Net Recognition Results on Isolated Words.* IEEE International Conference on Neural Networks (ICNN-87), San Diego, CA, June 21-24, 1987: Proceedings. Volume 4. pp 427-434. New York: Institute of Electrical and Electronics Engineers.

[Gold, B 88a] Gold, Bernard; Lippmann, Richard P. (1988). *A Neural Network for Isolated-Word Recognition.* IEEE International Conference on Acoustics, Speech, and Signal Processing (ICASSP-88), New York, April 11-14, 1988: Proceedings.

[Gold, C 63*] SEE: Tulving, E 63a. *Stimulus Information and Contextual Information as Determinants of Tachistoscopic Recognition of Words.*

[Goldberg, DE 88*] SEE: Das, R 88a. *Discrete-Time Parameter Estimation with Genetic Algorithms.*

[Goldberg, K 88a] Goldberg, Ken; Pearlmutter, Barak. (1988). Using a Neural Network to Learn the Dynamics of the CMU Direct-Drive Arm II. CMU-CS-88-160. Pittsburgh, PA: Carnegie Mellon University, Computer Science Department.

[Golden, RM 86*] SEE: Anderson, JA 86c. *Concepts in Distributed Systems.*

[Golden, RM 86a] Golden, Richard M. (1986). *A Developmental Neural Model of Visual Word Perception.* Cognitive Science. Volume 10. pp 241-276.

[Golden, RM 86b] Golden, Richard M. (1986). *Representing Causal Schemata in Connectionist Systems.* Cognitive Science Society, 8th, Amherst, MA, August 15-17, 1986: Proceedings. pp 13-22. Hillsdale, NJ: Lawrence Erlbaum.

[Golden, RM 86c] Golden, Richard M. (1986). *The "Brain-State-in-a-Box" Neural Model Is a Gradient Descent Algorithm.* Journal of Mathematical Psychology. Volume 30. pp 73-80.

[Golden, RM 88a] Golden, Richard M. (1988). *A Developmental Neural Model of Visual Word Perception.* In Waltz, David L. (ed.); Feldman, Jerome A. (ed.). Connectionist Models and Their Implications: Readings from Cognitive Science. pp 119-154. Norwood, NJ: Ablex.

[Golden, RM 88b] Golden, Richard M. (1988). *A Unified Framework for Connectionist Systems.* Biological Cybernetics. Volume 59. pp 109-120.

[Golden, RM 88c] Golden, Richard M. (1988). *Probabilistic Characterization of Neural Model Computations.* In Anderson, Dana Z. (ed.). Neural Information Processing Systems. pp 310-316. New York: American Institute of Physics.

[Golden, RM 88d] Golden, Richard M. (1988). *Relating Neural Networks to Traditional Engineering Approaches.* Conference of AI West, 1988: Proceedings.

[Goldman, D 73*] SEE: Homa, D 73a. *Prototype Abstraction and Classification of New Instances as a Function of the Number of Instances Defining the Prototype.*

[Goldstein, GD 62*] SEE: Yovitz, MC 62a. Self-Organizing Systems.

[Goldstein, JL 68a] Goldstein, Julius L.; Kiang, Nelson Y. S. (1968). *Neural Correlates of the Aural Combination Tone 2f(1)-f(2).* Proceedings of the IEEE. Volume 56. pp 981-992.

[Goldstein, Jr., MH 68*] SEE: Frishkopf, LS 68a. *Neural Coding in the Bullfrog's Auditory System: A Teleological Approach.*

[Goles, E 85a] Goles, E. (1985). *Dynamical Behaviour of Neural Networks.* SIAM Journal on Algebraic and Discrete Methods. Volume 6. pp 749-754.

[Goles, E 86a] Goles, E.; Vichniac, Gerard Y. (1986). *Invariants in Automata Networks.* Journal of Physics A: Mathematical and General. Volume 19. pp 961-965.

[Goles, E 86b] Goles, E.; Vichniac, Gerard Y. (1986). *Lyapunov Function for Parallel Neural Networks.* In Denker, John S. (ed.). Neural Networks for Computing, Snowbird, Utah, 1986. AIP Proceedings 151. pp 165-181. New York: American Institute of Physics.

[Gonzales, MEQ 87a] Gonzales, Maria Eunice Q. (1987). *On the Methodology of Connectionist Research Programmes in Vision.* IEEE International Conference on Neural Networks (ICNN-87), San Diego, CA, June 21-24, 1987: Proceedings. Volume 4. pp 309-316. New York: Institute of Electrical and Electronics Engineers.

[Gonzales, MEQ 88a] Gonzales, Maria Eunice Q.; French, Tony; Treffner, Paul J. (in press). A Naturalistic Approach to Mental Representation. New York: Wiley.

[Gooch, RP 81*] SEE: Widrow, B 81a. *Adpative Design of Digital Filters.*

[Gooch, RP 85*] SEE: Shynk, JJ 85a. *A Frequency-Domain Adaptive Pole-Zero Filter with Applications.*

[Goodall, SM 88*] SEE: D'Autrechy, CL 88a. *A General-Purpose Simulation Environment for Developing Connectionist Models.*

[Goode, BB 67*] SEE: Widrow, B 67a. *Adaptive Antenna Systems.*

[Goodman, G 86*] SEE: Takeda, M 86a. Neural Networks for Computation.

[Goodman, GO 81a] Goodman, George O.; McClelland, James L.; Gibbs, Jr, Raymond W. (1981). *The Role of Syntactic Context in Word Recognition.* Memory and Cognition. Volume 9. pp 580-586.

[Goodman, JW 84a] Goodman, Joseph W.; Leonberger, Frederick I.; Kung, Sun-Yuan; Athale, Ravindra A. (1984). *Optical Interconnections for VLSI Systems.* Proceedings of the IEEE. Volume 72. pp 850-866.

[Goodman, JW 87*] SEE: Bruck, J 87a. *A Generalized Convergence Theorem for Neural Networks and Its Applications in Combinatorial Optimization.*

[Goodman, JW 88*] SEE: Bruck, J 88a. *A Generalized Convergence Theorem for Neural Networks.*

[Goodman, JW 88*] SEE: Bruck, J 88e. *On the Power of Neural Networks for Solving Hard Problems.*

[Goodman, R 88*] SEE: Chiueh, T 88a. *A Neural Network Classifier Based on Coding Theory.*

[Goodman, RM 88*] SEE: Chiueh, T 88b. *High-Capacity Exponential Associative Memories.*

[Goodwin, JM 87*] SEE: Vidal, JJ 87a. *Implementing Neural Nets with Programmable Logic.*

[Goodwin, JM 87a] Goodwin, James M.; Rosen, Bruce E.; Vidal, Jacques J. (1987). *A Design for an Associative Spin Glass*

Processor. IEEE International Conference on Neural Networks (ICNN-87), San Diego, CA, June 21-24, 1987: Proceedings. Volume 3. pp 639-648. New York: Institute of Electrical and Electronics Engineers.

[Goodwin, JM 88*] SEE: Rosen, BE 88a. *Learning by State Recurrence Detection.*

[Goodwin, JM 88*] SEE: Rosen, BE 88b. Machine Operant Conditioning.

[Goodwin, JM 88*] SEE: Rosen, BE 88c. State Recurrence Learning.

[Goodwin, JM 88a] Goodwin, James M.; Rosen, Bruce E.; Vidal, Jacques J. (1988). Exploration of Learning in an Associative Magnetic Processor. Los Angeles, CA: University of California, Computer Science Department.

[Goodwin, JM 88b] Goodwin, James M.; Rosen, Bruce E.; Vidal, Jacques J. (1988). Progress on the Spin Chip Asociative Processor. Los Angeles, CA: University of California, Computer Science Department.

[Gorman, RP 88a] Gorman, R. Paul; Sejnowski, Terrence J. (1988). *Analysis of Hidden Units in a Layered network Trained to Classify Sonar Targets*. Neural Networks. Volume 1. pp 75-89.

[Gorman, RP 88b] Gorman, R. Paul; Sejnowski, Terrence J. (1988). *Learned Classification of Sonar Targets Using a Massively Parallel Network*. IEEE Transactions on Acoustics, Speech, and Signal Processing. Volume 36. pp 1135-1140.

[Gose, EE 66*] SEE: Mucciardi, AN 66a. *Evolutionary Pattern Recognition in Incomplete Nonlinear Multithreshold Networks.*

[Gose, EE 69*] SEE: Klopf, AH 69a. *An Evolutionary Pattern Recognition Network.*

[Gose, EE 69a] Gose, Earl E. (1969). *An Introduction to Biological and Mechanical Pattern Recognition*. In Watanbe, Satosi (ed.). Methodologies of Pattern Recognition. pp 203-252. New York: Academic Press.

[Gossard, DC 87*] SEE: Pabon, JA 87a. *A Methodology to Select Appropriate Learning Rate Parameters in Feed-Forward Networks.*

[Gossard, DC 88*] SEE: Pabon, JA 88a. *Connectionist Networks for Learning Coordinated Motion in Autonomous Systems.*

[Goto, Y 86*] SEE: Wallace, R 86a. *Progress in Robot Road-Following.*

[Gottlieb, A 83a] Gottlieb, Allan; Grishman, Ralph; Kruskal, Clyde P.; McAuliffe, Kevin P.; Rudolph, Larry; Snir, Marc. (1983). *The NYU Ultracomputer-Designing an MIMD Shared Memory Parallel Computer*. IEEE Transactions on Computers. Volume C-32. pp 175-189.

[Gouin, P 86*] SEE: Rimey, RD 86a. *Real-Time 3-D Object Classification Using a Learning System.*

[Goutos, GP 80*] SEE: Fairhurst, MC 80a. *An Evaluation of Cell Function in Relation to Reverberatory Modes in Cellular Networks.*

[Gove, RJ 87a] Gove, Robert John. (1987). *Integration of Symbolic and Multiple Digital Signal Processors with the Explorer / Odyssey for Image Processing and Understanding Applications*. pp 968-971. New York: Institute of Electrical and Electronics Engineers.

[Govind, G 88*] SEE: Ramamoorthy, PA 88b. *Signal Modeling and Prediction Using Neural Networks.*

[Graf, HP 86*] SEE: Jackel, LD 86a. *Artificial Neural Networks for Computing.*

[Graf, HP 86*] SEE: Hubbard, W 86a. *Electronic Neural Networks.*

[Graf, HP 86*] SEE: Jackel, LD 86b. *High Resolution Micro-Fabrication and Neural Networks.*

[Graf, HP 86a] Graf, Hans Peter; Jackel, Lawrence D.; Howard, Richard E.; Straughn, Brian L.; Denker, John S. (1986). *VLSI Implementation of a Neural Network Memory with Several Hundreds of Neurons*. In Denker, John S. (ed.). Neural Networks for Computing, Snowbird, Utah, 1986. AIP Proceedings 151. pp 182-187. New York: American Institute of Physics.

[Graf, HP 87*] SEE: Howard, RE 87a. *An Associative Memory Based on an Electronic Neural Network Architecture.*

[Graf, HP 87*] SEE: Jackel, LD 87b. *Electronic Neural Network Chips.*

[Graf, HP 87a] Graf, Hans Peter; Hubbard, Wayne E.; Jackel, Lawrence D.; de Vegvar, Paul G. N. (1987). *A CMOS Associative Memory Chip*. IEEE International Conference on Neural Networks (ICNN-87), San Diego, CA, June 21-24, 1987: Proceedings. Volume 3. pp 461-468. New York: Institute of Electrical and Electronics Engineers.

[Graf, HP 87b] Graf, Hans Peter; de Vegvar, Paul G. N. (1987). A CMOS Associative Memory Chip Based on Neural Networks. Holmdel, NJ: AT&T Bell Laboratories.

[Graf, HP 87c] Graf, Hans Peter; de Vegvar, Paul G. N. (1987). *A CMOS Implementation of a Neural Network Model*. In Lasleben, P. (ed.). Advanced Research in VLSI. pp 351-367. Cambridge, MA: MIT Press.

[Graf, HP 88*] SEE: Mackie, S 88b. *Implementations of Neural Network Models in Silicon.*

[Graf, HP 88*] SEE: Mackie, S 88c. *Microelectronic Implementations of Connectionist Neural Networks.*

[Graf, HP 88a] Graf, Hans Peter; Jackel, Lawrence D.; Hubbard, Wayne E. (March 1988). *VLSI Implementation of a Neural Network Model*. IEEE Computer Magazine. Volume 2. pp 41-49.

[Graham, N 71a] Graham, Norma; Nachmias, Jacob. (1971). *Detection of Grating Patterns Containing Two Spatial Frequencies: A Comparison of Single-Channel and Multiple-Channels Models*. Vision Research. Volume 11. pp 251-259.

[Grajski, KA 86a] Grajski, Kamil A.; Breiman, L.; Viana di Prisco, Gonzalo; Freeman, Walter J. (1986). *Classification of EEG Spatial Patterns with a Tree-Structured Methodology: CART*. IEEE Transactions on Biomedical Engineering. Volume BME-33. pp 1076-1086.

[Grajski, KA 86b] Grajski, Kamil A.; Freeman, Walter J. (1986). *Dynamic Model of Olfactory Discrimination*. In Denker, John S. (ed.). Neural Networks for Computing, Snowbird, Utah, 1986. AIP Proceedings 151. pp 188-193. New York: American Institute of Physics.

[Grajski, KA 87*] SEE: Freeman, WJ 87b. *Relation of Olfactory EEG to Behavior: Factor Analysis.*

[Grajski, KA 88a] Grajski, Kamil A. (1988). *Book Review*. Bulletin of Mathematical Biology. Volume 50. pp 414-419.

[Grajski, KA 88b] Grajski, Kamil A.; Freeman, Walter J. (in press). *Spatial EEG Correlates of Non-Associative and Associative Olfactory Learning in Rabbits*. Behavioral Neuroscience.

[Grajski, KA 88c] Grajski, Kamil A. (1988). *Tactile Information Processing: A Back-Propagation Model of Pattern Recognition in the Human Somatosensory System*. Neural Networks. Volume 1, Supplement 1. pp 253. Elmsford, NY: Pergamon.

[Granger, R 88*] SEE: Lynch, G 88a. *Simulation and Analysis of a Simple Cortical Network.*

[Granger, R 88a] Granger, Richard; Ambros-Ingerson, Jose; Lynch, Gary. (1988). *Derivation of Encoding Characteristics of Layer II Cerebral Cortex*. Journal of Cognitive Neuroscience. Volume 1. pp 64-91.

[Granger, R 88b] Granger, Richard; Ambros-Ingerson, Jose; Henry, Howard N.; Lynch, Gary. (1988). *Partitioning of Sensory Data by a Cortical Network*. In Anderson, Dana Z. (ed.). Neural Information Processing Systems. pp 317-337. New York: American Institute of Physics.

[Granger, R 89*] SEE: Lynch, G 89a. *Cortical Encoding of Memory: Hypotheses Derived from Analysis and Simulation of Physiological Learning Rules in Anatomical Structures.*

[Grant, PM 86a] Grant, P. M.; Sage, Jay P. (1986). A Comparison of Nonlinear Associative Memory and Matched-Filter Processing for Detecting Lines in Optical Images. Quarterly Technical Report. Lexington, MA: Massachusetts Institute of Technology, Lincoln Laboratory.

[Grant, PM 86b] Grant, P. M.; Sage, Jay P. (1986). *Comparison of Neural Network and Matched Filter Processing for Detecting Lines in Images.* In Denker, John S. (ed.). Neural Networks for Computing, Snowbird, Utah, 1986. AIP Proceedings 151. pp 194-199. New York: American Institute of Physics.

[Graziano, MJ 88a] Graziano, Michael J.; Noetzel, Andrew S. (1988). *Comparisons of Sigmoid Functions for Neural Network Convergence.* Conference on Information Sciences and Systems, Princeton, NJ, 1988: Proceedings.

[Green, CA 87*] SEE: Cheung, KF 87a. *Conventional and Composite Matched Filters with Error Correction: A Comparison.*

[Green, DM 60*] SEE: Marill, T 60a. *Statistical Recognition Functions and the Design of Pattern Recognizers.*

[Green, DM 63*] SEE: Marill, T 63a. *On the Effectiveness of Receptors in Recognition Systems.*

[Green, PE 87a] Green, Peter E.; Michalson, William R. (1987). *Real-Time Evidential Reasoning and Network Based Processing.* IEEE International Conference on Neural Networks (ICNN-87), San Diego, CA, June 21-24, 1987: Proceedings. Volume 2. pp 359-366. New York: Institute of Electrical and Electronics Engineers.

[Greene, PH 59a] Greene, Peter H. (1959). *An Approach to Computers that Perceive, Learn, and Reason.* Western Joint Computer Conference, San Francisco, CA, March 3-5, 1959: Proceedings. pp 181-186. New York: Institute of Radio Engineers.

[Greenwood, D 87*] SEE: Smith, T 87a. *Application of Genetic Algorithm to a Neural Network for Finding Optimal Paths.*

[Gremillion, M 87a] Gremillion, M.; Mandell, A. J.; Travis, Bryan J. (1987). *Neural Nets with Complex Structure: A Model of the Visual System.* IEEE International Conference on Neural Networks (ICNN-87), San Diego, CA, June 21-24, 1987: Proceedings. Volume 4. pp 235-246. New York: Institute of Electrical and Electronics Engineers.

[Grey, L 60*] SEE: Flores, I 60a. *Optimization of Reference Signals for Character Recognition Systems.*

[Griffin, RD 88*] SEE: Giles, CL 88a. *Encoding Geometric Invariances in Higher-Order Neural Networks.*

[Griffith, JS 63a] Griffith, J. S. (1963). *A Field Theory for Neural Nets, Part I: Derivation of the Field Equations.* Bulletin of Mathematical Biophysics. Volume 25. pp 111-120.

[Griffith, JS 63b] Griffith, J. S. (1963). *On the Stability of Brain-Like Structures.* Biophysical Journal. Volume 3. pp 299-308.

[Griffith, JS 65a] Griffith, J. S. (1965). *A Field Theory for Neural Nets, Part II: Properties of the Field Equations.* Bulletin of Mathematical Biophysics. Volume 27. pp 187-195.

[Griffith, JS 71a] Griffith, J. S. (1971). Mathematical Neurobiology. New York: Academic Press.

[Griffiths, LJ 67*] SEE: Widrow, B 67a. *Adaptive Antenna Systems.*

[Grillner, S 88a] Grillner, S.; Buchanan, J. T.; Lansner, Anders. (1988). *Simulation of the Segmental Burst Generating Network for Locomotion in Lamprey.* Neuroscience Letters. Volume 89. pp 31-35.

[Grimmelikhuijzen, CJP 88*] SEE: Bode, PM 88a. *Develpment of the Two-Part Pattern During Regeneration of the Head in*

Hydra.

[Grimson, WEL 75*] SEE: Caianiello, ER 75a. *Synthesis of Boolean Nets and Time Behavior of a General Mathematical Neuron.*

[Grinstein, G 86*] SEE: Hertz, JA 86a. *Memory Networks with Asymmetric Bonds.*

[Grishman, R 83*] SEE: Gottlieb, A 83a. *The NYU Ultracomputer-Designing an MIMD Shared Memory Parallel Computer.*

[Grissel, R 83a] Grissel, R. (ed.); Hodgson, J. P. E. (ed.); Yanowich, M. (ed.). (1983). Oscillations in Mathematical Biology. New York: Springer-Verlag.

[Grondin, RO 87*] SEE: Clark, LT 87a. *Comparison of a Pipelined "Best Match" Content Addressable Memory with Neural Networks.*

[Grondin, RO 87*] SEE: Shiue, LC 87a. *On Designing Fuzzy Learning Neural-Automata.*

[Grondin, RO 88*] SEE: Akers, LA 88a. *Limited Interconnectivity in Synthetic Neural Systems.*

[Groner, GF 63*] SEE: Widrow, B 63b. *Practical Applications for Adaptive Data-Processing Systems.*

[Groner, GF 63*] SEE: Talbert, LR 63a. A Real-Time Adaptive Speech-Recognition System.

[Groner, GF 66*] SEE: Koford, JS 66a. *The Use of an Adaptive Threshold Element to Design a Linear Optimal Pattern Classifier.*

[Gross, DJ 84a] Gross, D. J.; Mezard, M. (1984). *Simplest Spin Glass.* Nuclear Physics. Volume B240. pp 431-452.

[Gross, GW 88*] SEE: Kowalski, JM 88a. *Coupled Rotators as a Model for Spontaneous Activity in Neural Networks.*

[Gross, T 85a] Gross, Thomas; Kung, H. T.; Lam, Monica S.; Webb, Jon A. (1985). *Warp as a Machine for Low-Level Vision.* IEEE International Conference on Robotics and Automation, St. Louis, MO, March 25-28, 1985: Proceedings. pp 790-800. Silver Spring, MD: Institute of Electrical and Electronics Engineers.

[Gross, T 86*] SEE: Annaratone, M 86a. *Warp Architecture and Implementation.*

[Grossberg, S 67a] Grossberg, Stephen. (1967). *Nonlinear Difference-Differential Equations in Prediction and Learning Theory.* Proceedings of the National Academy of Sciences. Volume 58. pp 1329-1334.

[Grossberg, S 68a] Grossberg, Stephen. (1968). *A Prediction Theory for Some Nonlinear Functional-Differential Equations I. Learning of Lists.* Journal of Mathematical Analysis and Applications. Volume 21. pp 643-694.

[Grossberg, S 68b] Grossberg, Stephen. (1968). *A Prediction Theory for Some Nonlinear Functional-Differential Equations. II. Learning of Patterns.* Journal of Mathematical Analysis and Applications. Volume 22. pp 490-522.

[Grossberg, S 68c] Grossberg, Stephen. (1968). *Global Ratio Limit Theorems for Some Nonlinear Functional-Differential Equations, I.* Bulletin of the American Mathematical Society. Volume 74. pp 95-99.

[Grossberg, S 68d] Grossberg, Stephen. (1968). *Global Ratio Limit Theorems for Some Nonlinear Functional-Differential Equations, II.* Bulletin of the American Mathematical Society. Volume 74. pp 100-105.

[Grossberg, S 68e] Grossberg, Stephen. (1968). *Some Nonlinear Networks Capable of Learning a Spatial Pattern of Arbitrary Complexity.* Proceedings of the National Academy of Sciences. Volume 59. pp 368-372.

[Grossberg, S 68f] Grossberg, Stephen. (1968). *Some Physiological and Biochemical Consequences of Psychological Postulates.* Proceedings of the National Academy of Sciences. Volume 60. pp 758-765.

[Grossberg, S 69a] Grossberg, Stephen. (1969). *Enabling Fields: A Theory of Learning with Physiological Implications.* Journal of Mathematical Psychology. Number 6. pp 209-239.

[Grossberg, S 69b] Grossberg, Stephen. (1969). *Learning and Energy-Entropy Dependence in Some Nonlinear Functional-Differential Systems.* Bulletin of the American Mathematical Society. Volume 75. pp 1238-1242.

[Grossberg, S 69c] Grossberg, Stephen. (1969). *On Learning and Energy-Entropy Dependence in Recurrent and Nonrecurrent Signed Networks.* Journal of Statistical Physics. Number 1. pp 319-350.

[Grossberg, S 69d] Grossberg, Stephen. (1969). *On Learning, Information, Lateral Inhibition, and Transmitters.* Mathematical Biosciences. Number 4. pp 255-310.

[Grossberg, S 69e] Grossberg, Stephen. (1969). *On Learning of Spatiotemporal Patterns by Networks with Ordered Sensory and Motor Components: Excitatory Components of the Cerebellum.* Studies in Applied Mathematics. Number 48. pp 105-132.

[Grossberg, S 69f] Grossberg, Stephen. (1969). *On the Global Limits and Oscillations of a System of Nonlinear Differential Equations Describing a Flow on a Probabilistic Network.* Journal of Differential Equations. Volume 5. pp 531-563.

[Grossberg, S 69g] Grossberg, Stephen. (1969). *On the Production and Release of Chemical Transmitters and Related Topics in Cellular Control.* Journal of Theoretical Biology. Volume 22. pp 325-364.

[Grossberg, S 69h] Grossberg, Stephen. (1969). *On the Serial Learning of Lists.* Mathematical Biosciences. Number 4. pp 201-253.

[Grossberg, S 69i] Grossberg, Stephen. (1969). *On the Variational Systems of Some Nonlinear Difference-Differential Equations.* Journal of Differential Equations. Volume 6. pp 544-577.

[Grossberg, S 69j] Grossberg, Stephen. (1969). *Some Networks That Can Learn, Remember, and Reproduce Any Number of Complicated Space-Time Patterns, I.* Journal of Mathematics and Mechanics. Volume 19. pp 53-91.

[Grossberg, S 70a] Grossberg, Stephen. (1970). *Neural Pattern Discrimination.* Journal of Theoretical Biology. Number 27. pp 291-337.

[Grossberg, S 70b] Grossberg, Stephen; Pepe, James. (1970). *Schizophrenia: Possible Dependence of Associational Span, Bowing, and Primacy vs. Recency on Spiking Threshold.* Behavioral Science. Volume 15. pp 359-362.

[Grossberg, S 70c] Grossberg, Stephen. (1970). *Some Networks That Can Learn, Remember, and Reproduce Any Number of Complicated Space-Time Patterns, II.* Studies in Applied Mathematics. Volume 49. pp 135-166.

[Grossberg, S 71a] Grossberg, Stephen. (1971). *On the Dynamics of Operant Conditioning.* Journal of Theoretical Biology. Number 33. pp 225-255.

[Grossberg, S 71b] Grossberg, Stephen. (1971). *Pavlovian Pattern Learning by Nonlinear Neural Networks.* Proceedings of the National Academy of Sciences. Volume 68. pp 828-831.

[Grossberg, S 71c] Grossberg, Stephen; Pepe, James. (1971). *Spiking Threshold and Overarousal Effects in Serial Learning.* Journal of Statistical Physics. Number 3. pp 95-125.

[Grossberg, S 72a] Grossberg, Stephen. (1972). *A Neural Theory of Punishment and Avoidance, Part I: Qualitative Theory.* Mathematical Biosciences. Number 15. pp 39-67.

[Grossberg, S 72b] Grossberg, Stephen. (1972). *A Neural Theory of Punishment and Avoidance, Part II: Quantitative Theory.* Mathematical Biosciences. Number 15. pp 253-285.

[Grossberg, S 72c] Grossberg, Stephen. (1972). *Neural Expectation: Cerebellar and Retinal Analogs of Cells Fired by*

Learnable or Unlearned Pattern Classes. Kybernetik. Volume 10. pp 49-57.

[Grossberg, S 72d] Grossberg, Stephen. (1972). *Pattern Learning by Functional-Differential Neural Networks with Arbitrary Path Weights*. Delay and Functional Differential Equations and Their Applications. pp 121-160. New York: Academic Press.

[Grossberg, S 73a] Grossberg, Stephen. (1973). *Contour Enhancement, Short-Term Memory, and Constancies in Reverberating Neural Networks*. Studies in Applied Mathematics. Number 52. pp 217-257.

[Grossberg, S 74a] Grossberg, Stephen. (1974). *Classical and Instrumental Learning by Neural Networks*. In Rosen, R. (ed.); Snell, F. (ed.). Progress in Theoretical Biology. Volume 3. New York: Academic Press.

[Grossberg, S 75*] SEE: Ellias, SA 75a. *Pattern Formation, Contrast Control, and Oscillations in the Short Term Memory of Shunting On-Center Off-Surround Networks*.

[Grossberg, S 75a] Grossberg, Stephen. (1975). *A Neural Model of Attention, Reinforcement, and Discrimination Learning*. International Review of Neurobiology. Number 18. pp 263-327.

[Grossberg, S 75b] Grossberg, Stephen; Levine, Daniel S. (1975). *Some Developmental and Attentional Biases in the Contrast Enhancement and Short-Term Memory of Recurrent Neural Networks*. Journal of Theoretical Biology. Volume 53. pp 341-380.

[Grossberg, S 76*] SEE: Levine, DS 76a. *Visual Illusions in Neural Networks: Line Neutralization, Tilt After Effect, and Angle Expansion*.

[Grossberg, S 76a] Grossberg, Stephen. (1988). *Adaptive Pattern Classification and Universal Recoding: I. Parallel Development and Coding of Neural Feature Detectors*. In Anderson, James A. (ed.); Rosenfeld, Edward (ed.). Neurocomputing: Foundations of Research. pp 245-258. Cambridge, MA: MIT Press.

[Grossberg, S 76b] Grossberg, Stephen. (1976). *Adaptive Pattern Classification and Universal Recording, Part I: Parallel Development and Coding of Neural Feature Detectors*. Biological Cybernetics. Volume 23. pp 121-134.

[Grossberg, S 76c] Grossberg, Stephen. (1976). *Adaptive Pattern Classification and Universal Recording, Part II: Feedback, Expectation, Olfaction, and Illusions*. Biological Cybernetics. Volume 23. pp 187-202.

[Grossberg, S 76d] Grossberg, Stephen. (1976). *On the Development of Feature Detectors in the Visual Cortex with Applications to Learning and Reaction-Diffusion Systems*. Biological Cybernetics. Volume 21. pp 145-159.

[Grossberg, S 77a] Grossberg, Stephen. (1977). *Do We Know What We Want? Do We Know When We Want It?*. Brain Theory Newsletter. Volume 3. pp 16-18.

[Grossberg, S 77b] Grossberg, Stephen. (1977). *Pattern Formation by the Global Limits of a Nonlinear Competitive Interactin in n-Dimensions*. Journal of Mathematical Biology. Volume 4. pp 237-256.

[Grossberg, S 78a] Grossberg, Stephen. (1978). *A Theory of Human Memory: Self-Organization and Performance of Sensory-Motor Codes, Maps, and Plans*. Progress in Theoretical Biology. Volume 5. pp 233-374. New York: Academic Press.

[Grossberg, S 78b] Grossberg, Stephen. (1978). *A Theory of Visual Coding, Memory, and Development*. In Leeuwenberg, E. (ed.); Buffart, H. (ed.). Formal Theories of Visual Perception. New York: Wiley.

[Grossberg, S 78c] Grossberg, Stephen. (1978). *Behavioral Contrast in Short Term Memory: Serial Binary Memory Models or Parallel Continuous Memory Models?*. Journal of Mathematical Psychology. Volume 17. pp 199-219.

[Grossberg, S 78d] Grossberg, Stephen. (1978). *Communication, Memory and Development*. In Rosen, R. (ed.); Snell, F.

(ed.). Progress in Theoretical Biology. Volume 5. New York: Academic Press.

[Grossberg, S 78e] Grossberg, Stephen. (1978). *Competition, Decision, Consensus*. Journal of Mathematical Analysis and Applications. Volume 66. pp 470-493.

[Grossberg, S 78f] Grossberg, Stephen. (1978). *Decisions, Patterns, and Oscillations in the Dynamics of Competitive Systems with Applications to Volterra-Lotka Systems*. Journal of Theoretical Biology. Number 73. pp 101-130.

[Grossberg, S 78g] Grossberg, Stephen. (1978). *Do All Neural Networks Really Look Alike? A Comment on Anderson, Silverstein, Ritz and Jones*. Psychological Review. Volume 85. pp 592-596.

[Grossberg, S 80a] Grossberg, Stephen. (1980). *Biological Competition: Decision Rules, Pattern Formation and Oscillations*. Proceedings of the National Academy of Sciences. Volume 77. pp 2338-2342.

[Grossberg, S 80b] Grossberg, Stephen. (1980). *Direct Perception or Adaptive Resonance?*. Behavioral and Brain Sciences. Number 3. p 385.

[Grossberg, S 80c] Grossberg, Stephen. (1980). *How Does a Brain Build a Cognitive Code?*. Psychological Review. Volume 87. pp 1-51.

[Grossberg, S 80d] Grossberg, Stephen. (1988). *How Does a Brain Build a Cognitive Code?*. In Anderson, James A. (ed.); Rosenfeld, Edward (ed.). Neurocomputing: Foundations of Research. pp 349-400. Cambridge, MA: MIT Press.

[Grossberg, S 80e] Grossberg, Stephen. (1980). *Human and Computer Rules and Representations are not Equivalent*. Behavioral and Brain Sciences. Number 3. pp 136-138.

[Grossberg, S 80f] Grossberg, Stephen. (1980). *Intracellular Mechanisms of Adaptation and Self-Regulation in Self-Organizing Networks: The Role of Chemical Transducers*. Bulletin of Mathematical Biology. Volume 42. pp 365-396.

[Grossberg, S 81a] Grossberg, Stephen. (1981). *Adaptive Resonance in Development, Perception, and Cognition*. In Grossberg, Stephen (ed.). Mathematical Psychology and Psychophysiology. SIAM-AMS Proceedings, Vol 13. pp 107-156. Providence, RI: American Mathematical Society.

[Grossberg, S 81b] Grossberg, Stephen. (1981). *Psychophysiological Substrates of Schedule Interaction and Behavioral Contrast*. Mathematical Psychology and Psychophysiology. SIAM-AMS Proceedings, Vol 13. pp 157-186. Providence, RI: American Mathematical Society.

[Grossberg, S 82a] Grossberg, Stephen. (1982). *A Psychophysiological Theory of Reinforcement, Drive, Motivation, and Attention*. Journal of Theoretical Neurobiology. Volume 1. pp 286-369.

[Grossberg, S 82b] Grossberg, Stephen. (1982). *Associative and Competitive Principles of Learning and Development: The Temporal Unfolding and Stability of STM and LTM Patterns*. In Amari, Shun-ichi (ed.); Arbib, Michael A. (ed.). Competition and Cooperation in Neural Nets. pp 295-341. New York: Springer-Verlag.

[Grossberg, S 82c] Grossberg, Stephen. (1982). Studies of Mind and Brain: Neural Principles of Learning, Perception, Development, Cognition and Motor Control. Boston: D Reidel.

[Grossberg, S 82d] Grossberg, Stephen. (1982). *The Processing of Expected and Unexpected Events During Conditioning and Attention: A Psychophysiological Theory*. Psychological Review. Volume 89. pp 529-572.

[Grossberg, S 83*] SEE: Cohen, MA 83a. *Absolute Stability of Global Pattern Formation and Parallel Memory Storage by Competitive Neural Networks*.

[Grossberg, S 83*] SEE: Cohen, MA 83b. *Absolute Stability of Global Pattern Formation and Parallel Memory Storage by*

Competitive Neural Networks.

[Grossberg, S 83*] SEE: Carpenter, GA 83a. *A Neural Theory of Circadian Rhythms: The Gated Pacemaker.*

[Grossberg, S 83a] Grossberg, Stephen. (1983). *Neural Substrates of Binocular Form Perception: Filtering, Matching, Diffusion, and Resonance.* Synergetics of the Brain. Volume 23.

[Grossberg, S 83b] Grossberg, Stephen. (1983). *The Quantized Geometry of Visual Space: The Coherent Computation of Depth, Form, and Lightness.* Behavioral and Brain Sciences. Volume 6. pp 625-692.

[Grossberg, S 84*] SEE: Cohen, MA 84a. *Neural Dynamics of Brightness Perception: Features, Boundaries, Diffusion, and Resonance.*

[Grossberg, S 84*] SEE: Carpenter, GA 84a. *A Neural Theory of Circadian Rhythms: Aschoff's Rule in Diurnal and Nocturnal Mammals.*

[Grossberg, S 84*] SEE: Cohen, MA 84b. *Some Global Properties of Binocular Resonances: Disparity Matching, Filling-in, and Figure-Ground Synthesis.*

[Grossberg, S 84a] Grossberg, Stephen. (1984). *Neuroethology and Theoretical Neurobiology.* Behavioral and Brain Sciences. Volume 7. pp 388-390.

[Grossberg, S 84b] Grossberg, Stephen; Cohen, Michael A. (1984). *Some Global Properties of Binocular Resonances: Disparity Matching, Filling-In, and Figure-Ground Synthesis.* Figural Synthesis. pp 117-151. Hillsdale, NJ: Lawrence Erlbaum.

[Grossberg, S 84c] Grossberg, Stephen. (1984). *Some Psychophysiological and Pharmacological Correlates of a Developmental, Cognitive and Motivational Theory.* Brain and Information: Event Related Potentials. pp 58-151. New York: New York Academy of Sciences.

[Grossberg, S 84d] Grossberg, Stephen. (1984). *The Microscopic Analysis of Behavior: Toward a Synthesis of Instrumental, Perceptual, and Cognitive Ideas.* Behavioral and Brain Sciences. Volume 7. pp 594-595.

[Grossberg, S 84e] Grossberg, Stephen. (1984). *Unitization, Automaticity, Temporal Order, and Word Recognition.* Cognition and Brain Theory. Volume 7. pp 263-283.

[Grossberg, S 85*] SEE: Carpenter, GA 85a. *Category Learning and Adaptive Pattern Recognition: A Neural Network Model.*

[Grossberg, S 85*] SEE: Carpenter, GA 85b. *Neural Dynamics of Adaptive Pattern Recognition: Priming, Search, Attention, and Category Formation.*

[Grossberg, S 85*] SEE: Carpenter, GA 85c. *Neural Dynamics of Circadian Rhythms: The Mammalian Hypothalamic Pacemaker.*

[Grossberg, S 85a] Grossberg, Stephen. (1985). *Cognitive Self-Organization and Neural Modularity.* Behavioral and Brain Sciences. Volume 8. pp 18-19.

[Grossberg, S 85b] Grossberg, Stephen. (1985). *Four Frames Do Not Suffice.* Behavioral and Brain Sciences. Volume 8. pp 294-295.

[Grossberg, S 85c] Grossberg, Stephen; Mingolla, Ennio. (1985). *Neural Dynamics of Perceptual Grouping: Textures, Boundaries, and Emergent Segmentations.* Perception and Psychophysics. Volume 38. pp 141-171.

[Grossberg, S 85d] Grossberg, Stephen. (1985). *The Role of Learning in Sensory-Motor Control.* Behavioral and Brain Sciences. Volume 8. pp 155-157.

[Grossberg, S 85e] Grossberg, Stephen. (March 1985). *The Role of Learning in Sensory-Motor Control.* Behavioral and Brain

Sciences. Volume 8. pp 155-157.

[Grossberg, S 86*] SEE: Carpenter, GA 86b. *Absolutely Stable Learning of Recognition Codes by a Self-Organizing Neural Network.*

[Grossberg, S 86*] SEE: Carpenter, GA 86c. *Adaptation and Transmitter Gating in Vertebrate Photoreceptors.*

[Grossberg, S 86*] SEE: Carpenter, GA 86d. *Adaptive Resonance Theory: Stable Self-organization of Neural Recognition Codes in Response to Arbitrary Lists of Input Patterns.*

[Grossberg, S 86*] SEE: Carpenter, GA 86e. *Associative Learning, Adaptive Pattern Recognition, and Cooperative-Competitive Decision Making by Neural Networks.*

[Grossberg, S 86*] SEE: Carpenter, GA 86g. Computing with Neural Networks: The Role of Symmetry.

[Grossberg, S 86*] SEE: Carpenter, GA 86h. *Discovering Order in Chaos: Stable Self-Organization of Neural Recognition Codes.*

[Grossberg, S 86*] SEE: Cohen, MA 86b. *Neural Dynamics of Brightness Perception: Features, Boundaries, Diffusion and Resonance.*

[Grossberg, S 86*] SEE: Carpenter, GA 86i. *Neural Dynamics of Category Learning and Recognition: Attention, Memory Consolidation, and Amnesia.*

[Grossberg, S 86*] SEE: Carpenter, GA 86j. *Neural Dynamics of Category Learning and Recognition: Attention, Memory Consolidation, and Amnesia.*

[Grossberg, S 86*] SEE: Carpenter, GA 86k. *Neural Dynamics of Category Learning and Recognition: Structural Invariants, Evoked Potentials, and Reinforcement.*

[Grossberg, S 86*] SEE: Cohen, MA 86c. *Neural Dynamics of Speech and Language Coding: Developmental Programs, Perceptual Grouping, and Competition for Short-Term Memory.*

[Grossberg, S 86*] SEE: Carpenter, GA 86l. *Neural Theory of Circadian Rhythms: Aschoff's Rule in Diurnal and Nocturnal Mammals.*

[Grossberg, S 86*] SEE: Carpenter, GA 86m. *Neural Theory of Circadian Rhythms: Split Rhythms, After-Effects, and Motivational Interactions.*

[Grossberg, S 86*] SEE: Carpenter, GA 86n. *Neural Theory of Circadian Rhythms: The Gated Pacemaker.*

[Grossberg, S 86*] SEE: Banquet, J 86a. *Structure of Event-Related Potentials During Learning: An Experimental and Theoretical Analysis.*

[Grossberg, S 86*] SEE: Cohen, MA 86e. *Unitized Recognition Codes for Parts and Wholes: The Unique Cue in Configural Discriminations.*

[Grossberg, S 86a] Grossberg, Stephen (ed.). (1986). Adaptive Brain, Volumes I & II. New York: Elsevier.

[Grossberg, S 86b] Grossberg, Stephen. (1986). *Adaptive Compensation to Changes in the Oculomotor Plant.* Advances in the Biosciences: Adaptive Processes in Visual and Oculomotor Systems, Proceedings of a Conference in Asolimar, California, 16-20 June 1985. Volume 57. pp 341-345. Elmsford, NY: Pergamon.

[Grossberg, S 86c] Grossberg, Stephen. (1986). *Adaptive Self-Organization of Serial Order in Behavior: Speech, Language, and Motor Control.* In Grossberg, Stephen (ed.). Adaptive Brain, Volume II. pp 310-400. New York: Elsevier.

[Grossberg, S 86d] Grossberg, Stephen. (1986). *Associative and Competitive Principles of Learning and Development: The Temporal Unfolding and Stability of STM and LTM Patterns.* In Grossberg, Stephen (ed.). Adaptive Brain, Volume I. pp 448-485. New York: Elsevier.

[Grossberg, S 86e] Grossberg, Stephen. (March 1986). *Brain Metaphors, Theories, and Facts.* Behavioral and Brain Sciences. Volume 9. pp 97-98.

[Grossberg, S 86f] Grossberg, Stephen; Mingolla, Ennio. (1986). *Computer Simulation of Neural Networks for Perceptual Psychology.* Behavior Research Methods, Instruments, and Computers. Volume 18. pp 601-607.

[Grossberg, S 86g] Grossberg, Stephen. (1986). *Cooperative Self-Organization of Multiple Neural Systems During Adaptive Sensory-Motor Control.* In Guthrie, D. M. (ed.). Aims and Methods in Neuroethology. Manchester, England: Manchester University Press.

[Grossberg, S 86h] Grossberg, Stephen; Kuperstein, Michael. (1986). Neural Dynamics of Adaptive Sensory-Motor Control: Ballistic Eye Movements. Amsterdam: North-Holland.

[Grossberg, S 86i] Grossberg, Stephen; Stone, Gregory O. (1986). *Neural Dynamics of Attention Switching and Temporal Order Information in Short-Term Memory.* Memory and Cognition. Volume 14. pp 451-468.

[Grossberg, S 86j] Grossberg, Stephen; Mingolla, Ennio. (1986). *Neural Dynamics of Form Perception: Boundary Completion, Illusory Figures, and Neon Color Spreading.* In Grossberg, Stephen (ed.). Adaptive Brain, Volume II. pp 80-142. New York: Elsevier.

[Grossberg, S 86k] Grossberg, Stephen; Mingolla, Ennio. (1986). *Neural Dynamics of Perceptual Grouping: Textures, Boundaries, and Emergent Segmentations.* In Grossberg, Stephen (ed.). Adaptive Brain, Volume II. pp 143-210. New York: Elsevier.

[Grossberg, S 86l] Grossberg, Stephen; Stone, Gregory O. (1986). *Neural Dynamics of Word Recognition and Recall: Attentional Priming, Learning, and Resonance.* In Grossberg, Stephen (ed.). Adaptive Brain, Volume II. pp 401-455. New York: Elsevier.

[Grossberg, S 86m] Grossberg, Stephen; Stone, Gregory O. (1986). *Neural Dynamics of Word Recognition and Recall: Attentional Priming, Learning and Resonance.* Psychological Review. Volume 93. pp 46-74.

[Grossberg, S 86n] Grossberg, Stephen. (1986). *Processing of Expected and Unexpected Events During Conditioning and Attention: A Psychophysiological Theory.* In Grossberg, Stephen (ed.). Adaptive Brain, Volume I. pp 180-237. New York, NY: Elsevier.

[Grossberg, S 86o] Grossberg, Stephen. (1986). *Psychophysiological Theory of Reinforcement, Drive, Motivation and Attention.* In Grossberg, Stephen (ed.). Adaptive Brain, Volume I. pp 1-81. New York: Elsevier.

[Grossberg, S 86p] Grossberg, Stephen. (1986). *Quantized Geometry of Visual Space: The Coherent Computation of Depth, Form and Lightness.* In Grossberg, Stephen (ed.). Adaptive Brain, Volume II. pp 1-79. New York: Elsevier.

[Grossberg, S 86q] Grossberg, Stephen. (1986). *Some Psychophysiological and Pharmacological Correlates of a Developmental, Cognitive, and Motivational Theory.* In Grossberg, Stephen (ed.). Adaptive Brain, Volume I. pp 82-179. New York: Elsevier.

[Grossberg, S 86r] Grossberg, Stephen. (1986). *The Adaptive Self-Organization of Serial Order in Behavior: Speech, Language, and Motor Control.* In Schwab, Eileen C. (ed.); Nusbaum, Howard C. (ed.). Pattern Recognition by Humans and Machines: Speech Perception. Volume I. pp 187-294. New York: Academic Press.

[Grossberg, S 87*] SEE: Carpenter, GA 87c. *ART-2: Self-Organization of Stable Category Recognition Codes for Analog Input Patterns.*

[Grossberg, S 87*] SEE: Carpenter, GA 87d. *Invariant Pattern Recognition and Recall by an Attentive Self-Organizing ART Architecture in a Nonstationary World.*

[Grossberg, S 87*] SEE: Cohen, MA 87a. *Masking Fields: A Massively Parallel Architecture for Learning, Recognizing, and Predicting Multiple Groupings of Patterned Data.*

[Grossberg, S 87*] SEE: Cohen, MA 87b. *Masking Fields: A Massively Parallel Neural Architecture for Learning, Recognizing, and Predicting Multiple Groupings of Patterned Data.*

[Grossberg, S 87*] SEE: Carpenter, GA 87a. *A Massively Parallel Architecture for a Self-Organizing Neural Pattern Recognition Machine.*

[Grossberg, S 87*] SEE: Bullock, D 87a. *A Neural Network Architecture for Automatic Trajectory Formation and Coordination of Multiple Effectors During Variable-Speed Arm Movements.*

[Grossberg, S 87*] SEE: Carpenter, GA 87e. *Neural Networks: Introduction to the 1 December 1987 Issue of Applied Optics.*

[Grossberg, S 87*] SEE: Banquet, J 87a. *Probing Cognitive Processes Through the Structure of Event-Related Potentials During Learning: An Experimental and Theoretical Analysis.*

[Grossberg, S 87*] SEE: Cohen, MA 87c. *Recent Developments in a Neural Model of Real-Time Speech Analysis and Synthesis.*

[Grossberg, S 87*] SEE: Carpenter, GA 87f. *Technical Comments: Computing with Neural Networks.*

[Grossberg, S 87a] Grossberg, Stephen; Marshall, Jonathan. (1987). *A Computational Model of How Cortical Complex Cells Multiplex Information About Position, Contrast, Orientation, Spatial Frequency, Disparity.* IEEE International Conference on Neural Networks (ICNN-87), San Diego, CA, June 21-24, 1987: Proceedings. Volume 4. pp 203-214. New York: Institute of Electrical and Electronics Engineers.

[Grossberg, S 87b] Grossberg, Stephen; Schmajuk, Nestor A. (1987). *A Neural Network Architecture for Attentionally-Modulated Pavlovian Conditioning: Conditioned Reinforcement, Inhibition, and Opponent Processing.* IEEE International Conference on Neural Networks (ICNN-87), San Diego, CA, June 21-24, 1987: Proceedings. Volume 2. pp 795-804. New York: Institute of Electrical and Electronics Engineers.

[Grossberg, S 87c] Grossberg, Stephen; Todorvic, Dejan. (1987). *A Neural Network Architecture for Brightness Perception Under Constant and Variable Illumination Conditions.* IEEE International Conference on Neural Networks (ICNN-87), San Diego, CA, June 21-24, 1987: Proceedings. Volume 4. pp 185-192. New York: Institute of Electrical and Electronics Engineers.

[Grossberg, S 87d] Grossberg, Stephen; Mingolla, Ennio. (1987). *A Neural Network Architecture for Preattentive Vision: Multiple Scale Segmentation and Regularization.* IEEE International Conference on Neural Networks (ICNN-87), San Diego, CA, June 21-24, 1987: Proceedings. Volume 4. pp 177-184. New York: Institute of Electrical and Electronics Engineers.

[Grossberg, S 87e] Grossberg, Stephen. (1987). *Associative and Cooperative / Competitive Mechanisms in Neural Networks.* IEEE International Conference on Neural Networks (ICNN-87), San Diego, CA, June 21-24, 1987: Proceedings. Tutorial 1. San Diego: Institute of Electrical and Electronic Engineers.

[Grossberg, S 87f] Grossberg, Stephen; Levine, Daniel S. (1987). *Attentional Mechanisms in Neural Information Processing: Examples From Pavlovian Conditioning.* IEEE International Conference on Neural Networks (ICNN-87), San Diego, CA, June 21-24, 1987: Proceedings. Volume 4. pp 49-58. New York: Institute of Electrical and Electronics Engineers.

[Grossberg, S 87g] Grossberg, Stephen. (1987). *Competitive Learning: From Interactive Activation to Adaptive Resonance.* Cognitive Science. Volume 11. pp 23-63.

[Grossberg, S 87h] Grossberg, Stephen. (1987). *Cortical Dynamics of Three-Dimensional Form, Color, and Brightness Perception, I: Monocular Theory*. Perception and Psychophysics. Volume 41. pp 87-116.

[Grossberg, S 87i] Grossberg, Stephen. (1987). *Cortical Dynamics of Three-Dimensional Form, Color, and Brightness Perception, II: Binocular Theory*. Perception and Psychophysics. Volume 41. pp 117-158.

[Grossberg, S 87j] Grossberg, Stephen; Schmajuk, Nestor A. (1987). *Neural Dynamics of Attentionally Modulated Pavlovian Conditioning: Conditioned Reinforcement, Inhibition, and Opponent Processing*. Psychobiology. Volume 15. pp 195-240.

[Grossberg, S 87k] Grossberg, Stephen; Gutowski, William E. (1987). *Neural Dynamics of Decision Making Under Risk: Affective Balance and Cognitive-Emotional Interactions*. Psychological Review. Volume 94. pp 300-318.

[Grossberg, S 87l] Grossberg, Stephen; Gutowski, William E. (1987). *Neural Dynamics of Decision Making Under Risk: Affective Balance Theory*. IEEE International Conference on Neural Networks (ICNN-87), San Diego, CA, June 21-24, 1987: Proceedings. Volume 3. pp 31-38. New York: Institute of Electrical and Electronics Engineers.

[Grossberg, S 87m] Grossberg, Stephen; Mingolla, Ennio. (1987). *Neural Dynamics of Surface Perception: Boundary Webs, Illuminants, and Shape-From-Shading*. Computer Vision, Graphics, and Image Processing. Volume 37. pp 116-165.

[Grossberg, S 87n] Grossberg, Stephen. (1987). *Recent Developments in the Design of Real-Time Non-Linear Neural Network Architectures*. IEEE International Conference on Neural Networks (ICNN-87), San Diego, CA, June 21-24, 1987: Proceedings. Volume 1. pp 1-76. New York: Institute of Electrical and Electronics Engineers.

[Grossberg, S 87o] Grossberg, Stephen. (June 1987). *Stable Self-Organization of Sensory Recognition Codes: Is Chaos Necessary?*. Behavioral and Brain Sciences. Volume 10. pp 179-180.

[Grossberg, S 88*] SEE: Carpenter, GA 88a. *The ART of Adaptive Pattern Recognition by a Self-Organizing Neural Network.*

[Grossberg, S 88*] SEE: Bullock, D 88b. *Neural Dynamics of Planned Arm Movements: Emergent Invariants and Speed-Accuracy Properties During Trajectory Formation.*

[Grossberg, S 88*] SEE: Bullock, D 88c. *Neural Dynamics of Planned Arm Movements: Emergent Invariants and Speed-Accuracy Properties During Trajectory Formation.*

[Grossberg, S 88*] SEE: Bullock, D 88d. *The Vite Model: A Neural Command Circuit for Generating Arm and Articulator Trajectories.*

[Grossberg, S 88a] Grossberg, Stephen. (1988). *Competitive Learning: From Interactive Activation to Adaptive Resonance.* In Waltz, David L. (ed.); Feldman, Jerome A. (ed.). Connectionist Models and Their Implications: Readings from Cognitive Science. pp 243-284. Norwood, NJ: Ablex.

[Grossberg, S 88b] Grossberg, Stephen; Todorovic, Dejan. (1988). *Neural Dynamics of 1-D and 2-D Brightness Perception: A Unified Model of Classical and Recent Phenomena*. Perceptions and Psychophysics. Volume 43. pp 241-277.

[Grossberg, S 88c] Grossberg, Stephen; Todorovic, Dejan. (1988). *Neural Dynamics of 1-D and 2-D Brightness Perception: A Unified Model of Classical and Recent Phenomena*. Neural Networks and Natural Intelligence. pp 127-194. Cambridge, MA: MIT Press.

[Grossberg, S 88d] Grossberg, Stephen. (1988). Neural Networks and Natural Intelligence. Cambridge, MA: MIT Press.

[Grzywacx, NM 86a] Grzywacx, N. M.; Yuille, Alan L. (1986). *Motion Correspondence and Analog Networks*. In Denker, John S. (ed.). Neural Networks for Computing, Snowbird, Utah, 1986. AIP Proceedings 151. pp 200-205. New York: American Institute of Physics.

[Grzywacz, N 87a] Grzywacz, Norberto; Koch, Christof. (1987). *Functional Properties of Models for Direction Selectivity in the Retina*. Synapse. Volume 1. pp 417-434.

[Guckenheimer, J 82a] Guckenheimer, John. (1982). *Noise in Chaotic Systems*. Nature. Volume 298. pp 358-361.

[Guerin, A 88a] Guerin, Anne; Herault, Jeanny. (1988). *An Efficient Vector/Scalar Processing Structure for Adaptive Algorithms*. Signal Processing IV: Theories and Applications. pp 331-334. New York: Elsevier.

[Guest, CC 84a] Guest, Clark C.; Mirsalehi, M. M.; Gaylord, T. K. (1984). *EXCLUSIVE OR Processing (Binary Image Subtraction) Using Thick Fourier Holograms*. Applied Optics. Volume 23. pp 3444-3454.

[Guest, CC 87*] SEE: Kosko, B 87i. *Optical Bi-directional Associative Memories*.

[Guest, CC 87*] SEE: Te Kolste, RD 87a. *Optical Competitive Neural Network with Optical Feedback*.

[Guest, CC 87*] SEE: Yang, H 87a. *Performance of Backpropagation for Rotation Invariant Pattern Recognition*.

[Guest, CC 87a] Guest, Clark C.; TeKolste, Robert. (1987). *Designs and Devices for Optical Bidirectional Associative Memories*. Applied Optics. Volume 26. pp 5055-5060.

[Guevara, MR 83a] Guevara, Michael R.; Glass, Leon; Mackey, Michael C.; Shrier, Alvin. (1983). *Chaos in Neurobiology*. IEEE Transactions on Systems, Man, and Cybernetics. Volume SMC-13. pp 790-798.

[Guez, A 87*] SEE: Eilbert, JL 87a. *The Control of Behavior in a Hierarchical Neural Network: Modelling Attention Channels and Behavior Modes*.

[Guez, A 87*] SEE: Kam, M 87a. *On the Design of a Content-Addressable Memory via Binary Neural Networks*.

[Guez, A 87a] Guez, Allon; Eilbert, James L.; Kam, Moshe. (1987). *Neuromorphic Architecture for Adaptive Robot Control: A Preliminary Analysis*. IEEE International Conference on Neural Networks (ICNN-87), San Diego, CA, June 21-24, 1987: Proceedings. Volume 4. pp 567-572. New York: Institute of Electrical and Electronics Engineers.

[Guez, A 88a] Guez, Allon; Eilbert, James L.; Kam, Moshe. (1988). *Neural Network Architecture for Control*. IEEE Control Systems Magazine. pp 22-25.

[Guez, A 88b] Guez, Allon; Protopopsecu, Vladimir; Barhen, Jacob. (1988). *On the Stability, Storage Capacity, and Design of Nonlinear Continuous Neural Networks*. IEEE Transactions on Systems, Man and Cybernetics. SMC-18. pp 80-87.

[Guilfoyle, PS 84*] SEE: Rhodes, WT 84a. *Acoustooptic Algebraic Processing Architectures*.

[Gulati, S 87a] Gulati, S.; Iyengar, S. S.; Toomarian, N.; Protopopescu, Vladimir; Barhen, Jacob. (1987). *Nonlinear Neural Networks for Deterministic Scheduling*. IEEE International Conference on Neural Networks (ICNN-87), San Diego, CA, June 21-24, 1987: Proceedings. Volume 4. pp 745-752. New York: Institute of Electrical and Electronics Engineers.

[Gullapalli, V 88a] Gullapalli, Vijaykumar. (1988). A Stochastic Algorithm for Learning Real-Valued Functions via Reinforcement Feedback. COINS Technical Report 88-91. pp 1-38. Amherst, MA: University of Massachusetts, Computer and Information Science.

[Gullichsen, E 87a] Gullichsen, Eric; Chang, Ernest. (1987). *Pattern Classification by Neural Network: An Experimental System for Icon Recognition*. IEEE International Conference on Neural Networks (ICNN-87), San Diego, CA, June 21-24, 1987: Proceedings. Volume 4. pp 725-732. New York: Institute of Electrical and Electronics Engineers.

[Gulyas, B 88*] SEE: Orban, GA 88a. *Image Segregation by Motion: Cortical Mechanism and Implementation in Neural Networks*.

[Gunger, D 86a] Gunger, David; Skrzypek, Josef. (1986). *A Connectionist Architecture for Matching 3-D Models to Moving*

Edge Features. Intelligent Robots and Computer Vision. SPIE Volume 726. Bellingham, WA: Society of Photo-Optical Instrumentation Engineers.

[Gunger, D 87*] SEE: Paik, E 87a. *UCLA SFINX - A Neural Network Simulation Environment.*

[Gunger, D 87a] Gunger, David; Skrzypek, Josef. (1987). *UCLA PUNNS - A Neural Network Machine for Computer Vision.* Image Understanding and the Man-Machine Interface. SPIE Volume 758. pp 5-10. Bellingham, WA: Society of Photo-Optical Instrumentation Engineers.

[Gupta, A 86a] Gupta, Anoop. (1986). Parallelism in Production Systems. CMU-CS-86-122. Pittsburgh, PA: Carnegie Mellon University, Computer Science Department.

[Gupta, B 88*] SEE: Alspector, J 88b. *Performance of a Stochastic Learning Microchip.*

[Gupta, NK 73*] SEE: Widrow, B 73a. *Punish / Reward: Learning With a Critic in Adaptive Threshold Systems.*

[Gurney, KN 87*] SEE: Milligan, DK 87a. *RAM-Unit Learning Networks.*

[Gutfreund, H 85*] SEE: Amit, DJ 85a. *Spin-Glass Models of Neural Networks.*

[Gutfreund, H 85*] SEE: Amit, DJ 85b. *Storing Infinite Numbers of Patterns in a Spin-Glass Model of Neural Networks.*

[Gutfreund, H 87*] SEE: Meidan, M 87a. Electronic Analog Neural Network with Noise and Time Delays.

[Gutfreund, H 88a] Gutfreund, Hanoch. (1988). *Neural Networks with Hierarchically Correlated Patterns.* Physical Review A. Volume 37. pp 570-577.

[Gutfreund, H 88b] Gutfreund, Hanoch; Mezard, M. (1988). *Processing of Temporal Sequences in Neural Networks.* Physical Review Letters. Volume 61. pp 235-238.

[Gutowski, WE 87*] SEE: Grossberg, S 87k. *Neural Dynamics of Decision Making Under Risk: Affective Balance and Cognitive-Emotional Interactions.*

[Gutowski, WE 87*] SEE: Grossberg, S 87l. *Neural Dynamics of Decision Making Under Risk: Affective Balance Theory.*

[Gutschow, T 86*] SEE: Hecht-Nielsen, R 86b. Sensor Processing Using Artificial Neural Systems.

[Gutschow, T 87a] Gutschow, Todd; Hecht-Nielsen, Robert. *Pattern Recognition Using Artificial Neural Systems.* Principles of Neural Modeling.

[Gutzmann, KM 87a] Gutzmann, Kurt M. (1987). *Combinatorial Optimization Using a Continuous State Boltzmann Machine.* IEEE International Conference on Neural Networks (ICNN-87), San Diego, CA, June 21-24, 1987: Proceedings. Volume 3. pp 721-734. New York: Institute of Electrical and Electronics Engineers.

[Guyon, I 85*] SEE: Personnaz, L 85a. *Information Storage and Retrieval in Spin-Glass Like Neural Networks.*

[Guyon, I 86*] SEE: Personnaz, L 86a. *A Biologically Constrained Learning Mechanism in Networks of Formal Neurons.*

[Guyon, I 86*] SEE: Personnaz, L 86b. *Collective Computational Properties of Neural Networks: New Learning Mechanisms.*

[Guyon, I 86*] SEE: Personnaz, L 86c. *Designing a Neural Network Satisfying a Given Set of Constraints.*

[Guyon, I 86*] SEE: Personnaz, L 86e. *Neural Network Design for Efficient Information Retrieval.*

[Guyon, I 86*] SEE: Personnaz, L 86d. *Neural Network Design for Efficient Information Retrieval.*

[Guyon, I 86*] SEE: Personnaz, L 86f. *Simple Selectionist Learning Rule for Neural Networks.*

[Guyon, I 86a] Guyon, I.; Personnaz, L.; Siarry, P.; Dreyfus, G. (1986). *Engineering Applications of Spin Glass Concepts.* Proceedings of the Heidelberg Colloquium on Glassy Dynamics - Spin Glasses, Optimization and Neural Networks, June 9-13, 1986. Physics Volume 275. New York: Springer-Verlag.

[Guyon, I 87*] SEE: Personnaz, L 87a. *High-Order Neural Networks: Information Storage without Errors.*

[Guyon, I 88*] SEE: Dreyfus, G 88a. *High Order Neural Networks for Efficient Associative Memory Design.*

[Guyon, I 88a] Guyon, I.; Personnaz, L.; Dreyfus, G. (1988). *Of Points and Loops.* In Eckmiller, Rolf (ed.); von der Malsburg, Christoph (ed.). Neural Computers. NATO ASI Series, Volume F41. pp 261-269. Berlin: Springer-Verlag.

[Habibi, A 73a] Habibi, Ali. (1972). *Two-Dimensional Bayesian Estimate of Images.* Proceedings of the IEEE. Volume 60. pp 878-883.

[Habibi, A 86a] Habibi, Ali. (1986). *Image Processing and Understanding.* Handbook of Pattern Recognition and Image Processing. pp 169-189. New York: Academic Press.

[Hadeler, KP 74a] Hadeler, K. P. (1974). *On the Theory of Lateral Inhibition.* Kybernetik. Volume 14. pp 161-165.

[Haggerty, J 88*] SEE: Vidal, JJ 88b. *Synchronization in Neural Nets.*

[Haibt, LH 56*] SEE: Rochester, N 56b. *Tests on a Cell Assembly Theory of the Action of the Brain, Using a Large Digital Computer.*

[Haibt, LH 56*] SEE: Rochester, N 56a. *Tests on a Cell Assembly Theory of the Action of the Brain, Using a Large Digital Computer.*

[Haken, H 86*] SEE: Schoner, G 86a. *A Stochastic Theory of Phase Transitions in Human Hand Movement.*

[Haken, H 87*] SEE: Fuchs, A 87a. *Spatio-Temporal Analysis of Multi-channel Alpha EEG Map Series.*

[Haken, H 87a] Haken, Hermann (ed.). (1987). Computational Systems - Natural and Artificial. Heidelberg: Springer.

[Haken, H 88*] SEE: Fuchs, A 88a. *Computer Simulations of Pattern Recognition as a Dynamical Process of a Synergetic System.*

[Haken, H 88*] SEE: Banzhaf, W 88a. *A Network for Recognition and Classification of Continuous Patterns.*

[Haken, H 88a] Haken, Hermann. (1988). Information and Self-Organization: A Macroscopic Approach to Complex Systems. New York: Springer-Verlag.

[Haken, H 88b] Haken, Hermann. (1988). *Learning in Synergetic Systems for Pattern Recognition and Associative Action.* Zeitschrift fur Physik B; Condensed Matter. Volume 71. pp 521-526.

[Haken, H 88c] Haken, Hermann. (1988). *Nonequilibrium Phase Transitions in Pattern Recognition and Associative Memory.* Zeitschrift fur Physik B, Condensed Matter. Volume 70. pp 121-123. New York: Springer-Verlag.

[Haken, H 88d] Haken, Hermann. (1988). *Synergetics and Computers.* Journal of Computational and Applied Mathematics. Volume 22. pp 197-202.

[Haken, H 88e] Haken, Hermann. (1988). *Synergetics in Pattern Recognition and Associative Action.* Neural and Synergetic Computers. Series in Synergetics, Vol 42. pp 2-15. New York: Springer-Verlag.

[Hall, JL 79*] SEE: Schroeder, MR 79a. *Optimizing Digital Speech Coders by Exploiting Masking Properties of the Human Ear.*

[Halle, M 62a] Halle, M.; Stevens, K. (1962). *Speech Recognition: A Model and a Program for Research.* IRE Transactions on Information Theory. Volume IT-8. pp 155-159.

[Hallett, M 88*] SEE: Spitzer, AR 88a. A Method for Determining Optimal Interelectrode Spacing for Cerebral Topographic Mapping.

[Hallikainen, J 88a] Hallikainen, J.; Parkkinen, Jussi; Jaaskelainen, Timo. (1988). *Acousto-Optic Color Spectrometer.* Review of Scientific Instruments. Volume 59. pp 81-83.

[Haltsonen, S 78*] SEE: Jalanko, M 78a. *Application of Orthogonal Projection Principles to Simultaneous Phonemic Segmentation and Labeling of Continuous Speech.*

[Haltsonen, S 78a] Haltsonen, Seppo; Jalanko, Matti; Bry, Kalle-J; Kohonen, Teuvo. (1978). *Application of Novelty Filter to Segmentation of Speech.* IEEE International Conference on Acoustics, Speech, and Signal Processing (ICASSP-78), Tulsa, OK, April 10-12, 1978: Proceedings. pp 565-568. New York: Institute of Electrical and Electronics Engineers.

[Haltsonen, S 80*] SEE: Kohonen, T 80a. *A Thousand-Word Recognition System Based on the Learning Subspace Method and Redundant Hash Addressing.*

[Haltsonen, S 84*] SEE: Kohonen, T 84a. *On-Line Recognition of Spoken Words from a Large Vocabulary.*

[Hammerstrom, D 86*] SEE: Bailey, J 86a. How to Make a Billion Connections.

[Hammerstrom, D 86a] Hammerstrom, Daniel. (1986). A Connectivity Analysis of Recursive, Auto-Associative Connection Networks. CS/E-06-009. Beaverton, OR: Oregon Graduate Center, Computer Science/Engineering Department.

[Hammerstrom, D 86b] Hammerstrom, Daniel; Kravitz, Roy. (1986). A Design Methodology for High-Performance, General-Purpose VLSI. CS/E-86-008. Beaverton, OR: Oregon Graduate Center, Computer Science/Engineering Department.

[Hammerstrom, D 86c] Hammerstrom, Daniel. (1986). Computer vs the Brain. Beaverton, OR: Oregon Graduate Center, Computer Science/Engineering Department.

[Hammerstrom, D 86d] Hammerstrom, Daniel. (1986). Connectionist/Neural Network Bibliography. CS/E-86-010. Beaverton, OR: Oregon Graduate Center, Computer Science/Engineering Department.

[Hammerstrom, D 86e] Hammerstrom, Daniel; Bahr, Casey; Bailey, Jim; Baker, Tom; Beaver, Gary; Jagla, Kevin. (1986). The OGC Cognitive Architecture Project: Silicon Implementation of Connectionist/Neural Networks. Beaverton, OR: Oregon Graduate Center, Computer Science/Engineering Department.

[Hammerstrom, D 87a] Hammerstrom, Daniel. (1987). A Connectionist/Neural Network Bibliography, Volume II. CS/E-87-008. Beaverton, OR: Oregon Graduate Center, Computer Science/Engineering Department.

[Hammerstrom, D 87b] Hammerstrom, Daniel. (1987). *Connecting with the Human Mind: A New Approach to Building Computers Draws on the Research of Neuroscientists.* Visions. Volume 3. pp 13-14.

[Hammerstrom, D 88*] SEE: Bahr, C 88c. *Concurrent Neural Network Simulation: Two Examples within a Single, Integrated Neural Network Hardware Development Environment.*

[Hammerstrom, D 88*] SEE: May, N 88a. Fault Simulation of a Wafer-Scale Integrated Neural Network.

[Hammerstrom, D 88*] SEE: May, N 88b. Fltsim Detailed Description and Operation.

[Hammerstrom, D 88*] SEE: Rudnick, M 88a. *An Interconnect Structure for Wafer Scale Neurocomputers.*

[Hammerstrom, D 88*] SEE: Rudnick, M 88b. Physical Broadcast Structure.

[Hammerstrom, D 88a] Hammerstrom, Daniel. (1988). A Connectivity Analysis of a Class of Simple Associative Neural Networks. CS/E-86-009. Beaverton, OR: Oregon Graduate Center, Computer Science/Engineering Department.

[Hammerstrom, D 88b] Hammerstrom, Daniel. (1988). *The Connectivity Analysis of Simple Association.* In Anderson, Dana Z. (ed.). Neural Information Processing Systems. pp 338-347. New York: American Institute of Physics.

[Hammond, N 84a] Hammond, N.; Hinton, Geoffrey E.; Barnard, P.; MacLean, A. (1985). *Evaluating the Interface of a Document Processor: A Comparison of Expert Judgement and User Observation.* Human-Computer Interaction (Interact '84), London, September 4-7, 1984. pp 725-729. Amsterdam: Elsevier.

[Hampson, S 83a] Hampson, Steven; Kibler, Dennis. (1983). *A Boolean Complete Neural Model of Adaptive Behavior.* Biological Cybernetics. Volume 49. pp 9-19.

[Han, TS 75a] Han, Te Sun. (1975). *Linear Dependence Structure of the Entropy Space.* Information and Control. Volume 29. pp 337-368.

[Hancock, H 87*] SEE: Draper, JS 87a. *A Microcomputer Neural Net Benchmarked Against Standard Classification Techniques.*

[Hancock, JC 66*] SEE: Patrick, EA 66a. *Nonsupervised Sequential Classification and Recognition of Patterns.*

[Hancock, PJB 88*] SEE: Phillips, WA 88a. *On the Acquisition of Object Concepts from Sensory Data.*

[Handelman, DA 88*] SEE: Lane, SH 88a. *A Neural Network Computational Map Approach to Reflexive Motor Control.*

[Handelman, DA 88a] Handelman, David A.; Lane, Stephen H.; Gelfand, Jack J. (1988). *Integration of Knowledge-Based System and Neural Network Techniques for Robotic Control.* IFAC Workshop on Artificial Intelligence in Real-Time Control, Swansea, September, 1988.

[Hansen, FR 82a] Hansen, F. R.; Elliott, H. (1982). *Image Segmentation Using Simple Markov Field Models.* Computer Graphics and Image Processing 20. Volume 20. pp 101-132.

[Hanson, AR 78a] Hanson, Allen R.; Riseman, Edward M. (1978). *Segmentation of Natural Scenes.* Computer Vision Systems. pp 129-163. New York: Academic Press.

[Hanson, AR 87*] SEE: Riseman, EM 87a. *A Methodology for the Development of General Knowledge-Based Vision Systems.*

[Hanson, AR 87*] SEE: Arbib, MA 87d. Vision, Brain, and Cooperative Computation.

[Hanson, SJ 87a] Hanson, Stephen Jose; Kegl, Judy. (1987). *PARSNIP: A Connectionist Network that Learns Natural Language Grammar from Exposure to Natural Language Sentences.* Cognitive Science Society, 9th, Seattle, WA, July 16-18, 1987: Proceedings. pp 106-119. Hillsdale, NJ: Lawrence Erlbaum.

[Hanson, SJ 88a] Hanson, Stephen Jose; Burr, David J. (1988). *Minkowski-r Back-Propagation: Learning in Connectionist Models with Non-Euclidian Error Signals.* In Anderson, Dana Z. (ed.). Neural Information Processing Systems. pp 348-357. New York: American Institute of Physics.

[Hanson, WA 86*] SEE: Cruz, CA 86a. *Flow-of-Activation Processing: Parallel Associative Networks (PAN).*

[Hanson, WA 87*] SEE: Cruz, CA 87b. *Knowledge Processing Through Flow-of-Activation.*

[Hanson, WA 87*] SEE: Cruz, CA 87c. *Neural Network Emulation Hardware Design Considerations.*

[Hanson, WA 87b] Hanson, William A.; Cruz, Claude A.; Tam, Jason Y. (1987). *CONE - Computational Network Environment.* IEEE International Conference on Neural Networks (ICNN-87), San Diego, CA, June 21-24, 1987: Proceedings. Volume 3. pp 531-538. New York: Institute of Electrical and Electronics Engineers.

[Harley, TJ 65*] SEE: Abend, K 65a. *Classification of Binary Random Patterns.*

[Harley, WS 86*] SEE: Cohen, NJ 86a. *Skill Learning and Repetition Priming in Symmetry Detection: Parallel Studies of Human Subjects and Connectionist Models.*

[Harman, HH 76a] Harman, H. H. (1976). Modern Factor Analysis. Chicago: University of Chicago Press.

[Harmon, LD 59a] Harmon, Leon D. (1959). *Artificial Neuron.* Science. Volume 129. pp 962-963.

[Harmon, LD 61a] Harmon, Leon D. (1961). *Studies with Artificial Neurons, I: Properties and Functions of an Artificial.* Kybernetik. Volume 1. pp 89-101.

[Harmon, LD 66a] Harmon, Leon D.; Lewis, Edwin R. (1966). *Neural Modeling.* Physiological Review. Volume 46. pp 513-591.

[Harnish, RM 89*] SEE: Nadel, L 89a. Neural Connections, Mental Computation.

[Harper, P 87*] SEE: Samad, T 87b. *Associative Memory Storage Using a Variant of the Generalized Delta Rule.*

[Harris, CJ 81a] Harris, C. J.; Billings, S. A. (1981). Self-Tuning and Adaptive Control: Theory and Applications. London: Peter Peregrinus.

[Harrison, TD 86*] SEE: Prager, RW 86a. *Boltzmann Machines for Speech Recognition.*

[Harrison, TD 88a] Harrison, T. D.; Fallside, Frank. (1988). A Connectionist Structure for Phoneme Recognition. CUED/F-INFENG/TR.15. Cambridge: Cambridge University, Engineering Department.

[Hart, PE 68a] Hart, Peter E. (1968). *Condensed Nearest Neighbor Rule.* IEEE Transactions on Information Theory. Volume IT-14. pp 515-516.

[Hart, PE 72*] SEE: Duda, RO 72a. *Use of the Hough Transform to Detect Lines and Curves in Pictures.*

[Hart, PE 73*] SEE: Duda, RO 73a. Pattern Classification and Scene Analysis.

[Harth, EM 70*] SEE: Anninos, PA 70a. *Dynamics of Neural Structures.*

[Harth, EM 70a] Harth, Erich M.; Csermely, T. J.; Beek, B.; Lindsay, R. D. (1970). *Brain Functions and Neural Dynamics.* Journal of Theoretical Biology. Volume 26. pp 93-120.

[Harth, EM 74a] Harth, Erich M.; Tzanakou, E. (1974). *ALOPEX: A Stochastic Method for Determining Visual Receptive Fields.* Vision Research. Volume 14. pp 1475-1482.

[Harth, EM 78*] SEE: Finette, S 78a. *Anisotropic Connectivity and Cooperative Phenomena as a Basis for Orientation Sensitivity in the Visual Cortex.*

[Harth, EM 82*] SEE: Shaw, GL 82a. *Cooperativity in Brain Function: Assemblies of Approximately 30 Neurons.*

[Harth, EM 82a] Harth, Erich M. (1982). *Cybernetic Principles in Perception and Motor Control.* Progress in Cybernetics and Systems Research. Volume 9. pp 253-258.

[Harth, EM 82b] Harth, Erich M. (1982). *On the Spontaneous Emergence of Neuronal Schemata.* In Amari, Shun-ichi (ed.); Arbib, Michael A. (ed.). Competition and Cooperation in Neural Nets. pp 286-294. New York: Springer-Verlag.

[Harth, EM 83a] Harth, Erich M. (1983). *Order and Chaos in Neural Systems: An Approach to the Dynamics of Higher Brain Functions*. IEEE Transactions on Systems, Man and Cybernetics. Volume SMC-13. pp 782-789.

[Harth, EM 87*] SEE: Unnikrishnan, KP 87a. *Role of Feedback in Visual Perception*.

[Hartley, RL 87*] SEE: Szu, HH 87a. *Fast Simulated Annealing*.

[Hartley, RL 87*] SEE: Szu, HH 87e. *Nonconvex Optimization by Fast Simulated Annealing*.

[Hartley, RL 87*] SEE: Szu, HH 87f. *Nonconvex Optimization by Fast Simulated Annealing*.

[Hartley, RL 87a] Hartley, Ralph L.; Szu, Harold H. (1987). *A Comparison of the Computational Power of Neural Network Models*. IEEE International Conference on Neural Networks (ICNN-87), San Diego, CA, June 21-24, 1987: Proceedings. Volume 3. pp 15-22. New York: Institute of Electrical and Electronics Engineers.

[Hartline, DK 78a] Hartline, Daniel K. (1978). *Computer Models of Bursting Neural Networks*. Brain Theory Newsletter. Volume 3. pp 60-62.

[Hartman, H 86a] Hartman, H.; Vichniac, Gerard Y. (1986). *Inhomogeneous Celluar Automata (INCA)*. In Bienenstock, Elie L. (ed.). Disordered Systems and Biological Organization. pp 53-57. New York: Springer-Verlag.

[Hartmann, G 82a] Hartmann, Georg. (1982). *Recursive Features of Circular Receptive Fields*. Biological Cybernetics. Volume 43. pp 199-208.

[Hartmann, G 83a] Hartmann, Georg. (1983). *Processing of Continuous Lines and Edges by the Visual System*. Biological Cybernetics. Volume 47. pp 43-50.

[Hartmann, G 85a] Hartmann, Georg. (1985). *Hierarchical Contour Coding by the Visual Cortex*. Models of the Visual Cortex. pp 137-145. New York: Wiley.

[Hartmann, G 87a] Hartmann, Georg. (1987). *Recognition of Hierarchically Encoded Images by Technical and Biological Systems*. Biological Cybernetics. Volume 57. pp 73-84.

[Hartmann, G 88a] Hartmann, Georg. (1988). *Mapping Images to a Hierarchical Data Structure - A Way to Knowledge-based Pattern Recognition*. In Eckmiller, Rolf (ed.); von der Malsburg, Christoph (ed.). Neural Computers. NATO ASI Series, Volume F41. pp 91-100. Berlin: Springer-Verlag.

[Hartzell, HC 81a] Hartzell, H. Criss. (1981). *Mechanisms of Slow Synaptic Potentials*. Nature. Volume 291. pp 539-544.

[Hary, JM 82a] Hary, J. M.; Massaro, Dominic W. (1982). *Categorical Results do not Imply Categorical Perception*. Perception and Psycholphysics. Number 32. pp 409-418.

[Hassner, M 80a] Hassner, Martin; Sklansky, Jack. (1980). *The Use of Markov Random Fields as Models of Texture*. Computer Graphics and Image Processing. Volume 12.

[Hassoun, MH 88a] Hassoun, Mohamad H.; Clark, D. W. (1988). *An Adaptive Attentive Learning Algorithm for Single-Layer Neural Networks*. IEEE International Conference on Neural Networks (ICNN-88), San Diego, CA, July 24-27, 1988: Proceedings. Volume 1. pp 431-440. New York: Institute of Electrical and Electronics Engineers.

[Hassoun, MH 88b] Hassoun, Mohamad H.; Youssef, A. M. (in press). *High-Performance Recording Algorithm for Hopfield Model Associative Memories*. Optical Engineering.

[Hassoun, MH 88c] Hassoun, Mohamad H.; Spitzer, A. Robert. (in press). Pseudo-Unsupervised Neural Network Based Automated EMG Decomposition.

[Hassoun, MH 88d] Hassoun, Mohamad H. (1988). *Two-Level Neural Network for Deterministic Logic Processing*. Optical

Computing and Nonlinear Materials. SPIE Volume 881. pp 258-264. Bellingham, WA: Society of Photo-Optical Instrumentation Engineers.

[Hauptmann, AG 87a] Hauptmann, Alexander G.; Rudnicky, Alexander I. (1987). Talking to Computers: An Empirical Investigation. CMU-CS-87-186. Pittsburgh, PA: Carnegie Mellon University, Computer Science Department.

[Hawkins, HL 78*] SEE: Carr, TH 78a. *Perceptual Flexibility in Word Recognition: Strategies Affect Orthographic Computation But Not Lexical Access.*

[Hawkins, JK 60a] Hawkins, J. K. (1960). *A Magnetic Integrator for the Perceptron Program.* IRE International Convention Record. pp 88-95.

[Hawkins, JK 61a] Hawkins, J. K. (1961). *Self-Organizing Systems - A Review and Commentary.* Proceedings of the IRE. Volume 49. pp 31-48.

[Hay, JC 60a] Hay, J. C.; Martin, F. C.; Wightman, C. W. (1960). *The Mark I Perceptron - Design and Performance.* IRE International Convention Record. pp 78-87.

[Hayduk, JW 88a] Hayduk, John W.; Fong, Yu-Shan. (1988). Architecture for Fast Learning of Decision Regions in Neural Networks. Potsdam, NY: Clarkson University, Electrical and Computer Engineering Department.

[Hayes, JP 78a] Hayes, John P. (1978). *Associative Memories.* Computer Architecture and Organization. pp 388-400. New York: McGraw-Hill.

[Hayes, PJ 84*] SEE: Ballard, DH 84a. *Parallel Logical Inference.*

[Hayes-Roth, B 77a] Hayes-Roth, Barbara. (1977). *Evolution of Cognitive Structure and Processes.* Psychological Review. Volume 84. pp 260-278.

[Healy, LD 62*] SEE: Fischler, M 62a. *An Approach to General Pattern Recognition.*

[Healy, MJ 88a] Healy, Michael J. (1988). An Investigation of Knowledge Representation in a Neural Network. ETA-TR-94. Seattle, WA: Boeing Computer Services, Engineering and Scientific Services.

[Hearn, RH 75*] SEE: Widrow, B 75a. *Adaptive Noise Canceling: Principles and Applications.*

[Hebb, DO 49a] Hebb, Donald O. (1988). *The Organization of Behavior: The First Stage of Perception: Growth of the Assembly.* In Anderson, James A. (ed.); Rosenfeld, Edward (ed.). Neurocomputing: Foundations of Research. pp 45-56. Cambridge, MA: MIT Press.

[Hecht-Nielsen, R 81a] Hecht-Nielsen, Robert. (1981). *Neural Analog Information Processing.* Real-Time Signal Processing. SPIE Volume 298. pp 138-141. Bellingham, WA: Society of Photo-Optical Instrumentation Engineers.

[Hecht-Nielsen, R 82a] Hecht-Nielsen, Robert. (1982). *Neural Analog Processing.* Robotics and Industrial Inspection. SPIE Volume 360. pp 180-189. Bellingham, WA: Society of Photo-Optical Instrumentation Engineers.

[Hecht-Nielsen, R 83a] Hecht-Nielsen, Robert. (1983). *Book Review.* Journal of Mathematical Psychology. Volume 27. pp 335-340.

[Hecht-Nielsen, R 86a] Hecht-Nielsen, Robert. (1986). *Performance Limits of Optical, Electro-Optical, and Electronic Neurocomputers.* In Szu, Harold H. (ed.). Optical and Hybrid Computing. SPIE Volume 634. pp 277-306.

[Hecht-Nielsen, R 86b] Hecht-Nielsen, Robert; Gutschow, Todd. (1986). Sensor Processing Using Artificial Neural Systems. San Diego, CA: TRW, Rancho Carmel AI Center.

[Hecht-Nielsen, R 87*] SEE: Gutschow, T 87a. *Pattern Recognition Using Artificial Neural Systems.*

[Hecht-Nielsen, R 87a] Hecht-Nielsen, Robert. (1987). *Combinatorial Hypercompression.* IEEE International Conference on Neural Networks (ICNN-87), San Diego, CA, June 21-24, 1987: Proceedings. Volume 2. pp 455-462. New York: Institute of Electrical and Electronics Engineers.

[Hecht-Nielsen, R 87b] Hecht-Nielsen, Robert. (1987). *Counterpropagation Networks.* Applied Optics. Volume 26. pp 4979-4984.

[Hecht-Nielsen, R 87c] Hecht-Nielsen, Robert. (1987). *Counterpropagation Networks.* IEEE International Conference on Neural Networks (ICNN-87), San Diego, CA, June 21-24, 1987: Proceedings. Volume 2. pp 19-32. New York: Institute of Electrical and Electronics Engineers.

[Hecht-Nielsen, R 87d] Hecht-Nielsen, Robert. (1987). *History of Associative Networks.* Artificial Neural Networks, An Introductory Seminar and Exhibition, Cincinnati, OH, December 8, 1987. Cincinnati, OH: Institute of Advanced Manufacturing Sciences.

[Hecht-Nielsen, R 87e] Hecht-Nielsen, Robert. (1987). *Kolmogorov's Mapping Neural Network Existence Theorem.* IEEE International Conference on Neural Networks (ICNN-87), San Diego, CA, June 21-24, 1987: Proceedings. Volume 3. pp 11-14.

[Hecht-Nielsen, R 87f] Hecht-Nielsen, Robert. (1987). *Nearest Matched Filter Classification of Spatiotemporal Patterns.* Applied Optics. Volume 26. pp 1892-1899.

[Hecht-Nielsen, R 87g] Hecht-Nielsen, Robert. (1987). *Nearest Matched Filter Classification of Spatio-Temporal Patterns.* Artificial Neural System Design: Short Course. University of California, San Diego.

[Hecht-Nielsen, R 87h] Hecht-Nielsen, Robert. (1987). *Neurocomputer Applications.* IEEE International Conference on Neural Networks (ICNN-87), San Diego, CA, June 21-24, 1987: Proceedings. Tutorial. San Diego: Institute of Electrical and Electronics Engineers.

[Hecht-Nielsen, R 87i] Hecht-Nielsen, Robert. (1987). *Neurocomputing Applications Examples.* Artificial Neural Networks, An Introductory Seminar and Exhibition, Cincinnati, OH, December 8, 1987. Cincinnati, OH: Institute of Advanced Manufacturing Sciences.

[Hecht-Nielsen, R 88a] Hecht-Nielsen, Robert. (1988). *Applications of Counterpropagation Networks.* Neural Networks. Volume 1. pp 131-139. Elmsford, NY: Pergamon.

[Hecht-Nielsen, R 88b] Hecht-Nielsen, Robert. (1988). *Neurocomputer Applications.* In Eckmiller, Rolf (ed.); von der Malsburg, Christoph (ed.). Neural Computers. NATO ASI Series, Volume F41. pp 445-453. Berlin: Springer-Verlag.

[Hecht-Nielsen, R 88c] Hecht-Nielsen, Robert. (1988). *Neurocomputing: Picking the Human Brain.* IEEE Spectrum. pp 36-41.

[Hegstad, H 68*] SEE: Segundo, JP 68a. *Input-Output Relations in Computer-Simulated Nerve Cells: Influence of Statistical Properties, Strength, etc..*

[Heiligenberg, W 88*] SEE: Baldi, P 88c. *How Sensory Maps Could Enhance Resolution Through Ordered Arrangements of Broadly Tuned Receivers.*

[Heimer, L 68a] Heimer, Lennart. (1968). *The Tracing of Pathways in the Central Nervous System.* Proceedings of the IEEE. Volume 56. pp 950-959. New York: Institute of Electrical and Electronics Engineers.

[Heimfeld, S 88*] SEE: Koizumi, O 88a. *Plasticity in the Nervous System of Adult Hydra II. Conversion of Ganglion Cells of the Body Column into Epidermal Sensory Cells of the Hypostome.*

[Heisey, I 87a] Heisey, Inge; Skrzypek, Josef. (1987). *Color Constancy and Early Vision: A Connectionist Model.* IEEE

International Conference on Neural Networks (ICNN-87), San Diego, CA, June 21-24, 1987: Proceedings. Volume 4. pp 317-326. New York: Institute of Electrical and Electronics Engineers.

[Hemami, H 83a] Hemami, Hooshang; Stokes, Bradford T. (1983). *Four Neural Circuit Models and Their Role in the Organization of Voluntary Movement*. Biological Cybernetics. Volume 49. pp 69-77.

[Hendler, JA 85a] Hendler, J. A. (1985). *Integrating Marker-Passing and Problem Solving*. Cognitive Science Society, 7th, Irvine, CA, August 15-17, 1985: Proceedings. pp 130-139. Hillsdale, NJ: Lawrence Erlbaum.

[Hennessey, AK 87*] SEE: Farley, ET 87a. A Prototype Crossbar Switch for the Gamma Network.

[Hennessey, AK 87a] Hennessey, A. Kathleen; Walrath, J. (1987). The Gamma Network - A Physical and Logical Data Structure. Lubbock, TX: Texas Tech University.

[Hennessey, AK 87b] Hennessey, A. Kathleen; Farley, Earl. (1987). *The Gamma Network as an Interconnection Architecture for Neural Networks*. IEEE International Conference on Neural Networks (ICNN-87), San Diego, CA, June 21-24, 1987: Proceedings. Volume 3. pp 131-140. New York: Institute of Electrical and Electronics Engineers.

[Henon, M 86*] SEE: Coste, J 86a. *Invariant Cycles in the Random Mapping of n Integers into Themselves. Comparison with Kauffman Binary Network.*.

[Henry, HN 86a] Henry, Howard N. (1986). *A Neural Simulation of Classical Conditioning in Aplysia*. Cognitive Science Society, 8th, Amherst, MA, August 15-17, 1986: Proceedings. pp 661-668. Hillsdale, NJ: Lawrence Erlbaum.

[Henry, HN 88*] SEE: Granger, R 88b. *Partitioning of Sensory Data by a Cortical Network.*

[Henshaw, PD 86a] Henshaw, Philip D.; Todtenkopf, A. B. (1986). *Artificial Intelligence Applications of Fast Optical Memory Access*. In Szu, Harold H. (ed.). Optical and Hybrid Computing. SPIE Volume 634. pp 422-438. Bellingham, WA: Society of Photo-Optical Instrumentation Engineers.

[Herault, J 86a] Herault, Jeanny; Jutten, C. (1986). *Space or Time Adaptive Signal Processing by Neural Network Models*. In Denker, John S. (ed.). Neural Networks for Computing, Snowbird, Utah, 1986. AIP Proceedings 151. pp 206-211. New York: American Institute of Physics.

[Herault, J 88*] SEE: Guerin, A 88a. *An Efficient Vector/Scalar Processing Structure for Adaptive Algorithms.*

[Herault, J 88*] SEE: Jutten, C 88a. *Independent Components Analysis (INCA) Versus Principal Components Analysis.*

[Herault, J 88a] Herault, Jeanny. (1988). *Brain and Neurocomputing*. Signal Processing IV: Theories and Applications. pp 7-14. New York: Elsevier.

[Herault, J 88b] Herault, Jeanny; Jutten, Christian. (1988). *Independent Components Analysis in Neural Networks*. Systems with Learning and Memory Abilities. pp 483-497.

[Hertz, JA 86a] Hertz, J. A.; Grinstein, G.; Solla, S. A. (1986). *Memory Networks with Asymmetric Bonds*. In Denker, John S. (ed.). Neural Networks for Computing, Snowbird, Utah, 1986. AIP Proceedings 151. pp 212-218. New York: American Institute of Physics.

[Hertz, JA 87*] SEE: Cortes, C 87a. *Hierarchical Associative Networks.*

[Hertz, JA 87*] SEE: Nylen, M 87a. A Neural Network in the Visual Cortex.

[Hertz, JA 88*] SEE: Moller, P 88a. *A Cortical Network Model for Early Vision Processing.*

[Hertz, JA 88*] SEE: Krogh, A 88a. *Mean-field Analysis of Hierarchical Associative Networks with Magnetisation.*

[Hestenes, D 83a] Hestenes, David. *How the Brain Works: The Next Great Scientific Revolution.* Workshop on Maximum Entropy and Bayesian Methods in Applied Statistics, 3rd, August 1-4, 1983: Proceedings.

[Hestenes, D 87a] Hestenes, David. (1987). *How the Brain Works: The Next Great Scientific Revolution.* In Smith, C. R. (ed.); Erickson, G. J. (ed.). Maximum-Entropy and Bayesian Spectral Analysis and Estimation Problems. pp 173-205. Boston: D Reidel.

[Hestenes, D 88a] Hestenes, David. (1988). *Inductive Inference by Neural Networks.* In Erickson, G. J. (ed.); Smith, C. R. (ed.). Maximum-Entropy and Bayesian Methods in Science and Engineering. Volume 2. pp 19-32. Kluwer.

[Hewes, RP 88*] SEE: Miller, WT 88a. Real Time Experiments in Neural Network Based Learning Control During High Speed Nonrepetitive Robotic Operations.

[Hewett, TT 86a] Hewett, T. T. (1986). *The Electronic Spreadsheet as a Tool in Modeling and Simulating Neural Networks.* Modeling and Simulation on Microcomputers, 1986: Proceedings.

[Hewitt, C 75a] Hewitt, Carl. (1975). *Stereotypes as an Actor Approach Towards Solving the Problem of Procedural Attachment in Frame Theories.* Theoretical Issues in Natural Language Processing, An Interdisciplinary Workshop in Computational Linguistics, Psychology, Linguistics, Artificial Int.. pp 108-117. New Haven, CT: Yale University, Mathematical Social Sciences Board.

[Highleyman, WH 62a] Highleyman, W. H. (1962). *Linear Decision Functions, with Application to Pattern Recognition.* Proceedings of the IRE. Volume 50. pp 1501-1514.

[Hilborn, CG 69a] Hilborn, Charles G.; Lainiotis, Demetrios G. (1969). *Unsupervised Learning Minimum Risk Pattern Classification for Dependent Hypotheses and Dependent Measurements.* IEEE Transactions on Systems, Science, and Cybernetics. Volume SSC-5. pp 109-115.

[Hildebrant, RR 87*] SEE: Cervantes, JH 87a. *Comparison of Three Neuron-Based Computation Schemes.*

[Hildreth, E 86a] Hildreth, E.; Koch, Christof. (1986). The Analysis of Visual Motion: From Computational Theory to Neuronal Mechanisms. AI Memo #919. Cambridge, MA: Massachusetts Institute of Technology.

[Hillis, WD 82a] Hillis, W. Daniel. (1982). *New Computer Architectures and Their Relationships to Physics, or Why Computer Science is No Good.* International Journal of Theoretical Physics. Volume 21. pp 255-262.

[Hillis, WD 84a] Hillis, W. Daniel. (1984). *The Connection Machine: A Computer Architecture Based on Cellular Automata.* Physica. Volume 10-D. pp 213-228.

[Hiltz, FF 62a] Hiltz, F. F. (1962). *Analog Computer Simulation of a Neural Element.* IRE Transactions on Bio-Medical Electronics. Volume BME-9. pp 12-20.

[Hiltz, FF 63a] Hiltz, F. F. (1963). *Artificial Neuron.* Kybernetik. Volume 1. pp 231-236.

[Hinton, GE 79a] Hinton, Geoffrey E. (1979). *Book Review.* Journal of the Optical Society of America. Volume 69. pp 1492.

[Hinton, GE 79b] Hinton, Geoffrey E. (1979). *Imagery Without Arrays.* The Behavioral and Brain Sciences. Volume 2. pp 555-556.

[Hinton, GE 79c] Hinton, Geoffrey E. (1979). *Intentional Homunculi.* Contemporary Psychology. Volume 24. pp 746-748.

[Hinton, GE 79d] Hinton, Geoffrey E. (1979). *Some Demonstrations of the Effects of Structural Descriptions in Mental Imagery.* Cognitive Science. Volume 3. pp 231-250.

[Hinton, GE 80a] Hinton, Geoffrey E. (1980). *Inferring the Meaning of Direct Perception.* The Behavioral and Brain

Sciences. Volume 3. pp 387-388.

[Hinton, GE 80b] Hinton, Geoffrey E. (1980). Relaxation and its Role in Vision. PhD Dissertation. Edinburgh, Scotland: University of Edinburgh.

[Hinton, GE 81*] SEE: Anderson, JA 81b. *Models of Information Processing in the Brain.*

[Hinton, GE 81a] Hinton, Geoffrey E. (1981). *A Parallel Computation that Assigns Canonical Object-Based Frames of Reference.* International Joint Conference on Artificial Intelligence (IJCAI-81), 7th, Vancouver, August 24-28, 1981: Proceedings. Volume 2.

[Hinton, GE 81b] Hinton, Geoffrey E. (1981). *Implementing Semantic Networks in Parallel Hardware.* In Hinton, Geoffrey E. (ed.). Parallel Models of Associative Memory. pp 161-188. Hillsdale, NJ: Lawrence Erlbaum.

[Hinton, GE 81c] Hinton, Geoffrey E. (ed.); Anderson, James A. (ed.). (1981). Parallel Models of Associative Memory. Hillsdale, NJ: Lawrence Erlbaum.

[Hinton, GE 81d] Hinton, Geoffrey E. (1981). *Shape Representation in Parallel Systems.* International Joint Conference on Artificial Intelligence (IJCAI-81), 7th, Vancouver, August 24-28, 1981: Proceedings. Volume 2. pp 1088-1096.

[Hinton, GE 81e] Hinton, Geoffrey E. (1981). *Shape Representation in Parallel Systems.* International Joint Conference on Artificial Intelligence (IJCAI-81), 7th, Vancouver, August 24-28, 1981: Proceedings. Volume 2. pp 1088-1096. International Joint Conference on Artificial Intelligence.

[Hinton, GE 83*] SEE: Fahlman, SE 83a. *Massively Parallel Architectures for AI: Netl, Thistle, and Boltzmann Machines.*

[Hinton, GE 83*] SEE: Ballard, DH 83a. *Parallel Visual Computation.*

[Hinton, GE 83a] Hinton, Geoffrey E.; Sejnowski, Terrence J. (1983). *Analyzing Cooperative Computation.* Cognitive Science Society, 5th, Rochester, NY, May 18-20, 1983: Proceedings.

[Hinton, GE 83b] Hinton, Geoffrey E.; Sejnowski, Terrence J. (1983). *Optimal Perceptual Inference.* IEEE Computer Science Conference on Computer Vision and Pattern Recognition, Washington, DC, June, 1983: Proceedings. pp 448-453.

[Hinton, GE 84*] SEE: Hammond, N 84a. *Evaluating the Interface of a Document Processor: A Comparison of Expert Judgement and User Observation.*

[Hinton, GE 84a] Hinton, Geoffrey E.; Sejnowski, Terrence J.; Ackley, David H. (1984). Boltzmann Machines: Constraint Satisfaction Networks that Learn. CMU-CS-84-119. Pittsburgh, PA: Carnegie Mellon University, Computer Science Department.

[Hinton, GE 84b] Hinton, Geoffrey E. (1984). Distributed Representations. CMU-CS-84-157. Pittsburgh, PA: Carnegie Mellon University, Computer Science Department.

[Hinton, GE 84c] Hinton, Geoffrey E.; Smolensky, Paul. (1984). Parallel Computation and the Mass-Sprint Model of Motor Control. Technical Report 123. San Diego, CA: University of California, Center for Human Information Processing.

[Hinton, GE 85*] SEE: Ackley, DH 85a. *A Learning Algorithm for Boltzmann Machines.*

[Hinton, GE 85*] SEE: Ackley, DH 85b. *A Learning Algorithm for Boltzmann Machines.*

[Hinton, GE 85*] SEE: Rumelhart, DE 85b. Learning Internal Representations by Error Propagation.

[Hinton, GE 85*] SEE: Szeliski, R 85a. *Solving Random-Dot Stereograms Using the Heat Equation.*

[Hinton, GE 85*] SEE: Touretzky, DS 85a. *Symbols Among the Neurons: Details of a Connectionist Inference Architecture.*

[Hinton, GE 85a] Hinton, Geoffrey E. (1985). *Learning in Parallel Networks.* BYTE. Volume 10. pp 265-273.

[Hinton, GE 85b] Hinton, Geoffrey E.; Lang, Kevin J. (1985). *Shape Recognition and Illusory Conjunctions.* International Joint Conference on Artificial Intelligence (IJCAI-85), 9th, Los Angeles, CA, August 18-23, 1985: Proceedings. Volume 1. pp 252-259. Los Altos, CA: Morgan Kaufmann.

[Hinton, GE 85c] Hinton, Geoffrey E. (1985). *Three Frames Suffice.* The Behavioral and Brain Sciences. Volume 8. pp 296-297.

[Hinton, GE 86*] SEE: McClelland, JL 86b. *Appeal of Parallel Distributed Processing.*

[Hinton, GE 86*] SEE: Touretzky, DS 86a. A Distributed Connectionist Production System.

[Hinton, GE 86*] SEE: Plaut, DC 86a. Experiments on Back-Propagation.

[Hinton, GE 86*] SEE: Rumelhart, DE 86b. *General Framework for Parallel Distributed Processing.*

[Hinton, GE 86*] SEE: Pearlmutter, BA 86a. *G-Maximization: An Unsupervised Learning Procedure for Discovering Regularities.*

[Hinton, GE 86*] SEE: Rumelhart, DE 86d. *Learning Internal Representations by Error Propagation.*

[Hinton, GE 86*] SEE: Rumelhart, DE 86c. *Learning Internal Representations by Error Propagation.*

[Hinton, GE 86*] SEE: Rumelhart, DE 86f. *Learning Representations by Back-Propagating Errors.*

[Hinton, GE 86*] SEE: Rumelhart, DE 86e. *Learning Representations by Back-Propagating Errors.*

[Hinton, GE 86*] SEE: Sejnowski, TJ 86b. *Learning Symmetry Groups with Hidden Units: Beyond the Perceptron.*

[Hinton, GE 86*] SEE: Rumelhart, DE 86j. *Schemata and Sequential Thought Processes in PDP Models.*

[Hinton, GE 86a] Hinton, Geoffrey E. (1986). A Network that Induces a Distributed Representation Useful for Encoding Kinship Relations. Working Paper. Pittsburgh, PA: Carnegie Mellon University, Computer Science Department.

[Hinton, GE 86b] Hinton, Geoffrey E.; McClelland, James L.; Rumelhart, David E. (1986). *Distributed Representations.* In Rumelhart, David E. (ed.). Parallel Distributed Processing: Foundations. Volume I. pp 77-109. Cambridge, MA: MIT Press.

[Hinton, GE 86c] Hinton, Geoffrey E.; Sejnowski, Terrence J. (1986). *Learning and Relearning in Boltzmann Machines.* In Rumelhart, David E. (ed.). Parallel Distributed Processing: Foundations. Volume I. pp 282-317. Cambridge, MA: MIT Press.

[Hinton, GE 86d] Hinton, Geoffrey E. (1986). *Learning Distributed Representations of Concepts.* Cognitive Science Society, 8th, Amherst, MA, August 15-17, 1986: Proceedings. pp 1-12. Hillsdale, NJ: Lawrence Erlbaum.

[Hinton, GE 87*] SEE: Fahlman, SE 87a. *Connectionist Architectures for Artificial Intelligence.*

[Hinton, GE 87*] SEE: Sejnowski, TJ 87g. *Separating Figure from Ground with a Boltzmann Machine.*

[Hinton, GE 87a] Hinton, Geoffrey E. (1987). Connectionist Learning Procedures. CMU-CS-87-115. Pittsburgh, PA: Carnegie Mellon University.

[Hinton, GE 87b] Hinton, Geoffrey E.; Nowlan, Steven J. (1987). *How Learning Can Guide Evolution.* Complex Systems.

Volume 1. 495-502.

[Hinton, GE 87c] Hinton, Geoffrey E.; Sejnowski, Terrence J. (1987). *Neural Network Architectures for AI*. National Conference on Artificial Intelligence (AAAI-87), 6th, July 13-17, 1987: Proceedings. Tutorial MP2.

[Hinton, GE 87d] Hinton, Geoffrey E.; Plaut, David C. (1987). *Using Fast Weights to Deblur Old Memories*. Cognitive Science Society, 9th, Seattle, WA, July 16-18, 1987: Proceedings. pp 177-186. Hillsdale, NJ: Lawrence Erlbaum.

[Hinton, GE 88*] SEE: Ackley, DH 88a. *A Learning Algorithm for Boltzmann Machines*.

[Hinton, GE 88a] Hinton, Geoffrey E.; McClelland, James L. (1988). *Learning Representations by Recirculation*. In Anderson, Dana Z. (ed.). Neural Information Processing Systems. pp 358-366. New York: American Institute of Physics.

[Hintzman, DL 80a] Hintzman, D. L.; Ludlam, G. (1980). *Differential Forgetting of Prototypes and Old Instances: Simulation by an Exemplar-Based Classification Model*. Memory and Cognition. Volume 8. pp 378-382.

[Hirai, Y 75a] Hirai, Yuzo; Fukushima, Kunihiko. (1975). *A Model of Neural Network Extracting Binocular Parallax*. Biological Cybernetics. Volume 18. pp 19-29.

[Hirai, Y 78a] Hirai, Yuzo; Fukushima, Kunihiko. (1978). *An Inference upon the Neural Network Finding Binocular Correspondence*. Biological Cybernetics. Volume 31. pp 209-217.

[Hirai, Y 80a] Hirai, Yuzo. (1980). *A Template Matching Model for Pattern Recognition: Self-Organization of Templates and Template Matching by a Disinhibitory Neural Network*. Biological Cybernetics. Volume 38. pp 91-101.

[Hirai, Y 82a] Hirai, Yuzo. (1982). *A Learning Network Resolving Multiple Match in Associative Memory*. International Conference on Pattern Recognition, 6th, Munich, October 19-22, 1982: Proceedings. pp 1049-1052. Silver Spring, MD: Institute of Electrical and Electronics Engineers.

[Hirai, Y 83a] Hirai, Yuzo. (1983). *A Model of Human Associative Processor (HASP)*. IEEE Transactions on Systems, Man and Cybernetics. Volume SMC-13. pp 851-857.

[Hirai, Y 86a] Hirai, Yuzo. (1986). *On a Model of Human Associative Processor*. Transactions of the Institute of Electronics and Communications Engineers of Japan. Volume E 69. pp 549-557.

[Hirai, Y 86b] Hirai, Yuzo; Hiwatashi, Kenji. (1986). *Spatial Frequency Adaptation Under a Slit-Vision Condition*. Transactions of the Institute of Electronics and Communications Engineers of Japan. Volume E 69. pp 620-627.

[Hirai, Y 87a] Hirai, Yuzo; Ma, Qing. (1987). *Associate Networks Underlying Problem-Solving: Representation of the Process of Addition by a Model of an Associative Processor, HASP*. IEEE International Conference on Neural Networks (ICNN-87), San Diego, CA, June 21-24, 1987: Proceedings. Volume 2. pp 33-40. New York: Institute of Electrical and Electronics Engineers.

[Hirai, Y 87b] Hirai, Yuzo. (1987). *Representation of Control Structures by a Model of an Associative Processor, HASP*. IEEE International Conference on Neural Networks (ICNN-87), San Diego, CA, June 21-24, 1987: Proceedings. Volume 2. pp 447-454. New York: Institute of Electrical and Electronics Engineers.

[Hirai, Y 88a] Hirai, Yuzo; Ma, Qing. (1988). *Modeling the Process of Problem-Solving by Associative Networks Capable of Improving the Performance*. Biological Cybernetics. Volume 59. pp 353-365.

[Hirsch, MW 84a] Hirsch, Morris W. (1984). *The Dynamical Systems Approach to Differential Equations*. Bulletin (New Series) of the American Mathematical Society. Volume 11. pp 1-64.

[Hirsch, MW 85a] Hirsch, Morris W. (1985). *Systems of Differential Equations that Are Competitive or Cooperative II: Convergence Almost Everywhere*. SIAM Journal on Mathematical Analysis. Volume 16. pp 423-439.

[Hirsch, MW 87a] Hirsch, Morris W. (1987). *Convergence in Neural Nets.* IEEE International Conference on Neural Networks (ICNN-87), San Diego, CA, June 21-24, 1987: Proceedings. Volume 2. pp 115-126. New York: Institute of Electrical and Electronics Engineers.

[Hirst, G 85*] SEE: Selman, B 85a. *A Rule-Based Connectionist Parsing System.*

[Hiwatashi, K 86*] SEE: Hirai, Y 86b. *Spatial Frequency Adaptation Under a Slit-Vision Condition.*

[Ho, S 88*] SEE: Ramamoorthy, PA 88a. *A Neural Network Approach for Implementing Expert Systems.*

[Ho, Y 68a] Ho, Yu-Chi; Agrawala, Ashok K. (1968). *On Pattern Classification Algorithms: Introduction and Survey.* Proceedings of the IEEE. Volume 56. pp 2101-2114.

[Hochstein, S 87*] SEE: Kalay, A 87a. *Neural Patterns as Detection Units in Sensory Systems.*

[Hodgson, JPE 83*] SEE: Grissel, R 83a. Oscillations in Mathematical Biology.

[Hoff, ME 60*] SEE: Widrow, B 60b. *Adaptive Switching Circuits.*

[Hoff, ME 60*] SEE: Widrow, B 60c. *Adaptive Switching Circuits.*

[Hoff, ME 62a] Hoff, Marcian E. (1962). Learning Phenomena in Networks of Adaptive Switching Circuits. Technical Report 1554-1. Stanford, CA: Stanford University, Stanford Electronics Laboratory.

[Hoffman, A 62a] Hoffman, Allen. (1962). *The "Whirling Dervish", A Simulation Study in Learning and Recognition Systems.* IRE International Convention Record. pp 153-160.

[Hoffman, DD 87*] SEE: Richards, WA 87a. *Inferring Three-Dimensional Shapes from Two-Dimensional Silhouettes.*

[Hoffman, GW 86a] Hoffman, G. W.; Benson, M. W. (1986). *Neurons with Hysteresis Form a Network that Can Learn without Any Changes in Synaptic Connection Strengths.* In Denker, John S. (ed.). Neural Networks for Computing, Snowbird, Utah, 1986. AIP Proceedings 151. pp 219-226. New York: American Institute of Physics.

[Hoffmann, KP 88a] Hoffmann, K. P. (1988). *Innate and Learned Components in a Simple Visuo-Motor Reflex.* In Eckmiller, Rolf (ed.); von der Malsburg, Christoph (ed.). Neural Computers. NATO ASI Series, Volume F41. pp 371-380. Berlin: Springer-Verlag.

[Hofker, U 77*] SEE: Bunge, E 77b. *The AUROS Project: Automatic Recognition of Speakers by Computers.*

[Hofstadter, DR 81a] Hofstadter, Douglas R. (1981). *Metamagical Themas.* Scientific American. Volume 245. pp 22-43.

[Hofstadter, DR 83a] Hofstadter, Douglas R. (1983). *The Architecture of Jumbo.* International Machine Learning Workshop, Monticello, IL, June 22-24, 1983: Proceedings. pp 161-170.

[Hogg, T 84*] SEE: Huberman, BA 84a. *Adaptation and Self-Repair in Parallel Computing Structures.*

[Hogg, T 84a] Hogg, Tad; Huberman, Bernardo A. (1984). *Understanding Biological Computation: Reliable Learning and Recognition.* Proceedings of the National Academy of Sciences. Volume 81. pp 6871-6875.

[Hogg, T 87*] SEE: Shrager, J 87a. *Observation of Phase Transitions in Spreading Activation Networks.*

[Hogg, T 87*] SEE: Huberman, BA 87a. *Phase Transitions in Artificial Intelligence Systems.*

[Hogg, T 88*] SEE: Huberman, BA 88b. *The Behavior of Computational Ecologies.*

[Hogg, T 88*] SEE: Stornetta, WS 88a. *A Dynamical Approach to Temporal Pattern Processing.*

[Hogg, T 88a] Hogg, Tad. (1988). *Speech Recognition with Network-Based Classifiers.* A Commercial Assessment of the Applications of Neural Networks. Los Angeles, CA: Institute for International Research.

[Hohensee, B 87*] SEE: Krowitz, A 87a. *The State Space of Memory Clusters in the Hopfield Network.*

[Hohensee, WE 87*] SEE: Matheus, CJ 87a. *Learning in Artificial Neural Systems.*

[Holden, AV 75*] SEE: Rubio, JE 75a. *The Response of a Model Neurone to a White Noise Input.*

[Holland, JH 56*] SEE: Rochester, N 56a. *Tests on a Cell Assembly Theory of the Action of the Brain, Using a Large Digital Computer.*

[Holland, JH 56*] SEE: Rochester, N 56b. *Tests on a Cell Assembly Theory of the Action of the Brain, Using a Large Digital Computer.*

[Holland, JH 69a] Holland, John H. (1969). *Goal-Directed Pattern Recognition.* In Watanbe, Satosi (ed.). Methodologies of Pattern Recognition. pp 287-296. New York: Academic Press.

[Holland, JH 75a] Holland, John H. (1975). Adaptation in Natural and Artificial Systems. Ann Arbor: University of Michigan Press.

[Holland, JH 80a] Holland, John H. (1980). *Adaptive Algorithms for Discovering and Using General Patterns in Growing Knowledge-Bases.* International Journal of Policy Analysis and Information Systems. Volume 4. pp 217-240.

[Hollis, PW 87*] SEE: Takefuji, Y 87a. *Error Correcting System Based on Neural Circuits.*

[Hollis, PW 88*] SEE: Paulos, JJ 88a. *Neural Networks Using Analog Multipliers.*

[Holt, F 87*] SEE: Nguyen, DD 87a. *Stochastic Processing in a Neural Network Application.*

[Homa, D 73a] Homa, Donald; Cross, Joseph; Cornell, Don; Goldman, David; Shwartz, Steven. (1973). *Prototype Abstraction and Classification of New Instances as a Function of the Number of Instances Defining the Prototype.* Journal of Experimental Psychology. Volume 101. pp 116-122.

[Homa, D 81a] Homa, Donald; Sterling, Sharon; Trepel, Lawrence. (1981). *Limitations of Exemplar-Based Generalization and the Abstraction of Categorical Information.* Journal of Experimental Psychology: Human Learning and Memory. Volume 7. pp 418-439.

[Homma, T 88*] SEE: Atlas, LE 88a. *An Artificial Neural Network for Spatio-Temporal Bipolar Patterns: Application to Phoneme Classification.*

[Homma, T 88a] Homma, Toshiteru; Atlas, Les E.; Marks II, Robert Jackson. (1988). *An Artificial Neural Network for Spatio-Temporal Bipolar Patterns: Application to Phoneme Classification.* Neural Information Processing Systems. pp 31-40. New York: American Institute of Physics.

[Hon, E 63a] Hon, E.; Lee, S. (1963). *Noise Reduction in Fetal Electrocardiography.* American Journal of Obstetrics and Gynecology. Volume 87. pp 1087-1096.

[Hong, J 86*] SEE: Psaltis, D 86b. *Shift Invariance in Optical Associative Memories.*

[Hong, J 88*] SEE: Psaltis, D 88b. *Higher Order Associative Memories and Their Optical Implementations.*

[Honrubia, V 87*] SEE: Lau, CGY 87a. *Fast Component Threshold for Vestibular Nystagmus in the Rabbit.*

[Honrubia, V 87*] SEE: Elidan, J 87a. *The Neural Generators of the Vestibular Evoked Response.*

[Honrubia, V 87a] Honrubia, Vicente; Kuruvilla, Alexander; Mamikunian, Dawn; Eichel, James E. (1987). *Morphological Aspects of the Vestibular Nerve of the Squirrel Monkey.* Laryngoscope. Volume 97. pp 228-238.

[Honrubia, V 88*] SEE: Lau, CGY 88a. *Distributed Neural Information Processing in the Vestibulo-Ocular System.*

[Honrubia, V 88*] SEE: Baloh, RW 88a. *Eye Movements Induced by Linear Acceleration on a Parallel Swing.*

[Honrubia, V 89*] SEE: Elidan, J 89a. *The Firing Properties of Second-Order Vestibular Neurons in Correlation with the Far-Field Recorded Vestibular-Evoked Response.*

[Hood, G 86a] Hood, Greg. (1986). Artificial Organisms: A Neural Modeling Approach. Thesis Proposal. Pittsburgh, PA: Carnegie Mellon University.

[Hood, G 87a] Hood, Greg. (1987). *A Graphical Design System for Neural Networks.* IEEE International Conference on Neural Networks (ICNN-87), San Diego, CA, June 21-24, 1987: Proceedings. Volume 3. pp 453-460. New York: Institute of Electrical and Electronics Engineers.

[Hopfield, JJ 82a] Hopfield, John J. (1982). *Neural Networks and Physical Systems with Emergent Collective Computational Abilities.* Proceedings of the National Academy of Sciences. Volume 79. pp 2554-2558.

[Hopfield, JJ 82b] Hopfield, John J. (1988). *Neural Networks and Physical Systems with Emergent Collective Computational Abilities.* In Anderson, James A. (ed.); Rosenfeld, Edward (ed.). Neurocomputing: Foundations of Research. pp 460-464. Cambridge, MA: MIT Press.

[Hopfield, JJ 83a] Hopfield, John J.; Feinstein, D. I.; Palmer, R. G. (1983). *Unlearning Has a Stabilizing Effect in Collective Memories.* Nature. Volume 304. pp 158-159.

[Hopfield, JJ 84a] Hopfield, John J. (1984). *Neurons with Graded Response Have Collective Computational Properties Like Those of Two-State Neurons.* Proceedings of the National Academy of Sciences. Volume 81. pp 3088-3092.

[Hopfield, JJ 84b] Hopfield, John J. (1988). *Neurons With Graded Response Have Collective Computational Properties Like Those of Two-state Neurons.* In Anderson, James A. (ed.); Rosenfeld, Edward (ed.). Neurocomputing: Foundations of Research. pp 579-584. Cambridge, MA: MIT Press.

[Hopfield, JJ 85*] SEE: Gelperim, AE 85a. *The Logic of LIMAX Learning.*

[Hopfield, JJ 85a] Hopfield, John J.; Tank, David W. (1986). *Collective Computation with Continuous Variables.* Disordered Systems and Biological Organization. pp 155-170. Berlin Heidelberg: Springer-Verlag.

[Hopfield, JJ 85b] Hopfield, John J.; Tank, David W. (1985?). *Computing With Neural Circuits: A Model.* Science. Volume 233. pp 625-633.

[Hopfield, JJ 85c] Hopfield, John J.; Tank, David W. (1985). *Neural Computation of Decisions in Optimization Problems.* Biological Cybernetics. Volume 52. pp 141-152.

[Hopfield, JJ 86*] SEE: Platt, JC 86a. *Analog Decoding Using Neural Networks.*

[Hopfield, JJ 86*] SEE: Tank, DW 86a. *Simple Neural Optimization Networks: An A/D Converter, Signal Decision Circuit, and Linear Programming Circuit.*

[Hopfield, JJ 86a] Hopfield, John J.; Tank, David W. (1986). *Collective Computation with Continuous Variables.* In Bienenstock, Elie L. (ed.). Disordered Systems and Biological Organization. pp 155-170. New York: Springer-Verlag.

[Hopfield, JJ 86b] Hopfield, John J. (1986). *Physics, Biological Computation and Complimentarity.* The Lesson of Quantum

Theory. pp 295-314. Amsterdam: Elsevier.

[Hopfield, JJ 87*] SEE: Denker, JS 87a. *Automatic Learning, Rule Extraction, and Generalization.*

[Hopfield, JJ 87*] SEE: Tank, DW 87a. *Collective Computation in Neuronlike Circuits.*

[Hopfield, JJ 87*] SEE: Tank, DW 87b. *Concentrating Information in Time: Analog Neural Networks with Applications to Speech Recognition Problems.*

[Hopfield, JJ 87a] Hopfield, John J. (1987). *Learning Algorithms and Probability Distributions in Feed-Forward and Feed-Back Networks.* Proceedings of the National Academy of Sciences. Volume 84. pp 8429-8433.

[Hopfield, JJ 87b] Hopfield, John J. (1987). *Networks, Computations, Logic and Noise.* IEEE International Conference on Neural Networks (ICNN-87), San Diego, CA, June 21-24, 1987: Proceedings. Volume 1. pp 107-142. New York: Institute of Electrical and Electronics Engineers.

[Hopfield, JJ 87c] Hopfield, John J.; Tank, David W. (1987). *Technical Comments: Computing with Neural Networks.* Science. Volume 235. pp 1228-1229.

[Hoppensteadt, FC 81*] SEE: Keener, JP 81a. *Integrate-And-Fire Models of Nerve Membrane Response to Oscillatory Input.*

[Horn, B 75a] Horn, Berthold. (1975). *Obtaining Shape From Shading Information.* The Psychology of Computer Vision. pp 115-155. New York: McGraw-Hill.

[Horn, BKP 81a] Horn, B. K. P. (1981). *Hill-Shading and the Reflectance Map.* Proceedings of the IEEE. Volume 19. pp 14-47.

[Horn, D 88a] Horn, D.; Usher, M. (1988). *Capacities of Multiconnected Memory Models.* Journal de Physique. Volume 49. pp 389-395.

[Horn, D 88b] Horn, D.; Usher, M. (1988). Neural Networks with Dynamical Thresholds. TAUP-1648/88. Tel Aviv: Tel Aviv University, School of Physics and Astronomy.

[Horn, D 88c] Horn, D. (1988). Shift-Invariant Multiconnected Neural-Networks. TAUP 1686-88. Tel Aviv: Tel Aviv University, School of Physics and Astronomy.

[Horne, W 87a] Horne, William; Smith, Karl W.; McClung, Brian; Jamshidi, Mo. (1987). *Application-Oriented Connection Systems for Pattern Acquisition and Recognition.* IEEE International Conference on Neural Networks (ICNN-87), San Diego, CA, June 21-24, 1987: Proceedings. Volume 3. pp 141-148. New York: Institute of Electrical and Electronics Engineers.

[Hornik, K 88*] SEE: Baldi, P 88e. *Neural Networks and Principal Component Analysis: Learning from Examples Without Local Minima.*

[Hornik, K 88a] Hornik, Kurt; Halbert. (research). Multi-Layer Feedforward Networks Are Universal Approximators. San Diego, CA: University of California, Department of Economics.

[Horst, A 88*] SEE: Lehmann, D 88a. *Spontaneous and Information-Triggered Segments of Series of Human Brain Electric Field Maps.*

[Hoshimiya, N 79a] Hoshimiya, Nozomu; Yoshida, S.; Shogen, K.; Matsuo, Tadayuki. (1979). *Two-Terminal Electronic Circuit Neuron Model with Excitable Membrane V-I-t Characteristics.* Biological Cybernetics. Volume 35. pp 125-130.

[Hoshimiya, N 80*] SEE: Koga, K 80a. *A Hardware Neuron Model as a Component of the Neural Network.*

[Hoshimiya, N 82a] Hoshimiya, Nozomu; Futami, Ryoko. (1982). *A Preliminary Study of the Phoneme Recognition with Hardware*

Neural Network. Transactions of the Institute of Electronics and Communication Engineers of Japan. Volume E65. pp 895-902.

[Hoskins, J 87*] SEE: Jones, WP 87a. *Back-Propagation: A Generalized Delta Learning Rule*.

[Hosogi, S 88a] Hosogi, Shinya; Asakawa, Kazuo; Matsuo, Kazuhira. (1988). *Cerebellar Neural Network Model for Manipulator Control*. International Symposium on Bioelectronic and Molecular Electronic Devices, 2nd, Fujiyoshida, December 12-14, 1988: Proceedings.

[Hoss, W 87*] SEE: Niznik, CA 87a. *Modeling and Mathematical Formalism for Robotic Cerebellar Neural Network Pathway Information Measures*.

[Hostetler, LD 73*] SEE: Fukunaga, K 73a. *Optimization of k-Nearest-Neighbor Density Estimates*.

[Houk, JC 86*] SEE: Gielen, CCAM 86a. *Simple Changes in Reflex Threshold Cannot Explain All Aspects of Rapid Voluntary Movements*.

[Houk, JC 87*] SEE: Gielen, CCAM 87a. *A Model of the Motor Servo: Incorporating Nonlinear Spindle Receptor and Muscle Mechanical Properties*.

[Houk, JC 87a] Houk, James C. (1987). *Model of the Cerebellum as an Array of Adjustable Pattern Generators*. Cerebellum and Neuronal Plasticity. NATO ASI Series A148. pp 249-260. New York: Plenum.

[Houk, JC 88a] Houk, James C. (1988). *Control Strategies in Physiological Systems*. FASEB Journal. Volume 2. pp 97-107.

[Houk, JC 88b] Houk, James C. (1988). *Schema for Motor Control Utilizing a Network Model of the Cerebellum*. In Anderson, Dana Z. (ed.). Neural Information Processing Systems. pp 367-376. New York: American Institute of Physics.

[House, DH 85*] SEE: Arbib, MA 85c. Depth and Detours: An Essay on Visually Guided Behavior.

[House, DH 87*] SEE: Arbib, MA 87b. *Depth and Detours: An Essay on Visually Guided Behavior*.

[House, J 87a] House, James. (1987). *Neural Nets: The Dream Machines*. PC World. Volume 5. pp 285-291.

[Howard, RE 86*] SEE: Jackel, LD 86a. *Artificial Neural Networks for Computing*.

[Howard, RE 86*] SEE: Hubbard, W 86a. *Electronic Neural Networks*.

[Howard, RE 86*] SEE: Jackel, LD 86b. *High Resolution Micro-Fabrication and Neural Networks*.

[Howard, RE 86*] SEE: Graf, HP 86a. *VLSI Implementation of a Neural Network Memory with Several Hundreds of Neurons*.

[Howard, RE 87*] SEE: Denker, JS 87a. *Automatic Learning, Rule Extraction, and Generalization*.

[Howard, RE 87*] SEE: Jackel, LD 87a. *Building a Hierarchy with Neural Networks: An Example-Image Vector Quantization*.

[Howard, RE 87*] SEE: Jackel, LD 87b. *Electronic Neural Network Chips*.

[Howard, RE 87a] Howard, Richard E.; Schwartz, Daniel B.; Denker, John S.; Epworth, Roger W.; Graf, Hans Peter; Hubbard, Wayne E. (1987). *An Associative Memory Based on an Electronic Neural Network Architecture*. IEEE Transactions on Electron Devices. Volume ED-34. pp 1553-1556.

[Hoyle, G 73*] SEE: Willows, AOD 73a. *The Neuronal Basis of Behavior in Tritonia, III Neuronal Mechanism of a Fixed Action Pattern*.

[Hrechanyk, LM 82a] Hrechanyk, L. M.; Ballard, Dana H. (1982). *A Connectionist Model of Form Perception*. IEEE Workshop on

Computer Vision, Ringde, NH, August 23-25, 1982: Proceedings.

[Hsu, K 88a] Hsu, Ken; Brady, David; Psaltis, Demetri. (1988). *Experimental Demonstrations of Optical Neural Computers.* In Anderson, Dana Z. (ed.). Neural Information Processing Systems. pp 377-386. New York: American Institute of Physics.

[Hu, MJC 63*] SEE: Widrow, B 63b. *Practical Applications for Adaptive Data-Processing Systems.*

[Hu, MJC 63a] Hu, M. J. C. (1963). A Trainable Weather-Forecasting System. Technical Report 6759-1. Stanford, CA: Stanford University, Stanford Electronics Laboratory.

[Hu, ST 65a] Hu, S. T. (1965). Threshold Logic. Berkeley, CA: University of California Press.

[Hu, V 88*] SEE: Alspector, J 88c. *Stochastic Learning Networks and their Electronic Implementation.*

[Huang, J 87*] SEE: Sugie, N 87a. Localizing Sound Source by Incorporating Biological Auditory Mechanism.

[Huang, N 87a] Huang, Nai-Kuan. (1987). *A Learning Experiment on English Spelling Rules.* IEEE International Conference on Neural Networks (ICNN-87), San Diego, CA, June 21-24, 1987: Proceedings. Volume 2. pp 351-358. New York: Institute of Electrical and Electronics Engineers.

[Huang, N 88*] SEE: Chen, S 88a. A Case Study of Solving Optimization Problems Using Neural Network.

[Huang, N 88*] SEE: Xu, X 88a. A Generalized Neural Network Model.

[Huang, WY 87*] SEE: Lippmann, RP 87a. *A Framework for Speech Recognition Using Neural Networks.*

[Huang, WY 87a] Huang, William Y.; Lippmann, Richard P. (1987). *Comparisons Between Neural Net and Conventional Classifiers.* IEEE International Conference on Neural Networks (ICNN-87), San Diego, CA, June 21-24, 1987: Proceedings. Volume 4. pp 485-494. New York: Institute of Electrical and Electronics Engineers.

[Huang, WY 88a] Huang, William Y.; Lippmann, Richard P. (1988). *Neural Net and Traditional Classifiers.* In Anderson, Dana Z. (ed.). Neural Information Processing Systems. pp 387-396. New York: American Institute of Physics.

[Huang, WY 88b] Huang, William Y.; Lippmann, Richard P.; Nguyen, Thao K. P. (1988). *Neural Nets for Speech Recognition.* Meeting of the Acoustic Society of America, 115th, Seattle, WA, May 16-20, 1988.

[Hubbard, W 86a] Hubbard, W.; Schwartz, Daniel B.; Denker, John S.; Graf, Hans P.; Howard, R. E. (1986). *Electronic Neural Networks.* In Denker, John S. (ed.). Neural Networks for Computing, Snowbird, Utah, 1986. AIP Proceedings 151. pp 227-234. New York: American Institute of Physics.

[Hubbard, W 87*] SEE: Jackel, LD 87a. *Building a Hierarchy with Neural Networks: An Example-Image Vector Quantization.*

[Hubbard, WE 87*] SEE: Howard, RE 87a. *An Associative Memory Based on an Electronic Neural Network Architecture.*

[Hubbard, WE 87*] SEE: Graf, HP 87a. *A CMOS Associative Memory Chip.*

[Hubbard, WE 88*] SEE: Graf, HP 88a. *VLSI Implementation of a Neural Network Model.*

[Huberman, BA 84*] SEE: Hogg, T 84a. *Understanding Biological Computation: Reliable Learning and Recognition.*

[Huberman, BA 84a] Huberman, Bernardo A.; Hogg, Tad. (1984). *Adaptation and Self-Repair in Parallel Computing Structures.* Physical Review Letters. Volume 52. pp 1048-1051.

[Huberman, BA 85a] Huberman, Bernardo A. (1985). *Dynamics of Computing Structures.* Physica Scripta. Volume T9. pp

165–169.

[Huberman, BA 85b] Huberman, Bernardo A.; Kerszberg, Michel. (1985). *Ultradiffusion: The Relaxation of Hierarchical Systems*. Journal of Physics A, Mathematical and General. Volume 18. pp L331–L336.

[Huberman, BA 87*] SEE: Stornetta, WS 87a. *An Improved Three-Layer, Back Propagation Algorithm.*

[Huberman, BA 87*] SEE: Shrager, J 87a. *Observation of Phase Transitions in Spreading Activation Networks.*

[Huberman, BA 87a] Huberman, Bernardo A.; Hogg, Tad. (1987). *Phase Transitions in Artificial Intelligence Systems*. Artificial Intelligence. Volume 33. pp 155–171.

[Huberman, BA 88*] SEE: Stornetta, WS 88a. *A Dynamical Approach to Temporal Pattern Processing.*

[Huberman, BA 88a] Huberman, Bernardo A. (1988). *Asynchrony and Concurrency*. In Eckmiller, Rolf (ed.); von der Malsburg, Christoph (ed.). Neural Computers. NATO ASI Series, Volume F41. pp 455–465. Berlin: Springer-Verlag.

[Huberman, BA 88b] Huberman, Bernardo A.; Hogg, Tad. (1988). *The Behavior of Computational Ecologies*. The Ecology of Computation. pp 77–115. New York: Elsevier.

[Hudak, MJ 86a] Hudak, Michael J.; Marcellus, Daniel H. (1986). *Demon-Based Associative Memories*. Cybernetics and Systems. Volume 17. pp 249–275.

[Hughes, CS 87*] SEE: Tenorio, MF 87d. *Real Time Noisy Image Segmentation Using an Artificial Neural Network Model.*

[Hull, JJ 88*] SEE: Pawlicki, TF 88a. *Neural Network Models and Their Application to Handwritten Digit Recognition.*

[Hummel, RA 79*] SEE: Zucker, SW 79a. *Toward a Low-Level Description of Dot Clusters: Labeling Edge, Interior, and Noise Points.*

[Hummel, RA 83a] Hummel, Robert A.; Zucker, Steven W. (1983). *On the Foundations of Relaxation Labeling Processes*. IEEE Transactions on Pattern Analysis and Machine Intelligence. Volume PAMI-5. pp 267–287.

[Hummel, RA 86*] SEE: Rosenfeld, A 86a. *Scene Labeling by Relaxation Operations.*

[Hunt, BR 77a] Hunt, B. R. (1977). *Bayesian Methods in Nonlinear Digital Image Restoration*. IEEE Transactions on Computers. Volume C-26. pp 219–229.

[Hunt, DJ 85*] SEE: Keller, JM 85a. *Incorporating Fuzzy Membership Functions into the Perceptron Algorithm.*

[Hunter, L 88a] Hunter, Lawrence. (1988). *Artificial Neural Networks as Theories of Mind*. Neural Networks. Volume 1.

[Hunter, L 88b] Hunter, Lawrence. (September 19, 1988). *Melding Computers and Biology for Commercial Benefit*. New Science: AI Research Notes. pp 793–795.

[Hurlbert, AC 85a] Hurlbert, Anya C.; Poggio, Tomaso A. (1985). Spotlight on Attention. AI Memo #817. Cambridge, MA: Massachusetts Institute of Technology.

[Hurlbert, AC 88*] SEE: Poggio, TA 88a. *Learning a Color Algorithm from Examples.*

[Hurlbert, AC 88a] Hurlbert, Anya C.; Poggio, Tomaso A. (1988). *Synthesizing a Color Algorithm from Examples*. Science. Volume 239. pp 482–486.

[Hurson, A 81a] Hurson, Alireza. (November 1981). *An Associative Backend Machine for Data Base Management*. IEEE Computer Society Workshop on Computer Architecture for Pattern Analysis and Image Data Base Management. pp 225–230.

[Hutchinson, CE 84*] SEE: Scott, DE 84a. *An Application of the GMDH Algorithm to Economic Modeling.*

[Hutchinson, J 88*] SEE: Koch, C 88b. *Computing Motion Using Resistive Networks.*

[Hutchinson, J 88a] Hutchinson, James; Koch, Christof; Luo, Jin; Mead, Carver A. (March 1988). *Computing Motion Using Analog and Binary Resistive Networks.* Computer. pp 52-53.

[Hutchinson, JM 86a] Hutchinson, J. M.; Koch, Christof. (1986). *Simple Analog and Hybrid Networks for Surface Interpolation.* In Denker, John S. (ed.). Neural Networks for Computing, Snowbird, Utah, 1986. AIP Proceedings 151. pp 235-240. New York: American Institute of Physics.

[Hutchison, WR 87a] Hutchison, William R.; Stephens, Kenneth R. (1987). *Integration of Distributed and Symbolic Knowledge Representations.* IEEE International Conference on Neural Networks (ICNN-87), San Diego, CA, June 21-24, 1987: Proceedings. Volume 2. pp 395-398. New York: Institute of Electrical and Electronics Engineers.

[Hutchison, WR 87b] Hutchison, William R.; Stephens, Kenneth R. (1987). *The Airline Marketing Tactician (AMT): A Commercial Application of Adaptive Networking.* IEEE International Conference on Neural Networks (ICNN-87), San Diego, CA, June 21-24, 1987: Proceedings. Volume 4. pp 753-756. New York: Institute of Electrical and Electronics Engineers.

[Hutchison, WR 88*] SEE: Stephens, KR 88a. *Behavior Analysis and Intelligent Machines.*

[Huttunen, MO 73a] Huttunen, Matti O. (1973). *General Model for the Molecular Events in Synapses During Learning.* Perspectives in Biology and Medicine. Volume 17. pp 103-108.

[Hwang, JN 88*] SEE: Kung, S 88a. *An Algebraic Projection Analysis for Back-Propagation Learning.*

[Hwang, JN 88*] SEE: Kung, S 88b. *Digital VLSI Architectures for Artificial Neural Nets.*

[Hwang, JN 89*] SEE: Kung, S 89a. *Neural Net Architectures for Robotic Applications.*

[Hwang, JN 89a] Hwang, J. N.; Kung, Sun-Yuan. ((submitted)). *A Unifying Viewpoint of Multilayer Perceptrons and Hidden Markov Models.* IEEE International Symposium on Circuits and Systems, Portland, OR, May, 1989.

[Hytonen, T 87*] SEE: Jaaskelainen, T 87a. *Diffraction Regimes of Phase Gratings.*

[Ibaraki, T 68*] SEE: Yajima, S 68a. *On Autonomous Logic Nets of Threshold Elements.*

[Iberall, T 85*] SEE: Arbib, MA 85b. *Coordinated Control Programs for Movements of the Hand.*

[Iberall, T 86a] Iberall, Thea; Bingham, Geoffrey; Arbib, Michael A. (1986). *Opposition Space as a Structuring Concept for the Analysis of Skilled Hand Movements.* Generation and Modulation of Action Patterns. pp 158-173. Berlin: Springer-Verlag.

[Iberall, T 87*] SEE: Arbib, MA 87c. *Schemas That Integrate Vision and Touch for Hand Control.*

[Iberall, T 87a] Iberall, Thea. (1987). *A Ballpark Approach to Modelling Human Prehension.* IEEE International Conference on Neural Networks (ICNN-87), San Diego, CA, June 21-24, 1987: Proceedings. Volume 4. pp 535-544. New York: Institute of Electrical and Electronics Engineers.

[Iberall, T 87b] Iberall, Thea. (1987). *Grasp Planning for Human Prehension.* International Joint Conference on Artificial Intelligence (IJCAI-87), 10th, Milan, August 22-28, 1987: Proceedings. pp 1153-1156.

[Iberall, T 87c] Iberall, Thea. (1987). *The Nature of Human Prehension: Three Dextrous Hands in One.* IEEE International Conference on Robotics and Automation, Raleigh, NC, March 31-April 3, 1987: Proceedings. pp 396-401. New York:

Institute of Electrical and Electronics Engineers.

[Ikeda, S 84a] Ikeda, Saburo. (1984). *Nonlinear Prediction Models for River Flows and Typhoon Precipitation by Self-Organizing Methods.* In Farlow, Stanley J. (ed.). Self-Organizing Methods in Modeling: GMDH Type Algorithms. pp 149-168. New York: Marcel Dekker.

[Immes, L 88*] SEE: Becker, LA 88a. *Committing Facts to Long Term Memory.*

[Ineda, N 88*] SEE: Torioka, T 88a. *Pattern Separating Functioning of Two Layered Random Nerve Nets with Feedforward Inhibitory Connections.*

[Inui, T 88*] SEE: Miyake, S 88a. *Neural Computation Research for Vision - Simulation on a Parallel Processing Computer.*

[Irani, EA 88a] Irani, Erach A.; Matts, John P.; Long, John M.; Slagle, James R. (1988). Using Artificial Neural Nets for Statistical Discovery: Observations after Using Back-propagation, Expert Systems, and Multiple-Linear Regression. AI Memo. Minneapolis, MN: University of Minnesota, Computer Science Department.

[Irie, B 88a] Irie, Bunpei; Miyake, Sei. (1988). *Capabilities of Three-Layered Perceptrons.* IEEE International Conference on Neural Networks (ICNN-88), San Diego, CA, July 24-27, 1988: Proceedings. Volume 1. pp 641-648. New York: Institute of Electrical and Electronics Engineers.

[Irish, BW 87*] SEE: Parunak, HVD 87a. *Supplier-Consumer Nets for Material Handling.*

[Isenhour, TL 75*] SEE: Ritter, GL 75a. *An Algorithm for a Selective Nearest Neighbor Decision Rule.*

[Isham, V 81a] Isham, Valerie. (1981). *An Introduction to Spatial Point Processes and Markov Random Fields.* International Statistical Review. Volume 49. pp 21-43.

[Ishihara, S 86a] Ishihara, Satoshi. (1986). *Recent Advances in Optical Computing in Japan.* In Szu, Harold H. (ed.). Optical and Hybrid Computing. SPIE Volume 634. pp 31-50. Bellingham, WA: Society of Photo-Optical Instrumentation Engineers.

[Isobe, M 87*] SEE: Kawato, M 87a. *A Hierarchical Model for Voluntary Movement and Its Application to Robotics.*

[Isobe, M 88*] SEE: Kawato, M 88a. *Hierarchical Neural Network Model for Voluntary Movement with Application to Robotics.*

[Ito, M 67*] SEE: Eccles, JC 67a. The Cerebellum as a Neuronal Machine.

[Ito, M 79a] Ito, Masao. (1979). *Is the Cerebellum Really a Computer.* TINS. pp 122-126.

[Ito, T 83*] SEE: Fukushima, K 83c. *Neocognitron: A Neural Network Model for a Mechanism of Visual Pattern Recognition.*

[Ito, T 83*] SEE: Fukushima, K 83b. *Neocognitron: A Neural Network Model for a Mechanism of Visual Pattern Recognition.*

[Ivakhnenko, AG 71a] Ivakhnenko, A. G. (1971). *Polynomial Theory of Complex Systems.* IEEE Transactions on Systems, Man and Cybernetics. Volume SMC-1. pp 364-378.

[Ivakhnenko, AG 84a] Ivakhnenko, A. G. (1984). *Past, Present, and Future of GMDH.* In Farlow, Stanley J. (ed.). Self-Organizing Methods in Modeling: GMDH Type Algorithms. pp 105-120. New York: Marcel Dekker.

[Iverson, L 88*] SEE: Zucker, SW 88a. *Computational Networks in Early Vision: From Orientation Selection to Optical Flow.*

[Ivry, R 84*] SEE: Beck, J 84a. *The Perception of Transparency with Achromatic Colors.*

[Iyengar, SS 87*] SEE: Gulati, S 87a. *Nonlinear Neural Networks for Deterministic Scheduling.*

[Iyer, VK 88*] SEE: Ramamoorthy, PA 88b. *Signal Modeling and Prediction Using Neural Networks.*

[Jaaskelainen, T 87*] SEE: Parkkinen, J 87a. *Color Representation Using Statistical Pattern Recognition.*

[Jaaskelainen, T 87a] Jaaskelainen, Timo; Hytonen, T. (1987). *Diffraction Regimes of Phase Gratings.* Optics Communications. Volume 64. pp 19-22.

[Jaaskelainen, T 88*] SEE: Hallikainen, J 88a. *Acousto-Optic Color Spectrometer.*

[Jablon, N 85a] Jablon, Neil; Widrow, Bernard. (1985). *Adaptive Superresolution Phenomena in Aberrant Arrays.* Asilomar Conference on Circuits, Systems, and Computers, 19th, November 6-8, 1985: Conference Record. pp 309-314. Washington, DC: Institute of Electrical and Electronics Engineers.

[Jackel, LD 86*] SEE: Graf, HP 86a. *VLSI Implementation of a Neural Network Memory with Several Hundreds of Neurons.*

[Jackel, LD 86a] Jackel, Lawrence D.; Howard, R. E.; Graf, Hans P.; Straughn, Brian L.; Denker, John S. (1986). *Artificial Neural Networks for Computing.* Journal of Vacuum Science B. Volume 4. pp 61-63.

[Jackel, LD 86b] Jackel, Lawrence D.; Howard, R. E.; Graf, Hans P.; Denker, John S.; Straughn, Brian L. (1986). *High Resolution Micro-Fabrication and Neural Networks.* In Bienenstock, Elie L. (ed.). Disordered Systems and Biological Organization. pp 193-196. New York: Springer-Verlag.

[Jackel, LD 87*] SEE: Graf, HP 87a. *A CMOS Associative Memory Chip.*

[Jackel, LD 87a] Jackel, Lawrence D.; Howard, R. E.; Denker, John S.; Hubbard, W.; Solla, S. A. (1987). *Building a Hierarchy with Neural Networks: An Example-Image Vector Quantization.* Applied Optics. Volume 26. pp 5081-5084.

[Jackel, LD 87b] Jackel, Lawrence D.; Graf, Hans P.; Howard, R. E. (1987). *Electronic Neural Network Chips.* Applied Optics. Volume 26. pp 5077-5080.

[Jackel, LD 88*] SEE: Graf, HP 88a. *VLSI Implementation of a Neural Network Model.*

[Jackson, MD 75a] Jackson, Mark D.; McClelland, James L. (1975). *Sensory and Cognitive Determinants of Reading Speed.* Journal of Verbal Learning and Verbal Behavior. Volume 14. pp 565-574.

[Jackson, MD 79a] Jackson, Mark D.; McClelland, James L. (1979). *Processing Determinants of Reading Speed.* Journal of Experimental Psychology: General. Volume 108. pp 151-181.

[Jacobi, GT 62*] SEE: Yovitz, MC 62a. Self-Organizing Systems.

[Jacobsen, A 83*] SEE: Gilchrist, AL 83a. *The Classification and Integration of Edges as Critical to the Perception of Reflectance and Illumination.*

[Jacyna, GM 87a] Jacyna, Gary M.; Malaret, Erick R. (1987). *Classification Performance of an Auto-Associative Memory Based on Neural Networks.* IEEE International Conference on Neural Networks (ICNN-87), San Diego, CA, June 21-24, 1987: Proceedings. Volume 3. pp 329-338. New York: Institute of Electrical and Electronics Engineers.

[Jagla, K 86*] SEE: Hammerstrom, D 86e. The OGC Cognitive Architecture Project: Silicon Implementation of Connectionist/Neural Networks.

[Jagla, K 88*] SEE: Bahr, C 88c. *Concurrent Neural Network Simulation: Two Examples within a Single, Integrated Neural Network Hardware Development Environment.*

[Jain, AK 81a] Jain, Anil K. (1981). *Advances in Mathematical Models for Image Processing.* Proceedings of the IEEE. Volume 69. pp 502-528.

[Jain, AK 83*] SEE: Cross, GR 83a. *Markov Random Field Texture Models.*

[Jain, AK 86a] Jain, Anil K. (1986). *Cluster Analysis.* In Young, Tzay Y. (ed.); Fu, King-Sun (ed.). Handbook of Pattern Recognition and Image Processing. pp 33-59. New York: Academic Press.

[Jain, RC 86*] SEE: Knoll, TF 86a. *Recognizing Partially Visible Objects Using Feature Indexed Hypotheses.*

[Jalanko, M 78*] SEE: Haltsonen, S 78a. *Application of Novelty Filter to Segmentation of Speech.*

[Jalanko, M 78a] Jalanko, Matti; Haltsonen, Seppo; Bry, Kalle-J; Kohonen, Teuvo. (1978). *Application of Orthogonal Projection Principles to Simultaneous Phonemic Segmentation and Labeling of Continuous Speech.* International Joint Conference on Pattern Recognition, 4th, Kyoto, November 7-10, 1978: Proceedings. pp 1006-1008. New York: Institute of Electrical and Electronics Engineers.

[Jalanko, M 78b] Jalanko, Matti; Kohonen, Teuvo. (1978). *Application of the Subspace Method to Speech Recognition.* IEEE International Conference on Acoustics, Speech, and Signal Processing (ICASSP-78), Tulsa, OK, April 10-12, 1978: Proceedings. pp 561-564. New York: Institute of Electrical and Electronics Engineers.

[Jalanko, M 79*] SEE: Reuhkala, E 79a. *A Redundant Hash Addressing Method Adapted for the Postprocessing and Error-Correction of Computer-Recognized Speech.*

[Jalanko, M 79*] SEE: Kohonen, T 79a. *Spectral Classification of Phonemes by Learning Subspaces.*

[Jalanko, M 80*] SEE: Kohonen, T 80a. *A Thousand-Word Recognition System Based on the Learning Subspace Method and Redundant Hash Addressing.*

[James, M 85a] James, Mike. (1985). Classification Algorithms. New York: Wiley.

[Jamshidi, M 87*] SEE: Horne, W 87a. *Application-Oriented Connection Systems for Pattern Acquisition and Recognition.*

[Jang, J 88*] SEE: Lee, S 88a. *Optical Implementation of Associative Memory with Controlled Bit-Significance.*

[Jang, J 88a] Jang, Ju-Seog; Lee, Soo-Young; Shin, Sang-Yung. (1988). *An Optimization Network for Matrix Inversion.* In Anderson, Dana Z. (ed.). Neural Information Processing Systems. pp 397-401. New York: American Institute of Physics.

[Jang, J 88b] Jang, Ju-Seog; Shin, Sang-Yung; Lee, Soo-Young. (1988). *Optical Implementation of Quadratic Associative Memory with Outer-Product Storage.* Optics Letters. Volume 13. pp 693-695.

[Jang, J 88c] Jang, Ju-Seog; Jung, Su-Won; Lee, Soo-Young; Shin, Sang-Yung. (1988). *Optical Implementation of the Hopfield Model for Two-Dimensional Associative Memory.* Optics Letters. Volume 13. pp 248-250.

[Jannarone, R 88*] SEE: Takefuji, Y 88a. *Multinomial Conjunctoid Statistical Learning Machines.*

[Jansen, JKS 69*] SEE: Enger, PS 69a. *A Biological Model of the Excitation of a Second Order Sensory Neurone.*

[Jansen, JKS 69*] SEE: Walloe, L 69a. *A Computer-Simulated Model of a Second Order Sensory Neurone.*

[Janssens, R 88*] SEE: Tesauro, G 88c. *Scaling Relationships in Back-Propagation Learning: Dependence on Predicate Order.*

[Jaron, D 87*] SEE: Wolf, L 87a. *A Trainable Model for Near Periodic Signals.*

[Jarvis, RA 70a] Jarvis, R. A. (1970). *Adaptive Global Search in a Time Varying Environment Using a Probabilistic Automaton with Pattern Recognition Supervision.* IEEE Transactions on Systems Science and Cybernetics. Volume SSC-6. pp

209-217.

[Jarvis, RA 77a] Jarvis, R. A. (1977). *Teaching a Stochastic Automaton to Skillfully Play Hand / Eye Games.* Journal of Cybernetics and Information Science. Volume 1. pp 161-177.

[Jaynes, ET 68a] Jaynes, Edwin T. (1968). *Prior Probabilities.* IEEE Transactions on Systems Science and Cybernetics. Volume SSC-4. pp 227-241.

[Jaynes, ET 79a] Jaynes, Edwin T. (1979). *Where Do We Stand on Maximum Entropy?.* Maximum Entropy Formalism Conference, Massachusetts Institute of Technology, May 2-4, 1978: Proceedings. pp 15-118. Cambridge, MA: MIT Press.

[Jefferson, DR 85a] Jefferson, David R. (1985). *Virtual Time.* ACM Transactions on Programming Languages and Systems. Volume 7. pp 404-425.

[Jeffrey, W 86a] Jeffrey, W.; Rosner, R. (1986). *Neural Network Processing as a Tool for Function Optimization.* In Denker, John S. (ed.). Neural Networks for Computing, Snowbird, Utah, 1986. AIP Proceedings 151. pp 241-246. New York: American Institute of Physics.

[Jelinek, F 76a] Jelinek, Frederick. (1976). *Continuous Speech Recognition by Statistical Methods.* IEEE Proceedings. Volume 64. pp 532-556.

[Jelinek, F 83*] SEE: Bahl, LR 83a. *A Maximum Likelihood Approach to Continuous Speech Recognition.*

[Jenik, F 64a] Jenik, F. (1964). *Pulse Processing by Neuron Models.* Neural Theory and Modeling, Proceedings of the 1962 Ojai Symposium. pp 190-212. Stanford: Stanford University Press.

[Jenkins, BK 87*] SEE: Zhou, Y 87a. *A Novel Approach to Image Restoration Based on a Neural Network.*

[Jenkins, BK 88*] SEE: Zhou, Y 88b. *Image Restoration Using a Neural Network.*

[Jenkins, G 76*] SEE: Box, G 76a. Time Series Analysis: Forecasting and Control.

[Jensen, RV 87a] Jensen, Roderick V. (1987). *Classical Chaos.* American Scientist. Volume 75. pp 168-181.

[Jernigan, ME 82*] SEE: Wardell, RW 82a. *Modelling the Retina as a Stochastic Receptor Array.*

[Jesorsky, P 77*] SEE: Bunge, E 77b. *The AUROS Project: Automatic Recognition of Speakers by Computers.*

[Jespers, P 88*] SEE: Sirletti, B 88a. *An Algorithm for Pattern Recognition with VLSI Neural Networks.*

[Jespers, P 88*] SEE: Verleysen, M 88a. *A High-Storage Content-Addressable Memory and Its Learning Algorithm.*

[Jespers, P 88*] SEE: Verleysen, M 88b. *A Large VLSI Fully Interconnected Neural Network.*

[Jespers, P 88*] SEE: Verleysen, M 88c. *A Large VLSI Hopfield Network for Pattern Recognition Problems.*

[Jespers, P 88*] SEE: Verleysen, M 88g. *Neural Network or High-Storage Content-Addressable Memory: VLSI Circuit and Learning Algorithm.*

[Jespers, P 88*] SEE: Verleysen, M 88d. *A New CMOS Architecture for Neural Networks.*

[Jespers, P 88*] SEE: Verleysen, M 88e. *A New VLSI Architecture for Large Hopfield's Neural Networks.*

[Jespers, P 88*] SEE: Verleysen, M 88f. *A New VLSI Architecture for Neural Associative Memories.*

[Johannet, A 86*] SEE: Personnaz, L 86f. *Simple Selectionist Learning Rule for Neural Networks.*

[John, ER 78a] John, E. Roy; Schwartz, Eric L. (1978). *The Neurophysiology of Information Processing and Cognition.* Annual Review of Psychology. Volume 29. pp 1-29.

[Johnson, DS 79*] SEE: Garey, MR 79a. Computers and Intractability: A Guide to the Theory of NP-Completeness.

[Johnson, G 88a] Johnson, George. (May 20, 1988). *Scientists Identify "Gate" in Brain as Crucial to Memory.* New York Times. pp C1.

[Johnson, JL 88a] Johnson, John L. (1988). *Architectural Relationships Involving Symbolic Substitution.* Applied Optics. Volume 27. pp 529-533.

[Johnson Jr, CR 76*] SEE: Widrow, B 76b. *Stationary and Nonstationary Learning Characteristics of the LMS Adaptive Filter.*

[Johnson, K 86a] Johnson, Kenneth; Fong, Yu-Shan. (1986). *A Simplified Model of Retinal Function with Parallel Architecture and Integrated Space-Time Filtering for Low Level Visual Processing.* Visual Communications and Image Processing. SPIE Volume 707. pp 172-181. Bellingham, WA: Society of Photo-Optical Instrumentation Engineers.

[Johnson, K 87a] Johnson, Kenneth; Fong, Yu-Shan. (1987). *Two New Theories of Retinal Function.* IEEE International Conference on Neural Networks (ICNN-87), San Diego, CA, June 21-24, 1987: Proceedings. Volume 4. pp 97-106. New York: Institute of Electrical and Electronics Engineers.

[Johnson, M 87a] Johnson, Mark; Scanlon, Raymond D. (1987). *Experiences with a Feeling-Thinking Machine.* IEEE International Conference on Neural Networks (ICNN-87), San Diego, CA, June 21-24, 1987: Proceedings. Volume 2. pp 71-78. New York: Institute of Electrical and Electronics Engineers.

[Johnson, M 87b] Johnson, Mark; Scanlon, Raymond D. (1987). Non-Living Intelligence. ARCCB-TR-87009. Watervliet, NY: U S Army ARDEC.

[Johnson, M 88*] SEE: Scanlon, RD 88a. *Design of a Feeling-Thinking Machine.*

[Johnson, RC 86a] Johnson, R. Colin. (April 7, 1986). *Brain-Emulating Circuits Need 'Sleep' and 'Dreams'.* Electronic Engineering Times. pp 53 f.

[Johnson, RC 86b] Johnson, R. Colin. (December 22, 1986). *Carver Mead Joins Forces with Neural Net Startup.* Electronic Engineering Times. p 21.

[Johnson, RC 86c] Johnson, R. Colin. (March 24, 1986). *Nerve Nets Built On Chip.* Electronic Engineering Times. p 1.

[Johnson, RC 86d] Johnson, R. Colin. (December 1, 1986). *Neural Nets Inspire Computer Designer.* Electronic Engineering Times. pp 31 f.

[Johnson, RC 87*] SEE: Brown, C 87c. *Engineer A Black-Box Solution in the First "Real Cognizer" Contest.*

[Johnson, RC 87*] SEE: Brown, C 87e. *In Back-to-Basics Movement, VLSI Expert Calls for More Simplicity in Computation.*

[Johnson, RC 87*] SEE: Brown, C 87g. *Neural Net Pioneer Grossberg Extends ART-2 Architecture into Analog Domain.*

[Johnson, RC 87*] SEE: Brown, C 87j. *Neural Networks Get a Conference to Call Their Own.*

[Johnson, RC 87*] SEE: Brown, C 87l. *Theories of Neural Nets Expose the Limits of AI.*

[Johnson, RC 87a] Johnson, R. Colin. (1987). *Beyond Neural Networks - The Discipline of Intelligent Systems Design Emerges.* Electronic Engineering Times. Issue 464. pp PT34.

[Johnson, RC 87b] Johnson, R. Colin. (1987). *IEEE Puts Neural Nets into Focus.* Electronic Engineering Times. Issue 462.

pp 45-46.

[Johnson, RC 87c] Johnson, R. Colin. (March 30, 1987). *Japan Develops Neural Networks: Part 1, Academia.* Electronic Engineering Times. p 35 f.

[Johnson, RC 87d] Johnson, R. Colin. (March 23, 1987). *Japan's 6th Generation Project: Striking A Fine Balance.* Electronic Engineering Times. p 33 f.

[Johnson, RC 87e] Johnson, R. Colin. (August 10, 1987). *Learning Modules Permit Scalable Nerual Systems.* Electronic Engineering Times. pp 53, 73.

[Johnson, RC 87f] Johnson, R. Colin. (January 12, 1987). *Neural Net Developer Nestor Sprints From Startup Gate.* Electronic Engineering Times.

[Johnson, RC 87g] Johnson, R. Colin. (March 23, 1987). *Neural Net Speech System In Works.* Electronic Engineering Times. p 10.

[Johnson, RC 87h] Johnson, R. Colin. (1987). *Neural Nets to Arm NASA Space Robots?.* Electronic Engineering Times. Issue 457. pp 49-50.

[Johnson, RC 87i] Johnson, R. Colin. (April 6, 1987). *Neural Networks in Japan: Part 2.* Electronic Engineering Times.

[Johnson, RC 87j] Johnson, R. Colin. (1987). *Neural Networks Naive, Says Minsky.* Electronic Engineering Times. Issue 445. pp 41.

[Johnson, RC 87k] Johnson, R. Colin. (January 5, 1987). *Neural Systems Make Comeback After Nearly 50-Year Gestation.* Electronic Engineering Times. 23.

[Johnson, RC 88a] Johnson, R. Colin; Brown, Chappell. (1988). Cognizers: Neural Networks and Machines That Think. New York: Wiley.

[Johnson, RC 88b] Johnson, R. Colin. (1988). *NASA Unites Neural Nets, Fuzzy Logic.* Electronic Engineering Times. Issue 487. pp 39-41.

[Johnson, RC 88c] Johnson, R. Colin. (1988). *Nestor Frees Neural Nets from Hardware Chains.* Electronic Engineering Times. Issue 471. pp 33-34.

[Johnson, RC 88d] Johnson, R. Colin. (1988). *Networking the Brain.* Electronic Engineering Times. Issue 493. pp 41.

[Johnson, RC 88e] Johnson, R. Colin. (1988). *Neural Model: One Equation Does It All.* Electronic Engineering Times. Issue 479. pp 53-54.

[Johnson, RC 88f] Johnson, R. Colin. (1988). *Neural Net for HDTV.* Electronic Engineering Times. Issue 494. pp 51-52.

[Johnson, RC 88g] Johnson, R. Colin. (1988). *Neural Network Startups Proliferate Across the U.S..* The Scientist. Volume 2. pp 1.

[Johnson, RC 88h] Johnson, R. Colin. (1988). Neural Networks. Volumes 1-4. Portland, OR: Cognizer Connection.

[Johnson, RC 88i] Johnson, R. Colin. (1988). *Neural Networks Finally Get Real.* Electronic Engineering Times. Issue 512. pp 117-127.

[Johnson, RC 88j] Johnson, R. Colin. (1988). *Neural Optical Chip Modeled.* Electronic Engineering Times. pp 53.

[Johnson, RC 88k] Johnson, R. Colin. (1988). *Optics and Neural Nets: Marriage of Convenience.* Electronic Engineering

Times. Issue 469. pp 41-42.

[Johnson, RC 88l] Johnson, R. Colin. (1988). *Smart Thinking Shows at Neural Conference.* Electronic Engineering Times. Issue 499. pp 49.

[Johnson, RC 88m] Johnson, R. Colin. (1988). *The Fastest Software-only Neural Net?.* Electronic Engineering Times. Issue 502. pp 53.

[Johnston, A 88*] SEE: Dobson, VG 88b. *Differences in Performance of Binary Hetero-Associative Networks Using Link-Enabling vs. Link-Disabling Local Modification Rules.*

[Johnston, JC 73a] Johnston, James C.; McClelland, James L. (1973). *Visual Factors in Word Perception.* Perception and Psychophysics. Volume 14. pp 365-370.

[Johnston, JC 74a] Johnston, James C.; McClelland, James L. (1974). *Perception of Letters in Words: Seek Not and Ye Shall Find.* Science. Volume 184. pp 1192-1194.

[Johnston, JC 77*] SEE: McClelland, JL 77b. *The Role of Familiar Units in Perception of Words and Nonwords.*

[Johnston, JC 78a] Johnston, James C. (1978). *A Test of the Sophisticated Guessing Theory of Word Perception.* Cognitive Psychology. Volume 10. pp 123-153.

[Johnston, JC 80a] Johnston, James C.; McClelland, James L. (1980). *Experimental Tests of a Hierarchical Model of Word Identification.* Journal of Verbal Learning and Verbal Behavior. Volume 19. pp 503-524.

[Jones, KS 73a] Jones, Karen Sparck. (1973). *Index Term Weighting.* Information Storage and Retrieval. Volume 9. pp 619-633.

[Jones, RS 77*] SEE: Anderson, JA 77b. *Distinctive Features, Categorical Perception, and Probability Learning: Some Applications of a Neural Model.*

[Jones, RS 77*] SEE: Anderson, JA 77a. *Distinctive Features, Categorical Perception, and Probability Learning: Some Applications of a Neural Model.*

[Jones, WP 86a] Jones, William P. (1986). *On the Applied Use of Human Memory Models: The Memory Extender Personal Filing System.* International Journal of Man-Machine Studies. Volume 25. pp 191-228.

[Jones, WP 87a] Jones, William P.; Hoskins, Josiah. (1987). *Back-Propagation: A Generalized Delta Learning Rule.* Byte. Volume 12. pp 155-162.

[Jones, WP 88a] Jones, William P. (1988). *ANNA: An Adaptive Neural Network Anticipator for Human/Computer Interfacing.* Meeting of the International Neural Network Society (INNS), 1st, Boston, MA, September, 1988.

[Jong, T 88*] SEE: Tai, H 88a. *Neural Networks with Higher-Order Nonlinearity.*

[Jong, T 88a] Jong, T-L; Tai, H-M. (1988). *Hopfield Model with Complementary Binary Representations.* Electronics Letters. Volume 24. pp 1224-1225.

[Jordan, MI 86a] Jordan, Michael I. (1986). *Attractor Dynamics and Parallelism in a Connectionist Sequential Machine.* Cognitive Science Society, 8th, Amherst, MA, August 15-17, 1986: Proceedings. pp 531-546. Hillsdale, NJ: Lawrence Erlbaum.

[Jordan, MI 86b] Jordan, Michael I. (1986). *Introduction to Linear Algebra in Parallel Distributed Processing.* In Rumelhart, David E. (ed.). Parallel Distributed Processing: Foundations. Volume I. pp 365-422. Cambridge, MA: MIT Press.

[Jordan, MI 86c] Jordan, Michael I. (1986). Serial Order: A Parallel Distributed Processing Approach. ICS-8604. San Diego, CA: University of California, San Diego, Institute for Cognitive Science.

[Jordan, MI 87*] SEE: Barto, AG 87b. *Gradient Following Without Back-Propagation in Layered Networks.*

[Jordan, MI 88a] Jordan, Michael I.; Rosenbaum, David A. (1988). Action. COINS Technical Report 88-26. pp 1-68. Amherst, MA: University of Massachusetts, Computer and Information Science.

[Jordan, MI 88b] Jordan, Michael I. (1988). Supervised Learning and Systems with Excess Degrees of Freedom. COINS Technical Report 88-27. Amherst, MA: University of Massachusetts.

[Jorgensen, C 86a] Jorgensen, Chuck; Matheus, Christopher J. (1986). *Catching Knowledge in Neural Nets.* AI Expert. Volume 1. pp 30-41.

[Jorgensen, CC 87a] Jorgensen, C. C. (1987). *Neural Network Representation of Sensor Graphs in Autonomous Robot Path Planning.* IEEE International Conference on Neural Networks (ICNN-87), San Diego, CA, June 21-24, 1987: Proceedings. Volume 4. pp 507-515. New York: Institute of Electrical and Electronics Engineers.

[Jorgensen, CC 88*] SEE: Barhen, J 88b. *Concurrent Neuromorphic Algorithms for Autonomous Robots.*

[Joseph, RD 60a] Joseph, Roger David. (1960). *On Predicting Perceptron Performance.* IRE International Convention Record. pp 71-77.

[Josin, GM 77a] Josin, Gary M. (1987). *Self-Control in Neural Nets.* Biological Cybernetics. Volume 27. pp 185-187.

[Josin, GM 87a] Josin, Gary M. (1987). *Combinations of Neural Systems for Particular Application Situations.* IEEE International Conference on Neural Networks (ICNN-87), San Diego, CA, June 21-24, 1987: Proceedings. Volume 4. pp 517-524. New York: Institute of Electrical and Electronics Engineers.

[Josin, GM 87b] Josin, Gary M. (1987). *Neural Systems for Use as a Research Tool.* Computer Simulation in Brain Science. Cambridge: Cambridge University Press.

[Josin, GM 87c] Josin, Gary M. (1987). *Neural-Network Heuristics.* Byte. Volume 12. pp 183-192.

[Juang, BH 85a] Juang, B. H.; Rabiner, L. R.; Levinson, Stephen E.; Sondhi, M. M. (1985). *Recent Developments in the Application of Hidden Markov Models to Speaker-Independent Isolated Word Recognition.* IEEE International Conference on Acoustics, Speech, and Signal Processing (ICASSP-85), April, 1985: Proceedings.

[Judd, JS 87a] Judd, J. Stephen. (1987). Complexity of Connectionist Learning with Various Node Functions. COINS Technical Report 87-60. pp 1-23. Amherst, MA: University of Massachusetts, Computer and Information Science.

[Judd, JS 87b] Judd, J. Stephen. (1987). *Learning in Networks is Hard.* IEEE International Conference on Neural Networks (ICNN-87), San Diego, CA, June 21-24, 1987: Proceedings. Volume 2. pp 685-692. New York: Institute of Electrical and Electronics Engineers.

[Judd, JS 88a] Judd, J. Stephen. (1988). *On the Complexity of Loading Shallow Neural Networks.* Journal of Complexity. Volume 4. pp 177-192.

[Julesz, B 71a] Julesz, Bela. (1971). Foundations of Cyclopean Perception. Chicago: University of Chicago Press.

[Julesz, B 81a] Julesz, Bela. (1981). *Textrons, the Elements of Texture Perception and Their Interactions.* Nature. Volume 290. pp 91-97.

[Julesz, B 83*] SEE: Bergen, JR 83a. *Parallel Versus Serial Processing in Rapid Pattern Discrimination.*

[Julian, WH 87*] SEE: Cohen, MS 87a. *Neural Networks with a Hopf Bifurcation: Slowly Modulated Waves.*

[Jung, S 88*] SEE: Jang, J 88c. *Optical Implementation of the Hopfield Model for Two-Dimensional Associative Memory.*

[Juola, JF 71a] Juola, J. F.; Fischler, I.; Wood, C. T.; Atkinson, Reilly C. (1971). *Recognition Time for Information Stored in Long-Term Memory.* Perception and Psychophysics. Volume 10. pp 8-14.

[Jusczyk, PW 80a] Jusczyk, P. W. (ed.); Klein, R. M. (ed.). (1980). The Nature of Thought: Essays in Honor of D.O. Hebb. Hillsdale, NJ: Lawrence Erlbaum.

[Jutten, C 86*] SEE: Herault, J 86a. *Space or Time Adaptive Signal Processing by Neural Network Models.*

[Jutten, C 88*] SEE: Herault, J 88b. *Independent Components Analysis in Neural Networks.*

[Jutten, C 88a] Jutten, Christian; Herault, Jeanny. (1988). *Independent Components Analysis (INCA) Versus Principal Components Analysis.* Signal Processing IV: Theories and Applications. pp 643-646. New York: Elsevier.

[Kabrisky, M 87a] Kabrisky, Matthew; Rogers, Stephen K.; Mills, James P.; Kolesar, Edward. (1987). *A Multiplexed Multi-Electrode Semiconductor Brain Electrode Implant.* IEEE International Conference on Neural Networks (ICNN-87), San Diego, CA, June 21-24, 1987: Proceedings. Volume 3. pp 227-242. New York: Institute of Electrical and Electronics Engineers.

[Kabrisky, M 88*] SEE: Ruck, DW 88a. Multisensor Target Detection and Classification.

[Kadar, I 87a] Kadar, Ivan. (1987). *Robust Tracking Novelty Filters Based on Linear Models.* IEEE International Conference on Neural Networks (ICNN-87), San Diego, CA, June 21-24, 1987: Proceedings. Volume 4. pp 611-617. New York: Institute of Electrical and Electronics Engineers.

[Kahn, P 85a] Kahn, Philip. (1985). *Local Determination of a Moving Contrast Edge.* IEEE Transactions on Pattern Analysis and Machine Intelligence. Volume PAMI-7. pp 402-409.

[Kailath, T 74a] Kailath, T. (1974). *A View of Three Decades of Linear Filtering Theory.* IEEE Transactions on Information Theory. Volume IT-20. pp 145-181.

[Kak, AC 86a] Kak, A. C.; Roberts, B. A. (1986). *Reconstruction from Projections: Applications in Computerized Tomography.* In Young, Tzay Y. (ed.); Fu, King-Sun (ed.). Handbook of Pattern Recognition and Image Processing. pp 649-694. New York: Academic Press.

[Kalay, A 87a] Kalay, Ainat; Hochstein, Shaul; Rudolph, Larry. (1987). *Neural Patterns as Detection Units in Sensory Systems.* IEEE International Conference on Neural Networks (ICNN-87), San Diego, CA, June 21-24, 1987: Proceedings. Volume. pp 137-146. New York: Institute of Electrical and Electronics Engineers. .

[Kalita, J 87a] Kalita, Jugal; Shastri, Lokendra. (1987). *Generation of Simple Sentences in English Using the Connectionist Model of Computation.* Cognitive Science Society, 9th, Seattle, WA, July 16-18, 1987: Proceedings. pp 555-565. Hillsdale, NJ: Lawrence Erlbaum.

[Kalos, MH 75*] SEE: Bortz, AB 75a. *A New Algorithm for Monte Carlo Simulation of Ising Spin Systems.*

[Kam, M 87*] SEE: Guez, A 87a. *Neuromorphic Architecture for Adaptive Robot Control: A Preliminary Analysis.*

[Kam, M 87a] Kam, Moshe; Cheng, Roger; Guez, Allon. (1987). *On the Design of a Content-Addressable Memory via Binary Neural Networks.* IEEE International Conference on Neural Networks (ICNN-87), San Diego, CA, June 21-24, 1987: Proceedings. Volume 2. pp 513-522. New York: Institute of Electrical and Electronics Engineers.

[Kam, M 88*] SEE: Guez, A 88a. *Neural Network Architecture for Control.*

[Kamangar, FA 88a] Kamangar, Farhad A.; Smith, Jeffrey O. (1988). *A Neural Network Based Controller for Robotic*

Manipulators. Modeling and Simulation Conference, 19th, Pittsburgh, PA, May 5-6, 1988: Proceedings. Volume 19. pp 2043-2050. Research Triangle Park, NC: Instrument Society of America.

[Kamgar-Parsi, B 87*] SEE: Kamgar-Parsi, B 87a. *An Efficient Model of Neural Networks for Optimization.*

[Kamgar-Parsi, B 87a] Kamgar-Parsi, Behzad; Kamgar-Parsi, Behrooz. (1987). *An Efficient Model of Neural Networks for Optimization.* IEEE International Conference on Neural Networks (ICNN-87), San Diego, CA, June 21-24, 1987: Proceedings. Volume 3. pp 785-???. New York: Institute of Electrical and Electronics Engineers.

[Kamiyama, Y 88*] SEE: Usui, S 88a. *Physiological Engineering Model of the Retinal Horizontal Cell Layer.*

[Kamm, CA 87*] SEE: Landauer, TK 87a. *Teaching a Minimally Structured Back-Propagation Network to Recognize Speech Sounds.*

[Kammen, DM 87*] SEE: Daugman, JG 87a. *Image Statistics, Gases, and Visual Neural Primitives.*

[Kammen, DM 88a] Kammen, Daniel M.; Yuille, Alan L. (1988). *Spontaneous Symmetry-Breaking Energy Functions and the Emergence of Orientation Selective Cortical Cells.* Biological Cybernetics. Volume 59. pp 23-31.

[Kamp, Y 87*] SEE: Delsarte, P 87a. *Low Rank Matrices with a Given Sign Pattern.*

[Kampfner, RR 83a] Kampfner, Roberto R.; Conrad, Michael. (1983). *Computational Modeling of Evolutionary Learning Processes in the Brain.* Bulletin of Mathematical Biology. Volume 45. pp 931-968.

[Kampfner, RR 83b] Kampfner, Roberto R.; Conrad, Michael. (1983). *Sequential Behavior and Stability Properties of Enzymatic Neuron Networks.* Bulletin of Mathematical Biology. Volume 45. pp 969-980.

[Kampfner, RR 88*] SEE: Conrad, M 88b. *Neuronal Dynamics and Evolutionary Learning.*

[Kanade, T 86*] SEE: Wallace, R 86a. *Progress in Robot Road-Following.*

[Kanade, T 87*] SEE: Matthies, L 87a. Kalman Filter-Based Algorithms for Estimating Depth from Image Sequences.

[Kanal, LN 65*] SEE: Abend, K 65a. *Classification of Binary Random Patterns.*

[Kanal, LN 80a] Kanal, Laveen N. (1981). *Markov Mesh Models.* Image Modeling. pp 239-243. New York: Academic Press.

[Kanal, LN 86a] Kanal, Laveen N.; Dattatreya, G. R. (1986). *Problem Solving Methods for Pattern Recognition.* In Young, Tzay Y. (ed.); Fu, King-Sun (ed.). Handbook of Pattern Recognition and Image Processing. pp 143-168. New York: Academic Press.

[Kanal, LN 88*] SEE: Tsao, T 88a. *A Novel Architecture for Real Time Pick Up of 3D Motion and 3D Layout Information from the Flow of the Optic Array.*

[Kanatani, K 77*] SEE: Amari, S 77c. *Mathematical Foundation for Statistical Neurodynamics.*

[Kandel, A 86a] Kandel, A. (1986). Fuzzy Mathematical Techniques with Applications. Reading, MA: Addison-Wesley.

[Kandel, A 87a] Kandel, A. (ed.). (1987). Fuzzy Expert Systems. Reading, MA: Addison-Wesley.

[Kandel, ER 82a] Kandel, Eric R.; Schwartz, James H. (1982). *Molecularbiology of Learning: Modulation of Transmitter Release.* Science. Volume 218. pp 433-443.

[Kane, B 87a] Kane, Beverly. (August 3, 1987). *Neural Networking Research Gains New Life.* MacWeek. p 5.

[Kane, B 87b] Kane, Beverly. (1987). *Neural Networks Think Big.* Intelligent Systems Analyst. Volume 8. pp 1-9.

[Kanerva, P 84a] Kanerva, Pentti. (1984). Self-Propagating Search: A Unified Theory of Memory. Report Number CLSI-84-7. Stanford, CA: Stanford University, Center for the Study of Language and Information.

[Kanerva, P 86a] Kanerva, Pentti. (1986). *Parallel Structures in Human and Computer Memory.* In Denker, John S. (ed.). Neural Networks for Computing, Snowbird, Utah, 1986. AIP Proceedings 151. pp 247-258. New York: American Institute of Physics.

[Kanerva, P 88a] Kanerva, Pentti. (1988). Sparse Distributed Memory. Cambridge, MA: MIT Press.

[Kangas, J 87*] SEE: Kohonen, T 87e. *Microprocessor Implementation of a Large Vocabulary Speech Recognizer and Phonetic Typewriter for Finnish and Japanese.*

[Kangas, J 88*] SEE: Kohonen, T 88b. *Phonetic Typewriter for Finnish and Japanese.*

[Kanizsa, G 76a] Kanizsa, Gaetano. (1976). *Subjective Contours.* Scientific American. Volume 234. pp 48-52.

[Kanwal, JS 88a] Kanwal, Jagmeet S. (1988). *How the Catfish Tracks Its Prey: An Interactive "Pipelined" Processing System May Direct Foraging Via Reticulospinal Neurons.* Neural Information Processing Systems. pp 402-411. New York: American Institute of Physics.

[Kanwal, JS 88b] Kanwal, Jagmeet S.; Caprio, John. (1988). *Overlapping Taste and Tactile Maps of the Oropharynx in the Vagal Lobe of the Channel Catfish, Ictalurus punctatus.* Journal of Neurobiology. Volume 19. pp 211-222.

[Karayiannis, NB 88a] Karayiannis, Nicolaos B.; Venetsanopoulos, Anastasios N. (1988). *The Correlation Associative Memory Realizes Hebbian Learning.* Neural Networks. Volume 1.

[Karsai, G 88a] Karsai, Gabor; Ramaswamy, K.; Andersen, K.; Cook, G. E. (1988). Gas Tungsten Arc Weld Modeling Using a Mapping Network. Nashville, TN: Vanderbilt University, Department of Electrical Engineering.

[Karunasiri, RPG 87*] SEE: Bandara, KMSV 87a. *Interfacing Multispectral Sensors to Real Time Processors Based on Neural Network Models.*

[Karunasiri, RPG 88*] SEE: Coon, DD 88a. *Infrared Transient Sensing.*

[Kashyap, RL 70a] Kashyap, Rangasami L.; Blaydon, C. C.; Fu, King-Sun. (1970). *Stochastic Approximation.* Adaptive Learning and Pattern Recognition Systems: Theory and Applications. pp 339-354. New York: Academic Press.

[Kashyap, RL 82*] SEE: Chellappa, R 82a. *Digital Image Restoration Using Spatial Interaction Models.*

[Kashyap, RL 83a] Kashyap, Rangasami L.; Chellappa, Ramalingam. (1983). *Estimation and Choice of Neighbors in Spatial-Interaction Models of Images.* IEEE Transactions on Information Theory. Volume IT-29. pp 60-72.

[Katayama, K 88*] SEE: Tsutsumi, K 88a. *Neural Computation for Controlling the Configuration of 2-Dimentional Truss Structure.*

[Kato, K 88a] Kato, K.; Nakane, Kaoru. (1988). *An Associative Memory Model Based on a Quantification Method.* Neural Networks. Volume 1, Supplement 1. pp 188.

[Katsuno, I 71*] SEE: Suzuki, R 71a. *Dynamics of "Neuron Ring".*

[Katz, BF 88a] Katz, Bruce F. (1988). *A Connectionist Model of Chronic Pain.* ESPCI European Conference on Neural Networks (nEuro'88), 1st, Paris, June, 1988: Preceedings.

[Kaufman, H 67a] Kaufman, Howard. (1967). *An Experimental Investigation of Process Identification by Competitive Evolution.* IEEE Transactions on Systems, Sciences and Cybernetics. Volume SSC-3. pp 11-16.

[Kaufman, L 73a] Kaufman, Lloyd; Bacon, Joshua; Barroso, Felix. (1973). *Stereopsis Without Image Segregation*. Vision Research. Volume 13. pp 137-147.

[Kaufman, L 74a] Kaufman, Lloyd. (1974). Sight and Mind: An Introduction to Visual Perception. New York: Oxford University Press.

[Kaufman, SA 82*] SEE: Small, SL 82b. ISCON: An Interactive Simulator for Connectionist Networks.

[Kaufman, SA 86a] Kaufman, S. A. (1986). *Boolean Systems, Adaptive Automata, Evolution*. In Bienenstock, Elie L. (ed.). Disordered Systems and Biological Organization. pp 339-360. New York: Springer-Verlag.

[Kaunitz, J 75*] SEE: Widrow, B 75a. *Adaptive Noise Canceling: Principles and Applications*.

[Kautz, WH 61*] SEE: Myhill, J 61a. *On the Size of Weights Required for Linear-Input Switching Functions*.

[Kautz, WH 61a] Kautz, William H. (1961). *The Realization of Symmetric Switching Functions with Linear-Input Logical Elements*. IRE Transactions of Electronic Computers. Volume EC-10. pp 371-378.

[Kawahara, T 80a] Kawahara, Takuji. (1980). *Coupled Van der Pol Oscillators - A Model of Excitatory and Inhibitory Neural Interactions*. Biological Cybernetics. Volume 39. pp 37-43.

[Kawamoto, AH 84a] Kawamoto, Alan H.; Anderson, James A. (1984). *Lexical Access Using a Neural Network*. Cognitive Science Society, 6th, Boulder, CO, June 28-30, 1984: Proceedings. pp 204-213. Hillsdale, NJ: Lawrence Erlbaum.

[Kawamoto, AH 85a] Kawamoto, Alan H.; Anderson, James A. (1985). *A Neural Network Model of Multistable Perception*. Acta Psychologica. Volume 59. pp 35-65.

[Kawamoto, AH 85b] Kawamoto, Alan H. (1985). Dynamic Processes in the Resolution of Lexical Ambiguity: PhD Thesis. Providence, RI: Brown University.

[Kawamoto, AH 86*] SEE: McClelland, JL 86f. *Mechanisms of Sentence Processing: Assigning Roles to Constituents*.

[Kawamoto, AH 86a] Kawamoto, Alan H. (1986). *Solving Puzzles with a Connectionist Network Using a Hill-Climbing Heuristic*. Cognitive Science Society, 8th, Amherst, MA, August 15-17, 1986: Proceedings. pp 514-521. Hillsdale, NJ: Lawrence Erlbaum.

[Kawamura, S 87a] Kawamura, Sadao; Miyazaki, Fumio; Arimoto, Suguru. (1987). *Intelligent Control of Robot Motion Based on Learning Method*. IEEE International Symposium on Intelligent Control, Philadelphia, PA, January 18-20, 1987: Proceedings. New York: Institute of Electrical and Electronics Engineers.

[Kawamura, S 87b] Kawamura, Sadao; Yoshimi, T.; Miyazaki, F.; Arimoto, Suguru. (1987). *Technical Issues in Learning Control for Robot Motions*. International Conference on Advanced Robotics, 3rd, Versailles, October 13-15, 1987: Proceedings.

[Kawamura, S 88a] Kawamura, Sadao; Miyazaki, Fumio; Arimoto, Suguru. (1988). *Realization of Robot Motion Based on a Learning Method*. IEEE Transactions on Systems, Man and Cybernetics. Volume 18. pp 126-134.

[Kawano, I 68*] SEE: Yajima, S 68a. *On Autonomous Logic Nets of Threshold Elements*.

[Kawato, M 87a] Kawato, Mitsuo; Uno, Yoji; Isobe, Michiaki; Suzuki, Ryoji. (June 1987). *A Hierarchical Model for Voluntary Movement and Its Application to Robotics*. IEEE International Conference on Neural Networks (ICNN-87), San Diego, CA, June 21-24, 1987: Proceedings. Volume 4. pp 573-582. New York: Institute of Electrical and Electronics Engineers.

[Kawato, M 87b] Kawato, Mitsuo; Furukawa, Kazunori; Suzuki, Ryoji. (1987). *A Hierarchical Neural-Network Model for*

Control and Learning of Voluntary Movement. Biological Cybernetics. Volume 57. pp 169-185.

[Kawato, M 88a] Kawato, Mitsuo; Uno, Yoji; Isobe, Michiaki; Suzuki, Ryoji. (1988). *Hierarchical Neural Network Model for Voluntary Movement with Application to Robotics*. IEEE Control Systems Magazine. Volume 8. pp 8-16.

[Kaye, AR 73a] Kaye, A. R.; George, D. A. (1973). *Relationships Between LMS Compensators for Intersymbol Interference in Linear and Non-Linear Modulation Schemes*. IEEE Transactions on Information Theory. Volume IT-19. p 244.

[Kazantsev, VS 87*] SEE: Mazurov, VD 87a. *Solving of Optimization and Identification Problems by the Committee Methods*.

[Kazmierczak, H 63a] Kazmierczak, H.; Steinbuch, K. (1963). *Adaptive Systems in Pattern Recognition*. IEEE Transactions on Electronic Computers. Volume EC-12. pp 822-835.

[Keele, SW 68*] SEE: Posner, MI 68a. *On the Genesis of Abstract Ideas*.

[Keele, SW 70*] SEE: Posner, MI 70a. *Retention of Abstract Ideas*.

[Keeler, JD 86a] Keeler, James D. (1986). *Basins of Attraction of Neural Network Models*. In Denker, John S. (ed.). Neural Networks for Computing, Snowbird, Utah, 1986. AIP Proceedings 151. pp 259-264. New York: American Institute of Physics.

[Keeler, JD 87a] Keeler, James D. (1987). *Comparison of Information Capacity of Hebbian Neural Networks*. IEEE International Conference on Neural Networks (ICNN-87), San Diego, CA, June 21-24, 1987: Proceedings. Volume 3. pp 253-260. New York: Institute of Electrical and Electronics Engineers.

[Keeler, JD 87b] Keeler, James D. *Information Capacity of Outer-Product Neural Networks*. Physics Letters A. Volume 124. pp 53-58. New York: Elsevier.

[Keeler, JD 88a] Keeler, James D. (1988). *Capacity for Patterns and Sequences in Kanerva's SDM as Compared to Other Associative Memory Models*. In Anderson, Dana Z. (ed.). Neural Information Processing Systems. pp 412-421. New York: American Institute of Physics.

[Keeler, JD 88b] Keeler, James D. (1988). *Comparison Between Kanerva's SDM and Hopfield-type Neural Networks*. Cognitive Science. Volume 12. pp 299-329.

[Keener, JP 81a] Keener, James P.; Hoppensteadt, F. C.; Rinzel, John. (1981). *Integrate-And-Fire Models of Nerve Membrane Response to Oscillatory Input*. SIAM Journal on Applied Mathematics. Volume 41. pp 503-517.

[Keener, JP 83a] Keener, James P. (1983). *Analog Circuitry for the van der Pol and FitzHugh-Nagumo Equations*. IEEE Transactions on Systems, Man and Cybernetics. Volume SMC-13. pp 1010-1014.

[Kegl, J 87*] SEE: Hanson, SJ 87a. *PARSNIP: A Connectionist Network that Learns Natural Language Grammar from Exposure to Natural Language Sentences*.

[Kehoe, EJ 86a] Kehoe, E. James. (1986). *A Layered Network Model for Learning-to-Learn and Configuration in Classical Conditioning*. Cognitive Science Society, 8th, Amherst, MA, August 15-17, 1986: Proceedings. pp 154-175. Hillsdale, NJ: Lawrence Erlbaum.

[Keller, EL 85a] Keller, E. L. (ed.); Zee, D. S. (ed.). (1985). Asilomar Conference on Adaptive Processes in Visual and Oculomoter Systems, June 16-20, 1985: Proceedings. New York: Pergamon.

[Keller, HB 61a] Keller, Herbert B. (1961). *Finite Automata, Pattern Recognition and Perceptrons*. Journal of the Association for Computing Machinery. Volume 8. pp 1-20.

[Keller, JM 85a] Keller, James M.; Hunt, Douglas J. (1985). *Incorporating Fuzzy Membership Functions into the Perceptron*

Algorithm. IEEE Transactions on Pattern Analysis and Machine Intelligence. Volume PAMI-7. pp 693-699.

[Keller, R 87a] Keller, R.; Prater, M. (1987). *A Context-Driven World Model Appropriate for Training and Testing of Vision-Based Learning Networks*. IEEE International Conference Neural Networks (ICNN-87), San Diego, CA, June 21-24, 1987: Proceedings. Volume 4. pp 757-770. New York: Institute of Electrical and Electronics Engineers.

[Kellum, C 87a] Kellum, Charles. (1987). *An Adaptive Reference Monitor for Information System Security*. IEEE International Conference on Neural Networks (ICNN-87), San Diego, CA, June 21-24, 1987: Proceedings. Volume 4. pp 667-676. New York: Institute of Electrical and Electronics Engineers.

[Kelly, EJ 65a] Kelly, E. J.; Wishner, R. P. (1965). *Matched-Filter Theory for High-Velocity, Accelerating Targets*. IEEE Transactions on Military Electronics. Volume MIL-9. pp 56-69.

[Kelso, JAS 86*] SEE: Schoner, G 86a. *A Stochastic Theory of Phase Transitions in Human Hand Movement*.

[Kelso, JAS 88*] SEE: Schoner, G 88d. *Dynamic Pattern Generation in Behavioral and Neural Systems*.

[Kelso, JAS 88*] SEE: Schoner, G 88a. *A Dynamic Pattern Theory of Behavioral Change*.

[Kelso, JAS 88*] SEE: Schoner, G 88e. *Dynamic Patterns of Biological Coordination: Theoretical Strategy and New Results*.

[Kelso, JAS 88*] SEE: Schoner, G 88b. *A Synergetic Theory of Environmentally-Specified and Learned Patterns of Movement Coordination, I. Relative Phase Dynamics*.

[Kelso, JAS 88*] SEE: Schoner, G 88c. *A Synergetic Theory of Environmentally-Specified and Learned Patterns of Movement Coordination, II. Component Oscillator Dynamics*.

[Kemke, C 87a] Kemke, Christel. (1987). *Modelling Neural Networks by Means of Networks of Finite Automata*. IEEE International Conference on Neural Networks (ICNN-87), San Diego, CA, June 21-24, 1987: Proceedings. Volume 3. pp 23-30. New York: Institute of Electrical and Electronics Engineers.

[Kemke, C 88a] Kemke, Christel. (1988). *Der Neuere Konnektionismus, Ein Uberblick*. Informatik Spektrum. Volume 11. pp 143-162.

[Kemke, C 88b] Kemke, Christel. (1988). *Representing Neural Networks Models by Finite Automata*. ESPCI European Conference on Neural Networks (nEuro'88), 1st, Paris, June 6-9, 1988: Proceedings.

[Kennedy, MP 87a] Kennedy, Michael Peter; Chua, Leon O. (1987). *Circuit Theoretic Solutions for Neural Networks - An Old Approach to a New Problem*. IEEE International Conference on Neural Networks (ICNN-87), San Diego, CA, June 21-24, 1987: Proceedings. Volume 2. pp 169-176. New York: Institute of Electrical and Electronics Engineers.

[Kennedy, MP 87b] Kennedy, Michael Peter; Chua, Leon O. (1987). *Unifying the Tank and Hopfield Linear Programming Circuit and the Canonical Nonlinear Programming Circuit of Chua and Lin*. IEEE Transactions on Circuits and Systems. Volume CAS-34. pp 210-214.

[Kennedy, MP 88a] Kennedy, Michael Peter; Chua, Leon O. (1988). *Neural Networks for Nonlinear Programming*. IEEE Transactions on Circuits and Systems. Volume 35. pp 554-562.

[Kenny, SB 83*] SEE: Rosenbaum, DA 83a. *Hierarchical Control of Rapid Movement Sequences*.

[Kent, EW 78a] Kent, Ernest W. (January, 1978). *The Brains of Men and Machines, Part 1: Biological Models for Robotics*. BYTE. Volume 3. pp 11-106.

[Kent, EW 78b] Kent, Ernest W. (February, 1978). *The Brains of Men and Machines, Part 2: How the Brain Controls Outputs*. BYTE. Volume 3. pp 84-158.

[Kent, EW 78c] Kent, Ernest W. (March, 1978). *The Brains of Men and Machines, Part 3: How the Brain Analyzes Input.* BYTE. Volume 3. pp 74-108.

[Kent, EW 78d] Kent, Ernest W. (1978). *The Brains of Men and Machines, Part 4: The Machinery of Emotion and Choice.* BYTE. Volume 3. pp 66-89.

[Kent, RD 77a] Kent, R. D.; Minifie, F. D. (1977). *Coarticulation in Recent Speech Production Models.* Journal of Phonetics. Volume 5. pp 115-133.

[Kergosien, YL 86a] Kergosien, Y. L. (1986). *Local Versus Global Minima, Hysteresis, Multiple Meanings.* In Bienenstock, Elie L. (ed.). Disordered Systems and Biological Organization. pp 333-336. New York: Springer-Verlag.

[Kernighan, BW 73*] SEE: Lin, S 73a. *An Effective Heuristic Algorithm for the Traveling Salesman Problem.*

[Kerola, T 87*] SEE: Bagherzadeh, N 87a. *On Parallel Execution of the Traveling Salesman Problem on a Neural Network Model.*

[Kersten, D 87a] Kersten, Daniel; O'Toole, Alice J.; Sereno, Margaret E.; Knill, David C.; Anderson, James A. (1987). *Associative Learning of Scene Parameters from Images.* Applied Optics. Volume 26. pp 4999-5006.

[Kerszberg, M 85*] SEE: Huberman, BA 85b. *Ultradiffusion: The Relaxation of Hierarchical Systems.*

[Kerszberg, M 86*] SEE: Bergman, A 86a. *The Evolution of Data Processing Abilities in Competing Automata.*

[Kerszberg, M 86a] Kerszberg, Michel. (1986). *Emergence of Hierarchical Data Structures in Parallel Computation.* In Bienenstock, Elie L. (ed.). Disordered Systems and Biological Organization. pp 205-208. New York: Springer-Verlag.

[Kerszberg, M 87*] SEE: Bergman, A 87a. *Breeding Intelligent Automata.*

[Kesten, H 58a] Kesten, H. (1958). *Accelerated Stochastic Approximation.* Annals of Mathematical Statistics. Volume 29. pp 41-59.

[Khanna, SK 86*] SEE: Moopenn, A 86a. *Error Correction and Asymmetry in a Binary Memory Matrix.*

[Khanna, SK 87*] SEE: Thakoor, AP 87a. *Binary Synaptic Connections Based on Memory Switching in a-Si:H for Artificial Neural Networks.*

[Khanna, SK 87*] SEE: Thakoor, AP 87b. *Electronic Hardware Implementations of Neural Networks.*

[Khanna, SK 87*] SEE: Moopenn, A 87a. *A Neurocomputer Based on an Analog-Digital Hybrid Architecture.*

[Khanna, SK 87*] SEE: Lamb, JL 87a. *Resistive Synaptic Interconnects for Electronic Neural Networks.*

[Khanna, SK 88*] SEE: Moopenn, A 88a. *Programmable Synaptic Chip for Electronic Neural Networks.*

[Kiang, NYS 68*] SEE: Goldstein, JL 68a. *Neural Correlates of the Aural Combination Tone 2f(1)-f(2).*

[Kibler, D 83*] SEE: Hampson, S 83a. *A Boolean Complete Neural Model of Adaptive Behavior.*

[Kiech, EL 88*] SEE: Dietz, WE 88a. *Pattern-Based Fault Diagnosis Using Neural Networks.*

[Kiech, EL 89*] SEE: Dietz, WE 89a. *Neural Network Models Applied to Real-Time Fault Diagnosis.*

[Kienker, PK 86*] SEE: Sejnowski, TJ 86b. *Learning Symmetry Groups with Hidden Units: Beyond the Perceptron.*

[Kienker, PK 86a] Kienker, Paul K.; Sejnowski, Terrence J. (1986). *Separating Figure from Ground with a Parallel Network.* Perception. Volume 15. pp 197-216.

[Killion, TH 77*] SEE: Becker, CA 77a. *Interaction of Visual and Cognitive Effects in Word Recognition.*

[Kilmer, W 74a] Kilmer, William; Olinski, Melanie. (1974). *Model of a Plausible Learning Scheme for CA3 Hippocampus.* Kybernetik. Volume 16. pp 133-144.

[Kilmer, WL 69a] Kilmer, W. L.; McCulloch, Warren S.; Blum, J. (1969). *A Model of the Vertebrate Central Command System.* In Chaplin, G. B. B. (ed.); Gaines, Brian R. (ed.); Gedye, J. L. (ed.). International Journal of Man-Machine Studies. Volume 1. pp 279-309.

[Kilmer, WL 76*] SEE: Arbib, MA 76a. *Neural Models and Memory.*

[Kilmer, WL 78*] SEE: Morse, A 78a. *Neural Net Capable of Competitive and Cooperative Computation.*

[Kim, DO 75*] SEE: Pfeiffer, RR 75a. *Cochlear Nerve Filter Responses: Distribution Along the Cochlear Partition.*

[Kim, J 78a] Kim, J.; Mueller, C. W. (1978). Introduction to Factor Analysis: What It Is and How to Do It. pp 7-13. Beverly Hills, CA: Sage Publications.

[Kimball, O 85*] SEE: Schwartz, RM 85a. *Context-Dependent Modeling for Acoustic-Phonetic Recognition of Continuous Speech.*

[Kimme, C 75a] Kimme, C.; Sklansky, Jack; Ballard, Dana H. (1975). *Finding Circles by an Array of Accumulators.* Communications of the ACM. Volume 18. pp 120-122.

[Kimura, Y 88*] SEE: Shimohara, K 88a. *A Neural Network System with an Automatic Generation Mechanism for Distorted Patterns.*

[Kindrick, J 87*] SEE: Parunak, HVD 87a. *Supplier-Consumer Nets for Material Handling.*

[King, DJ 88*] SEE: Gallant, SI 88d. *Experiments with Sequential Associative Memories.*

[King, HT 86*] SEE: Annaratone, M 86a. *Warp Architecture and Implementation.*

[Kinoshita, J 87a] Kinoshita, June; Palevsky, Nicholas G. (May 1987). *Computing with Neural Networks.* High Technology.

[Kinoshita, J 89a] Kinoshita, June. (January, 1989). *Memory in a Neuron.* Scientific American. pp 28.

[Kinouchi, Y 87*] SEE: Nagashino, H 87a. *Self-Organizing Process of a Neural Network Model for Pattern Detection.*

[Kinser, JM 88a] Kinser, Jason M.; Caulfield, H. John; Shamir, Joseph. (1988). *Design for a Massive All-Optical Bidirectional Associative Memory: the Big BAM.* Applied Optics. Volume 27. pp 3442-3444.

[Kirby, KG 86a] Kirby, Kevin G.; Conrad, Michael. (1986). *Intraneuronal Dynamics as a Substrate for Evolutionary Learning.* Physica. Volume 22D. pp 205-215.

[Kirby, KG 88*] SEE: Conrad, M 88b. *Neuronal Dynamics and Evolutionary Learning.*

[Kirby, KG 88a] Kirby, Kevin G.; Conrad, Michael. (1988). *Bit-Vector Optimization Algorithms for Control of Learning in Neurons with Second-Messenger Dynamics.* IEEE International Conference on Neural Networks (ICNN-88), San Diego, CA, July 24-27, 1988: Proceedings. Volume 2. pp 55-62. New York: Institute of Electrical and Electronics Engineers.

[Kirkham, J 71*] SEE: Clema, JK 71a. *CONSIM (Conflict Simulator): Risk, Cost and Benefit in Political Simulations.*

[Kirkpatrick, S 78a] Kirkpatrick, Scott; Sherrington, David. (1978). *Infinite-Ranged Models of Spin-Glasses.* Physical Review B, Solid State. Volume 17. pp 4384-4403.

[Kirkpatrick, S 83a] Kirkpatrick, Scott; Gelatt, C. D.; Vecchi, Mario P. (1988). *Optimization by Simulated Annealing.* In Anderson, James A. (ed.); Rosenfeld, Edward (ed.). Neurocomputing: Foundations of Research. pp 554-568. Cambridge, MA: MIT Press.

[Kirkpatrick, S 83b] Kirkpatrick, Scott; Gelatt, C. D.; Vecchi, Mario P. (1983). *Optimization by Simulated Annealing.* Science. Volume 220. pp 671-680.

[Kishimoto, K 79a] Kishimoto, K.; Amari, Shun-ichi. (1979). *Existence and Stability of Local Excitations in Neural Fields.* Journal of Mathematical Biosciences. Volume 7. pp 303-318.

[Kitagawa, M 87*] SEE: Fukaya, M 87a. *Two-Level Neural Networks: Learning by Interaction with Environment.*

[Kitahashi, T 76*] SEE: Nagami, H 76a. *Characterization of Dynamical Behavior Associated with a Single Neural Equation.*

[Kitamura, S 88a] Kitamura, S.; Kurematsu, Y.; Nakai, Y. (1988). *Application of the Neural Network for the Trajectory Planning of a Biped Locomotive Robot.* Neural Networks. Volume 1, Supplement 1. pp 344.

[Kittler, J 80*] SEE: Devijver, PA 80a. On the Nearest Neighbor Rule.

[Kittler, J 82*] SEE: Devijver, PA 82a. Pattern Recognition: A Statistical Approach.

[Kittler, J 86a] Kittler, Josef. (1986). *Feature Selection and Extraction.* In Young, Tzay Y. (ed.); Fu, King-Sun (ed.). Handbook of Pattern Recognition and Image Processing. pp 60-84. New York: Academic Press.

[Klafter, J 86a] Klafter, Joseph; Shlesinger, Michael F. (1986). *On the Relationship Among Three Theories of Relaxation in Disordered Systems.* Proceedings of the National Academy of Sciences. Volume 83. pp 848-851.

[Klassen, M 88a] Klassen, Myungsook; Pao, Yoh Han; Chen, Victor. (1988). *Characteristics of the Functional Link Net: A Higher Order Delta Rule Net.* IEEE International Conference on Neural Networks (ICNN-88), San Diego, CA, July 24-27, 1988: Proceedings.

[Klatt, DH 80a] Klatt, Dennis H. (1980). *Scriber and LAFS: Two New Approaches to Speech Analysis.* In Lea, W. A. (ed.). Trends in Speech Recognition. Englewood Cliffs, NJ: Prentice-Hall.

[Kleene, SC 56a] Kleene, S. C. (1956). *Representation of Events in Nerve Nets and Finite Automata.* Automata Studies. pp 3-41. Princeton, NJ: Princeton University Press.

[Klein, RM 80*] SEE: Jusczyk, PW 80a. The Nature of Thought: Essays in Honor of D.O. Hebb.

[Klement, EP 81a] Klement, E. P. (1981). *Operations on Fuzzy Sets and Fuzzy Numbers Related to Triangular Norms.* International Symposium on Multiple-Valued Logic, 11th, Oklahoma City/Norman, OK, May 27-29, 1981: Proceedings. pp 218-225. New York: Institute of Electrical and Electronics Engineers.

[Klopf, AH 69a] Klopf, A. Harry; Gose, Earl E. (1969). *An Evolutionary Pattern Recognition Network.* IEEE Transactions on Systems Man and Cybernetics. Volume SSC-5. pp 247-250.

[Klopf, AH 72a] Klopf, A. Harry. (1972). Brain Function and Adaptive Systems: A Heterostatic Theory. AFCRL-72-0164, AD-742259. Bedford, MA: Air Force Cambridge Research Laboratories.

[Klopf, AH 79a] Klopf, A. Harry. (1979). *Goal-Seeking Systems from Goal-Seeking Components: Implications for AI.* Cognitive Brain Theory Newsletter. Number 3. p 2.

[Klopf, AH 82a] Klopf, A. Harry. (1982). The Hedonistic Neuron: A Theory of Memory Learning and Intelligence.

Washington, DC: Hemisphere.

[Klopf, AH 86a] Klopf, A. Harry. (1986). *Drive-Reinforcement Model of Single Neuron Function: An Alternate to the Hebbian Neuronal Model*. In Denker, John S. (ed.). Neural Networks for Computing, Snowbird, Utah, 1986. AIP Proceedings 151. pp 265-270. New York: American Institute of Physics.

[Klopf, AH 87a] Klopf, A. Harry. (1987). A Neuronal Model of Classical Conditioning. Interim Report for 10/79-9/87. Wright-Patterson Air Force Base, OH: AL/AFWAL.

[Klopf, AH 87b] Klopf, A. Harry. (1987). *Drive-Reinforcement Learning: A Real-Time Learning Mechanism for Unsupervised Learning*. IEEE International Conference on Neural Networks (ICNN-87), San Diego, CA, June 21-24, 1987: Proceedings. Volume 2. pp 441-446. New York: Institute of Electrical and Electronics Engineers.

[Klopf, AH 88a] Klopf, A. Harry. (1988). *A Neuronal Model of Classical Conditioning*. Psychobiology. Volume 16. pp 85-125.

[Knapp, AG 84a] Knapp, Andrew G.; Anderson, James A. (1984). *A Signal Averaging Model for Concept Formation*. Journal of Experimental Psychology: Learning, Memory, and Cognition. Volume 10. pp 616-637.

[Knapp, AG 84b] Knapp, Andrew G.; Anderson, James A. (1988). *Theory of Categorization Based on Distributed Memory Storage*. In Anderson, James A. (ed.); Rosenfeld, Edward (ed.). Neurocomputing: Foundations of Research. pp 588-610. Cambridge, MA: MIT Press.

[Knapp, AG 84c] Knapp, Andrew G.; Anderson, James A. (1984). *Theory of Categorization Based on Distributed Memory Storage*. Journal of Experimental Psychology: Learning, Memory, and Cognition. Volume 10. pp 616-637.

[Knight, BW 62*] SEE: Block, HD 62a. *Analysis of a Four-Layer Series-Coupled Perceptron, II*.

[Knight, BW 78*] SEE: Brodie, SE 78a. *The Response of the Limulus Retina to Moving Stimuli: A Prediction by Fourier Synthesis*.

[Knill, DC 87*] SEE: Kersten, D 87a. *Associative Learning of Scene Parameters from Images*.

[Knisley, JR 87*] SEE: Dress, WB 87a. *A Darwinian Approach to Artificial Neural Systems*.

[Knoll, TF 86a] Knoll, T. F.; Jain, R. C. (1986). *Recognizing Partially Visible Objects Using Feature Indexed Hypotheses*. IEEE Journal of Robotics and Automation. Volume RA-2. pp 3-13.

[Koch, C 81*] SEE: Poggio, TA 81a. Microelectronics in Nerve Cells: Dendritic Morphology and Information Processing.

[Koch, C 83a] Koch, Christof; Poggio, Tomaso A.; Torre, Vincent. (1983). *Nonlinear Interactions in a Dendritic Tree: Localization, Timing, and Role in Information Processing*. Proceedings of the National Academy of Sciences. Volume 80. pp 2799-2802.

[Koch, C 84*] SEE: Poggio, TA 84a. An Analog Model of Computation for the Ill-Posed Problems of Early Vision.

[Koch, C 84a] Koch, Christof; Poggio, Tomaso A. (1984). Biophysics of Computation: Neurons, Synapses and Membranes. AI Memo #795. Cambridge, MA: Massachusetts Institute of Technology.

[Koch, C 85*] SEE: Poggio, TA 85a. *Computational Vision and Regularization Theory*.

[Koch, C 85a] Koch, Christof; Marroquin, Jose Luis; Yuille, Alan L. (1985). Analog Neuronal Networks in Early Vision. AI Memo #751. Cambridge, MA: Massachusetts Institute of Technology.

[Koch, C 86*] SEE: Hildreth, E 86a. The Analysis of Visual Motion: From Computational Theory to Neuronal Mechanisms.

[Koch, C 86*] SEE: Hutchinson, JM 86a. *Simple Analog and Hybrid Networks for Surface Interpolation.*

[Koch, C 86a] Koch, Christof; Marroquin, Jose Luis; Yuille, Alan L. (1986). *Analog Neuronal Networks in Early Vision.* Proceedings of the National Academy of Sciences. Volume 83. pp 4263-4267.

[Koch, C 87*] SEE: Grzywacz, N 87a. *Functional Properties of Models for Direction Selectivity in the Retina.*

[Koch, C 87*] SEE: Ferster, D 87a. *Neuronal Connections Underlying Orientation Selectivity in Cat Visual Cortex.*

[Koch, C 87*] SEE: Poggio, TA 87a. *Synapses that Compute Motion.*

[Koch, C 87b] Koch, Christof. (1987). *The Action of the Corticofugal Pathway on Sensory Thalamic Nuclei: A Hypothesis.* Neuroscience. Volume 23. pp 399-406.

[Koch, C 88*] SEE: Sejnowski, TJ 88a. *Computational Neuroscience.*

[Koch, C 88*] SEE: Hutchinson, J 88a. *Computing Motion Using Analog and Binary Resistive Networks.*

[Koch, C 88a] Koch, Christof. (1988). *Computing Motion in the Presence of Discontinuities: Algorithm and Analog Networks.* In Eckmiller, Rolf (ed.); von der Malsburg, Christoph (ed.). Neural Computers. NATO ASI Series, Volume F41. pp 101-110. Berlin: Springer-Verlag.

[Koch, C 88b] Koch, Christof; Luo, Jin; Mead, Carver A.; Hutchinson, James. (1988). *Computing Motion Using Resistive Networks.* In Anderson, Dana Z. (ed.). Neural Information Processing Systems. pp 422-431. New York: American Institute of Physics.

[Koenderink, JJ 79a] Koenderink, Jan J.; van Doorn, Andrea J. (1979). *The Structure of Two-Dimensional Scalar Fields with Applications to Vision.* Biological Cybernetics. Volume 33. pp 151-158.

[Koenderink, JJ 86a] Koenderink, Jan J.; van Doorn, Andrea J. (1986). *Depth and Shape from Differential Perspective in the Presence of Bending Deformations.* Journal of the Optical Society of America A. Volume 3. pp 242-249.

[Koenderink, JJ 86b] Koenderink, Jan J.; van Doorn, Andrea J. (1986). *Dynamic Shape.* Biological Cybernetics. Volume 53. pp 383-396.

[Koenderink, JJ 86c] Koenderink, Jan J. (1986). *Optic Flow.* Vision Research. Volume 26. pp 161-180.

[Koenderink, JJ 87*] SEE: Toet, A 87a. *The Construction of a Simultaneous Functional Order in Nervous Systems I. Relevance of Signal Covariances and Signal Coincidences.*

[Koenderink, JJ 87*] SEE: Toet, A 87b. *The Construction of a Simultaneous Functional Order in Nervous Systems II. Computing Geometrical Structures.*

[Koenderink, JJ 87*] SEE: Toet, A 87c. *The Construction of a Simultaneous Functional Order in Nervous Systems III. The Influence of Environmental Constraints on the Resulting Functional Order.*

[Koenderink, JJ 87*] SEE: Richards, WA 87a. *Inferring Three-Dimensional Shapes from Two-Dimensional Silhouettes.*

[Koenderink, JJ 87*] SEE: van de Grind, WA 87a. *Influence of Contrast on Foveal and Peripheral Detection of Coherent Motion in Moving Random-Dot Patterns.*

[Koenderink, JJ 87a] Koenderink, Jan J. (1987). *An International Representation for Solid Shape Based on the Topological Properties of the Apparent Contour.* Image Understanding 1985-86. pp 257-285. Norwood, NJ: Ablex.

[Koenderink, JJ 87b] Koenderink, Jan J. (1987). *Color Atlas Theory.* Journal of the Optical Society of America A. Volume 4.

pp 1314-1321.

[Koenderink, JJ 87c] Koenderink, Jan J.; van Doorn, Andrea J. (1987). *Facts on Optic Flow.* Biological Cybernetics. Volume 56. pp 247-254.

[Koenderink, JJ 87d] Koenderink, Jan J.; van Doorn, Andrea J. (1987). *Representation of Local Geometry in the Visual System.* Biological Cybernetics. Volume 44. pp 367-375.

[Koenderink, JJ 88a] Koenderink, Jan J. (1988). *Design Principles for a Front-end Visual System.* In Eckmiller, Rolf (ed.); von der Malsburg, Christoph (ed.). Neural Computers. NATO ASI Series, Volume F41. pp 111-118. Berlin: Springer-Verlag.

[Koenderink, JJ 88b] Koenderink, Jan J. (1988). *Operational Significance of Receptive Field Assemblies.* Biological Cybernetics. Volume 58. pp 163-171.

[Koenderink, JJ 88c] Koenderink, Jan J. (1988). *Scale-Time.* Biological Cybernetics. Volume 58. pp 159-162.

[Koenderink, JJ 88d] Koenderink, Jan J.; Richards, Whitman. (1988). *Two-Dimensional Curvature Operators.* Journal of the Optical Society of America A. Volume 5. pp 1136-1141.

[Koford, JS 63*] SEE: Talbert, LR 63a. A Real-Time Adaptive Speech-Recognition System.

[Koford, JS 66a] Koford, J. S.; Groner, G. F. (1966). *The Use of an Adaptive Threshold Element to Design a Linear Optimal Pattern Classifier.* IEEE Transactions on Information Theory. Volume IT-12. pp 42-50.

[Koga, K 80a] Koga, Kazutoshi; Hoshimiya, Nozomu; Matsuo, Tadayuki. (1980). *A Hardware Neuron Model as a Component of the Neural Network.* Transactions of the Institute of Electronics and Communication Engineers of Japan. Volume E63. pp 70-71.

[Kohler, W 51a] Kohler, Wolfgang. (1951). *Discussion: Why the Mind Is in the Head.* Cerebral Mechanisms in Behavior, The Hixon Symposium. pp 57-83. New York: Wiley.

[Kohler, W 51b] Kohler, Wolfgang. (1951). *Relational Determination in Perception.* Cerebral Mechanisms in Behavior, The Hixon Symposium. pp 200-243. New York: Wiley.

[Kohli, CK 80*] SEE: Ajmera, RC 80a. *Retinal-Type Neuristor Sections.*

[Kohonen, T 72a] Kohonen, Teuvo. (1972). *Correlation Matrix Memories.* IEEE Transactions on Computers. Volume C-21. pp 353-359.

[Kohonen, T 72b] Kohonen, Teuvo. (1988). *Correlation Matrix Memories.* In Anderson, James A. (ed.); Rosenfeld, Edward (ed.). Neurocomputing: Foundations of Research. pp 174-180. Cambridge, MA: MIT Press.

[Kohonen, T 73a] Kohonen, Teuvo. (1973). *A New Model for Randomly Organized Associative Memory.* International Journal of Neuroscience. Volume 5. pp 27-29.

[Kohonen, T 73b] Kohonen, Teuvo; Ruohonen, Matti. (1973). *Representation of Associated Data by Matrix Operators.* IEEE Transactions on Computers. Volume C-22. pp 701-702.

[Kohonen, T 74a] Kohonen, Teuvo. (1974). *An Adaptive Associative Memory Principle.* IEEE Transactions on Computers. Volume C-23. pp 444-445.

[Kohonen, T 74b] Kohonen, Teuvo. (1974). *On the Implementation of Logic and Tables by Linear Networks.* International Symposium on Multiple-Valued Logic, Morgantown, WV, May 19-31, 1974: Proceedings. pp 493-504.

[Kohonen, T 75a] Kohonen, Teuvo. (1975). Where Does the Fundamental Information Processing Ability of Biological Systems

Come From?. Technology Report TKK-F-A238. Helsinki: Helsinki University of Technology.

[Kohonen, T 76a] Kohonen, Teuvo; Reuhkala, Erkki; Makisara, Kai; Vainio, L. (1976). *Associative Recall of Images*. Biological Cybernetics. Volume 22. pp 159-168.

[Kohonen, T 76b] Kohonen, Teuvo; Oja, Erkki. (1976). *Fast Adaptive Formation of Orthogonalizing and Associative Memory in Recurrent Networks of Neuron-Like Elements*. Biological Cybernetics. Volume 21. pp 85-95.

[Kohonen, T 77a] Kohonen, Teuvo. (1977). Associative Memory: A System-Theoretic Approach. New York: Springer-Verlag.

[Kohonen, T 78*] SEE: Haltsonen, S 78a. *Application of Novelty Filter to Segmentation of Speech.*

[Kohonen, T 78*] SEE: Jalanko, M 78a. *Application of Orthogonal Projection Principles to Simultaneous Phonemic Segmentation and Labeling of Continuous Speech.*

[Kohonen, T 78*] SEE: Jalanko, M 78b. *Application of the Subspace Method to Speech Recognition.*

[Kohonen, T 78a] Kohonen, Teuvo; Reuhkala, Erkki. (1978). *A Very Fast Associative Method for the Recognition and Correction of Misspelt Words, Based on Redundant Hash Addressing*. International Joint Conference on Pattern Recognition, 4th, Kyoto, November 7-10, 1978: Proceedings. pp 807-809. New York: Institute of Electrical and Electronics Engineers.

[Kohonen, T 79*] SEE: Reuhkala, E 79a. *A Redundant Hash Addressing Method Adapted for the Postprocessing and Error-Correction of Computer-Recognized Speech.*

[Kohonen, T 79a] Kohonen, Teuvo; Nemeth, Gabor; Bry, Kalle-J; Jalanko, Matti; Riittinen, Heikki. (1979). *Spectral Classification of Phonemes by Learning Subspaces*. IEEE International Conference on Acoustics, Speech, and Signal Processing (ICASSP-79), Washington, DC, April 2-4, 1979: Proceedings. pp 97-100. New York: Institute of Electrical and Electronics Engineers.

[Kohonen, T 80a] Kohonen, Teuvo; Riittinen, H.; Jalanko, Matti; Reuhkala, Erkki; Haltsonen, Seppo. (1980). *A Thousand-Word Recognition System Based on the Learning Subspace Method and Redundant Hash Addressing*. International Conference on Pattern Recognition, 5th, Miami Beach, FL, December 1-4, 1980: Proceedings. Volume 1. pp 158-165. New York: Institute of Electrical and Electronics Engineers.

[Kohonen, T 81a] Kohonen, Teuvo. (1981). *Automatic Formation of Topological Maps of Patterns in a Self-Organizing System*. In Oja, Erkki (ed.); Simula, O. (ed.). Scandinavian Conference on Image Analysis, 2nd, Helsinki, June 15-17, 1981: Proceedings. pp 214-220. Espoo: Pattern Recognition Society of Finland.

[Kohonen, T 81b] Kohonen, Teuvo. (1984). *Self-Organized Formation of Feature Maps*. Cybernetic Systems: Recogntion, Learning, Self-Organisation. pp 3-12. Letchworth: Research Studies Press.

[Kohonen, T 81c] Kohonen, Teuvo; Oja, Erkki; Lehtio, Pekka. (1981). *Storage and Processing of Information in Distributed Associative Memory Systems*. In Hinton, Geoffrey E. (ed.). Parallel Models of Associative Memory. pp 105-144. Hillsdale, NJ: Lawrence Erlbaum.

[Kohonen, T 82a] Kohonen, Teuvo. (1982). *A Simple Paradigm for the Self-Organized Formation of Structured Feature Maps*. In Amari, Shun-ichi (ed.); Arbib, Michael A. (ed.). Competition and Cooperation in Neural Nets. pp 248-266. New York: Springer-Verlag.

[Kohonen, T 82b] Kohonen, Teuvo. (1982). *Analysis of a Simple Self-Organizing Process*. Biological Cybernetics. Volume 44. pp 135-140.

[Kohonen, T 82c] Kohonen, Teuvo. (1982). *Clustering, Taxonomy, and Topological Maps of Patterns*. International Conference on Pattern Recognition, 6th, Munich, October 19-22, 1982: Proceedings. pp 114-128.

[Kohonen, T 82d] Kohonen, Teuvo. (1982). *Clustering, Taxonomy, and Topological Maps of Patterns.* International Conference on Pattern Recognition, 6th, Munich, October 19-22, 1982: Proceedings. pp 114-128. Silver Spring, MD: Institute of Electrical and Electronics Engineers.

[Kohonen, T 82e] Kohonen, Teuvo. (1982). *Self-Organized Formation of Topologically Correct Feature Maps.* Biological Cybernetics. Volume 43. pp 59-69.

[Kohonen, T 82f] Kohonen, Teuvo. (1988). *Self-organized Formation of Topologically Correct Feature Maps.* In Anderson, James A. (ed.); Rosenfeld, Edward (ed.). Neurocomputing: Foundations of Research. pp 511-522. Cambridge, MA: MIT Press.

[Kohonen, T 84a] Kohonen, Teuvo; Rittinen, Heikki; Reuhkala, Erkki; Haltsonen, Seppo. (1984). *On-Line Recognition of Spoken Words from a Large Vocabulary.* Information Sciences. Volume 33. pp 3-30.

[Kohonen, T 84b] Kohonen, Teuvo; Makisara, Kai; Saramaki, Tapio. (1984). *Phonotopic Maps - Insightful Representation of Phonological Features for Speech Recognition.* International Conference on Pattern Recognition, 7th, Montreal, July 30-August 2, 1984: Proceedings. Volume 1. pp 182-185. Silver Spring, MD: Institute of Electrical and Electronics Engineers.

[Kohonen, T 84c] Kohonen, Teuvo. (1984). Self Organization and Associative Memory. New York: Springer-Verlag.

[Kohonen, T 84d] Kohonen, Teuvo. (1984). *Self-Organizing Feature Maps and Abstractions.* Artificial Intelligence and Information-Control Systems of Robots, Proceedings of the Third International Conference, Czechoslovakia, June 11-15, 1984. pp 39-45. Amsterdam: Elsevier.

[Kohonen, T 85a] Kohonen, Teuvo; Oja, Erkki; Lehtio, Pekka. (1985). *A Model of Cortical Associative Memory.* In Levy, William B. (ed.); Anderson, James A. (ed.); Lehmkuhle, Stephen (ed.). Synaptic Modifications, Neuron Selectivity, and Nervous System Organization. pp 209-222. Hillsdale, NJ: Lawrence Erlbaum.

[Kohonen, T 86*] SEE: Torkkola, K 86a. *Microprocessor-Based Word Recognizer for a Large Vocabulary.*

[Kohonen, T 86*] SEE: Venta, O 86a. *A Non-Stochastic Method for the Correction of Sentences.*

[Kohonen, T 86a] Kohonen, Teuvo. (1986). *Dynamically Expanding Context, with Application to the Correction of Symbol Strings in the Recognition of Continuous Speech.* International Conference on Pattern Recognition, 8th, Paris, October 27-31, 1986: Proceedings. pp 1148-1151. Washington, DC: Institute of Electrical and Electronics Engineers.

[Kohonen, T 86b] Kohonen, Teuvo. (1986). Learning Vector Quantization for Pattern Recognition. TKK-F-A601. Helsinki: Helsinki University of Technology.

[Kohonen, T 86c] Kohonen, Teuvo; Makisara, Kai. (1986). *Representation of Sensory Information in Self-Organizing Feature Maps.* In Denker, John S. (ed.). Neural Networks for Computing, Snowbird, Utah, 1986. AIP Proceedings 151. pp 271-276. New York: American Institute of Physics.

[Kohonen, T 86d] Kohonen, Teuvo. (1986). *Representation of Sensory Information in Self-Organizing Feature Maps, and Relation of these Maps to Distributed Memory Networks.* Hybrid and Optical Computing. SPIE Volume 634. Bellingham, WA: Society of Photo-Optical Instrumentation Engineers.

[Kohonen, T 87a] Kohonen, Teuvo. (1987). *Adaptive, Associative, and Self-Organizing Functions in Neural Computing.* Applied Optics. Volume 26. pp 4910-4918.

[Kohonen, T 87b] Kohonen, Teuvo; Oja, Erkki. (1987). *Computing with Neural Networks.* Science. Volume 235. pp 1227.

[Kohonen, T 87c] Kohonen, Teuvo. (1987). Content-Addressable Memories. (2nd ed). Berlin: Springer-Verlag.

[Kohonen, T 87d] Kohonen, Teuvo. (1987). *Dynamically Expanding Context*. Journal of Intelligent Systems. Volume 1. pp 79-95.

[Kohonen, T 87e] Kohonen, Teuvo; Torkkola, Kari; Shozakai, Makoto; Kangas, Jari; Venta, Olli. (1987). *Microprocessor Implementation of a Large Vocabulary Speech Recognizer and Phonetic Typewriter for Finnish and Japanese*. European Conference on Speech Technology, Edinburgh, September 1987. Volume 2. pp 377-380.

[Kohonen, T 87f] Kohonen, Teuvo. (1987). *Self-Learning Rules by Dynamically Expanding Context*. IEEE International Conference on Neural Networks (ICNN-87), San Diego, CA, June 21-24, 1987: Proceedings. Volume 2. pp 3-10. New York: Institute of Electrical and Electronics Engineers.

[Kohonen, T 87g] Kohonen, Teuvo. (1987). *State of the Art in Neural Computing*. IEEE International Conference on Neural Networks (ICNN-87), San Diego, CA, June 21-24, 1987: Proceedings. Volume 1. pp 77-90. New York: Institute of Electrical and Electronics Engineers.

[Kohonen, T 87h] Kohonen, Teuvo; Oja, Erkki. (1987). *Technical Comments: Computing with Neural Networks*. Science. Volume 235. p 1227.

[Kohonen, T 88*] SEE: Venta, O 88a. *A Content-Addressing Software Method for the Emulation of Neural Networks*.

[Kohonen, T 88a] Kohonen, Teuvo. (1988). *An Introduction to Neural Computing*. Neural Networks. Volume 1. pp 3-16.

[Kohonen, T 88b] Kohonen, Teuvo; Torkkola, Kari; Shozakai, Makoto; Kangas, Jari; Venta, Olli. (1988). *Phonetic Typewriter for Finnish and Japanese*. IEEE International Conference on Acoustics, Speech, and Signal Processing (ICASSP-88), New York, NY, April 11-14, 1988: Proceedings. pp 607-610. New York: Institute of Electrical and Electronics Engineers.

[Kohonen, T 88c] Kohonen, Teuvo. (1988). Self-Organization and Associative Memory, Second Edition. Berlin: Springer-Verlag.

[Kohonen, T 88d] Kohonen, Teuvo. (March, 1988). *The "Neural" Phonetic Typewriter*. Computer. Volume 21. pp 11-22.

[Kohonen, T 88e] Kohonen, Teuvo. (1988). *The Role of Adaptive and Associative Circuits in Future Computer Designs*. In Eckmiller, Rolf (ed.); von der Malsburg, Christoph (ed.). Neural Computers. NATO ASI Series, Volume F41. pp 1-8. Berlin: Springer-Verlag.

[Kohout, L 80*] SEE: Bandler, W 80a. *Fuzzy Power Sets and Fuzzy Implication Operators*.

[Koi, H 84*] SEE: Fujita, S 84a. *Application of GMDH to Environmental System Modeling and Management*.

[Koizumi, O 86a] Koizumi, Osamu; Bode, Hans R. (1986). *Plasticity in the Nervous System of Adult Hydra I. The Position-Dependent Expression of FMRFamide-Like Immunoreactivity*. Developmental Biology. Volume 116. pp 407-421.

[Koizumi, O 88*] SEE: Bode, PM 88a. *Develpment of the Two-Part Pattern During Regeneration of the Head in Hydra*.

[Koizumi, O 88a] Koizumi, Osamu; Heimfeld, Shelly; Bode, Hans R. (1988). *Plasticity in the Nervous System of Adult Hydra II. Conversion of Ganglion Cells of the Body Column into Epidermal Sensory Cells of the Hypostome*. Developmental Biology. Volume 129. pp 358-371.

[Kokkinidis, M 84*] SEE: Anninos, PA 84b. *A Neural Net Model for Multiple Memory Domains*.

[Kokkinidis, M 84*] SEE: Anninos, PA 84c. *Noisy Neural Nets Exhibiting Memory Domains*.

[Kokkinidis, M 85a] Kokkinidis, Michael; Anninos, Photios A. (1985). *Noisy Neural Nets Exhibiting Epileptic Features*.

Journal of Theoretical Biology. Volume 113. pp 559-588.

[Kokkinidis, M 87*] SEE: Anninos, PA 87a. *A Mathematical Formalism for the Finite Difference Equation in Neural Nets with Chemical Markers.*

[Kolen, JF 88a] Kolen, John F. (1988). *Faster Learning Through a Probabilistic Approximation Algorithm.* International Conference on Neural Networks (ICNN-88), San Diego, CA, July 24-27, 1988: Proceedings.

[Kolen, JF 89*] SEE: Goel, A 89a. *Learning in Connectionist Networks: Has the Credit Assignment Problem Been Solved?.*

[Kolesar, E 87*] SEE: Kabrisky, M 87a. *A Multiplexed Multi-Electrode Semiconductor Brain Electrode Implant.*

[Kollias, S 88*] SEE: Anastassiou, D 88a. *Digital Image Halftoning Using Neural Networks.*

[Kollias, S 88a] Kollias, Stefanos; Anastassiou, Dimitris. (1988). *Adaptive Training of Multilayer Neural Networks Using a Least Squares Estimation Technique.* IEEE International Conference on Neural Networks (ICNN-88), San Diego, CA, July 24-27, 1988: Proceedings. Volume 1. pp 383-390. New York: Institute of Electrical and Electronics Engineers.

[Kollias, S 89a] Kollias, Stefanos; Tsai, Tu-Chih; Anastassiou, Dimitris. ((to appear)). *Image Halftoning and Reconstruction Using a Neural Network.* IEEE International Conference on Acoustics, Speech and Signal Processing (ICASSP-89), Glasgow, May 23-26, 1989: Proceedings.

[Kolodner, JL 87*] SEE: Lancaster, JS 87a. *Problem Solving in a Natural Task as a Function of Experience.*

[Kolodzy, PJ 87a] Kolodzy, Paul J. (1987). *Multidimensional Machine Vision Using Neural Networks.* IEEE International Conference on Neural Networks (ICNN-87), San Diego, CA, June 21-24, 1987: Proceedings. Volume 2. pp 747-758. New York: Institute of Electrical and Electronics Engineers.

[Kondo, T 84*] SEE: Tamura, H 84a. *On Revised Algorithms of GMDH With Applications.*

[Konheim, AG 63a] Konheim, Alan G. (1963). *A Geometric Convergence Theorem for the Perceptron.* Journal of the Society for Industrial and Applied Mathematics. Volume 11. pp 1-14.

[Kontoravdis, D 88*] SEE: Stafylopatis, A 88a. *Spatial Organization of Neural Networks: A Probabilistic Modeling Approach.*

[Kopec, GE 84a] Kopec, Gary E. (1984). *The Integrated Signal Processing System ISP.* IEEE Transactions on Acoustics, Speech, and Signal Processing. Volume ASSP-32. pp 842-851.

[Korn, AF 88a] Korn, Axel F. (1988). *Towards a Primal Sketch of Real World Scenes in Early Vision.* In Eckmiller, Rolf (ed.); von der Malsburg, Christoph (ed.). Neural Computers. NATO ASI Series, Volume F41. pp 119-128. Berlin: Springer-Verlag.

[Korn, H 87a] Korn, Henri; Burnod, Yves; Faber, Donald S. (1987). *Spontaneous Quantal Currents in a Central Neuron Match Predictions from Binomial Analysis of Evoked Responses.* Proceedings of the National Academy of Sciences. Volume 84. pp 5981-5985.

[Korpel, A 82a] Korpel, A. (1982). *Gabor: Frequency, Time and Memory.* Applied Optics. Volume 21. pp 3624-3632.

[Kosko, B 85a] Kosko, Bart; Limm, John S. (1985). *Vision as Causal Activation and Association.* Intelligent Robots and Computer Vision. SPIE Volume 579. pp 104-109. Bellingham, WA: Society of Photo-Optical Instrumentation Engineers.

[Kosko, B 86a] Kosko, Bart. (1986). *Counting with Fuzzy Sets.* IEEE Transactions on Pattern Analysis and Machine

Intelligence. Volume PAMI-8. pp 556-557.

[Kosko, B 86b] Kosko, Bart. (1986). *Differential Hebbian Learning.* In Denker, John S. (ed.). Neural Networks for Computing, Snowbird, Utah, 1986. AIP Proceedings 151. pp 277-282. New York: American Institute of Physics.

[Kosko, B 86c] Kosko, Bart. (1986). *Fuzzy Cognitive Maps.* International Journal on Man-Machine Studies. Volume 24. pp 65-75.

[Kosko, B 86d] Kosko, Bart. (1986). *Fuzzy Entropy and Conditioning.* Information Sciences. Volume 40. pp 165-174.

[Kosko, B 86e] Kosko, Bart. (1986). *Fuzzy Knowledge Combination.* International Journal of Intelligent Systems. Volume 1. pp 293-320.

[Kosko, B 87a] Kosko, Bart. (1987). *Adaptive Bidirectional Associative Memories.* Applied Optics. Volume 26. pp 4947-4960.

[Kosko, B 87b] Kosko, Bart. (1987). *Adaptive Inference in Fuzzy Knowledge Networks.* IEEE International Conference on Neural Networks (ICNN-87), San Diego, CA, June 21-24, 1987: Proceedings. Volume 2. pp 261-268. New York: Institute of Electrical and Electronics Engineers.

[Kosko, B 87c] Kosko, Bart. (1987). *Associative Memory.* IEEE International Conference on Neural Networks (ICNN-87), San Diego, CA, June 21-24, 1987: Proceedings. Tutorial 5. San Diego: Institute of Electrical and Electronics Engineers.

[Kosko, B 87d] Kosko, Bart. (1987). *Competitive Adaptive Bidirectional Associative Memories.* IEEE International Conference on Neural Networks (ICNN-87), San Diego, CA, June 21-24, 1987: Proceedings. Volume 2. pp 759-766. New York: Institute of Electrical and Electronics Engineers.

[Kosko, B 87e] Kosko, Bart. (1987). *Constructing an Associative Memory.* BYTE. Volume 12. pp 137-144.

[Kosko, B 87f] Kosko, Bart. (in press). *Fuzziness Versus Probability.* Statistical Science.

[Kosko, B 87g] Kosko, Bart. (1987). *Fuzzy Associative Memories.* Fuzzy Expert Systems. Reading, MA: Addison-Wesley.

[Kosko, B 87h] Kosko, Bart. (in press). Hidden Patterns in Combined Knowledge Networks. San Diego, CA: VERAC, Inc.

[Kosko, B 87i] Kosko, Bart; Guest, Clark C. (1987). *Optical Bi-directional Associative Memories.* Image Understanding. SPIE Volume 758. Bellingham, WA: Society of Photo-Optical Instrumentation Engineers.

[Kostyuk, PG 68a] Kostyuk, P. G.; Vasilenko, D. A. (1968). *Transformation of Cortical Motor Signals in Spinal Cord.* Proceedings of the IEEE. Volume 56. pp 1049-1058.

[Kotliar, BE 83a] Kotliar, Boris E.; Weinberger, Norman M. (1983). Neural Mechanism of Conditioning. Elmsford, NY: Pergamon.

[Kottas, J 86*] SEE: Warde, C 86a. *Hybrid Optical Inference Machines: Architectural Considerations.*

[Koutsougeras, C 88a] Koutsougeras, C.; Papachristou, C. A. (1988). *Learning Discrete Mappings - Athena's Approach.* Languages for Automation. pp 31-36. New York: Institute of Electrical and Electronics Engineers.

[Koutsougeras, C 88b] Koutsougeras, C.; Papachristou, C. A. (1988). *Training of a Neural Network for Pattern Classification Based on an Entropy Measure.* IEEE International Conference on Neural Networks (ICNN-88), San Diego, CA, July 24-27, 1988: Proceedings. Volume 1. pp 247-254. New York: Institute of Electrical and Electronics Engineers.

[Kowalski, JM 88a] Kowalski, J. M.; Ansari, Assad; Dawes, R.; Gross, G. W.; Prueitt, P. (1988). *Coupled Rotators as a*

Model for Spontaneous Activity in Neural Networks. Midcon/88, Dallas, TX, August 30-September 1, 1988: Conference Record.

[Kraft, DH 81*] SEE: Buell, DA 81a. *A Model for a Weighted Retrieval System.*

[Kraft III, LG 87*] SEE: Miller, WT 87b. *Application of a General Learning Algorithm to the Control of Robotic Manipulators.*

[Krasner, M 84*] SEE: Schwartz, RM 84a. *Improved Hidden Markov Modeling of Phonemes for Continuous Speech Recognition.*

[Krasner, M 85*] SEE: Schwartz, RM 85a. *Context-Dependent Modeling for Acoustic-Phonetic Recognition of Continuous Speech.*

[Kraul, C 87a] Kraul, Chris. (January 20, 1987). *Computer Firm is Closing in on Dream Machine.* San Diego Union.

[Kravitz, R 86*] SEE: Hammerstrom, D 86b. A Design Methodology for High-Performance, General-Purpose VLSI.

[Kretz, H 60*] SEE: Zemanek, H 60a. *A Model for Neurophysiological Functions.*

[Krieger, D 88a] Krieger, D.; Solomon, J.; Levitan, S.; Berger, T.; Barrioneuvo, G.; Sclabassi, Robert J. (1988). *A Neurobiologically Based Neural Net Model of the Hippocampus.* Modeling and Simulation Conference, 19th, Pittsburgh, PA, May 5-6, 1988: Proceedings. Volume 19. pp 2397-2402. Research Triangle Park, NC: Instrument Society of America.

[Krieger, DN 88*] SEE: Sclabassi, RJ 88a. *A Systems Theoretic Approach to the Study of CNS Function.*

[Kriener, B 77*] SEE: Bunge, E 77b. *The AUROS Project: Automatic Recognition of Speakers by Computers.*

[Krile, TF 86a] Krile, Thomas F.; Walkup, John F. (1986). *Optical Systems Research at Texas Tech University.* In Szu, Harold H. (ed.). Optical and Hybrid Computing. SPIE Volume 634. pp 457-474. Bellingham, WA: Society of Photo-Optical Instrumentation Engineers.

[Krishnan, G 87*] SEE: Walters, DKW 87c. *Color Separation Using General Purpose Computer Vision Algorithms.*

[Krishnan, G 87*] SEE: Walters, DKW 87d. *Perceptual Significance Hierarchy: A Computer Vision Theory for Color Separation.*

[Krishnan, G 88a] Krishnan, Ganapathy; Walters, Deborah K. W. (1988). *Segmenting Intersecting and Incomplete Boundaries.* Applications of Artificial Intelligence. SPIE Volume 937. pp 550-556. Bellingham, WA: Society of Photo-Optical Instrumentation Engineers.

[Krishnaprasad, PS 88*] SEE: Pati, YC 88a. *Neural Networks for Tactile Perception.*

[Kristan, WB 80a] Kristan, W. B. (1980). Information Processing in the Nervous System. pp 241-261. New York: Raven.

[Krivonogov, AI 87*] SEE: Mazurov, VD 87a. *Solving of Optimization and Identification Problems by the Committee Methods.*

[Krogh, A 87*] SEE: Cortes, C 87a. *Hierarchical Associative Networks.*

[Krogh, A 88a] Krogh, A.; Hertz, J. A. (1988). *Mean-field Analysis of Hierarchical Associative Networks with Magnetisation.* Journal of Physics A: Mathematical and General. Volume 21. pp 2211-2224.

[Krohn, K 65a] Krohn, Kenneth; Rhodes, John. (1965). *Nets of Threshold Elements.* Information and Control. Volume 8. pp 579-588.

[Krone, G 86a] Krone, G.; Mallot, Hanspeter A.; Palm, Gunther; Schuz, A. (1986). *Spatiotemporal Receptive Fields: A*

Dynamical Model Derived from Cortical Architectonics. Proceedings of the Royal Society of London Series B. Volume 226. pp 421-444.

[Krowitz, A 87a] Krowitz, A.; Rendell, L.; Hohensee, B. (1987). *The State Space of Memory Clusters in the Hopfield Network*. IEEE International Conference on Neural Networks (ICNN-87), San Diego, CA, June 21-24, 1987: Proceedings. Volume 3. pp 339-346. New York: Institute of Electrical and Electronics Engineers.

[Kruskal, CP 83*] SEE: Gottlieb, A 83a. *The NYU Ultracomputer-Designing an MIMD Shared Memory Parallel Computer*.

[Kubala, F 87*] SEE: Schwartz, RM 87a. *Rapid Speaker Adaptation Using a Probabilistic Spectral Mapping*.

[Kube, P 86a] Kube, Paul. (1986). *Using Frequency-and Orientation-Tuned Channels to Determine Surface Slant*. Cognitive Science Society, 8th, Amherst, MA, August 15-17, 1986: Proceedings. pp 235-244. Hillsdale, NJ: Lawrence Erlbaum.

[Kubota, M 88*] SEE: Alkon, DL 88b. *Regulation of Hermissenda K+ Channels by Cytoplasmic and Membrane-Associated C-Kinase*.

[Kuczewski, RM 87*] SEE: Myers, MH 87a. Application of New Artificial Neural System Information Processing Principles to Pattern Classification: Final Report, US Army Research Office.

[Kuczewski, RM 87a] Kuczewski, Robert M.; Myers, Michael H.; Crawford, William J. (1987). Exploration of Backward Error Propagation as a Self-Organizational Structure. San Diego, CA: TRW Mead.

[Kuczewski, RM 87b] Kuczewski, Robert M.; Myers, Michael H.; Crawford, William J. (1987). *Exploration of Backward Error Propagation as a Self-Organized Structure*. IEEE International Conference on Neural Networks (ICNN-87), San Diego, CA, June 21-24, 1987: Proceedings. Volume 2. pp 89-96. New York: Institute of Electrical and Electronics Engineers.

[Kuczewski, RM 87c] Kuczewski, Robert M. (1987). *Neural Network Approaches to Multi-Target Tracking*. IEEE International Conference on Neural Networks (ICNN-87), San Diego, CA, JUne 21-24, 1987: Proceedings. Volume 4. pp 619-634. New York: Institute of Electrical and Electronics Engineers.

[Kuczewski, RM 87d] Kuczewski, Robert M. (1987). Neural Network Approaches to Multi-Target Tracking. San Diego, CA: TRW Mead.

[Kuczewski, RM 87e] Kuczewski, Robert M.; Myers, Michael H.; Crawford, William J. (1987). *Neurocomputer Workstations and Processors: Approaches and Applications*. IEEE International Conference on Neural Networks (ICNN-87), San Diego, CA, June 21-24, 1987: Proceedings. Volume 3. pp 487-500. New York: Institute of Electrical and Electronics Engineers.

[Kuczewski, RM 87f] Kuczewski, Robert M.; Myers, Michael H.; Crawford, William J. (1987). Neurocomputer Workstations and Processors: Approaches and Applications. San Diego, CA: TRW Mead.

[Kudrycki, TP 88a] Kudrycki, Tomasz P. (1988). Neural Network Implementation of a Medical Diagnosis Expert System. Master's Thesis. Cincinnati, OH: University of Cincinnati, Electrical and Computer Engineering.

[Kuh, A 88a] Kuh, Anthony; Dickinson, Bradley W. (in press). *Information Capacity of Associative Memories*. IEEE Transactions on Information Theory.

[Kuh, A 88b] Kuh, Anthony. (1988). *Performance Measures for Associative Memories that Learn and Forget*. In Anderson, Dana Z. (ed.). Neural Information Processing Systems. pp 432-441. New York: American Institute of Physics.

[Kuklinski, T 88*] SEE: Ward, JR 88a. *A Model for Variability Effects in Hand-Printing, with Implications for the Design of On Line Character Recognition Systems*.

[Kumar, BVKV 86*] SEE: Montgomery, BL 86a. *Evaluation of the Use of the Hopfield Neural Network Model as a Nearest-Neighbor Algorithm.*

[Kumar, BVKV 87*] SEE: Lemmon, M 87a. *Neural Networks as Associative Memories.*

[Kumar, BVKV 88*] SEE: Lemmon, M 88a. *Emulating the Dynamics for a Class of Laterally Inhibited Neural Networks.*

[Kung, HT 84a] Kung, H. T. (1984). *Systolic Algorithms for the CMU Warp Processor.* International Conference on Pattern Recognition, 7th, Montreal, July 30-August 2, 1984: Proceedings. Volume 1. pp 570-577. Silver Spring, MD: Institute of Electrical and Electronics Engineers.

[Kung, HT 85*] SEE: Gross, T 85a. *Warp as a Machine for Low-Level Vision.*

[Kung, HT 85a] Kung, H. T.; Webb, Jon A. (1986). Mapping Image Processing Operations onto a Linear Systolic Machine. CMU-CS-86-137. pp 1-23. Pittsburgh, PA: Carnegie Mellon University, Computer Science Department.

[Kung, HT 85b] Kung, H. T. (1985). *Memory Requirements for Balanced Computer Architectures.* Journal of Complexity. Volume 1. pp 147-157.

[Kung, S 84*] SEE: Goodman, JW 84a. *Optical Interconnections for VLSI Systems.*

[Kung, S 88*] SEE: Vlontzos, JA 88a. *A Hierarchical System for Character Recognition with Stochastic Knowledge Representation.*

[Kung, S 88a] Kung, Sun-Yuan; Hwang, J. N. (1988). *An Algebraic Projection Analysis for Back-Propagation Learning.* IEEE International Conference on Neural Networks (ICNN-88), San Diego, CA, July 24-27, 1988: Proceedings. Volume 1. pp 363-370. New York: Institute of Electrical and Electronics Engineers.

[Kung, S 88b] Kung, Sun-Yuan; Hwang, J. N. (1988). *Digital VLSI Architectures for Artificial Neural Nets.* IEEE International Conference on Neural Networks (ICNN-88), San Diego, CA, July 24-27, 1988: Proceedings. Volume 2. pp 165-172. New York: Institute of Electrical and Electronics Engineers.

[Kung, S 89*] SEE: Hwang, JN 89a. *A Unifying Viewpoint of Multilayer Perceptrons and Hidden Markov Models.*

[Kung, S 89a] Kung, Sun-Yuan; Hwang, J. N. (in press). *Neural Net Architectures for Robotic Applications.* IEEE Journal of Robotics and Automation.

[Kuperstein, M 81a] Kuperstein, Michael; Whittington, Douglas A. (1981). *A Practical 24 Channel Microelectrode for Neural Recording in Vivo.* IEEE Transactions on Biomedical Engineering. Volume BME-28. pp 288-293.

[Kuperstein, M 86*] SEE: Grossberg, S 86h. Neural Dynamics of Adaptive Sensory-Motor Control: Ballistic Eye Movements.

[Kurata, K 81*] SEE: Nagano, T 81a. *A Self-Organizing Neural Network Model for the Development of Complex Cells.*

[Kurato, K 80*] SEE: Nagano, T 80a. *A Model of the Complex Cell Based on Recent Neurophysiological Findings.*

[Kurematsu, Y 88*] SEE: Kitamura, S 88a. *Application of the Neural Network for the Trajectory Planning of a Biped Locomotive Robot.*

[Kurogi, S 87a] Kurogi, Shuichi. (1987). *A Model of Neural Network for Spatiotemporal Pattern Recognition.* Biological Cybernetics. Volume 57. pp 103-114.

[Kurogi, S 88a] Kurogi, Shuichi. (1988). *Abilities and Limitations of a Neural Network Model for Spoken Word Recognition.* IEEE International Conference on Neural Networks (ICNN-88), San Diego, CA, July 24-27, 1988: Proceedings. Volume 2. pp 205-214. New York: Institute of Electrical and Electronics Engineers.

[Kurten, KE 87a] Kurten, K. E. (1987). *Phase Transitions in Quasirandom Neural Networks.* IEEE International Conference on Neural Networks (ICNN-87), San Diego, CA, June 21-24, 1987: Proceedings. Volume 2. pp 197-204. New York: Institute of Electrical and Electronics Engineers.

[Kuruvilla, A 87*] SEE: Honrubia, V 87a. *Morphological Aspects of the Vestibular Nerve of the Squirrel Monkey.*

[Kurz, B 76*] SEE: Stuller, JA 76a. *Two-Dimensional Markov Representations of Sampled Images.*

[Kushner, BG 86*] SEE: Athale, RA 86a. *Attentive Associative Architectures and Their Implications to Optical Computing.*

[Kussul, EM 82*] SEE: Gol'tsev, AD 82a. *Controlling the Motion of a Moving Robot Using a Neuroid Network.*

[Kwong, SK 86*] SEE: Yariv, A 86a. *Associative Memories Based on Message-Bearing Optical Modes in Phase-Conjugate Resonator.*

[Labos, E 86a] Labos, E. (1986). *Convergence, Cycling or Strange Motion in the Adaptive Synthesis of Neurons.* IIASA Conference on Modelling and Adaptive Control, Sopron, Hungary, July, 1986: Proceedings. Volume 105. pp 214-225. New York: Springer-Verlag.

[Labos, E 87a] Labos, E. (1987). *The Most Complicated Networks of Formal Neurons.* IEEE International Conference on Neural Networks (ICNN-87), San Diego, CA, June 21-24, 1987: Proceedings. Volume 3. pp 301-308. New York: Institute of Electrical and Electronics Engineers.

[Lachs, G 83a] Lachs, Gerard; Saia, Rosalie A.; Teich, Malvin C. (1983). *A Neural-Counting Model Based on Physiological Characteristics of the Peripheral Auditory System, Part IV, Application to Response.* IEEE Transactions on Systems, Man, and Cybernetics. Volume SMC-13. pp 964-972.

[Lachter, J 88a] Lachter, Joel; Bever, Thomas G. (1988). *The Relation Between Linguistic Structure and Associative Theories of Language Learning: A Constructive Critique of Some Connectionist Learning Models.* In Pinker, Steven (ed.); Mehler, Jacques (ed.). Connections and Symbols. pp 195-248. Cambridge, MA: MIT Press.

[Lacolle, B 86a] Lacolle, B. (1986). *Symbolic Computation Methods for Some Spin Glasses Problems.* In Bienenstock, Elie L. (ed.). Disordered Systems and Biological Organization. pp 133-148. New York: Springer-Verlag.

[Laden, B 87*] SEE: Lundell, J 87a. *A Dynamic Connectionist Model of Problem Solving.*

[LaGamma, EF 87*] SEE: Black, IB 87a. *Biochemistry of Information Storage in the Nervous System.*

[Lainiotis, DG 69*] SEE: Hilborn, CG 69a. *Unsupervised Learning Minimum Risk Pattern Classification for Dependent Hypotheses and Dependent Measurements.*

[Laird, JE 84a] Laird, John E. (1984). *Universal Sub-goaling.* CMU-CS-84-128. Pittsburgh, PA: Carnegie Mellon University, Computer Science Department.

[Lakshmivarahan, S 77*] SEE: Narendra, KS 77a. *Learning Automata: A Critique.*

[Lakshmivarahan, S 79a] Lakshmivarahan, S. (1979). e-Optimal Learning Algorithms: Non-Absorbing Barrier Type. Report EECS-7901. Norman, OK: University of Oklahoma, Department of Electrical Engineering.

[Lakshmivarahan, S 81a] Lakshmivarahan, S. (1981). Learning Algorithms and Applications. New York: Springer-Verlag.

[Lam, MS 85*] SEE: Gross, T 85a. *Warp as a Machine for Low-Level Vision.*

[Lam, MS 86*] SEE: Annaratone, M 86a. *Warp Architecture and Implementation.*

[Lam, T 88*] SEE: Ross, MD 88a. *Macular Bioaccelerometers on Earth and in Space.*

[Lam, T 88*] SEE: Ross, MD 88c. *The Neuroanatomical Substrate for Information Processing in Macular Endorgans.*

[Lamb, JL 86*] SEE: Thakoor, AP 86a. *Binary Synaptic Connections Based on Memory Switching in a SI:H.*

[Lamb, JL 87*] SEE: Thakoor, AP 87a. *Binary Synaptic Connections Based on Memory Switching in a-Si:H for Artificial Neural Networks.*

[Lamb, JL 87*] SEE: Daud, T 87a. *Neural Network Based Feed-Forward High Density Associative Memory.*

[Lamb, JL 87a] Lamb, J. L.; Thakoor, Anilkumar P.; Moopenn, Alexander; Khanna, S. K. (1987). *Resistive Synaptic Interconnects for Electronic Neural Networks.* Journal of Vacuum Science Technology. Volume 5. pp 1407-1411.

[Lambe, J 85a] Lambe, John; Moopenn, Alexander; Thakoor, Anilkumar P. (1985). Electronic Device Aspects of Neural Network Memories. NASA / CIT.

[Lambe, J 86*] SEE: Thakoor, AP 86a. *Binary Synaptic Connections Based on Memory Switching in a SI:H.*

[Lambe, J 86*] SEE: Moopenn, A 86a. *Error Correction and Asymmetry in a Binary Memory Matrix.*

[Lambe, J 87*] SEE: Thakoor, AP 87b. *Electronic Hardware Implementations of Neural Networks.*

[Lambe, J 87*] SEE: Moopenn, A 87b. *Electronic Implementation of Associative Memory Based on Neural Network Models.*

[Lancaster, JS 87a] Lancaster, Juliana S.; Kolodner, Janet L. (1987). *Problem Solving in a Natural Task as a Function of Experience.* Cognitive Science Society, 9th, Seattle, WA, July 16-18, 1987: Proceedings. pp 727-736. Hillsdale, NJ: Lawrence Erlbaum.

[Land, EH 77a] Land, Edwin H. (1977). *The Retinex Theory of Color Vision.* Scientific American. Volume 237. pp 108-128.

[Landa, J 87*] SEE: Szu, HH 87b. *Future Perspectives on Neural Network Research.*

[Landa, J 87a] Landa, Joseph; Scheff, Kim; Szu, Harold N. (1987). *Binocular Fusion Using Simulated Annealing.* IEEE International Conference on Neural Networks (ICNN-87), San Diego, CA, June 21-24, 1987: Proceedings. Volume 4. pp 327-334. New York: Institute of Electrical and Electronics Engineers.

[Landahl, HD 43a] Landahl, H. D.; McCulloch, Warren S.; Pitts, Walter H. (1943). *A Statistical Consequence of the Logical Calculus of Nervous Nets.* Bulletin of Mathematical Biophysics. Volume 5. pp 135-137.

[Landau, HJ 84*] SEE: Wyner, AD 84a. *Optimum Waveform Signal Sets with Amplitude and Energy Constraints.*

[Landauer, TK 87a] Landauer, T. K.; Kamm, C. A.; Singhal, S. (1987). *Teaching a Minimally Structured Back-Propagation Network to Recognize Speech Sounds.* Cognitive Science Society, 9th, Seattle, WA, July 16-18, 1987: Proceedings. pp 531-536. Hillsdale, NJ: Lawrence Erlbaum.

[Lando, JF 88*] SEE: Sanes, JN 88a. *Rapid Reorganization of Adult Rat Motor Cortex Somatic Representation Patterns after Motor Nerve Injury.*

[Lane, SH 88*] SEE: Handelman, DA 88a. *Integration of Knowledge-Based System and Neural Network Techniques for Robotic Control.*

[Lane, SH 88a] Lane, Stephen H.; Handelman, David A.; Gelfand, Jack J. (1988). *A Neural Network Computational Map Approach to Reflexive Motor Control.* IEEE Symposium on Intelligent Control, 3rd, Arlington, VA, August, 1988: Proceedings. New York: Institute of Electrical and Electronics Engineers.

[Lang, KJ 85*] SEE: Hinton, GE 85b. *Shape Recognition and Illusory Conjunctions.*

[Langenbacher, H 88*] SEE: Moopenn, A 88a. *Programmable Synaptic Chip for Electronic Neural Networks.*

[Langhofer, L 87*] SEE: Elidan, J 87a. *The Neural Generators of the Vestibular Evoked Response.*

[Langhofer, L 89*] SEE: Elidan, J 89a. *The Firing Properties of Second-Order Vestibular Neurons in Correlation with the Far-Field Recorded Vestibular-Evoked Response.*

[Langley, P 82*] SEE: Sleeman, D 82a. *Learning from Solution Paths: An Approach to the Credit Assignment Problem.*

[Lansner, A 83a] Lansner, Anders; Ekeberg, Oerjan. (1983). A Program Package Implementing a System of Interacting Adaptive Associative Networks. Report TRITA-NA-8302. Stockholm, Sweden: Royal Institute of Technology, Department of Numerical Analysis.

[Lansner, A 85a] Lansner, Anders. (1985). Prototype Extraction and Matching in Associative Nets. Report TRITA-NA-8516. Stockholm, Sweden: Royal Institute of Technology, Department of Numerical Analysis.

[Lansner, A 85b] Lansner, Anders; Ekeberg, Oerjan. (1985). *Reliability and Speed of Recall in an Associative Network.* IEEE Transactions on Pattern Analysis and Machine Intelligence. Volume PAMI-7. pp 490-498.

[Lansner, A 86a] Lansner, Anders. (1986). Investigations into the Pattern Processing Capabilities of Associative Nets. TRITA-NA-8601. Stockholm: Royal Institute of Technology, Numerical Analysis Department.

[Lansner, A 87a] Lansner, Anders; Ekeberg, Oerjan. (1987). *An Associative Network Solving the "4-Bit ADDER Problem".* IEEE International Conference on Neural Networks (ICNN-87), San Diego, CA, June 21-24, 1987: Proceedings. Volume 2. pp 549-556. New York: Institute of Electrical and Electronics Engineers.

[Lansner, A 88*] SEE: Grillner, S 88a. *Simulation of the Segmental Burst Generating Network for Locomotion in Lamprey.*

[Lapedes, A 86a] Lapedes, Alan; Farber, Robert. (1986). *Programming a Massively Parallel, Computation Universal System: Static Behavior.* In Denker, John S. (ed.). Neural Networks for Computing, Snowbird, Utah, 1986. AIP Proceedings 151. pp 283-298. New York: American Institute of Physics.

[Lapedes, A 87a] Lapedes, Alan; Farber, Robert. (1987). Nonlinear Signal Processing Using Neural Networks: Prediction and System Modelling. Preprint LA-UR-87-2662. Los Alamos, NM: Los Alamos National Laboratory.

[Lapedes, A 88a] Lapedes, Alan; Farber, Robert. (1988). *How Neural Networks Work.* In Anderson, Dana Z. (ed.). Neural Information Processing Systems. pp 442-456. New York: American Institute of Physics.

[Lapinski, RH 77*] SEE: Tweedy, JR 77a. *Semantic-Context Effects on Word Recognition: Influence of Varying the Proportion of Items Presented in an Appropriate Context.*

[Larimore, MG 76*] SEE: Widrow, B 76b. *Stationary and Nonstationary Learning Characteristics of the LMS Adaptive Filter.*

[Larochelle, S 80a] Larochelle, Serge; McClelland, James L.; Rodriguez, Elizabeth. (1980). *Context and the Allocation of Resources in Word Recognition.* Journal of Experimental Psychology: Human Perception and Performance. Volume 6. pp 686-694.

[Larsen, H 78*] SEE: Rush, S 78a. *A Practical Algorithm for Solving Dynamic Membrane Equations.*

[Larson, J 89*] SEE: Lynch, G 89a. *Cortical Encoding of Memory: Hypotheses Derived from Analysis and Simulation of Physiological Learning Rules in Anatomical Structures.*

[Lashley, KS 50a] Lashley, K. S. (1950). *In Search of the Engram.* Symposium of the Society for Experimental Biology. Volume 4. pp 454-482.

[Lashley, KS 50b] Lashley, K. S. (1988). *In Search of the Engram.* In Anderson, James A. (ed.); Rosenfeld, Edward (ed.).

Neurocomputing: Foundations of Research. pp 59-64. Cambridge, MA: MIT Press.

[Lau, CGY 87a] Lau, Clifford G. Y.; Honrubia, Vicente. (1987). *Fast Component Threshold for Vestibular Nystagmus in the Rabbit*. Journal of Comparative Physiology A. Volume 160. pp 585-592.

[Lau, CGY 87b] Lau, Clifford G. Y. (1987). *Research Challenges and Opportunities in Electronic Neural Systems.* IEEE International Conference on Neural Networks (ICNN-87), San Diego, CA, June 21-24, 1987: Proceedings. Volume 3. pp 201-204. New York: Institute of Electrical and Electronics Engineers.

[Lau, CGY 88a] Lau, Clifford G. Y.; Honrubia, Vicente. (1988). *Distributed Neural Information Processing in the Vestibulo-Ocular System.* In Anderson, Dana Z. (ed.). Neural Information Processing Systems. pp 457-466. New York: American Institute of Physics.

[Laughery, KR 69a] Laughery, Kenneth R. (1969). *Computer Simulation of Short-Term Memory: A Component Decay Model.* The Psychology of Learning and Motivation. Volume 3. pp 135-200.

[Lauria, FE 87*] SEE: Noos, FA 87a. *An Assembler for a Neural Network.*

[Law, PR 63*] SEE: Talbert, LR 63a. A Real-Time Adaptive Speech-Recognition System.

[Lawler, EL 86a] Lawler, E. L.; Lenstra, J. K.; Rinnooy Kan, A. H. G.; Shmoys, D. B. (1986). The Traveling Salesman Problem. New York: Wiley.

[Lazzaro, J 86*] SEE: Mueller, P 86a. *Machine for Neural Computation of Acoustical Patterns with Application to Real Time Speech Recognition.*

[Lazzaro, JP 86a] Lazzaro, John Paul. (1986). anaLOG: A Functional Simulator for VLSI Neural Systems. Report 5229:TR:86. Pasadena, CA: California Institute of Technology, Computer Science Department.

[Le Cun, Y 86a] Le Cun, Y. (1986). *Learning Process in an Asymmetric Threshold Network.* In Bienenstock, Elie L. (ed.). Disordered Systems and Biological Organization. pp 233-240. New York: Springer-Verlag.

[Le Cun, Y 87*] SEE: Soulie, FF 87a. *Evaluation of Network Architectures on Test Learning Tasks.*

[Le Texier, J 88*] SEE: Angeniol, B 88a. *Self-Organizing Feature Maps and the Travelling Salesman Problem.*

[Leake, B 76a] Leake, Barbara; Anninos, Photios A. (1976). *Effect of Connectivity on the Activity of Neural Net Models.* Journal of Theoretical Biology. Volume 58. pp 337-363.

[Lebow, WM 84a] Lebow, William M.; Mehra, Raman K.; Rice, Harbert; Toldalagi, Paul M. (1984). *Forecasting Applications of GMDH in Agricultural and Meteorological Time Series.* In Farlow, Stanley J. (ed.). Self-Organizing Methods in Modeling: GMDH Type Algorithms. pp 121-148. New York: Marcel Dekker.

[Lebowitz, JL 75*] SEE: Bortz, AB 75a. *A New Algorithm for Monte Carlo Simulation of Ising Spin Systems.*

[Leddy, B 87*] SEE: Bagherzadeh, N 87a. *On Parallel Execution of the Traveling Salesman Problem on a Neural Network Model.*

[Lederhendler, II 87a] Lederhendler, I. Izja; Alkon, Daniel L. (1987). *Associatively Reduced Withdrawal from Shadows in Hermissenda: A Direct Behavioral Analog of Photoreceptor Responses to Brief Light Steps.* Behavioral and Neural Biology. Volume 47. pp 227-249.

[Lee, C 83*] SEE: Brown, PF 83a. *Bayesian Adaptation in Speech Recognition.*

[Lee, D 88*] SEE: Pawlicki, TF 88a. *Neural Network Models and Their Application to Handwritten Digit Recognition.*

[Lee, D 88a] Lee, Dar-Shyang; Srihari, Sargur N.; Pawlicki, Thaddeus F. (1988). Experiments with Neural Network Models for Handwritten Digit Recognition. Buffalo, NY: State University of New York, Computer Science Department.

[Lee, EB 87*] SEE: Poliac, MO 87a. *A Crew Scheduling Problem.*

[Lee, HY 86*] SEE: Chen, HH 86a. *High Order Correlation Model for Associative Memory.*

[Lee, JN 84*] SEE: Fisher, AD 84a. *Associative Processor Architectures for Optical Computing.*

[Lee, JN 84*] SEE: Athale, RA 84a. *Optical Processing Using Outer-Product Concepts.*

[Lee, JN 85*] SEE: Fisher, AD 85a. *An Adaptive Optical Computing Element.*

[Lee, JN 86*] SEE: Fisher, AD 86a. *Implementations of Adaptive Associative Optical Computing Elements.*

[Lee, JN 87*] SEE: Fisher, AD 87b. *Optical Implementations of Associative Networks with Versatile Adaptive Learning Capabilities.*

[Lee, JSJ 87a] Lee, James S. J.; Nguyen, Dziem D.; Lin, C. (1987). Intelligent Recognition and Learning Control: Integrating Intelligent Processing and Neural Networks for Adaptive Moving Object Tracking. Seattle, WA: Boeing High Tech Center.

[Lee, JSJ 88*] SEE: Nguyen, DD 88a. *A New LMS-Based Algorithm for Rapid Adaptive Classification in Dynamic Environments: Theory and Preliminary Results.*

[Lee, JSJ 88a] Lee, James S. J.; Bezdek, James C.; Lin, C. (1988). *A Computer Vision Architecture for Intelligence Control of Neural Recognition in Dynamic Environments.* National Conference on Artificial Intelligence (AAAI-88), 7th, Minneapolis, MN, August 20, 1988: Proceedings.

[Lee, JSJ 88b] Lee, James S. J.; Bezdek, James C. (1988). *A Feature Projection Based Adaptive Pattern Recognition Network.* IEEE International Conference on Neural Networks (ICNN-88), San Diego, CA, July 24-27, 1988: Proceedings. Volume 1. pp 497-505. New York: Institute of Electrical and Electronics Engineers.

[Lee, K 85a] Lee, Kai-Fu. (1985). Incremental Network Generation in Template-Based Word Recognition. CMU-CS-85-181. Pittsburgh, PA: Carnegie Mellon University, Computer Science Department.

[Lee, K 86*] SEE: Shikano, K 86a. Speaker Adaptation Through Vector Quantization.

[Lee, K 86a] Lee, Kai-Fu. (1986). A Pattern Classification Approach to Evaluation Function Learning. CMU-CS-86-173. Pittsburgh, PA: Carnegie Mellon University, Computer Science Department.

[Lee, K 87a] Lee, Kai-Fu. (1987). Speaker-Independent Continuous Speech Recognition Using Hidden Markow Models. Thesis Proposal. Pittsburgh, PA: Carnegie Mellon University, Computer Science Department.

[Lee, K 88a] Lee, Kai-Fu. (1988). Large-Vocabulary Speaker-Independent Continuous Speech Recognition: The SPHINX System. CMU-CS-88-148. Pittsburgh, PA: Carnegie Mellon University, Computer Science Department.

[Lee, KS 86*] SEE: Farhat, NH 86b. *Optical Analog of Two-Dimensional Neural Networks and Their Application in Recognition of Radar Targets.*

[Lee, LS 87*] SEE: Stoll, HM 87a. A Continuous-Time Optical Neural Network.

[Lee, LS 87a] Lee, L. S.; Stoll, H. M.; Tackitt, M. C. (1987). A Continuous-Time Optical Neural Network Associative Memory. Palos Verdes Peninsula, CA: Northrop Research and Technology Center.

[Lee, M 88*] SEE: Pacheco, M 88a. *A Simple VLSI Architecture for Neurocomputing.*

[Lee, MH 82a] Lee, M. H.; Marudarajan, A. R. (1982). *A Computer Package for the Evaluation of Neuron Models Involving Large Uniform Networks.* International Journal of Man-Machine Studies. Volume 17. pp 189–210.

[Lee, S 63*] SEE: Hon, E 63a. *Noise Reduction in Fetal Electrocardiography.*

[Lee, S 88*] SEE: Jang, J 88b. *Optical Implementation of Quadratic Associative Memory with Outer-Product Storage.*

[Lee, S 88*] SEE: Jang, J 88c. *Optical Implementation of the Hopfield Model for Two-Dimensional Associative Memory.*

[Lee, S 88*] SEE: Jang, J 88a. *An Optimization Network for Matrix Inversion.*

[Lee, S 88a] Lee, Soo-Young; Jang, Ju-Seog; Shin, Sang-Yung; Shim, Chang-Sup. (1988). *Optical Implementation of Associative Memory with Controlled Bit-Significance.* Applied Optics. Volume 27. pp 1921-1923.

[Lee, SH 80*] SEE: Cederquist, J 80a. *Coherent Optical Feedback for the Analog Solution of Partial Differential Equations.*

[Lee, W 88*] SEE: Tenorio, MF 88a. *Self Organizing Neural Networks for the Identification Problem.*

[Lee, Y 86a] Lee, Yillbyung. (1986). A Neural Network Model of Frog Retina: A Discrete Time-Space Approach. COINS Technical Report 86-39. Amherst, MA: University of Massachusetts, Computer and Information Science.

[Lee, YC 86*] SEE: Chen, HH 86a. *High Order Correlation Model for Associative Memory.*

[Lee, YC 86*] SEE: Maxwell, T 86a. *Nonlinear Dynamics of Artificial Neural Systems.*

[Lee, YC 87*] SEE: Maxwell, T 87a. *Generalization in Neural Networks: The Contiguity Problem.*

[Lee, YC 87*] SEE: Sun, GZ 87a. *Learning Stereopsis with Neural Networks.*

[Lee, YC 88*] SEE: Sun, GZ 88a. *A Novel Net that Learns Sequential Decision Process.*

[Legendy, CR 67a] Legendy, C. R. (1967). *On the Scheme by Which the Human Brain Stores Information.* Mathematical Biosciences. Volume 1. pp 555-597.

[Lehar, S 87a] Lehar, Steve; Weaver, John. (1987). *A Developmental Approach to Neural Network Design.* IEEE International Conference on Neural Networks (ICNN-87), San Diego, CA, June 21-24, 1987: Proceedings. Volume 2. pp 97-104. New York: Institute of Electrical and Electronics Engineers.

[Lehky, SR 88a] Lehky, Sidney R.; Sejnowski, Terrence J. (1988). *Network Model of Shape-From-Shading: Neural Function Arises from Both Receptive and Projective Fields.* Nature. Volume 333. pp 452-454.

[Lehmann, D 87*] SEE: Fuchs, A 87a. *Spatio-Temporal Analysis of Multi-channel Alpha EEG Map Series.*

[Lehmann, D 87a] Lehmann, Dietrich; Ozaki, H.; Pal, I. (1987). *EEG Alpha Map Series: Brain Micro-States by Space-Oriented Adaptive Segmentation.* Electroencephalography and Clinical Neurophysiology. Volume 67. pp 271-288.

[Lehmann, D 87b] Lehmann, Dietrich; Brandeis, D.; Ozaki, H.; Pal, I. (1987). *Human Brain EEG Fields: Micro-States and Their Functional Significance.* Computational Systems - Natural and Artificial, Proceedings of the International Symposium on Synergetics at Schloss Elmau, Bavaria, May 4-9, 1987. pp 65–73. Heidelberg: Springer-Verlag.

[Lehmann, D 88a] Lehmann, Dietrich; Brandeis, D.; Horst, A.; Ozaki, H.; Pal, I. (1988). *Spontaneous and Information-Triggered Segments of Series of Human Brain Electric Field Maps.* In Anderson, Dana Z. (ed.). Neural Information Processing Systems. pp 467-473. New York: American Institute of Physics.

[Lehmkuhle, W 84*] SEE: Levy, WB 84a. Synaptic Change in the Nervous System.

[Lehtio, P 81*] SEE: Kohonen, T 81c. *Storage and Processing of Information in Distributed Associative Memory Systems.*

[Lehtio, P 85*] SEE: Kohonen, T 85a. *A Model of Cortical Associative Memory.*

[Leighton, R 87*] SEE: Wieland, A 87a. *Geometric Analysis of Neural Network Capabilities.*

[Leiserson, CE 85a] Leiserson, C. E. (1985). *Fat-Trees: Universal Networks for Hardware Efficient Supercomputing.* IEEE Transactions on Computers. Volume C-34. pp 892-901.

[Leith, EN 62a] Leith, Emmett N.; Upatnieks, Juris. (1962). *Reconstructed Wavefronts and Communication Theory.* Journal of the Optical Society of America. Volume 52. pp 1123-1130.

[Lemmon, M 87a] Lemmon, Michael; Kumar, B. V. K. Vijaya. (1987). *Neural Networks as Associative Memories.* IEEE International Conference on Neural Networks (ICNN-87), San Diego, CA, June 21-24, 1987: Proceedings. Volume 2. pp 213-222. New York: Institute of Electrical and Electronics Engineers.

[Lemmon, M 88a] Lemmon, Michael; Kumar, B. V. K. Vijaya. (in press). *Emulating the Dynamics for a Class of Laterally Inhibited Neural Networks.* Neural Networks.

[Lendaris, GG 88a] Lendaris, George G. (1988). *Conceptual Graph Knowledge Systems as Problem Context for Neural Networks.* IEEE International Conference on Neural Networks (ICNN-88), San Diego, CA, July 24-27, 1988: Proceedings. Volume 1. pp 133-140. New York: Institute of Electrical and Electronics Engineers.

[Lendaris, GG 88b] Lendaris, George G. (1988). *Neural Networks, Potential Assistants to Knowledge Engineers.* International Symposium on Knowledge Engineering, 3rd, Madrid, October 17-29, 1988.

[Lendaris, GG 88c] Lendaris, George G. (1988). *Representing Conceptual Graphs for Parallel Processing.* Conceptual Graphs Workshop (AAAI-88), 3rd, Minneapolis, MN, 1988: Proceedings.

[Lenstra, JK 86*] SEE: Lawler, EL 86a. The Traveling Salesman Problem.

[Leonberger, FI 84*] SEE: Goodman, JW 84a. *Optical Interconnections for VLSI Systems.*

[Leong, HM 88a] Leong, Harrison MonFook. (1988). *Optimization with Artificial Neural Network Systems: A Mapping Principle and a Comparison to Gradient Based Methods.* In Anderson, Dana Z. (ed.). Neural Information Processing Systems. pp 474-484. New York: American Institute of Physics.

[Lerner, RM 60a] Lerner, Robert M. (1960). *A Matched Filter Detection System for Complicated Doppler Shifted Signals.* IRE Transactions on Information Theory. Volume IT-6. pp 373-385.

[Lesser, VR 81a] Lesser, Victor R.; Corkill, Daniel D. (1981). *Functionally-Accurate, Cooperative Distributed Systems.* IEEE Transactions on Systems, Man, and Cybernetics. Volume SMC-11. pp 81-96.

[Lesser, VR 85*] SEE: Durfee, EH 85a. Coherent Cooperation among Communicating Problem Solvers.

[Lesser, VR 85*] SEE: Durfee, EH 85b. Increasing Coherence in a Distributed Problem Solving Network.

[Lester, BP 87a] Lester, Bruce P. (1987). *A Neural Network Simulation Language Based on Multi-Pascal.* IEEE International Conference on Neural Networks (ICNN-87), San Diego, CA, June 21-24, 1987: Proceedings. Volume 3. pp 347-354. New York: Institute of Electrical and Electronics Engineers.

[Lettvin, JY 59a] Lettvin, J. Y.; Maturana, H. R.; McCulloch, Warren S.; Pitts, W. H. (1959). *What the Frog's Eye Tells the Frog's Brain.* Proceedings of the IRE. Volume 47. pp 1940-1951.

[Leutkemeyer, K 82*] SEE: McGill, KC 82a. *On the Nature and Elimination of Stimulus Artifact in Nerve Signals Evoked and*

Recorded Using Surface Electrodes.

[Leven, SJ 87a] Leven, Samuel J.; Levine, Daniel S. (1987). *Effects of Reinforcement on Knowledge Retrieval and Evaluation.* IEEE International Conference on Neural Networks (ICNN-87), San Diego, CA, June 21-24, 1987: Proceedings. Volume 2. pp 269-278. New York: Institute of Electrical and Electronics Engineers.

[Levin, JA 76a] Levin, J. A. (1976). Proteus: An Activation Framework for Cognitive Process Models. Technical Report ISI/WP-2. California: Information Sciences Institute.

[Levine, DS 75*] SEE: Grossberg, S 75b. *Some Developmental and Attentional Biases in the Contrast Enhancement and Short-Term Memory of Recurrent Neural Networks.*

[Levine, DS 75a] Levine, Daniel S. (1975). Studies in the Transformation and Storage of Patterns in Reverberating Neural Networks. Ph.D. Dissertation. Cambridge, MA: Massachusetts Institute of Technology.

[Levine, DS 76a] Levine, Daniel S.; Grossberg, Stephen. (1976). *Visual Illusions in Neural Networks: Line Neutralization, Tilt After Effect, and Angle Expansion.* Journal of Theoretical Biology. Volume 61. pp 477-504.

[Levine, DS 79a] Levine, Daniel S. (1979). *Existence of a Limiting Pattern for a System of Non-Linear Equations Describing Interpopulation Competition.* Bulletin of Mathematical Biology. Volume 41. pp 617-628.

[Levine, DS 83a] Levine, Daniel S. (1983). *Neural Population Modeling and Psychology: A Review.* Mathematical Biosciences. Volume 66. pp 1-86.

[Levine, DS 85*] SEE: Cohen, MA 85a. *Unbounded Oscillatory Solutions for a System of Interactive Populations.*

[Levine, DS 86a] Levine, Daniel S. (1986). *A Neural Network Theory of Frontal Lobe Function.* Cognitive Science Society, 8th, Amherst, MA, August 15-17, 1986: Proceedings. pp 716-727. Hillsdale, NJ: Lawrence Erlbaum.

[Levine, DS 87*] SEE: Grossberg, S 87f. *Attentional Mechanisms in Neural Information Processing: Examples From Pavlovian Conditioning.*

[Levine, DS 87*] SEE: Leven, SJ 87a. *Effects of Reinforcement on Knowledge Retrieval and Evaluation.*

[Levinson, SE 81a] Levinson, Stephen E.; Liberman, M. Y. (April 1981). *Speech Recognition by Computer.* Scientific American. Volume 244. pp 64-76.

[Levinson, SE 83a] Levinson, Stephen E.; Rabiner, L. R.; Sondhi, M. M. (1983). *An Introduction to the Application of the Theory of Probabilistic Functions of a Markov Process to Automatic Speech Recognition.* Bell System Technical Journal. Volume 62. pp 1035-1074.

[Levinson, SE 85*] SEE: Juang, BH 85a. *Recent Developments in the Application of Hidden Markov Models to Speaker-Independent Isolated Word Recognition.*

[Levinson, SE 85a] Levinson, Stephen E. (1985). *Structural Methods in Automatic Speech Recognition.* Proceedings of the IEEE. Volume 73. pp 1625-1650.

[Levitan, S 88*] SEE: Krieger, D 88a. *A Neurobiologically Based Neural Net Model of the Hippocampus.*

[Levy, BC 87a] Levy, Bernard C. (1987). *Global Optimization with Stochastic Neural Networks.* IEEE International Conference on Neural Networks (ICNN-87), San Diego, CA, June 21-24, 1987: Proceedings. Volume 3. pp 681-690. New York: Institute of Electrical and Electronics Engineers.

[Levy, J 88*] SEE: Stenning, K 88a. *Knowledge-Rich Solutions to the Binding Problem: A Simulation of Some Human Computational Mechanisms.*

[Levy, P 56a] Levy, Paul. (1956). *A Special Problem of Brownian Motion, and a General Theory of Gaussian Random Functions*. Berkeley Symposium on Mathematical Statistics and Probability, 3rd, 1956: Proceedings. Volume 2. pp 133-175. University of California Press.

[Levy, WB 79a] Levy, William B.; Merrill, Walter; Rapaport, David; Stell, Mark. (1979). *Associative-Contingency Learning and Neuronal Networks*. Sistm Quarterly; Incorporating The Brain Theory Newsletter. Volume 11. pp 21-26.

[Levy, WB 84a] Levy, William B. (ed.); Anderson, James A. (ed.); Lehmkuhle, W. (ed.). (1984). Synaptic Change in the Nervous System. Hillsdale, NJ: Lawrence Erlbaum.

[Levy, WB 85a] Levy, William B. (1985). *Associative Changes at the Synapse: LTP in the Hippocampus*. In Levy, William B. (ed.); Anderson, James A. (ed.); Lehmkuhle, Stephen (ed.). Synaptic Modifications, Neuron Selectivity, and Nervous System Organization. pp 5-34. Hillsdale, NJ: Lawrence Erlbaum.

[Levy, WB 85b] Levy, William B.; Desmond, Nancy L. (1985). *The Rules of Elemental Synaptic Plasticity*. In Levy, William B. (ed.); Anderson, James A. (ed.); Lehmkuhle, Stephen (ed.). Synaptic Modifications, Neuron Selectivity, and Nervous System Organization. pp 105-122. Hillsdale, NJ: Lawrence Erlbaum.

[Levy, WB 87a] Levy, William B.; Burger, B. (1987). *Electrophysiological Observations Which Help Describe an Associative Synaptic Modification Rule*. IEEE International Conference on Neural Networks (ICNN-87), San Diego, CA, June 21-24, 1987: Proceedings. Volume 4. pp 11-16. New York: Institute of Electrical and Electronics Engineers.

[Lewis, ER 64a] Lewis, Edwin R. (1964). *An Electronic Model of the Neuron Based on the Dynamics of Potassium and Sodium Ion Fluxes*. Neural Theory and Modeling, Proceedings of the 1962 Ojai Symposium. pp 154-189. Stanford, CA: Stanford University Press.

[Lewis, ER 66*] SEE: Harmon, LD 66a. *Neural Modeling*.

[Lewis, ER 66a] Lewis, Edwin R. (1966). *Neuroelectric Potentials Derived from an Extended Version of the Hodgkin-Huxley Model*. Journal of Theoretical Biology. Volume 10. pp 125-158.

[Lewis, ER 68a] Lewis, Edwin R. (1968). *An Electronic Model of Neuroelectric Point Processes*. Kybernetik. Volume 5. pp 30-46.

[Lewis, ER 68b] Lewis, Edwin R. (1968). *Synchronization in Small Groups of Neurons: A Study with Electronic Models*. Cybernetic Problems in Bionics. pp 777-789. New York: Gordon and Breach.

[Lewis, ER 68c] Lewis, Edwin R. (1968). *The Iron Wire Model of the Neuron: A Review*. Cybernetic Problems in Bionics. pp 247-273. New York: Gordon and Breach.

[Lewis, ER 68d] Lewis, Edwin R. (1968). *Using Electronic Circuits to Model Simple Neuroelectric Interactions*. Proceedings of the IEEE. Volume 56. pp 931-949.

[Lewis, ER 77*] SEE: MacGregor, RJ 77a. Neural Modeling.

[Lewis, ER 83a] Lewis, Edwin R. (1983). *The Elements of Single Neurons: A Review*. IEEE Transactions on Systems, Man and Cybernetics. Volume SMC-13. pp 702-710.

[Lewis, PM 62a] Lewis, P. M. (1962). *The Characteristic Selection Problem in Recognition Systems*. IRE Transactions on Information Theory. Volume IT-8. pp 171-178.

[Lewis, PM 67a] Lewis, P. M.; Coates, C. L. (1967). Threshold Logic. New York: Wiley.

[Li, B 81a] Li, Bingxi. (1981). *Uniqueness and Stability of a Limit Cycle for a Third Order Dynamical System Arising in*

Neuron Modelling. Nonlinear Analysis: Theory, Methods & Applications. Volume 5. pp 13-19.

[Li, J 88*] SEE: Michel, AN 88a. *Qualitative Analysis of Neural Networks.*

[Li, JH 88a] Li, Jian Hua; Michel, Anthony N.; Porod, Wolfgang. (1988). *Qualitative Analysis and Synthesis of a Class of Neural Networks.* IEEE Transactions on Circuits and Systems. Volume 35. pp 976-986.

[Li, T 87a] Li, Tangqiu; Chun, Hon Wai. (1987). *Massively Parallel Network-Based Natural Language Parsing System.* IEEE International Conference on Computers and Applications, 2nd, June, 1987: Proceedings.

[Li, W 86*] SEE: Xu, J 86a. *The Dynamics of Large Scale Neuron-Glia Network and Its Relation to the Brain Functions (I).*

[Liao, PF 87*] SEE: Von Lehmen, A 87a. *Influence of Interconnection Weight Discretization and Noise in an Optoelectronic Neural Network.*

[Liao, PF 88*] SEE: Paek, EG 88a. *A Derivation of Neural Network Models for Associative Memory.*

[Liao, PF 88*] SEE: Von Lehmen, A 88a. *Factors Influencing Learning by Backpropagation.*

[Liberman, AM 57a] Liberman, Alvin M. (1957). *Some Results of Research on Speech Perception.* Journal of the Acoustical Society of America. Volume 29. pp 117-123.

[Liberman, F 79*] SEE: Cooper, LN 79a. *A Theory for the Acquisition and Loss of Neuron Specificity in Visual Cortex.*

[Liberman, MY 81*] SEE: Levinson, SE 81a. *Speech Recognition by Computer.*

[Licklider, JCR 52a] Licklider, J. C. R. (1952). *On the Process of Speech Perception.* Journal of the Acoustical Society of America. Volume 24. pp 590-594.

[Lieblich, I 78a] Lieblich, I.; Amari, Shun-ichi. (1978). *An Extended First Approximation Model for the Amygdaloid Kindling Phenomenon.* Biological Cybernetics. Volume 28. pp 129-135.

[Liederman, J 87*] SEE: Bullock, D 87b. *Reconciling Stable Asymmetry with Recovery of Function: An Adaptive Systems Perspective on Functional Plasticity.*

[Liesis, V 72*] SEE: Burattini, E 72a. *A Method of Analysis of the Models of Neural Nets.*

[Lim, JS 85*] SEE: Curtis, SR 85a. *Signal Reconstruction from Fourier Transform Sign Information.*

[Limm, JS 85*] SEE: Kosko, B 85a. *Vision as Causal Activation and Association.*

[Lin, C 87*] SEE: Lee, JSJ 87a. Intelligent Recognition and Learning Control: Integrating Intelligent Processing and Neural Networks for Adaptive Moving Object Tracking.

[Lin, C 88*] SEE: Lee, JSJ 88a. *A Computer Vision Architecture for Intelligence Control of Neural Recognition in Dynamic Environments.*

[Lin, CC 74a] Lin, C. C.; Segal, L. A. (1974). Mathematics Applied to Deterministic Problems in the Natural Sciences. New York: Macmillan.

[Lin, S 73a] Lin, S.; Kernighan, B. W. (1973). *An Effective Heuristic Algorithm for the Traveling Salesman Problem.* Operations Research. Volume 21. pp 498-516.

[Lin, T 88*] SEE: Chua, LO 88a. *A Neural Network Approach to Transform Image Coding.*

[Lindgren, N 65a] Lindgren, Nilo. (1965). *Machine Recognition of Human Language: Part 2, Theoretical Models of Speech*

Perception and Language. IEEE Spectrum. Volume 2. pp 45-59.

[Lindley, DV 87a] Lindley, D. V. (1987). *The Probability Approach to the Treatment of Uncertainty in Artificial Intelligence and Expert Systems*. Statistical Science. Volume 2. pp 17-24.

[Lindsay, I 88*] SEE: White, HJ 88a. *Digital and Analogue Holographic Associative Memories*.

[Lindsay, RD 70*] SEE: Harth, EM 70a. *Brain Functions and Neural Dynamics*.

[Lininger, DM 87*] SEE: Anderson, DZ 87b. *Dynamic Optical Interconnects: Volume Holograms as Optical Two-Port Operators*.

[Linsker, R 86a] Linsker, Ralph. (1986). *From Basic Network Principles to Neural Architecture: Emergence of Orientation Columns*. Proceedings of the National Academy of Sciences. Volume 83. pp 8779-8783.

[Linsker, R 86b] Linsker, Ralph. (1986). *From Basic Network Principles to Neural Architecture: Emergence of Orientation-Selective Cells*. Proceedings of the National Academy of Sciences. Volume 83. pp 8390-8394.

[Linsker, R 86c] Linsker, Ralph. (1986). *From Basic Network Principles to Neural Architecture: Emergence of Spatial-Opponent Cells*. Proceedings of the National Academy of Sciences. Volume 83. pp 7508-7512.

[Linsker, R 88a] Linsker, Ralph. (1988). *Self-Organization in a Perceptual Network*. Computer. Volume 21. pp 105-117.

[Linsker, R 88b] Linsker, Ralph. (1988). *Towards an Organizing Principle for a Layered Perceptual Network*. In Anderson, Dana Z. (ed.). Neural Information Processing Systems. pp 485-494. New York: American Institute of Physics.

[Lippincott, WL 87*] SEE: Fisher, AD 87b. *Optical Implementations of Associative Networks with Versatile Adaptive Learning Capabilities*.

[Lippman, RP 87a] Lippman, Richard P.; Gold, Bernard; Malpass, M. L. (1987). A Comparison of Hamming and Hopfield Neural Nets for Pattern Classification. Technical Report 769. Lexington, MA: Massachusetts Institute of Technology, Lincoln Laboratory.

[Lippmann, RP 87*] SEE: Huang, WY 87a. *Comparisons Between Neural Net and Conventional Classifiers*.

[Lippmann, RP 87*] SEE: Gold, B 87a. *Some Neural Net Recognition Results on Isolated Words*.

[Lippmann, RP 87a] Lippmann, Richard P.; Huang, William Y. (1987). *A Framework for Speech Recognition Using Neural Networks*. In Anderson, Dana Z. (ed.). Neural Information Processing Systems. New York: American Institute of Physics.

[Lippmann, RP 87b] Lippmann, Richard P. (1987). *An Introduction to Computing with Neural Nets*. IEEE ASSP Magazine. Volume 4. pp 4-22.

[Lippmann, RP 87c] Lippmann, Richard P. (1987). *Computing with Neural Nets*. IEEE Transactions on Acoustics, Speech and Signal Processing. Volume 4. pp 4-22.

[Lippmann, RP 87d] Lippmann, Richard P.; Gold, Ben. (1987). *Neural-Net Classifiers Useful for Speech Recognition*. IEEE International Conference on Neural Networks (ICNN-87), San Diego, CA, June 21-24, 1987: Proceedings. Volume 4. pp 417-426. New York: Institute of Electrical and Electronics Engineers.

[Lippmann, RP 88*] SEE: Huang, WY 88a. *Neural Net and Traditional Classifiers*.

[Lippmann, RP 88*] SEE: Huang, WY 88b. *Neural Nets for Speech Recognition*.

[Lippmann, RP 88*] SEE: Gold, B 88a. *A Neural Network for Isolated-Word Recognition*.

[Lippmann, RP 88*] SEE: Mann, J 88a. *A Self-Organizing Neural Net Chip.*

[Lippmann, RP 88a] Lippmann, Richard P. (1988). A Review of Studies Applying Neural Net Models to Speech Recognition Problems. Lexington, MA: Massachusetts Institute of Technology, Lincoln Laboratory.

[Lippmann, RP 88b] Lippmann, Richard P. (1988). *A Survey of Neural Network Models.* International Conference on Supercomputing, 3rd, Boston, MA, May 15-20, 1988: Proceedings.

[Lippmann, RP 88c] Lippmann, Richard P. (1988). *Neural Network Classifiers for Speech Recognition.* Lincoln Laboratory Journal. Volume 1. pp 107-124.

[Little, MJ 86a] Little, Michael J.; Bak, C. S. (1986). *Enhanced Memory Capacity of a Hopfield Neural Network.* Real Time Signal Processing. SPIE Volume 698. pp 150-156. Bellingham, WA: Society of Photo-Optical Instrumentation Engineers.

[Little, WA 75a] Little, W. A.; Shaw, Gordon L. (1975). *A Statistical Theory of Short and Long Term Memory.* Behavioral Biology. Volume 14. pp 115-133.

[Little, WA 75b] Little, W. A.; Shaw, Gordon L. (1988). *A Statistical Theory of Short and Long Term Memory.* In Anderson, James A. (ed.); Rosenfeld, Edward (ed.). Neurocomputing: Foundations of Research. pp 231-242. Cambridge, MA: MIT Press.

[Little, WA 78a] Little, W. A.; Shaw, Gordon L. (1978). *Analytic Study of the Storage Capacity of a Neural Network.* Mathematical Biosciences. Volume 39. pp 281-290.

[Littlestone, N 87a] Littlestone, Nick. (1987). *Learning in a Layered Network with Many Fixed-Function Hidden Nodes.* IEEE International Conference on Neural Networks (ICNN-87), San Diego, CA, June 21-24, 1987: Proceedings. Volume 2. pp 471-478. New York: Institute of Electrical and Electronics Engineers.

[Liu, CN 66*] SEE: Chow, CK 66b. *An Approach to Structure Adaptation in Pattern Recognition.*

[Lively, R 88a] Lively, Richard; Walters, Deborah K. W. (1988). *Integration of Detector Responses for Texture Segmentation.* Applications of Artificial Intelligence. SPIE Volume 937. pp 86-93. Bellingham, WA: Society of Photo-Optical Instrumentation Engineers.

[Llinas, R 79*] SEE: Pellionisz, AJ 79a. *Brain Modeling by Tensor Network Theory and Computer Simulation.*

[Llinas, R 82*] SEE: Pellionisz, AJ 82a. *Space-Time Representation in the Brain: The Cerebellum as a Predictive Space-Time Metric Tensor.*

[Llinas, R 82*] SEE: Pellionisz, AJ 82b. *Tensor Theory of Brain Function: The Cerebellum as a Space-Time Metric.*

[Loebner, EE 87a] Loebner, Egon E. (1987). *Concurrency Assurance in Vertebrate Retinas.* IEEE International Conference on Neural Networks (ICNN-87), San Diego, CA, June 21-24, 1987: Proceedings. Volume 4. pp 147-162. New York: Institute of Electrical and Electronics Engineers.

[Loftus, EF 75*] SEE: Collins, AM 75a. *A Spreading-Activation Theory of Semantic Processing.*

[Long, JM 88*] SEE: Irani, EA 88a. Using Artificial Neural Nets for Statistical Discovery: Observations after Using Back-propagation, Expert Systems, and Multiple-Linear Regression.

[Longuet-Higgins, HC 68a] Longuet-Higgins, H. C. (1968). *Holographic Model of Temporal Recall.* Nature. Volume 217. pp 104.

[Longuet-Higgins, HC 68b] Longuet-Higgins, H. C. (1968). *The Non-Local Storage of Temporal Information.* Proceedings of the

Royal Society of London, Series B. Volume 171. pp 327-334.

[Longuet-Higgins, HC 69*] SEE: Willshaw, DJ 69c. *The Holophone: Recent Developments.*

[Longuet-Higgins, HC 69*] SEE: Willshaw, DJ 69b. *Non-Holographic Associative Memory.*

[Longuet-Higgins, HC 69*] SEE: Willshaw, DJ 69a. *Non-Holographic Associative Memory.*

[Longuet-Higgins, HC 70*] SEE: Willshaw, DJ 70a. *Associative Memory Models.*

[Longuet-Higgins, HC 70a] Longuet-Higgins, H. C.; Willshaw, David J.; Buneman, O. P. (1970). *Theories of Associative Recall.* Quarterly Review of Biophysics. Volume 3. pp 223-244.

[Loos, H 88a] Loos, Hendricus (ed.); Anderson, Dana Z. (ed.). (1988). *Reflexive Associative Memories.* Neural Information Processing Systems. pp 495-504. New York: American Institute of Physics.

[Lopez, LR 88a] Lopez, Luis R. (1988). *Motion Sensitive Cellular Automata.* High Speed Computing. SPIE Volume 880. pp 2-10. Bellingham, WA: Society of Photo-Optical Instrumentation Engineers.

[Lorente de No, R 49a] Lorente de No, Rafael. (1949). *Cerebral Cortex: Architecture, Intracortical Connections, Motor Projections.* Physiology of the Nervous System. pp 288-330. New York: Oxford University Press.

[LoTurco, JJ 88a] LoTurco, Joseph J.; Coulter, Douglas A.; Alkon, Daniel L. (1988). *Enhancement of Synaptic Potentials in Rabbit CA1 Pyramidal Neurons Following Classical Conditioning.* Proceedings of the National Academy of Science. Volume 85. pp 1672-1676.

[Love, WR 87*] SEE: Gawronski, RP 87a. *Artina: A Flexible Computer Model of Retinal Processes for Moving Images.*

[Love, WR 87*] SEE: Gawronski, RR 87a. *Modelling of an Adaptive Neuronlike Network for the Construction of a Learning Control System.*

[Lowrance, JD 81*] SEE: Garvey, TD 81a. *An Inference Technique for Integrating Knowledge from Disparate Sources.*

[Lowry, SR 75*] SEE: Ritter, GL 75a. *An Algorithm for a Selective Nearest Neighbor Decision Rule.*

[Luce, PA 86*] SEE: Pisoni, DB 86a. *Speech Perception: Research, Theory, and the Principal Issues.*

[Ludlam, G 80*] SEE: Hintzman, DL 80a. *Differential Forgetting of Prototypes and Old Instances: Simulation by an Exemplar-Based Classification Model.*

[Lundell, J 87a] Lundell, James; Laden, Bernice. (1987). *A Dynamic Connectionist Model of Problem Solving.* Cognitive Science Society, 9th, Seattle, WA, July 16-18, 1987: Proceedings. pp 663-676. Hillsdale, NJ: Lawrence Erlbaum.

[Luo, J 88*] SEE: Hutchinson, J 88a. *Computing Motion Using Analog and Binary Resistive Networks.*

[Luo, J 88*] SEE: Koch, C 88b. *Computing Motion Using Resistive Networks.*

[Lutton, JL 86*] SEE: Bonomi, E 86a. *Statistical Mechanics: A General Approach to Combinatorial Optimization.*

[Lyman, J 87*] SEE: Reber, WL 87a. *An Artificial Neural System Design for Rotation and Scale Invariant Pattern Recognition.*

[Lynch, G 84a] Lynch, Gary; Baudry, Michel. (1984). *The Biochemistry of Memory: A New and Specific Hypothesis.* Science. Volume 224. pp 1057-1063.

[Lynch, G 88*] SEE: Granger, R 88a. *Derivation of Encoding Characteristics of Layer II Cerebral Cortex.*

[Lynch, G 88*] SEE: Granger, R 88b. *Partitioning of Sensory Data by a Cortical Network.*

[Lynch, G 88a] Lynch, Gary; Granger, Richard. (1988). *Simulation and Analysis of a Simple Cortical Network.* Psychology of Learning and Motivation. Volume 22.

[Lynch, G 89a] Lynch, Gary; Granger, Richard; Larson, John; Baudry, Michel. (1989). *Cortical Encoding of Memory: Hypotheses Derived from Analysis and Simulation of Physiological Learning Rules in Anatomical Structures.* In Nadel, Lynn (ed.). Neural Connections, Mental Computation. pp 181-224. Cambridge, MA: MIT Press.

[Lyon, RF 88a] Lyon, Richard F.; Mead, Carver A. (1988). *An Analog Electronic Cochlea.* IEEE Transactions on Acoustics, Speech, and Signal Processing. Volume 36. pp 1119-1134.

[Lyons, D 85*] SEE: Arbib, MA 85b. *Coordinated Control Programs for Movements of the Hand.*

[Lyons, D 85a] Lyons, Damian; Arbib, Michael A. (1985). A Task-Level Model of Distributed Computation for Sensory-Based Control of Complex Robot Systems. COINS Technical Report 85-30. Amherst, MA: University of Massachusetts, Computer Science Department.

[Lyons, D 87*] SEE: Arbib, MA 87c. *Schemas That Integrate Vision and Touch for Hand Control.*

[Ma, Q 87*] SEE: Hirai, Y 87a. *Associate Networks Underlying Problem-Solving: Representation of the Process of Addition by a Model of an Associative Processor, HASP.*

[Ma, Q 88*] SEE: Hirai, Y 88a. *Modeling the Process of Problem-Solving by Associative Networks Capable of Improving the Performance.*

[Ma, SN 87*] SEE: Coon, DD 87a. *Farey-Fraction Frequency Modulation in the Neuronlike Output of Silicon p-i-n Diodes at 4.2 K.*

[MacDonald, B 88a] MacDonald, Bruce. (1988). *Connecting to the Past.* In Anderson, Dana Z. (ed.). Neural Information Processing Systems. pp 505-514. New York: American Institute of Physics.

[MacGregor, RJ 74a] MacGregor, Ronald J.; Palasek, R. L. (1974). *Computer Simulation of Rhythmic Oscillation in Neuron Pools.* Kybernetik. Volume 16. pp 79-86.

[MacGregor, RJ 77a] MacGregor, Ronald J.; Lewis, Edwin R. (1977). Neural Modeling. New York: Plenum.

[MacGregor, RJ 87*] SEE: Gaudiano, P 87a. *Dynamic Activity and Memory Traces in Computer-Simulated Recurrently-Connected Neural Networks.*

[MacGregor, RJ 87a] MacGregor, Ronald J. (1987). Neural and Brain Modeling. New York: Academic Press.

[MacGregor, RJ 88a] MacGregor, Ronald J.; Tajchman, Gary. (1988). *Theory of Dynamic Similarity in Neuronal Systems.* Journal of Neurophysiology. Volume 60. pp 751-768.

[MacKay, DM 54a] MacKay, Donald M. (1954). *On Comparing the Brain with Machines.* American Scientist. Volume 42. pp 261-268.

[MacKay, DM 56a] MacKay, Donald M. (1956). *Towards an Information-Flow Model of Human Behaviour.* British Journal of Psychology. Volume 67. pp 30-43.

[Mackey, MC 77a] Mackey, Michael C. (1977). *Oscillation and Chaos in Physiological Control Systems.* Science. Volume 197. pp 287-289.

[Mackey, MC 83*] SEE: Guevara, MR 83a. *Chaos in Neurobiology.*

[Mackie, S 88a] Mackie, Stuart; Denker, John S. (1988). *A Digital Implementation of a Best Match Classifier*. IEEE Conference on Custom Integrated Circuits, Rochester, NY, May 16-19, 1988: Technical Digest. New York: Institute of Electrical and Electronics Engineers.

[Mackie, S 88b] Mackie, Stuart; Graf, Hans P.; Schwartz, Daniel B. (1988). *Implementations of Neural Network Models in Silicon*. In Eckmiller, Rolf (ed.); von der Malsburg, Christoph (ed.). Neural Computers. NATO ASI Series, Volume F41. pp 467-476. Berlin: Springer-Verlag.

[Mackie, S 88c] Mackie, Stuart; Graf, Hans P.; Schwartz, Daniel B.; Denker, John S. (1988). *Microelectronic Implementations of Connectionist Neural Networks*. In Anderson, Dana Z. (ed.). Neural Information Processing Systems. pp 515-523. New York: American Institute of Physics.

[MacLean, A 84*] SEE: Hammond, N 84a. *Evaluating the Interface of a Document Processor: A Comparison of Expert Judgement and User Observation.*

[MacLennan, BJ 87a] MacLennan, Bruce J. (1987). *Technology-Independent Design of Neurocomputers: The Universal Field Computer*. IEEE International Conference on Neural Networks (ICNN-87), San Diego, CA, June 21-24, 1987: Proceedings. Volume 3. pp 39-49. New York: Institute of Electrical and Electronics Engineers.

[MacLennan, BJ 88a] MacLennan, Bruce J. (March). *Field Computation: A Model of Massively Parallel Computation in Electronic, Optical, Molecular and Biological Systems*. Parallel Models of Intelligence: How Can Slow Components Think So Fast?. pp 180-183. Washington, DC: American Association for Artificial Intelligence.

[Macukow, B 87a] Macukow, Bohdan; Arsenault, Henri H. (1987). *Modification of the Threshold Condition for a Content-Addressable Memory Based on the Hopfield Model*. Applied Optics. Volume 26. pp 34-36.

[Macukow, B 87b] Macukow, Bohdan; Arsenault, Henri H. (1987). *Optical Associative Memory Model Based on Neural Networks Having Variable Interconnection Weights*. Applied Optics. Volume 26. pp 924-928.

[Macy, SA 87*] SEE: Shepanski, JF 87a. *Manual Training Technqiues of Autonomous Systems Based on Artificial Neural Networks.*

[Macy, SA 88*] SEE: Shepanski, JF 88b. *Teaching Artificial Neural Systems to Drive: Manual Training Techniques for Autonomous Systems.*

[Maddox, AB 87a] Maddox, Anthony B.; Pustejovsky, James. (1987). *Linguistic Descriptions of Visual Event Perceptions*. Cognitive Science Society, 9th, Seattle, WA, July 16-18, 1987: Proceedings. pp 442-454. Hillsdale, NJ: Lawrence Erlbaum.

[Maddox, J 87a] Maddox, John. (1987). *Simulating Memory by Numbers*. Nature. Volume 325. p 11.

[Maginu, K 88*] SEE: Amari, S 88b. *Statistical Neurodynamics of Associative Memory.*

[Maher, MA 87a] Maher, Mary Ann; Mead, Carver A. (1987). A Physical Charge-Controlled Model for MOS Transistors. Pasadena, CA: California Institute of Technology.

[Mahowald, MA 87*] SEE: Sivilotti, MA 87a. *Real-Time Visual Computations Using Analog CMOS Processing Arrays.*

[Mahowald, MA 87*] SEE: Sivilotti, MA 87b. *Real-time Visual Computations Using Analog CMOS Processing Arrays.*

[Mahowald, MA 88*] SEE: Mead, CA 88a. *A Silicon Model of Early Visual Processing.*

[Maitra, S 73*] SEE: Widrow, B 73a. *Punish / Reward: Learning With a Critic in Adaptive Threshold Systems.*

[Makhoul, J 84*] SEE: Schwartz, RM 84a. *Improved Hidden Markov Modeling of Phonemes for Continuous Speech Recognition.*

[Makhoul, J 85*] SEE: Schwartz, RM 85a. *Context-Dependent Modeling for Acoustic-Phonetic Recognition of Continuous Speech.*

[Makhoul, J 85a] Makhoul, John; Roucos, Salim; Gish, Herbert. (1985). *Vector Quantization in Speech Coding.* Proceedings of the IEEE. Volume 73. pp 1551-1588.

[Makisara, K 76*] SEE: Kohonen, T 76a. *Associative Recall of Images.*

[Makisara, K 84*] SEE: Kohonen, T 84b. *Phonotopic Maps - Insightful Representation of Phonological Features for Speech Recognition.*

[Makisara, K 86*] SEE: Kohonen, T 86c. *Representation of Sensory Information in Self-Organizing Feature Maps.*

[Makki, RZ 88*] SEE: Bou-Ghazale, S 88a. *Design and Implementation of an Associative Memory for Intensive Pattern Matching.*

[Malaret, ER 87*] SEE: Jacyna, GM 87a. *Classification Performance of an Auto-Associative Memory Based on Neural Networks.*

[Malina, W 81a] Malina, Witold. (1981). *On an Extended Fisher Criterion for Feature Selection.* IEEE Transactions on Pattern Analysis and Machine Intelligence. Volume PAMI-3. pp 611-614.

[Mallot, HA 85a] Mallot, Hanspeter A. (1985). *An Overall Description of Retinotopic Mapping in the Cat's Visual Cortex Areas 17, 18, and 19.* Biological Cybernetics. Volume 52. pp 45-51.

[Mallot, HA 86*] SEE: Krone, G 86a. *Spatiotemporal Receptive Fields: A Dynamical Model Derived from Cortical Architectonics.*

[Mallot, HA 87*] SEE: von Seelen, W 87a. *Characteristics of Neuronal Systems in the Visual Cortex.*

[Mallot, HA 87*] SEE: von Seelen, W 87b. *Properties of Cortical Systems - A Basis for Computer Vision.*

[Mallot, HA 88*] SEE: von Seelen, W 88a. *Parallelism and Redundancy in Neural Networks.*

[Mallot, HA 88a] Mallot, Hanspeter A. (1988). *Why Corticies? Neural Computation in the Vertebrate Visual System.* In Eckmiller, Rolf (ed.); von der Malsburg, Christoph (ed.). Neural Computers. NATO ASI Series, Volume F41. pp 129-138. Berlin: Springer-Verlag.

[Malone, JM 84a] Malone, James M. (1984). *Regression Without Models: Directions in the Search for Structure.* In Farlow, Stanley J. (ed.). Self-Organizing Methods in Modeling: GMDH Type Algorithms. pp 67-86. New York: Marcel Dekker.

[Malpass, ML 87*] SEE: Lippman, RP 87a. A Comparison of Hamming and Hopfield Neural Nets for Pattern Classification.

[Malpass, ML 87*] SEE: Gold, B 87a. *Some Neural Net Recognition Results on Isolated Words.*

[Mamikunian, D 87*] SEE: Honrubia, V 87a. *Morphological Aspects of the Vestibular Nerve of the Squirrel Monkey.*

[Mandell, AJ 87*] SEE: Gremillion, M 87a. *Neural Nets with Complex Structure: A Model of the Visual System.*

[Mandler, G 64*] SEE: Tulving, E 64a. *Interaction of Two Sources of Information in Tachistoscopic Word Recognition.*

[Mangis, JK 88*] SEE: Vogl, TP 88a. *Accelerating the Convergence of the Back-Propagation Method.*

[Mann, J 87*] SEE: Raffel, JI 87a. *A Generic Architecture for Wafer-Scale Neuromorphic Systems.*

[Mann, J 88a] Mann, James; Lippmann, Richard P.; Berger, Bob; Raffel, Jack I. (1988). *A Self-Organizing Neural Net*

Chip. IEEE Custom Integrated Circuits Conference, 1988: Proceedings. pp 10.3.1-10.3.5. New York: Institute of Electrical and Electronics Engineers.

[Mantey, PE 67*] SEE: Widrow, B 67a. *Adaptive Antenna Systems*.

[Marc, ME 86*] SEE: Axelrad, H 86a. *Recurrent Collateral Inhibition Simulated in a Simple Neuronal Automata Assembly*.

[Marcelja, S 80a] Marcelja, S. (1980). *Electrical Coupling of Photoreceptors in Retinal Network Models*. Biological Cybernetics. Volume 39. pp 15-20.

[Marcellus, DH 86*] SEE: Hudak, MJ 86a. *Demon-Based Associative Memories*.

[Marcus, CM 88a] Marcus, Charles M.; Westervelt, R. M. (1988). *Basins of Attraction for Electronic Neural Networks*. In Anderson, Dana Z. (ed.). Neural Information Processing Systems. pp 524-433. New York: American Institute of Physics.

[Marcus, CM 88b] Marcus, Charles M.; Westervelt, R. M. (in press). *Stability of Analog Neural Networks with Delay*. Physical Review A.

[Marcus, R 87a] Marcus, Robert. (1987). *A Connectionist Algorithm for Minimum Cost Network Flows*. IEEE International Conference on Neural Networks (ICNN-87), San Diego, CA, June 21-24, 1987: Proceedings. Volume 3. pp 735-740. New York: Institute of Electrical and Electronics Engineers.

[Maren, AJ 88a] Maren, A. J.; Minsky, V.; Ali, M. (1988). *A Multilayer Cooperative/Competitive Method for Creating Hierarchical Structures by Clustering Maximally-Related Nodes*. IEEE International Conference on Neural Networks (ICNN-88), San Diego, CA, July 24-27, 1988: Proceedings. Volume 2. pp 95-105. New York: Institute of Electrical and Electronics Engineers.

[Maren, AJ 88b] Maren, A. J.; Ali, M. (1988). *Hierarchical Scene Structure Representations to Facilitate Image Understanding*. International Conference on Industrial and Engineering Applications of Artificial Intelligence and Expert Systems, 1st, Tullahoma, TN, June 1-3, 1988. pp 657-667. Tullahoma, TN: University of Tennessee, Space Institute.

[Marill, T 60a] Marill, T.; Green, D. M. (1960). *Statistical Recognition Functions and the Design of Pattern Recognizers*. IRE Transactions on Electronic Computers. Volume EC-9. pp 472-477.

[Marill, T 63a] Marill, T.; Green, D. M. (1963). *On the Effectiveness of Receptors in Recognition Systems*. IEEE Transactions on Information Theory. Volume IT-9. pp 11-17.

[Marinaro, M 88*] SEE: Caianiello, ER 88a. *The Inverse Problem for Linear Boolean Nets*.

[Marko, H 70a] Marko, Hans; Giebel, H. (1970). *Recognition of Handwritten Characters with a System of Homogeneous Layers*. Nachrichtentechnische Zeitschrift. Volume 23. pp 455-459.

[Marko, H 74a] Marko, Hans. (1974). *A Biological Approach to Pattern Recognition*. IEEE Transactions on Systems, Man and Cybernetics. Volume SMC-4. pp 34-39.

[Marks II, RJ 87*] SEE: Cheung, KF 87a. *Conventional and Composite Matched Filters with Error Correction: A Comparison*.

[Marks II, RJ 87*] SEE: Cheung, KF 87b. *Neural Net Associative Memories Based on Convex Set Projections*.

[Marks II, RJ 87*] SEE: Cheung, KF 87c. *Synchronous vs Asynchronous Behavior of Hopfield's CAM Neural Net*.

[Marks II, RJ 87a] Marks II, Robert Jackson; Atlas, Les E.; Cheung, Kwan F. (1987). *A Class of Continuous Level Neural Nets*. Optics and the Information Age. SPIE Volume 813. pp 29-30. Bellingham, WA: Society of Photo-Optical

Instrumentation Engineers.

[Marks II, RJ 87b] Marks II, Robert Jackson. (1987). *Class of Continuous Level Associative Memory Neural Nets.* Applied Optics. Volume 26. pp 2005-2010.

[Marks II, RJ 87c] Marks II, Robert Jackson; Ritcey, James A.; Atlas, Les E.; Cheung, Kwan F. (1987). *Composite Matched Filter Output Partitioning.* Applied Optics. Volume 26. pp 2274-2278.

[Marks II, RJ 87d] Marks II, Robert Jackson; Atlas, Les E. (1987). *Composite Matched Filtering with Error Correction.* Optics Letters. Volume 12. pp 135-137.

[Marks II, RJ 88*] SEE: Atlas, LE 88a. *An Artificial Neural Network for Spatio-Temporal Bipolar Patterns: Application to Phoneme Classification.*

[Marks II, RJ 88*] SEE: Homma, T 88a. *An Artificial Neural Network for Spatio-Temporal Bipolar Patterns: Application to Phoneme Classification.*

[Marks II, RJ 88*] SEE: Oh, S 88a. *Effect of Clock Skew in Iterative Neural Network and Optical Feedback Processors.*

[Marks II, RJ 88a] Marks II, Robert Jackson; Atlas, Les E.; Oh, Seho; Cheung, Kwan F. (1988). *Architectures for a Continuous Level Neural Network Based on Alternating Orthogonal Projections.* Neural Network Models for Optical Computing. SPIE Volume 882. pp 90-92. Bellingham, WA: Society of Photo-Optical Instrumentation Engineers.

[Marks II, RJ 88b] Marks II, Robert Jackson; Atlas, Les E.; Oh, Seho. (1988). *Generalization in Layered Classification Neural Networks.* IEEE International Symposium on Circuits and Systems, Helsinki, June 7-9, 1988: Proceedings. pp 503-506. New York: Institute of Electrical and Electronics Engineers.

[Marks II, RJ 88c] Marks II, Robert Jackson; Atlas, Les E.; Choi, J. J.; Oh, Seho; Cheung, Kwan F.; Park, Dong Chul. (1988). *Nonlinearity Requirements for Correlation Based Associative Memories.* Optical Computing and Nonlinear Materials. SPIE Volume 881. pp 179-183. Bellingham, WA: Society of Photo-Optical Instrumentation Engineers.

[Marks II, RJ 88d] Marks II, Robert Jackson; Atlas, Les E.; Oh, Seho; Cheung, Kwan F. (1988). *Optical-Processor Architectures for Alternating-Projection Neural Networks.* Optics Letters. Volume 13. pp 533-535.

[Marks II, RJ 88e] Marks II, Robert Jackson; Atlas, Les E.; Choi, J. J.; Oh, Seho; Cheung, Kwan F.; Park, Dong Chul. (1988). *Performance Analysis of Associative Memories with Nonlinearities in the Correlation Domain.* Applied Optics. Volume 27. pp 2900-2904.

[Marks II, RJ 88f] Marks II, Robert Jackson; Atlas, Les E.; Park, Dong Chul; Oh, Seho. (1988). *The Effect of Stochastic Interconnects in Artificial Neural Network Classification.* IEEE International Conference on Neural Networks (ICNN-88), San Diego, CA, July 24-27, 1988: Proceedings. Volume 2. pp 437-442. New York: Institute of Electrical and Electronics Engineers.

[Marks II, RJ 88g] Marks II, Robert Jackson; Atlas, Les E.; Oh, Seho; Ritcey, James A. (1988). *The Performance of Convex Set Projection Based Neural Networks.* In Anderson, Dana Z. (ed.). Neural Information Processing Systems. pp 534-543. New York: American Institute of Physics.

[Marmarelis, VZ 83*] SEE: Curlander, JC 83a. *Processing of Visual Information in the Distal Neurons of the Vertebrate Retina.*

[Marom, E 85*] SEE: Dunning, GJ 85a. *All-Optical Associative Holographic Memory with Feedback Using Phase Conjugate Mirrors.*

[Marom, E 86*] SEE: Soffer, BH 86a. *Associative Holographic Memory with Feedback Using Phase-Conjugate Mirrors.*

[Marom, E 86*] SEE: Owechko, Y 86b. *Associative Memory in a Phase Conjugate Resonator Cavity Utilizing a Hologram.*

[Marom, E 86*] SEE: Owechko, Y 86a. *Associative Memory in a Phase Conjugate Resonator Cavity Utilizing a Hologram.*

[Marom, E 86*] SEE: Dunning, GJ 86a. *Optical Holographic Associative Memory Using a Phase Conjugate Resonator.*

[Marom, E 87*] SEE: Dunning, GJ 87a. *All-Optical Associative Memory with Shift Invariance and Multiple-Image Recall.*

[Marom, E 87*] SEE: Owechko, Y 87a. *Holographic Associative Memory with Nonlinearities in the Correlation Domain.*

[Marom, E 87*] SEE: Owechko, Y 87c. *Video Rate Optical Associative Memory with Shift Invariance.*

[Marple, SL 87a] Marple, S. Lawrence. (1987). Digital Spectral Analysis with Applications. Englewood Cliffs, NJ: Prentice-Hall.

[Marr, D 69a] Marr, David. (1969). *A Theory of Cerebellar Cortex.* Journal of Physiology. Volume 202. pp 437-470.

[Marr, D 70a] Marr, David. (1970). *A Theory of Cerebral Neocortex.* Proceedings of the Royal Society of London, Series B. Volume 176. pp 161-234.

[Marr, D 76a] Marr, David; Poggio, Tomaso A. (1988). *Cooperative Computation of Stereo Disparity.* In Anderson, James A. (ed.); Rosenfeld, Edward (ed.). Neurocomputing: Foundations of Research. pp 261-268. Cambridge, MA: MIT Press.

[Marr, D 76b] Marr, David; Poggio, Tomaso A. (1976). *Cooperative Computation of Stereo Disparity.* Science. Volume 194. pp 283-287.

[Marr, D 76c] Marr, David. (1976). From Computational Theory to Psychology and Neurophysiology. Working Paper 131. Cambridge, MA: Massachusetts Institute of Technology, AI Laboratory.

[Marr, D 77a] Marr, David; Poggio, Tomaso A. (1977). *From Understanding Computation to Understanding Neural Circuitry.* Neurosciences Research Progress Bulletin. Volume 15. pp 470-488.

[Marr, D 77b] Marr, David. (1977). Representing Visual Information. AIM-465. Cambridge, MA: Massachusetts Institute of Technology, AI Laboratory.

[Marr, D 78a] Marr, David; Palm, Gunther; Poggio, Tomaso A. (1978). *Analysis of a Cooperative Stereo Algorithm.* Biological Cybernetics. Volume 28. pp 223-239.

[Marr, D 78b] Marr, David; Nishihara, H. K. (1978). *Representation and Recognition of the Spatial Organization of Three-Dimensional Shapes.* Proceedings of the Royal Society of London, Series B. Volume 200. pp 269-294.

[Marr, D 79a] Marr, David; Poggio, Tomaso A. (1979). *A Computational Theory of Human Stereo Vision.* Proceedings of the Royal Society of London, Series B. Volume 204. pp 301-328.

[Marr, D 82a] Marr, David. (1982). Vision. New York: W H Freeman.

[Marrakchi, A 87*] SEE: Von Lehmen, A 87a. *Influence of Interconnection Weight Discretization and Noise in an Optoelectronic Neural Network.*

[Marrakchi, A 88*] SEE: Von Lehmen, A 88a. *Factors Influencing Learning by Backpropagation.*

[Marrian, CRK 87a] Marrian, C. R. K.; Peckerar, M. C. (1987). *Electronic 'Neural' Net Algorithm for Maximum Entropy Deconvolution.* IEEE International Conference on Neural Networks (ICNN-87), San Diego, CA, June 21-24, 1987: Proceedings. Volume 3. pp 749-758. New York: Institute of Electrical and Electronics Engineers.

[Marrian, CRK 88*] SEE: Culhane, AD 88a. A Neural Net Approach to Discrete Fourier Transforms.

[Marroquin, JL 85*] SEE: Koch, C 85a. Analog Neuronal Networks in Early Vision.

[Marroquin, JL 85a] Marroquin, Jose Luis. (1985). Probabilistic Solution of Inverse Problems. PhD Thesis. Cambridge, MA: Massachusetts Institute of Technology.

[Marroquin, JL 86*] SEE: Koch, C 86a. *Analog Neuronal Networks in Early Vision.*

[Marshall, J 87*] SEE: Grossberg, S 87a. *A Computational Model of How Cortical Complex Cells Multiplex Information About Position, Contrast, Orientation, Spatial Frequency, Disparity.*

[Marshall, JC 82*] SEE: Arbib, MA 82b. Neural Models of Language Processes.

[Marslen-Wilson, WD 78a] Marslen-Wilson, William D.; Welsh, Alan. (1978). *Processing Interactions and Lexical Access During Word Recognition in Continuous Speech.* Cognitive Psychology. Volume 10. pp 29-63.

[Marsocci, VA 82a] Marsocci, Velio A. (1982). *Electrical Network Modeling of Active Membranes of Nerves.* CRC Critical Reviews in Biomedical Engineering. Volume 8. pp 135-194.

[Martin, FC 60*] SEE: Hay, JC 60a. *The Mark I Perceptron - Design and Performance.*

[Martin, T 62*] SEE: Mueller, P 62a. *General Principles of Operations in Neuron Nets with Application to Acoustical Pattern Recognition.*

[Martindale, C 88a] Martindale, Colin. (1988). *Aesthetics, Psychobiology, and Cognition.* In Farley, F. (ed.); Neperud, R. (ed.). The Foundations of Aesthetics, Art, and Art Education. pp 7-42. New York: Praeger.

[Martindale, C 88b] Martindale, Colin; Moore, Kathleen. (1988). *Priming, Prototypicality, and Preference.* Journal of Experimental Psychology: Human Perception and Performance. Volume 14. pp 661-670.

[Martinez, TR 88a] Martinez, Tony R.; Vidal, Jacques J. (1988). *Adaptive Parallel Logic Networks.* Journal of Parallel and Distributed Computing. Volume 5. pp 26-57.

[Martland, D 87a] Martland, D. (1987). *Auto-Associative Pattern Storage Using Synchronous Boolean Networks.* IEEE International Conference on Neural Networks (ICNN-87), San Diego, CA, June 21-24, 1987: Proceedings. Volume 3. pp 355-366. New York: Institute of Electrical and Electronics Engineers.

[Martland, D 87b] Martland, D. (1987). *Behaviour of Autonomous, (Synchronous), Boolean Networks.* IEEE International Conference on Neural Networks (ICNN-87), San Diego, CA, June 21-24, 1987: Proceedings. Volume 2. pp 243-250. New York: Institute of Electrical and Electronics Engineers.

[Martland, D 88a] Martland, D. (in press). *Configuration Boolean Networks.* ESPCI European Conference on Neural Networks (nEuro'88), 1st, Paris, June 6-9, 1988: Proceedings.

[Marudarajan, AR 82*] SEE: Lee, MH 82a. *A Computer Package for the Evaluation of Neuron Models Involving Large Uniform Networks.*

[Maruyama, M 87*] SEE: Amari, S 87b. *On the Topological Representation of Signals in Self-Organizing Nerve Fields.*

[Mason, M 75a] Mason, Mildred. (1975). *Reading Ability and Letter Search Time: Effects of Orthographic Structure Defined by Single-Letter Positional Frequency.* Journal of Experimental Psychology: General. Volume 104. pp 146-166.

[Massaro, DW 79a] Massaro, Dominic W.; Venezky, Richard L.; Taylor, Glen A. (1979). *Orthographic Regularity, Positional Frequency, and Visual Processing of Letter Strings.* Journal of Experimental Psychology: General. Volume 108. pp 107-124.

[Massaro, DW 82*] SEE: Hary, JM 82a. *Categorical Results do not Imply Categorical Perception.*

[Massaro, DW 87a] Massaro, Dominic W.; Cohen, Michael M. (1987). *Process and Connectionist Models of Pattern Recognition.* Cognitive Science Society, 9th, Seattle, WA, July 16-18, 1987: Proceedings. pp 258-264. Hillsdale, NJ: Lawrence Erlbaum.

[Mastebroek, HAK 88*] SEE: Schuling, FH 88a. *An Adaptive and Heterodyne Filtering Procedure for the Imaging of Moving Objects.*

[Masterbroek, HAK 83*] SEE: Zaagman, WH 83a. *Adaptive Strategies in Fly Vision: On Their Image-Processing Qualities.*

[Matano, K 71*] SEE: Suzuki, R 71a. *Dynamics of "Neuron Ring".*

[Matheus, CJ 86*] SEE: Jorgensen, C 86a. *Catching Knowledge in Neural Nets.*

[Matheus, CJ 87a] Matheus, Christopher J.; Hohensee, William E. (1987). *Learning in Artificial Neural Systems.* Computer Intelligence. Volume 3. pp 283-294.

[Matsumoto, H 84*] SEE: Tsutsumi, K 84b. *Ring Neural Network qua a Generator of Rhythmic Oscillation with Period Control Mechanism.*

[Matsumoto, H 84*] SEE: Tsutsumi, K 84a. *A Synaptic Modification Algorithm in Consideration of the Generation of Rhythmic Oscillation in a Ring Neural Network.*

[Matsumoto, H 87*] SEE: Tsutsumi, K 87a. *Neural Computation and Learning Strategy for Manipulator Position Control.*

[Matsumoto, H 88*] SEE: Tsutsumi, K 88a. *Neural Computation for Controlling the Configuration of 2-Dimentional Truss Structure.*

[Matsuo, K 87a] Matsuo, Kazuhiro. (1987). *Self-Organization of Cooperative and Competitive Networks in Ecological Paradigm.* IEEE International Conference on Neural Networks (ICNN-87), San Diego, CA, June 21-24, 1987: Proceedings. Volume 3. pp 277-280. New York: Institute of Electrical and Electronics Engineers.

[Matsuo, K 88*] SEE: Hosogi, S 88a. *Cerebellar Neural Network Model for Manipulator Control.*

[Matsuo, T 79*] SEE: Hoshimiya, N 79a. *Two-Terminal Electronic Circuit Neuron Model with Excitable Membrane V-I-t Characteristics.*

[Matsuo, T 80*] SEE: Koga, K 80a. *A Hardware Neuron Model as a Component of the Neural Network.*

[Matsuo, T 84*] SEE: Takahashi, K 84a. *Integration of Multi-Microelectrode and Interface Circuits by Silicon Planar and Three-Dimensional Fabrication Technology.*

[Matsuzaki, K 86*] SEE: Wallace, R 86a. *Progress in Robot Road-Following.*

[Matthies, H 86a] Matthies, Hansjurgen (ed.). (1986). Learning and Memory: Mechanisms of Information Storage in the Nervous System. Oxford, England: Pergamon.

[Matthies, L 87a] Matthies, Larry; Szeliski, Richard; Kanade, Takeo. (1987). Kalman Filter-Based Algorithms for Estimating Depth from Image Sequences. CMU-CS-87-185. Pittsburgh, PA: Carnegie Mellon University, Computer Science Department.

[Matts, JP 88*] SEE: Irani, EA 88a. Using Artificial Neural Nets for Statistical Discovery: Observations after Using Back-propagation, Expert Systems, and Multiple-Linear Regression.

[Mattson, RL 59a] Mattson, R. L. (1959). *A Self-Organizing Logical System.* Eastern Joint Computer Conference, Boston, MA, December 1-3, 1959: Convention Record. pp 212-217.

[Mattson, RL 60a] Mattson, R. L. (1960). An Approach to Pattern Recognition Using Linear Threshold Devices. Report LMSD-702680. Lockheed Missiles and Space Company.

[Mattson, RL 62*] SEE: Fischler, M 62a. *An Approach to General Pattern Recognition.*

[Mattson, RL 62a] Mattson, R. L. (1962). The Analysis and Synthesis of Adaptive Systems Which Use Networks of Threshold Elements. Technical Report 1553-3. Stanford, CA: Stanford University, Stanford Electronics Laboratory.

[Mattson, RL 63a] Mattson, R. L.; Firschein, O.; Fischler, M. (1963). *An Experimental Investigation of a Class of Pattern Recognition Synthesis Algorithms.* IEEE Transactions on Electronic Computers. Volume EC-14.

[Mattson, RL 65a] Mattson, R. L.; Dammann, J. E. (1965). *A Technique for Determining and Coding Subclasses in Pattern Recognition Problems.* IBM Journal of Research and Development. Volume 9. pp 294-302.

[Maturana, HR 59*] SEE: Lettvin, JY 59a. *What the Frog's Eye Tells the Frog's Brain.*

[Maxwell, T 86*] SEE: Chen, HH 86a. *High Order Correlation Model for Associative Memory.*

[Maxwell, T 86a] Maxwell, Tom; Giles, C. Lee; Lee, Y. C.; Chen, H. H. (1986). *Nonlinear Dynamics of Artificial Neural Systems.* In Denker, John S. (ed.). Neural Networks for Computing, Snowbird, Utah, 1986. AIP Proceedings 151. pp 299-304. New York: American Institute of Physics.

[Maxwell, T 87*] SEE: Giles, CL 87a. *Learning Invariance, and Generalization in High-Order Neural Networks.*

[Maxwell, T 87a] Maxwell, Tom; Giles, C. Lee; Lee, Y. C. (1987). *Generalization in Neural Networks: The Contiguity Problem.* IEEE International Conference on Neural Networks (ICNN-87), San Diego, CA, June 21-24, 1987: Proceedings. Volume 2. pp 41-46. New York: Institute of Electrical and Electronics Engineers.

[Maxwell, T 88*] SEE: Giles, CL 88a. *Encoding Geometric Invariances in Higher-Order Neural Networks.*

[May, D 88a] May, David; Shepherd, Roger. (1988). *The Transputer.* In Eckmiller, Rolf (ed.); von der Malsburg, Christoph (ed.). Neural Computers. NATO ASI Series, Volume F41. pp 477-486. Berlin: Springer-Verlag.

[May, N 88a] May, Norm; Hammerstrom, Daniel. (1988). Fault Simulation of a Wafer-Scale Integrated Neural Network. CS/E-88-020. Beaverton, OR: Oregon Graduate Center.

[May, N 88b] May, Norm; Hammerstrom, Daniel. (1988). Fltsim Detailed Description and Operation. CS/E-88-021. Beaverton, OR: Oregon Graduate Center.

[May, RM 76a] May, Robert M. (1976). *Simple Mathematical Models with Very Complicated Dynamics.* Nature. Volume 261. pp 459-467.

[Mays, CH 63*] SEE: Talbert, LR 63a. A Real-Time Adaptive Speech-Recognition System.

[Mays, CH 63a] Mays, C. Hugh. (1963). Adaptive Threshold Logic. Technical Report 1557-1. Stanford, CA: Stanford University, Stanford Electronics Laboratory.

[Mays, CH 64a] Mays, C. Hugh. (1964). *Effects of Adaptation Parameters on Convergence Time and Tolerance for Adaptive Threshold Elements.* IEEE Transactions on Electronic Computers. Volume EC-13. pp 465-468.

[Mazaika, PK 87a] Mazaika, Paul K. (1987). *A Mathematical Model for the Boltzmann Machine.* IEEE International Conference on Neural Networks (ICNN), San Diego, CA, June 21-24, 1987: Proceedings. Volume 3. pp 157-164. New York: Institute of Electrical and Electronics Engineers.

[Mazurov, VD 87a] Mazurov, Vl D.; Krivonogov, Al I.; Kazantsev, Vl S. (1987). *Solving of Optimization and Identification*

Problems by the Committee Methods. Pattern Recognition. Volume 20. pp 371-378.

[McAulay, AD 87a] McAulay, Alastair D. (1987). *Engineering Design Neural Networks Using Split Inversion Learning*. IEEE International Conference on Neural Networks (ICNN-87), San Diego, CA, June 21-24, 1987: Proceedings. Volume 4. pp 635-642. New York: Institute of Electrical and Electronics Engineers.

[McAuliffe, KP 83*] SEE: Gottlieb, A 83a. *The NYU Ultracomputer-Designing an MIMD Shared Memory Parallel Computer*.

[McBeth, JH 87a] McBeth, J. Harold. (1987). Simulated Annealing. San Diego, CA: General Dynamics, Electronics Division.

[McClelland, JL 73*] SEE: Johnston, JC 73a. *Visual Factors in Word Perception*.

[McClelland, JL 74*] SEE: Johnston, JC 74a. *Perception of Letters in Words: Seek Not and Ye Shall Find*.

[McClelland, JL 75*] SEE: Jackson, MD 75a. *Sensory and Cognitive Determinants of Reading Speed*.

[McClelland, JL 75a] McClelland, James L. (1976). *Preliminary Letter Identification in the Perception of Words and Nonwords*. Journal of Experimental Psychology: Human Perception and Performance. Volume 2. pp 80-91.

[McClelland, JL 77a] McClelland, James L. (1977). *Letter and Configuration Information in Word Identification*. Journal of Verbal Learning and Verbal Behavior. Volume 16. pp 137-150.

[McClelland, JL 77b] McClelland, James L.; Johnston, James C. (1977). *The Role of Familiar Units in Perception of Words and Nonwords*. Perception and Psychophysics. Volume 22. pp 249-261.

[McClelland, JL 78a] McClelland, James L. (1978). *Cognitive Psychology: The Way We Were*. Contemporary Psychology. Volume 23. pp 860-861.

[McClelland, JL 78b] McClelland, James L. (1978). *Perception and Masking of Wholes and Parts*. Journal of Experimental Psychology: Human Perception and Performance. Volume 4. pp 210-223.

[McClelland, JL 78c] McClelland, James L. (1978). *Phenomenology of Perception*. Science. Volume 201. pp 899-900.

[McClelland, JL 79*] SEE: Jackson, MD 79a. *Processing Determinants of Reading Speed*.

[McClelland, JL 79a] McClelland, James L. (1979). *On the Time Relations of Mental Processes: An Examination of Systems of Processes in Cascade*. Psychological Review. Volume 86. pp 287-330.

[McClelland, JL 79b] McClelland, James L.; Miller, Jeff. (1979). *Structural Factors in Figure Perception*. Perception and Psychophysics. Volume 26. pp 221-229.

[McClelland, JL 80*] SEE: Larochelle, S 80a. *Context and the Allocation of Resources in Word Recognition*.

[McClelland, JL 80*] SEE: Johnston, JC 80a. *Experimental Tests of a Hierarchical Model of Word Identification*.

[McClelland, JL 81*] SEE: Goodman, GO 81a. *The Role of Syntactic Context in Word Recognition*.

[McClelland, JL 81a] McClelland, James L.; Rumelhart, David E. (1981). *An Interactive Activation Model of Context Effects in Letter Perception, Part 1: An Account of Basic Findings*. Psychological Review. Volume 88. pp 375-407.

[McClelland, JL 81b] McClelland, James L.; Rumelhart, David E. (1988). *An Interactive Activation Model of Context Effects in Letter Perception: Part 1, An Account of Basic Findings*. In Anderson, James A. (ed.); Rosenfeld, Edward (ed.). Neurocomputing: Foundations of Research. pp 404-436. Cambridge, MA: MIT Press.

[McClelland, JL 81c] McClelland, James L.; O'Regan, J. K. (1981). *Expectations Increase the Benefit Derived from Parafoveal*

Visual Information in Reading Words Aloud. Journal of Experimental Psychology: Human Perception and Performance. Volume 7. pp 634-644.

[McClelland, JL 81d] McClelland, James L.; O'Regan, J. K. (1981). *On Visual and Contextual Factors in Reading: A Reply to Rayner and Slowiaczek*. Journal of Experimental Psychology: Human Perception and Performance. Volume 7. pp 652-657.

[McClelland, JL 82*] SEE: Rumelhart, DE 82a. *An Interactive Activation Model of Context Effects in Letter Perception: Part 2*.

[McClelland, JL 83*] SEE: Elman, JL 83a. Speech Perception as a Cognitive Process: The Interactive Activation Model.

[McClelland, JL 85*] SEE: Rumelhart, DE 85c. *Levels Indeed! A Response to Broadbent*.

[McClelland, JL 85a] McClelland, James L.; Rumelhart, David E. (1985). *Distributed Memory and the Representation of General and Specific Information*. Journal of Experimental Psychology. Volume 114. pp 159-188.

[McClelland, JL 85b] McClelland, James L. (1985). *Putting Knowledge in its Place: A Scheme for Programming Parallel Processing Structures on the Fly*. Cognitive Science. Volume 9. pp 113-146.

[McClelland, JL 86*] SEE: Hinton, GE 86b. *Distributed Representations*.

[McClelland, JL 86*] SEE: Elman, JL 86a. *Exploiting Lawful Variability in the Speech Wave*.

[McClelland, JL 86*] SEE: Rumelhart, DE 86b. *General Framework for Parallel Distributed Processing*.

[McClelland, JL 86*] SEE: Rumelhart, DE 86g. *On Learning the Past Tenses of English Verbs*.

[Mcclelland, JL 86*] SEE: Rumelhart, DE 86h. Parallel Distributed Processing: Explorations in the Microstructure of Cognition.

[McClelland, JL 86*] SEE: Rumelhart, DE 86i. *PDP Models and General Issues in Cognitive Science*.

[McClelland, JL 86*] SEE: Rumelhart, DE 86j. *Schemata and Sequential Thought Processes in PDP Models*.

[McClelland, JL 86a] McClelland, James L.; Rumelhart, David E. (1986). *Amnesia and Distributed Memory*. In Rumelhart, David E. (ed.). Parallel Distributed Processing: Explorations in the Microstructure of Cognition. Volume II. pp 503-528. Cambridge, MA: MIT Press.

[McClelland, JL 86b] McClelland, James L.; Rumelhart, David E.; Hinton, Geoffrey E. (1986). *Appeal of Parallel Distributed Processing*. In Rumelhart, David E. (ed.). Parallel Distributed Processing: Explorations in the Microstructure of Cognition. Volume I, Foundations. pp 3-44. Cambridge, MA: MIT Press.

[McClelland, JL 86c] McClelland, James L.; Feldman, Jerome A.; Adelson, Beth; Bower, Gordon H.; McDermott, Drew. (1986). *Connectionist Models and Cognitive Science: Goals, Directions and Implications*. Connectionist Models and Cognitive Science: Goals, Directions and Implications: Report to the National Science Foundation. np. np.

[McClelland, JL 86d] McClelland, James L.; Rumelhart, David E. (1986). *Distributed Model of Human Learning and Memory*. In Rumelhart, David E. (ed.). Parallel Distributed Processing: Explorations in the Microstructure of Cognition. Volume II. pp 170-215. Cambridge, MA: MIT Press.

[McClelland, JL 86e] McClelland, James L.; Elman, Jeffrey L. (1986). *Interactive Processes in Speech Perception: The TRACE Model*. In Rumelhart, David E. (ed.). Parallel Distributed Processing: Explorations in the Microstructure of Cognition. Volume II. pp 58-121. Cambridge, MA: MIT Press.

[McClelland, JL 86f] McClelland, James L.; Kawamoto, Alan H. (1986). *Mechanisms of Sentence Processing: Assigning Roles to*

Constituents. In Rumelhart, David E. (ed.). Parallel Distributed Processing: Explorations in the Microstructure of Cognition. Volume II. pp 272-326. Cambridge, MA: MIT Press.

[McClelland, JL 86g] McClelland, James L. (1986). Parallel Distributed Processing and Role Assignment Constraints. Pittsburgh, PA: Carnegie Mellon University, Computer Science Department.

[McClelland, JL 86h] McClelland, James L.; Mozer, Michael C. (1986). *Perceptual Interactions in Two-Word Displays: Familiarity and Similarity Effects*. Journal of Experimental Psychology: Human Perception and Performance. Volume 12. pp 18-35.

[McClelland, JL 86i] McClelland, James L. (1986). *Programmable Blackboard Model of Reading*. In Rumelhart, David E. (ed.). Parallel Distributed Processing: Explorations in the Microstructure of Cognition. Volume II. pp 122-169. Cambridge, MA: MIT Press.

[McClelland, JL 86j] McClelland, James L. (1986). *Resource Requirements of Standard and Programmable Nets*. In Rumelhart, David E. (ed.). Parallel Distributed Processing: Explorations in the Microstructure of Cognition. Volume I, Foundations. pp 460-487. Cambridge, MA: MIT Press.

[McClelland, JL 86k] McClelland, James L.; Elman, Jeffrey L. (1986). *The TRACE Model of Speech Perception*. Cognitive Psychology. Volume 18. pp 1-86.

[McClelland, JL 87a] McClelland, James L. (1987). Explaining PDP: A Handbook of Models, Programs and Exercises. Cambridge, MA: MIT Press.

[McClelland, JL 87b] McClelland, James L.; St John, Mark F. (1987). *Reconstructive Memory for Sentences: A PDP Approach*. Ohio University Inference Conference, 1986. np. np.

[McClelland, JL 87c] McClelland, James L. (1987). *The Basis of Lawful Behavior: Rules or Connections?*. The Computers and Philosophy Newsletter. Volume 2. pp 10-16.

[McClelland, JL 87d] McClelland, James L. (1987). The Case for Interactionism in Language Processing. Technical Report AIP-2. Pittsburgh, PA: Carnegie Mellon University, Psychology Department.

[McClelland, JL 88*] SEE: Servan-Schreiber, D 88a. Encoding Sequential Structure in Simple Recurrent Networks.

[McClelland, JL 88*] SEE: St John, MF 88a. Learning and Applying Contextual Constraints in Sentence Comprehension.

[McClelland, JL 88*] SEE: Hinton, GE 88a. *Learning Representations by Recirculation*.

[McClelland, JL 88*] SEE: Cohen, JC 88a. On the Control of Automatic Processes: A Parallel Distributed Processing Model of the Stroop Effect.

[McClelland, JL 88a] McClelland, James L.; Rumelhart, David E. (1988). Explorations in Parallel Distributed Processing: A Handbook of Models, Programs, and Exercises. Cambridge, MA: MIT Press.

[McClelland, JL 88b] McClelland, James L. (1988). Parallel Distributed Processing: Implications for Cognition and Development. Technical Report AIP-47. Pittsburgh, PA: Carnegie Mellon University, Psychology Department.

[McClelland, JL 88c] McClelland, James L. (1988). *Putting Knowledge in its Place: A Scheme for Programming Parallel Processing Structures on the Fly*. In Waltz, David L. (ed.); Feldman, Jerome A. (ed.). Connectionist Models and Their Implications: Readings from Cognitive Science. pp 63-96. Norwood, NJ: Ablex.

[McClung, B 87*] SEE: Horne, W 87a. *Application-Oriented Connection Systems for Pattern Acquisition and Recognition*.

[McCool, JM 75*] SEE: Widrow, B 75a. *Adaptive Noise Canceling: Principles and Applications*.

[McCool, JM 75*] SEE: Widrow, B 75b. *The Complex LMS Algorithm.*

[McCool, JM 76*] SEE: Widrow, B 76a. *A Comparison of Adaptive Algorithms Based on the Methods of Steepest Descent and Random Search.*

[McCool, JM 76*] SEE: Widrow, B 76b. *Stationary and Nonstationary Learning Characteristics of the LMS Adaptive Filter.*

[McCool, JM 77*] SEE: Widrow, B 77a. *Comments on "An Adaptive Recursive LMS Filter".*

[McCool, JM 78*] SEE: Dentino, M 78a. *Adaptive Filtering in the Frequency Domain.*

[McCulloch, WS 43*] SEE: Landahl, HD 43a. *A Statistical Consequence of the Logical Calculus of Nervous Nets.*

[McCulloch, WS 43a] McCulloch, Warren S.; Pitts, Walter H. (1988). *A Logical Calculus of the Ideas Immanent in Nervous Activity.* In Anderson, James A. (ed.); Rosenfeld, Edward (ed.). Neurocomputing: Foundations of Research. pp 18-28. Cambridge, MA: MIT Press.

[McCulloch, WS 43b] McCulloch, Warren S.; Pitts, Walter H. (1943). *A Logical Calculus of the Ideas Immanent in Nervous Activity.* Bulletin of Mathematical Biophysics. Volume 5. pp 115-133.

[McCulloch, WS 47*] SEE: Pitts, WH 47b. *How We Know Universals: The Perception of Auditory and Visual Forms.*

[McCulloch, WS 47*] SEE: Pitts, WH 47a. *How We Know Universals: The Perception of Auditory and Visual Forms.*

[McCulloch, WS 59*] SEE: Lettvin, JY 59a. *What the Frog's Eye Tells the Frog's Brain.*

[McCulloch, WS 69*] SEE: Kilmer, WL 69a. *A Model of the Vertebrate Central Command System.*

[McDermott, D 86*] SEE: McClelland, JL 86c. *Connectionist Models and Cognitive Science: Goals, Directions and Implications.*

[McEliece, RJ 86a] McEliece, R. J.; Posner, E. C.; Rodemich, E. R.; Venkatesh, Santosh S. (1986). Capacity of the Hopfield Associative Memory. Pasadena, CA: California Institute of Technology.

[McEliece, RJ 87a] McEliece, R. J.; Posner, E. C.; Rodemich, E. R.; Venkatesh, Santosh S. (1987). *The Capacity of the Hopfield Associative Memory.* IEEE Transactions on Information Theory. Volume IT-33. pp 461-482.

[McGill, KC 82a] McGill, Kevin C.; Cummins, Kenneth L.; Dorfman, Leslie J.; Berlizot, Bruno B.; Leutkemeyer, Kelly; Widrow, Bernard. (1982). *On the Nature and Elimination of Stimulus Artifact in Nerve Signals Evoked and Recorded Using Surface Electrodes.* IEEE Transactions on Biomedical Engineering. Volume BME-29. pp 129-137.

[McGuire, BA 84a] McGuire, Barbara A.; Stevens, John K.; Sterling, Peter. (1984). *Microcircuitry of Bipolar Cells in Cat Retina.* Journal of Neuroscience. Volume 4. pp 2920-2938.

[McGuire, BA 86a] McGuire, Barbara A.; Stevens, John K.; Sterling, Peter. (1986). *Microcircuitry of Beta Ganglion Cells in Cat Retina.* Journal of Neuroscience. Volume 6. pp 907-918.

[McIntyre, J 88*] SEE: Mussa-Ivaldi, FA 88c. *Theoretical and Experimental Perspectives on Arm Trajectory Formation: A Distributed Model of Motor Redundancy.*

[McKoon, G 80a] McKoon, Gail; Ratcliff, Roger. (1980). *Priming in Item Recognition: The Organization of Propositions in Memory for Text.* Journal of Verbal Learning and Verbal Behavior. Volume 19. pp 369-386.

[McLaren, RW 70*] SEE: Mendel, JM 70b. *Reinforcement-Learning Control and Pattern Recognition Systems.*

[McLean, CR 85*] SEE: Albus, JS 85a. *Hierarchical Control for Robots, and Teleoperators.*

[McNaughton, BL 89a] McNaughton, B. L. (1989). *Neuronal Mechanisms for Spatial Computation and Information Storage.* In Nadel, Lynn (ed.); Cooper, Lynn A. (ed.); Culicover, Peter (ed.); Harnish, R. Michael (ed.). Neural Connections, Mental Computation. pp 284-350. Cambridge, MA: MIT Press.

[McNeill, M 87a] McNeill, Martin. (1987). *Neural Network Paradigms.* IEEE International Conference on Neural Networks (ICNN-87), San Diego, CA, June 21-24, 1987: Proceedings. Volume 2. pp 79-89. New York: Institute of Electrical and Electronics Engineers.

[Mead, CA 85*] SEE: Sivilotti, MA 85a. *A Novel Associative Memory Implemented Using Collective Computation.*

[Mead, CA 85a] Mead, Carver A.; Wawrzynek, John. (1985). *A New Discipline for CMOS Design: An Architecture for Sound Synthesis.* Chapel Hill Conference on Very Large Scale Integration, 1985. pp 87-104. Rockville, MD: Computer Science Press.

[Mead, CA 85b] Mead, Carver A. (1985). *A Sensitive Electronic Photoreceptor.* Chapel Hill Conference on Very Large Scale Integration, 1985. pp 463-471. Rockville, MD: Computer Science Press.

[Mead, CA 86*] SEE: Sivilotti, MA 86a. *VLSI Architectures for Implementation of Neural Networks.*

[Mead, CA 87*] SEE: Maher, MA 87a. A Physical Charge-Controlled Model for MOS Transistors.

[Mead, CA 87*] SEE: Sivilotti, MA 87b. *Real-time Visual Computations Using Analog CMOS Processing Arrays.*

[Mead, CA 87*] SEE: Sivilotti, MA 87a. *Real-Time Visual Computations Using Analog CMOS Processing Arrays.*

[Mead, CA 87a] Mead, Carver A. (1987). *Neural Hardware for Vision.* Engineering and Science. Volume 50. pp 2-7.

[Mead, CA 87b] Mead, Carver A. (1987). *Silicon Models of Neural Computation.* IEEE International Conference on Neural Networks (ICNN-87), San Diego, CA, June 21-24, 1987: Proceedings. Volume 1. pp 91-106. New York: Institute of Electrical and Electronics Engineers.

[Mead, CA 88*] SEE: Lyon, RF 88a. *An Analog Electronic Cochlea.*

[Mead, CA 88*] SEE: Hutchinson, J 88a. *Computing Motion Using Analog and Binary Resistive Networks.*

[Mead, CA 88*] SEE: Koch, C 88b. *Computing Motion Using Resistive Networks.*

[Mead, CA 88a] Mead, Carver A.; Mahowald, M. A. (1988). *A Silicon Model of Early Visual Processing.* Neural Networks. Volume 1. pp 91-97.

[Mead, CA 89a] Mead, Carver A. (1987). Analog VLSI and Neural Systems. New York: Addison-Wesley.

[Mehler, J 88*] SEE: Pinker, S 88a. Connections and Symbols.

[Mehr, D 87a] Mehr, David; Richfield, Steve. (1987). *Neural Net Application to Optical Character Recognition.* IEEE International Conference on Neural Networks (ICNN-87), San Diego, CA, June 21-24, 1987: Proceedings. Volume 4. pp 771-778. New York: Institute of Electrical and Electronics Engineers.

[Mehra, RK 84*] SEE: Lebow, WM 84a. *Forecasting Applications of GMDH in Agricultural and Meteorological Time Series.*

[Meidan, M 87a] Meidan, M.; Amit, Daniel J.; Gutfreund, Hanoch; Sompolinsky, H. (1987). Electronic Analog Neural Network with Noise and Time Delays. Jerusalem: Hebrew University, Racah Institute of Physics.

[Mel, BW 86a] Mel, Bartlett W. (1986). *A Connectionist Learning Model for 3-Dimensional Mental Rotation, Zoom, and Pan.* Cognitive Science Society, 8th, Amherst, MA, August 15-17, 1986: Proceedings. pp 562-571. Hillsdale, NJ:

Lawrence Erlbaum.

[Mel, BW 86b] Mel, Bartlett W. (1986). *A Connectionist Model for Simple Visual Functions*. Working Paper Number 70. Champaign-Urbana, IL: University of Illinois, Coordinated Sciences Laboratory, AI Group.

[Mel, BW 88a] Mel, Bartlett W. (1988). *MURPHY: A Robot that Learns by Doing*. In Anderson, Dana Z. (ed.). Neural Information Processing Systems. pp 544-553. New York: American Institute of Physics.

[Melkonian, DS 82a] Melkonian, D. S.; Mkrtchian, H. H.; Fanardjian, V. V. (1982). *Simulation of Learning Processes in Neuronal Networks of the Cerebellum*. Biological Cybernetics. Volume 45. pp 79-88.

[Mendel, JM 68a] Mendel, J. M.; Zapalac, J. J. (1968). *The Application of Techniques of Artificial Intelligence to Control System Design*. In Leondes, C. T. (ed.). Advances in Control Systems: Theory and Application. Volume 6. pp 2-94.

[Mendel, JM 70a] Mendel, J. M. (ed.); Fu, King-Sun (ed.). (1970). Adaptation, Learning, and Pattern Recognition Systems: Theory and Applications. New York: Academic Press.

[Mendel, JM 70b] Mendel, J. M.; McLaren, R. W. (1970). *Reinforcement-Learning Control and Pattern Recognition Systems*. In Mendel, J. M. (ed.); Fu, King-Sun (ed.). Adaptation, Learning and Pattern Recognition Systems: Theory and Applications. pp 287-317. New York: Academic Press.

[Mendel, JM 70c] Mendel, J. M. (1970). *Synthesis of Quasi-Optimal Switching Surfaces by Means of Training Techniques*. Adaptative, Learning, and Pattern Recognition Systems: Theory and Applications. pp 163-195. New York: Academic Press.

[Menzilcioglu, O 86*] SEE: Annaratone, M 86a. *Warp Architecture and Implementation*.

[Merat, FL 74a] Merat, Frank L. (1974). A Content Addressable Memory Based on Walsh Holography. Master's Thesis. Cleveland, OH: Case Western Reserve University.

[Merat, FL 75*] SEE: Pao, Y 75a. *Distributed Associative Memory for Patterns*.

[Mercer, RL 83*] SEE: Bahl, LR 83a. *A Maximum Likelihood Approach to Continuous Speech Recognition*.

[Merrill, W 79*] SEE: Levy, WB 79a. *Associative-Contingency Learning and Neuronal Networks*.

[Mesrobian, E 87a] Mesrobian, Edmond; Skrzypek, Josef. (1987). *A Connectionist Architecture for Computing Textural Segmentation*. Conference on Electro-Optic Imaging Systems and Devices. SPIE Proceedings Volume 755.

[Mesrobian, E 87b] Mesrobian, Edmond; Skrzypek, Josef. (1987). *Discrimination of Natural Textures: A Neural Network Architecture*. IEEE International Conference on Neural Networks (ICNN-87), San Diego, CA, June 21-24, 1987: Proceedings. Volume 4. pp 247-258. New York: Institute of Electrical and Electronics Engineers.

[Messner, RA 85a] Messner, Richard A.; Szu, Harold H. (1985). *An Image Processing Architecture for Real Time Generation of Scale and Rotation Invariant Patterns*. Computer Vision, Graphics, and Image Processing. Volume 31. pp 50-66.

[Messner, RA 86*] SEE: Szu, HH 86a. *Adaptive Invariant Novelty Filters*.

[Metelli, F 74a] Metelli, Fabio. (1974). *The Perceptron of Transparency*. Scientific American. Volume 230. pp 91-98.

[Metelli, F 85a] Metelli, Fabio; Da Pos, Osvaldo; Cavedon, Adele. (1985). *Balanced and Unbalanced, Complete and Partial Transparency*. Perception and Psychophysics. Volume 38. pp 354-366.

[Meyer, G 88*] SEE: Ross, MD 88a. *Macular Bioaccelerometers on Earth and in Space*.

[Meyer, G 88*] SEE: Ross, MD 88c. *The Neuroanatomical Substrate for Information Processing in Macular Endorgans.*

[Meyer, GE 83a] Meyer, Glenn E.; Senecal, Mary. (1983). *The Illusion of Transparency and Chromatic Subjective Contours.* Perception and Psychophysics. Volume 34. pp 58-64.

[Meystel, A 86a] Meystel, A. (1986). *Cognitive Control of Autonomous Mobile Robots: Nested Hierarchical Intelligent Module.* Cognitive Science Society, 8th, Amherst, MA, August 15-17, 1986: Proceedings. pp 747-764. Hillsdale, NJ: Lawrence Erlbaum.

[Mezard, M 84*] SEE: Gross, DJ 84a. *Simplest Spin Glass.*

[Mezard, M 84a] Mezard, M.; Parisi, Giorgio; Sourlas, N.; Toulouse, G.; Virasoro, M. A. (1984). *Nature of the Spin-Glass Phase.* Physical Review Letters. Volume 52. pp 1156-1159.

[Mezard, M 84b] Mezard, M.; Parisi, Giorgio; Sourlas, N.; Toulouse, G.; Virasoro, M. A. (1984). *Replica Symmetry Breaking and the Nature of the Spin Glass Phase.* Journal de Physique. Volume 45. pp 843-854.

[Mezard, M 85a] Mezard, M.; Virasoro, M. A. (1985). *The Microstructure of Ultrametricity.* Journal de Physique. Volume 46. pp 1293-1307.

[Mezard, M 86a] Mezard, M. (1986). *On the Statistical Physics of Spin Glasses.* In Bienenstock, Elie L. (ed.). Disordered Systems and Biological Organization. pp 119 f. New York: Springer-Verlag.

[Mezard, M 88*] SEE: Gutfreund, H 88b. *Processing of Temporal Sequences in Neural Networks.*

[Miceli, WJ 86a] Miceli, William J. (1986). *Neural Processing Systems.* In Szu, Harold H. (ed.). Optical and Hybrid Computing. SPIE Volume 634. pp 349-351. Bellingham, WA: Society of Photo-Optical Instrumentation Engineers.

[Michael, C 87a] Michael, Chester. (June 15, 1987). *Dress Rehearsal for Real-Time Artificial Neural Networks.* Electronic Products. pp 19-28.

[Michalson, WR 87*] SEE: Green, PE 87a. *Real-Time Evidential Reasoning and Network Based Processing.*

[Michel, AN 88*] SEE: Li, JH 88a. *Qualitative Analysis and Synthesis of a Class of Neural Networks.*

[Michel, AN 88a] Michel, A. N.; Farrell, J. A.; Li, J. (1988). *Qualitative Analysis of Neural Networks.* IEEE International Symposium on Circuits and Systems, Helsinki, June 6-9, 1988. pp 989-992. New York: Institute of Electrical and Electronics Engineers.

[Michel, AN 88b] Michel, A. N.; Farrell, J. A.; Porod, Wolfgang. (1988). *Stability Results for Neural Networks.* In Anderson, Dana Z. (ed.). Neural Information Processing Systems. pp 554-563. New York: American Institute of Physics.

[Michie, D 68a] Michie, Donald; Chambers, R. A. (1968). *BOXES: An Experiment in Adaptive Control.* Machine Intelligence. Volume 2. pp 137-152.

[Michie, D 68b] Michie, Donald; Chambers, R. A. (1968). *Boxes as a Model of Pattern-Formation.* In Waddington, C. H. (ed.). Towards a Theoretical Biology. pp 206-215. Edinburgh: Edinburgh University Press.

[Michie, D 86a] Michie, Donald. (1986). On Machine Intelligence. New York: Wiley.

[Miikkulainen, R 87a] Miikkulainen, Risto. (1987). Self-Organizing Process Based on Lateral Inhibition and Weight Redistribution. Technical Report UCLA-AI-87-16. Los Angeles, CA: University of California, Computer Science Department.

[Miikkulainen, R 88a] Miikkulainen, Risto; Dyer, Michael G. (in press). *Encoding Input/Output Representations in*

Connectionist Cognitive Systems. Connectionist Models Summer School, 1988: Proceedings.

[Miikkulainen, R 88b] Miikkulainen, Risto; Dyer, Michael G. (1988). Forming Global Representations with Extended Backpropagation. Technical Report UCLA-AI-88-7. Los Angeles, CA: University of California, Computer Science Department.

[Millecchia, R 87*] SEE: Brown, PB 87a. *Analog Memory for Continuous-Voltage, Discrete-Time Implementations of Neural Networks*.

[Miller, GA 54a] Miller, George A.; Bruner, Jerome S.; Postman, Leo. (1954). *Familiarity of Letter Sequences and Tachistoscopic Identification*. Journal of General Psychology. Volume 50. pp 129-139.

[Miller, J 79*] SEE: McClelland, JL 79b. *Structural Factors in Figure Perception*.

[Miller, JP 83*] SEE: Selverston, AI 83a. *Neural Mechanisms for the Production of Cyclic Motor Patterns*.

[Miller, RK 87a] Miller, Richard K. (1987). Neural Networks: Implementing Associative Memory Models in Neurocomputers. Madison, GA: SEAI Technical Publications.

[Miller, RK 88a] Miller, Richard K. (1988). Survey on Artificial Neural Systems. Survey Report Number 11. Madison GA: Future Technology Surveys, Inc.

[Miller, RN 80*] SEE: Rinzel, J 80a. *Numerical Calculation of Stable and Unstable Periodic Solutions to the Hodgkin-Huxley Equations*.

[Miller, WC 67a] Miller, W. C. (1967). A Modified Mean Square Error Criterion for Use in Unsupervised Learning. TR 6778-2. Stanford, CA: Stanford University, Stanford Electronics Laboratory.

[Miller, WT 86a] Miller, W. Thomas. (1986). *A Nonlinear Learning Controller for Robotic Manipulators*. Intelligent Robots and Computer Vision. SPIE Volume 726. pp 416-423. Bellingham, WA: Society of Photo-Optical Instrumentation Engineers.

[Miller, WT 87*] SEE: Glanz, FH 87a. *Shape Recognition Using a CMAC Based Learning*.

[Miller, WT 87a] Miller, W. Thomas. (1987). *A Learning Controller for Nonrepetitive Robotic Operations*. Workshop on Space Telerobotics: Proceedings. Volume 11. pp 273-280. Pasadena, CA: Jet Propulsion Laboratory.

[Miller, WT 87b] Miller, W. Thomas; Glanz, Filson H.; Kraft III, L. Gordon. (1987). *Application of a General Learning Algorithm to the Control of Robotic Manipulators*. International Journal of Robotics Research. Volume 6. pp 84-98.

[Miller, WT 87c] Miller, W. Thomas. (1987). *Sensor-Based Control of Robotic Manipulators Using a General Learning Algorithm*. IEEE Journal of Robotics and Automation. Volume RA-3. pp 157-165.

[Miller, WT 88*] SEE: Glanz, FH 88a. Deconvolution Using a CMAC Neural Network.

[Miller, WT 88a] Miller, W. Thomas; Hewes, Robert P. (1988). Real Time Experiments in Neural Network Based Learning Control During High Speed Nonrepetitive Robotic Operations. Durham, NH: University of New Hampshire, Electrical and Computer Engineering.

[Miller, WT 88b] Miller, W. Thomas. (1988). Real Time Learned Sensor Processing and Motor Control for a Robot with Vision. Durham, NH: University of New Hampshire, Electrical and Computer Engineering.

[Milligan, DK 87a] Milligan, D. K.; Gurney, K. N. (1987). *RAM-Unit Learning Networks*. IEEE International Conference on Neural Networks (ICNN-87), San Diego, CA, June 21-24, 1987: Proceedings. Volume 2. pp 557-566. New York:

Institute of Electrical and Electronics Engineers.

[Mills, JP 87*] SEE: Kabrisky, M 87a. *A Multiplexed Multi-Electrode Semiconductor Brain Electrode Implant.*

[Mills, JP 88*] SEE: Ruck, DW 88a. Multisensor Target Detection and Classification.

[Milnor, J 86a] Milnor, J. (1986). *Directional Entropies of Celluar Automaton-Maps.* In Bienenstock, Elie L. (ed.). Disordered Systems and Biological Organization. pp 113-115. New York: Springer-Verlag.

[Milsum, J 66a] Milsum, John. (1966). Biological Control Systems Analysis. New York: McGraw-Hill.

[Mimo, A 87*] SEE: Chun, HW 87a. *A Massively Parallel Model of Schema Selection.*

[Mimo, A 87*] SEE: Chun, HW 87b. *A Model of Schema Selection Using Marker Passing and Connectionist Spreading Activation.*

[Minamitani, H 82a] Minamitani, Haruyuki. (1982). *A Neural Network Model of Pain Mechanisms: Functional Properties of the Network Cells Providing the Power Law.* IEEE Transactions on Systems, Man and Cybernetics. Volume SMC-12. pp 907-913.

[Mine, H 70a] Mine, Hiashi; Osaki, Shunji. (1970). Markovian Decision Processes. New York: Elsevier.

[Mingolla, E 85*] SEE: Grossberg, S 85c. *Neural Dynamics of Perceptual Grouping: Textures, Boundaries, and Emergent Segmentations.*

[Mingolla, E 86*] SEE: Grossberg, S 86f. *Computer Simulation of Neural Networks for Perceptual Psychology.*

[Mingolla, E 86*] SEE: Grossberg, S 86j. *Neural Dynamics of Form Perception: Boundary Completion, Illusory Figures, and Neon Color Spreading.*

[Mingolla, E 86*] SEE: Grossberg, S 86k. *Neural Dynamics of Perceptual Grouping: Textures, Boundaries, and Emergent Segmentations.*

[Mingolla, E 86a] Mingolla, Ennio; Todd, J. T. (1986). *Perception of Solid Shape from Shading.* Biological Cybernetics. Volume 53. pp 137-151.

[Mingolla, E 87*] SEE: Grossberg, S 87m. *Neural Dynamics of Surface Perception: Boundary Webs, Illuminants, and Shape-From-Shading.*

[Mingolla, E 87*] SEE: Grossberg, S 87d. *A Neural Network Architecture for Preattentive Vision: Multiple Scale Segmentation and Regularization.*

[Mingolla, E 87a] Mingolla, Ennio. (1987). *Neural Networks for Computational Vision.* IEEE International Conference on Neural Networks (ICNN-87), San Diego, CA, June 21-24, 1987: Proceedings. Tutorial 9. San Diego: Institute of Electrical and Electronics Engineers.

[Minifie, FD 77*] SEE: Kent, RD 77a. *Coarticulation in Recent Speech Production Models.*

[Minnick, RC 61a] Minnick, Robert C. (1961). *Linear-Input Logic.* IRE Transactions of Electronic Computers. Volume EC-10. pp 6-16.

[Minsky, ML 54a] Minsky, Marvin L. (1954). Theory of Neural-Analog Reinforcement Systems and its Application to the Brain-Model Problem. (PhD Thesis). Princeton, NJ: Princeton University.

[Minsky, ML 60a] Minsky, Marvin L.; Selfridge, Oliver G. (1960). *Learning in Random Nets.* In Cherry, Colin (ed.). Information Theory, London Symposium, 4th, Held at the Royal Institution, London, August 29-September 2, 1960.

London: Butterworths.

[Minsky, ML 61a] Minsky, Marvin L. (1963). *Steps Toward Artificial Intelligence.* Computers and Thought. pp 406-450. New York: McGraw-Hill.

[Minsky, ML 61b] Minsky, Marvin L. (1961). *Steps Toward Artificial Intelligence.* Proceedings of the IRE. Volume 49. pp 8-30.

[Minsky, ML 69a] Minsky, Marvin L.; Papert, Seymour A. (1969). Perceptrons: An Introduction to Computational Geometry. Cambridge, MA: MIT Press.

[Minsky, ML 69b] Minsky, Marvin L.; Papert, Seymour A. (1988). *Perceptrons: Introduction.* In Anderson, James A. (ed.); Rosenfeld, Edward (ed.). Neurocomputing: Foundations of Research. pp 161-170. Cambridge, MA: MIT Press.

[Minsky, ML 75a] Minsky, Marvin L. (1975). *A Framework for Representing Knowledge.* The Psychology of Computer Vision. New York: McGraw-Hill.

[Minsky, ML 77a] Minsky, Marvin L. (1977). *Plain Talk About Neurodevelopmental Epistemology.* International Joint Conference on Artificial Intelligence (IJCAI-77), 5th, Cambridge, MA, August 22-25, 1977: Proceedings. Volume 2. pp 1083-1092. International Joint Conference on Artificial Intelligence.

[Minsky, ML 80a] Minsky, Marvin L. (1980). *K-Lines: A Theory of Memory.* Cognitive Science. Volume 4. pp 117-133.

[Minsky, ML 86a] Minsky, Marvin L. (1985). The Society of Mind. New York: Simon and Schuster.

[Minsky, ML 88a] Minsky, Marvin L.; Papert, Seymour A. (1988). Perceptrons: Expanded Edition. Cambridge, MA: MIT Press.

[Minsky, V 88*] SEE: Maren, AJ 88a. *A Multilayer Cooperative/Competitive Method for Creating Hierarchical Structures by Clustering Maximally-Related Nodes.*

[Minton, S 85a] Minton, Steven. (1985). A Game-Playing Program that Learns by Analyzing Examples. CMU-CS-85-130. Pittsburgh, PA: Carnegie Mellon University, Computer Science Department.

[Mirsalehi, MM 84*] SEE: Guest, CC 84a. *EXCLUSIVE OR Processing (Binary Image Subtraction) Using Thick Fourier Holograms.*

[Mishchenko, EF 62*] SEE: Pontryagin, LS 62a. The Mathematical Theory of Optimal Processes.

[Mishkin, M 87a] Mishkin, Mortimer; Appenzeller, Tim. (1987). *The Anatomy of Memory.* Scientific American. Volume 256. pp 80-89.

[Mitchell, CE 81a] Mitchell, Charles E.; Friesen, W. Otto. (1981). *A Neuromime System for Neural Circuit Analysis.* Biological Cybernetics. Volume 40. pp 127-137.

[Mitchell, TM 82*] SEE: Sleeman, D 82a. *Learning from Solution Paths: An Approach to the Credit Assignment Problem.*

[Mitchison, G 83*] SEE: Crick, F 83a. *The Function of Dream Sleep.*

[Mitra, SK 72*] SEE: Rao, CR 72a. *Generalized Inverse of a Matrix and Its Applications.*

[Mix, DF 81*] SEE: Overman, KC 81a. *A Novel View of Fourier Analysis.*

[Miyahara, S 86*] SEE: Farhat, NH 86b. *Optical Analog of Two-Dimensional Neural Networks and Their Application in Recognition of Radar Targets.*

[Miyajima, S 83*] SEE: Nagano, T 83a. *A Neural Network Model for the Developmental Process of Hypercomplex Cells.*

[Miyake, S 75*] SEE: Fukushima, K 75a. *A Self-Organizing Multilayer Neural Network.*

[Miyake, S 77*] SEE: Fukushima, K 77a. *Self-Organizing Neural Network with a Function of Associative Memory - Feedback Type Cognitron -*

[Miyake, S 78*] SEE: Fukushima, K 78a. *Self-Organizing Neural Network with a Function of Associative Memory: Feedback Type Cognitron.*

[Miyake, S 80*] SEE: Fukushima, K 80b. *Neocognitron: Self-Organizing Network Capable of Position-Invariant Recognition of Patterns.*

[Miyake, S 80*] SEE: Fukushima, K 80c. *Neocognitron: Self-Organizing Network Capable of Position-Invariant Recognition of Patterns.*

[Miyake, S 82*] SEE: Fukushima, K 82a. *Neocognitron: A New Algorithm for Pattern Recognition Tolerant of Deformation and Shifts in Position.*

[Miyake, S 82*] SEE: Fukushima, K 82b. *Neocognitron: A Self-Organizing Neural Network Model for a Mechanism of Visual Pattern Recognition.*

[Miyake, S 83*] SEE: Fukushima, K 83c. *Neocognitron: A Neural Network Model for a Mechanism of Visual Pattern Recognition.*

[Miyake, S 83*] SEE: Fukushima, K 83b. *Neocognitron: A Neural Network Model for a Mechanism of Visual Pattern Recognition.*

[Miyake, S 83a] Miyake, Sei; Fukushima, Kunihiko. (1983). *A Neural Network Model for the Mechanism of Feature Extraction - A Self-Organizing Network with Feedback Inhibition.* Transactions of the Institute of Electronics and Communication Engineers of Japan. Volume E65. pp 500-507.

[Miyake, S 84*] SEE: Fukushima, K 84b. *A Neural Network Model for the Mechanism of Feature Extraction.*

[Miyake, S 86a] Miyake, Sei; Fukushima, Kunihiko. (1986). *Neural Network Model for the Mechanism of Pattern Information Processing.* In Denker, John S. (ed.). Neural Networks for Computing, Snowbird, Utah, 1986. AIP Proceedings 151. pp 305-308. New York: American Institute of Physics.

[Miyake, S 88*] SEE: Irie, B 88a. *Capabilities of Three-Layered Perceptrons.*

[Miyake, S 88a] Miyake, Sei; Inui, Toshio. (1988). *Neural Computation Research for Vision - Simulation on a Parallel Processing Computer.* European Meeting on Cybernetics and Systems Research, 9th, Vienna, April 5-8, 1988: Proceedings.

[Miyake, S 88b] Miyake, Sei. (1988). *Neurocomputer.* International Workshop on Future Electron Devices - Three Dimensional Integration, 5th, Miyagi-Zao, May 30-June 1, 1988: Proceedings.

[Miyake, S 88c] Miyake, Sei; Fukushima, Kunihiko. ((to appear)). *Self-Organizing Neural Network with the Mechanism of Feedback Information Processing.* US-Japan Joint Seminar: Proceedings.

[Miyata, Y 87a] Miyata, Yoshiro. (1987). *Organization of Action Sequences in Motor Learning: A Connectionist Approach.* Cognitive Science Society, 9th, Seattle, WA, July 16-18, 1987: Proceedings. pp 496-507. Hillsdale, NJ: Lawrence Erlbaum.

[Miyazaki, F 87*] SEE: Kawamura, S 87a. *Intelligent Control of Robot Motion Based on Learning Method.*

[Miyazaki, F 87*] SEE: Kawamura, S 87b. *Technical Issues in Learning Control for Robot Motions.*

[Miyazaki, F 88*] SEE: Kawamura, S 88a. *Realization of Robot Motion Based on a Learning Method.*

[Mize, AS 87*] SEE: Draper, JS 87a. *A Microcomputer Neural Net Benchmarked Against Standard Classification Techniques.*

[Mizuno, H 78a] Mizuno, Hisashi. (1978). *A Self-Organizing Multi-Layered Neural Network Realizing a Function in the Human Recognition and Memory Mechanisms.* IEEE International Conference on Cybernetics and Society, 1978: Proceedings. pp 1039-1043.

[Mizuno, H 80a] Mizuno, Hisashi. (1980). *A Self-Organizing Multi-Layered Neural Network Realizing a Funciton in the Human Recognition and Memory Mechanisms.* Transactions of the Institute of Electronics and Communication Engineers of Japan. Volume E63. pp 132-133.

[Mjolsness, E 85a] Mjolsness, Eric. (1985). Neural Networks, Pattern Recognition, and Fingerprint Hallucination. Report 5198:TR:85. Pasadena, CA: California Institute of Technology.

[Mjolsness, E 86a] Mjolsness, Eric; Sharp, David H. (1986). *Preliminary Analysis of Recursively Generated Networks.* In Denker, John S. (ed.). Neural Networks for Computing, Snowbird, Utah, 1986. AIP Proceedings 151. pp 309-314. New York: American Institute of Physics.

[Mjolsness, E 87a] Mjolsness, Eric. (1987). *Control of Attention in Neural Networks.* IEEE International Conference on Neural Networks (ICNN-87), San Diego, CA, June 21-24, 1987: Proceedings. Volume 2. pp 567-574. New York: Institute of Electrical and Electronics Engineers.

[Mjolsness, E 87b] Mjolsness, Eric; Sharp, David H.; Alpert, Bardley K. (1987). *Recursively Generated Neural Networks.* IEEE International Conference on Neural Networks (ICNN-87), San Diego, CA, June 21-24, 1987: Proceedings. Volume 3. pp 165-172. New York: Institute of Electrical and Electronics Engineers.

[Mjolsness, E 88a] Mjolsness, Eric; Gindi, Gene R.; Anandan, P. (1988). Optimization in Model Matching and Perceptual Organization: A First Look. YALEU/DCS/RR-634. New Haven, CT: Yale University, Department of Computer Science.

[Mjolsness, E 88b] Mjolsness, Eric; Sharp, David H.; Alpert, Bradley K. (1988). Scaling, Machine Learning, and Genetic Neural Nets. LA-UR-88-142. Los Alamos, NM: Los Alamos National Laboratory.

[Mkrtchian, HH 82*] SEE: Melkonian, DS 82a. *Simulation of Learning Processes in Neuronal Networks of the Cerebellum.*

[Mokhoff, N 87a] Mokhoff, Nicholas. (March 15, 1987). *Parallelism Breeds a New Class of Supercomputers.* Computer Design. pp 53-64.

[Moll, KL 71a] Moll, Kenneth L.; Daniloff, Raymond G. (1971). *Investigation of the Timing of Velar Movements during Speech.* Journal of the Acoustical Society of America. Volume 50. pp 678-684.

[Moller, P 87*] SEE: Nylen, M 87a. A Neural Network in the Visual Cortex.

[Moller, P 87a] Moller, Per. (1987). Statistical Mechanics of Visual Processing: A Model for Layer IVc of the Primary Visual Cortex of Primates. Nordita-87/44 S (preprint). Copenhagen: Nordisk Institut for Teoretisk Atomfysik.

[Moller, P 88a] Moller, Per; Nylen, M.; Hertz, J. A. (1988). *A Cortical Network Model for Early Vision Processing.* In Eckmiller, Rolf (ed.); von der Malsburg, Christoph (ed.). Neural Computers. NATO ASI Series, Volume F41. pp 139-148. Berlin: Springer-Verlag.

[Molnar, CE 67a] Molnar, Charles E.; Ornstein, Severo M.; Anne, Antharvedi. (1967). *The CHASM: A Macromodular Computer for Analyzing Neuron Models.* Joint Computer Conference, Atlantic City, NJ, April 18-20, 1967: AFIPS Proceedings. Volume 30. pp 393-401. New York: American Federation of Information Processing Societies.

[Molnar, CE 68a] Molnar, Charles E.; Pfeiffer, Russell R. (1968). *Interpretation of Spontaneous Spike Discharge Patterns of*

Neurons in the Cochlear Nucleus. Proceedings of the IEEE. Volume 56. pp 993-1004.

[Molnar, DO 87a] Molnar, Daniel O. (1987). *Modeling Accuracy Improvements for the Group Method of Data Handling.* IEEE International Conference on Neural Networks (ICNN-87), San Diego, CA, June 21-24, 1987: Proceedings. Volume 4. pp 839-846. New York: Institute of Electrical and Electronics Engineers.

[Montalvo, FS 75a] Montalvo, F. S. (1975). *Consensus versus Competition in Neural Networks: A Comparative Analysis of Three Models.* International Journal of Man-Machine Studies. Volume 7. pp 333-346.

[Montgomery, BL 86a] Montgomery, Bruce L.; Kumar, B. V. K. Vijaya. (1986). *Evaluation of the Use of the Hopfield Neural Network Model as a Nearest-Neighbor Algorithm.* Applied Optics. Volume 25. pp 3759-3766.

[Montroll, EW 83a] Montroll, Elliott W.; Shlesinger, Michael F. (1983). *Maximum Entropy Formalism, Fractals, Scaling Phenomena, and 1/f Noise: A Tale of Tails.* Journal of Statistical Physics. Volume 32. pp 209-230.

[Moody, J 87*] SEE: Baum, EB 87a. *New "Neural" Algorithms for Associative Memory.*

[Moody, J 87a] Moody, John. (1987). *Perspectives on Associative Memories.* IEEE International Conference on Neural Networks (ICNN-87), San Diego, CA, June 21-24, 1987: Proceedings. Volume 3. pp 59-62. New York: Institute of Electrical and Electronics Engineers.

[Moody, J 88*] SEE: Baum, EB 88a. *Internal Representations for Associative Memory.*

[Moody, J 88a] Moody, John; Darken, Christian. (1988). Fast Learning in Networks of Locally-Tuned Processing Units. YALEU/DCS/RR-654. New Haven, CT: Yale University, Department of Computer Science.

[Moody, J 88b] Moody, John; Darken, Christian. (1988). Learning with Localized Receptive Fields. YALEU/DCS/RR-649. New Haven, CT: Yale University, Department of Computer Science.

[Moopenn, A 85*] SEE: Lambe, J 85a. Electronic Device Aspects of Neural Network Memories.

[Moopenn, A 86*] SEE: Thakoor, AP 86a. *Binary Synaptic Connections Based on Memory Switching in a SI:H.*

[Moopenn, A 86a] Moopenn, Alexander; Khanna, S. K.; Lambe, John; Thakoor, Anilkumar P. (1986). *Error Correction and Asymmetry in a Binary Memory Matrix.* In Denker, John S. (ed.). Neural Networks for Computing, Snowbird, Utah, 1986. AIP Proceedings 151. pp 315-320. New York: American Institute of Physics.

[Moopenn, A 87*] SEE: Thakoor, AP 87a. *Binary Synaptic Connections Based on Memory Switching in a-Si:H for Artificial Neural Networks.*

[Moopenn, A 87*] SEE: Thakoor, AP 87b. *Electronic Hardware Implementations of Neural Networks.*

[Moopenn, A 87*] SEE: Daud, T 87a. *Neural Network Based Feed-Forward High Density Associative Memory.*

[Moopenn, A 87*] SEE: Lamb, JL 87a. *Resistive Synaptic Interconnects for Electronic Neural Networks.*

[Moopenn, A 87a] Moopenn, Alexander; Thakoor, Anilkumar P.; Duong, T.; Khanna, S. K. (1987). *A Neurocomputer Based on an Analog-Digital Hybrid Architecture.* IEEE International Conference on Neural Networks (ICNN-87), San Diego, CA, June 21-24, 1987: Proceedings. Volume 3. pp 479-486. New York: Institute of Electrical and Electronics Engineers.

[Moopenn, A 87b] Moopenn, Alexander; Lambe, John; Thakoor, Anilkumar P. (1987). *Electronic Implementation of Associative Memory Based on Neural Network Models.* IEEE Transactions on Systems, Man and Cybernetics. Volume SMC-17. pp 325-331.

[Moopenn, A 88*] SEE: Thakoor, AP 88a. Electronic Neurocomputers.

[Moopenn, A 88a] Moopenn, Alexander; Langenbacher, H.; Tahkoor, Anilkumar P.; Khanna, S. K. (1988). *Programmable Synaptic Chip for Electronic Neural Networks.* In Anderson, Dana Z. (ed.). <u>Neural Information Processing Systems</u>. pp 564-572. New York: American Institute of Physics.

[Moore, GP 66a] Moore, George P.; Perkel, Donald H.; Segundo, Jose P. (1966). *Statistical Analysis and Functional Interpretation of Neuronal Spike Data.* <u>Annual Review of Physiology</u>. Volume 28. pp 493-522.

[Moore, GP 68*] SEE: Segundo, JP 68a. *Input-Output Relations in Computer-Simulated Nerve Cells: Influence of Statistical Properties, Strength, etc..*

[Moore, JW 68a] Moore, John W. (1968). *Specifications for Nerve Membrane Models.* <u>Proceedings of the IEEE</u>. Volume 56. pp 895-905.

[Moore, JW 85*] SEE: Schmajuk, NA 85a. *Real-Time Attentional Models for Classical Conditioning and the Hippocampus.*

[Moore, JW 85a] Moore, John W.; Desmond, John E.; Berthier, Neil E.; Blazis, Diane E. J.; Sutton, Richard S.; Barto, Andrew G. (1985). *Connectionistic Learning in Real Time: Sutton-Barto Adaptive Element and Classical Conditioning of the Rabbit Nictitating Membrane Response.* <u>Cognitive Science Society, 7th, Irvine, CA, August 15-17, 1985: Proceedings</u>.

[Moore, JW 86*] SEE: Schmajuk, NA 86a. *A Real-Time Attentional-Associative Network for Classical Conditioning of the Rabbit's NMR.*

[Moore, JW 86*] SEE: Blazis, DEJ 86a. *Simulation of the Classically Conditioned Nictitating Membrane Response by a Neuron-Like Adaptive Element.*

[Moore, JW 87*] SEE: Schmajuk, NA 87a. <u>Two Attentional Models of Classical Conditioning: Variations in CS Effectiveness Revisited</u>.

[Moore, JW 88*] SEE: Desmond, JE 88a. *Adaptive Timing in Neural Networks: The Conditioned Response.*

[Moore, JW 88*] SEE: Schmajuk, NA 88a. *The Hippocampus and the Classically Conditioned Nictitating Membrane Response: A Real-Time Attentinal-Associative Model.*

[Moore, K 88*] SEE: Martindale, C 88b. *Priming, Prototypicality, and Preference.*

[Moore, T 87*] SEE: Wolf, L 87a. *A Trainable Model for Near Periodic Signals.*

[Moore, WR 88a] Moore, Will R. (1988). *Conventional Fault-tolerance and Neural Computers.* In Eckmiller, Rolf (ed.); von der Malsburg, Christoph (ed.). <u>Neural Computers</u>. NATO ASI Series, Volume F41. pp 29-37. Berlin: Springer-Verlag.

[Morasso, P 86a] Morasso, P.; Mussa-Ivaldi, F. A. (1986). *The Role of Physical Constraints in Natural and Artificial Manipulation.* <u>IEEE International Conference on Robotics and Automation, San Francisco, CA, April 7-10, 1986: Proceedings</u>. pp 1901-1905. Washington, DC: Institute of Electrical and Electronics Engineers.

[Morasso, P 88*] SEE: Mussa-Ivaldi, FA 88b. *Kinematic Networks: A Distributed Model for Representing and Regularizing Motor Redundancy.*

[Morasso, P 88*] SEE: De Rossi, D 88a. *Skin & Muscles: The Basic Wet-Ware for Sensory-Motor Control.*

[Morasso, P 88a] Morasso, P.; Zaccaria, R.; Mussa Ivaldi, F. A. (1988). *A Parallel Distributed Architecture for Motor Control.* <u>Neural Networks</u>. Volume 1. pp 348.

[Morgan, NH 87*] SEE: Gevins, AS 87a. *"Ignorance-Based" Neural-Network Signal Processing in Brain Research.*

[Morgan, NH 88*] SEE: Gevins, AS 88a. *Applications of Neural-Network (NN) Signal Processing in Brain Research.*

[Morikawa, T 72*] SEE: Nakajima, M 72a. *Automatic Character Reading Using a Holographic Data Processing Technique.*

[Morishita, I 70a] Morishita, Iwao. (1970). *Analysis of an Adaptive Threshold Logic Unit.* IEEE Transactions on Computers. Volume C-19. pp 1181-1192.

[Morishita, I 72a] Morishita, Iwao; Yajima, Akio. (1972). *Analysis and Simulation of Networks of Mutually Inhibiting Neurons.* Kybernetik. Volume 11. pp 154-165.

[Morishita, I 77*] SEE: Tokura, T 77a. *Analysis and Simulation of Double Layer Neural Networks with Mutually Inhibiting Interconnections.*

[Morris, LR 77a] Morris, L. Robert. (1977). *Automatic Generation of Time Efficient Digital Signal Processing Software.* IEEE Transactions on Acoustics, Speech, and Signal Processing. Volume ASSP-25. pp 74-79.

[Morris, RJT 88a] Morris, R. J. T.; Rubin, L. D.; Wong, W. S. (1988). *A Decentralized Tunable Short Term Neural Network Memory and Application to Tracking.* IEEE International Conference on Neural Networks (ICNN-88), San Diego, CA, July 24-27, 1988: Proceedings. Volume 1. pp 199-206. New York: Institute of Electrical and Electronics Engineers.

[Morse, A 78a] Morse, A.; Kilmer, W. L. (1978). *Neural Net Capable of Competitive and Cooperative Computation.* Biological Cybernetics. Volume 30. pp 1-6.

[Morse, KG 87a] Morse, Kingsley G. (1987). *Neuroscience as a Possible Research Standard for Neural Networks.* IEEE International Conference on Neural Networks (ICNN-87), San Diego, CA, June 21-24, 1987: Proceedings. Volume 4. pp 827-830. New York: Institute of Electrical and Electronics Engineers.

[Moses, HE 74a] Moses, H. E.; Quesada, A. F. (1974). *The Power Spectrum of the Mellin Transformation with Applications to Scaling of Physical Quantities.* Journal of Mathematical Physics. Volume 15. pp 748-752.

[Mosseri, R 86a] Mosseri, R.; Sadoc, J. F. (1986). *Order and Defects in Geometrically Frustrated Systems.* In Bienenstock, Elie L. (ed.). Disordered Systems and Biological Organization. pp 149-152. New York: Springer-Verlag.

[Mosteller, F 51*] SEE: Bush, RR 51a. *A Mathematical Model for Simple Learning.*

[Mosteller, F 51*] SEE: Bush, RR 51b. *A Mathematical Model for Simple Learning.*

[Mosteller, F 51*] SEE: Bush, RR 51d. *A Model for Stimulus Generalization and Discrimination.*

[Mosteller, F 51*] SEE: Bush, RR 51c. *A Model for Stimulus Generalization and Discrimination.*

[Mosteller, F 55*] SEE: Bush, RR 55a. Stochastic Models for Learning.

[Mottishaw, P 87*] SEE: Gardner, E 87a. *Zero Temperature Parallel Dynamics for Infinite Range Spin Glasses and Neural Networks.*

[Mountcastle, VB 67a] Mountcastle, Vernon B. (1967). *The Problem of Sensing and the Neural Coding of Sensory Events.* The Neurosciences: A Study Program. pp 393-408. New York: Rockefeller University Press.

[Mountcastle, VB 79a] Mountcastle, Vernon B. (1979). *An Organizing Principle for Cerebral Function.* The Neurosciences: Fourth Study Program. pp 21-42. Cambridge, MA: MIT Press.

[Moussouris, J 74a] Moussouris, John. (1974). *Gibbs and Markov Random Systems with Constraints.* Journal of Statistical Physics. Volume 10. pp 11-33.

[Movshon, JA 85a] Movshon, J. Anthony; Adelson, Edward H.; Gizzi, Martin S.; Newsome, William T. (1985). *The Analysis of Moving Visual Patterns.* In Chagas, Carlos (ed.); Gattass, Ricardo (ed.). Pattern Recognition Mechanisms. pp 118-151. New York: Springer-Verlag.

[Mozer, MC 81*] SEE: Anderson, JA 81a. *Categorization and Selective Neurons.*

[Mozer, MC 84a] Mozer, Michael C. (1984). Inductive Information Retrieval Using Parallel Distributed Computation. ICS-8406. San Diego, CA: University of California, San Diego, Institute for Cognitive Science.

[Mozer, MC 86*] SEE: McClelland, JL 86h. *Perceptual Interactions in Two-Word Displays: Familiarity and Similarity Effects.*

[Mozer, MC 86a] Mozer, Michael C. (1986). Early Parallel Processing in Reading: A Connectionist Approach. ICS-8611. San Diego, CA: University of California, San Diego, Institute for Cognitive Science.

[Mozer, MC 86b] Mozer, Michael C. (1986). RAMBOT: A Connectionist Expert System that Learns by Example. ICS-8610. San Diego, CA: University of California, San Diego, Institute for Cognitive Science.

[Mozer, MC 87a] Mozer, Michael C. (1987). *RAMBOT: A Connectionist Expert System that Learns by Example.* IEEE International Conference on Neural Networks (ICNN-87), San Diego, CA, June 21-24, 1987: Proceedings. Volume 2. pp 693-700. New York: Institute of Electrical and Electronics Engineers.

[Mozurkewich, G 84*] SEE: Chamberlin, RV 84a. *Time Decay of the Remanent Magnetization in Spin-Glasses.*

[Mucciardi, AN 66a] Mucciardi, Anthony N.; Gose, Earl E. (1966). *Evolutionary Pattern Recognition in Incomplete Nonlinear Multithreshold Networks.* IEEE Transactions on Electronic Computers. Volume EC-15. pp 257-261.

[Mucciardi, AN 84*] SEE: Barron, RL 84a. *Adaptive Learning Networks: Development and Application in the United States of Algorithms Related to GMDH.*

[Mueller, CW 78*] SEE: Kim, J 78a. Introduction to Factor Analysis: What It Is and How to Do It.

[Mueller, P 62a] Mueller, P.; Martin, T.; Putzrath, F. (1962). *General Principles of Operations in Neuron Nets with Application to Acoustical Pattern Recognition.* Biological Prototypes and Synthetic Systems, Proceedings. Volume 1. pp 192-212. New York: Plenum.

[Mueller, P 86a] Mueller, P.; Lazzaro, J. (1986). *Machine for Neural Computation of Acoustical Patterns with Application to Real Time Speech Recognition.* In Denker, John S. (ed.). Neural Networks for Computing, Snowbird, Utah, 1986. AIP Proceedings 151. pp 321-326. New York: American Institute of Physics.

[Mueller, P 87a] Mueller, P.; Spiro, P.; Blackman, D.; Furman, R. E. (1987). *Neural Computation of Visual Images.* IEEE International Conference on Neural Networks (ICNN-87), San Diego, CA, June 21-24, 1987: Proceedings. Volume 4. pp 75-88. New York: Institute of Electrical and Electronics Engineers.

[Muir, A 81a] Muir, A. (1981). *The Construction of Homogeneous Nets.* Bulletin of Mathematical Biology. Volume 43. pp 415-426.

[Muir, A 83a] Muir, A.; Warner, M. W. (1983). *The Dynamics of Symmetric Nets.* Bulletin of Mathematical Biology. Volume 45. pp 781-792.

[Muir, A 84a] Muir, A.; Warner, M. W. (1984). *Lattice Valued Relations and Automata.* Discrete Applied Mathematics. Volume 7. pp 65-78.

[Muller, FJ 76a] Muller, F. J.; Taylor, W. K. (1976). *A Parallel Processing Model for the Study of Possible Pattern Recognition Mechanisms in the Brain.* Pattern Recognition. Volume 8. pp 47-52.

[Munro, PW 82*] SEE: Bienenstock, EL 82a. *A Theory for the Development of Neuron Selectivity: Orientation Selectivity and Binocular Interactions in the Visual Cortex.*

[Munro, PW 82*] SEE: Bienenstock, EL 82b. *Theory for the Development of Neuron Selectivity: Orientation Specificity and Binocular Interaction in Visual Cortex.*

[Munro, PW 85*] SEE: Cooper, LN 85a. *Neuron Selectivity: Single Neuron and Neuron Networks.*

[Munro, PW 86a] Munro, Paul W. (1986). *Self-Supervised Learning: A Scheme for Discovery of "Natural" Categories by Single Units.* Cognitive Science Society, 8th, Amherst, MA, August 15-17, 1986: Proceedings. pp 765-772. Hillsdale, NJ: Lawrence Erlbaum.

[Munro, PW 86b] Munro, Paul W. (1986). *State-Dependent Factors Influencing Neural Plasticity: A Partial Account of the Critical Period.* In Rumelhart, David E. (ed.). Parallel Distributed Processing: Explorations in the Microstructure of Cognition. Volume 2. pp 471-502. Cambridge, MA: MIT Press.

[Munro, PW 87*] SEE: Cottrell, GW 87a. Image Compression by Back Propagation: An Example of Extensional Programming.

[Munro, PW 87*] SEE: Cottrell, GW 87b. *Learning Internal Representations from Gray-Scale Images: An Example of Extensional Programming.*

[Munro, PW 87a] Munro, Paul W. (1987). *A Dual Back-Propagation Scheme for Scalar Reward Learning.* Cognitive Science Society, 9th, Seattle, WA, July 16-18, 1987: Proceedings. pp 165-176. Hillsdale, NJ: Lawrence Erlbaum.

[Munro, PW 88*] SEE: Cosic, C 88a. *Learning to Represent and Understand Locative Prepositional Phrases.*

[Munro, PW 88a] Munro, Paul W.; Anderson, James A. (1988). *Tools for Connectionist Modeling: The Dynamical Systems Methodology.* Behavior Research Methods, Instruments, and Computers. Volume 20. 276-281.

[Murakami, K 81a] Murakami, Kenji; Aibara, Tsunehiro. (1981). *Construction of a Distributed Associative Memory on the Basis of Bayes Discriminant Rule.* IEEE Transactions on Pattern Analysis and Machine Intelligence. Volume PAMI-3. pp 210-214.

[Murakami, K 82a] Murakami, Kenji; Aibara, Tsunehiro. (1982). *Optimal Association with Partly Missing Key Vectors.* Biological Cybernetics. Volume 44. pp 151-155.

[Murakami, K 88a] Murakami, Kenji; Yamajo, Yasuhiro; Aibara, Tsunehiro. (1988). *Construction of Associative Memory from Incomplete Memorized Information Using Space Reduction Method.* Transactions of the Institute of Electronics, Information and Communication Engineers of Japan. Volume J71-D. pp 151-157.

[Muraki, S 86a] Muraki, Shigeru; Sugie, Noboru. (1986). *Speech Prosthesis System for Laryngectomees Using "Electropalatograph" and a Small Keyboard.* Electronics and Communications in Japan. Volume 69. pp 81-90.

[Murdock, BB 83a] Murdock, Bennet B. (1983). *A Distributed Memory Model for Serial-Order Information.* Psychological Review. Volume 90. pp 316-338.

[Murdock Jr, BB 82a] Murdock Jr, Bennet B. (1982). *A Theory for the Storage and Retrieval of Item and Associative Information.* Psychological Review. Volume 89. pp 609-626.

[Murphy, GL 85*] SEE: Cohen, B 85a. *Models of Concepts.*

[Murphy, GL 86*] SEE: Anderson, JA 86b. *Concepts in Connectionist Models.*

[Murphy, GL 86*] SEE: Anderson, JA 86c. *Concepts in Distributed Systems.*

[Murphy, GL 86*] SEE: Anderson, JA 86e. *Psychological Concepts in Parallel Systems.*

[Murray, AF 87a] Murray, Alan F.; Smith, Anthony V. W. (1987). *Asynchronous Arithmetic for VLSI Neural Systems.* Electronics Letters. Volume 23. pp 642-643.

[Murray, AF 88*] SEE: Butler, ZF 88a. *VLSI Bit-Serial Neural Networks.*

[Murray, AF 88a] Murray, Alan F.; Smith, Anthony V. W. (1988). A Novel Computational and Signalling Method for VLSI Neural Networks. Edinburgh: University of Edinburgh, Department of Electrical Engineering.

[Murray, AF 88b] Murray, Alan F.; Smith, Anthony V. W. (1988). *Asynchronous VLSI Neural Networks Using Pulse-Stream Arithmetic.* IEEE Journal of Solid-State Circuits. Volume 23. pp 688-697.

[Murray, AF 88c] Murray, Alan F.; Smith, Anthony V. W.; Butler, Zoe F. (1988). *Bit-Serial Neural Networks.* In Anderson, Dana Z. (ed.). Neural Information Processing Systems. pp 573-583. New York: American Institute of Physics.

[Murray, AF 88d] Murray, Alan F.; Smith, Anthony V. W.; Tarassenko, Lionel. (1988). *Fully-Programmable Analogue VLSI Devices for the Implementation of Neural Networks.* Workshop on AI, July, 1988.

[Murray, AF 88e] Murray, Alan F.; Butler, Zoe F.; Smith, Anthony V. W. (1988). *VLSI Neural Networks.* IEE Colloquium, London, February, 1988.

[Murray, W 81*] SEE: Gill, PE 81a. Practical Optimization.

[Mussa Ivaldi, FA 88*] SEE: Morasso, P 88a. *A Parallel Distributed Architecture for Motor Control.*

[Mussa Ivaldi, FA 88*] SEE: De Rossi, D 88a. *Skin & Muscles: The Basic Wet-Ware for Sensory-Motor Control.*

[Mussa-Ivaldi, FA 86*] SEE: Morasso, P 86a. *The Role of Physical Constraints in Natural and Artificial Manipulation.*

[Mussa-Ivaldi, FA 88a] Mussa-Ivaldi, F. A. (1988). *Do Neurons in the Motor Cortex Encode Movement Direction? An Alternative Hypothesis.* Neuroscience Letters. Volume 9. pp 106-111.

[Mussa-Ivaldi, FA 88b] Mussa-Ivaldi, F. A.; Morasso, P.; Zaccaria, R. (1988). *Kinematic Networks: A Distributed Model for Representing and Regularizing Motor Redundancy.* Biological Cybernetics. Volume 60. pp 1-16.

[Mussa-Ivaldi, FA 88c] Mussa-Ivaldi, F. A.; McIntyre, J.; Bizzi, E. (1988). *Theoretical and Experimental Perspectives on Arm Trajectory Formation: A Distributed Model of Motor Redundancy.* Biological and Artificial Intelligence Systems. pp 563-577. Leiden: Escom.

[Myers, HJ 82*] SEE: Cruz, CA 82a. Associative Networks.

[Myers, HJ 83*] SEE: Cruz, CA 83a. Associative Networks, II.

[Myers, MH 86a] Myers, Michael H. (1986). *Some Speculations on Artificial Neural System Technology.* IEEE 1986 National Aerospace and Electronics Conference (NAECON 1986), 38th: Proceedings. Volume 4. pp 1298-1302. New York: Institute of Electrical and Electronics Engineers.

[Myers, MH 87*] SEE: Kuczewski, RM 87a. Exploration of Backward Error Propagation as a Self-Organizational Structure.

[Myers, MH 87*] SEE: Kuczewski, RM 87b. *Exploration of Backward Error Propagation as a Self-Organized Structure.*

[Myers, MH 87*] SEE: Kuczewski, RM 87e. *Neurocomputer Workstations and Processors: Approaches and Applications.*

[Myers, MH 87*] SEE: Kuczewski, RM 87f. Neurocomputer Workstations and Processors: Approaches and Applications.

[Myers, MH 87a] Myers, Michael H.; Kuczewski, Robert M.; Crawford, William J. (1987). Application of New Artificial Neural

System Information Processing Principles to Pattern Classification: Final Report, US Army Research Office. AD-A180-442. TRW Military Electronics and Avionics Division.

[Myhill, J 61a] Myhill, J.; Kautz, William H. (1961). *On the Size of Weights Required for Linear-Input Switching Functions*. IRE Transactions on Electronic Computers. Volume EC-10. pp 288-290.

[Nachmias, J 71*] SEE: Graham, N 71a. *Detection of Grating Patterns Containing Two Spatial Frequencies: A Comparison of Single-Channel and Multiple-Channels Models*.

[Nadal, JP 88*] SEE: Dreyfus, G 88a. *High Order Neural Networks for Efficient Associative Memory Design*.

[Nadel, L 89a] Nadel, Lynn (ed.); Cooper, Lynn A. (ed.); Culicover, Peter (ed.); Harnish, R. Michael (ed.). (1989). Neural Connections, Mental Computation. Cambridge, MA: MIT Press.

[Nagami, H 76a] Nagami, Hitoshi; Kitahashi, Tadahiro. (1976). *Characterization of Dynamical Behavior Associated with a Single Neural Equation*. Mathematical Biosciences. Volume 32. pp 221-237.

[Nagano, T 73a] Nagano, Takashi. (1973). *A Learning Machine with New Structure and Function*. IEEE Transactions on Systems, Man and Cybernetics. Volume SMC-3. pp 466-472.

[Nagano, T 80a] Nagano, Takashi; Kurato, Koji. (1980). *A Model of the Complex Cell Based on Recent Neurophysiological Findings*. Biological Cybernetics. Volume 38. pp 103-105.

[Nagano, T 81a] Nagano, Takashi; Kurata, Koji. (1981). *A Self-Organizing Neural Network Model for the Development of Complex Cells*. Biological Cybernetics. Volume 40. pp 195-200.

[Nagano, T 83a] Nagano, Takashi; Miyajima, Shinichi. (1983). *A Neural Network Model for the Developmental Process of Hypercomplex Cells*. IEEE Transactions on Systems, Man and Cybernetics. Volume SMC-13. pp 847-851.

[Nagashino, H 81a] Nagashino, Hirofumi; Tamura, Hiroshi; Ushita, Tomiyuki. (1981). *Relations Between Initial Conditions and Periodic Firing Modes in Reciprocal Inhibition Neural Networks*. Transactions of the Institute of Electronics and Communication Engineers of Japan. Volume E64. pp 378-385.

[Nagashino, H 87a] Nagashino, Hirofumi; Kinouchi, Yohsuke; Ushita, Tomiyuki. (1987). *Self-Organizing Process of a Neural Network Model for Pattern Detection*. IEEE International Conference on Neural Networks (ICNN-87), San Diego, CA, June 21-24, 1987: Proceedings. Volume 2. pp 105-114. New York: Institute of Electrical and Electronics Engineers.

[Nagata, S 88*] SEE: Watanabe, N 88a. *Mobile Robot Control by Neural Networks and Their Associated Learning Algorithm*.

[Nagel, RN 80*] SEE: Albus, JS 80a. *A Measurement and Control Model for Adaptive Robots*.

[Nagel, RN 81*] SEE: Albus, JS 81b. *Theory and Practice of Hierarchical Control*.

[Nagendran, S 87*] SEE: Caelli, TM 87a. *Fast Edge-Only Matching Techniques for Robot Pattern Recognition*.

[Nagumo, J 62a] Nagumo, Jin-ichi; Arimoto, Suguru; Yoshizawa, Shuji. (1962). *An Active Pulse Transmission Line Simulating Nerve Axon*. Proceedings of the IRE. Volume 50. pp 2061-2070.

[Nagumo, J 67a] Nagumo, Jin-ichi; Noda, Atsuhiko. (1967). *A Learning Method for System Identification*. IEEE Transactions on Automatic Control. Volume AC-12. pp 282-289.

[Nagumo, J 81*] SEE: Osada, H 81a. *Theoretical Study on Cantor Function-Like Response Characteristics of a Neuron Model*.

[Nagumo, J 82*] SEE: Yoshizawa, S 82a. *Pulse Sequences Generated by a Degenerate Analog Neuron Model*.

[Nagy, G 66a] Nagy, George; Shelton, G. L. (1966). *Self-Corrective Character Recognition System.* IEEE Transactions on Information Theory. Volume IT-12. pp 215-222.

[Nagy, G 68a] Nagy, George. (1968). *The State of the Art in Pattern Recognition.* Proceedings of the IEEE. Volume 56. pp 836-862.

[Nahi, NE 72a] Nahi, N. E.; Assefi, T. (1972). *Bayesian Recursive Image Estimation.* IEEE Transactions on Computers. Volume C-21. pp 734-738.

[Naito, S 88*] SEE: Alkon, DL 88b. *Regulation of Hermissenda K+ Channels by Cytoplasmic and Membrane-Associated C-Kinase.*

[Naito, S 88a] Naito, Shigetaka; Bank, Barry; Alkon, Daniel L. (1988). *Transient and Persistent Depolarization-Induced Changes of Protein Phosphorylation in a Molluscan Nervous System.* Journal of Neurochemistry. Volume 50. pp 704-711.

[Nakai, Y 88*] SEE: Kitamura, S 88a. *Application of the Neural Network for the Trajectory Planning of a Biped Locomotive Robot.*

[Nakajima, M 72a] Nakajima, M.; Morikawa, T.; Sakurai, K. (1972). *Automatic Character Reading Using a Holographic Data Processing Technique.* Applied Optics. Volume 11. pp 362-371.

[Nakane, K 88*] SEE: Kato, K 88a. *An Associative Memory Model Based on a Quantification Method.*

[Nakano, K 72a] Nakano, Kaoru. (1972). *Associatron: A Model of Associative Memory.* IEEE Transactions on Systems, Man and Cybernetics. Volume SMC-2. pp 380-388.

[Nakano, R 88*] SEE: Saito, K 88a. Medical Diagnostic Expert System Based on PDP Model.

[Nakashima, Y 88*] SEE: Bode, PM 88a. *Develpment of the Two-Part Pattern During Regeneration of the Head in Hydra.*

[Narendra, KS 74a] Narendra, Kumpati S.; Thathachar, M. A. L. (1974). *Learning Automata: A Survey.* IEEE Transactions on Systems, Man and Cybernetics. Volume SMC-4. pp 323-334.

[Narendra, KS 77a] Narendra, Kumpati S.; Lakshmivarahan, S. (1977). *Learning Automata: A Critique.* Journal of Cybernetics and Information Science. Volume 1. pp 53-65.

[Narendra, KS 83a] Narendra, Kumpati S.; Wheeler, R. M. (1983). *An N-Player Sequential Stochastic Game with Identical Payoffs.* IEEE Transactions on Systems, Man and Cybernetics. Volume SMC-13. pp 1154-1158.

[Narendra, KS 86*] SEE: Barto, AG 86a. *Cooperativity in Networks of Pattern Recognizing Stochastic Learning Automata.*

[Narendra, PM 75*] SEE: Fukunaga, K 75a. *A Branch and Bound Algorithm for Computing k-Nearest Neighbors.*

[Nasrabadi, NM 88*] SEE: Clifford, SP 88a. *Integration of Stereo Vision and Optical Flow Using Markov Random Fields.*

[Nasrabadi, NM 88a] Nasrabadi, Nasser M.; Feng, Yushu. (1988). *Vector Quantization of Images Based Upon the Kohonen Self-Organizing Feature Maps.* IEEE International Conference on Neural Networks (ICNN-88), San Diego, CA, July 24-27, 1988: Proceedings. Volume 1. pp 101-108. New York: Institute of Electrical and Electronics Engineers.

[Nass, MM 75a] Nass, Menasche M.; Cooper, Leon N. (1975). *A Theory for the Development of Feature Detecting Cells in the Visual Cortex.* Biological Cybernetics. Volume 19. pp 1-18.

[Navlakha, JK 88*] SEE: Feild Jr, WB 88a. On Choosing Non-Random Input Exemplars to Form Stable States in a Hopfield Neural Net.

[Navlakha, JK 88*] SEE: Feild Jr, WB 88b. Transputer Implementation of Hopfield Neural Network.

[Navon, D 86a] Navon, David. (1986). Visibility or Disability: Notes on Attention. ICS-8606. San Diego, CA: University of California, San Diego, Institute for Cognitive Science.

[Neely, JH 76a] Neely, James H. (1976). *Semantic Priming and Retrieval from Lexical Memory: Evidence for Facilitatory and Inhibitory Processes.* Memory and Cognition. Volume 4. pp 648-654.

[Neely, JH 77a] Neely, James H. (1977). *Semantic Priming and Retrieval from Lexical Memory: Roles of Inhibitionless Spreading Activation and Limited-Capacity Attention.* Journal of Experimental Psychology: General. Volume 106. pp 226-254.

[Negrini, R 88a] Negrini, R.; Sami, M. G.; Scarabottolo, N.; Stefanelli, R. (1988). *Fault-tolerance in Imaging-oriented Systolic Arrays.* In Eckmiller, Rolf (ed.); von der Malsburg, Christoph (ed.). Neural Computers. NATO ASI Series, Volume F41. pp 39-50. Berlin: Springer-Verlag.

[Neisser, U 60*] SEE: Selfridge, OG 60a. *Pattern Recognition by Machine.*

[Nelken, I 87a] Nelken, Izzy; Oxley, Don. (1987). *Parallel Gaussian Elimination on an Optically Interconnected Data Flow Computer.* Mathematics and Computers in Simulation. Volume 29. pp 515-529.

[Nelken, I 88a] Nelken, Izzy. (1988). *Improving the Learning Rate of Neural Networks.* Annual Meeting of the International Neural Networks Society, 1st, Boston, MA, September, 1988: Proceedings.

[Nelson, ME 87*] SEE: Bower, JM 87a. Piriform (Olfactory) Cortex Model on the Hypercube.

[Nelson, TJ 83a] Nelson, T. J. (1983). *A Neural Network Model for Cognitive Activity.* Biological Cybernetics. Volume 49. pp 79-88.

[Nemeth, G 79*] SEE: Kohonen, T 79a. *Spectral Classification of Phonemes by Learning Subspaces.*

[Nemhauser, GL 88a] Nemhauser, George L.; Wolsey, Laurence A. (1988). Integer and Combinatorial Optimizations. New York: Wiley.

[Newcomb, RW 79*] SEE: Dimopoulos, N 79a. *Modeling Networks of Morishita Neurons With Application to the Cerebellum.*

[Newcomb, RW 80*] SEE: Ajmera, RC 80a. *Retinal-Type Neuristor Sections.*

[Newcomb, RW 80*] SEE: Dimopoulos, N 80a. *Stability Properties of a Class of Large Scale Neural Networks.*

[Newcomb, RW 83*] SEE: Chande, A 83a. *Neural-Type Robotics: An Overview.*

[Newcomb, RW 84*] SEE: Nizkik, CA 84a. *Computer Networking Capacity in Robotic Neural Systems.*

[Newcomb, RW 85*] SEE: El-Leithy, N 85a. *Overview of Neural-Type Electronics.*

[Newcomb, RW 87*] SEE: El-Leithy, N 87a. *A Basic MOS Neural-Type Junction: A Perspective on Neural-Type Microsystems.*

[Newcomb, RW 87*] SEE: El-Leithy, N 87b. *A Semistate Model for Neural-Type Junction Circuits.*

[Newcomb, RW 87a] Newcomb, R. W.; El-Leithy, N. (1987). *Semistate Description of an MOS Neural-Type Cell.* Mathematical Theory Networks and Systems (MTNS-87), Phoenix, AZ, June 15-19, 1987: Proceedings.

[Newell, A 55a] Newell, Allen. (1955). *The Chess Machine: An Example of Dealing with a Complex Task by Adaptation.* Western Joint Computer Conference, 1955: Proceedings. pp 101-108. New York: Institute of Radio Engineers.

[Newell, A 57a] Newell, Allen; Shaw, J. C.; Simon, Herbert A. (1957). *Empirical Explorations of the Logic Theory Machine:*

A Case Study in Heuristic. Western Joint Computer Conference, Los Angeles, CA, February 26-28, 1957: Proceedings. pp 218-230. New York: Institute of Radio Engineers.

[Newell, A 80a] Newell, Allen. (1980). *Physical Symbol Systems*. Cognitive Science. Volume 4. pp 135-183.

[Newman, CM 88a] Newman, Charles M. (1988). *Memory Capacity in Neural Network Models: Rigorous Lower Bounds*. Neural Networks. Volume 1. pp 223-238.

[Newport, JP 85a] Newport, John Paul. (1985). *What Bell Laboratories is Learning from Slugs*. Fortune. Volume 111. p 117.

[Newquist, HP 86a] Newquist, Harvey P. *Chips for Brains: Computers Built with Neural Networks*.

[Newsome, WT 85*] SEE: Movshon, JA 85a. *The Analysis of Moving Visual Patterns*.

[Ngai, JY 87a] Ngai, John Y.; Seitz, Charles L. (1987). A Framework for Adaptive Routing. Report 5246:TR:87. Pasadena, CA: California Institute of Technology, Computer Science Department.

[Nguyen, DD 87*] SEE: Lee, JSJ 87a. Intelligent Recognition and Learning Control: Integrating Intelligent Processing and Neural Networks for Adaptive Moving Object Tracking.

[Nguyen, DD 87a] Nguyen, Dziem D.; Holt, Fred. (1987). *Stochastic Processing in a Neural Network Application*. IEEE International Conference on Neural Networks (ICNN-87), San Diego, CA, June 21-24, 1987: Proceedings. Volume 3. pp 281-292. New York: Institute of Electrical and Electronics Engineers.

[Nguyen, DD 88a] Nguyen, Dziem D.; Lee, James S. J. (1988). *A New LMS-Based Algorithm for Rapid Adaptive Classification in Dynamic Environments: Theory and Preliminary Results*. IEEE International Conference on Neural Networks (ICNN-88), San Diego, CA, July 24-27, 1988: Proceedings. Volume 1. pp 455-463. New York: Institute of Electrical and Electronics Engineers.

[Nguyen, HG 88*] SEE: Blackburn, MR 88a. An Artificial Neural System for Autonomous Undersea Vehicles.

[Nguyen, HT 87*] SEE: Yager, RR 87a. Fuzzy Sets and Applications: Selected Papers by L A Zadeh.

[Nguyen, TKP 88*] SEE: Huang, WY 88b. *Neural Nets for Speech Recognition*.

[Nicolic, ZJ 66a] Nicolic, Z. J.; Fu, King-Sun. (1966). *An Algorithm for Learning without External Supervision and its Application to Learning Control Systems*. IEEE Transactions on Automatic Control. Volume AC-11. pp 414-422.

[Nicolis, JS 84a] Nicolis, John S. (1984). *The Role of Chaos in Reliable Information Processing*. Journal of the Franklin Institute. Volume 317. pp 289-307.

[Niez, J 86*] SEE: Peretto, P 86a. *Collective Properties of Neural Networks*.

[Niez, J 86*] SEE: Peretto, P 86b. *Long Term Memory Storage Capacity of Multiconnected Neural Networks*.

[Niez, J 86*] SEE: Peretto, P 86c. *Stochastic Dynamics of Neural Networks*.

[Nigam, A 80*] SEE: Feldman, JA 80a. *A Model and Proof Technique for Message-Based Systems*.

[Niki, K 81a] Niki, Kazuhisa; Saburi, Masahisa. (1981). *A Neuronal Model with Synaptic Long-Term Memories*. Transactions of the Institute of Electronics and Communication Engineers of Japan. Volume E64. pp 940-947. Institute of Electronics and Communication Engineers of Japan.

[Nilsson, NJ 64*] SEE: Block, HD 64a. *Determination and Detection of Features in Patterns*.

[Nilsson, NJ 65a] Nilsson, Nils J. (1965). Learning Machines: Foundations of Trainable Pattern-Classifying Systems. New York: McGraw-Hill.

[Nilsson, NJ 67*] SEE: Rosen, CA 67a. *An Intelligent Automaton.*

[Nilsson, NJ 67a] Nilsson, Nils J.; Raphael, Bertram. (1967). *Preliminary Design of an Intelligent Robot.* Computer and Information Sciences-II. pp 235-259. New York: Academic Press.

[Nilsson, NJ 71*] SEE: Fikes, RE 71a. *STRIPS: A New Approach to the Application of Theorem Proving to Problem Solving.*

[Niranjan, M 88a] Niranjan, M.; Fallside, Frank. (1988). Neural Networks and Radial Basis Functions in Classifying Static Speech Patterns. CUED/F-INFENG/TR.22. Cambridge: Cambridge University, Engineering Department.

[Nishida, K 84a] Nishida, K. (1984). *A Full Associative Memory Chip for Data-Flow Computer.* Bulletin of the Electrotechnical Laboratory (Japan). Volume 48. pp 713-715.

[Nishihara, HK 78*] SEE: Marr, D 78b. *Representation and Recognition of the Spatial Organization of Three-Dimensional Shapes.*

[Nitta, J 81*] SEE: Toko, K 81a. *Application of Bifurcation Theory to Nerve Excitations.*

[Nizkik, CA 84a] Nizkik, Carol A.; Newcomb, R. W. (1984). *Computer Networking Capacity in Robotic Neural Systems.* Computer Communications. Volume 7. pp 85-91.

[Niznik, CA 83a] Niznik, Carol A. (1983). *The Neural Path Probabilistic Delay Model.* IEEE Transactions on Systems, Man and Cybernetics. Volume SMC-13. pp 1014-1018.

[Niznik, CA 87a] Niznik, Carol A.; Hoss, Wayne; Watts, Charles. (1987). *Modeling and Mathematical Formalism for Robotic Cerebellar Neural Network Pathway Information Measures.* IEEE International Conference on Neural Networks (ICNN-87), San Diego, CA, June 21-24, 1987: Proceedings. Volume 4. pp 583-592. New York: Institute of Electrical and Electronics Engineers.

[Noda, A 67*] SEE: Nagumo, J 67a. *A Learning Method for System Identification.*

[Noest, AJ 86a] Noest, Andre J. (1986). *Percolation and Frustration in Neural Networks.* In Bienenstock, Elie L. (ed.). Disordered Systems and Biological Organization. pp 381-384. New York: Springer-Verlag.

[Noest, AJ 88a] Noest, Andre J. (1988). *Phasor Neural Networks.* In Anderson, Dana Z. (ed.). Neural Information Processing Systems. pp 584-591. New York: American Institute of Physics.

[Noetzel, AS 88*] SEE: Graziano, MJ 88a. *Comparisons of Sigmoid Functions for Neural Network Convergence.*

[Noetzel, AS 88*] SEE: Silverman, RH 88a. *Time-Sequential Self-Organization of Hierarchical Neural Networks.*

[Nolfi, S 87*] SEE: Parisi, D 87a. *Connectionist Modeling of Syntactic Constraints on Sentence Processing.*

[Nolfi, S 88a] Nolfi, Stefano; Parisi, Domenico. (1988). *Learning to Understand Sentences in a Connectionist Network.* IEEE International Conference on Neural Networks (ICNN-88), San Diego, CA, July 24-27, 1988: Proceedings. Volume 2. pp 215-219. New York: Institute of Electrical and Electronics Engineers.

[Nomura, J 84a] Nomura, Junji. (1984). *A Method for Predicting Sales Amount by the Use of IWSM and GMDH.* In Farlow, Stanley J. (ed.). Self-Organizing Methods in Modeling: GMDH Type Algorithms. pp 215-224. New York: Marcel Dekker.

[Noos, FA 87a] Noos, Francesco A.; Lauria, Francesco E. (1987). *An Assembler for a Neural Network.* IEEE International

Conference on Neural Networks (ICNN-87), San Diego, CA, June 21-24, 1987: Proceedings. Volume 4. pp 601-610. New York: Institute of Electrical and Electronics Engineers.

[Norman, DA 79a] Norman, Donald A.; Bobrow, Daniel G. (1979). *Descriptions: An Intermediate Stage in Memory Retrieval.* Cognitive Psychology. Volume 11. pp 107-123.

[Norman, DA 81a] Norman, Donald A. (1981). *A Psychologist Views Human Processing: Human Errors and Other Phenomena Suggest Processing Mechanisms.* International Joint Conference on Artificial Intelligence (IJCAI-81), 7th, Vancouver, August 24-28, 1981: Proceedings. Volume 2. pp 1097-1101. International Joint Conference on Artificial Intelligence.

[Norman, DA 82*] SEE: Rumelhart, DE 82b. *Simulating a Skilled Typist: A Study of Skilled Cognitive-Motor Performance.*

[Norman, DA 86a] Norman, Donald A. (1986). *Reflections on Cognition and Parallel Distributed Processing.* In Rumelhart, David E. (ed.). Parallel Distributed Processing: Explorations in the Microstructure of Cognition. Volume 2. pp 531-546. Cambridge, MA: MIT Press.

[Norman, MF 74a] Norman, M. F. (1974). *Markovian Learning Processes.* SIAM Review. Volume 16. pp 143-162.

[Norrod, FE 87a] Norrod, Forrest E.; O'Neill, Michael D.; Gat, Erann. (1987). *Feedback-Induced Sequentiality in Neural Networks.* IEEE International Conference on Neural Networks (ICNN-87), San Diego, CA, June 21-24, 1987: Proceedings. Volume 2. pp 251-260. New York: Institute of Electrical and Electronics Engineers.

[Novick, DG 87*] SEE: Douglas, SA 87a. *Consistency and Variation in Spatial Reference.*

[Novikoff, A 62a] Novikoff, A. (1963). *On Convergence Proofs for Perceptrons.* Mathematical Theory of Automata: Proceedings. Volume 12. pp 615-622. Brooklyn, NY: Polytechnic Press of the Polytechnic Institute of Brooklyn.

[Nowlan, SJ 86*] SEE: Plaut, DC 86a. Experiments on Back-Propagation.

[Nowlan, SJ 87*] SEE: Hinton, GE 87b. *How Learning Can Guide Evolution.*

[Nusbaum, HC 86*] SEE: Schwab, EC 86b. Pattern Recognition by Humans and Machines.

[Nusbaum, HC 86*] SEE: Schwab, EC 86a. Pattern Recognition by Humans and Machines.

[Nusbaum, HC 86a] Nusbaum, Howard C.; Schwab, Eileen C. (1986). *The Role of Attention and Active Processing in Speech Perception.* In Schwab, Eileen C. (ed.); Nusbaum, Howard C. (ed.). Pattern Recognition by Humans and Machines. Volume 1, Speech Perception. pp 113-158. New York: Academic Press.

[Nygaard, K 69*] SEE: Walloe, L 69a. *A Computer-Simulated Model of a Second Order Sensory Neurone.*

[Nylen, M 87a] Nylen, M.; Moller, Per; Hertz, J. A. (1987). A Neural Network in the Visual Cortex. Nordita-87/41 S (preprint). Copenhagen: Nordisk Institut for Teoretisk Atomfysik.

[Nylen, M 88*] SEE: Moller, P 88a. *A Cortical Network Model for Early Vision Processing.*

[O'Keefe, J 89a] O'Keefe, John. (1989). *Computations the Hippocampus Might Perform.* In Nadel, Lynn (ed.); Cooper, Lynn A. (ed.); Culicover, Peter (ed.); Harnish, R. Michael (ed.). Neural Connections, Mental Computation. pp 225-284. Cambridge, MA: MIT Press.

[O'Neill, MD 87*] SEE: Norrod, FE 87a. *Feedback-Induced Sequentiality in Neural Networks.*

[O'Regan, JK 81*] SEE: McClelland, JL 81c. *Expectations Increase the Benefit Derived from Parafoveal Visual Information in Reading Words Aloud.*

[O'Regan, JK 81*] SEE: McClelland, JL 81d. *On Visual and Contextual Factors in Reading: A Reply to Rayner and Slowiaczek.*

[O'Toole, AJ 87*] SEE: Kersten, D 87a. *Associative Learning of Scene Parameters from Images.*

[Obusek, CJ 71*] SEE: Warren, RM 71a. *Speech Perception and Phonemic Restorations.*

[Oettinger, AG 52a] Oettinger, Anthony G. (1952). *Programming A Digital Computer to Learn.* Philosophical Magazine. Volume 43, Series 7. pp 1243-1263.

[Offner, F 40a] Offner, Franklin; Weinberg, Alvin M.; Young, Gale. (1940). *Nerve Conduction Theory: Some Mathematical Consequences of Bernstein's Model.* Bulletin of Mathematical Biophysics. Volume 2. pp 89-103.

[Ogielski, AT 85a] Ogielski, Andrew T.; Stein, D. L. (1985). *Dynamics on Ultrametric Spaces.* Physical Review Letters. Volume 55. pp 1634-1637.

[Ogmen, H 87a] Ogmen, Haluk; Gagne, Simon. (1987). *Dynamic Pattern Matching with Filling-In.* IEEE International Conference on Systems, Man and Cybernetics, 1987: Proceedings. pp 733-738. New York: Institute of Electrical and Electronics Engineers.

[Ogmen, H 87b] Ogmen, Haluk; Gagne, Simon. (1987). *Early Spatio-Temporal Integration by Gated Competitive Networks.* IEEE International Conference on Neural Networks (ICNN-87), San Diego, CA, June 21-24, 1987: Proceedings. Volume 2. pp 805-812. New York: Institute of Electrical and Electronics Engineers.

[Ogmen, H 87c] Ogmen, Haluk; Gagne, Simon. (April 1987). *Low Level Encoding of Motion by Asynchronous Shunting Networks.* Multinational Instrumentation Conference and Exposition (Miconex-87), Winnipeg, April, 1987: Proceedings.

[Ogmen, H 87d] Ogmen, Haluk; Gagne, Simon. (1987). *Phototransduction in Invertebrates.* Biological Cybernetics. Volume 56. pp 27-35.

[Oguztoreli, MN 75a] Oguztoreli, M. N. (1975). *On the Activities in a Continuous Neural Network.* Biological Cybernetics. Volume 18. pp 41-48.

[Oguztoreli, MN 79a] Oguztoreli, M. N. (1979). *Activity Analysis of Neural Networks.* Biological Cybernetics. Volume 34. pp 159-169.

[Oguztoreli, NN 83a] Oguztoreli, N. N. (1983). *Modeling and Simulation of Vertebrate Primary Visual System: Basic Network.* IEEE Transactions on Systems, Man and Cybernetics. Volume SMC-13. pp 766-782.

[Oguztoreli, NN 86a] Oguztoreli, N. N.; Caelli, Terry M. (1986). *An Inverse Problem in Neural Processing.* Biological Cybernetics. Volume 53. pp 239-245.

[Oh, S 84a] Oh, Se-Young. (1984). *A Walsh-Hadamard Based Distributed Storage Device for the Associative Search of Information.* IEEE Transactions on Pattern Analysis and Machine Intelligence. Volume PAMI-6. pp 617-623.

[Oh, S 84b] Oh, Se-Young. (1984). *Cross-talk Minimization for a Walsh-Hadamard Based Associative Memory.* International Conference on Pattern Recognition, 7th, Montreal, July 30-August 2, 1984: Proceedings. Volume 1. pp 355-357.

[Oh, S 87*] SEE: Cheung, KF 87b. *Neural Net Associative Memories Based on Convex Set Projections.*

[Oh, S 88*] SEE: Marks II, RJ 88a. *Architectures for a Continuous Level Neural Network Based on Alternating Orthogonal Projections.*

[Oh, S 88*] SEE: Marks II, RJ 88f. *The Effect of Stochastic Interconnects in Artificial Neural Network Classification.*

[Oh, S 88*] SEE: Marks II, RJ 88b. *Generalization in Layered Classification Neural Networks.*

[Oh, S 88*] SEE: Marks II, RJ 88c. *Nonlinearity Requirements for Correlation Based Associative Memories.*

[Oh, S 88*] SEE: Marks II, RJ 88d. *Optical-Processor Architectures for Alternating-Projection Neural Networks.*

[Oh, S 88*] SEE: Marks II, RJ 88e. *Performance Analysis of Associative Memories with Nonlinearities in the Correlation Domain.*

[Oh, S 88*] SEE: Marks II, RJ 88g. *The Performance of Convex Set Projection Based Neural Networks.*

[Oh, S 88a] Oh, Seho; Atlas, Les E.; Marks II, Robert Jackson; Park, Dong Chul. (1988). *Effect of Clock Skew in Iterative Neural Network and Optical Feedback Processors.* IEEE International Conference on Neural Networks (ICNN-88), San Diego, CA, July 24-27, 1988: Proceedings. Volume 2. pp 429-435. New York: Institute of Electrical and Electronics Engineers.

[Ohashi, K 84a] Ohashi, Kenichi. (1984). *GMDH Forecasting of United States Interest Rates.* In Farlow, Stanley J. (ed.). Self-Organizing Methods in Modeling: GMDH Type Algorithms. pp 199-214. New York: Marcel Dekker.

[Ohnishi, N 87*] SEE: Sugie, N 87a. Localizing Sound Source by Incorporating Biological Auditory Mechanism.

[Oja, E 76*] SEE: Kohonen, T 76b. *Fast Adaptive Formation of Orthogonalizing and Associative Memory in Recurrent Networks of Neuron-Like Elements.*

[Oja, E 79*] SEE: Cooper, LN 79a. *A Theory for the Acquisition and Loss of Neuron Specificity in Visual Cortex.*

[Oja, E 81*] SEE: Kohonen, T 81c. *Storage and Processing of Information in Distributed Associative Memory Systems.*

[Oja, E 85*] SEE: Kohonen, T 85a. *A Model of Cortical Associative Memory.*

[Oja, E 87*] SEE: Kohonen, T 87b. *Computing with Neural Networks.*

[Oja, E 87*] SEE: Kohonen, T 87h. *Technical Comments: Computing with Neural Networks.*

[Oja, E 87a] Oja, Erkki; Parkkinen, Jussi. (1987). *Texture Subspaces.* Pattern Recognition Theory and Applications. NATO ASI Series, Volume F30. pp 21-33. New York: Springer-Verlag.

[Okabe, Y 87*] SEE: Fukaya, M 87a. *Two-Level Neural Networks: Learning by Interaction with Environment.*

[Okajima, K 86a] Okajima, Kenji. (1986). *A Mathematical Model of the Primary Visual Cortex and Hypercolumn.* Biological Cybernetics. Volume 54. pp 107-114.

[Okajima, K 87a] Okajima, Kenji; Tanaka, Shigeru; Fujiwara, S. (1987). *A Heteroassociative Memory Network with Feedback Connection.* IEEE International Conference on Neural Networks (ICNN-87), San Diego, CA, June 21-24, 1987: Proceedings. Volume 2. pp 711-718. New York: Institute of Electrical and Electronics Engineers.

[Okajima, K 88a] Okajima, Kenji. (1988). *Towards a Model for Visual Information Processing.* Symposium on Future Electron Devices, 7th, Tokyo, November 1-2, 1988. Volume IV-5. pp 91-95.

[Okuda, M 74a] Okuda, Masami; Yoshida, Akihiko; Takahashi, Koichi. (1974). *A Dynamical Behaviour of Active Regions in Randomly Connected Neural Networks.* Journal of Theoretical Biology. Volume 48. pp 51-73.

[Olafsson, S 88a] Olafsson, Sverrir; Abu-Mostafa, Yaser S. (1988). *The Capacity of Multilevel Threshold Functions.* IEEE Transactions on Pattern Analysis and Machine Intelligence. Volume 10. pp 277-281.

[Oldham, WJB 87*] SEE: Rogers, CH 87a. *Isolated Word Recognition with an Artificial Neural Network.*

[Olinski, M 74*] SEE: Kilmer, W 74a. *Model of a Plausible Learning Scheme for CA3 Hippocampus.*

[Omidvar, M 88*] SEE: Cheung, JY 88a. *Mathematical Analysis of Learning Behavior of Neuronal Models.*

[Omohundro, SM 87a] Omohundro, Stephen M. (1987). Efficient Algorithms with Neural Network Behavior. UIUCDCS-R-87-1331. pp 1-75. Urbana-Champaign, IL: University of Illinois, Department of Computer Science.

[Oonuki, M 76a] Oonuki, Makoto. (1976). *Firing Probability of a Neural Network: First-order Differential Equation.* Journal of Theoretical Biology. Volume 58. pp 59-81.

[Oppenheim, AV 81a] Oppenheim, Alan V. (1981). *The Importance of Phase in Signals.* Proceedings of the IEEE. Volume 69. pp 527-541.

[Oppenheim, AV 85*] SEE: Curtis, SR 85a. *Signal Reconstruction from Fourier Transform Sign Information.*

[Or, W 88*] SEE: Ross, MD 88c. *The Neuroanatomical Substrate for Information Processing in Macular Endorgans.*

[Orbach, R 84*] SEE: Chamberlin, RV 84a. *Time Decay of the Remanent Magnetization in Spin-Glasses.*

[Orban, GA 88*] SEE: Van Hulle, MM 88b. *Correlative Binding in Higher Order Network for Sensorial Representation.*

[Orban, GA 88*] SEE: Van Hulle, MM 88c. *Entropy Driven Artificial Neuronal Networks and Sensorial Representation: A Proposal.*

[Orban, GA 88*] SEE: Van Hulle, MM 88a. *A Field Approach to Higher Order ANNs for Sensorial Representation.*

[Orban, GA 88*] SEE: Devos, MR 88a. *Self-Adapting Back-Propagation.*

[Orban, GA 88*] SEE: Van Hulle, MM 88d. *Sensory ANNs and Symbolic Decomposition.*

[Orban, GA 88a] Orban, G. A.; Gulyas, B. (1988). *Image Segregation by Motion: Cortical Mechanism and Implementation in Neural Networks.* In Eckmiller, Rolf (ed.); von der Malsburg, Christoph (ed.). Neural Computers. NATO ASI Series, Volume F41. pp 149-158. Berlin: Springer-Verlag.

[Ornstein, SM 67*] SEE: Molnar, CE 67a. *The CHASM: A Macromodular Computer for Analyzing Neuron Models.*

[Osada, H 81a] Osada, Hiroyasu; Yoshizawa, Shuji; Nagumo, Jin-ichi. (1981). *Theoretical Study on Cantor Function-Like Response Characteristics of a Neuron Model.* Transactions of the Institute of Electronics and Communication Engineers of Japan. Volume E64. pp 566-573.

[Osada, H 82*] SEE: Yoshizawa, S 82a. *Pulse Sequences Generated by a Degenerate Analog Neuron Model.*

[Osaki, S 70*] SEE: Mine, H 70a. Markovian Decision Processes.

[Osborne, NN 82a] Osborne, Neville N. (ed.). (1983). Dale's Principle and Communication Between Neurons. New York: Pergamon.

[Osborne, NN 86a] Osborne, Neville N. (ed.); Chader, Gerald J. (ed.). (1986). Progress in Retinal Research. Volume 5.

[Ostgaard, MA 62a] Ostgaard, M. A.; Butsch, L. M. (September, 1962). *Adaptive and Self-Organizing Flight Control Systems.* Aerospace Engineering. Volume 21. pp 80-81.

[Ott, E 81a] Ott, Edward. (October 1981). *Strange Attractors and Chaotic Motions of Dynamical Systems.* Reviews of Modern Physics. Volume 53, Number 4, Part 1. pp 655-671. American Physical Society.

[Ovchinnikov, S 87*] SEE: Yager, RR 87a. Fuzzy Sets and Applications: Selected Papers by L A Zadeh.

[Overman, KC 81a] Overman, Kelly C.; Mix, Dwight F. (1981). *A Novel View of Fourier Analysis.* Proceedings of IEEE.

Volume 69. pp 1372-1373.

[Overton, KJ 87a] Overton, Kenneth J. (1987). *Robot Tactile Sensing and Schema Control.* In Arbib, Michael A. (ed.); Hanson, Allen R. (ed.). Vision, Brain and Cooperative Computation. pp 511-526. Cambridge, MA: MIT Press.

[Owechko, Y 85*] SEE: Dunning, GJ 85a. *All-Optical Associative Holographic Memory with Feedback Using Phase Conjugate Mirrors.*

[Owechko, Y 86*] SEE: Soffer, BH 86a. *Associative Holographic Memory with Feedback Using Phase-Conjugate Mirrors.*

[Owechko, Y 86*] SEE: Dunning, GJ 86a. *Optical Holographic Associative Memory Using a Phase Conjugate Resonator.*

[Owechko, Y 86a] Owechko, Yuri; Marom, E.; Soffer, Bernard H.; Dunning, G. J. (1986). *Associative Memory in a Phase Conjugate Resonator Cavity Utilizing a Hologram.* Neural Network Models for Optical Computing. SPIE Volume 700. pp 296-300. Bellingham, WA: Society of Photo-Optical Instrumentation Engineers.

[Owechko, Y 86b] Owechko, Yuri; Marom, E.; Soffer, Bernard H.; Dunning, G. J. (1986). *Associative Memory in a Phase Conjugate Resonator Cavity Utilizing a Hologram.* Optical Computing. SPIE Volume 700. pp 296-300. Bellingham, WA: Society of Photo-Optical Instrumentation Engineers.

[Owechko, Y 87*] SEE: Dunning, GJ 87a. *All-Optical Associative Memory with Shift Invariance and Multiple-Image Recall.*

[Owechko, Y 87a] Owechko, Yuri; Dunning, G. J.; Marom, E.; Soffer, Bernard H. (1987). *Holographic Associative Memory with Nonlinearities in the Correlation Domain.* Applied Optics. Volume 26. pp 1900-1910.

[Owechko, Y 87b] Owechko, Yuri. (1987). *Optoelectronic Resonator Neural Networks.* Applied Optics. Volume 26. pp 5104-5111.

[Owechko, Y 87c] Owechko, Yuri; Dunning, G. J.; Soffer, Bernard H.; Marom, E. (1987). *Video Rate Optical Associative Memory with Shift Invariance.* IEEE International Conference on Neural Networks (ICNN-87), San Diego, CA, June 21-24, 1987: Proceedings. Volume 3. pp 557-564. New York: Institute of Electrical and Electronics Engineers.

[Owechko, Y 88a] Owechko, Yuri; Soffer, Bernard H.; Dunning, G. J. (1988). *Optoelectronic Neural Networks Based on Holographically Interconnected Image Processors.* Neural Network Models for Optical Computing. SPIE Volume 882. pp 143-153. Bellingham, WA: Society of Photo-Optical Instrumentation Engineers.

[Owen, WG 83*] SEE: Torre, V 83a. *The Dynamics of Electrically Interacting Cells.*

[Oxley, D 87*] SEE: Nelken, I 87a. *Parallel Gaussian Elimination on an Optically Interconnected Data Flow Computer.*

[Oyster, JM 86a] Oyster, J. Michael; Broadwell, Walter; Vicuna, Fernando. (1986). Associative Network Applications to Robot Vision. Report G320-2777. Los Angeles, CA: IBM Los Angeles Scientific Center.

[Oyster, JM 87a] Oyster, J. Michael; Skrzypek, Josef. (1987). *Computing Shape with Neural Networks: A Proposal.* IEEE International Conference on Neural Networks (ICNN-87), San Diego, CA, June 21-24, 1987: Proceedings. Volume 4. pp 335-344. New York: Institute of Electrical and Electronics Engineers.

[Ozaki, H 87*] SEE: Lehmann, D 87a. *EEG Alpha Map Series: Brain Micro-States by Space-Oriented Adaptive Segmentation.*

[Ozaki, H 87*] SEE: Lehmann, D 87b. *Human Brain EEG Fields: Micro-States and Their Functional Significance.*

[Ozaki, H 88*] SEE: Lehmann, D 88a. *Spontaneous and Information-Triggered Segments of Series of Human Brain Electric Field Maps.*

[Pabon, JA 87a] Pabon, Jahir A.; Gossard, David C. (1987). *A Methodology to Select Appropriate Learning Rate Parameters in*

Feed-Forward Networks. Conference on Aerospace Applications of Artificial Intelligence (AAAIC), 3rd, Dayton, OH, October 5-9, 1987: Proceedings.

[Pabon, JA 88a] Pabon, Jahir A.; Gossard, David C. (in press). *Connectionist Networks for Learning Coordinated Motion in Autonomous Systems.* National Conference on Artificial Intelligence (AAAI-88), 7th, St. Paul, MN, August 22-26, 1988.

[Pacheco, M 88a] Pacheco, M.; Bavan, S.; Lee, M.; Treleaven, P. C. (1988). *A Simple VLSI Architecture for Neurocomputing.* Annual Meeting of the International Neural Networks Society, 1st, Boston, MA, September, 1988: Proceedings.

[Packard, NH 86*] SEE: Crutchfield, JP 86a. *Chaos.*

[Paek, EG 85*] SEE: Farhat, NH 85a. *Optical Implementation of the Hopfield Model.*

[Paek, EG 85*] SEE: Farhat, NH 85b. *Optical Implementation of the Hopfield Model.*

[Paek, EG 87*] SEE: Von Lehmen, A 87a. *Influence of Interconnection Weight Discretization and Noise in an Optoelectronic Neural Network.*

[Paek, EG 88*] SEE: Von Lehmen, A 88a. *Factors Influencing Learning by Backpropagation.*

[Paek, EG 88a] Paek, Eung Gi; Liao, Paul F.; Gharavi, Hamid. (in press). *A Derivation of Neural Network Models for Associative Memory.* IEEE Circuits and Systems.

[Paek, EG 88b] Paek, Eung Gi; Von Lehmen, A. (in press). *Holographic Associative Memory for Word-Break Recognition.* Optics Letters.

[Paek, EG 88c] Paek, Eung Gi; Von Lehmen, A. (in press). *Real-Time Holographic Associative Memory for Identifying Words in a Continuous Letter String.* Optical Engineering.

[Page, EW 87*] SEE: Tagliarini, GA 87a. *Solving Constraint Satisfaction Problems with Neural Networks.*

[Page, EW 88*] SEE: Tagliarini, GA 88a. *A Neural-Network Solution to the Concentrator Assignment Problem.*

[Page, EW 88a] Page, Edward W.; Tagliarini, Gene A. (1988). *Algorithm Development for Neural Networks.* SPIE Symposium on Innovative Science and Technology, Los Angeles, CA, January, 1988.

[Paget, G 88*] SEE: Cochet, Y 88a. *Reseaux Neuronzux Pour Bases D'Images.*

[Paget, G 88*] SEE: Cochet, Y 88b. *ZZENN (Zig Zag Epigenetic Neural Networks), A New Approach to Connectionist Machine Learning.*

[Paget, G 88*] SEE: Cochet, Y 88c. *ZZENN: Zig Zag Epigenetic Neural Networks and Their Use in Information Systems.*

[Paik, E 87a] Paik, Eugene; Gunger, David; Skrzypek, Josef. (1987). *UCLA SFINX - A Neural Network Simulation Environment.* IEEE International Conference on Neural Networks (ICNN-87), San Diego, CA, June 21-24, 1987: Proceedings. Volume 3. pp 367-376. New York: Institute of Electrical and Electronics Engineers.

[Pal, I 87*] SEE: Lehmann, D 87a. *EEG Alpha Map Series: Brain Micro-States by Space-Oriented Adaptive Segmentation.*

[Pal, I 87*] SEE: Lehmann, D 87b. *Human Brain EEG Fields: Micro-States and Their Functional Significance.*

[Pal, I 88*] SEE: Lehmann, D 88a. *Spontaneous and Information-Triggered Segments of Series of Human Brain Electric Field Maps.*

[Palasek, RL 74*] SEE: MacGregor, RJ 74a. *Computer Simulation of Rhythmic Oscillation in Neuron Pools.*

[Palevsky, NG 87*] SEE: Kinoshita, J 87a. *Computing with Neural Networks.*

[Palm, G 78*] SEE: Marr, D 78a. *Analysis of a Cooperative Stereo Algorithm.*

[Palm, G 80a] Palm, Gunther. (1982). *How Useful are Associative Memories?.* Biomathematics in 1980. pp 145-153. Amsterdam: North-Holland.

[Palm, G 80b] Palm, Gunther. (1980). *On Associative Memory.* Biological Cybernetics. Volume 36. pp 19-31.

[Palm, G 81a] Palm, Gunther. (1981). *Towards a Theory of Cell Assemblies.* Biological Cybernetics. Volume 39. pp 181-194.

[Palm, G 82a] Palm, Gunther. (1982). Neural Assemblies: An Alternative Approach to Artificial Intelligence. New York: Springer-Verlag.

[Palm, G 84a] Palm, Gunther. (1984). *Local Synaptic Modification Can Lead to Organized Connectivity Patterns in Associative Memory.* Synergetics, From Microscopic to Macroscopic Order, Proceedings of the International Symposium on Synergetics at Berlin, July 4-8, 1983. Series in Synergetics v 22. pp 229-242. New York: Springer-Verlag.

[Palm, G 84b] Palm, Gunther; Bonhoeffer, Tobias. (1984). *Parallel Processing for Associative and Neuronal Networks.* Biological Cybernetics. Volume 51. pp 201-204.

[Palm, G 86*] SEE: Krone, G 86a. *Spatiotemporal Receptive Fields: A Dynamical Model Derived from Cortical Architectonics.*

[Palm, G 86a] Palm, Gunther. (1986). *Associative Networks and Cell Assemblies.* In Palm, Gunther (ed.); Aertsen, A. (ed.). Brain Theory. pp 211-228. Berlin: Springer-Verlag.

[Palm, G 87a] Palm, Gunther. (1987). *Associative Memory and Threshold Control in Neural Networks.* In Casti, John L. (ed.); Karlqvist, Anders (ed.). Real Brains, Artificial Minds. pp 165-179. Amsterdam: North-Holland.

[Palm, G 87b] Palm, Gunther. (1987). *Comments on the Paper "Threshold Regulation - A Key to the Understanding of the Combined Dynamics of EEG and Event-Related Potentials" by Elbert and Rockstroh.* Journal of Psychophysiology. Volume 4. pp 335-337.

[Palm, G 87c] Palm, Gunther. (1987). *Computing with Neural Networks.* Science. Volume 235. pp 1227-1228.

[Palm, G 87d] Palm, Gunther. (1987). *Technical Comments: Computing with Neural Networks.* Science. Volume 235. pp 1227-1228.

[Palm, G 88a] Palm, Gunther. (1988). *On the Asymptotic Information Storage Capacity of Neural Networks.* In Eckmiller, Rolf (ed.); von der Malsburg, Christoph (ed.). Neural Computers. NATO ASI Series, Volume F41. pp 269-280. Berlin: Springer-Verlag.

[Palmer, RG 83*] SEE: Hopfield, JJ 83a. *Unlearning Has a Stabilizing Effect in Collective Memories.*

[Palmer, RG 84a] Palmer, R. G.; Stein, D. L.; Abrahams, Elihu; Anderson, P. W. (1984). *Models of Hierarchically Constrained Dynamics for Glassy Relaxation.* Physical Review Letters. Volume 53. pp 958-961.

[Palmieri, G 68a] Palmieri, G.; Wanke, E. (1968). *A Pattern Recognition Machine.* Kybernetik. Volume 4. pp 69-80.

[Pandya, AS 87*] SEE: Unnikrishnan, KP 87a. *Role of Feedback in Visual Perception.*

[Pao, Y 75a] Pao, Yoh-Han; Merat, Frank L. (1975). *Distributed Associative Memory for Patterns.* IEEE Transactions on

Systems, Man and Cybernetics. Volume SMC-5. pp 620-625.

[Pao, Y 78a] Pao, Yoh-Han; Schultz, W. L. (1978). *An Associative Memory Technique for the Recognition of Patterns.* International Joint Conference on Pattern Recognition, 4th, Kyoto, November 7-10, 1978: Proceedings.

[Pao, YH 88*] SEE: Klassen, M 88a. *Characteristics of the Functional Link Net: A Higher Order Delta Rule Net.*

[Papachristou, CA 88*] SEE: Koutsougeras, C 88a. *Learning Discrete Mappings - Athena's Approach.*

[Papachristou, CA 88*] SEE: Koutsougeras, C 88b. *Training of a Neural Network for Pattern Classification Based on an Entropy Measure.*

[Papadopoullos, MS 79a] Papadopoullos, Maximos S.; Anninos, Photios A.; Zenone, Silvio. (1979). *Memory Topology of Neural Nets.* Journal of Theoretical Biology. Volume 80. pp 505-535.

[Papert, SA 60a] Papert, Seymour A. (1961). *Some Mathematical Models of Learning.* Information Theory, London Symposium, 4th, Held at the Royal Institution, London, August 29th-September 2nd, 1960. pp 242-251. Belfast: Universities Press.

[Papert, SA 69*] SEE: Minsky, ML 69a. Perceptrons: An Introduction to Computational Geometry.

[Papert, SA 69*] SEE: Minsky, ML 69b. *Perceptrons: Introduction.*

[Papert, SA 88*] SEE: Minsky, ML 88a. Perceptrons: Expanded Edition.

[Parberry, I 88a] Parberry, Ian. (1988). A Primer on the Complexity Theory of Neural Networks (Preliminary Version). CS-88-39. University Park, PA: Pennsylvania State University, Department of Computer Science.

[Parberry, I 88b] Parberry, Ian; Schnitger, Georg. (1988). *Parallel Computation with Threshold Functions.* Journal of Computer and System Sciences. Volume 36. pp 278-302.

[Parberry, I 88c] Parberry, Ian; Schnitger, Georg. (in press). *Relating Boltzmann Machines to Conventional Models of Computation.* Neural Networks.

[Parga, N 86a] Parga, N.; Virasoro, M. A. (1986). *The Ultrametric Organization of Memories in a Neural Network.* Journal de Physique. Volume 47. pp 1857-1864.

[Parhami, B 73a] Parhami, Behrooz. (1973). *Associative Memories and Processors: An Overview and Selected Bibliography.* Proceedings of the IEEE. Volume 61. pp 722-730.

[Parisi, D 87a] Parisi, Domenico; Nolfi, Stefano. (1987). *Connectionist Modeling of Syntactic Constraints on Sentence Processing.* IEEE International Conference on Neural Networks (ICNN-87), San Diego, CA, June 21-24, 1987: Proceedings. Volume 2. pp 507-512. New York: Institute of Electrical and Electronics Engineers.

[Parisi, D 88*] SEE: Nolfi, S 88a. *Learning to Understand Sentences in a Connectionist Network.*

[Parisi, G 84*] SEE: Mezard, M 84a. *Nature of the Spin-Glass Phase.*

[Parisi, G 84*] SEE: Mezard, M 84b. *Replica Symmetry Breaking and the Nature of the Spin Glass Phase.*

[Parisi, G 86a] Parisi, Giorgio. (1986). *Asymmetric Neural Networks and the Process of Learning.* Journal of Physics A: Mathematical and General. Volume 19. pp L675-680.

[Park, CH 86*] SEE: Psaltis, D 86a. *Nonlinear Discriminant Functions and Associative Memories.*

[Park, CH 88*] SEE: Psaltis, D 88b. *Higher Order Associative Memories and Their Optical Implementations.*

[Park, DC 88*] SEE: Oh, S 88a. *Effect of Clock Skew in Iterative Neural Network and Optical Feedback Processors.*

[Park, DC 88*] SEE: Marks II, RJ 88f. *The Effect of Stochastic Interconnects in Artificial Neural Network Classification.*

[Park, DC 88*] SEE: Marks II, RJ 88c. *Nonlinearity Requirements for Correlation Based Associative Memories.*

[Park, DC 88*] SEE: Marks II, RJ 88e. *Performance Analysis of Associative Memories with Nonlinearities in the Correlation Domain.*

[Parker, DB 85a] Parker, David B. (1985). Learning-Logic: Casting the Cortex of the Human Brain in Silicon. TR-47. Cambridge, MA: MIT, Center for Computational Research in Economics and Management.

[Parker, DB 86a] Parker, David B. (1986). *Comparison of Algorithms for Neuron-Like Cells.* In Denker, John S. (ed.). Neural Networks for Computing, Snowbird, Utah, 1986. AIP Proceedings 151. pp 327-332. New York: American Institute of Physics.

[Parker, DB 87a] Parker, David B. (1987). *Optimal Algorithms for Adaptive Networks: Second Order Back Propagation, Second Order Direct Propagaion and Second Order Hebbian Learning.* IEEE International Conference on Neural Networks (ICNN-87), San Diego, CA, June 21-24, 1987: Proceedings. Volume 2. pp 593-600. New York: Institute of Electrical and Electronics Engineers.

[Parkkinen, J 87*] SEE: Oja, E 87a. *Texture Subspaces.*

[Parkkinen, J 87a] Parkkinen, Jussi; Jaaskelainen, Timo. (1987). *Color Representation Using Statistical Pattern Recognition.* Applied Optics. Volume 26. pp 4240-4245.

[Parkkinen, J 88*] SEE: Hallikainen, J 88a. *Acousto-Optic Color Spectrometer.*

[Parks, M 87a] Parks, Mike. (1987). *Characterization of the Boltzmann Machine Learning Rate.* IEEE International Conference on Neural Networks (ICNN-87), San Diego, CA, June 21-24, 1987: Proceedings. Volume 3. pp 715-720. New York: Institute of Electrical and Electronics Engineers.

[Parnas, I 79a] Parnas, I.; Segev, I. (1979). *A Mathematical Model for Conduction of Action Potentials Along Bifurcating Axons.* Journal of Physiology. Volume 295. pp 323-343.

[Parthasarathy, K 87*] SEE: Gindi, GR 87a. *Winner-Take-All Networks and Associative Memory: Analysis and Optical Realization.*

[Parunak, HVD 87a] Parunak, H. Van Dyke; Kindrick, James; Irish, Bruce W. (1987). *Supplier-Consumer Nets for Material Handling.* IEEE International Conference on Neural Networks (ICNN-87), San Diego, CA, June 21-24, 1987: Proceedings. Volume 4. pp 787-794. New York: Institute of Electrical and Electronics Engineers.

[Parzen, E 62a] Parzen, E. (1962). *On Estimation of a Probability Density Function and Mode.* Annals of Mathematical Statistics. Volume 33. pp 1065-1076.

[Patarnello, S 87*] SEE: Carnevali, P 87a. *Exhaustive Thermodynamical Analysis of Boolean Learning Networks.*

[Patarnello, S 87a] Patarnello, S.; Carnevali, Paolo. (1987). *Learning Networks of Neurons with Boolean Logic.* Europhysics Letters. Volume 4. pp 503-508.

[Patarnello, S 88a] Patarnello, Stefano; Carnevali, Paolo. (1988). *Learning Networks of Neurons with Boolean Logic.* In Eckmiller, Rolf (ed.); von der Malsburg, Christoph (ed.). Neural Computers. NATO ASI Series, Volume F41. pp 281-289. Berlin: Springer-Verlag.

[Patel, JS 87*] SEE: Von Lehmen, A 87a. *Influence of Interconnection Weight Discretization and Noise in an Optoelectronic*

Neural Network.

[Patel, JS 88*] SEE: Von Lehmen, A 88a. *Factors Influencing Learning by Backpropagation.*

[Pati, YC 88a] Pati, Y. C.; Friedman, D.; Krishnaprasad, P. S.; Yao, C. T.; Peckerar, M. C.; Yang, R. (1988). *Neural Networks for Tactile Perception.* IEEE International Conference on Robots and Automation, Philadelphia, PA, April 25-29, 1988: Proceedings. pp 292-297. New York: Institute of Electrical and Electronics Engineers.

[Patrick, EA 66a] Patrick, Edward A.; Hancock, J. C. (1966). *Nonsupervised Sequential Classification and Recognition of Patterns.* IEEE Transactions on Information Theory. Volume IT-12. pp 362-372.

[Patrick, EA 70*] SEE: Fischer, FP 70a. *A Preprocessing Algorithm for Nearest Neighbor Decision Rules.*

[Patrick, EA 86a] Patrick, Edward A.; Fattu, James M. (1986). Artificial Intelligence with Statistical Pattern Recognition. Englewood Cliffs, NJ: Prentice-Hall.

[Paulos, JJ 88a] Paulos, John J.; Hollis, Paul W. (1988). *Neural Networks Using Analog Multipliers.* IEEE International Symposium on Circuits and Systems, Helsinki, June 7-9, 1988: Proceedings. pp 499-502. New York: Institute of Electrical and Electronics Engineers.

[Pawlicki, TF 88*] SEE: Lee, D 88a. Experiments with Neural Network Models for Handwritten Digit Recognition.

[Pawlicki, TF 88a] Pawlicki, Thaddeus F.; Lee, Dar-Shyang; Hull, Jonathan J.; Srihari, Sargur N. (1988). *Neural Network Models and Their Application to Handwritten Digit Recognition.* IEEE International Conference on Neural Networks (ICNN-88), San Diego, CA, July 24-27, 1988: Proceedings. Volume 2. pp 63-70. New York: Institute of Electrical and Electronics Engineers.

[Pawlicki, TF 88b] Pawlicki, Thaddeus F. (1988). NORA: Neural-network Object Recognition Architecture. Buffalo, NY: State University of New York, Computer Science Department.

[Paxman, RG 85*] SEE: Smith, WE 85a. *Image Reconstruction from Coded Data: I. Reconstruction Algorithms and Experimental Results.*

[Pearl, J 85a] Pearl, Judea. (1985). *Bayesian Networks: A Model of Self-Activated Memory for Evidential Reasoning.* Cognitive Science Society, 7th, Irvine, CA, August 15-17, 1985: Proceedings. pp 329-334.

[Pearl, J 85b] Pearl, Judea. (1985). *How to Do with Probabilities What People Say You Can't.* Conference on Artificial Intelligence Applications, 2nd, Miami Beach, FL, December 11-13, 1985: Proceedings. Washington: IEEE Computer Society Press.

[Pearl, J 86a] Pearl, Judea. (1986). *Fusion, Propagation, and Structuring in Belief Networks.* Artificial Intelligence. Volume 29. pp 241-288.

[Pearl, J 86b] Pearl, Judea. (1986). *Fusion, Propagation, and Structuring in Belief Networks.* Artificial Intelligence. Volume 29. pp 241-288.

[Pearl, J 86c] Pearl, Judea. (1986). *On Evidential Reasoning in a Hierarchy of Hypotheses.* Artificial Intelligence. Volume 28. pp 9-15.

[Pearl, J 86d] Pearl, Judea. (1986). *On Evidential Reasoning in a Hierarchy of Hypotheses.* Artificial Intelligence. Volume 28. pp 9-15.

[Pearl, J 87*] SEE: Geffner, H 87a. *On the Probabilistic Semantics of Connectionist Networks.*

[Pearl, J 87a] Pearl, Judea. (1987). *Distributed Revision of Composite Beliefs.* Artificial Intelligence. Volume 33. pp

173-215.

[Pearlmutter, B 88*] SEE: Goldberg, K 88a. Using a Neural Network to Learn the Dynamics of the CMU Direct-Drive Arm II.

[Pearlmutter, BA 86a] Pearlmutter, B. A.; Hinton, Geoffrey E. (1986). *G-Maximization: An Unsupervised Learning Procedure for Discovering Regularities*. In Denker, John S. (ed.). Neural Networks for Computing, Snowbird, Utah, 1986. AIP Proceedings 151. pp 333-338. New York: American Institute of Physics.

[Pearson, JC 86*] SEE: Silverman, DJ 86a. *Associative Recall Properties of the Trion Model of Cortical Organization.*

[Pearson, JC 87*] SEE: Edelman, GM 87a. *Computer Simulation of Somatotopic Map Plasticity.*

[Pearson, JC 87a] Pearson, John C.; Finkel, Leif H.; Edelman, Gerald M. (1987). *Plasticity in the Organization of Adult Cerebral Cortical Maps: A Computer Simulation Based on Neuronal Group Selection*. Journal of Neuroscience. Volume 7. pp 4209-4223.

[Pearson, JC 88*] SEE: Gelfand, JJ 88a. *Multisensor Integration in Biological Systems.*

[Pearson, JC 88a] Pearson, John C.; Gelfand, Jack J.; Sullivan, W. E.; Peterson, Richard M.; Spence, Clay D. (1988). *Neural Network Approach to Sensory Fusion*. Sensor Fusion. SPIE 931. pp 103-108. Bellingham, WA: Society of Photo-Optical Instrumentation Engineers.

[Peckerar, MC 87*] SEE: Marrian, CRK 87a. *Electronic 'Neural' Net Algorithm for Maximum Entropy Deconvolution.*

[Peckerar, MC 88*] SEE: Culhane, AD 88a. A Neural Net Approach to Discrete Fourier Transforms.

[Peckerar, MC 88*] SEE: Pati, YC 88a. *Neural Networks for Tactile Perception.*

[Peelle, HA 81a] Peelle, Howard A. (1981). *Teaching a Topic in Cybernetics with APL: An Introduction to Neural Net Modeling*. In Anderson, William L. (ed.); Smith, David G. (ed.). Quote Quad. Volume 12. pp 235-239.

[Peleg, S 80a] Peleg, Shmuel. (1980). *A New Probabilistic Relaxation Scheme*. IEEE Transactions on Pattern Analysis and Machine Intelligence. Volume PAMI-2. pp 362-369.

[Pellegrin, D 86a] Pellegrin, D. (1986). *Dynamics of Random Boolean Networks*. In Bienenstock, Elie L. (ed.). Disordered Systems and Biological Organization. pp 67 f. New York: Springer-Verlag.

[Pellionisz, AJ 73a] Pellionisz, Andras J.; Szentagothai, John. (1973). *Dynamic Single Unit Simulation of a Realistic Cerebellar Network Model*. Brain Research. Volume 49. pp 83-99.

[Pellionisz, AJ 74a] Pellionisz, Andras J.; Szentagothai, John. (1974). *Dynamic Single Unit Simulation of a Realistic Cerebellar Network Model. II. Purkinje Cell Activity within the Basic Circuit and Modified by Inhibitory Sys.*. Brain Research. Volume 68. pp 19-40.

[Pellionisz, AJ 78a] Pellionisz, Andras J. (1978). *Synthesis of Fragmented Data on Neuronal Systems: A Computer Model of Cerebellum*. Progress in Cybernetics and Systems Research. Volume 3. pp 411-427.

[Pellionisz, AJ 79a] Pellionisz, Andras J.; Llinas, R. (1979). *Brain Modeling by Tensor Network Theory and Computer Simulation*. Neuroscience. Volume 4. pp 323-348.

[Pellionisz, AJ 82a] Pellionisz, Andras J.; Llinas, R. (1982). *Space-Time Representation in the Brain: The Cerebellum as a Predictive Space-Time Metric Tensor*. Neuroscience. Volume 7. pp 2949-2970.

[Pellionisz, AJ 82b] Pellionisz, Andras J.; Llinas, R. (1982). *Tensor Theory of Brain Function: The Cerebellum as a Space-Time Metric*. In Amari, Shun-ichi (ed.); Arbib, Michael A. (ed.). Competition and Cooperation in Neural

Nets. pp 394-417. New York: Springer-Verlag.

[Pellionisz, AJ 83a] Pellionisz, Andras J. (1983). *Brain Theory: Connecting Neurobiology to Robotics.* Journal of Theoretical Neurobiology. Volume 2. pp 185-211.

[Pellionisz, AJ 84a] Pellionisz, Andras J. (1984). *Coordination:A Vector-Matrix Description of Transformations of Overcomplete CNS Coordinates and Tensorial Solution Using the Moore-Penrose Generalized Inverse.* Journal of Theoretical Biology. Volume 110. pp 353-375.

[Pellionisz, AJ 86a] Pellionisz, Andras J. (1986). *Old Dogmas and New Axioms in Brain Theory.* Behavioral and Brain Sciences. Volume 9. pp 103-104.

[Pellionisz, AJ 86b] Pellionisz, Andras J. (1986). *Tensor Network Theory and its Application in Computer Modeling of the Metaorganization of Sensorimotor Hierarchies of Gaze.* In Denker, John S. (ed.). Neural Networks for Computing, Snowbird, Utah, 1986. AIP Proceedings 151. pp 339-344. New York: American Institute of Physics.

[Pellionisz, AJ 87*] SEE: Daunicht, WJ 87a. *Spatial Arrangement of the Vestibular and the Oculomotor System in the Rat.*

[Pellionisz, AJ 87a] Pellionisz, Andras J. (1987). *Sensorimotor Operations: A Ground for the Co-Evolution of Brain Theory with Neurobotics and Neurocomputers.* IEEE International Conference on Neural Networks (ICNN-87), San Diego, CA, June 21-24, 1987: Proceedings. Volume 4. pp 593-600. New York: Institute of Electrical and Electronics Engineers.

[Pellionisz, AJ 88a] Pellionisz, Andras J. (1988). *Tensor Geometry: A Language of Brains and Neurocomputers. Generalized Coordinates in Neuroscience & Robotics.* In Eckmiller, Rolf (ed.); von der Malsburg, Christoph (ed.). Neural Computers. NATO ASI Series, Volume F41. pp 381-391. Berlin: Springer-Verlag.

[Pemberton, JC 87*] SEE: Vidal, JJ 87a. *Implementing Neural Nets with Programmable Logic.*

[Peng, J 87*] SEE: Becker, LA 87a. *Network Processing of Hierarchical Knowledge for Classification and Diagnosis.*

[Peng, J 87*] SEE: Becker, LA 87b. *Using Activation Networks for Analogical Ordering of Consideration: One Method for Integrating Connectionist and Symbolic Processing.*

[Penla, N 87*] SEE: Somani, AK 87a. *Compact Neural Network.*

[Penla, N 88*] SEE: Somani, AK 88b. Reduced Interconnection Neural Networks.

[Pentland, AP 86a] Pentland, Alex P. (1986). Perceptual Organization and the Representation of Natural Form. Technical Note Number 357. Menlo Park, CA: SRI International.

[Penz, PA 86a] Penz, P. Andrew; Wiggins, R. (1986). *Digital Signal Processor Accelerators for Neural Network Simulations.* In Denker, John S. (ed.). Neural Networks for Computing, Snowbird, Utah, 1986. AIP Proceedings 151. pp 345-355. New York: American Institute of Physics.

[Penz, PA 87*] SEE: Collins, DR 87a. *Optical Neurocomputers Implementation Using Deformable Mirror Arrays.*

[Penz, PA 87a] Penz, P. Andrew. (1987). *The Closeness Code: An Integer to Binary Vector Transformation Suitable for Neural Network Architectures.* IEEE International Conference on Neural Networks (ICNN-87), San Diego, CA, June 21-24, 1987: Proceedings. Volume 3. pp 515-522. New York: Institute of Electrical and Electronics Engineers.

[Pepe, J 70*] SEE: Grossberg, S 70b. *Schizophrenia: Possible Dependence of Associational Span, Bowing, and Primacy vs. Recency on Spiking Threshold.*

[Pepe, J 71*] SEE: Grossberg, S 71c. *Spiking Threshold and Overarousal Effects in Serial Learning.*

[Perera, AGU 86*] SEE: Coon, DD 86a. *Spectral Information Coding by Infrared Photoreceptors.*

[Perera, AGU 87*] SEE: Coon, DD 87a. *Farey-Fraction Frequency Modulation in the Neuronlike Output of Silicon p-i-n Diodes at 4.2 K.*

[Perera, AGU 87*] SEE: Coon, DD 87b. *Parallel Asynchronous Focal Plane Array Processing.*

[Perera, AGU 87*] SEE: Coon, DD 87c. *Semiconductor Electronic Concepts for Neural Network Emulation.*

[Perera, AGU 88*] SEE: Coon, DD 88b. *Integrate-and-Fire Coding in Hodgkin-Huxley Circuits Employing Silicon Diodes.*

[Perera, AGU 88*] SEE: Coon, DD 88c. *Neuron-Like Transient Phenomena in Silicon p-i-n Structures.*

[Perera, AGU 88*] SEE: Coon, DD 88d. *New Hardware for Massive Neural Networks.*

[Peretto, P 84a] Peretto, Pierre. (1984). *Collective Properties of Neural Networks: A Statistical Physics Approach.* Biological Cybernetics. Volume 50. pp 51-62.

[Peretto, P 86a] Peretto, Pierre; Niez, Jean-Jacques. (1986). *Collective Properties of Neural Networks.* In Bienenstock, Elie L. (ed.). Disordered Systems and Biological Organization. pp 171-186. New York: Springer-Verlag.

[Peretto, P 86b] Peretto, Pierre; Niez, Jean-Jacques. (1986). *Long Term Memory Storage Capacity of Multiconnected Neural Networks.* Biological Cybernetics. Volume 54. pp 53-63.

[Peretto, P 86c] Peretto, Pierre; Niez, Jean-Jacques. (1986). *Stochastic Dynamics of Neural Networks.* IEEE Transactions on Systems, Man and Cybernetics. Volume SMC-16. pp 73-83.

[Perez, R 73*] SEE: Glass, L 73a. *Perception of Random Dot Interference Patterns.*

[Perkel, DH 64a] Perkel, Donald H. (1964). A Digital-Computer Model of Nerve Cell Functioning. Report RM-4132-NIH. Santa Monica, CA: Rand Corporation.

[Perkel, DH 66*] SEE: Moore, GP 66a. *Statistical Analysis and Functional Interpretation of Neuronal Spike Data.*

[Perkel, DH 68*] SEE: Segundo, JP 68a. *Input-Output Relations in Computer-Simulated Nerve Cells: Influence of Statistical Properties, Strength, etc..*

[Perkel, DH 76a] Perkel, Donald H. (1976). *A Computer Program for Simulating a Network of Interacting Neurons, Part I: Organization and Physiological Assumptions.* Computers and Biomedical Research. Volume 9. pp 31-43.

[Perkel, DH 76b] Perkel, Donald H.; Smith, Mark S. (1976). *A Computer Program for Simulating a Network of Interacting Neurons, Part II: Programming Aspects.* Computers and Biomedical Research. Volume 9. pp 45-66.

[Perkel, DH 76c] Perkel, Donald H. (1976). *A Computer Program for Simulating a Network of Interacting Neurons, Part III: Applications.* Computers and Biomedical Research. Volume 9. pp 67-74.

[Personnaz, L 85a] Personnaz, L.; Guyon, I.; Dreyfus, G. (1985). *Information Storage and Retrieval in Spin-Glass Like Neural Networks.* Journal de Physique, Lettres. Volume 46. pp 359-365.

[Personnaz, L 86*] SEE: Guyon, I 86a. *Engineering Applications of Spin Glass Concepts.*

[Personnaz, L 86a] Personnaz, L.; Guyon, I.; Dreyfus, G.; Toulouse, G. (1986). *A Biologically Constrained Learning Mechanism in Networks of Formal Neurons.* Journal of Statistical Physics. Volume 43. pp 411-422.

[Personnaz, L 86b] Personnaz, L.; Guyon, I.; Dreyfus, G. (1986). *Collective Computational Properties of Neural Networks:*

New Learning Mechanisms. Physical Review A. Volume 34. pp 4217-4228.

[Personnaz, L 86c] Personnaz, L.; Guyon, I.; Dreyfus, G. (1986). *Designing a Neural Network Satisfying a Given Set of Constraints.* In Denker, John S. (ed.). Neural Networks for Computing, Snowbird, Utah, 1986. AIP Proceedings 151. pp 356-359. New York: American Institute of Physics.

[Personnaz, L 86d] Personnaz, L.; Guyon, I.; Dreyfus, G. (1986). *Neural Network Design for Efficient Information Retrieval.* In Bienenstock, Elie L. (ed.). Disordered Systems and Biological Organization. pp 227-232. New York: Springer-Verlag.

[Personnaz, L 86e] Personnaz, L.; Guyon, I.; Dreyfus, G. (1986). *Neural Network Design for Efficient Information Retrieval.* Series Disordered Systems and Biological Organization. NATO ASI Series, Volume F20. pp 227-231. Berlin Heidelberg: Springer-Verlag.

[Personnaz, L 86f] Personnaz, L.; Guyon, I.; Johannet, A.; Dreyfus, G.; Toulouse, G. (1986). *Simple Selectionist Learning Rule for Neural Networks.* In Denker, John S. (ed.). Neural Networks for Computing, Snowbird, Utah, 1986. AIP Proceedings 151. pp 360-363. New York: American Institute of Physics.

[Personnaz, L 87a] Personnaz, L.; Guyon, I.; Dreyfus, G. (1987). *High-Order Neural Networks: Information Storage without Errors.* Europhysics Letters. Volume 4. pp 863-867.

[Personnaz, L 88*] SEE: Dreyfus, G 88a. *High Order Neural Networks for Efficient Associative Memory Design.*

[Personnaz, L 88*] SEE: Guyon, I 88a. *Of Points and Loops.*

[Pertile, G 70*] SEE: Anninos, PA 70a. *Dynamics of Neural Structures.*

[Peters, SL 87a] Peters, Sandra L.; Shapiro, Stuart C. (1987). *A Representation for Natural Category Systems.* Cognitive Science Society, 9th, Seattle, WA, July 16-18, 1987: Proceedings. pp 379-390. Hillsdale, NJ: Lawrence Erlbaum.

[Peterson, D 70a] Peterson, David. (1970). *Some Convergence Properties of a Nearest Neighbor Decision Rule.* IEEE Transactions on Information Theory. Volume IT-16. pp 26-31.

[Peterson, I 87a] Peterson, Ivars. (1987). *Robots with a Lot of Nerve: Eye-hand Coordination May Help a Robot Learn to See.* Science News. Volume 131. pp 362-363.

[Peterson, RM 88*] SEE: Pearson, JC 88a. *Neural Network Approach to Sensory Fusion.*

[Petrie, T 70*] SEE: Baum, LE 70a. *A Maximization Technique Occurring in the Statistical Analysis of Probabilistic Functions of Markov Chains.*

[Petsche, T 87a] Petsche, Thomas; Dickinson, Bradley W. (1987). *A Locally Interconnected Neural Network for Deconvolution.* Conference on Information Sciences and Systems, 21st, Baltimore, MD, March 25-27, 1987: Proceedings. pp 381-386. Baltimore, MD: Johns Hopkins University.

[Petsche, T 88a] Petsche, Thomas; Dickinson, Bradley W. (1988). *A Trellis-Structured Neural Network.* In Anderson, Dana Z. (ed.). Neural Information Processing Systems. pp 592-601. New York: American Institute of Physics.

[Petsche, T 88b] Petsche, Thomas. (1988). Topics in Neural Networks. PhD Dissertation. Princeton, NJ: Princeton University, Department of Electrical Engineering.

[Pettigrew, JD 85a] Pettigrew, J. D. (1985). *Some Constraints Operating on the Synaptic Modifications Underlying Binocular Competition in the Developing Visual Cortex.* In Levy, William B. (ed.); Anderson, James A. (ed.); Lehmkuhle, Stephen (ed.). Synaptic Modifications, Neuron Selectivity, and Nervous System Organization. pp 79-88. Hillsdale, NJ: Lawrence Erlbaum.

[Pettit, FR 87a] Pettit, Francis R. (1987). *Post-Digital Electronics*. Encyclopedia of Physical Science and Technology. Volume 11. pp 195-205.

[Pettus, RO 87*] SEE: Zhang, W 87a. *NPN Relations and Cognitive Map Development*.

[Pfaffelhuber, E 75a] Pfaffelhuber, E. (1975). *Correlation Memory Models: A First Approximation in a General Learning Scheme*. Biological Cybernetics. Volume 18. pp 217-223.

[Pfeiffer, RR 68*] SEE: Molnar, CE 68a. *Interpretation of Spontaneous Spike Discharge Patterns of Neurons in the Cochlear Nucleus*.

[Pfeiffer, RR 75a] Pfeiffer, Russell R.; Kim, D. O. (1975). *Cochlear Nerve Filter Responses: Distribution Along the Cochlear Partition*. Journal of the Acoustics Society of America. Volume 58. pp 867-869.

[Pham, KM 88a] Pham, K. M.; Degoulet, P. (1988). *MOSAIC: A Macro-Connectionist Organization System for Artificial Intelligence Computation*. IEEE International Conference on Neural Networks (ICNN-88), San Diego, CA, July 24-27, 1988: Proceedings. Volume 2. pp 533-540. New York: Institute of Electrical and Electronics Engineers.

[Phillips, WA 88a] Phillips, W. A.; Hancock, P. J. B.; Willson, N. J.; Smith, L. S. (1988). *On the Acquisition of Object Concepts from Sensory Data*. In Eckmiller, Rolf (ed.); von der Malsburg, Christoph (ed.). Neural Computers. NATO ASI Series, Volume F41. pp 159-168. Berlin: Springer-Verlag.

[Pickard, DK 77a] Pickard, D. K. (1977). *A Curious Binary Lattice Process*. Journal of Applied Probability. Volume 14. pp 717-731.

[Pierce, WH 61*] SEE: Widrow, B 61a. *Birth, Life, and Death in Microelectronic Systems*.

[Pilpel, S 87a] Pilpel, Shairy. (March 1987). *A Balanced Scalable Parallel Processor*. VLSI Systems Design. pp 80-88.

[Pineda, FJ 87a] Pineda, Fernando J. (1987). *Generalization of Back-Propagation to Recurrent Neural Networks*. Physical Review Letters. Volume 59. pp 2229-2232.

[Pineda, FJ 88a] Pineda, Fernando J. (in press). *Dynamics and Architecture in Neural Computation*. Journal of Complexity.

[Pineda, FJ 88b] Pineda, Fernando J. (1988). *Generalization of Backpropagation to Recurrent and Higher Order Neural Networks*. In Anderson, Dana Z. (ed.). Neural Information Processing Systems. pp 602-611. New York: American Institute of Physics.

[Pinette, B 85*] SEE: Sutton, RS 85a. *Learning of World Models by Connectionist Networks*.

[Pinker, S 88a] Pinker, Steven (ed.); Mehler, Jacques (ed.). (1988). Connections and Symbols. Cambridge, MA: MIT Press.

[Pinker, S 88b] Pinker, Steven. (1988). *On Language and Connectionism: Analysis of a Parallel Distributed Processing Model of Language Acquisition*. In Pinker, Steven (ed.); Mehler, Jacques (ed.). Connections and Symbols. pp 73-194. Cambridge, MA: MIT Press.

[Piske, UAW 63*] SEE: Steinbuch, K 63a. *Learning Matrices and Their Applications*.

[Pisoni, DB 81a] Pisoni, David B. (1981). *In Defense of Segmental Representations in Speech Perception*. Journal of the Acoustical Society of America. Volume 69. pp S1,S32.

[Pisoni, DB 86a] Pisoni, David B.; Luce, Paul A. (1986). *Speech Perception: Research, Theory, and the Principal Issues*. In Schwab, Eileen C. (ed.); Nusbaum, Howard C. (ed.). Pattern Recognition by Humans and Machines: Speech Perception. Volume I. pp 1-50. New York: Academic Press.

[Pitney, GA 87*] SEE: Smith, T 87a. *Application of Genetic Algorithm to a Neural Network for Finding Optimal Paths*.

[Pitts, WH 43*] SEE: McCulloch, WS 43a. *A Logical Calculus of the Ideas Immanent in Nervous Activity.*

[Pitts, WH 43*] SEE: McCulloch, WS 43b. *A Logical Calculus of the Ideas Immanent in Nervous Activity.*

[Pitts, WH 43*] SEE: Landahl, HD 43a. *A Statistical Consequence of the Logical Calculus of Nervous Nets.*

[Pitts, WH 47a] Pitts, Walter H.; McCulloch, Warren S. (1947). *How We Know Universals: The Perception of Auditory and Visual Forms.* Bulletin of Mathematical Biophysics. Volume 9. pp 127-147.

[Pitts, WH 47b] Pitts, Walter H.; McCulloch, Warren S. (1988). *How We Know Universals: The Perception of Auditory and Visual Forms.* In Anderson, James A. (ed.); Rosenfeld, Edward (ed.). Neurocomputing: Foundations of Research. pp 32-41. Cambridge, MA: MIT Press.

[Pitts, WH 59*] SEE: Lettvin, JY 59a. *What the Frog's Eye Tells the Frog's Brain.*

[Platt, JC 86a] Platt, John C.; Hopfield, John J. (1986). *Analog Decoding Using Neural Networks.* In Denker, John S. (ed.). Neural Networks for Computing, Snowbird, Utah, 1986. AIP Proceedings 151. pp 364-369. New York: American Institute of Physics.

[Platt, JC 88a] Platt, John C.; Barr, Alan H. (1988). *Constrained Differential Optimization.* In Anderson, Dana Z. (ed.). Neural Information Processing Systems. pp 612-621. New York: American Institute of Physics.

[Platt, JC 88b] Platt, John C.; Barr, Alan H. (1988). Constrained Differential Optimization for Neural Networks. CS-TR-88-17. Pasadena, CA: California Institute of Technology, Computer Science Department.

[Platzman, LK 83*] SEE: Bartholdi, JJ 83a. *A Minimal Technology Routing System for Meals on Wheels.*

[Platzman, LK 84*] SEE: Bartholdi, JJ 84a. Heuristics Based on Space-Filling Curves for Combinatorial Problems in the Plane.

[Platzman, LK 86*] SEE: Bartholdi, JJ 86a. *Retrieval Strategies for a Carousel Conveyor.*

[Plaut, DC 84a] Plaut, David C. (1984). Visual Recognition of Simple Objects by a Connection Network. TR-143. Rochester, NY: University of Rochester, Computer Science Department.

[Plaut, DC 86*] SEE: Derthick, MA 86a. *Is Distributed Connectionism Compatible with the Physical Symbol System Hypothesis.*

[Plaut, DC 86a] Plaut, David C.; Nowlan, Steven J.; Hinton, Geoffrey E. (1986). Experiments on Back-Propagation. CMU-CS-86-126. Pittsburgh, PA: Carnegie Mellon University, Computer Science Department.

[Plaut, DC 87*] SEE: Hinton, GE 87d. *Using Fast Weights to Deblur Old Memories.*

[Plumbley, MD 88a] Plumbley, M. D.; Fallside, Frank. (1988). An Information-Theoretic Approach to Unsupervised Connectionist Models. CUED/F-INFENG/TR.7. Cambridge: Cambridge University, Engineering Department.

[Poggio, TA 75a] Poggio, Tomaso A. (1975). *On Optimal Nonlinear Associative Recall.* Biological Cybernetics. Volume 19. pp 201-209.

[Poggio, TA 76*] SEE: Marr, D 76a. *Cooperative Computation of Stereo Disparity.*

[Poggio, TA 76*] SEE: Marr, D 76b. *Cooperative Computation of Stereo Disparity.*

[Poggio, TA 77*] SEE: Marr, D 77a. *From Understanding Computation to Understanding Neural Circuitry.*

[Poggio, TA 78*] SEE: Marr, D 78a. *Analysis of a Cooperative Stereo Algorithm.*

[Poggio, TA 79*] SEE: Marr, D 79a. *A Computational Theory of Human Stereo Vision.*

[Poggio, TA 81a] Poggio, Tomaso A.; Koch, Christof; Torre, Vincent. (1981). Microelectronics in Nerve Cells: Dendritic Morphology and Information Processing. AI Memo 650. Cambridge, MA: Massachusetts Institute of Technology.

[Poggio, TA 82a] Poggio, Tomaso A. (1982). Visual Algorithms. AI Memo 683. Cambridge, MA: Massachusetts Institute of Technology.

[Poggio, TA 83*] SEE: Koch, C 83a. *Nonlinear Interactions in a Dendritic Tree: Localization, Timing, and Role in Information Processing.*

[Poggio, TA 84*] SEE: Koch, C 84a. Biophysics of Computation: Neurons, Synapses and Membranes.

[Poggio, TA 84a] Poggio, Tomaso A.; Koch, Christof. (1984). An Analog Model of Computation for the Ill-Posed Problems of Early Vision. AI Memo 783. Cambridge, MA: Massachusetts Institute of Technology.

[Poggio, TA 84b] Poggio, Tomaso A.; Torre, Vincent. (1984). Ill-Posed Problems and Regularization Analysis in Early Vision. AI Memo 773. Cambridge, MA: Massachusetts Institute of Technology.

[Poggio, TA 84c] Poggio, Tomaso A. (1984). Vision by Man and Machine: How the Brain Processes Visual Information May Be Suggested by Studies in Computer Vision (and Vice Versa). AI Memo 776. Cambridge, MA: Massachusetts Institute of Technology.

[Poggio, TA 85*] SEE: Hurlbert, AC 85a. Spotlight on Attention.

[Poggio, TA 85a] Poggio, Tomaso A.; Torre, Vincent; Koch, Christof. (1985). *Computational Vision and Regularization Theory.* Nature. Volume 317. pp 314-319.

[Poggio, TA 87a] Poggio, Tomaso A.; Koch, Christof. (1987). *Synapses that Compute Motion.* Scientific American. Volume 256. pp 46-71.

[Poggio, TA 88*] SEE: Hurlbert, AC 88a. *Synthesizing a Color Algorithm from Examples.*

[Poggio, TA 88a] Poggio, Tomaso A.; Hurlbert, Anya C. (1988). *Learning a Color Algorithm from Examples.* In Anderson, Dana Z. (ed.). Neural Information Processing Systems. pp 622-631. New York: American Institute of Physics.

[Poliac, MO 87a] Poliac, M. O.; Lee, E. B.; Slagle, James R.; Wick, M. R. (1987). *A Crew Scheduling Problem.* IEEE International Conference on Neural Networks (ICNN-87), San Diego, CA, June 21-24, 1987: Proceedings. Volume 4. pp 779-786. New York: Institute of Electrical and Electronics Engineers.

[Pollack, JB 82a] Pollack, Jordan B. (1982). *Natural Language Processing Using Spreading Activation and Lateral Inhibition.* Cognitive Science Society, 4th, Ann Arbor, MI, August 4-6, 1982: Proceedings. Hillsdale, NJ: Lawrence Erlbaum.

[Pollack, JB 84a] Pollack, Jordan B.; Waltz, David L. (1984). *Parallel Interpretation of Natural Language.* International Conference on Fifth Generation Computer Systems, Tokyo, November 6-9, 1984: Proceedings. pp 686-691. Amsterdam: North-Holland.

[Pollack, JB 85*] SEE: Waltz, DL 85a. *Massively Parallel Parsing: A Strongly Interactive Model of Natural Language Interpretation.*

[Pollack, JB 86a] Pollack, Jordan B.; Waltz, David L. (1986). *Interpretation of Natural Language: A Potential Application of Parallelism.* Byte. pp 189-198.

[Pollack, JB 87a] Pollack, Jordan B. (1987). *Cascaded Back-Propagation on Dynamic Connectionist Networks.* Cognitive

Science Society, 9th, Seattle, WA, July 16-18, 1987: Proceedings. pp 391-404. Hillsdale, NJ: Lawrence Erlbaum.

[Pollack, JB 88*] SEE: Waltz, DL 88c. *Massively Parallel Parsing: A Strongly Interactive Model of Natural Language Interpretation.*

[Pollatsek, A 86*] SEE: Trehub, A 86a. *Seeing-More-Than-is-There: A Probe of Retinoid Networks.*

[Pont, MJ 88a] Pont, M. J.; Damper, R. I. (1988). *A Neural Model of Infant Speech Perception.* FASE Symposium (SPEECH-88), 7th, Edinburgh, August 22-26, 1988.

[Pont, MJ 88b] Pont, M. J. (1988). *A Neural Model of the Mechanisms Underlying Infant Perception of Voice-Onset Time.* Neural Networks. Volume 1, Supplement 1.

[Pontryagin, LS 62a] Pontryagin, L. S.; Boltyanskil, V. G.; Gamkrelidze, R. V.; Mishchenko, E. F. (1962). The Mathematical Theory of Optimal Processes. New York: Wiley.

[Porod, W 88*] SEE: Li, JH 88a. *Qualitative Analysis and Synthesis of a Class of Neural Networks.*

[Porod, W 88*] SEE: Michel, AN 88b. *Stability Results for Neural Networks.*

[Port, O 86a] Port, Otis. (June 2, 1986). *Computers That Come Awfully Close to Thinking.* Business Week. pp 92-95.

[Port, RL 88*] SEE: Sclabassi, RJ 88b. *Nonlinear Systems Analysis of the Hippocampal Perforant Path-Dentate Projection. I. Theoretical and Interpretational Considerations.*

[Posner, EC 86*] SEE: McEliece, RJ 86a. Capacity of the Hopfield Associative Memory.

[Posner, EC 87*] SEE: McEliece, RJ 87a. *The Capacity of the Hopfield Associative Memory.*

[Posner, MI 68a] Posner, Michael I.; Keele, Steven W. (1968). *On the Genesis of Abstract Ideas.* Journal of Experimental Psychology. Volume 77. pp 353-363.

[Posner, MI 70a] Posner, Michael I.; Keele, Steven W. (1970). *Retention of Abstract Ideas.* Journal of Experimental Psychology. Volume 83. pp 304-308.

[Post, PB 69a] Post, Paul B. (1969). *A Lifelike Model for Association Relevance.* International Joint Conference on Artificial Intelligence (IJCAI-69), 1st, Washington, DC, May 7-9, 1969: Proceedings. pp 271-280.

[Post, SD 87a] Post, Stephen D. (1987). *A Bipartite Connectionist Model to Represent N-ary Boolean and Linear Constraints.* IEEE International Conference on Neural Networks (ICNN-87), San Diego, CA, June 21-24, 1987: Proceedings. Volume 3. pp 87-94. New York: Institute of Electrical and Electronics Engineers.

[Postman, L 51*] SEE: Bruner, JS 51a. *Expectation and the Perception of Color.*

[Postman, L 54*] SEE: Miller, GA 54a. *Familiarity of Letter Sequences and Tachistoscopic Identification.*

[Potu, B 88a] Potu, B.; Ramamoorthy, P. A. (1988). *A Fully Digital Architecture for Multi-State Neural Networks.* Meeting of the International Neural Network Society (INNS), 1st, Boston, MA, September 6-10, 1988: Proceedings.

[Prade, H 85a] Prade, Henri. (1985). *A Computational Approach to Approximate and Plausible Reasoning with Applications to Expert Systems.* IEEE Transactions on Pattern Analysis and Machine Intelligence. Volume PAMI-7. pp 260-283.

[Prager, JM 80a] Prager, John M. (1980). *Extracting and Labeling Boundary Segments in Natural Scenes.* IEEE Transactions on Pattern Analysis and Machine Intelligence. Volume PAMI-2. pp 16-27.

[Prager, MH 84a] Prager, Michael H. (1984). *An SAS Program for Simplified GMDH Models.* In Farlow, Stanley J. (ed.).

Self-Organizing Methods in Modeling: GMDH Type Algorithms. pp 291-316. New York: Marcel Dekker.

[Prager, MH 84b] Prager, Michael H.; Saila, Saul B. (1984). *Predictive GMDH Models of Shrimp Catches: Some Practical Considerations*. In Farlow, Stanley J. (ed.). Self-Organizing Methods in Modeling: GMDH Type Algorithms. pp 179-198. New York: Marcel Dekker.

[Prager, RW 86a] Prager, R. W.; Harrison, T. D.; Fallside, Frank. (1986). *Boltzmann Machines for Speech Recognition*. Computer Speech and Language. Volume 1. pp 3-27. New York: Academic Press.

[Prager, RW 87a] Prager, R. W.; Fallside, Frank. (1987). *A Comparison of the Boltzmann Machine and the Back Propagation Network as Recognizers of Static Speech patterns*. Computer Speech and Language. Volume 22. pp 179-183. New York: Academic Press.

[Prager, RW 88a] Prager, R. W.; Fallside, Frank. (1988). The Modified Kanerva Model for Automatic Speech Recognition. CUED/F-INFENG/TR.6. Cambridge: Cambridge University, Engineering Department.

[Prata, A 85*] SEE: Farhat, NH 85b. *Optical Implementation of the Hopfield Model*.

[Prata, A 85*] SEE: Farhat, NH 85a. *Optical Implementation of the Hopfield Model*.

[Prater, M 87*] SEE: Keller, R 87a. *A Context-Driven World Model Appropriate for Training and Testing of Vision-Based Learning Networks*.

[Prazdny, K 83*] SEE: Beck, J 83a. *A Theory of Textural Segmentation*.

[Prazdny, K 84*] SEE: Beck, J 84a. *The Perception of Transparency with Achromatic Colors*.

[Prazdny, K 85a] Prazdny, K. (1985). *On the Nature of Inducing Forms Generating Perceptions of Illusory Contours*. Perception and Psychophysics. Volume 37. pp 237-242.

[Prazdny, S 81*] SEE: Beck, J 81a. *Highlights and the Perception of Glossiness*.

[Preparata, FP 81a] Preparata, Franco P.; Vuillemin, Jean. (1981). *The Cube-Connected Cycles: A Versatile Network for Parallel Computation*. Communications of the Association for Computing Machinery. Volume 24. pp 300-309.

[Preston Jr, K 86a] Preston Jr, Kendall. (1986). *Cellular Logic Arrays for Image Processing*. Handbook of Pattern Recognition and Image Processing. pp 395-436. New York: Academic Press.

[Pribram, KH 69a] Pribram, Karl H. (1969). *The Neurophysiology of Remembering*. Scientific American. Volume 220. pp 73-86.

[Pribram, KH 86a] Pribram, Karl H. (1986). *The Cognitive Revolution and Mind/Brain Issues*. American Psychologist. Volume 41. pp 507-520.

[Probert, TH 84*] SEE: Brooks, HA 84a. *Let's Ask GMDH What Effect the Environment Has on Fisheries*.

[Probst, D 86*] SEE: de Mori, R 86a. *Computer Recognition of Speech*.

[Prohorov, S 83*] SEE: Soucek, B 83a. *Event-Train Correlation and Real-Time Microcomputer Systems*.

[Protopopescu, V 87*] SEE: Barhen, J 87a. *Concurrent Neuromorphic Algorithms for Optimization of the Computational Load of a Hypercube Supercomputer*.

[Protopopescu, V 87*] SEE: Gulati, S 87a. *Nonlinear Neural Networks for Deterministic Scheduling*.

[Protopopescu, V 87*] SEE: Barhen, J 87b. *Optimization of the Computational Load of a Hypercube Supercomputer Onboard a*

Mobile Robot.

[Protopopsecu, V 88*] SEE: Guez, A 88b. *On the Stability, Storage Capacity, and Design of Nonlinear Continuous Neural Networks.*

[Provence, JD 87a] Provence, John D. (1987). *Neural Network Implementation for Maximum-Likelihood Sequence Estimation of Binary Signals in Gaussian Noise.* IEEE International Conference on Neural Networks (ICNN-87), San Diego, CA, June 21-24, 1987: Proceedings. Volume 3. pp 703-714. New York: Institute of Electrical and Electronics Engineers.

[Prueitt, P 88*] SEE: Kowalski, JM 88a. *Coupled Rotators as a Model for Spontaneous Activity in Neural Networks.*

[Psaltis, D 76*] SEE: Casasent, DP 76a. *Position, Rotation, and Scale Invariant Optical Correlation.*

[Psaltis, D 84a] Psaltis, Demetri; Farhat, Nabil H. (1984). *A New Approach to Optical Information Processing Based on the Hopfield Model.* Digest of the 13th Congress of the ICO, Sapporo, Japan. Volume ICO-13. p 24.

[Psaltis, D 85*] SEE: Abu-Mostafa, YS 85a. *Computation Power of Parallelism in Optical Architectures.*

[Psaltis, D 85*] SEE: Farhat, NH 85a. *Optical Implementation of the Hopfield Model.*

[Psaltis, D 85*] SEE: Farhat, NH 85b. *Optical Implementation of the Hopfield Model.*

[Psaltis, D 85a] Psaltis, Demetri; Farhat, Nabil H. (1985). *Optical Information Processing Based on an Associative Memory Model of Neural Nets with Thresholding and Feedback.* Optics Letters. Volume 10. pp 98-100.

[Psaltis, D 86*] SEE: Venkatesh, SS 86b. Linear Maps with Point Rules: Applications to Pattern Classification and Associative Memory.

[Psaltis, D 86a] Psaltis, Demetri; Park, Cheol Hoon. (1986). *Nonlinear Discriminant Functions and Associative Memories.* In Denker, John S. (ed.). Neural Networks for Computing, Snowbird, Utah, 1986. AIP Proceedings 151. pp 370-375. New York: American Institute of Physics.

[Psaltis, D 86b] Psaltis, Demetri; Hong, J.; Venkatesh, Santosh S. (1986). *Shift Invariance in Optical Associative Memories.* Optical Computing. SPIE Volume 625. pp 189-195. Bellingham, WA: Society of Photo-Optical Instrumentation Engineers.

[Psaltis, D 87*] SEE: Wagner, K 87a. *Multilayer Optical Learning Networks.*

[Psaltis, D 87*] SEE: Wagner, K 87b. *Nonlinear Etalons in Competitive Optical Learning Networks.*

[Psaltis, D 87*] SEE: Abu-Mostafa, YS 87a. *Optical Neural Computers.*

[Psaltis, D 87a] Psaltis, Demetri; Wagner, Kelvin; Brady, David. (1987). *Learning in Optical Neural Computers.* IEEE International Conference on Neural Networks (ICNN-87), San Diego, CA, June 21-24, 1987: Proceedings. Volume 3. pp 549-556. New York: Institute of Electrical and Electronics Engineers.

[Psaltis, D 87b] Psaltis, Demetri; Sideris, Athanasios; Yamamura, Alan. (1987). *Neural Controllers.* IEEE International Conference on Neural Networks (ICNN-87), San Diego, CA, June 21-24, 1987: Proceedings. Volume 4. pp 551-558. New York: Institute of Electrical and Electronics Engineers.

[Psaltis, D 88*] SEE: Hsu, K 88a. *Experimental Demonstrations of Optical Neural Computers.*

[Psaltis, D 88a] Psaltis, Demetri; Brady, David; Wagner, Kelvin. (1988). *Adaptive Optical Networks Using Photorefractive Crystals.* Applied Optics. Volume 27. pp 1752-1759.

[Psaltis, D 88b] Psaltis, Demetri; Park, Cheol Hoon; Hong, John. (1988). *Higher Order Associative Memories and Their*

Optical Implementations. Neural Networks. Volume 1. pp 149-163.

[Puleo P, F 81a] Puleo P, Francisco. (1981). *Simulation of a Neural Network*. International Conference on Cybernetics and Society, Atlanta, GA, October 26-28, 1981: Proceedings. pp 251-256. New York: Institute of Electrical and Electronics Engineers.

[Pustejovsky, J 87*] SEE: Maddox, AB 87a. *Linguistic Descriptions of Visual Event Perceptions*.

[Putzrath, F 62*] SEE: Mueller, P 62a. *General Principles of Operations in Neuron Nets with Application to Acoustical Pattern Recognition*.

[Pylyshyn, ZW 80a] Pylyshyn, Zenon W. (1980). *Computation and Cognition: Issues in the Foundations of Cognitive Science*. The Behavioral and Brain Sciences. Volume 3. pp 111-169.

[Pylyshyn, ZW 84a] Pylyshyn, Zenon W. (1984). *Why "Computing" Requires Symbols*. Cognitive Science Society, 6th, Boulder, CO, June 28-30, 1984: Proceedings. pp 71-73. Hillsdale, NJ: Lawrence Erlbaum.

[Pylyshyn, ZW 86a] Pylyshyn, Zenon W. (1986). *Cognitive Science and the Study of Cognition and Language*. In Schwab, Eileen C. (ed.); Nusbaum, Howard C. (ed.). Pattern Recognition by Humans and Machines: Speech Perception. Volume I. pp 295-314. New York: Academic Press.

[Pylyshyn, ZW 88*] SEE: Fodor, JA 88a. *Connectionism and Cognitive Architecture: A Critical Analysis*.

[Qian, N 88a] Qian, Ning; Sejnowski, Terrence J. (1988). *Electrodiffusion Model of Electrical Conduction in Neuronal Processes*. Cellular Mechanisms of Conditioning and Behavioral Plasticity. pp 237-244. New York: Plenum.

[Quesada, AF 74*] SEE: Moses, HE 74a. *The Power Spectrum of the Mellin Transformation with Applications to Scaling of Physical Quantities*.

[Quillian, MR 68a] Quillian, M. Ross. (1968). *Semantic Memory*. Semantic Information Processing. pp 227-270. Cambridge, MA: MIT Press.

[Quillian, MR 69*] SEE: Collins, AM 69a. *Retrieval Time from Semantic Memory*.

[Rabin, DE 86*] SEE: Zipser, D 86c. *P3: A Parallel Network Simulating System*.

[Rabiner, LR 83*] SEE: Levinson, SE 83a. *An Introduction to the Application of the Theory of Probabilistic Functions of a Markov Process to Automatic Speech Recognition*.

[Rabiner, LR 85*] SEE: Juang, BH 85a. *Recent Developments in the Application of Hidden Markov Models to Speaker-Independent Isolated Word Recognition*.

[Rabinowitz, P 55*] SEE: Cole, KS 55a. *Automatic Computation of Nerve Excitation*.

[Rafelski, J 84*] SEE: Clark, JW 84a. *Self-Organization of Neural Networks*.

[Raffel, JI 87a] Raffel, Jack I.; Mann, James; Berger, Robert; Soares, Antonio; Gilbert, Sheldon. (1987). *A Generic Architecture for Wafer-Scale Neuromorphic Systems*. IEEE International Conference on Neural Networks (ICNN-87), San Diego, CA, June 21-24, 1987: Proceedings. Volume 3. pp 501-513. New York: Institute of Electrical and Electronics Engineers.

[Raffel, JI 88*] SEE: Mann, J 88a. *A Self-Organizing Neural Net Chip*.

[Raffel, JI 88a] Raffel, Jack I. (1988). Electronic Implementation of Neuromorphic Systems. Lexington, MA: Massachusetts Institute of Technology, Lincoln Laboratory.

[Raffel, JI 88b] Raffel, Jack I. (1988). The Application of Wafer-Scale Technology to Neuromorphic Systems. Lexington, MA: Massachusetts Institute of Technology, Lincoln Laboratory.

[Raibert, MH 78a] Raibert, M. H. (1978). *A Model for Sensorimotor Control and Learning.* Biological Cybernetics. Volume 29. pp 29-36.

[Rall, W 60a] Rall, Wilfrid. (1960). *Membrane Potential Transients and Membrane Time Constant of Motoneurons.* Experimental Neurology. Volume 2. pp 503-532.

[Ramamoorthy, PA 88*] SEE: Potu, B 88a. *A Fully Digital Architecture for Multi-State Neural Networks.*

[Ramamoorthy, PA 88a] Ramamoorthy, P. A.; Ho, S. (1988). *A Neural Network Approach for Implementing Expert Systems.* Meeting of the International Neural Network Society (INNS), 1st, Boston, MA, September 6-10, 1988: Proceedings.

[Ramamoorthy, PA 88b] Ramamoorthy, P. A.; Govind, G.; Iyer, V. K. (1988). *Signal Modeling and Prediction Using Neural Networks.* Meeting of the International Neural Network Society (INNS), 1st, Boston, MA, September 6-10, 1988: Proceedings.

[Raman, S 75*] SEE: Anninos, PA 75a. *Derivation of a Mathematical Equation for the EEG and the General Solution within the Brain and in Space.*

[Ramaswamy, K 88*] SEE: Karsai, G 88a. Gas Tungsten Arc Weld Modeling Using a Mapping Network.

[Ramesham, R 87*] SEE: Daud, T 87a. *Neural Network Based Feed-Forward High Density Associative Memory.*

[Ramirez, RW 85a] Ramirez, Robert W. (1985). The FFT: Fundamentals and Concepts. Englewood Cliffs, NJ: Prentice-Hall.

[Randall, DL 68*] SEE: Dill, JC 68a. *PLEXUS: An On-Line System for Modeling Neural Networks.*

[Rao, CR 72a] Rao, C. Radhakrishna; Mitra, Sujit Kumar. (1972). *Generalized Inverse of a Matrix and Its Applications.* Berkeley Symposium on Mathematical Statistics and Probability, 6th, 1972: Proceedings. Volume 1. pp 601-620. University of California Press.

[Rapaport, D 79*] SEE: Levy, WB 79a. *Associative-Contingency Learning and Neuronal Networks.*

[Raphael, B 67*] SEE: Nilsson, NJ 67a. *Preliminary Design of an Intelligent Robot.*

[Rapoport, A 48*] SEE: Shimbel, A 48a. *A Statistical Approach to the Theory of the Central Nervous System.*

[Rapoport, A 51a] Rapoport, Anatol. (1951). *Addition and Multiplication Theorems for the Input of Two Neurons Converging on a Third.* Bulletin of Mathematical Biophysics. Volume 13. pp 179-188.

[Rapoport, A 52a] Rapoport, Anatol. (1952). *Ignition Phenomena in Random Nets.* Bulletin of Mathematical Biophysics. Volume 14. pp 35-44.

[Rashevsky, N 36a] Rashevsky, N. (1936). *Physico-Mathematical Aspects of Excitation and Conduction in Nerves.* Cold Spring Harbor Symposia on Quantitative Biology. Volume IV. pp 90-97. Johnson Reprint Corporation.

[Rasnow, B 87*] SEE: Carman, GJ 87a. *Optimal Mapping of the Configuration Space of a Network of Neurons.*

[Ratcliff, R 80*] SEE: McKoon, G 80a. *Priming in Item Recognition: The Organization of Propositions in Memory for Text.*

[Ratcliff, R 81a] Ratcliff, Roger. (1981). *Parallel-Processing Mechanisms and Processing of Organized Information in Human Memory.* In Hinton, Geoffrey E. (ed.). Parallel Models of Associative Memory. pp 265-284. Hillsdale, NJ: Lawrence Erlbaum.

[Ratcliff, R 87a] Ratcliff, Roger. (1987). *Order Information and Distributed Memory Models*. Cognitive Science Society, 9th, Seattle, WA, July 16-18, 1987: Proceedings. pp 474-486. Hillsdale, NJ: Lawrence Erlbaum.

[Ratliff, F 78*] SEE: Brodie, SE 78a. *The Response of the Limulus Retina to Moving Stimuli: A Prediction by Fourier Synthesis*.

[Rauch-Hindin, WB 88a] Rauch-Hindin, Wendy B. (1988). *Problems with Paradigms*. AI Expert. Volume 3. pp 55-60.

[Razavi, H 88*] SEE: Bou-Ghazale, S 88a. *Design and Implementation of an Associative Memory for Intensive Pattern Matching*.

[Reber, WL 87a] Reber, William L.; Lyman, John. (1987). *An Artificial Neural System Design for Rotation and Scale Invariant Pattern Recognition*. IEEE International Conference on Neural Networks (ICNN-87), San Diego, CA, June 21-24, 1987: Proceedings. Volume 4. pp 277-284. New York: Institute of Electrical and Electronics Engineers.

[Recce, M 88a] Recce, M.; Treleaven, P. C. (1988). *Parallel Architectures for Neural Computers*. In Eckmiller, Rolf (ed.); von der Malsburg, Christoph (ed.). Neural Computers. NATO ASI Series, Volume F41. pp 487-495. Berlin: Springer-Verlag.

[Reddy, R 86*] SEE: Shikano, K 86a. Speaker Adaptation Through Vector Quantization.

[Reder, SM 76*] SEE: Estes, WK 76a. *Serial Position Functions for Letter Identification at Brief and Extended Exposure Durations*.

[Redol, RA 88*] SEE: Almeida, LB 88b. *Backpropagation in Perceptrons with Feedback*.

[Reece, P 87a] Reece, Peter. (January, 1987). *Perceptrons and Neural Nets*. AI Expert. Volume 2. pp 50-57.

[Reed, AV 76a] Reed, Adam V. (1976). *List Length and the Time Course of Recognition in Immediate Memory*. Memory and Cognition. Volume 4. pp 16-30.

[Reedy, RE 88*] SEE: Shimabukuro, RL 88a. *Dual-Polarity Nonvolatile MOS Analogue Memory (MAM) Cell for Neural-Type Circuitry*.

[Reeke Jr, GN 82*] SEE: Edelman, GM 82a. *Selective Networks Capable of Representative Transformations, Limited Generalizations, and Associative Memory*.

[Reeke Jr, GN 84a] Reeke Jr, George N.; Edelman, Gerald M. (1984). *Selective Networks and Recognition Automata*. Computer Culture: The Scientific, Intellectual, and Social Impact of the Computer. Volume 426. pp 181-201.

[Reeke Jr, GN 86a] Reeke Jr, George N.; Edelman, Gerald M. (1986). *Recognition Automata Based on Selective Neural Networks*. Structure and Dynamics of Nucleic Acid, Proteins and Membranes. pp 329-353. New York: Plenum.

[Reeke Jr, GN 87a] Reeke Jr, George N.; Edelman, Gerald M. (1987). *Selective Neural Networks and Their Implications for Recognition Automata*. International Journal of Supercomputer Applications. Volume 1. pp 44-89.

[Reeke Jr, GN 88a] Reeke Jr, George N.; Edelman, Gerald M. (1988). *Real Brains and Artificial Intelligence*. Daedalus. Volume 117. pp 143-173.

[Reeke Jr, GN 89*] SEE: Finkel, LH 89a. *A Population Approach to the Neural Basis of Perceptual Categorization*.

[Reese, R 88*] SEE: Cuykendall, R 88a. *Scaling Techniques for Neural Algorithms*.

[Reese, R 89*] SEE: Cuykendall, R 89a. *Scaling the Neural TSP Algorithm*.

[Reese, WD 75*] SEE: Altes, RA 75a. *Doppler Tolerant Classification of Distributed Targets - a Bionic Sonar*.

[Regan, D 84a] Regan, D. (ed.); Shapley, R. M. (ed.); Sperkreijse, H. (ed.). (1986). Systems Approach in Vision: Proceedings of a Workshop, Amsterdam, The Netherlands, 27-29 August 1984. New York: Pergamon.

[Reggia, JA 85a] Reggia, James A. (1985). *Virtual lateral Inhibition in Parallel Activation Models of Associative Memory.* International Joint Conference on Artificial Intelligence (IJCAI-85), 9th, Los Angeles, CA, August 18-23, 1985: Proceedings.

[Reggia, JA 87*] SEE: Seidman, SB 87a. *Analysis of a Competition-Based Connectionist Model.*

[Reggia, JA 87a] Reggia, James A. (1987). *Properties of a Competition-Based Activation Mechanism in Neuromimetic Network Models.* IEEE International Conference on Neural Networks (ICNN-87), San Diego, CA, June 21-24, 1987: Proceedings. Volume 2. pp 131-138. New York: Institute of Electrical and Electronics Engineers.

[Reggia, JA 88*] SEE: Wang, PY 88a. *Analysis of Competition-Based Spreading Activation in Connectionist Models.*

[Reggia, JA 88*] SEE: D'Autrechy, CL 88a. *A General-Purpose Simulation Environment for Developing Connectionist Models.*

[Reggia, JA 88a] Reggia, James A.; Sutton III, Granger G. (1988). *Self-Processing Networks and Their Biomedical Implications.* IEEE International Conference on Neural Networks (ICNN-88), San Diego, CA, July 24-27, 1988: Proceedings. New York: Institute of Electrical and Electronics Engineers.

[Reicher, GM 69a] Reicher, Gerald M. (1969). *Perceptual Recognition as a Function of Meaningfulness of Stimulus Material.* Journal of Experimental Psychology. Volume 81. pp 275-280.

[Reilly, DL 82*] SEE: Cooper, LN 82a. Self Organizing General Pattern Class Separator and Identifier.

[Reilly, DL 82a] Reilly, Douglas L.; Cooper, Leon N.; Elbaum, Charles. (1982). *A Neural Model for Category Learning.* Biological Cybernetics. Volume 45. pp 35-41.

[Reilly, DL 86*] SEE: Rimey, RD 86a. *Real-Time 3-D Object Classification Using a Learning System.*

[Reilly, DL 87a] Reilly, Douglas L.; Scofield, Christopher L.; Elbaum, Charles; Cooper, Leon N. (1987). *Learning System Architectures Composed of Multiple Learning Modules.* IEEE International Conference on Neural Networks (ICNN-87), San Diego, CA, June 21-24, 1987: Proceedings. Volume 2. pp 495-504. New York: Institute of Electrical and Electronics Engineers.

[Reilly, DL 88*] SEE: Scofield, CL 88b. *Pattern Class Degeneracy in an Unrestricted Storage Density Memory.*

[Renals, S 88*] SEE: Rohwer, R 88b. *Training Recurrent Networks.*

[Renals, S 88*] SEE: Rohwer, R 88c. *Unstable Connectionist Networks in Speech Recognition.*

[Renals, S 88a] Renals, Steve; Rohwer, Richard; Terry, Mark. (1988). *A Comparison of Speech Recognition Front Ends Using a Connectionist Classifier.* Federation of Acoustical Societies of Europe (FASE-88), August 22-26, 1988: Proceedings.

[Rendell, L 87*] SEE: Krowitz, A 87a. *The State Space of Memory Clusters in the Hopfield Network.*

[Rentschler, I 87*] SEE: Caelli, TM 87b. *Visual Pattern Recognition in Humans.*

[Reuhkala, E 76*] SEE: Kohonen, T 76a. *Associative Recall of Images.*

[Reuhkala, E 78*] SEE: Kohonen, T 78a. *A Very Fast Associative Method for the Recognition and Correction of Misspelt Words, Based on Redundant Hash Addressing.*

[Reuhkala, E 79a] Reuhkala, Erkki; Jalanko, Matti; Kohonen, Teuvo. (1979). *A Redundant Hash Addressing Method Adapted for the Postprocessing and Error-Correction of Computer-Recognized Speech.* IEEE International Conference on Acoustics, Speech, and Signal Processing (ICASSP-79), Washington, DC, April 2-4, 1979: Proceedings. pp 591-594. New York: Institute of Electrical and Electronics Engineers.

[Reuhkala, E 80*] SEE: Kohonen, T 80a. *A Thousand-Word Recognition System Based on the Learning Subspace Method and Redundant Hash Addressing.*

[Reuhkala, E 84*] SEE: Kohonen, T 84a. *On-Line Recognition of Spoken Words from a Large Vocabulary.*

[Rhodes, J 65*] SEE: Krohn, K 65a. *Nets of Threshold Elements.*

[Rhodes, WT 84a] Rhodes, William T.; Guilfoyle, Peter S. (1984). *Acoustooptic Algebraic Processing Architectures.* Proceedings of the IEEE. Volume 72. pp 820-830.

[Ricart, R 87a] Ricart, Richard. (1987). *Neural Network Based Architectures for Aerospace Applications.* Workshop on Space Operations Automation & Robotics, 1st, Houston, TX, August 5-7, 1987: Proceedings.

[Ricart, R 88a] Ricart, Richard. (1988). *Backward Conditioning: A Neural Network Model which Exhibits Both Excitatory and Inhibitory Conditioning.* Neural Networks. Volume 1, Supplement 1.

[Ricciardi, LM 67*] SEE: Caianiello, ER 67a. *Reverberations and Control of Neural Networks.*

[Rice, H 84*] SEE: Lebow, WM 84a. *Forecasting Applications of GMDH in Agricultural and Meteorological Time Series.*

[Richards, W 88*] SEE: Koenderink, JJ 88d. *Two-Dimensional Curvature Operators.*

[Richards, WA 87a] Richards, W. A.; Koenderink, Jan J.; Hoffman, D. D. (1987). *Inferring Three-Dimensional Shapes from Two-Dimensional Silhouettes.* Journal of the Optical Society of America, A. Volume 4. pp 1168-1175.

[Richardson, WH 72a] Richardson, William Hadley. (1972). *Bayesian-Based Iterative Method of Image Restoration.* Journal of the Optical Society of America. Volume 62. pp 55-59.

[Richer, I 68*] SEE: Dill, JC 68a. *PLEXUS: An On-Line System for Modeling Neural Networks.*

[Richfield, S 87*] SEE: Mehr, D 87a. *Neural Net Application to Optical Character Recognition.*

[Richfield, S 87a] Richfield, Steve. (1987). *A Logarithmic Vector Processor for Neural Net Applications.* IEEE International Conference on Neural Networks (ICNN-87), San Diego, CA, June 21-24, 1987: Proceedings. Volume 4. pp 733-740. New York: Institute of Electrical and Electronics Engineers.

[Richfield, S 87b] Richfield, Steve. (1987). *High-Speed Learning in Deductive-Reasoning Systems.* IEEE International Conference on Neural Networks (ICNN-87), San Diego, CA, June 21-24, 1987: Proceedings. Volume 2. pp 575-584. New York: Institute of Electrical and Electronics Engineers.

[Richter, J 82a] Richter, Jacob; Ullman, Shimon. (1982). *A Model for the Temporal Organization of X- and Y-Type Receptive Fields in the Primate Retina.* Biological Cybernetics. Volume 43. pp 127-145.

[Ridgway, WC 62a] Ridgway, W. C. (1962). An Adaptive Logic System with Generalizing Properties. Technical Report 1556-1. Stanford, CA: Stanford University, Stanford Electronics Laboratory.

[Riecken, ME 88*] SEE: Allen, RB 88a. *Interacting and Communicating Connectionist Agents.*

[Rietman, E 88a] Rietman, Ed. (1988). Experiments in Artificial Neural Networks. Blue Ridge Summit, PA: Tab.

[Rigler, AK 88*] SEE: Vogl, TP 88a. *Accelerating the Convergence of the Back-Propagation Method.*

[Riittinen, H 79*] SEE: Kohonen, T 79a. *Spectral Classification of Phonemes by Learning Subspaces.*

[Riittinen, H 80*] SEE: Kohonen, T 80a. *A Thousand-Word Recognition System Based on the Learning Subspace Method and Redundant Hash Addressing.*

[Riittinen, H 86*] SEE: Torkkola, K 86a. *Microprocessor-Based Word Recognizer for a Large Vocabulary.*

[Riley, MS 84*] SEE: Smolensky, P 84a. Harmony Theory: Problem Solving, Parallel Cognitive Models, and Thermal Plastics.

[Rimey, RD 86a] Rimey, Raymond D.; Gouin, Philip; Scofield, Christopher L.; Reilly, Douglas L. (1987). *Real-Time 3-D Object Classification Using a Learning System.* Intelligent Robots and Computer Vision. SPIE Volume 726. pp 552-557. Bellingham, WA: Society of Photo-Optical Instrumentation Engineers.

[Rinnooy Kan, AHG 86*] SEE: Lawler, EL 86a. The Traveling Salesman Problem.

[Rinott, Y 88*] SEE: Baldi, P 88a. *Asymptotic Normality of Some Graph Related Statistics.*

[Rinott, Y 88*] SEE: Baldi, P 88g. *On Normal Approximations of Distributions in Terms of Depenency Graphs.*

[Rinzel, J 80a] Rinzel, John; Miller, Robert N. (1980). *Numerical Calculation of Stable and Unstable Periodic Solutions to the Hodgkin-Huxley Equations.* Mathematical Biosciences. Volume 49. pp 27-59.

[Rinzel, J 81*] SEE: Keener, JP 81a. *Integrate-And-Fire Models of Nerve Membrane Response to Oscillatory Input.*

[Rips, LJ 73a] Rips, Lance J.; Shoben, Edward J.; Smith, Edward E. (1973). *Semantic Distance and the Verification of Semantic Relations.* Journal of Verbal Learning and Verbal Behavior. Volume 12. pp 1-20.

[Riseman, EM 71a] Riseman, Edward M. (1971). *Logical Networks for Feature Extraction.* IEEE Transactions on Systems, Man and Cybernetics. Volume SMC-1. pp 43-55.

[Riseman, EM 78*] SEE: Hanson, AR 78a. *Segmentation of Natural Scenes.*

[Riseman, EM 87a] Riseman, Edward M.; Hanson, Allen R. (1987). *A Methodology for the Development of General Knowledge-Based Vision Systems.* In Arbib, Michael A. (ed.); Hanson, Allen R. (ed.). Vision, Brain, and Cooperative Computation. pp 285-328. Cambridge, MA: MIT Press.

[Rissland, EL 84*] SEE: Selfridge, OG 84a. Adaptive Control of Ill-Defined Systems.

[Ritcey, JA 87*] SEE: Marks II, RJ 87c. *Composite Matched Filter Output Partitioning.*

[Ritcey, JA 87*] SEE: Cheung, KF 87a. *Conventional and Composite Matched Filters with Error Correction: A Comparison.*

[Ritcey, JA 88*] SEE: Marks II, RJ 88g. *The Performance of Convex Set Projection Based Neural Networks.*

[Ritter, GL 75a] Ritter, G. L.; Woodruff, H. B.; Lowry, S. R.; Isenhour, T. L. (1975). *An Algorithm for a Selective Nearest Neighbor Decision Rule.* IEEE Transactions on Information Theory. Volume IT-21. pp 665-669.

[Ritter, H 86*] SEE: Buhmann, J 86c. *Physicists Explore Human and Artificial Intelligence.*

[Ritter, H 86a] Ritter, Helge; Schulten, Klaus. (1986). *On the Stationary State of Kohonen's Self-Organizing Sensory Mapping.* Biological Cybernetics. Volume 54. pp 99-106.

[Ritter, H 86b] Ritter, Helge; Schulten, Klaus. (1986). *Topology Conserving Mappings for Learning Motor Tasks.* In Denker, John S. (ed.). Neural Networks for Computing, Snowbird, Utah, 1986. AIP Proceedings 151. pp 376-380. New York: American Institute of Physics.

[Ritter, H 87a] Ritter, Helge; Schulten, Klaus. (1987). *Planning a Dynamic Trajectory via Path Finding in Discretized Phase Space*. WOPPLOT 86: Parallel Processing-Logic, Organization and Technology. Volume 253. pp 29-39. Berlin: Springer-Verlag.

[Ritter, H 88a] Ritter, Helge; Schulten, Klaus. (in press). *Convergence Properties of Kohonen's Topology Conserving Maps: Fluctuations, Stability and Dimension Selection*. Biological Cybernetics.

[Ritter, H 88b] Ritter, Helge; Schulten, Klaus. (1988). *Extending Kohonen's Self-Organizing Mapping Algorithm to Learn Ballistic Movements*. In Eckmiller, Rolf (ed.); von der Malsburg, Christoph (ed.). Neural Computers. NATO ASI Series, Volume F41. pp 393-406. Berlin: Springer-Verlag.

[Ritter, H 88c] Ritter, Helge; Schulten, Klaus. (1988). *Kohonen's Self-Organizing Maps: Exploring their Computational Capabilities*. IEEE International Conference on Neural Networks (ICNN-88), San Diego, CA, July 24-27, 1988: Proceedings. Volume 1. pp 109-116. New York: Institute of Electrical and Electronics Engineers.

[Rittinen, H 84*] SEE: Kohonen, T 84a. *On-Line Recognition of Spoken Words from a Large Vocabulary.*

[Ritz, SA 77*] SEE: Anderson, JA 77b. *Distinctive Features, Categorical Perception, and Probability Learning: Some Applications of a Neural Model.*

[Ritz, SA 77*] SEE: Anderson, JA 77a. *Distinctive Features, Categorical Perception, and Probability Learning: Some Applications of a Neural Model.*

[Ritz, SA 77*] SEE: Anderson, JA 77d. *Vowel Pre-Processing with a Neurally Based Model.*

[Roach, AH 87*] SEE: Black, IB 87a. *Biochemistry of Information Storage in the Nervous System.*

[Robert, F 86a] Robert, Francois. (1986). *Basic Results for the Behavior of Discrete Iterations*. In Bienenstock, Elie L. (ed.). Disordered Systems and Biological Organization. pp 33-47. New York: Springer-Verlag.

[Robert, Y 86a] Robert, Yves; Tchuente, M. (1986). *On Some Dynamical Properties of Monotone Networks*. In Bienenstock, Elie L. (ed.). Disordered Systems and Biological Organization. pp 49-52. New York: Springer-Verlag.

[Roberts, BA 86*] SEE: Kak, AC 86a. *Reconstruction from Projections: Applications in Computerized Tomography.*

[Roberts, LC 60a] Roberts, Lawrence C. (1960). *Pattern Recognition with an Adaptive Network*. IRE International Convention Record. pp 66-70.

[Roberts, LG 65a] Roberts, L. G.; Vanderburgh, Jr, Alexander. (1965). *Machine Perception of Three-Dimensional Solids*. Optical and Electro-Optical Information Processing. pp 159-197. Cambridge, MA: MIT Press.

[Robinson, AJR 87a] Robinson, A. J. R.; Fallside, Frank. (1987). The Utility Driven Dynamic Error Propagation Network. CUED/F-INFENG/TR.1. Cambridge: Cambridge University, Engineering Department.

[Robinson, AJR 88a] Robinson, A. J. R.; Fallside, Frank. (1988). A Dynamic Connectionist Model for Phoneme Recognition: Preliminary Results. CUED/F-INFENG/TR.14. Cambridge: Cambridge University, Engineering Department.

[Robinson, AJR 88b] Robinson, A. J. R.; Fallside, Frank. (1988). *Static and Dynamic Error Propagation Networks with Application to Speech Coding*. In Anderson, Dana Z. (ed.). Neural Information Processing Systems. pp 632-641. New York: American Institute of Physics.

[Robinson, DA 68a] Robinson, David A. (1968). *The Electrical Properties of Metal Microelectrodes*. Proceedings of the IEEE. Volume 56. pp 1065-1071.

[Robinson, DA 68b] Robinson, David A. (1968). *The Oculomotor Control System: A Review*. Proceedings of the IEEE. Volume

56. pp 1032-1049.

[Robinson, DA 83*] SEE: Cannon, SC 83a. *A Proposed Neural Network for the Integrator of the Oculomotor System.*

[Robinson, DA 85*] SEE: Cannon, SC 85a. *An Improved Neural-Network Model for the Neural Integrator of the Oculomotor System: More Realistic Neuron Behavior.*

[Robinson, GB 88*] SEE: Sclabassi, RJ 88b. *Nonlinear Systems Analysis of the Hippocampal Perforant Path-Dentate Projection. I. Theoretical and Interpretational Considerations.*

[Rochester, N 56a] Rochester, N.; Holland, John H.; Haibt, L. H.; Duda, W. L. (1956). *Tests on a Cell Assembly Theory of the Action of the Brain, Using a Large Digital Computer.* IRE Transactions on Information Theory. Volume IT-2. pp 80-93.

[Rochester, N 56b] Rochester, N.; Holland, John H.; Haibt, L. H.; Duda, W. L. (1988). *Tests on a Cell Assembly Theory of the Action of the Brain, Using a Large Digital Computer.* In Anderson, James A. (ed.); Rosenfeld, Edward (ed.). Neurocomputing: Foundations of Research. pp 68-80. Cambridge, MA: MIT Press.

[Rock, I 80a] Rock, Irvin. (1980). *Difficulties with a Direct Theory of Perception.* The Behavioral and Brain Sciences. Volume 3. pp 398-399.

[Rodemich, ER 86*] SEE: McEliece, RJ 86a. Capacity of the Hopfield Associative Memory.

[Rodemich, ER 87*] SEE: McEliece, RJ 87a. *The Capacity of the Hopfield Associative Memory.*

[Rodrigues, J 51*] SEE: Bruner, JS 51a. *Expectation and the Perception of Color.*

[Rodriguez, E 80*] SEE: Larochelle, S 80a. *Context and the Allocation of Resources in Word Recognition.*

[Rogers, CH 87a] Rogers, Charles H.; Oldham, William J. B. (1987). *Isolated Word Recognition with an Artificial Neural Network.* IEEE International Conference on Neural Networks (ICNN-87), San Diego, CA, June 21-24, 1987: Proceedings. Volume 4. pp 435-442. New York: Institute of Electrical and Electronics Engineers.

[Rogers, M 87a] Rogers, Michael. (July 20, 1987). *Mimicking the Human Mind: Radical New Computers Teach Themselves.* Newsweek. pp 52-53.

[Rogers, SK 87*] SEE: Kabrisky, M 87a. *A Multiplexed Multi-Electrode Semiconductor Brain Electrode Implant.*

[Rogers, SK 88*] SEE: Ruck, DW 88a. Multisensor Target Detection and Classification.

[Rohrbaugh, J 70*] SEE: Eriksen, CW 70a. *Visual Masking in Multielement Displays.*

[Rohwer, R 87a] Rohwer, Richard; Forrest, Bruce M. (1987). *Training Time-Dependence in Neural Networks.* IEEE International Conference on Neural Networks (ICNN-87), San Diego, CA, June 21-24, 1987: Proceedings. Volume 2. pp 701-710. New York: Institute of Electrical and Electronics Engineers.

[Rohwer, R 88*] SEE: Renals, S 88a. *A Comparison of Speech Recognition Front Ends Using a Connectionist Classifier.*

[Rohwer, R 88a] Rohwer, Richard. (1988). *Connectionist Methods in Speech Recognition.* Federation of Acoustical Societies of Europe (FASE-88), August 22-26, 1988: Proceedings.

[Rohwer, R 88b] Rohwer, Richard; Renals, Steve. (1988). *Training Recurrent Networks.* ESPCI European Conference on Neural Networks (nEURO'88), 1st, Paris, June 6-9, 1988: Proceedings.

[Rohwer, R 88c] Rohwer, Richard; Renals, Steve; Terry, Mark. (April 1988). *Unstable Connectionist Networks in Speech*

Recognition. IEEE International Conference on Acoustics, Speech, and Signal Processing (ICASSP-88), New York, NY, April 11-14, 1988: Proceedings.

[Roll, G 87*] SEE: Strugala, M 87a. *An Associative Processor as a Design Rule Check Accelerator.*

[Rolls, ET 89a] Rolls, Edmund T. (1989). *Principles Underlying the Representation and Storage of Information in Neuronal Networks in the Primate Hippocampus and Cerebral Cortex.* In Zornetzer, S. F. (ed.); Davis, J. L. (ed.); Lau, C. (ed.). An Introduction to Neural and Electronic Networks. New York: Academic Press.

[Roney, KJ 80a] Roney, Kathleen J.; Shaw, Gordon L. (1980). *Analytic Study of Assemblies of Neurons in Memory Storage.* Mathematical Biosciences. Volume 51. pp 25-41.

[Rosch, E 75a] Rosch, Eleanor. (1975). *Cognitive Representations of Semantic Categories.* Journal of Experimental Psychology: General. Volume 104. pp 192-233.

[Rose, C 88*] SEE: Winters, JH 88a. *On Parallel Networks for Optimum Classification.*

[Rose, C 89*] SEE: Winters, JH 89a. *Minimum Distance Automata in Parallel Networks for Optimum Classification.*

[Rosen, BE 87*] SEE: Goodwin, JM 87a. *A Design for an Associative Spin Glass Processor.*

[Rosen, BE 88*] SEE: Goodwin, JM 88a. Exploration of Learning in an Associative Magnetic Processor.

[Rosen, BE 88*] SEE: Goodwin, JM 88b. Progress on the Spin Chip Asociative Processor.

[Rosen, BE 88a] Rosen, Bruce E.; Goodwin, James M.; Vidal, Jacques J. (1988). *Learning by State Recurrence Detection.* In Anderson, Dana Z. (ed.). Neural Information Processing Systems. pp 642-651. New York: American Institute of Physics.

[Rosen, BE 88b] Rosen, Bruce E.; Goodwin, James M.; Vidal, Jacques J. (1988). Machine Operant Conditioning. Los Angeles, CA: University of California, Computer Science Department.

[Rosen, BE 88c] Rosen, Bruce E.; Goodwin, James M.; Vidal, Jacques J. (1988). State Recurrence Learning. Los Angeles, CA: University of California, Computer Science Department.

[Rosen, CA 67a] Rosen, Charles A.; Nilsson, Nils J. (March 1967). *An Intelligent Automaton.* IEEE International Convention Record. pp 50-55.

[Rosenbaum, DA 80a] Rosenbaum, David A. (1980). *Human Movement Initiation: Specification of Arm, Direction, and Extent.* Journal of Experimental Psychology: General. Volume 109. pp 444-474.

[Rosenbaum, DA 83a] Rosenbaum, David A.; Kenny, Sandra B.; Derr, Marcia A. (1983). *Hierarchical Control of Rapid Movement Sequences.* Journal of Experimental Psychology: Human Perception and Performance. Volume 9. pp 86-102.

[Rosenbaum, DA 88*] SEE: Jordan, MI 88a. Action.

[Rosenberg, CR 86*] SEE: Sejnowski, TJ 86d. NETtalk: A Parallel Network that Learns to Read Aloud.

[Rosenberg, CR 86*] SEE: Sejnowski, TJ 86c. *NETtalk: A Parallel Network that Learns to Read Aloud.*

[Rosenberg, CR 86a] Rosenberg, Charles R.; Sejnowski, Terrence J. (1986). *The Spacing Effect on NETtalk, a Massively-Parallel Network.* Cognitive Science Society, 8th, Amherst, MA, August 15-17, 1986: Proceedings. pp 72-89. Hillsdale, NJ: Lawrence Erlbaum.

[Rosenberg, CR 87*] SEE: Sejnowski, TJ 87c. *Connectionist Models of Learning.*

[Rosenberg, CR 87*] SEE: Sejnowski, TJ 87f. *Parallel Networks that Learn to Pronounce English Text.*

[Rosenberg, CR 87a] Rosenberg, Charles R. (1987). *Revealing the Structure of NETtalk's Internal Representations.* Cognitive Science Society, 9th, Seattle, WA, July 16-18, 1987: Proceedings. pp 537-554. Hillsdale, NJ: Lawrence Erlbaum.

[Rosenberg, CR 88a] Rosenberg, Charles R.; Blelloch, Guy. (1988). *An Implementation of Network Learning on the Connection Machine.* In Waltz, David L. (ed.); Feldman, Jerome A. (ed.). Connectionist Models and Their Implications: Readings from Cognitive Science. pp 329-340. Norwood, NJ: Ablex.

[Rosenberg, IG 86a] Rosenberg, I. G. (1986). *Spin Glass and Pseudo Boolean Optimization.* In Bienenstock, Elie L. (ed.). Disordered Systems and Biological Organization. pp 327-332. New York: Springer-Verlag.

[Rosenblatt, F 57a] Rosenblatt, Frank. (1957). The Perceptron: A Perceiving and Recognizing Automaton, Project PARA. Repcrt 85-460-1. Ithaca, NY: Cornell University, Cornell Aeronautics Laboratory.

[Rosenblatt, F 58a] Rosenblatt, Frank. (1958). *The Perceptron: A Probabilistic Model for Information Storage and Organization in the Brain.* Psychological Review. Volume 65. pp 386-408.

[Rosenblatt, F 58b] Rosenblatt, Frank. (1988). *The Perceptron: A Probabilistic Model for Information Storage and Organization in the Brain.* In Anderson, James A. (ed.); Rosenfeld, Edward (ed.). Neurocomputing: Foundations of Research. pp 92-114. Cambridge, MA: MIT Press.

[Rosenblatt, F 60a] Rosenblatt, Frank. (1960). *Perceptron Simulation Experiments.* IRE Proceedings. Volume 48. pp 301-309.

[Rosenblatt, F 61a] Rosenblatt, Frank. (1961). Principles of Neurodynamics: Perceptrons and the Theory of Brain Mechanisms. Buffalo, NY: Cornell University, Aeronautical Laboratory.

[Rosenblatt, F 62*] SEE: Block, HD 62a. *Analysis of a Four-Layer Series-Coupled Perceptron, II.*

[Rosenblatt, F 62a] Rosenblatt, Frank. (1962). Principles of Neurodynamics: Perceptrons and the Theory of Brain Mechanisms. Washington, DC: Spartan.

[Rosenblatt, F 67a] Rosenblatt, Frank. (1967). *Recent Work on Theoretical Models of Biological Memory.* Computer and Information Sciences-II. pp 33-57. New York: Academic Press.

[Rosenblith, WA 68*] SEE: Weiss, TF 68a. *Scanning the Issue: Special Issue on Studies of Neural Elements and Systems.*

[Rosenfeld, A 77a] Rosenfeld, Azriel. (1977). *Iterative Methods in Image Analysis.* IEEE Computer Society Conference on Pattern Recognition and Image Processing, Troy, NY, June 6-8, 1977: Proceedings. pp 14-18. New York: Institute of Electrical and Electronics Engineers.

[Rosenfeld, A 78a] Rosenfeld, Azriel. (1978). *Iterative Methods in Image Analysis.* Pattern Recognition. Volume 10. pp 181-187.

[Rosenfeld, A 81*] SEE: Danker, AJ 81a. *Blob Detection by Relaxation.*

[Rosenfeld, A 81*] SEE: Davis, LS 81a. *Cooperating Processes for Low-Level Vision: A Survey.*

[Rosenfeld, A 83*] SEE: Beck, J 83a. *A Theory of Textural Segmentation.*

[Rosenfeld, A 84a] Rosenfeld, Azriel (ed.). (1984). Multi-Resolution Image Processing & Analysis. New York: Springer-Verlag.

[Rosenfeld, A 86a] Rosenfeld, Azriel; Hummel, Robert A.; Zucker, Steven W. (1976). *Scene Labeling by Relaxation*

Operations. IEEE Transactions on Systems, Man and Cybernetics. Volume SMC-6. pp 420-433.

[Rosenfeld, E 87a] Rosenfeld, Edward. (1987). *Neurocomputing - A New Industry*. IEEE International Conference on Neural Networks (ICNN-87), San Diego, CA, June 21-24, 1987: Proceedings. Volume 4. pp 831-838. New York: Institute of Electrical and Electronics Engineers.

[Rosenfeld, R 87a] Rosenfeld, Ronald; Touretzky, David S. (1987). *Four Capacity Models for Coarse-Coded Symbol Memories*. CMU-CS-87-182. Pittsburgh, PA: Carnegie Mellon University, Computer Science Department.

[Rosenfeld, R 87b] Rosenfeld, Ronald; Touretzky, David S. (1987). Scaling Properties of Coarse-Coded Symbol Memories. Working Paper.

[Rosenfeld, R 88a] Rosenfeld, Ronald; Touretzky, David S. (1988). *Scaling Properties of Coarse-Coded Symbol Memories*. In Anderson, Dana Z. (ed.). Neural Information Processing Systems. pp 652-661. New York: American Institute of Physics.

[Rosenzweig, MR 76a] Rosenzweig, Mark R. (ed.); Bennett, Edward L. (ed.). (1976). Neural Mechanisms of Learning and Memory. Cambridge, MA: MIT Press.

[Rosner, R 86*] SEE: Jeffrey, W 86a. *Neural Network Processing as a Tool for Function Optimization*.

[Ross, MD 88a] Ross, M. D.; Cutler, L.; Meyer, G.; Vaziri, P.; Lam, T. (1988). *Macular Bioaccelerometers on Earth and in Space*. In Hwang, J. C. (ed.). Basic and Applied Aspects of Vestibular Function. 219-229. Hong Kong: Hong Kong University Press.

[Ross, MD 88b] Ross, Muriel D. (1988). *Morphological Evidence for Parallel Processing of Information in Rat Macula*. Acta Oto-larynologica. Volume 106. pp 213-218.

[Ross, MD 88c] Ross, Muriel D.; Cutler, L.; Meyer, G.; Vaziri, P.; Lam, T.; Or, W. (1988). *The Neuroanatomical Substrate for Information Processing in Macular Endorgans*. IEEE International Conference on Engineering in Medicine and Biology, 10th, New Orleans, November 4-7, 1988: Proceedings. Volume 10. pp 1065-1066. New York: Institute of Electrical and Electronics Engineers.

[Rossler, J 87a] Rossler, J.; Varela, F. J. (1987). *Exact Results for the Average Dynamic Behavior of some Non-Linear Neural Networks*. Biological Cybernetics. Volume 57. pp 249-256.

[Rostykus, PL 88*] SEE: Somani, AK 88a. Evaluation of a Hypercube Based Compact Neural Network.

[Rotenberg, AR 71a] Rotenberg, A. R. (1971). *Behavior of Homogeneous Statistical Ensemble of Finite Automata*. Avtomatika I Telemekhanika. Volume 9. pp 84-92.

[Roucos, S 84*] SEE: Schwartz, RM 84a. *Improved Hidden Markov Modeling of Phonemes for Continuous Speech Recognition*.

[Roucos, S 85*] SEE: Schwartz, RM 85a. *Context-Dependent Modeling for Acoustic-Phonetic Recognition of Continuous Speech*.

[Roucos, S 85*] SEE: Makhoul, J 85a. *Vector Quantization in Speech Coding*.

[Rovnyak, SM 87*] SEE: Stirk, CW 87a. *Effects of System Noise on an Optical Implementation of an Additive Lateral Inhibitory Network*.

[Rovnyak, SM 88a] Rovnyak, S. M.; Stirk, Charles W.; Athale, Ravindra A. (1988). *Optical Implementation of Winner-Take-All Networks: Noise Considerations*. Neural Network Models for Optical Computing. SPIE Volume 882. pp 137-142. Bellingham, WA: Society of Photo-Optical Instrumentation Engineers.

[Rowe, B 87a] Rowe, Bruce. (March 9, 1987). *New Computer Learns on the Job*. San Diego Business Journal.

[Roweth, D 87*] SEE: Forrest, BM 87a. *Implementing Neural Network Models on Parallel Computers.*

[Roy, AE 60a] Roy, Archie E. (1960). *On a Method of Storing Information.* The Bulletin of Mathematical Biophysics. Volume 22. pp 139-168.

[Roy, AE 62a] Roy, Archie E. (1962). *On a method of Storing Information: II. A Further Study of Model Properties.* The Bulletin of Mathematical Biophysics. Volume 24. pp 39-69.

[Rozonder, LI 69a] Rozonder, L. I. (1969). *Random Logical Nets I.* Avtomatika I Telemekhanika. Volume 5. pp 137-147.

[Rozonder, LI 69b] Rozonder, L. I. (1969). *Random Logical Nets II.* Avtomatika I Telemekhanika. Volume 6. pp 99-109.

[Rozonder, LI 69c] Rozonder, L. I. (1969). *Random Logical Nets III.* Avtomatika I Telemekhanika. Volume 7. pp 127-136.

[Rubin, LD 88*] SEE: Morris, RJT 88a. *A Decentralized Tunable Short Term Neural Network Memory and Application to Tracking.*

[Rubin, V 83a] Rubin, Vivienne; Willett, Peter. (1983). *A Comparison of Some Hierarchial Monothetic Divisive Clustering Algorithms for Structure-Property Correlation.* Analytica Chimica Acta. Volume 151. pp 161-166.

[Rubio, JE 75a] Rubio, J. E.; Holden, Arun V. (1975). *The Response of a Model Neurone to a White Noise Input.* Biological Cybernetics. Volume 19. pp 191-195.

[Ruck, DW 88a] Ruck, Dennis W.; Rogers, Steven K.; Kabrisky, Matthew; Mills, James P. (1988). Multisensor Target Detection and Classification. Wright-Patterson Air Force Base, OH: Air Force Institute of Technology.

[Rudjord, T 66*] SEE: Andersen, P 66a. *Rhythmic Activity in a Simulated Neuronal Network.*

[Rudnick, M 88a] Rudnick, Mike; Hammerstrom, Daniel. (in press). *An Interconnect Structure for Wafer Scale Neurocomputers.* In Touretzky, David S. (ed.); Hinton, Geoffrey E. (ed.); Sejnowski, Terrence J. (ed.). Proceedings of the 1988 Connectionist Models Summer School. Los Altos, CA: Morgan Kaufmann.

[Rudnick, M 88b] Rudnick, Mike; Hammerstrom, Daniel. (1988). Physical Broadcast Structure. CS/E-88-018. Beaverton, OR: Oregon Graduate Center.

[Rudnicky, AI 87*] SEE: Hauptmann, AG 87a. Talking to Computers: An Empirical Investigation.

[Rudolph, L 83*] SEE: Gottlieb, A 83a. *The NYU Ultracomputer-Designing an MIMD Shared Memory Parallel Computer.*

[Rudolph, L 87*] SEE: Kalay, A 87a. *Neural Patterns as Detection Units in Sensory Systems.*

[Rumelhart, DE 70a] Rumelhart, David E. (1970). *A Multicomponent Theory of the Perception of Briefly Exposed Visual Displays.* Journal of Mathematical Psychology. Volume 7. pp 191-218.

[Rumelhart, DE 75a] Rumelhart, David E. (1975). *Notes on a Schema for Stories.* Representation and Understanding, Studies in Cognitive Science. pp 211-236. New York: Academic Press.

[Rumelhart, DE 81*] SEE: McClelland, JL 81a. *An Interactive Activation Model of Context Effects in Letter Perception, Part 1: An Account of Basic Findings.*

[Rumelhart, DE 81*] SEE: McClelland, JL 81b. *An Interactive Activation Model of Context Effects in Letter Perception: Part 1, An Account of Basic Findings.*

[Rumelhart, DE 82a] Rumelhart, David E.; McClelland, James L. (1982). *An Interactive Activation Model of Context Effects in Letter Perception: Part 2.* Psychological Review. Volume 89. pp 60-94.

[Rumelhart, DE 82b] Rumelhart, David E.; Norman, Donald A. (1982). *Simulating a Skilled Typist: A Study of Skilled Cognitive-Motor Performance*. Cognitive Science. Volume 6. pp 1-36.

[Rumelhart, DE 85*] SEE: McClelland, JL 85a. *Distributed Memory and the Representation of General and Specific Information*.

[Rumelhart, DE 85a] Rumelhart, David E.; Zipser, David. (1985). *Feature Discovery by Competitive Learning*. Cognitive Science. Volume 9. pp 75-112.

[Rumelhart, DE 85b] Rumelhart, David E.; Hinton, Geoffrey E.; Williams, Ronald J. (1985). Learning Internal Representations by Error Propagation. Report 8506. San Diego, CA: University of California, San Diego, Institute for Cognitive Science.

[Rumelhart, DE 85c] Rumelhart, David E.; McClelland, James L. (1985). *Levels Indeed! A Response to Broadbent*. Journal of Experimental Psychology: General. Volume 114. pp 193-197.

[Rumelhart, DE 86*] SEE: McClelland, JL 86a. *Amnesia and Distributed Memory*.

[Rumelhart, DE 86*] SEE: McClelland, JL 86b. *Appeal of Parallel Distributed Processing*.

[Rumelhart, DE 86*] SEE: McClelland, JL 86d. *Distributed Model of Human Learning and Memory*.

[Rumelhart, DE 86*] SEE: Hinton, GE 86b. *Distributed Representations*.

[Rumelhart, DE 86a] Rumelhart, David E.; Zipser, David. (1986). *Feature Discovery by Competitive Learning*. In Rumelhart, David E. (ed.). Parallel Distributed Processing: Explorations in the Microstructure of Cognition. Volume I, Foundations. pp 151-193. Cambridge, MA: MIT Press.

[Rumelhart, DE 86b] Rumelhart, David E.; Hinton, Geoffrey E.; McClelland, James L. (1986). *General Framework for Parallel Distributed Processing*. In Rumelhart, David E. (ed.). Parallel Distributed Processing: Explorations in the Microstructure of Cognition. Volume I, Foundations. pp 45-76. Cambridge, MA: MIT Press.

[Rumelhart, DE 86c] Rumelhart, David E.; Hinton, Geoffrey E.; Williams, Ronald J. (1986). *Learning Internal Representations by Error Propagation*. In Rumelhart, David E. (ed.). Parallel Distributed Processing: Explorations in the Microstructure of Cognition. Volume I, Foundations. pp 318-362. Cambridge, MA: MIT Press.

[Rumelhart, DE 86d] Rumelhart, David E.; Hinton, Geoffrey E.; Willams, Ronald J. (1988). *Learning Internal Representations by Error Propagation*. In Anderson, James A. (ed.); Rosenfeld, Edward (ed.). Neurocomputing: Foundations of Research. pp 675-695. Cambridge, MA: MIT Press.

[Rumelhart, DE 86e] Rumelhart, David E.; Hinton, Geoffrey E.; Williams, Ronald J. (1986). *Learning Representations by Back-Propagating Errors*. Nature. Volume 323. pp 533-536.

[Rumelhart, DE 86f] Rumelhart, David E.; Hinton, Geoffrey E.; Willams, Ronald J. (1988). *Learning Representations by Back-Propagating Errors*. In Anderson, James A. (ed.); Rosenfeld, Edward (ed.). Neurocomputing: Foundations of Research. pp 696-700. Cambridge, MA: MIT Press.

[Rumelhart, DE 86g] Rumelhart, David E.; McClelland, James L. (1986). *On Learning the Past Tenses of English Verbs*. In Rumelhart, David E. (ed.). Parallel Distributed Processing: Explorations in the Microstructure of Cognition. Volume II. pp 216-271. Cambridge, MA: MIT Press.

[Rumelhart, DE 86h] Rumelhart, David E. (ed.); Mcclelland, James L. (ed.). (1986). Parallel Distributed Processing: Explorations in the Microstructure of Cognition. Cambridge, MA: MIT Press.

[Rumelhart, DE 86i] Rumelhart, David E.; McClelland, James L. (1986). *PDP Models and General Issues in Cognitive Science*.

Parallel Distributed Processing: Explorations in the Microstructure of Cognition. Volume I, Foundations. pp 110-146. Cambridge, MA: MIT Press.

[Rumelhart, DE 86j] Rumelhart, David E.; Smolensky, Paul; McClelland, James L.; Hinton, Geoffrey E. (1986). *Schemata and Sequential Thought Processes in PDP Models*. In Rumelhart, David E. (ed.). Parallel Distributed Processing: Explorations in the Microstructure of Cognition. Volume II. pp 7-57. Cambridge, MA: MIT Press.

[Rumelhart, DE 88*] SEE: McClelland, JL 88a. Explorations in Parallel Distributed Processing: A Handbook of Models, Programs, and Exercises.

[Rumelhart, DE 88a] Rumelhart, David E.; Zipser, David. (1988). *Feature Discovery by Competitive Learning*. In Waltz, David L. (ed.); Feldman, Jerome A. (ed.). Connectionist Models and Their Implications: Readings from Cognitive Science. pp 205-242. Norwood, NJ: Ablex.

[Ruohonen, M 73*] SEE: Kohonen, T 73b. *Representation of Associated Data by Matrix Operators*.

[Rush, S 78a] Rush, Stanley; Larsen, Hugh. (1978). *A Practical Algorithm for Solving Dynamic Membrane Equations*. IEEE Transactions on Biomedical Engineering. Volume BME-25. pp 389-392.

[Russo, Jr, LS 88*] SEE: Deshmukh, VD 88b. *Computer Simulation of Event Related Potentials During Voluntary Cognition*.

[Rutherford, DA 79*] SEE: Braae, M 79a. *Theoretical and Linguistic Aspects of the Fuzzy Logic Controller*.

[Ryan, TW 87*] SEE: Winter, CL 87b. *TIN: A Trainable Inference Network*.

[Ryan, TW 87*] SEE: Winter, CL 87a. TIN: A Trainable Inference Network.

[Ryan, TW 87a] Ryan, Thomas W.; Winter, C. L.; Turner, Charles J. (1987). *Dynamic Control of an Artificial Neural System: The Property Inheritance Network*. Applied Optics. Volume 26. pp 4961-4971.

[Ryan, TW 87b] Ryan, Thomas W.; Winter, C. L. (1987). *Variations on Adaptive Resonance*. IEEE International Conference on Neural Networks (ICNN-87), San Diego, CA, June 21-24, 1987: Proceedings. Volume 2. pp 767-776. New York: Institute of Electrical and Electronics Engineers.

[Ryan, TW 88a] Ryan, Thomas W. (1988). *The Resonance Correlation Network*. IEEE International Conference on Neural Networks (ICNN-88), San Diego, CA, July 24-27, 1988: Proceedings. Volume 1. pp 673-680.

[Ryan, TW 88b] Ryan, Thomas W.; Cotter, Charles A. (1988). Vector Quantization Training by a Self-Organizing Neural Network. Tucson, AZ: Science Applications International Corporation.

[Sabbah, D 81*] SEE: Ballard, DH 81a. *On Shapes*.

[Sabbah, D 81a] Sabbah, Daniel. (1981). *Design of a Highly Parallel Visual Recognition System*. International Joint Conference on Artificial Intelligence (IJCAI-81), 7th, Vancouver, August 24-28, 1981: Proceedings. Volume 2. pp 722-727. International Joint Conference on Artificial Intelligence.

[Sabbah, D 85a] Sabbah, Daniel. (1985). *Computing with Connections in Visual Recognition of Origami Objects*. Cognitive Science. Volume 9. pp 25-50.

[Sabbah, D 88a] Sabbah, Daniel. (1988). *Computing with Connections in Visual Recognition of Origami Objects*. In Waltz, David L. (ed.); Feldman, Jerome A. (ed.). Connectionist Models and Their Implications: Readings from Cognitive Science. pp 155-180. Norwood, NJ: Ablex.

[Saburi, M 81*] SEE: Niki, K 81a. *A Neuronal Model with Synaptic Long-Term Memories*.

[Sadoc, JF 86*] SEE: Mosseri, R 86a. *Order and Defects in Geometrically Frustrated Systems*.

[Sage, JP 86*] SEE: Grant, PM 86b. *Comparison of Neural Network and Matched Filter Processing for Detecting Lines in Images.*

[Sage, JP 86*] SEE: Grant, PM 86a. A Comparison of Nonlinear Associative Memory and Matched-Filter Processing for Detecting Lines in Optical Images.

[Sage, JP 86a] Sage, Jay P.; Thompson, K.; Withers, R. S. (1986). *Artificial Neural Network Integrated Circuit Based on NMOS/CCD Principles.* In Denker, John S. (ed.). Neural Networks for Computing, Snowbird, Utah, 1986. AIP Proceedings 151. pp 381-385. New York: American Institute of Physics.

[Sage, JP 87b] Sage, Jay P. (1987). International Solid State Circuits Conference, February, 1987. Panel Discussion Participation.

[Sage, JP 88a] Sage, Jay P.; Withers, Richard S. (1988). *Analog Nonvolatile Memory for Neural Network Implementations.* Electrochemical Society Meeting, Chicago, IL, October 14, 1988: Proceedings.

[Saia, RA 83*] SEE: Lachs, G 83a. *A Neural-Counting Model Based on Physiological Characteristics of the Peripheral Auditory System, Part IV, Application to Response.*

[Saila, SB 84*] SEE: Prager, MH 84b. *Predictive GMDH Models of Shrimp Catches: Some Practical Considerations.*

[Saito, K 88a] Saito, Kazumi; Nakano, Ryohei. (1988). Medical Diagnostic Expert System Based on PDP Model. NTT Communications and Information Processing Laboratories.

[Sakakibara, M 88*] SEE: Usui, S 88a. *Physiological Engineering Model of the Retinal Horizontal Cell Layer.*

[Sakitt, B 82a] Sakitt, Barbara; Barlow, H. B. (1982). *A Model for the Economical Encoding of the Visual Image in Cerebral Cortex.* Biological Cybernetics. Volume 43. pp 97-108.

[Sakurai, K 72*] SEE: Nakajima, M 72a. *Automatic Character Reading Using a Holographic Data Processing Technique.*

[Sakurai, M 88a] Sakurai, Miyoko. (February 8, 1988). *Fujitsu Breeding Neural-Computer-Driven Robot.* Electronic Engineering Times. pp 33-34.

[Sakurai, M 88b] Sakurai, Miyoko. (May 23, 1988). *Recognition System Relies on Neurals.* Electronic Engineering Times. Issue 487. pp 39.

[Salas, PF 87*] SEE: Vidal, JJ 87b. *Real-Time Parallel Processing of Pixel Arrays Using Adaptive Logic.*

[Salasoo, A 85a] Salasoo, Aita; Shiffrin, Richard M.; Feustel, Timothy C. (1985). *Building Permanent Memory Codes: Codification and Repetition Effects in Word Identification.* Journal of Experimental Psychology: General. Volume 114. pp 50-77.

[Salazar, JM 85*] SEE: Babloyantz, A 85a. *Evidence of Chaotic Dynamics of Brain Activity During the Sleep Cycle.*

[Salehi, JA 88*] SEE: Vecchi, MP 88a. *Neuromorphic Networks Based on Sparse Optical Orthogonal Codes.*

[Salton, G 83a] Salton, G.; Fox, E. A.; Wu, H. (1983). *Extended Boolean Information Retrieval.* Communications of the ACM. Volume 26. pp 1022-1036.

[Salu, Y 83a] Salu, Yehuda. (1983). *Computer Simulations of Learning in Neural Systems.* Computers and Biomedical Research. Volume 16. pp 176-189.

[Salu, Y 84a] Salu, Yehuda. (1984). *Learning in Neural Networks, Models and Simulations.* IEEE International Conference on Systems, Man and Cybernetics, Halifax, Nova Scotia, October 12-14, 1987: Proceedings. pp 325-327.

[Samad, T 87a] Samad, Tariq. (1987). *A Connectionist Network that Learns to Process Some (Very) Simple Sentences.* Eastern States Conference on Linguistics (ESCOL-86), 3rd: Proceedings Linguistics. Ohio State University, Department of Linguistics.

[Samad, T 87b] Samad, Tariq; Harper, Paul. (1987). *Associative Memory Storage Using a Variant of the Generalized Delta Rule.* IEEE International Conference on Neural Networks (ICNN-87), San Diego, CA, June 21-24, 1987: Proceedings. Volume 3. pp 173-184. New York: Institute of Electrical and Electronics Engineers.

[Samad, T 87c] Samad, Tariq. (1987). *Refining and Redefining the Back-Propagation Learning Rule for Connectionist Networks.* IEEE International Conference on Systems, Man and Cybernetics, 1987: Proceedings. New York: Institute of Electrical and Electronics Engineers.

[Samad, T 88a] Samad, Tariq. (1988). *Back-Propagation is Significantly Faster if the Expected Value of the Source Unit is Used for Update.* Neural Networks. Volume 1.

[Samad, T 88b] Samad, Tariq. (1988). *Towards Connectionist Rule-Based Systems.* IEEE International Conference on Neural Networks (ICNN-88), San Diego, CA, July 24-27, 1988: Proceedings. Volume 2. pp 525-532.

[Sambur, MR 78a] Sambur, M. R. (1978). *Adaptive Noise Canceling for Speech Signals.* IEEE Transactions on Acoustics, Speech and Signal Processing. Volume ASSP-26. pp 419-423.

[Sami, MG 88*] SEE: Negrini, R 88a. *Fault-tolerance in Imaging-oriented Systolic Arrays.*

[Sampsell, JB 87*] SEE: Collins, DR 87a. *Optical Neurocomputers Implementation Using Deformable Mirror Arrays.*

[Samuel, AG 86a] Samuel, Arthur G. (1986). *The Role of the Lexicon in Speech Perception.* In Schwab, Eileen C. (ed.); Nusbaum, Howard C. (ed.). Pattern Recognition by Humans and Machines. Volume I, Speech Perception. pp 89-112. New York: Academic Press.

[Sanderson, AC 80a] Sanderson, Arthur C. (1980). *Input-Output Analysis of an IPFM Neural Model: Effects of Spike Regularity and Record Length.* IEEE Transactions on Biomedical Engineering. Volume BME-27. pp 120-131.

[Sanderson, AC 83a] Sanderson, Arthur C. (ed.); Zeevi, Yehoshua Y. (ed.). (1983). *Special Issue on Neural and Sensory Information Processing.* IEEE Transactions on Systems, Man and Cybernetics. Volume SMC-13.

[Sandini, G 83*] SEE: Torre, V 83a. *The Dynamics of Electrically Interacting Cells.*

[Sanes, JN 88a] Sanes, Jerome N.; Suner, Selim; Lando, James F.; Donoghue, John P. (1988). *Rapid Reorganization of Adult Rat Motor Cortex Somatic Representation Patterns after Motor Nerve Injury.* Proceedings of the National Academy of Sciences. Volume 85. pp 2003-2007.

[Sankar, PV 86*] SEE: Sklansky, J 86a. *Biomedical Image Analysis.*

[Sanots, E 87*] SEE: Charniak, E 87a. *A Connectionist Context-Free Parser Which is not Context-Free, But Then it is not Really Connectionist Either.*

[Sanz, J 88*] SEE: Bruck, J 88b. *A Study on Neural Networks.*

[Saramaki, T 84*] SEE: Kohonen, T 84b. *Phonotopic Maps - Insightful Representation of Phonological Features for Speech Recognition.*

[Saridis, GN 77a] Saridis, G. N. (1977). Self-Organizing Control of Stochastic Systems. New York: Marcel Dekker.

[Sasiela, RJ 86a] Sasiela, R. J. (1986). *Forgetting as a Way to Improve Neural-Net Behavior.* In Denker, John S. (ed.). Neural Networks for Computing, Snowbird, Utah, 1986. AIP Proceedings 151. pp 386-391. New York: American

Institute of Physics.

[Satyanarayana, S 87*] SEE: Tsividis, Y 87b. *Analogue Circuits for Variable-Synapse Electronic Neural Networks.*

[Satyanarayana, S 88*] SEE: Alspector, J 88c. *Stochastic Learning Networks and their Electronic Implementation.*

[Satyanarayana, S 88a] Satyanarayana, Srinagesh; Suyama, K.; Tsividis, Y. (1988). *Analog MOS Circuit Techniques in the VLSI Implementation of Neural Networks.* Neural Networks. Volume 1, Supplement. pp 406.

[Saund, E 86a] Saund, Eric. (1986). *Abstraction and Representation of Continuous Variables in Connectionist Networks.* National Conference on Artificial Intelligence (AAAI-86), 5th, Philadelphia, PA, August 11-15, 1986: Proceedings.

[Savely, RT 87a] Savely, Robert T. (1987). *The Implementation of Neural Network Technology.* IEEE International Conference on Neural Networks (ICNN-87), San Diego, CA, June 21-24, 1987: Proceedings. Volume 4. pp 477-484. New York: Institute of Electrical and Electronics Engineers.

[Sawada, R 82a] Sawada, R.; Sugie, Noboru. (1982). *Self-Organization of Neural Nets with Competitive and Cooperative Interaction.* In Amari, Shun-ichi (ed.); Arbib, Michael A. (ed.). Competition and Cooperation in Neural Nets. pp 238-247. New York: Springer-Verlag.

[Sawusch, JR 86a] Sawusch, James R. (1986). *Auditory and Phonetic Coding of Speech.* In Schwab, Eileen C. (ed.); Nusbaum, Howard C. (ed.). Pattern Recognition by Humans and Machines. Volume I, Speech Perception. pp 51-88. New York: Academic Press.

[Sayegh, S 88a] Sayegh, Samir; Tenorio, Manoel F. (in press). Expert System and Neural Network Approaches for Classical Guitar Fingering.

[Saylor, J 86a] Saylor, J.; Stork, David G. (1986). *Parallel Analog Neural Networks for Tree Searching.* In Denker, John S. (ed.). Neural Networks for Computing, Snowbird, Utah, 1986. AIP Proceedings 151. pp 392-397. New York: American Institute of Physics.

[Scalettar, R 88a] Scalettar, R.; Zee, A. (1988). *Emergence of Grandmother Memory in Feed Forward Networks: Learning with Noise and Forgetfulness.* In Waltz, David L. (ed.); Feldman, Jerome A. (ed.). Connectionist Models and Their Implications: Readings from Cognitive Science. pp 309-328. Norwood, NJ: Ablex.

[Scalettar, R 88b] Scalettar, R.; Zee, A. (1988). *Perception of Left and Right by a Feed Forward Net.* Biological Cybernetics. Volume 58. pp 193-201.

[Scanlon, RD 87*] SEE: Johnson, M 87a. *Experiences with a Feeling-Thinking Machine.*

[Scanlon, RD 87*] SEE: Johnson, M 87b. Non-Living Intelligence.

[Scanlon, RD 88a] Scanlon, Raymond D.; Johnson, Mark. (1988). *Design of a Feeling-Thinking Machine.* Army Conference on Applied Mathematics and Computing, 5th, West Point, NY, June 15-18, 1987: Proceedings. ARCCB-TR-88013. Watervliet, NY: U S Army ARDEC.

[Scarabottolo, N 88*] SEE: Negrini, R 88a. *Fault-tolerance in Imaging-oriented Systolic Arrays.*

[Schalk, TB 81*] SEE: Doddington, GR 81a. *Speech Recognition: Turning Theory into Practice.*

[Schank, RC 80a] Schank, Roger C. (1980). *Language and Memory.* Cognitive Science. Volume 4. pp 243-284.

[Schank, RC 84a] Schank, Roger C. (1984). Cognitive Computer: On Language, Learning and AI. Reading, MA: Addison-Wesley.

[Schatzman, M 86a] Schatzman, M. (1986). *Self Organizing Mathematical Models: Non-Linear Evolution Equations with a*

Convolution Term. In Bienenstock, Elie L. (ed.). Disordered Systems and Biological Organization. pp 385-388. New York: Springer-Verlag.

[Scheff, K 87*] SEE: Landa, J 87a. *Binocular Fusion Using Simulated Annealing*.

[Scheff, K 87a] Scheff, Kim; Szu, Harold H. (1987). *1-D Optical Cauchy Machine Infinite Film Spectrum Search*. IEEE International Conference on Neural Networks (ICNN-87), San Diego, CA, June 21-24, 1987: Proceedings. Volume 3. pp 673-679. New York: Institute of Electrical and Electronics Engineers.

[Scheff, K 87b] Scheff, Kim; Willey, Jefferson; Szu, Harold H. (1987). *Simulation Results of Multiple Layer Difference Learning*. IEEE International Conference on Neural Networks (ICNN-87), San Diego, CA, June 21-24, 1987: Proceedings. Volume 2. pp 679-684. New York: Institute of Electrical and Electronics Engineers.

[Scheibel, AB 82*] SEE: Shaw, GL 82a. *Cooperativity in Brain Function: Assemblies of Approximately 30 Neurons*.

[Scheidler, W 87*] SEE: Caelli, TM 87b. *Visual Pattern Recognition in Humans*.

[Schindler, M 87a] Schindler, Max. (January 22, 1987). *Neural Nets Learn to Outsmart a Salesman*. Electronic Design. pp 42-45.

[Schindler, M 87b] Schindler, Max (ed.). (January 22, 1987). *Neural Nets Learn to Outsmart a Salesman*. Electronic Design.

[Schmajuk, NA 85a] Schmajuk, Nestor A.; Moore, John W. (1986). *Real-Time Attentional Models for Classical Conditioning and the Hippocampus*. Physiological Psychology. Volume 13. pp 278-290.

[Schmajuk, NA 86a] Schmajuk, Nestor A.; Moore, John W. (1986). *A Real-Time Attentional-Associative Network for Classical Conditioning of the Rabbit's NMR*. Cognitive Science Society, 8th, Amherst, MA, August 15-17, 1986: Proceedings. pp 794-807. Hillsdale, NJ: Lawrence Erlbaum.

[Schmajuk, NA 87*] SEE: Grossberg, S 87j. *Neural Dynamics of Attentionally Modulated Pavlovian Conditioning: Conditioned Reinforcement, Inhibition, and Opponent Processing*.

[Schmajuk, NA 87*] SEE: Grossberg, S 87b. *A Neural Network Architecture for Attentionally-Modulated Pavlovian Conditioning: Conditioned Reinforcement, Inhibition, and Opponent Processing*.

[Schmajuk, NA 87a] Schmajuk, Nestor A.; Moore, John W. (1987). Two Attentional Models of Classical Conditioning: Variations in CS Effectiveness Revisited. COINS Technical Report 87-29. Amherst, MA: University of Massachusetts, Computer Science Department.

[Schmajuk, NA 88a] Schmajuk, Nestor A.; Moore, John W. (1988). *The Hippocampus and the Classically Conditioned Nictitating Membrane Response: A Real-Time Attentinal-Associative Model*. Psychobiology. Volume 16. pp 20-35.

[Schmajuk, NA 88b] Schmajuk, Nestor A. (1988). *The Hippocampus and the Control of Information Storage in the Brain*. Competition and Cooperation in Neural Nets II. New York: Springer-Verlag.

[Schmitt, FO 76a] Schmitt, Francis O.; Dev, Parvati; Smith, Barry H. (1976). *Electrotonic Processing of Information by Brain Cells*. Science. Volume 193. pp 114-120.

[Schneider, W 85a] Schneider, W. (1985). *Toward a Model of Attention and the Development of Automatic Processing*. Attention and Performance. Volume 11. pp 475-492.

[Schneider, W 86*] SEE: von der Malsburg, C 86a. *A Neural Cocktail-Party Processor*.

[Schnitger, G 88*] SEE: Parberry, I 88b. *Parallel Computation with Threshold Functions*.

[Schnitger, G 88*] SEE: Parberry, I 88c. *Relating Boltzmann Machines to Conventional Models of Computation.*

[Schoenfeld, R 56*] SEE: Berman, M 56a. *Invariants in Experimental Data on Linear Kinetics and the Formulation of Models.*

[Schoner, G 86a] Schoner, G.; Haken, Hermann; Kelso, J. A. S. (1986). *A Stochastic Theory of Phase Transitions in Human Hand Movement.* Biological Cybernetics. Volume 53. pp 247-257.

[Schoner, G 88a] Schoner, G.; Kelso, J. A. S. (in press). *A Dynamic Pattern Theory of Behavioral Change.* Journal of Theoretical Biology.

[Schoner, G 88b] Schoner, G.; Kelso, J. A. S. (1988). *A Synergetic Theory of Environmentally-Specified and Learned Patterns of Movement Coordination, I. Relative Phase Dynamics.* Biological Cybernetics. Volume 58. pp 71-80.

[Schoner, G 88c] Schoner, G.; Kelso, J. A. S. (1988). *A Synergetic Theory of Environmentally-Specified and Learned Patterns of Movement Coordination, II. Component Oscillator Dynamics.* Biological Cybernetics. Volume 58. pp 81-89.

[Schoner, G 88d] Schoner, G.; Kelso, J. A. S. (1988). *Dynamic Pattern Generation in Behavioral and Neural Systems.* Science. Volume 239. pp 1513-1520.

[Schoner, G 88e] Schoner, G.; Kelso, J. A. S. (1988). *Dynamic Patterns of Biological Coordination: Theoretical Strategy and New Results.* Dynamic Patterns in Complex Systems. pp 77-102. Singapore: World Scientific.

[Schroeder, MR 79a] Schroeder, M. R.; Atal, B.; Hall, J. L. (1979). *Optimizing Digital Speech Coders by Exploiting Masking Properties of the Human Ear.* Journal of the Acoustical Society of America. Volume 66. pp 1647-1652.

[Schuling, FH 88a] Schuling, F. H.; Mastebroek, H. A. K.; Zaagman, Willem H. (1988). *An Adaptive and Heterodyne Filtering Procedure for the Imaging of Moving Objects.* In Anderson, Dana Z. (ed.). Neural Information Processing Systems. pp 662-673. New York: American Institute of Physics.

[Schuling, FH 88b] Schuling, F. H. (1988). Processing of Moving Images in Natural and Artificial Visual Systems. Druk: Krips Repro Meppel.

[Schulten, K 86*] SEE: Buhmann, J 86b. *Influence of Noise on the Behavior of an Autoassociative Neural Network.*

[Schulten, K 86*] SEE: Ritter, H 86a. *On the Stationary State of Kohonen's Self-Organizing Sensory Mapping.*

[Schulten, K 86*] SEE: Buhmann, J 86c. *Physicists Explore Human and Artificial Intelligence.*

[Schulten, K 86*] SEE: Buhmann, J 86a. *A Physiological Neural Network as an Autoassociative Memory.*

[Schulten, K 86*] SEE: Divko, R 86a. *Stochastic Spin Models for Pattern Recognition.*

[Schulten, K 86*] SEE: Ritter, H 86b. *Topology Conserving Mappings for Learning Motor Tasks.*

[Schulten, K 87*] SEE: Buhmann, J 87a. *Influence of Noise on the Function of a "Physiological" Neural Network.*

[Schulten, K 87*] SEE: Ritter, H 87a. *Planning a Dynamic Trajectory via Path Finding in Discretized Phase Space.*

[Schulten, K 88*] SEE: Ritter, H 88a. *Convergence Properties of Kohonen's Topology Conserving Maps: Fluctuations, Stability and Dimension Selection.*

[Schulten, K 88*] SEE: Ritter, H 88b. *Extending Kohonen's Self-Organizing Mapping Algorithm to Learn Ballistic Movements.*

[Schulten, K 88*] SEE: Ritter, H 88c. *Kohonen's Self-Organizing Maps: Exploring their Computational Capabilities.*

[Schulten, K 88*] SEE: Buhmann, J 88a. *Storing Sequences of Biased Patterns in Neural Networks with Stochastic Dynamics.*

[Schultz, WL 78*] SEE: Pao, Y 78a. *An Associative Memory Technique for the Recognition of Patterns.*

[Schulz, M 88a] Schulz, Mario; Darianian, Mohsen; Tavangarian, Djamshid; Waldschmidt, Klaus. (1988). *An Associative Co-Processor Architecture with the UNIX Operating System for the Acceleration of CAD Tools.* In Luque, E. (ed.). Mind and Microcomputers and Their Applications. International Society for Mini and Microcomputers (ISMM).

[Schulze, E 87*] SEE: von Seelen, W 87b. *Properties of Cortical Systems - A Basis for Computer Vision.*

[Schuz, A 86*] SEE: Krone, G 86a. *Spatiotemporal Receptive Fields: A Dynamical Model Derived from Cortical Architectonics.*

[Schvaneveldt, RW 77*] SEE: Tweedy, JR 77a. *Semantic-Context Effects on Word Recognition: Influence of Varying the Proportion of Items Presented in an Appropriate Context.*

[Schwab, EC 86*] SEE: Nusbaum, HC 86a. *The Role of Attention and Active Processing in Speech Perception.*

[Schwab, EC 86a] Schwab, Eileen C. (ed.); Nusbaum, Howard C. (ed.). (1986). Pattern Recognition by Humans and Machines. Volume 2, Visual Perception. Orlando, FL: Academic Press.

[Schwab, EC 86b] Schwab, Eileen C. (ed.); Nusbaum, Howard C. (ed.). (1986). Pattern Recognition by Humans and Machines. Volume 1, Speech Perception. Orlando, FL: Academic Press.

[Schwartz, DB 86*] SEE: Hubbard, W 86a. *Electronic Neural Networks.*

[Schwartz, DB 87*] SEE: Howard, RE 87a. *An Associative Memory Based on an Electronic Neural Network Architecture.*

[Schwartz, DB 87*] SEE: Denker, JS 87a. *Automatic Learning, Rule Extraction, and Generalization.*

[Schwartz, DB 88*] SEE: Mackie, S 88b. *Implementations of Neural Network Models in Silicon.*

[Schwartz, DB 88*] SEE: Mackie, S 88c. *Microelectronic Implementations of Connectionist Neural Networks.*

[Schwartz, EL 78*] SEE: John, ER 78a. *The Neurophysiology of Information Processing and Cognition.*

[Schwartz, EL 80a] Schwartz, Eric L. (1980). *Computational Anatomy and Functional Architecture of Striate Cortex: A Spatial Mapping Approach to Perceptual Coding.* Vision Research. Volume 20. pp 645-669.

[Schwartz, EL 83a] Schwartz, Eric L. (1983). *Research Note - Cortical Mapping and Perceptual Invariance: A Reply to Cavanagh.* Vision Research. Volume 23. pp 831-835.

[Schwartz, EL 87*] SEE: Yeshurun, Y 87a. *An Ocular Dominance Column Map as a Data Structure for Stereo Segmentation.*

[Schwartz, EL 87a] Schwartz, Eric L.; Yeshurun, Yehezkel. (1987). *Towards a Non-Network Approach to "Neural Modeling": Some Basic Issues of Measurement, Simulation, and Computational Significance of Brain Maps.* IEEE International Conference on Neural Networks (ICNN-87), San Diego, CA, June 21-24, 1987: Proceedings. Volume 4. pp 225-234. New York: Institute of Electrical and Electronics Engineers.

[Schwartz, JH 82*] SEE: Kandel, ER 82a. *Molecularbiology of Learning: Modulation of Transmitter Release.*

[Schwartz, RG 88*] SEE: Tenorio, MF 88b. *Unrestricted English Text-to-Speech Conversion with a Distributed Adaptive Network.*

[Schwartz, RG 88a] Schwartz, Richard G.; Tom, M. Daniel; Tenorio, Manoel F. (1988). *Evaluating a Connectionist Network as a Simulation of Speech Development.* Meeting of the International Neural Network Society (INNS), 1st, Boston, MA, September 6-10, 1988.

[Schwartz, RM 84a] Schwartz, R. M.; Chow, Y. L.; Roucos, Salim; Krasner, M.; Makhoul, John. (1984). *Improved Hidden Markov Modeling of Phonemes for Continuous Speech Recognition.* IEEE International Conference on Acoustics, Speech, and Signal Processing (ICASSP-84), April, 1984: Proceedings. pp 35.6.1-4. New York: Institute of Electrical and Electronics Engineers.

[Schwartz, RM 85a] Schwartz, R. M.; Chow, Y. L.; Kimball, O.; Roucos, Salim; Krasner, M.; Makhoul, John. (1985). *Context-Dependent Modeling for Acoustic-Phonetic Recognition of Continuous Speech.* IEEE International Conference on Acoustics, Speech, and Signal Processing (ICASSP-85), April, 1985: Proceedings. pp 1205-1208. New York: Institute of Electrical and Electronics Engineers.

[Schwartz, RM 87a] Schwartz, R. M.; Chow, Y. L.; Kubala, F. (1987). *Rapid Speaker Adaptation Using a Probabilistic Spectral Mapping.* IEEE International Conference on Acoustics, Speech, and Signal Processing (ICASSP-87), Dallas, TX, April 6-9, 1987: Proceedings.

[Schwartz, T 88*] SEE: Wasserman, PD 88a. *Expert Focus, Neural Networks, Part 2: What are They and Why Is Everybody So Interested in Them Now?.*

[Schwartz, TJ 86a] Schwartz, Tom J. (September 1, 1986). *IBM Research Yields Artificial Neural Net Workstation.* Electronic Engineering Times. pp 67 f.

[Schwartz, TJ 86b] Schwartz, Tom J. (October 20, 1986). *TRW Unit Uses Neural Networks for Voice Recognition.* Electronic Engineering Times. p 55.

[Schwartz, TJ 87b] Schwartz, Tom J. (April 13, 1987). *Hecht-Nielsen Unveils First New Neural Net Architecture Since '84.* Electronic Engineering Times.

[Schwartz, TJ 87c] Schwartz, Tom J. (March 2, 1987). *Neural Net Leaders Strut Their Stuff at Federal Workshop.* Electronic Engineering Times.

[Schwartz, TJ 87d] Schwartz, Tom J. (March 2, 1987). *Neural Net Software for PC AT.* Electronic Engineering Times. p 27f.

[Schwartz, TJ 87e] Schwartz, Tom J. (July 27, 1987). *Report Outlines Commercialization of Neural Nets.* Electronic Engineering Times. p 76.

[Schwartz, TJ 88a] Schwartz, Tom J. (1988). *Board Creates Neural AT.* Electronic Engineering Times. Issue 471. pp 33.

[Sclabassi, RJ 88*] SEE: Krieger, D 88a. *A Neurobiologically Based Neural Net Model of the Hippocampus.*

[Sclabassi, RJ 88*] SEE: Berger, TW 88a. *Nonlinear Systems Analysis of the Hippocampal Perforant Path-Dentate Projection. II. Effects of Random Impulse Train Stimulation.*

[Sclabassi, RJ 88*] SEE: Berger, TW 88b. *Nonlinear Systems Analysis of the Hippocampal Perforant Path-Dentate Projection. III. Comparison of Random Train and Paired Impulse Stimulation.*

[Sclabassi, RJ 88a] Sclabassi, Robert J.; Krieger, Donald N.; Berger, Theodore W. (1988). *A Systems Theoretic Approach to the Study of CNS Function.* Annals of Biomedical Engineering. Volume 16. pp 17-34.

[Sclabassi, RJ 88b] Sclabassi, Robert J.; Eriksson, Jan L.; Port, Richard L.; Robinson, Gilbert B.; Berger, Theodore W. (1988). *Nonlinear Systems Analysis of the Hippocampal Perforant Path-Dentate Projection. I. Theoretical and Interpretational Considerations.* Journal of Neurophysiology. Volume 60. pp 1066-1076.

[Scofield, CL 85*] SEE: Cooper, LN 85a. *Neuron Selectivity: Single Neuron and Neuron Networks.*

[Scofield, CL 86*] SEE: Rimey, RD 86a. *Real-Time 3-D Object Classification Using a Learning System.*

[Scofield, CL 87*] SEE: Reilly, DL 87a. *Learning System Architectures Composed of Multiple Learning Modules.*

[Scofield, CL 88a] Scofield, Christopher L. (1988). *A Mean Field Theory of Layer IV of Visual Cortex and Its Application to Artificial Neural Networks.* In Anderson, Dana Z. (ed.). Neural Information Processing Systems. pp 683-692. New York: American Institute of Physics.

[Scofield, CL 88b] Scofield, Christopher L.; Reilly, Douglas L.; Elbaum, Charles; Cooper, Leon N. (1988). *Pattern Class Degeneracy in an Unrestricted Storage Density Memory.* In Anderson, Dana Z. (ed.). Neural Information Processing Systems. pp 674-682. New York: American Institute of Physics.

[Scott, AC 75a] Scott, Alwyn C. (1975). *The Electrophysics of a Nerve Fiber.* Reviews of Modern Physics. Volume 47. pp 487-533.

[Scott, DE 84a] Scott, Donald E.; Hutchinson, Charles E. (1984). *An Application of the GMDH Algorithm to Economic Modeling.* In Farlow, Stanley J. (ed.). Self-Organizing Methods in Modeling: GMDH Type Algorithms. pp 243-256. New York: Marcel Dekker.

[Scott, PD 79a] Scott, P. D. (1979). *Conditional Control Transfer Mechanisms in the Neocortex: 1. The Basic Model.* Journal of Theoretical Biology. Volume 81. pp 423-452.

[Sears, RW 65a] Sears, R. W. (1965). *Adaptive Representation for Pattern Recognition.* IEEE Transactions on Systems, Science, and Cybernetics. Volume SMC-1. pp 59-65.

[Sebestyen, GS 61a] Sebestyen, G. S. (1961). *Recognition of Membership in Classes.* IRE Transactions on Information Theory. Volume IT-7. p 44 f.

[Sebestyen, GS 62a] Sebestyen, G. S. (1963). Decision-making Processes in Pattern Recognition. New York: Macmillan.

[Segal, LA 74*] SEE: Lin, CC 74a. Mathematics Applied to Deterministic Problems in the Natural Sciences.

[Segall, Z 86a] Segall, Zary; Snyder, Larry. (September 1986). *Workshop on Performance Efficient Parallel Programming.* CMU-CS-86-180.

[Segev, I 79*] SEE: Parnas, I 79a. *A Mathematical Model for Conduction of Action Potentials Along Bifurcating Axons.*

[Segundo, JP 66*] SEE: Moore, GP 66a. *Statistical Analysis and Functional Interpretation of Neuronal Spike Data.*

[Segundo, JP 68a] Segundo, Jose P.; Perkel, Donald H.; Wyman, H.; Hegstad, H.; Moore, George P. (1968). *Input-Output Relations in Computer-Simulated Nerve Cells: Influence of Statistical Properties, Strength, etc..* Kybernetik. Volume 4. pp 157-171.

[Segundo, JP 79*] SEE: Brillinger, DR 79a. *Empirical Examination of the Threshold Model of Neuron Firing.*

[Seibert, M 88*] SEE: Waxman, AM 88b. *The Neural Analog Diffusion-Enhancement Layer (NADEL) and Early Visual Processing.*

[Seidman, SB 87a] Seidman, Stephen B.; Reggia, James A.; Wang, Pearl Y. (1987). *Analysis of a Competition-Based Connectionist Model.* IEEE International Conference on Neural Networks (ICNN-87), San Diego, CA, June 21-24, 1987: Proceedings. Volume 2. pp 139-146. New York: Institute of Electrical and Electronics Engineers.

[Seidman, SB 88*] SEE: Wang, PY 88a. *Analysis of Competition-Based Spreading Activation in Connectionist Models.*

[Seitz, CL 87*] SEE: Ngai, JY 87a. A Framework for Adaptive Routing.

[Sejnowski, TJ 81a] Sejnowski, Terrence J. (1981). *Skeleton Filters in the Brain.* In Hinton, Geoffrey E. (ed.). Parallel Models of Associative Memory. pp 189-212. Hillsdale, NJ: Lawrence Erlbaum.

[Sejnowski, TJ 83*] SEE: Hinton, GE 83a. *Analyzing Cooperative Computation.*

[Sejnowski, TJ 83*] SEE: Fahlman, SE 83a. *Massively Parallel Architectures for AI: Netl, Thistle, and Boltzmann Machines.*

[Sejnowski, TJ 83*] SEE: Hinton, GE 83b. *Optimal Perceptual Inference.*

[Sejnowski, TJ 83*] SEE: Ballard, DH 83a. *Parallel Visual Computation.*

[Sejnowski, TJ 84*] SEE: Hinton, GE 84a. Boltzmann Machines: Constraint Satisfaction Networks that Learn.

[Sejnowski, TJ 85*] SEE: Ackley, DH 85b. *A Learning Algorithm for Boltzmann Machines.*

[Sejnowski, TJ 85*] SEE: Ackley, DH 85a. *A Learning Algorithm for Boltzmann Machines.*

[Sejnowski, TJ 86*] SEE: Hinton, GE 86c. *Learning and Relearning in Boltzmann Machines.*

[Sejnowski, TJ 86*] SEE: Kienker, PK 86a. *Separating Figure from Ground with a Parallel Network.*

[Sejnowski, TJ 86*] SEE: Cohen, NJ 86a. *Skill Learning and Repetition Priming in Symmetry Detection: Parallel Studies of Human Subjects and Connectionist Models.*

[Sejnowski, TJ 86*] SEE: Rosenberg, CR 86a. *The Spacing Effect on NETtalk, a Massively-Parallel Network.*

[Sejnowski, TJ 86a] Sejnowski, Terrence J. (1986). *Higher-Order Boltzmann Machines.* In Denker, John S. (ed.). Neural Networks for Computing, Snowbird, Utah, 1986. AIP Proceedings 151. pp 398-403. New York: American Institute of Physics.

[Sejnowski, TJ 86b] Sejnowski, Terrence J.; Kienker, Paul K.; Hinton, Geoffrey E. (1986). *Learning Symmetry Groups with Hidden Units: Beyond the Perceptron.* Physica D. Volume 22. pp 260-275.

[Sejnowski, TJ 86c] Sejnowski, Terrence J.; Rosenberg, Charles R. (1988). *NETtalk: A Parallel Network that Learns to Read Aloud.* In Anderson, James A. (ed.); Rosenfeld, Edward (ed.). Neurocomputing: Foundations of Research. pp 663-672. Cambridge, MA: MIT Press.

[Sejnowski, TJ 86d] Sejnowski, Terrence J.; Rosenberg, Charles R. (1986). NETtalk: A Parallel Network that Learns to Read Aloud. JHU/EECS-86/01. Baltimore, MD: Johns Hopkins University, Electrical Engineering and Computer Science.

[Sejnowski, TJ 86e] Sejnowski, Terrence J. (1986). *Open Questions About Computation in Cerebral Cortex.* In Rumelhart, David E. (ed.). Parallel Distributed Processing: Explorations in the Microstructure of Cognition. Volume II. pp 372-389. Cambridge, MA: MIT Press.

[Sejnowski, TJ 87*] SEE: Hinton, GE 87c. *Neural Network Architectures for AI.*

[Sejnowski, TJ 87a] Sejnowski, Terrence J. (1987). *Computational Models and the Development of Topographic Projections.* Trends in Neurosciences. Volume 10. pp 304-305.

[Sejnowski, TJ 87b] Sejnowski, Terrence J. (1987). *Computing with Connections.* Mathematical Psychology. Volume 31. pp 203-210.

[Sejnowski, TJ 87c] Sejnowski, Terrence J.; Rosenberg, Charles R. (1987). *Connectionist Models of Learning.* Perspectives in Memory Research and Training. Cambridge, MA: MIT Press.

[Sejnowski, TJ 87d] Sejnowski, Terrence J. (1987). *Hopfield, Boltzman Machines, and Vision.* Artificial Neural Networks, An Introductory Seminar and Exhibition, Cincinnati, OH, December 8, 1987. Cincinnati, OH: Institute of Advanced Manufacturing Sciences.

[Sejnowski, TJ 87e] Sejnowski, Terrence J. (1987). *Learning Procedures in Multilayered Networks.* Artificial Neural

Networks, An Introductory Seminar and Exhibition, Cincinnati, OH, December 8, 1987. Cincinnati, OH: Institute of Advanced Manufacturing Sciences.

[Sejnowski, TJ 87f] Sejnowski, Terrence J.; Rosenberg, Charles R. (1987). *Parallel Networks that Learn to Pronounce English Text*. Complex Systems. Volume 1. pp 145-168.

[Sejnowski, TJ 87g] Sejnowski, Terrence J.; Hinton, Geoffrey E. (1987). *Separating Figure from Ground with a Boltzmann Machine*. In Arbib, Michael A. (ed.); Hanson, Allen R. (ed.). Vision, Brain, and Cooperative Computation. pp 703-724. Cambridge, MA: MIT Press.

[Sejnowski, TJ 88*] SEE: Tesauro, G 88a. *A 'Neural' Network that Learns to Play Backgammon.*

[Sejnowski, TJ 88*] SEE: Gorman, RP 88a. *Analysis of Hidden Units in a Layered network Trained to Classify Sonar Targets.*

[Sejnowski, TJ 88*] SEE: Qian, N 88a. *Electrodiffusion Model of Electrical Conduction in Neuronal Processes.*

[Sejnowski, TJ 88*] SEE: Gorman, RP 88b. *Learned Classification of Sonar Targets Using a Massively Parallel Network.*

[Sejnowski, TJ 88*] SEE: Ackley, DH 88a. *A Learning Algorithm for Boltzmann Machines.*

[Sejnowski, TJ 88*] SEE: Lehky, SR 88a. *Network Model of Shape-From-Shading: Neural Function Arises from Both Receptive and Projective Fields.*

[Sejnowski, TJ 88a] Sejnowski, Terrence J.; Koch, Christof; Churchland, Patricia Smith. (1988). *Computational Neuroscience*. Science. Volume 241. pp 1299-1306.

[Sejnowski, TJ 88b] Sejnowski, Terrence J. (1988). *Neural Network Learning Algorithms*. In Eckmiller, Rolf (ed.); von der Malsburg, Christoph (ed.). Neural Computers. NATO ASI Series, Volume F41. pp 291-300. Berlin: Springer-Verlag.

[Sejnowski, TJ 88c] Sejnowski, Terrence J. (1988). *Neural Populations Revealed*. Nature. Volume 332. pp 308.

[Sejnowski, TJ 89*] SEE: Churchland, PS 89a. *Neural Representation and Neural Computation.*

[Selfridge, OG 55a] Selfridge, Oliver G. (1955). *Pattern Recognition and Modern Computers*. Western Joint Computer Conference, 1955: Proceedings. pp 91-93. New York: Institute of Radio Engineers.

[Selfridge, OG 58a] Selfridge, Oliver G. (1988). *Pandemonium: A Paradigm for Learning*. In Anderson, James A. (ed.); Rosenfeld, Edward (ed.). Neurocomputing: Foundations of Research. pp 117-122. Cambridge, MA: MIT Press.

[Selfridge, OG 59a] Selfridge, Oliver G. (1959). *Pandemonium: A Paradigm for Learning*. Mechanization of Thought Processes. pp 511-526. London: Her Majesty's Stationery Office.

[Selfridge, OG 60*] SEE: Minsky, ML 60a. *Learning in Random Nets.*

[Selfridge, OG 60a] Selfridge, Oliver G.; Neisser, Ulric. (1960). *Pattern Recognition by Machine*. Scientific American. Volume 203. pp 60-68.

[Selfridge, OG 84a] Selfridge, Oliver G. (ed.); Rissland, E. L. (ed.); Arbib, Michael A. (ed.). (1984). Adaptive Control of Ill-Defined Systems. Volume 16. New York: Plenum.

[Selke, W 80*] SEE: Fisher, ME 80a. *Infinitely Many Commensurate Phases in a Simple Ising Model.*

[Selke, W 81*] SEE: Fisher, ME 81a. *Low Temperature Analysis of the Axial Next-Nearest Neighbor Ising Model Near its Multiphase Point.*

[Selman, B 85a] Selman, Bart; Hirst, Graeme. (1985). *A Rule-Based Connectionist Parsing System*. Cognitive Science Society, 7th, Irvine, CA, August 15-17, 1985: Proceedings. pp 212-219. Hillsdale, NJ: Lawrence Erlbaum.

[Selman, B 85b] Selman, Bart. (1985). Rule-Based Processing in a Connectionist System for Natural Language Understanding. CSRI Technical Report 168. Toronto, Ontario: University of Toronto, Computers Systems Research Institute.

[Selverston, AI 80a] Selverston, Allen I. (1980). *Are Central Pattern Generators Understandable?*. The Behavioral and Brain Sciences. Volume 3. pp 535-571.

[Selverston, AI 83a] Selverston, Allen I.; Miller, John P.; Wadepuhl, Martin. (1983). *Neural Mechanisms for the Production of Cyclic Motor Patterns*. IEEE Transactions on Systems, Man and Cybernetics. Volume SMC-13. pp 749-757.

[Senecal, M 83*] SEE: Meyer, GE 83a. *The Illusion of Transparency and Chromatic Subjective Contours*.

[Sereno, ME 87*] SEE: Kersten, D 87a. *Associative Learning of Scene Parameters from Images*.

[Sereno, ME 87a] Sereno, Margaret E. (1987). *Implementing Stages of Motion Analysis in Neural Networks*. Cognitive Science Society, 9th, Seattle, WA, July 16-18, 1987: Proceedings. pp 405-416. Hillsdale, NJ: Lawrence Erlbaum.

[Serra, J 82a] Serra, J. (1982). Image Analysis and Mathematical Morphology. New York: Academic Press.

[Serra, J 88a] Serra, J. (1988). Image Analysis and Mathematical Morphology. Volume 2. New York: Academic Press.

[Servan-Schreiber, D 88a] Servan-Schreiber, David; Cleeremans, Axel; McClelland, James L. (1988). Encoding Sequential Structure in Simple Recurrent Networks. CMU-CS-88-183. Pittsburgh, PA: Carnegie Mellon University, Computer Science Department.

[Sethi, IK 81a] Sethi, I. K. (1981). *Fast Algorithm for Recognizing Nearest Neighbors*. IEEE Transactions on Systems, Man and Cybernetics. Volume SMC-11. pp 245-248.

[Shae, ZY 87*] SEE: Farhat, NH 87a. *Architectures and Methodologies for Self-Organization and Stochastic Learning in Opto-Electronic Analogs of Neural Nets*.

[Shafer, G 86*] SEE: Shenoy, P 86a. *Propagating Belief Functions with Local Computations*.

[Shamir, J 88*] SEE: Kinser, JM 88a. *Design for a Massive All-Optical Bidirectional Associative Memory: the Big BAM*.

[Shamma, S 83*] SEE: Cannon, SC 83a. *A Proposed Neural Network for the Integrator of the Oculomotor System*.

[Shamma, SA 87a] Shamma, Shihab A. (1987). *Neural Networks for Speech Processing and Recognition*. IEEE International Conference on Neural Networks (ICNN-87), San Diego, CA, June 21-24, 1987: Proceedings. Volume 4. pp 397-406. New York: Institute of Electrical and Electronics Engineers.

[Shannon, CE 48a] Shannon, Claude E. (1948). *A Mathematical Theory of Communication*. The Bell System Technical Journal. Volume 27. pp 379-656.

[Shapiro, SC 87*] SEE: Peters, SL 87a. *A Representation for Natural Category Systems*.

[Shapley, RM 84*] SEE: Regan, D 84a. Systems Approach in Vision: Proceedings of a Workshop, Amsterdam, The Netherlands, 27-29 August 1984.

[Sharkey, NE 86a] Sharkey, N. E.; Sutcliffe, R. F. E.; Wobcke, W. R. (1986). *Mixing Binary and Continuous Connection Schemes for Knowledge Access*. National Conference on Artificial Intelligence (AAAI-86), 5th, Philadelphia, PA, August 11-15, 1986: Proceedings. pp 262-266.

[Sharkey, NE 88a] Sharkey, N. E. (1988). *A PDP System for Goal-Plan Decisions.* In Trappl, Robert (ed.). Cybernetics and Systems '88. pp 1031-1038. Kluwer.

[Sharkey, NE 88b] Sharkey, N. E. (in press). *Neural Network Learning Techniques.* In McTear, M. (ed.). An Introduction to Cognitive Science.

[Sharkey, NE 89a] Sharkey, N. E. (in press). *A PDP Learning Approach to Natural Language Understanding.* In Aleksander, Igor (ed.). Neural Computing Architectures: The Design of Brainlike Machines. London: Kogan Page.

[Sharkey, NE 89b] Sharkey, N. E. (1989). *Fast Connectionist Learning: Words and Case.* Artificial Intelligence Review. Volume 3. pp 33-47.

[Sharov, AM 81*] SEE: Elinson, MI 81a. *Creation of Efficient Recognition Systems and the Problems of Microelectronics.*

[Sharp, DH 86*] SEE: Mjolsness, E 86a. *Preliminary Analysis of Recursively Generated Networks.*

[Sharp, DH 87*] SEE: Cowan, JD 87a. Neural Nets.

[Sharp, DH 87*] SEE: Mjolsness, E 87b. *Recursively Generated Neural Networks.*

[Sharp, DH 88*] SEE: Cowan, JD 88a. *Neural Nets and Artificial Intelligence.*

[Sharp, DH 88*] SEE: Mjolsness, E 88b. Scaling, Machine Learning, and Genetic Neural Nets.

[Shastri, L 82*] SEE: Small, SL 82b. ISCON: An Interactive Simulator for Connectionist Networks.

[Shastri, L 82*] SEE: Small, SL 82c. *Toward Connectionist Parsing.*

[Shastri, L 84a] Shastri, Lokendra; Feldman, Jerome A. (1984). Semantic Networks and Neural Nets. TR-133. Rochester, NY: University of Rochester, Department of Computer Science.

[Shastri, L 85a] Shastri, Lokendra; Feldman, Jerome A. (1985). *Evidential Reasoning in Semantic Networks: A Formal Theory.* International Joint Conference on Artificial Intelligence (IJCAI-85), 9th, Los Angeles, August 18-23, 1985: Proceedings. pp 467-474.

[Shastri, L 85b] Shastri, Lokendra. (1985). Evidential Reasoning in Semantic Networks: A Formal Theory and its Parallel Implementation: PhD Thesis. Technical Report TR-166. Rochester, NY: University of Rochester.

[Shastri, L 87*] SEE: Kalita, J 87a. *Generation of Simple Sentences in English Using the Connectionist Model of Computation.*

[Shastri, L 87*] SEE: Watrous, RL 87a. *Learned Phonetic Discrimination Using Connectionist Networks.*

[Shastri, L 87*] SEE: Watrous, RL 87b. *Learning Acoustic Features from Speech Data Using Connectionist Networks.*

[Shastri, L 87*] SEE: Watrous, RL 87d. *Learning Phonetic Features Using Connectionist Networks: An Experiment in Speech Recognition.*

[Shastri, L 87a] Shastri, Lokendra. (1987). *A Connectionist Encoding of Semantic Networks.* Cognitive Science Society, 9th, Seattle, WA, July 16-18, 1987: Proceedings. pp 143-154. Hillsdale, NJ: Lawrence Erlbaum.

[Shaw, GL 75*] SEE: Little, WA 75b. *A Statistical Theory of Short and Long Term Memory.*

[Shaw, GL 75*] SEE: Little, WA 75a. *A Statistical Theory of Short and Long Term Memory.*

[Shaw, GL 77a] Shaw, Gordon L. (1977). *Memory Storage Capacity and the Role of Assemblies in Neural Networks.* Brain Theory

Newsletter. Volume 3. pp 35-38.

[Shaw, GL 78*] SEE: Little, WA 78a. *Analytic Study of the Storage Capacity of a Neural Network.*

[Shaw, GL 78a] Shaw, Gordon L. (1978). *Analytical Versus Computer Studies in Modeling Neural Networks.* Brain Theory Newsletter. Volume 3. pp 85-86.

[Shaw, GL 78b] Shaw, Gordon L. (1978). *Space-Time Correlations of Neuronal Firing Related to Memory Storage Capacity.* Brain Research Bulletin. Volume 3. pp 107-113.

[Shaw, GL 80*] SEE: Roney, KJ 80a. *Analytic Study of Assemblies of Neurons in Memory Storage.*

[Shaw, GL 82a] Shaw, Gordon L.; Harth, Erich M.; Scheibel, Arnold B. (1982). *Cooperativity in Brain Function: Assemblies of Approximately 30 Neurons.* Experimental Neurology. Volume 77. pp 324-358.

[Shaw, GL 86*] SEE: Silverman, DJ 86a. *Associative Recall Properties of the Trion Model of Cortical Organization.*

[Shaw, JC 57*] SEE: Newell, A 57a. *Empirical Explorations of the Logic Theory Machine: A Case Study in Heuristic.*

[Shaw, RS 86*] SEE: Crutchfield, JP 86a. *Chaos.*

[Shekhar, S 88*] SEE: Dutta, S 88a. *Bond Rating: A Non-Conservative Application of Neural Networks.*

[Shekhar, S 88*] SEE: Dutta, S 88b. *Using Neural Networks for Generalization Problems.*

[Shekhar, S 89*] SEE: Dutta, S 89a. *An Artificial Intelligence Approach to Predicting Bond Ratings.*

[Shelton, GL 66*] SEE: Nagy, G 66a. *Self-Corrective Character Recognition System.*

[Sheng, CL 69a] Sheng, C. L. (1969). Threshold Logic. New York: Academic Press.

[Shenoy, P 86a] Shenoy, P.; Shafer, G. (1986). *Propagating Belief Functions with Local Computations.* IEEE Expert. Volume 1. pp 43-52.

[Shepanski, JF 87a] Shepanski, John F.; Macy, S. A. (1987). *Manual Training Technqiues of Autonomous Systems Based on Artificial Neural Networks.* IEEE International Conference on Neural Networks (ICNN-87), San Diego, CA, June 21-24, 1987: Proceedings. Volume 4. pp 697-704. New York: Institute of Electrical and Electronics Engineers.

[Shepanski, JF 88a] Shepanski, John F. *Fast Learning in Artificial Neural Systems: Multilayer Perceptron Training Using Optimal Estimation.* IEEE International Conference on Neural Networks (ICNN-88), San Diego, CA, July 24-27, 1988: Proceedings.

[Shepanski, JF 88b] Shepanski, John F.; Macy, S. A. (1988). *Teaching Artificial Neural Systems to Drive: Manual Training Techniques for Autonomous Systems.* In Anderson, Dana Z. (ed.). Neural Information Processing Systems. pp 693-700. New York: American Institute of Physics.

[Shepard, RN 89a] Shepard, Roger N. (1989). *Internal Representation of Universal Regularities: A Challenge for Connectionism.* In Nadel, Lynn (ed.); Cooper, Lynn A. (ed.); Culicover, Peter (ed.); Harnish, R. Michael (ed.). Neural Connections, Mental Computation. pp 104-134. Cambridge, MA: MIT Press.

[Shepherd, GM 79a] Shepherd, Gordon M. (1979). The Synaptic Organization of the Brain. Oxford: Oxford University Press.

[Shepherd, R 88*] SEE: May, D 88a. *The Transputer.*

[Sherman, H 60a] Sherman, H. (1960). *A Quasi-Topological Method for the Recognition of Line Patterns.* International

Conference on Information Processing, Paris, June 15-20, 1959: Proceedings. pp 232-238. Paris: UNESCO.

[Sherrington, D 78*] SEE: Kirkpatrick, S 78a. *Infinite-Ranged Models of Spin-Glasses.*

[Sherwood, DL 87a] Sherwood, David L. (1987). *Bias and Anticipation in Neural Networks.* IEEE International Conference on Neural Networks (ICNN-87), San Diego, CA, June 21-24, 1987: Proceedings. Volume 3. pp 185-190. New York: Institute of Electrical and Electronics Engineers.

[Shiavi, RG 86a] Shiavi, Richard G.; Bourne, John R. (1986). *Methods of Biological Signal Processing.* In Young, Tzay Y. (ed.); Fu, King-Sun (ed.). Handbook of Pattern Recognition and Image Processing. pp 546-568. New York: Academic Press.

[Shibata, T 86a] Shibata, Tsutomu; Sugihara, Kokichi; Sugie, Noboru. (1986). *Recovering Three-Dimensional Structure and Motion of Jointed Objects from Orthographically Projected Optical Flow.* Systems and Computers in Japan. Volume 17. pp 60-69.

[Shiffrin, RM 85*] SEE: Salasoo, A 85a. *Building Permanent Memory Codes: Codification and Repetition Effects in Word Identification.*

[Shiffron, RM 68*] SEE: Atkinson, RC 68a. *Human Memory: A Proposed System and its Control Processes.*

[Shikano, K 86a] Shikano, Kiyohiro; Lee, Kai-Fu; Reddy, Raj. (1986). Speaker Adaptation Through Vector Quantization. CMU-CS-86-160. Pittsburgh, PA: Carnegie Mellon University, Computer Science Department.

[Shim, C 88*] SEE: Lee, S 88a. *Optical Implementation of Associative Memory with Controlled Bit-Significance.*

[Shimabukuro, R 88*] SEE: Shoemaker, PA 88a. *A Modifiable Weight Circuit for Use in Adaptive Neuromorphic Networks.*

[Shimabukuro, RL 88a] Shimabukuro, R. L.; Reedy, R. E.; Garcia, G. A. (1988). *Dual-Polarity Nonvolatile MOS Analogue Memory (MAM) Cell for Neural-Type Circuitry.* Electronics Letters. Volume 24. pp 1231-1232.

[Shimbel, A 48a] Shimbel, Alfonso; Rapoport, Anatol. (1948). *A Statistical Approach to the Theory of the Central Nervous System.* Bulletin of Mathematical Biophysics. Volume 10. pp 41-55.

[Shimohara, K 88*] SEE: Uchiyama, T 88a. *Boltzmann Machines in which Weights Converge to Global Optimal Solutions.*

[Shimohara, K 88a] Shimohara, Katsunori; Tokunaga, Yukio; Uchiyama, Tadasu; Kimura, Y. (1988). *A Neural Network System with an Automatic Generation Mechanism for Distorted Patterns.* Neural Networks. Volume 1, Supplement 1. pp 52.

[Shimohara, K 88b] Shimohara, Katsunori; Uchiyama, Tadasu; Tokunaga, Yukio. (1988). *Back-Propagation Networks for Event-Driven Temporal Sequence Processing.* IEEE International Conference on Neural Networks (ICNN-88), San Diego, CA, July 24-27, 1988: Proceedings. Volume 1. pp 665-672. New York: Institute of Electrical and Electronics Engineers.

[Shin, S 88*] SEE: Lee, S 88a. *Optical Implementation of Associative Memory with Controlled Bit-Significance.*

[Shin, S 88*] SEE: Jang, J 88b. *Optical Implementation of Quadratic Associative Memory with Outer-Product Storage.*

[Shin, S 88*] SEE: Jang, J 88c. *Optical Implementation of the Hopfield Model for Two-Dimensional Associative Memory.*

[Shin, S 88*] SEE: Jang, J 88a. *An Optimization Network for Matrix Inversion.*

[Shingai, R 78a] Shingai, Ryuuzo. (1978). *A Computer Use for Establishing and Proving Conjectures in Uniform Neural Networks.* Brain Theory Newsletter. Volume 3. pp 99-101.

[Shirai, Y 73a] Shirai, Yoshiaki. (1973). *A Context Sensitive Line Finder for Recognition of Polyhedra.* Artificial

Intelligence. Volume 4. pp 95-119.

[Shiue, LC 87a] Shiue, L. C.; Grondin, R. O. (1987). *On Designing Fuzzy Learning Neural-Automata.* IEEE International Conference on Neural Networks (ICNN-87), San Diego, CA, June 21-24, 1987: Proceedings. Volume 2. pp 299-308. New York: Institute of Electrical and Electronics Engineers.

[Shlesinger, MF 83*] SEE: Montroll, EW 83a. *Maximum Entropy Formalism, Fractals, Scaling Phenomena, and l/f Noise: A Tale of Tails.*

[Shlesinger, MF 86*] SEE: Klafter, J 86a. *On the Relationship Among Three Theories of Relaxation in Disordered Systems.*

[Shmoys, DB 86*] SEE: Lawler, EL 86a. The Traveling Salesman Problem.

[Shoben, EJ 73*] SEE: Rips, LJ 73a. *Semantic Distance and the Verification of Semantic Relations.*

[Shoemaker, PA 88a] Shoemaker, P. A.; Shimabukuro, R. (1988). *A Modifiable Weight Circuit for Use in Adaptive Neuromorphic Networks.* International Neural Network Society (INNS), 1st, Boston, MA, September 6-10, 1988: Proceedings. Volume 1, Supplement 1.

[Shogen, K 79*] SEE: Hoshimiya, N 79a. *Two-Terminal Electronic Circuit Neuron Model with Excitable Membrane V-I-t Characteristics.*

[Shortliffe, EH 75a] Shortliffe, Edward H.; Buchanan, Bruce G. (1975). *A Model of Inexact Reasoning in Medicine.* Mathematical Biosciences. Volume 23. pp 351-379.

[Shozakai, M 87*] SEE: Kohonen, T 87e. *Microprocessor Implementation of a Large Vocabulary Speech Recognizer and Phonetic Typewriter for Finnish and Japanese.*

[Shozakai, M 88*] SEE: Kohonen, T 88b. *Phonetic Typewriter for Finnish and Japanese.*

[Shrager, J 87a] Shrager, Jeff; Hogg, Tad; Huberman, Bernardo A. (1987). *Observation of Phase Transitions in Spreading Activation Networks.* Science. Volume 236. pp 1092-1094.

[Shrier, A 83*] SEE: Guevara, MR 83a. *Chaos in Neurobiology.*

[Shrier, S 87a] Shrier, Stefan; Barron, Roger L.; Gilstrap, Lewey O. (1987). *Polynomial and Neural Networks: Analogies and Engineering Applications.* IEEE International Conference on Neural Networks (ICNN-87), San Diego, CA, June 21-24, 1987: Proceedings. Volume 2. pp 431-439. New York: Institute of Electrical and Electronics Engineers.

[Shrivastava, Y 87a] Shrivastava, Yash; Dasgupta, Soura. (1987). *Convergence Issues in Perceptron Based Adaptive Neural Network Models.* Allerton Conference, 1987: Proceedings. pp 1133-1141.

[Shustek, LJ 75*] SEE: Friedman, JH 75a. *An Algorithm for Finding Nearest Neighbors.*

[Shwartz, S 73*] SEE: Homa, D 73a. *Prototype Abstraction and Classification of New Instances as a Function of the Number of Instances Defining the Prototype.*

[Shynk, JJ 85a] Shynk, John J.; Gooch, Richard P.; Widrow, Bernard. (1985). *A Frequency-Domain Adaptive Pole-Zero Filter with Applications.* Asilomar Conference on Circuits, Systems, and Computers, 19th, November 6-8, 1985: Conference Record. pp 299-303. Washington, DC: Institute of Electrical and Electronics Engineers.

[Siarry, P 86*] SEE: Guyon, I 86a. *Engineering Applications of Spin Glass Concepts.*

[Sibain, P 86a] Sibain, P. (1986). *Firing Response of a Neural Model with Threshold Modulations and Neural Dynamics.* In Denker, John S. (ed.). Neural Networks for Computing, Snowbird, Utah, 1986. AIP Proceedings 151. pp 404-407.

New York: American Institute of Physics.

[Sibul, LH 87a] Sibul, Leon H. (ed.). (1987). Adaptive Signal Processing. New York: Institute of Electrical and Electronics Engineers.

[Sideris, A 87*] SEE: Psaltis, D 87b. *Neural Controllers.*

[Siegel, MA 87*] SEE: Taber, WR 87a. *Estimation of Expert Weights Using Fuzzy Cognitive Maps.*

[Siegel, RM 88a] Siegel, Ralph M. (1988). *Discovering Structure from Motion in Monkey, Man and Machine.* In Anderson, Dana Z. (ed.). Neural Information Processing Systems. pp 701-708. New York: American Institute of Physics.

[Sietsma, J 88a] Sietsma, J.; Dow, R. J. F. (1988). *Neural Net Pruning - Why and How.* IEEE International Conference on Neural Networks (ICNN-88), San Diego, CA, July 24-27, 1988: Proceedings. Volume 1. pp 325-333. New York: Institute of Electrical and Electronics Engineers.

[Silverman, DJ 86a] Silverman, D. J.; Shaw, Gordon L.; Pearson, John C. (1986). *Associative Recall Properties of the Trion Model of Cortical Organization.* Biological Cybernetics. Volume 53. pp 259-271.

[Silverman, RH 88a] Silverman, Ronald H.; Noetzel, Andrew S. (1988). *Time-Sequential Self-Organization of Hierarchical Neural Networks.* In Anderson, Dana Z. (ed.). Neural Information Processing Systems. pp 709-714. New York: American Institute of Physics.

[Silverstein, JW 76a] Silverstein, Jack W. (1976). *Asymptotics Applied to a Neural Network.* Biological Cybernetics. Volume 22. pp 73-84.

[Silverstein, JW 77*] SEE: Anderson, JA 77b. *Distinctive Features, Categorical Perception, and Probability Learning: Some Applications of a Neural Model.*

[Silverstein, JW 77*] SEE: Anderson, JA 77a. *Distinctive Features, Categorical Perception, and Probability Learning: Some Applications of a Neural Model.*

[Silverstein, JW 77*] SEE: Anderson, JA 77d. *Vowel Pre-Processing with a Neurally Based Model.*

[Simmes, SD 87a] Simmes, Stephen D. (1987). *Programming Parallel Computers Using Energy Minimization Algorithms.* IEEE International Conference on Neural Networks (ICNN-87), San Diego, CA, June 21-24, 1987: Proceedings. Volume 2. pp 205-212. New York: Institute of Electrical and Electronics Engineers.

[Simmes, SD 88a] Simmes, Stephen D.; Turner, Charles J. (1988). Associative Memories on the Connection Machine. Tucson, AZ: Science Applications International Corporation.

[Simmons, GJ 64a] Simmons, G. J. (1964). *Application of an Associatively Addressed, Distributed Memory.* Joint Computer Conference, Spring 1964. Volume 25. pp 493-513. New York: American Federation of Information Processing Societies.

[Simon, HA 57*] SEE: Newell, A 57a. *Empirical Explorations of the Logic Theory Machine: A Case Study in Heuristic.*

[Simon, HA 76a] Simon, Herbert A. (1976). *The Information-Storage System Called "Human Memory".* Neural Mechanisms of Learning and Memory. pp 79-96. Cambridge, MA: MIT Press.

[Simpson, PK 87a] Simpson, Patrick K. (in press). *A Review of Artificial Neural Systems II: Paradigms, Applications, and Implementations.* CRC Critical Reviews in Artificial Intelligence.

[Simpson, PK 87b] Simpson, Patrick K. (in press). *A Survey of Artificial Neural Systems I: Foundations.* CRC Critical Reviews in Artificial Intelligence.

[Simpson, PK 88*] SEE: Taber, WR 88d. *The Recognition of Orca Calls with a Neural Network.*

[Singer, A 87a] Singer, Alexander; Donoghue, John P. (November, 1987). *A Computer Simulation of Cerebral Neocortex: Computational Capabilities of Nonlinear Neural Networks.* IEEE Conference on Neural Information Processing, Denver, CO, 1987: Proceedings.

[Singer, A 88a] Singer, Alexander; Donoghue, John P. (1988). *A Computer Simulation of Cerebral Neocortex: Computational Capabilities of Nonlinear Neural Networks.* In Anderson, Dana Z. (ed.). Neural Information Processing Systems. pp 715-729. New York: American Institute of Physics.

[Singer, W 85a] Singer, W. (1985). *Hebbian Modification of Synaptic Transmission as a Common Mechanism in Experience-Dependent Maturation of Cortical Functions.* In Levy, William B. (ed.); Anderson, James A. (ed.); Lehmkuhle, Stephen (ed.). Synaptic Modifications, Neuron Selectivity, and Nervous System Organization. pp 35-64. Hillsdale, NJ: Lawrence Erlbaum.

[Singer, W 87*] SEE: Frohn, H 87a. *Addendum: A Self-Organizing Neural Network Sharing Features of the Mammalian Visual System.*

[Singer, W 88*] SEE: von der Malsburg, C 88c. *Principles of Cortical Network Organization.*

[Singhal, S 87*] SEE: Landauer, TK 87a. *Teaching a Minimally Structured Back-Propagation Network to Recognize Speech Sounds.*

[Singleton, RC 62a] Singleton, Richard C. (1962). *A Test for Linear Separability as Applied to Self-Organizing Machines.* Self Organizing Systems. pp 503-524. Washington, DC: Spartan.

[Sirletti, B 88*] SEE: Verleysen, M 88a. *A High-Storage Content-Addressable Memory and Its Learning Algorithm.*

[Sirletti, B 88*] SEE: Verleysen, M 88b. *A Large VLSI Fully Interconnected Neural Network.*

[Sirletti, B 88*] SEE: Verleysen, M 88c. *A Large VLSI Hopfield Network for Pattern Recognition Problems.*

[Sirletti, B 88*] SEE: Verleysen, M 88g. *Neural Network or High-Storage Content-Addressable Memory: VLSI Circuit and Learning Algorithm.*

[Sirletti, B 88*] SEE: Verleysen, M 88d. *A New CMOS Architecture for Neural Networks.*

[Sirletti, B 88*] SEE: Verleysen, M 88e. *A New VLSI Architecture for Large Hopfield's Neural Networks.*

[Sirletti, B 88*] SEE: Verleysen, M 88f. *A New VLSI Architecture for Neural Associative Memories.*

[Sirletti, B 88a] Sirletti, Bruno; Verleysen, Michel; Vandemeulebroecke, Andre; Jespers, Paul. (1988). *An Algorithm for Pattern Recognition with VLSI Neural Networks.* Meeting of the International Neural Network Society (INNS), 1st, Boston, MA, September 6-10, 1988: Proceedings.

[Sivilotti, MA 85a] Sivilotti, Massimo A.; Emerling, Michael R.; Mead, Carver A. (1985). *A Novel Associative Memory Implemented Using Collective Computation.* Chapel Hill Conference on Very Large Scale Integration, 1985: Proceedings. pp 329-342. Rockville, MD: Computer Science Press.

[Sivilotti, MA 86a] Sivilotti, Massimo A.; Emerling, Michael R.; Mead, Carver A. (1986). *VLSI Architectures for Implementation of Neural Networks.* In Denker, John S. (ed.). Neural Networks for Computing, Snowbird, Utah, 1986. AIP Proceedings 151. pp 408-413. New York: American Institute of Physics.

[Sivilotti, MA 87a] Sivilotti, Massimo A.; Mahowald, Michelle A.; Mead, Carver A. (1987). *Real-Time Visual Computations Using Analog CMOS Processing Arrays.* Conference on Advanced Research in VLSI, March, 1987: Proceedings. pp

295-312.

[Sivilotti, MA 87b] Sivilotti, Massimo A.; Mahowald, Michelle A.; Mead, Carver A. (1988). *Real-time Visual Computations Using Analog CMOS Processing Arrays*. In Anderson, James A. (ed.); Rosenfeld, Edward (ed.). Neurocomputing: Foundations of Research. pp 703-712. Cambridge, MA: MIT Press.

[Skarda, CA 87a] Skarda, Christine A.; Freeman, Walter J. (1987). *How Brains Make Chaos in Order to Make Sense of the World*. Behavioral and Brain Sciences. Volume 10. pp 161-173.

[Sklansky, J 75*] SEE: Kimme, C 75a. *Finding Circles by an Array of Accumulators.*

[Sklansky, J 80*] SEE: Hassner, M 80a. *The Use of Markov Random Fields as Models of Texture.*

[Sklansky, J 81a] Sklansky, Jack; Wassel, Gustav N. (1981). Pattern Classifiers and Trainable Machines. Berlin: Springer-Verlag.

[Sklansky, J 86a] Sklansky, Jack; Sankar, P. V.; Walter, R. J. (1986). *Biomedical Image Analysis*. In Young, Tzay Y. (ed.); Fu, King-Sun (ed.). Handbook of Pattern Recognition and Image Processing. pp 629-648. New York: Academic Press.

[Skouras, A 84*] SEE: Anninos, PA 84a. *A Computer Model for Learning Processes and the Role of the Cerebral Commissures.*

[Skouras, A 84*] SEE: Anninos, PA 84c. *Noisy Neural Nets Exhibiting Memory Domains.*

[Skrzypek, J 86*] SEE: Gunger, D 86a. *A Connectionist Architecture for Matching 3-D Models to Moving Edge Features.*

[Skrzypek, J 87*] SEE: Heisey, I 87a. *Color Constancy and Early Vision: A Connectionist Model.*

[Skrzypek, J 87*] SEE: Oyster, JM 87a. *Computing Shape with Neural Networks: A Proposal.*

[Skrzypek, J 87*] SEE: Mesrobian, E 87a. *A Connectionist Architecture for Computing Textural Segmentation.*

[Skrzypek, J 87*] SEE: Mesrobian, E 87b. *Discrimination of Natural Textures: A Neural Network Architecture.*

[Skrzypek, J 87*] SEE: Gunger, D 87a. *UCLA PUNNS - A Neural Network Machine for Computer Vision.*

[Skrzypek, J 87*] SEE: Paik, E 87a. *UCLA SFINX - A Neural Network Simulation Environment.*

[Slack, JM 85*] SEE: Aitkenhead, AM 85a. Issues in Cognitive Modeling: A Reader.

[Slagle, JR 87*] SEE: Poliac, MO 87a. *A Crew Scheduling Problem.*

[Slagle, JR 88*] SEE: Irani, EA 88a. Using Artificial Neural Nets for Statistical Discovery: Observations after Using Back-propagation, Expert Systems, and Multiple-Linear Regression.

[Sleeman, D 82a] Sleeman, Derek; Langley, Pat; Mitchell, Tom M. (1982). *Learning from Solution Paths: An Approach to the Credit Assignment Problem*. AI Magazine. Volume 3. pp 48-52.

[Small, SL 80a] Small, Steven L. (1980). Word Expert Parsing: A Theory of Distributed Word-Based Natural Language Understanding. Report TR954. College Park, MD: University of Maryland, Department of Computer Science.

[Small, SL 82a] Small, Steven L. (1982). *Exploded Connections: Unchunking Schematic Knowledge*. Cognitive Science Society, 4th, Ann Arbor, MI, August 4-6, 1982: Proceedings. Hillsdale, NJ: Lawrence Erlbaum.

[Small, SL 82b] Small, Steven L.; Shastri, Lokendra; Brucks, M.; Kaufman, S. A. (1982). ISCON: An Interactive Simulator for Connectionist Networks. Report TR109. Rochester, NY: University of Rochester, Department of Computer Science.

[Small, SL 82c] Small, Steven L.; Cottrell, Garrison W.; Shastri, Lokendra. (1982). *Toward Connectionist Parsing.* National Conference on Artificial Intelligence (AAAI-82), 1st, Pittsburgh, PA, August 18-20, 1982: Proceedings. pp 247-250.

[Small, SL 83*] SEE: Cottrell, GW 83a. *A Connectionist Scheme for Modeling Word Sense Disambiguation.*

[Small, SL 83a] Small, Steven L. (1983). *A Computer Model of Sentence Comprehension.* Technical Report. Rochester, NY: University of Rochester, Department of Computer Science.

[Small, SL 84*] SEE: Cottrell, GW 84a. *Viewing Parsing as Word Sense Discrimination: A Connectionist Approach.*

[Smallwood, J 88*] SEE: Alkon, DL 88b. *Regulation of Hermissenda K+ Channels by Cytoplasmic and Membrane-Associated C-Kinase.*

[Smillie, J 84a] Smillie, John. (1984). *Competitive and Cooperative Tridiagonal Systems of Differential Equations.* SIAM Journal on Applied Mathematics. Volume 15. pp 530-534.

[Smith, AB 87a] Smith, Andrew B. (1987). *A Parallel PDP Network Simulator.* IEEE International Conference on Neural Networks (ICNN-87), San Diego, CA, June 21-24, 1987: Proceedings. Volume 3. pp 377-384. New York: Institute of Electrical and Electronics Engineers.

[Smith, AVW 87*] SEE: Murray, AF 87a. *Asynchronous Arithmetic for VLSI Neural Systems.*

[Smith, AVW 88*] SEE: Murray, AF 88b. *Asynchronous VLSI Neural Networks Using Pulse-Stream Arithmetic.*

[Smith, AVW 88*] SEE: Murray, AF 88c. *Bit-Serial Neural Networks.*

[Smith, AVW 88*] SEE: Murray, AF 88d. *Fully-Programmable Analogue VLSI Devices for the Implementation of Neural Networks.*

[Smith, AVW 88*] SEE: Murray, AF 88a. A Novel Computational and Signalling Method for VLSI Neural Networks.

[Smith, AVW 88*] SEE: Murray, AF 88e. *VLSI Neural Networks.*

[Smith, BH 76*] SEE: Schmitt, FO 76a. *Electrotonic Processing of Information by Brain Cells.*

[Smith, D 87*] SEE: Gallant, SI 87b. *Random Cells: An Idea Whose Time Has Come and Gone... And Come Again?.*

[Smith, DR 62a] Smith, D. R.; Davidson, C. H. (1962). *Maintained Activity in Neural Nets.* Journal of Association of Computing Machinery. Volume 9. pp 268-278.

[Smith, DR 67*] SEE: Brousil, JK 67a. *A Threshold Logic Network for Shape Invariance.*

[Smith, EE 71*] SEE: Aderman, D 71a. *Expectancy as a Determinant of Functional Units in Perceptual Recognition.*

[Smith, EE 73*] SEE: Rips, LJ 73a. *Semantic Distance and the Verification of Semantic Relations.*

[Smith, EE 75*] SEE: Spoehr, KT 75a. *The Role of Orthographic and Phonotactic Rules in Perceiving Letter Patterns.*

[Smith, FW 63*] SEE: Widrow, B 63b. *Practical Applications for Adaptive Data-Processing Systems.*

[Smith, FW 64*] SEE: Widrow, B 64b. *Pattern-Recognizing Control Systems.*

[Smith, FW 64*] SEE: Widrow, B 64c. *Pattern-Recognizing Control Systems.*

[Smith, FW 66a] Smith, Fred W. (1966). *A Trainable Non-Linear Function Generator.* IEEE Transactions on Automatic Control. Volume AC-11. pp 212-218.

[Smith, FW 66b] Smith, Fred W. (1966). *Design of Quasi-Optimal Minimum-Time Controllers.* IEEE Transactions on Automatic Control. Volume AC-11. pp 71-77.

[Smith, HL 86a] Smith, Hal L. (1986). *Periodic Competitive Differential Equations and the Discrete Dynamics of Competitive Maps.* Journal of Differential Equations. Volume 64. pp 165-194.

[Smith, HL 86b] Smith, Hal L. (1986). *Periodic Solutions of Periodic Competitive and Cooperative Systems.* SIAM Journal on Applied Mathematics. Volume 17. pp 1289-1318.

[Smith, JO 88*] SEE: Kamangar, FA 88a. *A Neural Network Based Controller for Robotic Manipulators.*

[Smith, KW 87*] SEE: Horne, W 87a. *Application-Oriented Connection Systems for Pattern Acquisition and Recognition.*

[Smith, LS 88*] SEE: Phillips, WA 88a. *On the Acquisition of Object Concepts from Sensory Data.*

[Smith, MS 76*] SEE: Perkel, DH 76b. *A Computer Program for Simulating a Network of Interacting Neurons, Part II: Programming Aspects.*

[Smith, PT 76a] Smith, Philip T.; Baker, Robert G. (1976). *The Influence of English Spelling Patterns on Pronunciation.* Journal of Verbal Learning and Verbal Behavior. Volume 15. pp 267-285.

[Smith, RG 86a] Smith, Robert G.; Freed, Michael A.; Sterling, Peter. (1986). *Microcircuitry of the Dark-Adapted Cat Retina: Functional Architecture of the Rod-Cone Network.* Journal of Neuroscience. Volume 6. pp 3505-3517.

[Smith, RG 87*] SEE: Sterling, P 87a. *Microcircuitry of the On-Beta Ganglion Cell in Daylight, Twilight, and Starlight.*

[Smith, RG 87*] SEE: Freed, MA 87a. *Rod Bipolar Array in the Cat Retina: Pattern of Input from Rods and GABA-Accumulating Amacrine Cells.*

[Smith, RG 88*] SEE: Sterling, P 88a. *Architecture of Rod and Cone Circuits to the On-Beta Ganglion Cell.*

[Smith, SR 88*] SEE: Wheeler, BC 88a. *Neural Network Models of an Escape Response.*

[Smith, T 87a] Smith, T.; Pitney, G. A.; Greenwood, D. (1987). *Application of Genetic Algorithm to a Neural Network for Finding Optimal Paths.* IEEE International Conference on Neural Networks (ICNN-87), San Diego, CA, June 21-24, 1987: Proceedings. Volume 4. pp 505-506. New York: Institute of Electrical and Electronics Engineers.

[Smith, VW 88*] SEE: Butler, ZF 88a. *VLSI Bit-Serial Neural Networks.*

[Smith, WE 85a] Smith, Warren E.; Paxman, Richard G.; Barrett, Harrison H. (1985). *Image Reconstruction from Coded Data: I. Reconstruction Algorithms and Experimental Results.* Journal of the Optical Society of America A. Volume 2. pp 491-500.

[Smolensky, P 83a] Smolensky, Paul. (1983). Harmony Theory: A Mathematical Framework for Stochastic Parallel Processing. ICS-8306. San Diego, CA: University of California, San Diego, Institute for Cognitive Science.

[Smolensky, P 83b] Smolensky, Paul. (1983). *Schema Selection and Stochastic Inference in Modular Environments.* National Conference on Artificial Intelligence (AAAI-83), 2nd, Washington, DC, August 22-26, 1983: Proceedings. pp 378-382. Hillsdale, NJ: Lawrence Erlbaum.

[Smolensky, P 83c] Smolensky, Paul. (1983). *Schema Selection and Stochastic Inference in Modular Environments.* National Conference on Artificial Intelligence (AAAI-83), 2nd, Washington, DC, August 22-26, 1983: Proceedings. pp

378-382. Washington, DC: American Association for Artificial Intelligence.

[Smolensky, P 84*] SEE: Hinton, GE 84c. Parallel Computation and the Mass-Sprint Model of Motor Control.

[Smolensky, P 84a] Smolensky, Paul; Riley, Mary S. (1984). Harmony Theory: Problem Solving, Parallel Cognitive Models, and Thermal Plastics. ICS-8404. San Diego, CA: University of California, San Diego, Institute for Cognitive Science.

[Smolensky, P 84b] Smolensky, Paul. (1984). *The Mathematical Role of Self-Consistency in Parallel Computation.* Cognitive Science Society, 6th, Boulder, CO, June 28-30, 1984: Proceedings. Hillsdale, NJ: Lawrence Erlbaum.

[Smolensky, P 86*] SEE: Rumelhart, DE 86j. *Schemata and Sequential Thought Processes in PDP Models.*

[Smolensky, P 86a] Smolensky, Paul. (1986). *Information Processing in Dynamical Systems: Foundations of Harmony Theory.* In Rumelhart, David E. (ed.). Parallel Distributed Processing: Explorations in the Microstructure of Cognition. Volume I, Foundations. pp 194-281. Cambridge, MA: MIT Press.

[Smolensky, P 86b] Smolensky, Paul. (1986). *Neural and Conceptual Interpretation of PDP Models.* In Rumelhart, David E. (ed.). Parallel Distributed Processing: Explorations in the Microstructure of Cognition. Volume II. pp 390-431. Cambridge, MA: MIT Press.

[Smolensky, P 87a] Smolensky, Paul. (1987). *On the Connectionist Reduction of Conscious Rule Interpretation.* Cognitive Science Society, 9th, Seattle, WA, July 16-18, 1987: Proceedings. pp 187-194. Hillsdale, NJ: Lawrence Erlbaum.

[Smolensky, P 87b] Smolensky, Paul. (1987). On the Proper Treatment of Connectionism. CU-CS-395-87. Boulder, CO: University of Colorado, Department of Computer Science.

[Smolensky, P 89a] Smolensky, Paul. (1989). *Connectionist Modeling: Neural Computation/Mental Connections.* In Nadel, Lynn (ed.); Cooper, Lynn A. (ed.); Culicover, Peter (ed.); Harnish, R. Michael (ed.). Neural Connections, Mental Computation. pp 49-67. Cambridge, MA: MIT Press.

[Smythe, EJ 87a] Smythe, Erich J. (1987). *Detection of Format Transitions by Connectionist Network.* IEEE International Conference on Neural Networks (ICNN-87), San Diego, CA, June 21-24, 1987: Proceedings. Volume 4. pp 495-504. New York: Institute of Electrical and Electronics Engineers.

[Snir, M 83*] SEE: Gottlieb, A 83a. *The NYU Ultracomputer-Designing an MIMD Shared Memory Parallel Computer.*

[Snyder, L 86*] SEE: Segall, Z 86a. *Workshop on Performance Efficient Parallel Programming.*

[Snyder, W 88*] SEE: Bilbro, GL 88a. *Image Segmentation with Neurocomputers.*

[Soares, A 87*] SEE: Raffel, JI 87a. *A Generic Architecture for Wafer-Scale Neuromorphic Systems.*

[Soffer, BH 85*] SEE: Dunning, GJ 85a. *All-Optical Associative Holographic Memory with Feedback Using Phase Conjugate Mirrors.*

[Soffer, BH 86*] SEE: Owechko, Y 86b. *Associative Memory in a Phase Conjugate Resonator Cavity Utilizing a Hologram.*

[Soffer, BH 86*] SEE: Owechko, Y 86a. *Associative Memory in a Phase Conjugate Resonator Cavity Utilizing a Hologram.*

[Soffer, BH 86*] SEE: Dunning, GJ 86a. *Optical Holographic Associative Memory Using a Phase Conjugate Resonator.*

[Soffer, BH 86a] Soffer, Bernard H.; Dunning, G. J.; Owechko, Yuri; Marom, E. (1986). *Associative Holographic Memory with Feedback Using Phase-Conjugate Mirrors.* Optics Letters. Volume 11. pp 118-120.

[Soffer, BH 87*] SEE: Dunning, GJ 87a. *All-Optical Associative Memory with Shift Invariance and Multiple-Image Recall.*

[Soffer, BH 87*] SEE: Owechko, Y 87a. *Holographic Associative Memory with Nonlinearities in the Correlation Domain.*

[Soffer, BH 87*] SEE: Owechko, Y 87c. *Video Rate Optical Associative Memory with Shift Invariance.*

[Soffer, BH 88*] SEE: Owechko, Y 88a. *Optoelectronic Neural Networks Based on Holographically Interconnected Image Processors.*

[Solla, S 87*] SEE: Denker, JS 87a. *Automatic Learning, Rule Extraction, and Generalization.*

[Solla, SA 86*] SEE: Hertz, JA 86a. *Memory Networks with Asymmetric Bonds.*

[Solla, SA 86a] Solla, S. A.; Sorkin, G. B.; White, S. R. (1986). *Configuration Space Analysis for Optimization Problems.* In Bienenstock, Elie L. (ed.). Disordered Systems and Biological Organization. pp 283-294. New York: Springer-Verlag.

[Solla, SA 87*] SEE: Jackel, LD 87a. *Building a Hierarchy with Neural Networks: An Example-Image Vector Quantization.*

[Solnick, B 84a] Solnick, Bennett; Davis, Thomas L.; Sterling, Peter. (1984). *Numbers of Specific Types of Neuron in Layer IVab of Cat Striate Cortex.* Proceedings of the National Academy of Sciences. Volume 81. pp 3898-3900.

[Solomon, J 88*] SEE: Krieger, D 88a. *A Neurobiologically Based Neural Net Model of the Hippocampus.*

[Solomonoff, RJ 57a] Solomonoff, R. J. (1957). *An Inductive Inference Machine.* IRE National Convention Record. pp 56-62.

[Somani, AK 87a] Somani, Arun K.; Penla, Narong. (1987). *Compact Neural Network.* IEEE International Conference on Neural Networks (ICNN-87), San Diego, CA, June 21-24, 1987: Proceedings. Volume 3. pp 191-200. New York: Institute of Electrical and Electronics Engineers.

[Somani, AK 88a] Somani, Arun K.; Rostykus, Phyllis L. (1988). Evaluation of a Hypercube Based Compact Neural Network. Seattle, WA: University of Washington, Department of Electrical Engineering.

[Somani, AK 88b] Somani, Arun K.; Penla, Narong. (1988). Reduced Interconnection Neural Networks. Seattle, WA: University of Washington, Department of Electrical Engineering.

[Sompolinsky, H 85*] SEE: Amit, DJ 85a. *Spin-Glass Models of Neural Networks.*

[Sompolinsky, H 87*] SEE: Meidan, M 87a. Electronic Analog Neural Network with Noise and Time Delays.

[Sondhi, MM 83*] SEE: Levinson, SE 83a. *An Introduction to the Application of the Theory of Probabilistic Functions of a Markov Process to Automatic Speech Recognition.*

[Sondhi, MM 85*] SEE: Juang, BH 85a. *Recent Developments in the Application of Hidden Markov Models to Speaker-Independent Isolated Word Recognition.*

[Sorkin, GB 86*] SEE: Solla, SA 86a. *Configuration Space Analysis for Optimization Problems.*

[Sornborger, AT 88a] Sornborger, Andrew T. *Attention Based Neural Network.* Meeting of the International Neural Network Society (INNS), 1st, Boston, MA, September 6-10, 1988: Proceedings.

[Soucek, B 83a] Soucek, Branko; Prohorov, Sergej. (1983). *Event-Train Correlation and Real-Time Microcomputer Systems.* Microprocessing and Microprogramming. Volume 11. pp 23-29.

[Soucek, B 88a] Soucek, Branko; Soucek, Marina. (1988). Neural and Massively Parallel Computers: The Sixth Generation. New York: Wiley.

[Soucek, M 88*] SEE: Soucek, B 88a. Neural and Massively Parallel Computers: The Sixth Generation.

[Soules, G 70*] SEE: Baum, LE 70a. *A Maximization Technique Occurring in the Statistical Analysis of Probabilistic Functions of Markov Chains.*

[Soulie, FF 86*] SEE: Bienenstock, EL 86a. Disordered Systems and Biological Organization.

[Soulie, FF 86a] Soulie, F. Fogelman. (1986). *Lyapunov Functions and Their Use in Automata Networks.* In Bienenstock, Elie L. (ed.). Disordered Systems and Biological Organization. pp 85-100. New York: Springer-Verlag.

[Soulie, FF 87a] Soulie, F. Fogelman; Gallinari, P.; Le Cun, Y.; Thiria, S. (1987). *Evaluation of Network Architectures on Test Learning Tasks.* IEEE International Conference on Neural Networks (ICNN-87), San Diego, CA, June 21-24, 1987: Proceedings. Volume 2. pp 653-660. New York: Institute of Electrical and Electronics Engineers.

[Sourlas, N 84*] SEE: Mezard, M 84a. *Nature of the Spin-Glass Phase.*

[Sourlas, N 84*] SEE: Mezard, M 84b. *Replica Symmetry Breaking and the Nature of the Spin Glass Phase.*

[Sparrow, C 80a] Sparrow, C. (1980). *Bifurcation and Chaotic Behaviour in Simple Feedback Systems.* Journal of Theoretical Biology. Volume 83. pp 93-105.

[Specht, DF 63*] SEE: Widrow, B 63b. *Practical Applications for Adaptive Data-Processing Systems.*

[Specht, DF 67a] Specht, Donald F. (1967). *Generation of Polynomial Discriminant Functions for Pattern Recognition.* IEEE Transactions on Electronic Computers. Volume EC-16. pp 308-319.

[Specht, DF 67b] Specht, Donald F. (1967). *Vectorcardiographic Diagnosis Using the Polynomial Discriminant Method of Pattern Recognition.* IEEE Transactions on Bio-Medical Engineering. Volume BME-14. pp 90-95.

[Specht, DF 88a] Specht, Donald F. (July 1988). *Probabilistic Neural Networks for Classification, Mapping, or Associative Memory.* IEEE International Conference on Neural Networks, July 1988, San Diego, California (Preprint).

[Spence, CD 88*] SEE: Gelfand, JJ 88a. *Multisensor Integration in Biological Systems.*

[Spence, CD 88*] SEE: Pearson, JC 88a. *Neural Network Approach to Sensory Fusion.*

[Spencer, EG 86a] Spencer, E. G. (1986). *Programmable Bistable Switches and Resistors for Neural Networks.* In Denker, John S. (ed.). Neural Networks for Computing, Snowbird, Utah, 1986. AIP Proceedings 151. pp 414-425. New York: American Institute of Physics.

[Sperkreijse, H 84*] SEE: Regan, D 84a. Systems Approach in Vision: Proceedings of a Workshop, Amsterdam, The Netherlands, 27-29 August 1984.

[Sperling, G 81a] Sperling, George. (1981). *Mathematical Models of Binocular Vision.* Mathematical Psychology and Psychophysiology. SIAM-AMS Proceedings Volume 13. pp 281-300. Providence, RI: American Mathematical Society.

[Sperling, G 87a] Sperling, George. (1987). *A Neuro-Biological Review.* IEEE International Conference on Neural Networks (ICNN-87), San Diego, CA, June 21-24, 1987: Proceedings. Tutorial. New York: Institute of Electrical and Electronics Engineers.

[Spinelli, DN 70a] Spinelli, D. N. (1970). *OCCAM: A Computer Model for a Content-Addressable Memory in the Central Nervous Systems.* Biology of Memory. pp 293-306. New York: Academic Press.

[Spinelli, DN 76*] SEE: Arbib, MA 76a. *Neural Models and Memory.*

[Spiro, P 87*] SEE: Mueller, P 87a. *Neural Computation of Visual Images.*

[Spitzer, AR 88*] SEE: Hassoun, MH 88c. Pseudo-Unsupervised Neural Network Based Automated EMG Decomposition.

[Spitzer, AR 88a] Spitzer, A. Robert; Cohen, Leonardo; Fabrikant, Judy; Hallett, Mark. (in press). A Method for Determining Optimal Interelectrode Spacing for Cerebral Topographic Mapping.

[Spitzer, F 71a] Spitzer, Frank. (1971). *Markov Random Fields and Gibbs Ensembles*. American Mathematical Monthly. Volume 78. pp 142-154.

[Spoehr, KT 75a] Spoehr, Kathryn T.; Smith, Edward E. (1975). *The Role of Orthographic and Phonotactic Rules in Perceiving Letter Patterns*. Journal of Experimental Psychology: Human Perception and Performance. Volume 104. pp 21-34.

[Spohrer, JC 83*] SEE: Brown, PF 83a. *Bayesian Adaptation in Speech Recognition.*

[Spragins, J 66a] Spragins, John. (1966). *Learning Without a Teacher*. IEEE Transactions on Information Theory. Volume IT-12. pp 223-230.

[Squires, KC 77*] SEE: Squires, NK 77a. *Bisensory Stimulation: Inferring Decision-Related Processes from the P300 Component.*

[Squires, NK 77a] Squires, Nancy K.; Donchin, Emanuel; Squires, Kenneth C. (1977). *Bisensory Stimulation: Inferring Decision-Related Processes from the P300 Component*. Journal of Experimental Psychology: Human Perception and Performance. Volume 3. pp 299-315.

[Srihari, SN 88*] SEE: Lee, D 88a. Experiments with Neural Network Models for Handwritten Digit Recognition.

[Srihari, SN 88*] SEE: Pawlicki, TF 88a. *Neural Network Models and Their Application to Handwritten Digit Recognition.*

[St Jacques, J 85*] SEE: Abu-Mostafa, YS 85b. *Information Capacity of the Hopfield Model.*

[St John, MF 87*] SEE: McClelland, JL 87b. *Reconstructive Memory for Sentences: A PDP Approach.*

[St John, MF 88a] St John, Mark F.; McClelland, James L. (1988). Learning and Applying Contextual Constraints in Sentence Comprehension. Technical Report AIP-39. Pittsburgh, PA: Carnegie Mellon University, Computer Science Department.

[Stafeev, VI 80*] SEE: Batalina, MA 80a. *Neuristor Line Made of S-Type Germanium Diodes.*

[Stafylopatis, A 87*] SEE: Gelenbe, E 87a. *Temporal Behavior of Neural Networks.*

[Stafylopatis, A 88a] Stafylopatis, Andreas; Dikaiakos, M.; Kontoravdis, D. (1988). *Spatial Organization of Neural Networks: A Probabilistic Modeling Approach*. In Anderson, Dana Z. (ed.). Neural Information Processing Systems. pp 740-749. New York: American Institute of Physics.

[Standley, DL 88*] SEE: Wyatt Jr, JL 88b. *Circuit Design Criteria for Stable Lateral Inhibition Neural Networks.*

[Standley, DL 88*] SEE: Wyatt Jr, JL 88a. *A Method for the Design of Stable Lateral Inhibition Networks that is Robust in the Presence of Circuit Parasitics.*

[Standley, DL 88a] Standley, D. L.; Wyatt Jr, John L. (1988). *Stability Theorem for Lateral Inhibition Networks that is Robust in the Presence of Circuit Parasitics*. IEEE International Conference on Neural Networks (ICNN-88), San Diego, CA, July 24-27, 1988: Proceedings. pp 27-36. New York: Institute of Electrical and Electronics Engineers.

[Stanley, JC 76a] Stanley, J. C. (1976). *Simulation Studies of a Temporal Sequence Memory Model*. Biological Cybernetics. Volume 24. pp 121-137.

[Stark, L 59a] Stark, Lawrence; Baker, Frank. (1959). *Stability and Oscillations in a Neurological Servomechanism.*

Journal of Neurophysiology. Volume 22. pp 156-164.

[Stark, L 67*] SEE: Cook, G 67a. *Derivation of a Model for the Human Eye-Positioning Mechanism.*

[Starmer, CF 87a] Starmer, C. F. (1987). *Characterizing Synaptic Plasticity with an Activity Dependent Model.* IEEE International Conference on Neural Networks (ICNN-87), San Diego, CA, June 21-24, 1987: Proceedings. Volume 4. pp 3-10. New York: Institute of Electrical and Electronics Engineers.

[Stearns, SD 85*] SEE: Widrow, B 85a. Adaptive Signal Processing.

[Steels, L 87a] Steels, Luc. (1987). *Self-Organization Through Selection.* IEEE International Conference on Neural Networks (ICNN-87), San Diego, CA, June 21-24, 1987: Proceedings. Volume 2. pp 55-62. New York: Institute of Electrical and Electronics Engineers.

[Steels, L 88a] Steels, Luc. (1988). *Steps Toward Common Sense.* European Conference on Artificial Intelligence (ECAI-88): Proceedings. pp 49-54. London: Pitman.

[Steenstrup, M 88*] SEE: Vichniac, GY 88a. *A Neural Network for the Optimization of Communications Network Design.*

[Stefanelli, R 88*] SEE: Negrini, R 88a. *Fault-tolerance in Imaging-oriented Systolic Arrays.*

[Stefik, M 81a] Stefik, Mark. (1981). *Planning and Meta-Planning (MOLGEN: Part 2).* Artificial Intelligence. Volume 16. pp 141-169.

[Stefik, M 81b] Stefik, Mark. (1981). *Planning with Constraints (MOLGEN: Part 1).* Artificial Intelligence. Volume 16. pp 111-139.

[Stein, DL 84*] SEE: Palmer, RG 84a. *Models of Hierarchically Constrained Dynamics for Glassy Relaxation.*

[Stein, DL 85*] SEE: Ogielski, AT 85a. *Dynamics on Ultrametric Spaces.*

[Stein, RB 72a] Stein, R. B. (1972). *The Stochastic Properties of Spike Trains Recorded from Nerve Cells.* Stochastic Point Processes: Satistical Analysis, Theory, and Applications. pp 700-731. New York: Wiley.

[Steinbuch, K 61a] Steinbuch, K. (1961/1963). *Die Lernmatrix.* Kybernetik. Volume 1. pp 36-45.

[Steinbuch, K 63*] SEE: Kazmierczak, H 63a. *Adaptive Systems in Pattern Recognition.*

[Steinbuch, K 63a] Steinbuch, K.; Piske, U. A. W. (1963). *Learning Matrices and Their Applications.* IEEE Transactions on Electronic Computers. Volume EC-12. pp 846-862.

[Steinbuch, K 63b] Steinbuch, K.; Zendeh, F. (1963). *Self-Correcting Translator Circuits.* Information Processing 1962, Proceedings of IFIP Congress 62; Organized by the International Federation for Information Processing, Munich, Aug 27-Sept 1,1962. pp 359-366. Amsterdam: North-Holland.

[Steinbuch, K 65a] Steinbuch, K.; Widrow, Bernard. (1965). *A Critical Comparison of Two Kinds of Adaptive Classification Networks.* IEEE Transactions on Electronic Computers. Volume EC-14. pp 737-740.

[Stell, M 79*] SEE: Levy, WB 79a. *Associative-Contingency Learning and Neuronal Networks.*

[Stenning, K 88a] Stenning, Keith; Levy, Joe. (1988). *Knowledge-Rich Solutions to the Binding Problem: A Simulation of Some Human Computational Mechanisms.* Knowledge-Based Systems. Volume 1. pp 143-152.

[Stent, GS 73a] Stent, Gunther S. (1973). *A Physiological Mechanism for Hebb's Postulate of Learning.* Proceedings of the National Academy of Sciences. Volume 70. pp 997-1001.

[Stent, GS 78*] SEE: Friesen, WO 78a. *Neural Circuits for Generating Rhythmic Movements.*

[Stephens, KR 87*] SEE: Hutchison, WR 87b. *The Airline Marketing Tactician (AMT): A Commercial Application of Adaptive Networking.*

[Stephens, KR 87*] SEE: Hutchison, WR 87a. *Integration of Distributed and Symbolic Knowledge Representations.*

[Stephens, KR 88a] Stephens, Kenneth R.; Hutchison, William R. (in press). *Behavior Analysis and Intelligent Machines.* Cybernetic.

[Sterling, P 84*] SEE: McGuire, BA 84a. *Microcircuitry of Bipolar Cells in Cat Retina.*

[Sterling, P 84*] SEE: Solnick, B 84a. *Numbers of Specific Types of Neuron in Layer IVab of Cat Striate Cortex.*

[Sterling, P 86*] SEE: McGuire, BA 86a. *Microcircuitry of Beta Ganglion Cells in Cat Retina.*

[Sterling, P 86*] SEE: Smith, RG 86a. *Microcircuitry of the Dark-Adapted Cat Retina: Functional Architecture of the Rod-Cone Network.*

[Sterling, P 87*] SEE: Freed, MA 87a. *Rod Bipolar Array in the Cat Retina: Pattern of Input from Rods and GABA-Accumulating Amacrine Cells.*

[Sterling, P 87a] Sterling, Peter; Cohen, Ethan; Freed, Michael A.; Smith, Robert G. (1987). *Microcircuitry of the On-Beta Ganglion Cell in Daylight, Twilight, and Starlight.* Neuroscience Research. Supplement 6. pp S269-S285.

[Sterling, P 88*] SEE: Freed, MA 88a. *The On-Alpha Ganglion Cell of the Cat Retina and Its Presynaptic Cell Types.*

[Sterling, P 88a] Sterling, Peter; Freed, Michael A.; Smith, Robert G. (1988). *Architecture of Rod and Cone Circuits to the On-Beta Ganglion Cell.* Journal of Neuroscience. Volume 8. pp 623-642.

[Sterling, S 81*] SEE: Homa, D 81a. *Limitations of Exemplar-Based Generalization and the Abstraction of Categorical Information.*

[Sternberg, RJ 86a] Sternberg, Robert J. (1986). *Inside Intelligence.* American Scientist. Volume 74. pp 137-143.

[Sternberg, S 69a] Sternberg, Saul. (1969). *Memory-Scanning: Mental Processes Revealed by Reaction-Time Experiments.* American Scientist. Volume 57. pp 421-457.

[Stevens, CF 68a] Stevens, Charles F. (1968). *Synaptic Physiology.* Proceedings of the IEEE. Volume 56. pp 916-930.

[Stevens, JK 84*] SEE: McGuire, BA 84a. *Microcircuitry of Bipolar Cells in Cat Retina.*

[Stevens, JK 85a] Stevens, John K. (1985). *Reverse Engineering the Brain.* BYTE. Volume 10. pp 287-299.

[Stevens, JK 86*] SEE: McGuire, BA 86a. *Microcircuitry of Beta Ganglion Cells in Cat Retina.*

[Stevens, K 62*] SEE: Halle, M 62a. *Speech Recognition: A Model and a Program for Research.*

[Stevens, ME 61a] Stevens, M. E. (1961). Automatic Character Recognition: A State-of-the-Art Report. Technical Note Number 112. Frederick, MD: National Bureau of Standards.

[Stewart, CV 87a] Stewart, Charles V.; Dyer, Charles R. (1987). *A Connectionist Model for Stereo Vision.* IEEE International Conference on Neural Networks (ICNN-87), San Diego, CA, June 21-24, 1987: Proceedings. Volume 4. pp 215-224. New York: Institute of Electrical and Electronics Engineers.

[Stewart, CV 87b] Stewart, Charles V.; Dyer, Charles R. (1987). Local Constraint Integration in a Connectionist Model of

Stereo Vision. Technical Report #726. Madison, WI: University of Wisconsin, Computer Sciences Department.

[Stewart, CV 88a] Stewart, Charles V.; Dyer, Charles R. (1988). Simulation of a Connectionist Stereo Algorithm on a Shared-Memory Multiprocessor. Technical Report #760. Madison, WI: University of Wisconsin, Computer Sciences Department.

[Stewart, CV 88b] Stewart, Charles V.; Dyer, Charles R. (1988). The Trinocular General Support Algorithm: A Three Camera Stereo Algorithm for Overcoming Binocular Matching Errors. Technical Report #768. Madison, WI: University of Wisconsin, Computer Sciences Department.

[Stiles, GS 85a] Stiles, G. S.; Denq, Dong-Lih. (1985). *On the Effect of Noise on the Moore-Penrose Generalized Inverse Associative Memory.* IEEE Transactions on Pattern Analysis and Machine Intelligences. Volume PAMI-7. pp 358-360.

[Stiney, M 87*] SEE: Brown, PB 87a. *Analog Memory for Continuous-Voltage, Discrete-Time Implementations of Neural Networks.*

[Stirk, CW 87a] Stirk, Charles W.; Rovnyak, S. M.; Athale, Ravindra A. (1987). *Effects of System Noise on an Optical Implementation of an Additive Lateral Inhibitory Network.* IEEE International Conference on Neural Networks (ICNN-87), San Diego, CA, June 21-24, 1987: Proceedings. Volume 3. pp 615-624. New York: Institute of Electrical and Electronics Engineers.

[Stirk, CW 88*] SEE: Athale, RA 88a. *Compact Architectures for Adaptive Neural Nets.*

[Stirk, CW 88*] SEE: Rovnyak, SM 88a. *Optical Implementation of Winner-Take-All Networks: Noise Considerations.*

[Stokes, BT 83*] SEE: Hemami, H 83a. *Four Neural Circuit Models and Their Role in the Organization of Voluntary Movement.*

[Stoll, HM 87*] SEE: Lee, LS 87a. A Continuous-Time Optical Neural Network Associative Memory.

[Stoll, HM 87a] Stoll, H. M.; Lee, L. S. (1987). A Continuous-Time Optical Neural Network. Palos Verdes Peninsula, CA: Northrop Research and Technology Center.

[Stone, GO 86*] SEE: Grossberg, S 86i. *Neural Dynamics of Attention Switching and Temporal Order Information in Short-Term Memory.*

[Stone, GO 86*] SEE: Grossberg, S 86l. *Neural Dynamics of Word Recognition and Recall: Attentional Priming, Learning, and Resonance.*

[Stone, GO 86*] SEE: Grossberg, S 86m. *Neural Dynamics of Word Recognition and Recall: Attentional Priming, Learning and Resonance.*

[Stone, GO 86a] Stone, Gregory O. (1986). *Analysis of the Delta Rule and Learning of Statistical Associations.* In Rumelhart, David E. (ed.). Parallel Distributed Processing: Explorations in the Microstructure of Cognition. Volume I, Foundations. pp 444-459. Cambridge, MA: MIT Press.

[Stoner, WW 87*] SEE: Brill, MH 87a. *Retinal Model with Adaptive Contrast Sensitivity and Resolution.*

[Stonham, TJ 76*] SEE: Fairhurst, MC 76a. *A Classification System for Alpha-Numeric Characters Based on Learning Network Techniques.*

[Stonham, TJ 87*] SEE: Austin, J 87c. The ADAM Associative Memory.

[Stonham, TJ 87*] SEE: Austin, J 87b. *Distributed Associative Memory for Use in Scene Analysis.*

[Stork, DG 85a] Stork, David G.; Falk, David S.; Brill, Dieter R. (1985). Seeing the Light: Optics in Nature,

Photography, Color, Vision, and Holography. New York: Harper and Row.

[Stork, DG 86*] SEE: Saylor, J 86a. *Parallel Analog Neural Networks for Tree Searching.*

[Stork, DG 87*] SEE: Cohen, MA 87c. *Recent Developments in a Neural Model of Real-Time Speech Analysis and Synthesis.*

[Stork, DG 87a] Stork, David G. (1987). *Counterpropagation Networks: Adaptive Hierarchical Networks for 'Near Optimal' Mappings.* Synapse Connection. Volume 1. pp 9-11, 17.

[Stork, DG 87b] Stork, David G. (1987). *Exaptation, Preadaptation, Evolution and Biological Neural Networks.* Synapse Connection. Volume 1. pp 2-5.

[Stork, DG 88a] Stork, David G. (1988). *A Brief Introduction to the Operation of an Adaptive Resonance Network.* Synapse Connection. Volume 2. pp 1-7.

[Stork, DG 88b] Stork, David G. (1988). *Complex Systems-Operational Approaches in Neurobiology, Physics, and Computers.* American Journal of Physics. Volume 56. pp 573-574.

[Stork, DG 88c] Stork, David G. (1988). *Recirculation Networks.* Synapse Connection. Volume 2. pp 5-6.

[Stork, DG 88d] Stork, David G. (1988). *Stability of Neural Networks for Optimization.* Synapse Connection. Volume 2. pp 12-13.

[Stornetta, WS 87a] Stornetta, W. Scott; Huberman, Bernardo A. (1987). *An Improved Three-Layer, Back Propagation Algorithm.* IEEE International Conference on Neural Networks (ICNN-87), San Diego, CA, June 21-24, 1987: Proceedings. Volume 2. pp 637-644. New York: Institute of Electrical and Electronics Engineers.

[Stornetta, WS 88a] Stornetta, W. Scott; Hogg, Tad; Huberman, Bernardo A. (1988). *A Dynamical Approach to Temporal Pattern Processing.* In Anderson, Dana Z. (ed.). Neural Information Processing Systems. pp 750-759. New York: American Institute of Physics.

[Straughan, JH 54*] SEE: Estes, WK 54a. *Analysis of a Verbal Conditioning Situation in Terms of Statistical Learning Theory.*

[Straughn, BL 86*] SEE: Jackel, LD 86a. *Artificial Neural Networks for Computing.*

[Straughn, BL 86*] SEE: Jackel, LD 86b. *High Resolution Micro-Fabrication and Neural Networks.*

[Straughn, BL 86*] SEE: Graf, HP 86a. *VLSI Implementation of a Neural Network Memory with Several Hundreds of Neurons.*

[Stromberg, W 86a] Stromberg, W.; Farr, T. (1986). *A Fourier-Based Textual Feature Extraction Procedure.* IEEE Transactions on Geoscience and Remote Sensing. Volume GE-24. pp 722-731.

[Strong, GW 87a] Strong, Gary W.; Whitehead, Bruce A. (1987). *A Factored Architecture to Solve Illusory Conjunctions in Neural Networks.* IEEE International Conference on Neural Networks (ICNN-87), San Diego, CA, June 21-24, 1987: Proceedings. Volume 2. pp 463-470. New York: Institute of Electrical and Electronics Engineers.

[Stroud, N 87*] SEE: Forrest, BM 87a. *Implementing Neural Network Models on Parallel Computers.*

[Stroud, N 88*] SEE: Gardner, E 88a. *Training with Noise: Application to Word and Text Storage.*

[Strugala, M 87a] Strugala, M.; Tavangarian, Djamshid; Waldschmidt, K.; Roll, G. (1987). *An Associative Processor as a Design Rule Check Accelerator.* VLSI and Computers, Proceedings of the 1st International Conference on Computer Technology, Systems and Applications, Hamburg, May 11-15, 1987. pp 426-431. New York: Institute of Electrical and Electronics Engineers.

[Studdert-Kennedy, M 80a] Studdert-Kennedy, Michael. (1980). *Speech Perception*. Language and Speech. Volume 23. pp 45-66.

[Stuller, JA 76a] Stuller, J. A.; Kurz, B. (1976). *Two-Dimensional Markov Representations of Sampled Images*. IEEE Transactions on Communications. Volume COM-24. pp 1148-1152.

[Suen, CY 86a] Suen, Ching Y. (1986). *Character Recognition by Computer and Applications*. In Young, Tzay Y. (ed.); Fu, King-Sun (ed.). Handbook of Pattern Recognition and Image Processing. pp 569-587. New York: Academic Press.

[Sugie, N 82*] SEE: Sawada, R 82a. *Self-Organization of Neural Nets with Competitive and Cooperative Interaction.*

[Sugie, N 82a] Sugie, Noboru. (1982). *Neural Models of Brightness Perception and Retinal Rivalry in Binocular Vision.* Biological Cybernetics. Volume 43. pp 13-21.

[Sugie, N 86*] SEE: Shibata, T 86a. *Recovering Three-Dimensional Structure and Motion of Jointed Objects from Orthographically Projected Optical Flow.*

[Sugie, N 86*] SEE: Muraki, S 86a. *Speech Prosthesis System for Laryngectomees Using "Electropalatograph" and a Small Keyboard.*

[Sugie, N 87a] Sugie, Noboru; Huang, Jie; Ohnishi, Noboru. (1987). Localizing Sound Source by Incorporating Biological Auditory Mechanism. Nagoya: Nagoya University, Electrical Engineering Department.

[Sugihara, K 86*] SEE: Shibata, T 86a. *Recovering Three-Dimensional Structure and Motion of Jointed Objects from Orthographically Projected Optical Flow.*

[Sullivan, M 85a] Sullivan, Michael; Cohen, Paul R. (1985). *An Endorsement-Based Plan Recognition Program.* International Joint Conference on Artificial Intelligence (IJCAI-85), 9th, Los Angeles, CA, August 18-23, 1985: Proceedings. Volume 1. pp 475-479. Los Altos, CA: Morgan Kaufmann.

[Sullivan, WE 88*] SEE: Gelfand, JJ 88a. *Multisensor Integration in Biological Systems.*

[Sullivan, WE 88*] SEE: Pearson, JC 88a. *Neural Network Approach to Sensory Fusion.*

[Sun, GZ 86*] SEE: Chen, HH 86a. *High Order Correlation Model for Associative Memory.*

[Sun, GZ 87a] Sun, G. Z.; Chen, H. H.; Lee, Y. C. (1987). *Learning Stereopsis with Neural Networks.* IEEE International Conference on Neural Networks (ICNN-87), San Diego, CA, June 21-24, 1987: Proceedings. Volume 4. pp 345-356. New York: Institute of Electrical and Electronics Engineers.

[Sun, GZ 88a] Sun, G. Z.; Lee, Y. C.; Chen, H. H. (1988). *A Novel Net that Learns Sequential Decision Process.* In Anderson, Dana Z. (ed.). Neural Information Processing Systems. pp 760-766. New York: American Institute of Physics.

[Suner, S 88*] SEE: Sanes, JN 88a. *Rapid Reorganization of Adult Rat Motor Cortex Somatic Representation Patterns after Motor Nerve Injury.*

[Sung, FP 83*] SEE: Cooper, DB 83a. *Multiple-Window Parallel Adaptive Boundary Finding in Computer Vision.*

[Sutcliffe, RFE 86*] SEE: Sharkey, NE 86a. *Mixing Binary and Continuous Connection Schemes for Knowledge Access.*

[Sutton III, GG 88*] SEE: D'Autrechy, CL 88a. *A General-Purpose Simulation Environment for Developing Connectionist Models.*

[Sutton III, GG 88*] SEE: Reggia, JA 88a. *Self-Processing Networks and Their Biomedical Implications.*

[Sutton, JP 88a] Sutton, Jeffrey P.; Beis, J. S.; Trainor, L. E. H. (1988). *Hierarchical Model of Memory and Memory Loss*. Journal of Physics A: Mathematics General. Volume 21. pp 4443-4454.

[Sutton, RS 78a] Sutton, Richard S. (1978). *Single Channel Theory: A Neuronal Theory of Learning*. Brain Theory Newsletter. Volume 3. pp 72-75.

[Sutton, RS 81*] SEE: Barto, AG 81a. *Associative Search Network: A Reinforcement Learning Associative Memory*.

[Sutton, RS 81*] SEE: Barto, AG 81b. Goal Seeking Components for Adaptive Intelligence: An Initial Assessment.

[Sutton, RS 81*] SEE: Barto, AG 81c. *Landmark Learning: An Illustration of Associative Search*.

[Sutton, RS 81a] Sutton, Richard S.; Barto, Andrew G. (1981). *An Adaptive Network that Constructs and Uses an Internal Model of its World*. Cognition and Brain Theory. Volume 4. pp 217-246.

[Sutton, RS 81b] Sutton, Richard S.; Barto, Andrew G. (1981). *Toward A Modern Theory of Adaptive Networks: Expectation and Prediction*. Psychological Review. Volume 88. pp 135-170.

[Sutton, RS 82*] SEE: Barto, AG 82a. Adaptive Neurons-Like Elements that Can Solve Difficult Learning Control Problems.

[Sutton, RS 82*] SEE: Barto, AG 82b. *Simulation of Anticipatory Responses in Classical Conditioning by a Neuron-Like Adaptive Element*.

[Sutton, RS 82*] SEE: Barto, AG 82c. *Synthesis of Non-Linear Control Surfaces by a Layered Associative Search Network*.

[Sutton, RS 83*] SEE: Barto, AG 83a. Neural Problem Solving.

[Sutton, RS 83*] SEE: Barto, AG 83b. *Neuron-like Adaptive Elements that Can Solve Difficult Learning Control Problems*.

[Sutton, RS 83*] SEE: Barto, AG 83c. *Neuron-like Elements That Can Solve Difficult Learning and Control Problems*.

[Sutton, RS 84*] SEE: Barto, AG 84b. Simulation Experiments with Goal-Seeking Adaptive Elements.

[Sutton, RS 84a] Sutton, Richard S. (1984). Temporal Aspects of Credit Assignment in Reinforcement Learning: PhD Thesis. Amherst, MA: University of Massachusetts.

[Sutton, RS 84b] Sutton, Richard S. (1984). Temporal Credit Assignment in Reinforcement Learning. COINS Technical Report 84-02. Amherst, MA: University of Massachusetts, Computer and Information Science.

[Sutton, RS 85*] SEE: Moore, JW 85a. *Connectionistic Learning in Real Time: Sutton-Barto Adaptive Element and Classical Conditioning of the Rabbit Nictitating Membrane Response*.

[Sutton, RS 85*] SEE: Barto, AG 85d. *Neural Problem Solving*.

[Sutton, RS 85a] Sutton, Richard S.; Pinette, B. (1985). *Learning of World Models by Connectionist Networks*. Cognitive Science Society, 7th, Irvine, CA, August 15-17, 1985: Proceedings.

[Sutton, RS 86a] Sutton, Richard S. (1986). *Two Problems with Back-Propagation and Other Steepest-Descent Learning Procedures for Networks*. Cognitive Science Society, 8th, Amherst, MA, August 15-17, 1986: Proceedings. pp 823-831. Hillsdale, NJ: Lawrence Erlbaum.

[Sutton, RS 87a] Sutton, Richard S.; Barto, Andrew G. (in press). *A Temporal-Difference Model of Classical Conditioning*. Cognitive Science Society, 9th, Seattle, WA, July 16-18, 1987: Proceedings.

[Suyama, K 88*] SEE: Satyanarayana, S 88a. *Analog MOS Circuit Techniques in the VLSI Implementation of Neural Networks*.

[Suzuji, H 88a] Suzuji, Hisashi; Arimoto, Suguru. (1988). *Self-Organization of Associative Database and Its Applications*. In Anderson, Dana Z. (ed.). Neural Information Processing Systems. pp 767-774. New York: American Institute of Physics.

[Suzuki, H 88a] Suzuki, Hisashi; Arimoto, Suguru. (1988). *Self-Organizing Model for Pattern Learning and Its Application to Robot Eyesight*. Conference on Artificial Intelligence Applications, 4th, San Diego, CA, March 14-18, 1988: Proceedings. pp 236-241. Washington, DC: Institute of Electrical and Electronics Engineers.

[Suzuki, H 88b] Suzuki, Hisashi; Arimoto, Suguru. (1988). *Self-Organizing Model of Learning and Its Application to Eyesight System of Robot*. International Workshop on Artificial Intelligence for Industrial Applications 1988. pp 411-416. New York: Institute of Electrical and Electronics Engineers.

[Suzuki, H 88c] Suzuki, Hisashi; Arimoto, Suguru. (1988). *Visual Control of Autonomous Mobile Robot Based on Self-Organizing Model for Pattern Learning*. Journal of Robotic Systems. Volume 5. New York: Wiley.

[Suzuki, J 76a] Suzuki, Jouji. (1976). *Speech Processing by Splicing of Autocorrelation Function*. IEEE International Conference on Acoustics, Speech, and Signal Processing (ICASSP-76), April, 1976: Proceedings. pp 713-716. New York: Institute of Electrical and Electronics Engineers.

[Suzuki, J 77a] Suzuki, Jouji. (1977). *Speech Processing System by the use of Short-Time Cross-Correlation Function*. IEEE International Conference on Acoustics, Speech, and Signal Processing (ICASSP-77), Hartford, CT, May 9-11, 1977: Proceedings. pp 24-27. New York: Institute of Electrical and Electronics Engineers.

[Suzuki, R 67a] Suzuki, Ryoji. (1967). *Mathematical Analysis and Application of Iron-Wire Neuron Model*. IEEE Transactions on Bio-Medical Engineering. Volume BME-14. pp 114-124.

[Suzuki, R 71a] Suzuki, Ryoji; Katsuno, Ikuko; Matano, Keiichi. (1971). *Dynamics of "Neuron Ring"*. Kybernetik. Volume 8. pp 39-45.

[Suzuki, R 87*] SEE: Kawato, M 87a. *A Hierarchical Model for Voluntary Movement and Its Application to Robotics*.

[Suzuki, R 87*] SEE: Kawato, M 87b. *A Hierarchical Neural-Network Model for Control and Learning of Voluntary Movement*.

[Suzuki, R 88*] SEE: Kawato, M 88a. *Hierarchical Neural Network Model for Voluntary Movement with Application to Robotics*.

[Swain, PH 86a] Swain, Philip H. (1986). *Remote Sensing*. In Young, Tzay Y. (ed.); Fu, King-Sun (ed.). Handbook of Pattern Recognition and Image Processing. pp 613-628. New York: Academic Press.

[Szeliski, R 85a] Szeliski, Richard; Hinton, Geoffrey E. (1985). *Solving Random-Dot Stereograms Using the Heat Equation*. IEEE Computer Society Conference on Computer Vision and Pattern Recognition, San Francisco, CA, June 19-23, 1985: Proceedings. pp 284-288. Silver Spring, MD: Institute of Electrical and Electronics Engineers.

[Szeliski, R 86a] Szeliski, Richard. (1986). Cooperative Algorithms for Solving Random Dot Stereograms. CMU-CS-86-133. Pittsburgh, PA: Carnegie Mellon University, Computer Science Department.

[Szeliski, R 87*] SEE: Matthies, L 87a. Kalman Filter-Based Algorithms for Estimating Depth from Image Sequences.

[Szentagothai, J 67*] SEE: Eccles, JC 67a. The Cerebellum as a Neuronal Machine.

[Szentagothai, J 68a] Szentagothai, John. (1968). *Structuro-Functional Considerations of the Cerebella Neuron Network*. Proceedings of the IEEE. Volume 56. pp 960-968.

[Szentagothai, J 73*] SEE: Pellionisz, AJ 73a. *Dynamic Single Unit Simulation of a Realistic Cerebellar Network Model*.

[Szentagothai, J 74*] SEE: Pellionisz, AJ 74a. *Dynamic Single Unit Simulation of a Realistic Cerebellar Network Model*.

II. Purkinje Cell Activity within the Basic Circuit and Modified by Inhibitory Sys..

[Szentagothai, J 74a] Szentagothai, John; Arbib, Michael A. (1974). *Conceptual Models of Neural Organization.* Neuroscience Research Program Bulletin. Volume 12. pp 307-510.

[Szu, HH 85*] SEE: Messner, RA 85a. *An Image Processing Architecture for Real Time Generation of Scale and Rotation Invariant Patterns.*

[Szu, HH 85a] Szu, Harold H. (1985). *Globally Connected Network Models for Computing Using Fine-Grained Processing Elements.* Proceedings of the International Conference on Lasers 1985. pp 92-97.

[Szu, HH 86*] SEE: Athale, RA 86b. *Attentive Associative Memory and Its Optical Implementation.*

[Szu, HH 86*] SEE: Athale, RA 86c. *Optical Implementation of Associative Memory with Controlled Nonlinearity in the Correlation Domain.*

[Szu, HH 86a] Szu, Harold H.; Messner, Richard A. (1986). *Adaptive Invariant Novelty Filters.* Proceedings of the IEEE. Volume 74. pp 518-519.

[Szu, HH 86b] Szu, Harold H. (1986). *Computed Tomography for Optical Computing.* In Szu, Harold H. (ed.). Optical and Hybrid Computing. SPIE Volume 634. pp 480-484. Bellingham, WA: Society of Photo-Optical Instrumentation Engineers.

[Szu, HH 86c] Szu, Harold H. (1986). *Fast Simulated Annealing.* In Denker, John S. (ed.). Neural Networks for Computing, Snowbird, Utah, 1986. AIP Proceedings 151. pp 420-425. New York: American Institute of Physics.

[Szu, HH 86d] Szu, Harold H. (1986). *Holographic Coordinate Transformations and Optical Computing.* In Szu, Harold H. (ed.). Optical and Hybrid Computing. SPIE Volume 634. pp 480-484. Bellingham, WA: Society of Photo-Optical Instrumentation Engineers.

[Szu, HH 86e] Szu, Harold H. (1986). *Iterative Restoration Algorithms for Nonlinear Constraint Computing.* In Szu, Harold H. (ed.). Optical and Hybrid Computing. SPIE Volume 634. pp 112-119. Bellingham, WA: Society of Photo-Optical Instrumentation Engineers.

[Szu, HH 86f] Szu, Harold H. (1986). *Nonlinear Signal Processing Using Fiber-Optics Neurograms.* In Szu, Harold H. (ed.). Optical and Hybrid Computing. SPIE Volume 634. pp 120-133. Bellingham, WA: Society of Photo-Optical Instrumentation Engineers.

[Szu, HH 86g] Szu, Harold H. (ed.). (1986). Optical and Hybrid Computing. SPIE Volume 634. Bellingham, WA: Society of Photo-Optical Instrumentation Engineers.

[Szu, HH 86h] Szu, Harold H. (1986). *Three Layers of Vector Outer Product Neural Networks for Optical Pattern Recognition.* In Szu, Harold H. (ed.). Optical and Hybrid Computing. SPIE Volume 634. pp 312-330. Bellingham, WA: Society of Photo-Optical Instrumentation Engineers.

[Szu, HH 87*] SEE: Scheff, K 87a. *1-D Optical Cauchy Machine Infinite Film Spectrum Search.*

[Szu, HH 87*] SEE: Hartley, RL 87a. *A Comparison of the Computational Power of Neural Network Models.*

[Szu, HH 87*] SEE: Scheff, K 87b. *Simulation Results of Multiple Layer Difference Learning.*

[Szu, HH 87a] Szu, Harold H.; Hartley, Ralph L. (1987). *Fast Simulated Annealing.* Physics Letters A. Volume 122. pp 157-162. New York: Elsevier.

[Szu, HH 87b] Szu, Harold H.; Landa, Joseph. (1987). *Future Perspectives on Neural Network Research.* IEEE International

Conference on Neural Networks (ICNN-87), San Diego, CA, June 21-24, 1987: Proceedings. Volume 1. pp 159-???. New York: Institute of Electrical and Electronics Engineers.

[Szu, HH 87c] Szu, Harold H. (1987). *Non-Convex Optimization.* Real Time Signal Processing IX. SPIE Volume 698. pp 59-65. Bellingham, WA: Society of Photo-Optical Instrumentation Engineers.

[Szu, HH 87d] Szu, Harold H. (1987). *Non-Convex Optimization.* SPIE. Bellingham, WA: Society of Photo-Optical Instrumentation Engineers.

[Szu, HH 87e] Szu, Harold H.; Hartley, Ralph L. (1987). *Nonconvex Optimization by Fast Simulated Annealing.* Proceedings of the IEEE. Volume 75. pp 1538-1540.

[Szu, HH 87f] Szu, Harold H.; Hartley, Ralph L. (1987). *Nonconvex Optimization by Fast Simulated Annealing.* Proceedings of the IEEE. Volume 75. pp 1538-1540.

[Szu, HH 87g] Szu, Harold H.; Caulfield, H. John. (1987). *Optical Expert Systems.* Applied Optics. Volume 26. pp 1943-1947.

[Szu, HN 87*] SEE: Landa, J 87a. *Binocular Fusion Using Simulated Annealing.*

[Tabatabai, A 88*] SEE: Troudet, T 88a. *An Adaptive Neural Net Approach to the Segmentation of Mixed Gray-Level and Binary Pictures.*

[Taber, WR 87a] Taber, W. R.; Siegel, M. A. (1987). *Estimation of Expert Weights Using Fuzzy Cognitive Maps.* IEEE International Conference on Neural Networks (ICNN-87), San Diego, CA, June 21-24, 1987: Proceedings. Volume 2. pp 319-326. New York: Institute of Electrical and Electronics Engineers.

[Taber, WR 88*] SEE: Deich, RO 88a. *Applications of Neural Networks to Underwater Acoustic Signatures: Fuzzy Decision Making and the Neuron Ring.*

[Taber, WR 88a] Taber, W. R.; Deich, R. O. (in press). *A Comparison of Feedforward Neural Networks: Fuzzy Operators and Acoustic Ship Signatures.* Journal of Intelligent Systems.

[Taber, WR 88b] Taber, W. R.; Deich, R. O. (1988). *Fuzzy Sets and Neural Networks.* Joint Technology Workshop on Neural Networks and Fuzzy Logic, 1st, Houston, TX, May 2-3, 1988: Proceedings.

[Taber, WR 88c] Taber, W. R.; Deich, R. O. (1988). *Risk Weighting in Knowledge Bases Using Fuzzy Cognitive Maps.* Artificial Intelligence and Advanced Computer Technology Conference, 4th, Long Beach, CA, May 4-6, 1988: Proceedings.

[Taber, WR 88d] Taber, W. R.; Deich, R. O.; Simpson, Patrick K.; Fagg, A. H. (1988). *The Recognition of Orca Calls with a Neural Network.* International Workship on Fuzzy System Applications, Iizuka City, Japan, August 21-24, 1988: Proceedings.

[Tabor, L 86*] SEE: Cohen, NJ 86a. *Skill Learning and Repetition Priming in Symmetry Detection: Parallel Studies of Human Subjects and Connectionist Models.*

[Tackitt, MC 87*] SEE: Lee, LS 87a. A Continuous-Time Optical Neural Network Associative Memory.

[Tagliaferri, R 88*] SEE: Caianiello, ER 88a. *The Inverse Problem for Linear Boolean Nets.*

[Tagliarini, GA 87a] Tagliarini, Gene A.; Page, Edward W. (1987). *Solving Constraint Satisfaction Problems with Neural Networks.* IEEE International Conference on Neural Networks (ICNN-87), San Diego, CA, June 21-24, 1987: Proceedings. Volume 3. pp 741-748. New York: Institute of Electrical and Electronics Engineers.

[Tagliarini, GA 88*] SEE: Page, EW 88a. *Algorithm Development for Neural Networks.*

[Tagliarini, GA 88a] Tagliarini, Gene A.; Page, Edward W. (1988). *A Neural-Network Solution to the Concentrator Assignment Problem.* In Anderson, Dana Z. (ed.). Neural Information Processing Systems. pp 775-782. New York: American Institute of Physics.

[Tahkoor, AP 88*] SEE: Moopenn, A 88a. *Programmable Synaptic Chip for Electronic Neural Networks.*

[Tai, H 88*] SEE: Jong, T 88a. *Hopfield Model with Complementary Binary Representations.*

[Tai, H 88a] Tai, H-M; Jong, T-L. (1988). *Neural Networks with Higher-Order Nonlinearity.* Electronics Letters. Volume 24. pp 1225-1226.

[Tajchman, G 88*] SEE: MacGregor, RJ 88a. *Theory of Dynamic Similarity in Neuronal Systems.*

[Takahashi, K 74*] SEE: Okuda, M 74a. *A Dynamical Behaviour of Active Regions in Randomly Connected Neural Networks.*

[Takahashi, K 84a] Takahashi, Kouro; Matsuo, Tadayuki. (1984). *Integration of Multi-Microelectrode and Interface Circuits by Silicon Planar and Three-Dimensional Fabrication Technology.* Sensors and Actuators. Volume 5. pp 89-99.

[Takeda, M 86a] Takeda, M.; Goodman, G. (1986). Neural Networks for Computation. EE Preprint. Stanford, CA: Stanford University, Electrical Engineering Department.

[Takefuji, Y 87*] SEE: Foo, YS 87a. *A Neural Network Scheme for 3-Dimensional Pattern Recognition and Construction.*

[Takefuji, Y 87a] Takefuji, Yoshiyasu; Hollis, Paul W.; Foo, Yoon-Pin Simon; Cho, Yong B. (1987). *Error Correcting System Based on Neural Circuits.* IEEE International Conference on Neural Networks (ICNN-87), San Diego, CA, June 21-24, 1987: Proceedings. Volume 3. pp 293-300. New York: Institute of Electrical and Electronics Engineers.

[Takefuji, Y 88*] SEE: Foo, YS 88a. *Integer Linear Programming Neural Networks for Job-Shop Scheduling.*

[Takefuji, Y 88*] SEE: Foo, YS 88b. *Stochastic Neural Networks for Solving Job-Shop Scheduling: Part 1. Problem Representation.*

[Takefuji, Y 88*] SEE: Foo, YS 88c. *Stochastic Neural Networks for Solving Job-Shop Scheduling: Part 2. Architecture and Simulations.*

[Takefuji, Y 88a] Takefuji, Yoshiyasu; Jannarone, Robert; Cho, Yong B.; Chen, Tatung. (1988). *Multinomial Conjunctoid Statistical Learning Machines.* International Symposium on Computer Architecture, 15th, Honolulu, HI, May 30-June 2, 1988: Proceedings. pp 12-16. New York: Institute of Electrical and Electronics Engineers.

[Takeuchi, A 78*] SEE: Amari, S 78a. *Mathematical Theory on Formation of Category Detecting Nerve Cells.*

[Takeuchi, A 79*] SEE: Amari, S 79a. *Theory of Self-Organizing Nerve Nets.*

[Takeuchi, A 79a] Takeuchi, Akikazu; Amari, Shun-ichi. (1979). *Formation of Topographic Maps and Columnar Microstructures in Nerve Fields.* Biological Cybernetics. Volume 35. pp 63-72.

[Takeuchi, Y 78a] Takeuchi, Yasuhiro; Adachi, Norihiko; Tokumaru, Hidekatsu. (1978). *The Stability of Generalized Volterra Equations.* Journal of Mathematical Analysis and Applications. Volume 62. pp 453-473.

[Talbert, LR 63*] SEE: Widrow, B 63b. *Practical Applications for Adaptive Data-Processing Systems.*

[Talbert, LR 63a] Talbert, Lee R.; Groner, G. F.; Koford, J. S.; Brown, P. J.; Law, P. R.; Mays, C. Hugh. (1963). A Real-Time Adaptive Speech-Recognition System. Technical Report Number 6760-1. Stanford, CA: Stanford University, Stanford Electronics Laboratory.

[Talbert, LR 70a] Talbert, Lee R. (1970). *The Sum-Line Extrapolative Algorithm and its Application to Statistical*

Classification Problems. IEEE Transactions on Systems, Science, and Cybernetics. Volume SSC-6. pp 229-239.

[Tam, JY 85*] SEE: Cruz, CA 85a. NEP: An Emulation-Assist Processor for Parallel Associative Networks.

[Tam, JY 86*] SEE: Cruz, CA 86a. *Flow-of-Activation Processing: Parallel Associative Networks (PAN).*

[Tam, JY 87*] SEE: Hanson, WA 87b. *CONE - Computational Network Environment.*

[Tam, JY 87*] SEE: Cruz, CA 87b. *Knowledge Processing Through Flow-of-Activation.*

[Tam, JY 87*] SEE: Cruz, CA 87c. *Neural Network Emulation Hardware Design Considerations.*

[Tamura, H 81*] SEE: Nagashino, H 81a. *Relations Between Initial Conditions and Periodic Firing Modes in Reciprocal Inhibition Neural Networks.*

[Tamura, H 84a] Tamura, Hiroyuki; Kondo, Tadashi. (1984). *On Revised Algorithms of GMDH With Applications.* In Farlow, Stanley J. (ed.). Self-Organizing Methods in Modeling: GMDH Type Algorithms. pp 225-242. New York: Marcel Dekker.

[Tamura, PN 77a] Tamura, P. N.; Wyant, J. C. (1977). *Matrix Multiplication Using Coherent Optical Techniques.* In Casasent, David P. (ed.); Sawchuk, Alexander A. (ed.). Optical Information Processing: Real-Time Devices and Novel Techniques. SPIE Volume 83. p 97 f. Palo Verde Hills, CA: Society of Photo-Optical Instrumentation Engineers.

[Tan, M 88a] Tan, Ming; Eshelman, Larry. (1988). The Impact of Noise on Learning. CMU-CS-88-144. Pittsburgh, PA: Carnegie Mellon University, Computer Science Department.

[Tanaka, S 87*] SEE: Okajima, K 87a. *A Heteroassociative Memory Network with Feedback Connection.*

[Tanaka, S 88a] Tanaka, Shigeru. (1988). *Theory of Self-Organization of Cortical Maps.* Society of Instrument and Control Engineers of Japan (SICE-88), Narashino, August, 1988: Proceedings. pp 1069-1072.

[Tank, DW 85*] SEE: Hopfield, JJ 85a. *Collective Computation with Continuous Variables.*

[Tank, DW 85*] SEE: Hopfield, JJ 85b. *Computing With Neural Circuits: A Model.*

[Tank, DW 85*] SEE: Gelperim, AE 85a. *The Logic of LIMAX Learning.*

[Tank, DW 85*] SEE: Hopfield, JJ 85c. *Neural Computation of Decisions in Optimization Problems.*

[Tank, DW 86*] SEE: Hopfield, JJ 86a. *Collective Computation with Continuous Variables.*

[Tank, DW 86a] Tank, David W.; Hopfield, John J. (1986). *Simple Neural Optimization Networks: An A/D Converter, Signal Decision Circuit, and Linear Programming Circuit.* IEEE Transactions on Circuits and Systems. Volume CAS-33.

[Tank, DW 87*] SEE: Hopfield, JJ 87c. *Technical Comments: Computing with Neural Networks.*

[Tank, DW 87a] Tank, David W.; Hopfield, John J. (1987). *Collective Computation in Neuronlike Circuits.* Scientific American. Volume 257. pp 104-114.

[Tank, DW 87b] Tank, David W.; Hopfield, John J. (1987). *Concentrating Information in Time: Analog Neural Networks with Applications to Speech Recognition Problems.* IEEE International Conference on Neural Networks (ICNN-87), San Diego, CA, June 21-24, 1987: Proceedings. Volume 4. pp 455-468. New York: Institute of Electrical and Electronics Engineers.

[Tao, KM 87a] Tao, K. Mike. (1987). *Neural Nets and Computer Vision Systems.* IEEE International Conference on Neural

Networks (ICNN-87), San Diego, CA, June 21-24, 1987: Proceedings. Volume 4. pp 643-650. New York: Institute of Electrical and Electronics Engineers.

[Tapia, RA 78a] Tapia, R. A.; Thompson, J. R. (1978). Non-Parametric Probability Density Estimation. Baltimore, MD: Johns Hopkins University.

[Tarassenko, L 88*] SEE: Murray, AF 88d. *Fully-Programmable Analogue VLSI Devices for the Implementation of Neural Networks.*

[Tarrant, R 86*] SEE: Gass, W 86a. *A Parallel Signal Processor System.*

[Tavangarian, D 87*] SEE: Strugala, M 87a. *An Associative Processor as a Design Rule Check Accelerator.*

[Tavangarian, D 88*] SEE: Schulz, M 88a. *An Associative Co-Processor Architecture with the UNIX Operating System for the Acceleration of CAD Tools.*

[Tavangarian, D 88a] Tavangarian, Djamshid. (1988). *Concept of a Flag-Oriented Fully Parallel Associative Processor Based on Flag-Algebra.* GI/ITG Symposium "Architecture of Computing Systems", 10th, Paderborn, 1988.

[Taylor, GA 79*] SEE: Massaro, DW 79a. *Orthographic Regularity, Positional Frequency, and Visual Processing of Letter Strings.*

[Taylor, WK 76*] SEE: Muller, FJ 76a. *A Parallel Processing Model for the Study of Possible Pattern Recognition Mechanisms in the Brain.*

[Tchuente, M 86*] SEE: Robert, Y 86a. *On Some Dynamical Properties of Monotone Networks.*

[Tchuente, M 86a] Tchuente, M. (1986). *Dynamics and Self Organization in One-Dimensional Arrays.* In Bienenstock, Elie L. (ed.). Disordered Systems and Biological Organization. pp 21-32. New York: Springer-Verlag.

[Te Kolste, RD 87a] Te Kolste, Robert D.; Guest, Clark C. (1987). *Optical Competitive Neural Network with Optical Feedback.* IEEE International Conference on Neural Networks (ICNN-87), San Diego, CA, June 21-24, 1987: Proceedings. Volume 3. pp 625-630. New York: Institute of Electrical and Electronics Engineers.

[Tebelski, J 88*] SEE: Derthick, MA 88a. *'Ensemble' Boltzmann Units have Collective Computational Properties like those of Hopfield and Tank Neurons.*

[Teich, MC 83*] SEE: Lachs, G 83a. *A Neural-Counting Model Based on Physiological Characteristics of the Peripheral Auditory System, Part IV, Application to Response.*

[TeKolste, R 87*] SEE: Guest, CC 87a. *Designs and Devices for Optical Bidirectional Associative Memories.*

[Tenorio, MF 87a] Tenorio, Manoel F. (in press). *Extended Fuzzy Cognitive Maps (EFCM) Engines for the Diagnosis of Electro-Mechanical and Electronic Equipment.* ISMM International Conference on Computer Applications in Design, Simulation and Analysis.

[Tenorio, MF 87b] Tenorio, Manoel F. (in press). *Knowledge Acquisition Tool for the Diagnosis of Electro-Mechanical and Electronic Equipment.* ESD/SMI Expert Systems Conference and Exposition, 2nd.

[Tenorio, MF 87c] Tenorio, Manoel F. (in press). *Parallel Stochastic Computing.* ISMM International Conference on Computer Applications in Design, Simulation and Analysis.

[Tenorio, MF 87d] Tenorio, Manoel F.; Hughes, Craig S. (1987). *Real Time Noisy Image Segmentation Using an Artificial Neural Network Model.* IEEE International Conference on Neural Networks (ICNN-87), San Diego, CA, June 21-24, 1987: Proceedings. Volume 4. pp 357-364. New York: Institute of Electrical and Electronics Engineers.

[Tenorio, MF 87e] Tenorio, Manoel F. (in press). *Simulating Nerves to Improve Cochlear Implants for the Hearing Impaired.* ISMM International Conference on Computer Applications in Design, Simulation and Analysis.

[Tenorio, MF 87f] Tenorio, Manoel F. (in press). *Using A Visual Programming Language to Simulate Adaptive Networks.* ISMM International Conference on Computer Applications in Design, Simulation and Analysis.

[Tenorio, MF 87g] Tenorio, Manoel F. (1987). *Using Neural Networks to Improve Cochlear Implant Speech Perception.* In Anderson, Dana Z. (ed.). Neural Information Processing Systems. pp 783-793. New York: American Institute of Physics.

[Tenorio, MF 87h] Tenorio, Manoel F. (in press). *Using SARCENet for Optimal Solutions of Scheduling Problems.* ESD/SMI Expert Systems Conference and Exposition, 2nd.

[Tenorio, MF 87i] Tenorio, Manoel F. (in press). *Using the Hyperbase Intelligent Processor for Fault Analysis and Detection.* ESD/SMI Expert Systems Conference and Exposition, 2nd.

[Tenorio, MF 87j] Tenorio, Manoel F. (in press). *Visual Programming as a User Interfaace for Programming Neural Networks.* ESD/SMI Expert Systems Conference and Exposition, 2nd.

[Tenorio, MF 88*] SEE: Schwartz, RG 88a. *Evaluating a Connectionist Network as a Simulation of Speech Development.*

[Tenorio, MF 88*] SEE: Sayegh, S 88a. Expert System and Neural Network Approaches for Classical Guitar Fingering.

[Tenorio, MF 88a] Tenorio, Manoel F.; Lee, Wei-Tsih. (1988). *Self Organizing Neural Networks for the Identification Problem.* Neural Information Processing Systems. New York: American Institute of Physics.

[Tenorio, MF 88b] Tenorio, Manoel F.; Schwartz, Richard G.; Tom, M. Daniel. (in press). *Unrestricted English Text-to-Speech Conversion with a Distributed Adaptive Network.* Pattern Recognition Letters.

[Termini, S 72*] SEE: De Luca, A 72a. *A Definition of a Nonprobabilistic Entropy in the Setting of Fuzzy Sets Theory.*

[Terry, M 88*] SEE: Renals, S 88a. *A Comparison of Speech Recognition Front Ends Using a Connectionist Classifier.*

[Terry, M 88*] SEE: Rohwer, R 88c. *Unstable Connectionist Networks in Speech Recognition.*

[Tesauro, G 86a] Tesauro, Gerald. (1986). *Simple Neural Models of Classical Conditioning.* Biological Cybernetics. Volume 55. pp 187-200.

[Tesauro, G 87a] Tesauro, Gerald. (1987). *Scaling Relationships in Back-Propagation Learning: Dependence on Training Set Size.* Complex Systems. Volume 1. pp 368-372.

[Tesauro, G 88a] Tesauro, Gerald; Sejnowski, Terrence J. (1988). *A 'Neural' Network that Learns to Play Backgammon.* In Anderson, Dana Z. (ed.). Neural Information Processing Systems. pp 794-803. New York: American Institute of Physics.

[Tesauro, G 88b] Tesauro, Gerald. (1988). *A Plausible Neural Circuit for Classical Conditioning Without Synpatic Plasticity.* Proceedings of the National Academy of Sciences. Volume 85. pp 2830-2833.

[Tesauro, G 88c] Tesauro, Gerald; Janssens, Robert. (1988). *Scaling Relationships in Back-Propagation Learning: Dependence on Predicate Order.* Complex Systems. Volume 2. pp 39-44.

[Thackara, JI 81*] SEE: Warde, C 81a. *Optical Information Processing Characteristics of the Microchannel Spatial Light Modulator.*

[Thackara, JI 83*] SEE: Warde, C 83a. *Operating Modes of the Microchannel Spatial Light Modulator.*

[Thakoor, AP 85*] SEE: Lambe, J 85a. Electronic Device Aspects of Neural Network Memories.

[Thakoor, AP 86*] SEE: Moopenn, A 86a. *Error Correction and Asymmetry in a Binary Memory Matrix.*

[Thakoor, AP 86a] Thakoor, Anilkumar P.; Lamb, J. L.; Moopenn, Alexander; Lambe, John. (1986). *Binary Synaptic Connections Based on Memory Switching in a SI:H.* In Denker, John S. (ed.). Neural Networks for Computing, Snowbird, Utah, 1986. AIP Proceedings 151. pp 426-431. New York: American Institute of Physics.

[Thakoor, AP 87*] SEE: Moopenn, A 87b. *Electronic Implementation of Associative Memory Based on Neural Network Models.*

[Thakoor, AP 87*] SEE: Daud, T 87a. *Neural Network Based Feed-Forward High Density Associative Memory.*

[Thakoor, AP 87*] SEE: Moopenn, A 87a. *A Neurocomputer Based on an Analog-Digital Hybrid Architecture.*

[Thakoor, AP 87*] SEE: Lamb, JL 87a. *Resistive Synaptic Interconnects for Electronic Neural Networks.*

[Thakoor, AP 87a] Thakoor, Anilkumar P.; Lamb, J. L.; Moopenn, Alexander; Khanna, S. K. (1987). *Binary Synaptic Connections Based on Memory Switching in a-Si:H for Artificial Neural Networks.* Materials Research Society Symposia Proceedings. Volume 95. pp 627-632.

[Thakoor, AP 87b] Thakoor, Anilkumar P.; Moopenn, Alexander; Lambe, John; Khanna, S. K. (1987). *Electronic Hardware Implementations of Neural Networks.* Applied Optics. Volume 26. pp 5085-5092.

[Thakoor, AP 88a] Thakoor, Anilkumar P.; Moopenn, Alexander. (1988). Electronic Neurocomputers. NASA Tech Briefs Volume 12. pp 12-13.

[Thathachar, MAL 74*] SEE: Narendra, KS 74a. *Learning Automata: A Survey.*

[Thiria, S 87*] SEE: Soulie, FF 87a. *Evaluation of Network Architectures on Test Learning Tasks.*

[Thomas, G 87*] SEE: Anninos, PA 87c. *A Neural Net Model for the Release of Prolactin During Pseudopregnancy in the Rat.*

[Thomason, MG 86a] Thomason, Michael G. (1986). *Syntactic Pattern Recognition: Stochastic Languages.* In Young, Tzay Y. (ed.); Fu, King-Sun (ed.). Handbook of Pattern Recognition and Image Processing. pp 119-142. New York: Academic Press.

[Thompson, JR 78*] SEE: Tapia, RA 78a. Non-Parametric Probability Density Estimation.

[Thompson, K 86*] SEE: Sage, JP 86a. *Artificial Neural Network Integrated Circuit Based on NMOS/CCD Principles.*

[Tillery, SI 87a] Tillery, Stephen I.; Combs, Nathan. (1987). *The Problem of Internal Representations in Neural Nets: Concepts, Implications and a Proposed Metric.* IEEE International Conference on Neural Networks (ICNN-87), San Diego, CA, June 21-24, 1987: Proceedings. Volume 2. pp 585-592. New York: Institute of Electrical and Electronics Engineers.

[Titchener, PF 81*] SEE: Widrow, B 81a. *Adpative Design of Digital Filters.*

[Titchener, PF 87*] SEE: Widrow, B 87b. *Fundamental Relations Between the LMS Algorithm and the DFT.*

[Todd, JT 86*] SEE: Mingolla, E 86a. *Perception of Solid Shape from Shading.*

[Todorovic, D 87*] SEE: Bullock, D 87b. *Reconciling Stable Asymmetry with Recovery of Function: An Adaptive Systems Perspective on Functional Plasticity.*

[Todorovic, D 88*] SEE: Grossberg, S 88c. *Neural Dynamics of 1-D and 2-D Brightness Perception: A Unified Model of*

Classical and Recent Phenomena.

[Todorovic, D 88*] SEE: Grossberg, S 88b. *Neural Dynamics of 1-D and 2-D Brightness Perception: A Unified Model of Classical and Recent Phenomena.*

[Todorvic, D 87*] SEE: Grossberg, S 87c. *A Neural Network Architecture for Brightness Perception Under Constant and Variable Illumination Conditions.*

[Todtenkopf, AB 86*] SEE: Henshaw, PD 86a. *Artificial Intelligence Applications of Fast Optical Memory Access.*

[Toet, A 87a] Toet, A.; Blom, J.; Koenderink, Jan J. (1987). *The Construction of a Simultaneous Functional Order in Nervous Systems I. Relevance of Signal Covariances and Signal Coincidences.* Biological Cybernetics. Volume 57. pp 115-125.

[Toet, A 87b] Toet, A.; Blom, J.; Koenderink, Jan J. (1987). *The Construction of a Simultaneous Functional Order in Nervous Systems II. Computing Geometrical Structures.* Biological Cybernetics. Volume 57. pp 127-136.

[Toet, A 87c] Toet, A.; Blom, J.; Koenderink, Jan J. (1987). *The Construction of a Simultaneous Functional Order in Nervous Systems III. The Influence of Environmental Constraints on the Resulting Functional Order.* Biological Cybernetics. Volume 57. pp 331-340.

[Toko, K 81a] Toko, Kiyoshi; Nitta, Junsaku; Urahama, Kiichi; Yamafuji, Kaoru. (1981). *Application of Bifurcation Theory to Nerve Excitations.* Transactions of the Institute of Electronics and Communication Engineers of Japan. Volume E64. pp 707-714.

[Tokumaru, H 78*] SEE: Takeuchi, Y 78a. *The Stability of Generalized Volterra Equations.*

[Tokunaga, Y 88*] SEE: Shimohara, K 88b. *Back-Propagation Networks for Event-Driven Temporal Sequence Processing.*

[Tokunaga, Y 88*] SEE: Uchiyama, T 88a. *Boltzmann Machines in which Weights Converge to Global Optimal Solutions.*

[Tokunaga, Y 88*] SEE: Shimohara, K 88a. *A Neural Network System with an Automatic Generation Mechanism for Distorted Patterns.*

[Tokura, T 77a] Tokura, T.; Morishita, Iwao. (1977). *Analysis and Simulation of Double Layer Neural Networks with Mutually Inhibiting Interconnections.* Biological Cybernetics. Volume 22. pp 83-92.

[Toldalagi, PM 84*] SEE: Lebow, WM 84a. *Forecasting Applications of GMDH in Agricultural and Meteorological Time Series.*

[Tolle, H 86*] SEE: Ersu, E 86a. *Acceleration of Learning by Variable Generalization for on-Line Self-Organizing Control.*

[Tolle, H 87*] SEE: Ersu, E 87a. *Learning Control Structures with Neuron-Like Associative Memory Systems.*

[Tolle, H 88*] SEE: Ersu, E 88a. *Hierarchical Learning Control: An Approach with Neuron-Like Associative Memories.*

[Tom, MD 88*] SEE: Schwartz, RG 88a. *Evaluating a Connectionist Network as a Simulation of Speech Development.*

[Tom, MD 88*] SEE: Tenorio, MF 88b. *Unrestricted English Text-to-Speech Conversion with a Distributed Adaptive Network.*

[Tomboulian, S 88a] Tomboulian, Sherryl. (1988). *Introduction to a System for Implementing Neural Net Connections on SIMD Architectures.* In Anderson, Dana Z. (ed.). Neural Information Processing Systems. pp 804-813. New York: American Institute of Physics.

[Tomita, M 85*] SEE: Carbonell, JG 85a. New Approaches to Machine Translation.

[Tomita, T 68a] Tomita, Tsuneo. (1968). *Electrical Response of Single Photoreceptors.* Proceedings of the IEEE. Volume

56. pp 1015-1023.

[Tomlin, RS 87*] SEE: Douglas, SA 87a. *Consistency and Variation in Spatial Reference.*

[Tong, RM 87*] SEE: Yager, RR 87a. Fuzzy Sets and Applications: Selected Papers by L A Zadeh.

[Toomarian, N 87*] SEE: Barhen, J 87a. *Concurrent Neuromorphic Algorithms for Optimization of the Computational Load of a Hypercube Supercomputer.*

[Toomarian, N 87*] SEE: Gulati, S 87a. *Nonlinear Neural Networks for Deterministic Scheduling.*

[Toomarian, N 87*] SEE: Barhen, J 87b. *Optimization of the Computational Load of a Hypercube Supercomputer Onboard a Mobile Robot.*

[Torioka, T 79a] Torioka, Toyoshi. (1979). *Pattern Separability in a Random Neural Net with Inhibitory Connections.* Biological Cybernetics. Volume 34. pp 53-62.

[Torioka, T 79b] Torioka, Toyoshi. (1979). *Pattern Separability of a Tightly Coupled Random Nerve Net with Inhibitory Connections.* Transactions of the Institute of Electronics and Communication Engineers of Japan. Volume E62. pp 864-870.

[Torioka, T 80a] Torioka, Toyoshi. (1980). *Further Consideration on Pattern Separability in a Random Neural Net with Inhibitory Connections.* Biological Cybernetics. Volume 36. pp 203-212.

[Torioka, T 80b] Torioka, Toyoshi. (1980). *Property of the Degree of Pattern Separation in a Two-Layer Random Neural Net with Inhibitory Connections.* Transactions of the Institute of Electronics and Communication Engineers of Japan. Volume E63. pp 590-591.

[Torioka, T 88a] Torioka, Toyoshi; Ineda, N. (1988). *Pattern Separating Functioning of Two Layered Random Nerve Nets with Feedforward Inhibitory Connections.* IEEE Transactions on Systems, Man, and Cybernetics. Volume 18. pp 358-366.

[Torkkola, K 86a] Torkkola, Kari; Riittinen, H.; Kohonen, Teuvo. (1986). *Microprocessor-Based Word Recognizer for a Large Vocabulary.* International Conference on Pattern Recognition, 8th, Paris, October 27-31, 1986: Proceedings. pp 814-816. Washington, DC: Institute of Electrical and Electronics Engineers.

[Torkkola, K 87*] SEE: Kohonen, T 87e. *Microprocessor Implementation of a Large Vocabulary Speech Recognizer and Phonetic Typewriter for Finnish and Japanese.*

[Torkkola, K 88*] SEE: Kohonen, T 88b. *Phonetic Typewriter for Finnish and Japanese.*

[Toronto, AF 64*] SEE: Warner, HR 64a. *Experience with Bayes Theorem for Computer Diagnosis of Congenital Heart Disease.*

[Torras i Genis, C 85a] Torras i Genis, Carme. (1985). Temporal-Pattern Learning in Neural Models. Number 63. Berlin: Springer-Verlag.

[Torras i Genis, C 86a] Torras i Genis, Carme. (1986). *Entrainment in Pacemakers Characterized by a V-Shaped PRC.* Journal of Mathematical Biology. Volume 24. pp 291-312.

[Torras i Genis, C 86b] Torras i Genis, Carme. (1986). *Neural Network Model with Rhythm-Assimilation Capacity.* IEEE Transactions on Systems, Man and Cybernetics. Volume SMC-16. pp 680-693.

[Torras i Genis, C 87a] Torras i Genis, Carme. (1987). *On the Relationship Between Two Models of NeuralEntrainment.* Biological Cybernetics. Volume 57. pp 313-319.

[Torras i Genis, C 88a] Torras i Genis, Carme. (1988). *Exploring Three Possibilities in Network Design: Spontaneous Node*

Activity, Node Plasticity and Temporal Coding. In Eckmiller, Rolf (ed.); von der Malsburg, Christoph (ed.). Neural Computers. NATO ASI Series, Volume F41. pp 301-310. Berlin: Springer-Verlag.

[Torre, V 81*] SEE: Poggio, TA 81a. Microelectronics in Nerve Cells: Dendritic Morphology and Information Processing.

[Torre, V 83*] SEE: Koch, C 83a. *Nonlinear Interactions in a Dendritic Tree: Localization, Timing, and Role in Information Processing.*

[Torre, V 83a] Torre, Vincent; Owen, W. Geoffrey; Sandini, Giulio. (1983). *The Dynamics of Electrically Interacting Cells.* IEEE Transactions on Systems, Man and Cybernetics. Volume SMC-13. pp 757-765.

[Torre, V 84*] SEE: Poggio, TA 84b. Ill-Posed Problems and Regularization Analysis in Early Vision.

[Torre, V 85*] SEE: Poggio, TA 85a. *Computational Vision and Regularization Theory.*

[Toulouse, G 84*] SEE: Mezard, M 84a. *Nature of the Spin-Glass Phase.*

[Toulouse, G 84*] SEE: Mezard, M 84b. *Replica Symmetry Breaking and the Nature of the Spin Glass Phase.*

[Toulouse, G 86*] SEE: Personnaz, L 86a. *A Biologically Constrained Learning Mechanism in Networks of Formal Neurons.*

[Toulouse, G 86*] SEE: Personnaz, L 86f. *Simple Selectionist Learning Rule for Neural Networks.*

[Touretzky, DS 81*] SEE: Fahlman, SE 81a. *Cancellation in a Parallel Semantic Network.*

[Touretzky, DS 85a] Touretzky, David S.; Hinton, Geoffrey E. (1985). *Symbols Among the Neurons: Details of a Connectionist Inference Architecture.* International Joint Conference on Artificial Intelligence (IJCAI-85), 9th, Los Angeles, CA, August 18-23, 1985: Proceedings. Volume 1. pp 238-243. Los Altos, CA: Morgan Kaufmann.

[Touretzky, DS 86a] Touretzky, David S.; Hinton, Geoffrey E. (1986). A Distributed Connectionist Production System. CMU-CS-86-172. Pittsburgh, PA: Carnegie Mellon University, Computer Science Department.

[Touretzky, DS 86b] Touretzky, David S. (1986). *BoltzCONS: Reconciling Connectionism with the Recursive Nature of Stacks and Trees.* Cognitive Science Society, 8th, Amherst, MA, August 15-17, 1986: Proceedings. pp 522-530. Hillsdale, NJ: Lawrence Erlbaum.

[Touretzky, DS 86c] Touretzky, David S. (1986). *Representing and Transforming Recursive Objects in a Neural Network, or "Trees DO Grow on Boltzmann Machines".* IEEE International Conference on Systems, Man, and Cybernetics, Atlanta, GA, October 14-17, 1986: Proceedings. Volume 1. pp 12-16. New York: Institute of Electrical and Electronics Engineers.

[Touretzky, DS 87*] SEE: Rosenfeld, R 87a. *Four Capacity Models for Coarse-Coded Symbol Memories.*

[Touretzky, DS 87*] SEE: Rosenfeld, R 87b. Scaling Properties of Coarse-Coded Symbol Memories.

[Touretzky, DS 87a] Touretzky, David S.; Geva, Shai. (1987). *A Distributed Connectionist Representation for Concept Structures.* Cognitive Science Society, 9th, Seattle, WA, July 16-18, 1987: Proceedings. pp 155-164. Hillsdale, NJ: Lawrence Erlbaum.

[Touretzky, DS 87b] Touretzky, David S. (1987). *Representing Conceptual Structures in a Neural Network.* IEEE International Conference on Neural Networks (ICNN-87), San Diego, CA, June 21-24, 1987: Proceedings. Volume 2. pp 279-286. New York: Institute of Electrical and Electronics Engineers.

[Touretzky, DS 87c] Touretzky, David S.; Derthick, Mark A. (1987). *Symbol Structures in Connectionist Networks: Five Properties and Two Architectures.* Intellectual Leverage: CompCon, 32nd, San Francisco, CA, February 23-27,

1987. pp 30-34.

[Touretzky, DS 88*] SEE: Rosenfeld, R 88a. *Scaling Properties of Coarse-Coded Symbol Memories.*

[Trainor, LEH 88*] SEE: Sutton, JP 88a. *Hierarchical Model of Memory and Memory Loss.*

[Travis, BJ 86a] Travis, Bryan J. (1986). *Layered Neural Network Model Applied to the Auditory System.* In Denker, John S. (ed.). Neural Networks for Computing, Snowbird, Utah, 1986. AIP Proceedings 151. pp 432-439. New York: American Institute of Physics.

[Travis, BJ 87*] SEE: Gremillion, M 87a. *Neural Nets with Complex Structure: A Model of the Visual System.*

[Travis, BJ 87a] Travis, Bryan J. (1987). *A Computational Model of the Auditory System.* IEEE International Conference on Neural Networks (ICNN-87), San Diego, CA, June 21-24, 1987: Proceedings. Volume 4. pp 67-74. New York: Institute of Electrical and Electronics Engineers.

[Treffner, PJ 87a] Treffner, Paul J. (1987). *Ecological Connectionism and Animal-Environment Mutuality.* IEEE International Conference on Neural Networks (ICNN-87), San Diego, CA, June 21-24, 1987: Proceedings. Volume 2. pp 813-???. New York: Institute of Electrical and Electronics Engineers.

[Treffner, PJ 88*] SEE: Gonzales, MEQ 88a. A Naturalistic Approach to Mental Representation.

[Trehub, A 77a] Trehub, Arnold. (1977). *Neuronal Models for Cognitive Processes: Networks for Learning, Perception, and Imagination.* Journal of Theoretical Biology. Volume 65. pp 141-169.

[Trehub, A 86a] Trehub, Arnold; Pollatsek, Alexander. (1986). *Seeing-More-Than-is-There: A Probe of Retinoid Networks.* Cognitive Science Society, 8th, Amherst, MA, August 15-17, 1986: Proceedings. pp 572-584. Hillsdale, NJ: Lawrence Erlbaum.

[Trehub, A 87a] Trehub, Arnold. (1987). *Visual-Cognitive Neuronal Networks.* In Arbib, Michael A. (ed.); Hanson, Allen R. (ed.). Vision, Brain, and Cooperative Computation. pp 623-664. Cambridge, MA: MIT Press.

[Treisman, AM 80a] Treisman, Anne M.; Gelade, Garry. (1980). *A Feature-Integration Theory of Attention.* Cognitive Psychology. Volume 12. pp 97-137.

[Treleaven, PC 88*] SEE: Recce, M 88a. *Parallel Architectures for Neural Computers.*

[Treleaven, PC 88*] SEE: Pacheco, M 88a. *A Simple VLSI Architecture for Neurocomputing.*

[Trenary, R 87a] Trenary, Robert; Conrad, Michael. (1987). *A Neuron Model of a Memory System for Autonomous Exploration of an Environment.* Intelligent Autonomous Systems. pp 601-609. Amsterdam: North-Holland.

[Trepel, L 81*] SEE: Homa, D 81a. *Limitations of Exemplar-Based Generalization and the Abstraction of Categorical Information.*

[Trick, MA 86*] SEE: Bartholdi, JJ 86b. *Stable Matching with Preferences Derived From a Psychological Model.*

[Tromp, JW 86*] SEE: Colvin, ME 86a. *Model for Cortical Function.*

[Troudet, T 88a] Troudet, Thierry; Tabatabai, Ali. (1988). *An Adaptive Neural Net Approach to the Segmentation of Mixed Gray-Level and Binary Pictures.* IEEE International Conference on Neural Networks (ICNN-88), San Diego, CA, July 24-27, 1988: Proceedings. Volume 1. pp 585-592. New York: Institute of Electrical and Electronics Engineers.

[Trussell, HJ 80a] Trussell, H. J. (1980). *The Relationship Between Image Restoration by the Maximum A Posteriori Method and a Maximum Entropy Method.* IEEE Transactions on Acoustics, Speech, and Signal Processing. Volume ASSP-28. pp

114-117.

[Tsai, T 89*] SEE: Kollias, S 89a. *Image Halftoning and Reconstruction Using a Neural Network.*

[Tsai, WT 88*] SEE: Chen, S 88a. A Case Study of Solving Optimization Problems Using Neural Network.

[Tsai, WT 88*] SEE: Xu, X 88b. Constructing Associative Memories Using Neural Networks.

[Tsai, WT 88*] SEE: Xu, X 88a. A Generalized Neural Network Model.

[Tsai, WT 88*] SEE: Xu, X 88c. On "Information Capacity of the Hopfield Model".

[Tsao, T 88a] Tsao, Thomas; Kanal, Laveen N. (1988). *A Novel Architecture for Real Time Pick Up of 3D Motion and 3D Layout Information from the Flow of the Optic Array.* NATO ASI Series, Real-Time Object Measurement and Classification. NATO ASI Series Volume F42. pp 267-280. New York: Springer-Verlag.

[Tsetlin, ML 71*] SEE: Gelfand, IM 71a. *Mathematical Modeling of Mechanisms of the Central Nervous System.*

[Tse-Yun, F 81a] Tse-Yun, Feng. (1981). *A Survey of Interconnection Networks (Concurrent Computing).* IEEE Computer. Volume 14. pp 12-27.

[Tsividis, Y 87a] Tsividis, Y. (1987). *Analog MOS Integrated Circuits - Certain New Ideas, Trends, and Obstacles.* IEEE Journal of Solid-State Circuits. Volume SC-22. pp 317-321.

[Tsividis, Y 87b] Tsividis, Y.; Satyanarayana, Srinagesh. (1987). *Analogue Circuits for Variable-Synapse Electronic Neural Networks.* Electronics Letters. Volume 23. pp 1313-1314.

[Tsividis, Y 87c] Tsividis, Y.; Anastassiou, Dimitris. (1987). *Switched-Capacitor Neural Networks.* Electronics Letters. Volume 23. pp 958-959.

[Tsividis, Y 88*] SEE: Satyanarayana, S 88a. *Analog MOS Circuit Techniques in the VLSI Implementation of Neural Networks.*

[Tsotsos, JK 87a] Tsotsos, John K. (1987). *Representational Axes and Temporal Cooperative Processes.* In Arbib, Michael A. (ed.); Hanson, Allen R. (ed.). Vision, Brain, and Cooperative Computation. pp 361-418. Cambridge, MA: MIT Press.

[Tsukahara, N 72a] Tsukahara, N. (1972). *The Properties of the Cerebello-Pontine Reverberating Circuit.* Brain Research. Volume 40. pp 67-71.

[Tsutsumi, K 84a] Tsutsumi, Kazuyoshi; Matsumoto, Haruya. (1984). *A Synaptic Modification Algorithm in Consideration of the Generation of Rhythmic Oscillation in a Ring Neural Network.* Biological Cybernetcs. Volume 50. pp 419-430.

[Tsutsumi, K 84b] Tsutsumi, Kazuyoshi; Matsumoto, Haruya. (1984). *Ring Neural Network qua a Generator of Rhythmic Oscillation with Period Control Mechanism.* Biological Cybernetcs. Volume 51. pp 181-194.

[Tsutsumi, K 87a] Tsutsumi, Kazuyoshi; Matsumoto, Haruya. (1987). *Neural Computation and Learning Strategy for Manipulator Position Control.* IEEE International Conference on Neural Networks (ICNN-87), San Diego, CA, June 21-24, 1987: Proceedings. Volume 4. pp 525-534. New York: Institute of Electrical and Electronics Engineers.

[Tsutsumi, K 88a] Tsutsumi, Kazuyoshi; Katayama, Koichi; Matsumoto, Haruya. (1988). *Neural Computation for Controlling the Configuration of 2-Dimentional Truss Structure.* IEEE International Conference on Neural Networks (ICNN-88), San Diego, CA, July 24-27, 1988: Proceedings. Volume 2. pp 575-586. New York: Institute of Electrical and Electronics Engineers.

[Tsypkin, YZ 68a] Tsypkin, Ya Z. (1968). *Self-Learning - What Is It?.* IEEE Transactions on Automatic Control. Volume

AC-13. pp 608-612.

[Tsypkin, YZ 71a] Tsypkin, Y. Z. (1971). Adaptation and Learning in Automatic Systems. New York: Academic Press.

[Tulving, E 63a] Tulving, Endel; Gold, Cecille. (1963). *Stimulus Information and Contextual Information as Determinants of Tachistoscopic Recognition of Words*. Journal of Experimental Psychology. Volume 66. pp 319-327.

[Tulving, E 64a] Tulving, Endel; Mandler, George; Baumal, Ruth. (1964). *Interaction of Two Sources of Information in Tachistoscopic Word Recognition*. Canadian Journal of Psychology. Volume 18. pp 62-71.

[Tulving, E 85a] Tulving, Endel. (1985). *How Many Memory Systems are there?*. American Psychologist. Volume 40. pp 385-398.

[Turing, AM 56a] Turing, A. M. (1956). *Can a Machine Think?*. The World of Mathematics. Volume 4. pp 2099-2123. New York: Simon and Schuster.

[Turner, CJ 87*] SEE: Ryan, TW 87a. *Dynamic Control of an Artificial Neural System: The Property Inheritance Network*.

[Turner, CJ 87*] SEE: Winter, CL 87a. TIN: A Trainable Inference Network.

[Turner, CJ 87*] SEE: Winter, CL 87b. *TIN: A Trainable Inference Network*.

[Turner, CJ 88*] SEE: Simmes, SD 88a. Associative Memories on the Connection Machine.

[Tweedy, JR 77a] Tweedy, James R.; Lapinski, Robert H.; Schvaneveldt, Roger W. (1977). *Semantic-Context Effects on Word Recognition: Influence of Varying the Proportion of Items Presented in an Appropriate Context*. Memory and Cognition. Volume 5. pp 84-89.

[Tyc-Dumont, S 87*] SEE: Bras, H 87a. *The Dendrites of Single Brain-Stem Motoneurons Intracellularly Labelled with Horseradish Peroxidase in the Cat: An Ultrastructural Analysis*.

[Tyc-Dumont, S 87*] SEE: Bras, H 87b. *The Dendrites of Single Brain-Stem Motoneurons Intracellularly Labelled with Horseradish Peroxidase in the Cat: Morphological and Electrical*.

[Tyc-Dumont, S 87*] SEE: Bras, H 87c. *The Mammalian Central Neurone is a Complex Computing Device*.

[Tyman, M 85*] SEE: Zargham, MR 85a. *Neural Petri Nets*.

[Tzanakou, E 74*] SEE: Harth, EM 74a. *ALOPEX: A Stochastic Method for Determining Visual Receptive Fields*.

[Uchiyama, T 88*] SEE: Shimohara, K 88b. *Back-Propagation Networks for Event-Driven Temporal Sequence Processing*.

[Uchiyama, T 88*] SEE: Shimohara, K 88a. *A Neural Network System with an Automatic Generation Mechanism for Distorted Patterns*.

[Uchiyama, T 88a] Uchiyama, Tadasu; Shimohara, Katsunori; Tokunaga, Yukio. (1988). *Boltzmann Machines in which Weights Converge to Global Optimal Solutions*. IEEE International Conference on Neural Networks (ICNN-88), San Diego, CA, July 24-27, 1988: Proceedings. Volume 1. pp 665-672. New York: Institute of Electrical and Electronics Engineers.

[Uhr, L 61a] Uhr, Leonard; Vossler, C. (1961). *A Pattern Recognition Program That Generates, Evaluates, and Adjusts to its Own Operators*. Western Joint Computer Conference, Los Angeles, CA, May 9-11, 1961: Proceedings. pp 555-569.

[Uhr, L 86a] Uhr, Leonard. (1986). *Parallel Architectures for Image Processing, Computer Vision, and Pattern Perception*. Handbook of Pattern Recognition and Image Processing. pp 437-469. New York: Academic Press.

[Ullman, S 79a] Ullman, Shimon. (1979). *Relaxation and Constrained Optimization by Local Processes.* Computer Graphics and Image Processing. Volume 10. pp 115-125.

[Ullman, S 82*] SEE: Richter, J 82a. *A Model for the Temporal Organization of X- and Y-Type Receptive Fields in the Primate Retina.*

[Ullmann, JR 62a] Ullmann, J. R. (1962). *A Consistency Technique for Pattern Association.* IRE Transactions on Information Theory. Volume IT-8. pp S74-S81.

[Unnikrishnan, KP 87a] Unnikrishnan, K. P.; Pandya, A. S.; Harth, Erich M. (1987). *Role of Feedback in Visual Perception.* IEEE International Conference on Neural Networks (ICNN-87), San Diego, CA, June 21-24, 1987: Proceedings. Volume 4. pp 259-268. New York: Institute of Electrical and Electronics Engineers.

[Uno, Y 87*] SEE: Kawato, M 87a. *A Hierarchical Model for Voluntary Movement and Its Application to Robotics.*

[Uno, Y 88*] SEE: Kawato, M 88a. *Hierarchical Neural Network Model for Voluntary Movement with Application to Robotics.*

[Upatnieks, J 62*] SEE: Leith, EN 62a. *Reconstructed Wavefronts and Communication Theory.*

[Urahama, K 81*] SEE: Toko, K 81a. *Application of Bifurcation Theory to Nerve Excitations.*

[Usher, M 88*] SEE: Horn, D 88a. *Capacities of Multiconnected Memory Models.*

[Usher, M 88*] SEE: Horn, D 88b. Neural Networks with Dynamical Thresholds.

[Ushita, T 81*] SEE: Nagashino, H 81a. *Relations Between Initial Conditions and Periodic Firing Modes in Reciprocal Inhibition Neural Networks.*

[Ushita, T 87*] SEE: Nagashino, H 87a. *Self-Organizing Process of a Neural Network Model for Pattern Detection.*

[Usui, S 88a] Usui, Shiro; Kamiyama, Yoshimi; Sakakibara, Manabu. (1988). *Physiological Engineering Model of the Retinal Horizontal Cell Layer.* IEEE International Conference on Neural Networks (ICNN-88), San Diego, CA, July 24-27, 1988: Proceedings. Volume 2. pp 87-93. New York: Institute of Electrical and Electronics Engineers.

[Utgoff, PE 85a] Utgoff, Paul E.; Barto, Andrew G. (1986). Learning Efficient Recognizers for Analytically Derived Concepts. COINS Technical Report 86-45. Amherst, MA: University of Massachusetts, Department of Computer Science.

[Utsunomiya, T 80*] SEE: Aihara, K 80a. *A Macroscopic Analysis on Large Scale Networks on Neuronlike Elements with Refractory.*

[Uttal, WR 69a] Uttal, William R. (1969). *Masking of Alphabetic Character Recognition by Dynamic Visual Noise (DVN).* Perception and Psychophysics. Volume 6. pp 121-128.

[Uttley, AM 56a] Uttley, Albert M. (1956). *Conditional Probability Machines and Conditioned Reflexes.* Automata Studies. pp 253-275. Princeton, NJ: Princeton University Press.

[Uttley, AM 59a] Uttley, Albert M. (1959). *Conditional Probability Computing in a Nervous System.* Mechanization of Thought Processes. London: Her Majesty's Stationery Office.

[Uttley, AM 59b] Uttley, Albert M. (1959). *The Design of Conditional Probability Computers.* Information and Control. Volume 2. pp 1-24.

[Uttley, AM 66a] Uttley, Albert M. (1966). *The Transmission of Information and the Effect of Local Feedback in Theoretical and Neural Networks.* Brain Research. Volume 2. pp 21-50.

[Uttley, AM 70a] Uttley, Albert M. (1970). *The Informon: A Network for Adaptive Pattern Recognition.* Journal of Theoretical Biology. Volume 27. pp 31-67.

[Uttley, AM 75a] Uttley, Albert M. (1975). *The Informon in Classical Conditioning.* Journal of Theoretical Biology. Volume 49. pp 355-376.

[Uttley, AM 76a] Uttley, Albert M. (1976). *A Two-Pathway Informon Theory of Conditioning and Adaptive Pattern Recognition.* Brain Research. Volume 102. pp 23-35.

[Uttley, AM 76b] Uttley, Albert M. (1976). *Neurophysiological Predictions of a Two-Pathway Informon Theory of Neural Conditioning.* Brain Research. Volume 102. pp 55-70.

[Uttley, AM 76c] Uttley, Albert M. (1976). *Simulation Studies of Learning in an Informon Network.* Brain Research. Volume 102. pp 35-53.

[Uttley, AM 79a] Uttley, Albert M. (1979). Information Transmission in the Nervous System. London: Academic Press.

[Vaid, A 88*] SEE: Zhou, Y 88b. *Image Restoration Using a Neural Network.*

[Vainio, L 76*] SEE: Kohonen, T 76a. *Associative Recall of Images.*

[van Campenhout, JM 81a] van Campenhout, Jan M.; Cover, Thomas M. (1981). *Maximum Entropy and Conditional Probability.* IEEE Transactions on Information Theory. Volume IT-27. pp 483-489.

[van de Grind, WA 87a] van de Grind, W. A.; Koenderink, Jan J.; van Doorn, Andrea J. (1987). *Influence of Contrast on Foveal and Peripheral Detection of Coherent Motion in Moving Random-Dot Patterns.* Journal of the Optical Society of America A. Volume 4. pp 1643-1652.

[van Dijk, BW 87*] SEE: Freeman, WJ 87e. *Spatial Patterns of Visual Cortical Fast EEG During Conditioned Reflex in a Rhesus Monkey.*

[van Doorn, AJ 79*] SEE: Koenderink, JJ 79a. *The Structure of Two-Dimensional Scalar Fields with Applications to Vision.*

[van Doorn, AJ 86*] SEE: Koenderink, JJ 86a. *Depth and Shape from Differential Perspective in the Presence of Bending Deformations.*

[van Doorn, AJ 86*] SEE: Koenderink, JJ 86b. *Dynamic Shape.*

[van Doorn, AJ 87*] SEE: Koenderink, JJ 87c. *Facts on Optic Flow.*

[van Doorn, AJ 87*] SEE: van de Grind, WA 87a. *Influence of Contrast on Foveal and Peripheral Detection of Coherent Motion in Moving Random-Dot Patterns.*

[van Doorn, AJ 87*] SEE: Koenderink, JJ 87d. *Representation of Local Geometry in the Visual System.*

[Van Essen, DC 87*] SEE: Anderson, CH 87a. *Shifter Circuits: A Computational Strategy for Dyanmic Aspects of Visual Proccssing.*

[Van Essen, DC 88*] SEE: DeYoe, EA 88a. *Concurrent Processing Streams in Monkey Visual Cortex.*

[van Heerden, PJ 63a] van Heerden, P. J. (1963). *A New Optical Method of Storing and Retrieving Information.* Applied Optics. Volume 2. pp 387-392.

[van Heerden, PJ 63b] van Heerden, P. J. (1963). *Theory of Optical Information Storage in Solids.* Applied Optics. Volume 2. pp 393-400.

[van Heerden, PJ 70a] van Heerden, P. J. (1970). *Models for the Brain*. Nature. Volume 225. pp 177-178.

[van Hemmen, JL 82a] van Hemmen, J. L. (1982). *Classical Spin-Glass Model*. Physical Review Letters. Volume 49. pp 409-412.

[Van Hulle, MM 88a] Van Hulle, M. M.; Orban, G. A. (1988). *A Field Approach to Higher Order ANNs for Sensorial Representation*. ESPCI European Conference on Neural Networks (nEURO'88), 1st, Paris, June 6-9, 1988: Proceedings.

[Van Hulle, MM 88b] Van Hulle, M. M.; Orban, G. A. (1988). *Correlative Binding in Higher Order Network for Sensorial Representation*. Neuro-Nimes '88, Nimes, November 15-17, 1988: Proceedings.

[Van Hulle, MM 88c] Van Hulle, M. M.; Orban, G. A. (in press). *Entropy Driven Artificial Neuronal Networks and Sensorial Representation: A Proposal*. Journal of Parallel and Distributed Computing.

[Van Hulle, MM 88d] Van Hulle, M. M.; Orban, G. A. (1988). *Sensory ANNs and Symbolic Decomposition*. DECUS Europe Symposium, Cannes, September 5-9, 1988: Proceedings.

[van Roggen, W 81*] SEE: Fahlman, SE 81a. *Cancellation in a Parallel Semantic Network*.

[Vandemeulebroecke, A 88*] SEE: Sirletti, B 88a. *An Algorithm for Pattern Recognition with VLSI Neural Networks*.

[Vandemeulebroecke, A 88*] SEE: Verleysen, M 88a. *A High-Storage Content-Addressable Memory and Its Learning Algorithm*.

[Vandemeulebroecke, A 88*] SEE: Verleysen, M 88g. *Neural Network or High-Storage Content-Addressable Memory: VLSI Circuit and Learning Algorithm*.

[VanderBrug, GJ 80*] SEE: Albus, JS 80a. *A Measurement and Control Model for Adaptive Robots*.

[Varela, FJ 87*] SEE: Rossler, J 87a. *Exact Results for the Average Dynamic Behavior of some Non-Linear Neural Networks*.

[Vasilenko, DA 68*] SEE: Kostyuk, PG 68a. *Transformation of Cortical Motor Signals in Spinal Cord*.

[Vaziri, P 88*] SEE: Ross, MD 88a. *Macular Bioaccelerometers on Earth and in Space*.

[Vaziri, P 88*] SEE: Ross, MD 88c. *The Neuroanatomical Substrate for Information Processing in Macular Endorgans*.

[Veasy, LG 64*] SEE: Warner, HR 64a. *Experience with Bayes Theorem for Computer Diagnosis of Congenital Heart Disease*.

[Vecchi, MP 83*] SEE: Kirkpatrick, S 83b. *Optimization by Simulated Annealing*.

[Vecchi, MP 83*] SEE: Kirkpatrick, S 83a. *Optimization by Simulated Annealing*.

[Vecchi, MP 88a] Vecchi, Mario P.; Salehi, Jawad A. (1988). *Neuromorphic Networks Based on Sparse Optical Orthogonal Codes*. In Anderson, Dana Z. (ed.). Neural Information Processing Systems. pp 814-823. New York: American Institute of Physics.

[Veelenturf, LPJ 81a] Veelenturf, L. P. J. (1981). *An Automata-Theoretical Approach to Developing Learning Neural Networks*. Cybernetics and Systems. Volume 12. pp 179-202.

[Venetsanopoulos, AN 88*] SEE: Karayiannis, NB 88a. *The Correlation Associative Memory Realizes Hebbian Learning*.

[Venezky, RL 79*] SEE: Massaro, DW 79a. *Orthographic Regularity, Positional Frequency, and Visual Processing of Letter Strings*.

[Venkatesh, SS 86*] SEE: McEliece, RJ 86a. Capacity of the Hopfield Associative Memory.

[Venkatesh, SS 86*] SEE: Psaltis, D 86b. *Shift Invariance in Optical Associative Memories.*

[Venkatesh, SS 86a] Venkatesh, Santosh S. (1986). *Epsilon Capacity of Neural Networks.* In Denker, John S. (ed.). Neural Networks for Computing, Snowbird, Utah, 1986. AIP Proceedings 151. pp 440-445. New York: American Institute of Physics.

[Venkatesh, SS 86b] Venkatesh, Santosh S.; Psaltis, Demetri. (1986). Linear Maps with Point Rules: Applications to Pattern Classification and Associative Memory. PhD Thesis. Pasadena, CA: California Institute of Technology.

[Venkatesh, SS 87*] SEE: McEliece, RJ 87a. *The Capacity of the Hopfield Associative Memory.*

[Venkatesh, SS 87*] SEE: Baldi, P 87a. *Number of Stable Points for Spin-Glasses and Neural Networks of Higher Orders.*

[Venkatesh, SS 87a] Venkatesh, Santosh S. (1987). *Computation with Neural Networks.* IEEE International Conference on Engineering in Medicine and Biology, 9th, 1987: Proceedings. pp 1364-1365. New York: Institute of Electrical and Electronics Engineers.

[Venkatesh, SS 88*] SEE: Baldi, P 88h. *On Properties of Networks of Neuron-Like Elements.*

[Venta, O 86a] Venta, Olli; Kohonen, Teuvo. (1986). *A Non-Stochastic Method for the Correction of Sentences.* International Conference on Pattern Recognition, 8th, Paris, October 27-31, 1986: Proceedings. pp 1214-1217. New York: Institute of Electrical and Electronics Engineers.

[Venta, O 86b] Venta, Olli. (1986). *N-Gram Driven Search for Sentences in a Syntactic Network.* IEEE International Conference on Acoustics, Speech, and Signal Processing (ICASSP-86), Tokyo, April 7-11, 1986: Proceedings. pp 1145-1148. New York: Institute of Electrical and Electronics Engineers.

[Venta, O 87*] SEE: Kohonen, T 87e. *Microprocessor Implementation of a Large Vocabulary Speech Recognizer and Phonetic Typewriter for Finnish and Japanese.*

[Venta, O 88*] SEE: Kohonen, T 88b. *Phonetic Typewriter for Finnish and Japanese.*

[Venta, O 88a] Venta, Olli; Kohonen, Teuvo. (1988). *A Content-Addressing Software Method for the Emulation of Neural Networks.* IEEE International Conference on Neural Networks (ICNN-88), San Diego, CA, July 24-27, 1988: Proceedings. Volume 1. pp 191-198. New York: Institute of Electrical and Electronics Engineers.

[Verber, CM 84a] Verber, Carl M. (1984). *Integrated-Optical Approaches to Numerical Optical Processing.* Proceedings of the IEEE. Volume 72. pp 942-953.

[Verleysen, M 88*] SEE: Sirletti, B 88a. *An Algorithm for Pattern Recognition with VLSI Neural Networks.*

[Verleysen, M 88a] Verleysen, Michel; Sirletti, Bruno; Vandemeulebroecke, Andre; Jespers, Paul. (in press). *A High-Storage Content-Addressable Memory and Its Learning Algorithm.* IEEE Circuit and System.

[Verleysen, M 88b] Verleysen, Michel; Sirletti, Bruno; Jespers, Paul. (1988). *A Large VLSI Fully Interconnected Neural Network.* Symposium on VLSI Circuits, Tokyo, August, 1988: Proceedings.

[Verleysen, M 88c] Verleysen, Michel; Sirletti, Bruno; Jespers, Paul. (1988). *A Large VLSI Hopfield Network for Pattern Recognition Problems.* Meeting of the International Neural Network Society (INNS), 1st, Boston, MA, September 6-10, 1988: Proceedings.

[Verleysen, M 88d] Verleysen, Michel; Sirletti, Bruno; Jespers, Paul. (1988). *A New CMOS Architecture for Neural Networks.* Workshop on VLSI for Artificial Intelligence, Oxford, July, 1988: Proceedings.

[Verleysen, M 88e] Verleysen, Michel; Sirletti, Bruno; Jespers, Paul. (1988). *A New VLSI Architecture for Large*

Hopfield's Neural Networks. ESSCIRC-88, Manchester, September, 1988: Proceedings.

[Verleysen, M 88f] Verleysen, Michel; Sirletti, Bruno; Jespers, Paul. (1988). *A New VLSI Architecture for Neural Associative Memories*. nEuro'88, Paris, June, 1988: Proceedings.

[Verleysen, M 88g] Verleysen, Michel; Sirletti, Bruno; Vandemeulebroecke, Andre; Jespers, Paul. (in press). *Neural Network or High-Storage Content-Addressable Memory: VLSI Circuit and Learning Algorithm*. IEEE Journal of Solid-State Circuits.

[Verveen, AA 68a] Verveen, A. A. (1968). *Fluctuation Phenomena in Nerve Membrane*. Proceedings of the IEEE. Volume 56. pp 906-916.

[Vetterli, M 87*] SEE: Widrow, B 87b. *Fundamental Relations Between the LMS Algorithm and the DFT*.

[Viana Di Prisco, G 84a] Viana Di Prisco, Gonzalo. (1984). *Hebb Synaptic Plasticity*. Progress in Neurobiology. Volume 22. pp 89-102.

[Viana di Prisco, G 86*] SEE: Grajski, KA 86a. *Classification of EEG Spatial Patterns with a Tree-Structured Methodology: CART*.

[Viana Di Prisco, G 86*] SEE: Freeman, WJ 86a. *Relation of Olfactory EEG to Behavior: Time Series Analysis*.

[Vichniac, GY 86*] SEE: Hartman, H 86a. *Inhomogeneous Celluar Automata (INCA)*.

[Vichniac, GY 86*] SEE: Goles, E 86a. *Invariants in Automata Networks*.

[Vichniac, GY 86*] SEE: Goles, E 86b. *Lyapunov Function for Parallel Neural Networks*.

[Vichniac, GY 86a] Vichniac, Gerard Y. (1986). *Cellular Automata Models of Disorder and Organization*. In Bienenstock, Elie L. (ed.). Disordered Systems and Biological Organization. pp 3 f. New York: Springer-Verlag.

[Vichniac, GY 88a] Vichniac, Gerard Y.; Gardner Lepp, M.; Steenstrup, M. (1988). *A Neural Network for the Optimization of Communications Network Design*. Neural Networks. Volume 1.

[Victor, J 86a] Victor, Jesse (ed.). *Bell Labs Models Parallel Processor on Neural Networks*. Mini-Micro Systems, August 1986. pp 43-49.

[Victor, J 86b] Victor, Jesse. (August, 1986). *Bell Labs Models Parallel Processor on Neural Networks*. Mini-Micro Systems. pp 43-49.

[Vicuna, F 86*] SEE: Oyster, JM 86a. Associative Network Applications to Robot Vision.

[Vidal, JJ 87*] SEE: Goodwin, JM 87a. *A Design for an Associative Spin Glass Processor*.

[Vidal, JJ 87a] Vidal, Jacques J.; Pemberton, Joseph C.; Goodwin, James M. (1987). *Implementing Neural Nets with Programmable Logic*. IEEE International Conference on Neural Networks (ICNN-87), San Diego, CA, June 21-24, 1987: Proceedings. Volume 3. pp 539-548. New York: Institute of Electrical and Electronics Engineers.

[Vidal, JJ 87b] Vidal, Jacques J.; Salas, Paul F. (1987). *Real-Time Parallel Processing of Pixel Arrays Using Adaptive Logic*. Real Time Image Processing: Concepts and Technologies. SPIE Volume 860. pp 26-30. Bellingham, WA: Society of Photo-Optical Instrumentation Engineers.

[Vidal, JJ 87c] Vidal, Jacques J. (1987). *Silicon Brains: Whither Neuromimetic Computer Architectures*. IEEE International Conference on Computer Design, VLSI in Computers, Rochester, NY, 1987: Proceedings. pp 17-20. New York: Institute of Electrical and Electronics Engineers.

[Vidal, JJ 88*] SEE: Martinez, TR 88a. *Adaptive Parallel Logic Networks.*

[Vidal, JJ 88*] SEE: Goodwin, JM 88a. Exploration of Learning in an Associative Magnetic Processor.

[Vidal, JJ 88*] SEE: Rosen, BE 88a. *Learning by State Recurrence Detection.*

[Vidal, JJ 88*] SEE: Rosen, BE 88b. Machine Operant Conditioning.

[Vidal, JJ 88*] SEE: Goodwin, JM 88b. Progress on the Spin Chip Asociative Processor.

[Vidal, JJ 88*] SEE: Rosen, BE 88c. State Recurrence Learning.

[Vidal, JJ 88a] Vidal, Jacques J. (1988). *Implementing Neural Nets with Programmable Logic.* IEEE Transactions on Acoustics, Speech, and Signal Processing. Volume 36. pp 1180-1189.

[Vidal, JJ 88b] Vidal, Jacques J.; Haggerty, John. (1988). *Synchronization in Neural Nets.* In Anderson, Dana Z. (ed.). Neural Information Processing Systems. pp 824-829. New York: American Institute of Physics.

[Vijaykumar, R 86a] Vijaykumar, Rukmini; Arbib, Michael A. (1986). A Task-Level Robot Planner for Assembly Operations. COINS Technical Report 86-31. Amherst, MA: University of Massachusetts, Department of Computer Science.

[Vilain, MB 85a] Vilain, M. B. (1985). *The Restricted Language Architecture of a Hybrid Representation System.* International Joint Conference on Artificial Intelligence (IJCAI-85), 9th, Los Angeles, CA, August 18-23, 1985: Proceedings. Los Altos, CA: Morgan Kaufman.

[Virasoro, MA 84*] SEE: Mezard, M 84a. *Nature of the Spin-Glass Phase.*

[Virasoro, MA 84*] SEE: Mezard, M 84b. *Replica Symmetry Breaking and the Nature of the Spin Glass Phase.*

[Virasoro, MA 85*] SEE: Mezard, M 85a. *The Microstructure of Ultrametricity.*

[Virasoro, MA 86*] SEE: Parga, N 86a. *The Ultrametric Organization of Memories in a Neural Network.*

[Virasoro, MA 86a] Virasoro, M. A. (1986). *Ultrametricity, Hopfield Model and all that.* In Bienenstock, Elie L. (ed.). Disordered Systems and Biological Organization. pp 197-204. New York: Springer-Verlag.

[Vittoz, EA 85a] Vittoz, Eric A. (1985). *Micropower Techniques.* Design of MOS VLSI Circuits for Telecommunications. pp 104-144. Englewood Cliffs, NJ: Prentice-Hall.

[Vlasov, AN 80*] SEE: Batalina, MA 80a. *Neuristor Line Made of S-Type Germanium Diodes.*

[Vlontzos, JA 88a] Vlontzos, J. A.; Kung, Sun-Yuan. (1988). *A Hierarchical System for Character Recognition with Stochastic Knowledge Representation.* IEEE International Conference on Neural Networks (ICNN-88), San Diego, CA, July 24-27, 1988: Proceedings. Volume 1. pp 601-608. New York: Institute of Electrical and Electronics Engineers.

[Voevodsky, J 87a] Voevodsky, John. (1987). *PLATO / ARISTOTLE: A Neural Net Knowledge Processor.* IEEE International Conference on Neural Networks (ICNN-87), San Diego, CA, June 21-24, 1987: Proceedings. Volume 2. pp 399-410. New York: Institute of Electrical and Electronics Engineers.

[Vogl, TP 88a] Vogl, T. P.; Mangis, J. K.; Rigler, A. K.; Zink, W. T.; Alkon, Daniel L. (1988). *Accelerating the Convergence of the Back-Propagation Method.* Biological Cybernetics. Volume 59. pp 257-263.

[Vold, TG 87a] Vold, Terje G. (1987). *Bayesian Image Processing, Gibbs Distributions, and Neural Networks.* IEEE International Conference on Neural Networks (ICNN-87), San Diego, CA, June 21-24, 1987: Proceedings. Volume 3. pp

691-702. New York: Institute of Electrical and Electronics Engineers.

[von der Malsburg, C 73a] von der Malsburg, Christoph. (1988). *Self-organization of Orientation Sensitive Cells in the Striate Cortex.* In Anderson, James A. (ed.); Rosenfeld, Edward (ed.). Neurocomputing: Foundations of Research. pp 212-228. Cambridge, MA: MIT Press.

[von der Malsburg, C 73b] von der Malsburg, Christoph. (1973). *Self-Organization of Orientation Sensitive Cells in the Striate Cortex.* Kybernetik. Volume 14. pp 85-100.

[von der Malsburg, C 75*] SEE: Willshaw, DJ 75a. *How Patterned Neural Connections Can be Set Up by Self-Organization.*

[von der Malsburg, C 76*] SEE: Willshaw, DJ 76b. *How Patterned Neural Connections Can be Set Up by Self-Organization.*

[von der Malsburg, C 77a] von der Malsburg, Christoph; Willshaw, David J. (1977). *How to Label Nerve Cells so that they ca Interconnect in an Ordered Fashion.* Proceedings of the National Academy of Sciences. Volume 74. pp 5176-5178.

[von der Malsburg, C 79*] SEE: Willshaw, DJ 79a. *A Marker Induction Mechanism for the Establishment of Ordered Neural Mapping: Its Application to the Retinotectal Connections.*

[von der Malsburg, C 79a] von der Malsburg, Christoph. (1979). *Development of Ocularity Domains and Growth Behaviour of Axon Terminals.* Biological Cybernetics. Volume 32. pp 49-62.

[von der Malsburg, C 81a] von der Malsburg, Christoph; Willshaw, David J. (1981). *Differential Equations for the Development of Topological Nerve Fibre Projections.* Mathematical Psychology and Psychophysiology. SIAM-AMS Proceedings Volume 13. pp 39-47. Providence, RI: American Mathematical Society.

[von der Malsburg, C 81b] von der Malsburg, Christoph. (1981). The Correlation Theory of Brain Function. Report 81-2. Gottingen: Max-Plank Institute, Department of Neuro-Biology.

[von der Malsburg, C 82a] von der Malsburg, Christoph; Cowan, Jack D. (1982). *Outline of a Theory for the Ontogenesis of Iso-Orientation Domains in Visual Cortex.* Biological Cybernetics. Volume 45. pp 49-56.

[von der Malsburg, C 85*] SEE: Cowan, JD 85a. *A Proposed Mechanism for the Origin and Development of Iso-Orientation Columns.*

[von der Malsburg, C 86a] von der Malsburg, Christoph; Schneider, W. (1986). *A Neural Cocktail-Party Processor.* Biological Cybernetics. Volume 54. pp 29-40.

[von der Malsburg, C 86b] von der Malsburg, Christoph; Bienenstock, Elie L. (1986). *Statistical Coding and Short-Term Synaptic Plasticity: A Scheme for Knowledge Representation in the Brain.* Disordered Systems and Biological Organization. NATO ASI Series, Volume F20. pp 247-272. Berlin: Springer-Verlag.

[von der Malsburg, C 87*] SEE: Bienenstock, EL 87a. *A Neural Network for Invariant Pattern Recognition.*

[von der Malsburg, C 87a] von der Malsburg, Christoph; Bienenstock, Elie L. (1987). *A Neural Network for the Retrieval of Superimposed Connection Patterns.* Europhysics Letters. Volume 3. pp 1243-1249.

[von der Malsburg, C 87b] von der Malsburg, Christoph. (1987). *Synaptic Plasticity as Basis of Brain Organization.* The Neural and Molecular Bases of Learning. pp 411-432. New York: Wiley.

[von der Malsburg, C 88a] von der Malsburg, Christoph. (1988). *Goal and Architecture of Neural Computers.* In Eckmiller, Rolf (ed.); von der Malsburg, Christoph (ed.). Neural Computers. NATO ASI Series, Volume F41. pp 23-28. Berlin: Springer-Verlag.

[von der Malsburg, C 88b] von der Malsburg, Christoph. (1988). *Pattern Recognition by Labeled Graph Matching.* Neural

Networks. Volume 1. pp 141-148.

[von der Malsburg, C 88c] von der Malsburg, Christoph; Singer, W. (1988). *Principles of Cortical Network Organization.* Neurobiology of Neocortex. pp 69-99. New York: Wiley.

[von Foerster, H 62*] SEE: Ashby, WR 62a. *Instability of Pulse Activity in a Net with Threshold.*

[von Foerster, H 62a] von Foerster, Heinz (ed.); Zopf, George W. (ed.). (1962). Principles of Self-Organization. Elmsford, NY: Pergamon.

[Von Lehmen, A 87a] Von Lehmen, A.; Paek, Eung Gi; Liao, Paul F.; Marrakchi, A.; Patel, J. S. (in press). *Influence of Interconnection Weight Discretization and Noise in an Optoelectronic Neural Network.* Optics Letters.

[Von Lehmen, A 88*] SEE: Paek, EG 88b. *Holographic Associative Memory for Word-Break Recognition.*

[Von Lehmen, A 88*] SEE: Paek, EG 88c. *Real-Time Holographic Associative Memory for Identifying Words in a Continuous Letter String.*

[Von Lehmen, A 88a] Von Lehmen, A.; Paek, Eung Gi; Liao, Paul F.; Marrakchi, A.; Patel, J. S. (1988). *Factors Influencing Learning by Backpropagation.* IEEE International Conference on Neural Networks (ICNN-88), San Diego, CA, July 24-27, 1988: Proceedings. Volume 1. pp 335-341. New York: Institute of Electrical and Electronics Engineers.

[von Neumann, J 56a] von Neumann, John. (1956). *Probabilistic Logics and the Synthesis of Reliable Organisms from Unreliable Components.* Automata Studies. pp 43-98. Princeton: Princeton University Press.

[von Neumann, J 58a] von Neumann, John. (1988). *The Computer and the Brain.* In Anderson, James A. (ed.); Rosenfeld, Edward (ed.). Neurocomputing: Foundations of Research. pp 83-88. Cambridge, MA: MIT Press.

[von Seelen, W 87a] von Seelen, Werner; Mallot, Hanspeter A.; Giannakopoulos, F. (1987). *Characteristics of Neuronal Systems in the Visual Cortex.* Biological Cybernetics. Volume 56. pp 37-49.

[von Seelen, W 87b] von Seelen, Werner; Mallot, Hanspeter A.; Giannakopoulos, E.; Schulze, E. (1987). *Properties of Cortical Systems - A Basis for Computer Vision.* IEEE International Conference on Neural Networks (ICNN-87), San Diego, CA, June 21-24, 1987: Proceedings. Volume 4. pp 89-96. New York: Institute of Electrical and Electronics Engineers.

[von Seelen, W 88a] von Seelen, Werner; Mallot, Hanspeter A. (1988). *Parallelism and Redundancy in Neural Networks.* In Eckmiller, Rolf (ed.); von der Malsburg, Christoph (ed.). Neural Computers. NATO ASI Series, Volume F41. pp 51-60. Berlin: Springer-Verlag.

[Vossler, C 61*] SEE: Uhr, L 61a. *A Pattern Recognition Program That Generates, Evaluates, and Adjusts to its Own Operators.*

[Vuillemin, J 81*] SEE: Preparata, FP 81a. *The Cube-Connected Cycles: A Versatile Network for Parallel Computation.*

[Wada, M 82a] Wada, Mitsuo. (1982). *Cyclic Sequences Generated by Neuron Model.* Transactions of the Institute of Electronics and Communication Engineers of Japan. Volume E65. pp 135-142.

[Wada, M 82b] Wada, Mitsuo. (1982). *Stimulus-Response Characteristics of a Mathematical Neuron Model.* Transactions of the Institute of Electronics and Communication Engineers of Japan. Volume E65. pp 595-602.

[Wadepuhl, M 83*] SEE: Selverston, AI 83a. *Neural Mechanisms for the Production of Cyclic Motor Patterns.*

[Wagner, GP 84*] SEE: Dammasch, IE 84a. *On the Properties of Randomly Connected McCulloch-Pitts Networks: Differences*

Between Input-Constant and Input-Variant Networks.

[Wagner, GP 86*] SEE: Dammasch, IE 86b. *Self-Stabilization of Neuronal Networks, I. The Compensation Algorithm for Synaptogenesis.*

[Wagner, GP 88*] SEE: Dammasch, IE 88c. *Self-Stabilization of Neuronal Networks, II. Stability Conditions for Synaptogenesis.*

[Wagner, K 87*] SEE: Psaltis, D 87a. *Learning in Optical Neural Computers.*

[Wagner, K 87a] Wagner, Kelvin; Psaltis, Demetri. (1987). *Multilayer Optical Learning Networks.* Applied Optics. Volume 26. pp 5061-5076.

[Wagner, K 87b] Wagner, Kelvin; Psaltis, Demetri. (1987). *Nonlinear Etalons in Competitive Optical Learning Networks.* IEEE International Conference on Neural Networks (ICNN-87), San Diego, CA, June 21-24, 1987: Proceedings. Volume 3. pp 585-594. New York: Institute of Electrical and Electronics Engineers.

[Wagner, K 88*] SEE: Psaltis, D 88a. *Adaptive Optical Networks Using Photorefractive Crystals.*

[Waibel, A 86a] Waibel, Alex. (1986). *Suprasegmentals in Very Large Vocabulary Word Recognition.* In Schwab, Eileen C. (ed.); Nusbaum, Howard C. (ed.). Pattern Recognition by Humans and Machines: Speech Perception. Volume I. pp 159-186. New York: Academic Press.

[Waibel, AH 87*] SEE: Watrous, RL 87a. *Learned Phonetic Discrimination Using Connectionist Networks.*

[Walach, E 84*] SEE: Widrow, B 84a. *On the Statistical Efficiency of the LMS Algorithm with Nonstationary Inputs.*

[Walach, E 84a] Walach, Eugene; Widrow, Bernard. (1984). *The Least Mean Fourth (LMF) Adaptive Algorithm and Its Family.* IEEE Transactions on Information Theory. Volume IT-30. pp 275-283.

[Waldron, I 68*] SEE: Wilson, DM 68a. *Models for the Generation of the Motor Output Pattern in Flying Locusts.*

[Waldron, MB 71a] Waldron, Manjula Bhushan. (1971). Syneuristor: A Device and Concept. Report 6560-22. pp 1-110. Stanford, CA: Stanford University, Information Systems Laboratory.

[Waldrop, MM 88a] Waldrop, M. Mitchell. (1988). *Soar: A Unified Theory of Cogition?.* Science. Volume 241. pp 296-298.

[Waldrop, MM 88b] Waldrop, M. Mitchell. (1988). *Toward a Unified Theory of Cognition.* Science. Volume 241. pp 27-29.

[Waldschmidt, K 87*] SEE: Strugala, M 87a. *An Associative Processor as a Design Rule Check Accelerator.*

[Waldschmidt, K 88*] SEE: Schulz, M 88a. *An Associative Co-Processor Architecture with the UNIX Operating System for the Acceleration of CAD Tools.*

[Walker, CC 62*] SEE: Ashby, WR 62a. *Instability of Pulse Activity in a Net with Threshold.*

[Walker, MR 88*] SEE: Akers, LA 88a. *Limited Interconnectivity in Synthetic Neural Systems.*

[Walkup, JF 86*] SEE: Krile, TF 86a. *Optical Systems Research at Texas Tech University.*

[Wallace, DJ 86*] SEE: Bruce, AD 86a. *Learning and Memory Properties in Fully Connected Networks.*

[Wallace, DJ 87*] SEE: Bruce, AD 87a. *Dynamics and Statistical Mechanics of the Hopfield Model.*

[Wallace, DJ 87*] SEE: Forrest, BM 87a. *Implementing Neural Network Models on Parallel Computers.*

[Wallace, DJ 88*] SEE: Gardner, E 88a. *Training with Noise: Application to Word and Text Storage.*

[Wallace, R 86a] Wallace, R.; Matsuzaki, K.; Goto, Y.; Crisman, J.; Webb, Jon A.; Kanade, Takeo. (1986). *Progress in Robot Road-Following.* IEEE International Conference on Robotics and Automation, San Francisco, CA, April 7-10, 1986: Proceedings. Volume 3. pp 1615-1621. New York: Institute of Electrical and Electronics Engineers.

[Walloe, L 69*] SEE: Enger, PS 69a. *A Biological Model of the Excitation of a Second Order Sensory Neurone.*

[Walloe, L 69a] Walloe, Lars; Jansen, J. K. S.; Nygaard, K. (1969). *A Computer-Simulated Model of a Second Order Sensory Neurone.* Kybernetik. Volume 6. pp 130-141.

[Walrath, J 87*] SEE: Hennessey, AK 87a. The Gamma Network - A Physical and Logical Data Structure.

[Walsh, JL 23a] Walsh, J. L. (1923). *A Closed Set of Normal Orthogonal Functions.* American Journal of Mathematics. Volume 45.

[Walter, RJ 86*] SEE: Sklansky, J 86a. *Biomedical Image Analysis.*

[Walters, DKW 87a] Walters, Deborah K. W. (1987). *Automated Fake Color Separation: Combining Computer Vision and Computer Graphics.* Applications of Artificial Intelligence V. SPIE Volume 786. pp 541-548. Bellingham, WA: Society of Photo-Optical Instrumentation Engineers.

[Walters, DKW 87b] Walters, Deborah K. W. (1987). *Biological vs Artificial Representations of Variables in Fine-Grain Parallel Systems.* IEEE Computer Society Conference on Computer Vision and Pattern Recognition, Miami Beach, FL, November 30-December 2, 1987: Proceedings. pp 298-300. New York: Institute of Electrical and Electronics Engineers.

[Walters, DKW 87c] Walters, Deborah K. W.; Krishnan, Ganapathy. (1987). *Color Separation Using General Purpose Computer Vision Algorithms.* Conference on Artificial Intelligence Applications, 3rd, Kissimmee, FL, February 23-27, 1987: Proceedings. pp 188-193. New York: Institute of Electrical and Electronics Engineers.

[Walters, DKW 87d] Walters, Deborah K. W.; Krishnan, Ganapathy. (1987). *Perceptual Significance Hierarchy: A Computer Vision Theory for Color Separation.* National Conference on Artificial Intelligence (AAAI-87), 6th, July 13-17, 1987: Proceedings. Volume 2. pp 767-771. Washington, DC: American Association for Artificial Intelligence.

[Walters, DKW 87e] Walters, Deborah K. W. (1987). *Properties of Connectionist Variable Representations.* Cognitive Science Society, 9th, Seattle, WA, July 16-18, 1987: Proceedings. pp 265-273. Hillsdale, NJ: Lawrence Erlbaum.

[Walters, DKW 87f] Walters, Deborah K. W. (1987). *Response Mapping Functions: Classification and Analysis of Connectionist Representations.* IEEE International Conference on Neural Networks (ICNN-87), San Diego, CA, June 21-24, 1987: Proceedings. Volume 3. pp 79-86. New York: Institute of Electrical and Electronics Engineers.

[Walters, DKW 87g] Walters, Deborah K. W. (1987). *RHO-Space: A Neural Network for the Detection and Representation of Oriented Edges.* Cognitive Science Society, 9th, Seattle, WA, July 16-18, 1987: Proceedings. pp 455-460. Hillsdale, NJ: Lawrence Erlbaum.

[Walters, DKW 87h] Walters, Deborah K. W. (1987). *RHO-Space: A Neural Network for the Detection and Representation of Oriented Edges.* Cognitive Science Society, 9th, Seattle, WA, July 16-18, 1987: Proceedings. pp 455-460. Hillsdale, NJ: Lawrence Erlbaum.

[Walters, DKW 87i] Walters, Deborah K. W. (1987). *Selection of Image Primitives for General-Purpose Visual Processing.* Computer Vision, Graphics, and Image Processing. Volume 37. pp 261-298.

[Walters, DKW 87j] Walters, Deborah K. W. (1987). *The Representation of Variables in Connectionist Networks.* IEEE International Conference on Computer Vision, 1st, London, June 8-11, 1987: Proceedings. pp 698-702. New York:

Institute of Electrical and Electronics Engineers.

[Walters, DKW 88*] SEE: Lively, R 88a. *Integration of Detector Responses for Texture Segmentation.*

[Walters, DKW 88*] SEE: Krishnan, G 88a. *Segmenting Intersecting and Incomplete Boundaries.*

[Waltz, DL 75a] Waltz, David L. (1975). *Understanding Line Drawings of Scenes with Shadows.* The Psychology of Computer Vision. pp 19-91. New York: McGraw-Hill.

[Waltz, DL 78a] Waltz, David L. (1978). *A Parallel Model for Low-Level Vision.* Computer Vision Systems. pp 175-186. New York: Academic Press.

[Waltz, DL 84*] SEE: Pollack, JB 84a. *Parallel Interpretation of Natural Language.*

[Waltz, DL 85a] Waltz, David L.; Pollack, Jordan B. (1985). *Massively Parallel Parsing: A Strongly Interactive Model of Natural Language Interpretation.* Cognitive Science. Volume 9. pp 51-74.

[Waltz, DL 86*] SEE: Pollack, JB 86a. *Interpretation of Natural Language: A Potential Application of Parallelism.*

[Waltz, DL 88a] Waltz, David L.; Feldman, Jerome A. (1988). *Connectionist Models and Their Implications.* In Waltz, David L. (ed.); Feldman, Jerome A. (ed.). Connectionist Models and Their Implications: Readings from Cognitive Science. pp 1-12. Norwood, NJ: Ablex.

[Waltz, DL 88b] Waltz, David L. (ed.); Feldman, Jerome A. (ed.). (1988). Connectionist Models and Their Implications: Readings from Cognitive Science. Norwood, NJ: Ablex.

[Waltz, DL 88c] Waltz, David L.; Pollack, Jordan B. (1988). *Massively Parallel Parsing: A Strongly Interactive Model of Natural Language Interpretation.* In Waltz, David L. (ed.); Feldman, Jerome A. (ed.). Connectionist Models and Their Implications: Readings from Cognitive Science. pp 181-204. Norwood, NJ: Ablex.

[Wang, PY 87*] SEE: Seidman, SB 87a. *Analysis of a Competition-Based Connectionist Model.*

[Wang, PY 88a] Wang, Pearl Y.; Seidman, Stephen B.; Reggia, James A. (1988). *Analysis of Competition-Based Spreading Activation in Connectionist Models.* International Journal of Man-Machine Studies. Volume 28. pp 77-97.

[Wanke, E 68*] SEE: Palmieri, G 68a. *A Pattern Recognition Machine.*

[Ward, JR 88a] Ward, J. R.; Kuklinski, T. (1988). *A Model for Variability Effects in Hand-Printing, with Implications for the Design of On Line Character Recognition Systems.* IEEE Transactions on Systems, Man, and Cybernetics. Volume 18. pp 438-451.

[Warde, C 81a] Warde, Cardinal; Weiss, A. M.; Fisher, Arthur D.; Thackara, J. I. (1981). *Optical Information Processing Characteristics of the Microchannel Spatial Light Modulator.* Applied Optics. Volume 20. pp 2066-2074.

[Warde, C 83a] Warde, Cardinal; Thackara, J. I. (1983). *Operating Modes of the Microchannel Spatial Light Modulator.* Optical Engineering. Volume 22. p 695-703.

[Warde, C 86a] Warde, Cardinal; Kottas, James. (1986). *Hybrid Optical Inference Machines: Architectural Considerations.* Applied Optics. Volume 25. pp 940-947.

[Wardell, RW 82a] Wardell, R. W.; Jernigan, M. E. (1982). *Modelling the Retina as a Stochastic Receptor Array.* IEEE International Conference on Cybernetics and Society, Seattle, WA, October 28-30, 1982: Proceedings. pp 171-175. New York: Institute of Electrical and Electronics Engineers.

[Warden, WH 83*] SEE: Bartholdi, JJ 83a. *A Minimal Technology Routing System for Meals on Wheels.*

[Warner, HR 64a] Warner, Homer R.; Toronto, Alan F.; Veasy, L. George. (1964). *Experience with Bayes Theorem for Computer Diagnosis of Congenital Heart Disease.* Computers in Medicine and Biology. Volume 115. pp 558-567. New York: New York Academy of Sciences.

[Warner, MW 81a] Warner, M. W. (1981). *Tolerance Nets.* Bulletin of Mathematical Biology. Volume 43. pp 641-650.

[Warner, MW 83*] SEE: Muir, A 83a. *The Dynamics of Symmetric Nets.*

[Warner, MW 84*] SEE: Muir, A 84a. *Lattice Valued Relations and Automata.*

[Warren, RM 70a] Warren, Richard M. (1970). *Perceptual Restoration of Missing Speech Sounds.* Science. Volume 167. pp 392-393.

[Warren, RM 71a] Warren, Richard M.; Obusek, Charles J. (1971). *Speech Perception and Phonemic Restorations.* Perception and Psychophysics. Volume 9. pp 358-362.

[Wassel, GN 81*] SEE: Sklansky, J 81a. Pattern Classifiers and Trainable Machines.

[Wasserman, PD 88a] Wasserman, Philip D.; Schwartz, Tom. (1988). *Expert Focus, Neural Networks, Part 2: What are They and Why Is Everybody So Interested in Them Now?.* IEEE Expert. Volume 3. pp 10-15.

[Watanabe, N 88a] Watanabe, Nobuo; Nagata, Shigemi; Asakawa, Kazuo. (1988). *Mobile Robot Control by Neural Networks and Their Associated Learning Algorithm.* International Conference on Industrial Electronics, Control, and Instrumentation (IECON-88): Proceedings. pp 1054-1060.

[Waterman, DA 70a] Waterman, D. A. (1970). *Generalization Learning Techniques for Automating the Learning of Heuristics.* Artificial Intelligence. Volume 1. pp 121-170.

[Watrous, RL 87a] Watrous, Raymond L.; Shastri, L.; Waibel, A. H. (1987). *Learned Phonetic Discrimination Using Connectionist Networks.* European Conference on Speech Technology, Edinburgh, September 1987: Proceedings. pp 377-380.

[Watrous, RL 87b] Watrous, Raymond L.; Shastri, Lokendra. (1987). *Learning Acoustic Features from Speech Data Using Connectionist Networks.* Cognitive Science Society, 9th, Seattle, WA, July 16-18, 1987: Proceedings. pp 518-530. Hillsdale, NJ: Lawrence Erlbaum.

[Watrous, RL 87c] Watrous, Raymond L. (1987). *Learning Algorithms for Connectionist Networks: Applied Gradient Methods of Nonlinear Optimization.* IEEE International Conference on Neural Networks (ICNN-87), San Diego, CA, June 21-24, 1987: Proceedings. Volume 2. pp 619-628. New York: Institute of Electrical and Electronics Engineers.

[Watrous, RL 87d] Watrous, Raymond L.; Shastri, Lokendra. (1987). *Learning Phonetic Features Using Connectionist Networks: An Experiment in Speech Recognition.* IEEE International Conference on Neural Networks (ICNN-87), San Diego, CA, June 21-24, 1987: Proceedings. Volume 4. pp 381-388. New York: Institute of Electrical and Electronics Engineers.

[Watrous, RL 88a] Watrous, Raymond L. (1988). *Connectionist Speech Recognition Using the Temporal Flow Model.* IEEE Arden House Workshop on Automatic Speech Recognition, 1988.

[Watson, M 87a] Watson, Mark. (1987). *A Neural Network Model for Phoneme Recognition Using the Generalized Delta Rule for Connection Strength Modification.* IEEE International Conference on Neural Networks (ICNN-87), San Diego, CA, June 21-24, 1987: Proceedings. Volume 4. pp 389-396. New York: Institute of Electrical and Electronics Engineers.

[Watts, C 87*] SEE: Niznik, CA 87a. *Modeling and Mathematical Formalism for Robotic Cerebellar Neural Network Pathway Information Measures.*

[Wawrzynek, J 85*] SEE: Mead, CA 85a. *A New Discipline for CMOS Design: An Architecture for Sound Synthesis.*

[Waxman, AM 88a] Waxman, Allen M.; Wu, Jian; Bergholm, Fredrik. (1988). *Convected Activation Profiles and the Measurement of Visual Motion.* IEEE Computer Society Conference on Computer Vision and Pattern Recognition, Ann Arbor, MI, June 5-9, 1988: Proceedings. pp 717-723. New York: Institute of Electrical and Electronics Engineers.

[Waxman, AM 88b] Waxman, Allen M.; Seibert, Michael; Cunningham, Robert; Wu, Jian. (1988). *The Neural Analog Diffusion-Enhancement Layer (NADEL) and Early Visual Processing.* Visual Communications and Image Processing. SPIE Volume 1001. pp 1093-1102. Bellingham, WA: Society of Photo-Optical Instrumentation Engineers.

[Weaver, J 87*] SEE: Lehar, S 87a. *A Developmental Approach to Neural Network Design.*

[Webb, JA 85*] SEE: Kung, HT 85a. Mapping Image Processing Operations onto a Linear Systolic Machine.

[Webb, JA 85*] SEE: Gross, T 85a. *Warp as a Machine for Low-Level Vision.*

[Webb, JA 86*] SEE: Wallace, R 86a. *Progress in Robot Road-Following.*

[Wechsler, H 87a] Wechsler, Harry; Zimmerman, George Lee. (1987). *Fault-Tolerant Recognition Using DAM's.* IEEE International Conference on Neural Networks (ICNN-87), San Diego, CA, June 21-24, 1987: Proceedings. Volume 2. pp 719-726. New York: Institute of Electrical and Electronics Engineers.

[Wechsler, H 88*] SEE: Char, JM 88a. *Distributed and Fault-Tolerant Computation for Retrieval Tasks Using Distributed Associative Memories.*

[Wechsler, H 88a] Wechsler, Harry; Zimmerman, George Lee. (1988). *Invariant Object Recognition Using a Distributed Associative Memory.* In Anderson, Dana Z. (ed.). Neural Information Processing Systems. pp 830-839. New York: American Institute of Physics.

[Weems, C 82a] Weems, C. (1982). *TITANIC: A VLSI Based Content Addressable Parallel Array Processor.* International Conference on Computer Communication, 6th, London, September 7-10, 1982: Proceedings. pp 236-239.

[Weinberg, AM 40*] SEE: Offner, F 40a. *Nerve Conduction Theory: Some Mathematical Consequences of Bernstein's Model.*

[Weinberg, AM 40a] Weinberg, Alvin M. (1940). *The Equivalence of the Conduction Theories of Rashevsky and Rushton.* Bulletin of Mathematical Biophysics. Volume 2. pp 61-64.

[Weinberger, NM 83*] SEE: Kotliar, BE 83a. Neural Mechanism of Conditioning.

[Weisbuch, G 86*] SEE: Bienenstock, EL 86a. Disordered Systems and Biological Organization.

[Weisbuch, G 86a] Weisbuch, Gerard; d'Humieres, D. (1986). *Determining the Dynamic Landscape of Hopfield Networks.* In Bienenstock, Elie L. (ed.). Disordered Systems and Biological Organization. pp 187-192. New York: Springer-Verlag.

[Weisbuch, G 88a] Weisbuch, Gerard; Atlan, Henri. (1988). *Control of the Immune Response.* In Eckmiller, Rolf (ed.); von der Malsburg, Christoph (ed.). Neural Computers. NATO ASI Series, Volume F41. pp 497-501. Berlin: Springer-Verlag.

[Weiss, AM 81*] SEE: Warde, C 81a. *Optical Information Processing Characteristics of the Microchannel Spatial Light Modulator.*

[Weiss, N 70*] SEE: Baum, LE 70a. *A Maximization Technique Occurring in the Statistical Analysis of Probabilistic Functions of Markov Chains.*

[Weiss, TF 68a] Weiss, Thomas F. (ed.); Rosenblith, Walter A. (ed.). (1968). *Scanning the Issue: Special Issue on Studies of Neural Elements and Systems.* Proceedings of the IEEE. Volume 56. pp 891-894. New York: Institute of

Electrical and Electronics Engineers.

[Weisstein, N 68a] Weisstein, N. (1968). *A Rashevsky-Landahl Neural Net: Simulation of Metacontrast.* Psychological Review. Number 75. pp 494-521.

[Wellekens, CJ 87*] SEE: Bourland, H 87a. *Multilayer Perceptrons and Automatic Speech Recognition.*

[Wellekens, CJ 88*] SEE: Bourland, H 88a. Links Between Markov Models and Multilayer Perceptrons.

[Wellekens, CJ 88a] Wellekens, C. J. (1988). *Connectionism at the Philips Research Laboratory Brussels.* International Conference on "Connectionism in Perspective", Zurich, October 10-13, 1988.

[Wellekens, CJ 88b] Wellekens, C. J.; Bourlard, H. (1988). Les Applications des Modeles Connexionistes a la Reconnaissance de la Parole. Manuscript M 269. Brussels: Philips Research Laboratory.

[Welsh, A 78*] SEE: Marslen-Wilson, WD 78a. *Processing Interactions and Lexical Access During Word Recognition in Continuous Speech.*

[Werbos, PJ 82a] Werbos, Paul J. (1982). *Applications of Advances in Nonlinear Sensitivity Analysis.* System Modeling and Optimization. pp 762-770. New York: Springer-Verlag.

[Werbos, PJ 87a] Werbos, Paul J. (1987). *Building and Understanding Adaptive Systems: A Statistical/Numerical Approach to Factory Automation and Brain Research.* IEEE Transactions on Systems, Man and Cybernetics. Volume SMC-17. pp 7-20.

[Werbos, PJ 88a] Werbos, Paul J. (1988). *Backpropagation: Past and Future.* IEEE International Conference on Neural Networks (ICNN-88), San Diego, CA, July 24-27, 1988: Proceedings. Volume 1. pp 343-353. New York: Institute of Electrical and Electronics Engineers.

[Werling, R 85*] SEE: Clema, JK 85a. *Expert Systems and Machine Learning.*

[Werling, R 86*] SEE: Clema, JK 86a. *Knowledge Bases and Machine Learning.*

[Werner, G 75a] Werner, Gerhard. (1975). Feature Extraction by Neurons and Behavior. Cambridge, MA: MIT Press.

[Wesseling, D 77*] SEE: Bunge, E 77b. *The AUROS Project: Automatic Recognition of Speakers by Computers.*

[Westervelt, GL 86*] SEE: Babcock, KL 86a. *Complex Dynamics in Simple Neural Circuits.*

[Westervelt, RM 88*] SEE: Marcus, CM 88a. *Basins of Attraction for Electronic Neural Networks.*

[Westervelt, RM 88*] SEE: Marcus, CM 88b. *Stability of Analog Neural Networks with Delay.*

[Westlake, PR 70a] Westlake, Philip R. (1970). *The Possibilities of Neural Holographic Processes within the Brain.* Kybernetik. Volume 7. pp 129-153.

[Wheatley, TE 80*] SEE: Albus, JS 80a. *A Measurement and Control Model for Adaptive Robots.*

[Wheeler, BC 88a] Wheeler, B. C.; Smith, S. R. (1988). *Neural Network Models of an Escape Response.* Neural Networks. Volume 1, Supplement 1. pp 281.

[Wheeler, RM 83*] SEE: Narendra, KS 83a. *An N-Player Sequential Stochastic Game with Identical Payoffs.*

[White, H 87a] White, Halbert. (1987). *Some Asymptotic Results for Back-Propagation.* IEEE International Conference on Neural Networks (ICNN-87), San Diego, CA, June 21-24, 1987: Proceedings. Volume 3. pp 261-266. New York:

Institute of Electrical and Electronics Engineers.

[White, H 87b] White, Halbert. (1987). Some Asymptotic Results for Learning in Single Hidden Layer Feedforward Network Models. #87-13. San Diego, CA: University of California, Department of Economics.

[White, H 88*] SEE: Gallant, AR 88a. There Exists a Neural Network That Does Not Make Avoidable Mistakes.

[White, H 88a] White, Halbert. (1988). Economic Prediction Using Neural Networks: The Case of IBM Daily Stock Returns. #88-20. San Diego, CA: University of California, Department of Economics.

[White, H 88b] White, Halbert. (1988). The Case for Conceptual and Operational Separation of Network Architectures and Learning Mechanisms. #88-21. San Diego, CA: University of California, Department of Economics.

[White, HJ 88a] White, H. J.; Aldridge, N. B.; Lindsay, I. (1988). *Digital and Analogue Holographic Associative Memories*. Optical Engineering. Volume 27.

[White, HJ 88b] White, H. J. (1988). *Experimental Results from a Simple Neural Network*. Proceedings. SPIE Volume 963.

[White, HJ 88c] White, H. J.; Wright, W. A. (1988). *Holographic Implementation of a Discrete Hopfield Neural Network Model*. Proceedings. SPIE Volume 882.

[White, HJ 88d] White, H. J.; Wright, W. A. (1988). *Holographic Implementation of a Hopfield Model with Discrete Weightings*. Applied Optics. Volume 27. pp 331.

[White, M 88*] SEE: Bilbro, GL 88a. *Image Segmentation with Neurocomputers*.

[White, RL 81a] White, Roy L.; Gardner, Daniel. (1981). *Self-Inhibition Alters Firing Patterns of Neurons in Aplysia Buccal Ganglia*. Brain Reseach. Volume 209. pp 77-93.

[White, SR 86*] SEE: Solla, SA 86a. *Configuration Space Analysis for Optimization Problems*.

[Whitehead, BA 87*] SEE: Strong, GW 87a. *A Factored Architecture to Solve Illusory Conjunctions in Neural Networks*.

[Whitfield, IC 68a] Whitfield, I. C. (1968). *Centrifugal Control Mechanisms*. Proceedings of the IEEE. Volume 56. pp 1005-1008.

[Whitley, D 88a] Whitley, Darrell. (1988). Applying Genetic Algorithms to Neural Net Learning. Technical Report CS-88-128. Fort Collins, CO: Colorado State University.

[Whittington, DA 81*] SEE: Kuperstein, M 81a. *A Practical 24 Channel Microelectrode for Neural Recording in Vivo*.

[Wick, MR 87*] SEE: Poliac, MO 87a. *A Crew Scheduling Problem*.

[Wickelgren, WA 69a] Wickelgren, W. A. (1969). *Context-Sensitive Coding, Associative Memory, and Serial Order in (speech) Behavior*. Psychological Review. Volume 76. pp 1-15.

[Wickelgren, WA 79a] Wickelgren, W. A. (1979). *Chunking and Consolidation: A Theoretical Synthesis of Semantic Networks, Configuring in Conditioning, S-R Versus Cognitive Learning, etc.*. Psychological Review. Volume 86. pp 44-60.

[Widrow, B 59a] Widrow, Bernard. (1959). *Adaptive Sampled-Data Systems-A Statistical Theory of Adaptation*. WESCON Convention Record. pp 74-85.

[Widrow, B 60a] Widrow, Bernard. (1961). *Adaptive Sampled-Data Systems*. Automatic and Remote Control, Proceedings of the First International Congress of the International Federation of Automatic Control, Moscow, 1960. Volume I. pp 423-429. London: Butterworths.

[Widrow, B 60b] Widrow, Bernard; Hoff, Marcian E. (1988). *Adaptive Switching Circuits.* In Anderson, James A. (ed.); Rosenfeld, Edward (ed.). Neurocomputing: Foundations of Research. pp 126-134. Cambridge, MA: MIT Press.

[Widrow, B 60c] Widrow, Bernard; Hoff, Marcian E. (1960). *Adaptive Switching Circuits.* WESCON Convention Record. pp 96-104.

[Widrow, B 60d] Widrow, Bernard. (1960). An Adaptive "Adaline" Neuron Using Chemical "Memistors". Technical Report Number 1553-2. Stanford, CA: Stanford University, Stanford Electronics Laboratory.

[Widrow, B 61a] Widrow, Bernard; Pierce, W. H.; Angell, James B. (1961). *Birth, Life, and Death in Microelectronic Systems.* IRE Transactions on Military Electronics. MIL-5.

[Widrow, B 62a] Widrow, Bernard. (1962). *Generalization and Information Storage in Networks of Adaline "Neurons".* In Yovitz, M. C. (ed.); Jacobi, George T. (ed.); Goldstein, Gordon D. (ed.). Self-Organizing Systems. pp 435-461. Washington, DC: Spartan.

[Widrow, B 62b] Widrow, Bernard. (1962). *Rate of Adaptation in Control Systems.* ARS Journal. Volume 32. pp 1378-1385.

[Widrow, B 62c] Widrow, Bernard; Angell, James B. (1962). *Reliable, Trainable Networks for Computing and Control.* Aerospace Engineering. Volume 21. pp 78-123.

[Widrow, B 63a] Widrow, Bernard. (1963). *A Statistical Theory of Adaptation.* Adaptive Control Systems, Proceedings of a Symposium held in Garden City, Long Island, NY, October, 1960. pp 97-122. New York: Macmillan.

[Widrow, B 63b] Widrow, Bernard; Groner, G. F.; Hu, M. J. C.; Smith, Fred W.; Specht, D. F.; Talbert, Lee R. (1963). *Practical Applications for Adaptive Data-Processing Systems.* WESCON 63: Western Electronic Show and Convention, San Francisco, CA, August 20-23, 1987. Volume 11.4. pp 1-14.

[Widrow, B 64a] Widrow, Bernard. (1964). *Pattern Recognition and Adaptive Control.* Applications and Industry. pp 269-277. New York: Institute of Electrical and Electronics Engineers.

[Widrow, B 64b] Widrow, Bernard; Smith, Fred W. (1964). *Pattern-Recognizing Control Systems.* Computer and Information Sciences (COINS) Symposium Proceedings. pp 288-317. Washington, DC: Spartan.

[Widrow, B 64c] Widrow, Bernard; Smith, Fred W. (1964). *Pattern-Recognizing Control Systems.* In Tou, Julius T. (ed.); Wilcox, R. H. (ed.). Computer and Information Sciences: Collected Papers in Learning, Adaptation, and Control in Information Systems. pp 318-343. Washington, DC: Spartan.

[Widrow, B 65*] SEE: Steinbuch, K 65a. *A Critical Comparison of Two Kinds of Adaptive Classification Networks.*

[Widrow, B 66a] Widrow, Bernard. (1966). *"Bootstrap Learning" in Threshold Logic Systems.* Congress of the International Federation of Automatic Control, 3rd: Proceedings. Volume 1. pp 14D.1-14D.10. London: International Federation of Automatic Control.

[Widrow, B 66b] Widrow, Bernard. (1966). Adaptive Filters I: Fundamentals. SU-SEL-66-126. Stanford, CA: Stanford University, Stanford Electronics Laboratory.

[Widrow, B 67a] Widrow, Bernard; Mantey, P. E.; Griffiths, L. J.; Goode, B. B. (1967). *Adaptive Antenna Systems.* Proceedings of the IEEE. Volume 55. pp 2143-2159.

[Widrow, B 71a] Widrow, Bernard. (1971). *Adaptive Filters.* Aspects of Network and System Theory. pp 563-586. New York: Holt, Rinehart, and Winston.

[Widrow, B 73a] Widrow, Bernard; Gupta, Narendra K.; Maitra, Sidhartha. (1973). *Punish / Reward: Learning With a Critic in Adaptive Threshold Systems.* IEEE Transactions on Systems, Man and Cybernetics. Volume SMC-3. pp 455-465.

[Widrow, B 73b] Widrow, Bernard. (1973). *The "Rubber-Mask" Technique - I. Pattern Storage and Recognition.* Pattern Recognition. Volume 5. pp 175-197.

[Widrow, B 73c] Widrow, Bernard. (1973). *The "Rubber-Mask" Technique - II. Pattern Storage and Recognition.* Pattern Recognition. Volume 5. pp 199-211.

[Widrow, B 75a] Widrow, Bernard; Glover, John R.; McCool, John M.; Kaunitz, John; Williams, Charles S.; Hearn, Robert H. (1975). *Adaptive Noise Canceling: Principles and Applications.* Proceedings of the IEEE. Volume 63. pp 1692-1716.

[Widrow, B 75b] Widrow, Bernard; McCool, John M.; Ball, Michael. (1975). *The Complex LMS Algorithm.* Proceedings of the IEEE. Volume 63. pp 719-720.

[Widrow, B 76a] Widrow, Bernard; McCool, John M. (1976). *A Comparison of Adaptive Algorithms Based on the Methods of Steepest Descent and Random Search.* IEEE Transactions on Antennas and Propagation. Volume AP-24. pp 615-637.

[Widrow, B 76b] Widrow, Bernard; McCool, John M.; Larimore, Michael G.; Johnson Jr, C. Richard. (1976). *Stationary and Nonstationary Learning Characteristics of the LMS Adaptive Filter.* Proceedings of the IEEE. Volume 64. pp 1151-1162.

[Widrow, B 77a] Widrow, Bernard; McCool, John M. (1977). *Comments on "An Adaptive Recursive LMS Filter".* Proceedings of the IEEE. Volume 65. pp 1402-1404.

[Widrow, B 78*] SEE: Dentino, M 78a. *Adaptive Filtering in the Frequency Domain.*

[Widrow, B 79a] Widrow, Bernard. (1979). *A Review of Adaptive Antennas.* IEEE International Conference on Acoustics, Speech, and Signal Processing (ICASSP-79), Washington, DC, April 2-4, 1979: Proceedings. pp 273-278. New York: Institute of Electrical and Electronics Engineers.

[Widrow, B 80a] Widrow, Bernard. (1980). *On the Speed of Convergence of Adaptive Algorithms.* Asilomar Conference on Circuits, Systems, and Computers, 14th, 1980: Conference Record. pp 239-242. New York: Institute of Electrical and Electronics Engineers.

[Widrow, B 81*] SEE: Ferrara, Jr, ER 81a. *Multichannel Adaptive Filtering for Signal Enhancement.*

[Widrow, B 81*] SEE: Ferrara, Jr, ER 81b. *The Time-Sequenced Adaptive Filter.*

[Widrow, B 81a] Widrow, Bernard; Titchener, Paul F.; Gooch, Richard P. (1981). *Adpative Design of Digital Filters.* IEEE International Conference on Acoustics, Speech, and Signal Processing (ICASSP-81), Atlanta, GA, April 1, 1981: Proceedings. Volume 1. pp 243-246. New York: Institute of Electrical and Electronics Engineers.

[Widrow, B 82*] SEE: McGill, KC 82a. *On the Nature and Elimination of Stimulus Artifact in Nerve Signals Evoked and Recorded Using Surface Electrodes.*

[Widrow, B 84*] SEE: Walach, E 84a. *The Least Mean Fourth (LMF) Adaptive Algorithm and Its Family.*

[Widrow, B 84a] Widrow, Bernard; Walach, Eugene. (1984). *On the Statistical Efficiency of the LMS Algorithm with Nonstationary Inputs.* IEEE Transactions on Information Theory. Volume IT-30. pp 211-221.

[Widrow, B 85*] SEE: Jablon, N 85a. *Adaptive Superresolution Phenomena in Aberrant Arrays.*

[Widrow, B 85*] SEE: Shynk, JJ 85a. *A Frequency-Domain Adaptive Pole-Zero Filter with Applications.*

[Widrow, B 85a] Widrow, Bernard; Stearns, Samuel D. (1985). Adaptive Signal Processing. New York: Prentice-Hall.

[Widrow, B 87a] Widrow, Bernard. (1987). *Adaline and Madaline - 1963.* IEEE International Conference on Neural Networks

(ICNN-87), San Diego, CA, June 21-24, 1987: Proceedings. Volume 1. pp 143-158. New York: Institute of Electrical and Electronics Engineers.

[Widrow, B 87b] Widrow, Bernard; Baudrenghien, Philippe; Vetterli, Martin; Titchener, Paul F. (1987). *Fundamental Relations Between the LMS Algorithm and the DFT*. IEEE Transactions on Circuits and Systems. Volume CAS-34. pp 814-820.

[Widrow, B 87c] Widrow, Bernard; Winter, Rodney G.; Baxter, Robert A. (1987). *Learning Phenomena in Layered Neural Networks*. IEEE International Conference on Neural Networks (ICNN-87), San Diego, CA, June 21-24, 1987: Proceedings. Volume 2. pp 411-430. New York: Institute of Electrical and Electronics Engineers.

[Widrow, B 88a] Widrow, Bernard; Winter, Rodney G.; Baxter, Robert A. (1988). *Layered Neural Nets for Pattern Recognition*. IEEE Transactions on Acoustics, Speech, and Signal Processing. Volume 36. pp 1109-1118.

[Widrow, B 88b] Widrow, Bernard; Winter, Rodney. (March, 1988). *Neural Nets for Adaptive Filtering and Adaptive Pattern Recognition*. Computer. Volume 21. pp 25-39.

[Wieland, A 87a] Wieland, Alexis; Leighton, Russell. (1987). *Geometric Analysis of Neural Network Capabilities*. IEEE International Conference on Neural Networks (ICNN-87), San Diego, CA, June 21-24, 1987: Proceedings. Volume 3. pp 385-392. New York: Institute of Electrical and Electronics Engineers.

[Wiesel, TN 83*] SEE: Gilbert, CD 83a. *Clustered Intrinsic Connections in Cat Visual Cortex*.

[Wiggins, R 86*] SEE: Penz, PA 86a. *Digital Signal Processor Accelerators for Neural Network Simulations*.

[Wightman, CW 60*] SEE: Hay, JC 60a. *The Mark I Perceptron - Design and Performance*.

[Wigstrom, H 73a] Wigstrom, H. (1973). *A Neuron Model with Learning Capability and its Relation to Mechanisms of Association*. Kybernetik. Volume 12. pp 204-215.

[Wigstrom, H 74a] Wigstrom, Holger. (1974). *A Model of a Neural Network with Recurrent Inhibition*. Kybernetik. Volume 16. pp 103-112.

[Wilamowski, BM 75a] Wilamowski, B. M.; Czarnul, Z.; Bialko, M. (1975). *Novel Inductorless Neuristor Line*. Electronics Letters. Volume 11. pp 355-356.

[Wilby, WPL 60*] SEE: Gabor, D 60a. *A Universal Non-Linear Filter, Predictor, and Simulator which Optimizes Itself by a Learning Process*.

[Wilczek, F 87*] SEE: Baum, EB 87a. *New "Neural" Algorithms for Associative Memory*.

[Wilczek, F 88*] SEE: Baum, EB 88a. *Internal Representations for Associative Memory*.

[Wilczek, F 88*] SEE: Baum, EB 88d. *Supervised Learning of Probability Distributions by Neural Networks*.

[Wiles, J 87a] Wiles, Janet. (1987). *Analysis of a Distributed Associative Memory Model*. IEEE International Conference on Neural Networks (ICNN-87), San Diego, CA, June 21-24, 1987: Proceedings. Volume 3. pp 393-402. New York: Institute of Electrical and Electronics Engineers.

[Wilks, RS 87a] Wilks, Robert S. (1987). *A Leading Question*. Artificial Neural Networks, An Introductory Seminar and Exhibition, Cincinnati, OH, December 8, 1987. Cincinnati, OH: Institute of Advanced Manufacturing Sciences.

[Will, CA 87a] Will, Craig A. (1987). *Neural Network Architectures and Their Implications for Next Generation Information Systems*. pp 32-38. New York: Association for Computing Machinery.

[Willams, RJ 86*] SEE: Rumelhart, DE 86d. *Learning Internal Representations by Error Propagation*.

[Willams, RJ 86*] SEE: Rumelhart, DE 86f. *Learning Representations by Back-Propagating Errors.*

[Willcox, CR 88a] Willcox, Charles R. (in press). *Exponential Storage and Retrieval in Hierarchical Neural Networks.* Journal of Physics A.

[Willett, P 83*] SEE: Rubin, V 83a. *A Comparison of Some Hierarchial Monothetic Divisive Clustering Algorithms for Structure-Property Correlation.*

[Willey, J 87*] SEE: Scheff, K 87b. *Simulation Results of Multiple Layer Difference Learning.*

[Williams, CS 75*] SEE: Widrow, B 75a. *Adaptive Noise Canceling: Principles and Applications.*

[Williams, G 87a] Williams, Gregg. (1987). *An Introduction to Relaxation Methods.* BYTE. Volume 12. pp 111-124.

[Williams, RJ 83a] Williams, Ronald J. (1983). Unit Activation Rules for Cognitive Network Models. ICS-8303. San Diego, CA: University of California, San Diego, Institute for Cognitive Science.

[Williams, RJ 85*] SEE: Rumelhart, DE 85b. Learning Internal Representations by Error Propagation.

[Williams, RJ 85a] Williams, Ronald J. (1985). Feature Discovery Through Error-Correction Learning. ICS-8501. San Diego, CA: University of California, San Diego, Institute for Cognitive Science.

[Williams, RJ 85b] Williams, Ronald J. (1985). Inference of Spatial Relations by Self-Organizing Networks. ICS-8502. San Diego, CA: University of California, San Diego, Institute for Cognitive Science.

[Williams, RJ 86*] SEE: Rumelhart, DE 86c. *Learning Internal Representations by Error Propagation.*

[Williams, RJ 86*] SEE: Rumelhart, DE 86e. *Learning Representations by Back-Propagating Errors.*

[Williams, RJ 86a] Williams, Ronald J. (1986). *Inverting a Connectionist Network Mapping by Back-Propagation of Error.* Cognitive Science Society, 8th, Amherst, MA, August 15-17, 1986: Proceedings. pp 859-865. Hillsdale, NJ: Lawrence Erlbaum.

[Williams, RJ 86b] Williams, Ronald J. (1986). *Logic of Activation Functions.* In Rumelhart, David E. (ed.). Parallel Distributed Processing: Explorations in the Microstructure of Cognition. Volume I, Foundations. pp 423-443. Cambridge, MA: MIT Press.

[Williams, RJ 86c] Williams, Ronald J. (1986). Reinforcement Learning in Connectionist Networks: A Mathematical Analysis. ICS-8605. San Diego, CA: University of California, San Diego, Institute for Cognitive Science.

[Williams, RJ 87a] Williams, Ronald J. (1987). *A Class of Gradient-Estimating Algorithms for Reinforcement Learning in Neural Networks.* IEEE International Conference on Neural Networks (ICNN-87), San Diego, CA, June 21-24, 1987: Proceedings. Volume 2. pp 601-608. New York: Institute of Electrical and Electronics Engineers.

[Williams, RJ 88a] Williams, Ronald J. (1988). *On the Use of Backpropagation in Associative Reinforcement Learning.* IEEE International Conference on Neural Networks (ICNN-88), San Diego, CA, July 24-27, 1988: Proceedings. Volume 1. pp 263-270. New York: Institute of Electrical and Electronics Engineers.

[Williams, RJ 88b] Williams, Ronald J. (1988). Toward a Theory of Reinforcement-Learning Connectionist Systems. Technical Report NU-CCS-88-3. Boston: Northeastern University, College of Computer Science.

[Williams, T 87a] Williams, Tom. (March 1, 1987). *Optics and Neural Nets: Trying to Model the Human Brain.* Computer Design.

[Williams, WJ 83*] SEE: Fuller, MS 83a. *A Continuous Information Theoretic Approach to the Analysis of Cutaneous Receptor*

Neurons.

[Willows, AOD 73a] Willows, A. O. D.; Dorsett, D. A.; Hoyle, G. (1973). *The Neuronal Basis of Behavior in Tritonia, III Neuronal Mechanism of a Fixed Action Pattern.* Journal of Neurobiology. Volume 4. pp 255-285.

[Willshaw, DJ 69a] Willshaw, David J.; Buneman, O. P.; Longuet-Higgins, H. C. (1969). *Non-Holographic Associative Memory.* Nature. Volume 222. pp 960-962.

[Willshaw, DJ 69b] Willshaw, David J.; Buneman, O. P.; Longuet-Higgins, H. C. (1988). *Non-Holographic Associative Memory.* In Anderson, James A. (ed.); Rosenfeld, Edward (ed.). Neurocomputing: Foundations of Research. pp 153-156. Cambridge, MA: MIT Press.

[Willshaw, DJ 69c] Willshaw, David J.; Longuet-Higgins, H. C. (1969). *The Holophone: Recent Developments.* Machine Intelligence. Volume 4. pp 349-357.

[Willshaw, DJ 70*] SEE: Longuet-Higgins, HC 70a. *Theories of Associative Recall.*

[Willshaw, DJ 70a] Willshaw, David J.; Longuet-Higgins, H. C. (1970). *Associative Memory Models.* Machine Intelligence. pp 351-359. New York: Elsevier.

[Willshaw, DJ 72a] Willshaw, David J.; Buneman, O. P. (1972). *Parallel and Serial Methods of Pattern Matching.* Machine Intelligence. Volume 7. pp 357-367. New York: Wiley.

[Willshaw, DJ 75a] Willshaw, David J.; von der Malsburg, Christoph. (1975). *How Patterned Neural Connections Can be Set Up by Self-Organization.* Self-Organization in Neurogenesis. pp 431-445.

[Willshaw, DJ 76a] Willshaw, David J. (1972). *A Simple Network Capable of Inductive Generalization.* Proceedings of the Royal Society, Series B. Volume 182. pp 233-247.

[Willshaw, DJ 76b] Willshaw, David J.; von der Malsburg, Christoph. (1979). *How Patterned Neural Connections Can be Set Up by Self-Organization.* Proceedings of the Royal Society, Series B. Volume 194. pp 431-445.

[Willshaw, DJ 77*] SEE: von der Malsburg, C 77a. *How to Label Nerve Cells so that they can Interconnect in an Ordered Fashion.*

[Willshaw, DJ 79a] Willshaw, David J.; von der Malsburg, Christoph. (1979). *A Marker Induction Mechanism for the Establishment of Ordered Neural Mapping: Its Application to the Retinotectal Connections.* Philosophical Transactions of the Royal Society, Series B. Volume 287. pp 203-243.

[Willshaw, DJ 81*] SEE: von der Malsburg, C 81a. *Differential Equations for the Development of Topological Nerve Fibre Projections.*

[Willshaw, DJ 81a] Willshaw, David J. (1981). *Holography, Associative Memory, and Inductive Generalization.* In Hinton, Geoffrey E. (ed.). Parallel Models of Associative Memory. pp 83-104. Hillsdale, NJ: Lawrence Erlbaum.

[Willshaw, DJ 87*] SEE: Durbin, R 87a. *An Analogue Approach to the Traveling Salesman Problem Using an Elastic Net Method.*

[Willson, NJ 87a] Willson, N. J. (1987). A Neural Net for Size and Position Invariant Pattern Recognition. Technical Report CCCN-1. Stirling: Stirling University, Departments of Computer Science and Psychology.

[Willson, NJ 88*] SEE: Phillips, WA 88a. *On the Acquisition of Object Concepts from Sensory Data.*

[Willwacher, G 82a] Willwacher, Gerd. (1982). *Storage of a Temporal Pattern Sequence in a Network.* Biological Cybernetics. Volume 43. pp 115-126.

[Wilson, DL 72a] Wilson, Dennis L. (1972). *Asymptotic Properties of Nearest Neighbor Rules Using Edited Data.* IEEE Transactions on Systems, Man and Cybernetics. Volume SMC-2. pp 408-421.

[Wilson, DM 68a] Wilson, Donald M.; Waldron, I. (1968). *Models for the Generation of the Motor Output Pattern in Flying Locusts.* Proceedings of the IEEE. Volume 56. pp 1058-1064.

[Wilson, GV 87*] SEE: Forrest, BM 87a. *Implementing Neural Network Models on Parallel Computers.*

[Wilson, HR 73a] Wilson, H. R.; Cowan, Jack D. (1973). *A Mathematical Theory of the Functional Dynamics of Cortical and Thalamic Nervous Tissue.* Kybernetik. Volume 13. pp 55-80.

[Wilson, HR 75a] Wilson, Hugh R. (1975). *A Synaptic Model for Spatial Frequency Adaptation.* Journal of Theoretical Biology. Volume 50. pp 327-352.

[Wilson, HR 79a] Wilson, Hugh R.; Bergen, James R. (1979). *A Four Mechanism Model for Threshold Spatial Vision.* Vision Research. Volume 19. pp 19-32.

[Wilson, MA 87*] SEE: Bower, JM 87a. Piriform (Olfactory) Cortex Model on the Hypercube.

[Wilson, MA 87a] Wilson, Matthew A.; Bower, James M. (1987). *A Computer Simulation of Olfactory Cortex with Functional Implications for Storage and Retrieval of Olfactory Information.* In Anderson, Dana Z. (ed.). Neural Information Processing Systems. New York: American Institute of Physics.

[Wilson, MA 88*] SEE: Bower, JM 88a. *A Computer Simulation of Olfactory Cortex with Functional Implications for Storage and Retrieval of Olfactory Information.*

[Windecker, RC 78a] Windecker, Richard C. (1978). *Stochastic Combinational Networks.* Information Sciences. Volume 16. pp 185-234.

[Windecker, RC 88a] Windecker, Richard C. (1988). *Learning in Networks of Nondeterministic Adaptive Logic Elements.* In Anderson, Dana Z. (ed.). Neural Information Processing Systems. pp 840-849. New York: American Institute of Physics.

[Winston, JV 84*] SEE: Clark, JW 84a. *Self-Organization of Neural Networks.*

[Winston, PH 75a] Winston, Patrick Henry. (1975). *Learning Structural Descriptions from Examples.* The Psychology of Computer Vision. pp 157-209. New York: McGraw-Hill.

[Winter, CL 87*] SEE: Ryan, TW 87a. *Dynamic Control of an Artificial Neural System: The Property Inheritance Network.*

[Winter, CL 87*] SEE: Ryan, TW 87b. *Variations on Adaptive Resonance.*

[Winter, CL 87a] Winter, C. L.; Ryan, Thomas W.; Turner, Charles J. (1987). TIN: A Trainable Inference Network. Tucson, AZ: Science Applications International Corporation.

[Winter, CL 87b] Winter, C. L.; Ryan, Thomas W.; Turner, Charles J. (1987). *TIN: A Trainable Inference Network.* IEEE International Conference on Neural Networks (ICNN-87), San Diego, CA, June 21-24, 1987: Proceedings. Volume 2. pp 777-786. New York: Institute of Electrical and Electronics Engineers.

[Winter, CL 88a] Winter, C. L. (1988). *An Adaptive Network that Flees Pursuit.* Neural Networks.

[Winter, CL 88b] Winter, C. L. (1988). An Associative Memory with Heterogeneous Input Fields. Tucson, AZ: Science Applications International Corporation.

[Winter, R 88*] SEE: Widrow, B 88b. *Neural Nets for Adaptive Filtering and Adaptive Pattern Recognition.*

[Winter, RG 87*] SEE: Widrow, B 87c. *Learning Phenomena in Layered Neural Networks.*

[Winter, RG 88*] SEE: Widrow, B 88a. *Layered Neural Nets for Pattern Recognition.*

[Winters, JH 88a] Winters, Jack H.; Rose, Christopher. (1988). *On Parallel Networks for Optimum Classification.* Neural Networks. Volume 1. pp 416.

[Winters, JH 88b] Winters, Jack H. ((to appear)). *Superresolution for Ultrasonic Imaging in Air Using Neural Networks.* IEEE International Conference on Neural Networks (ICNN-88), San Diego, CA, July 24-27, 1988: Proceedings.

[Winters, JH 89a] Winters, Jack H.; Rose, Christopher. (in press). *Minimum Distance Automata in Parallel Networks for Optimum Classification.* Neural Networks.

[Wishner, RP 65*] SEE: Kelly, EJ 65a. *Matched-Filter Theory for High-Velocity, Accelerating Targets.*

[Withers, RS 86*] SEE: Sage, JP 86a. *Artificial Neural Network Integrated Circuit Based on NMOS/CCD Principles.*

[Withers, RS 88*] SEE: Sage, JP 88a. *Analog Nonvolatile Memory for Neural Network Implementations.*

[Wittner, BS 87*] SEE: Denker, JS 87a. *Automatic Learning, Rule Extraction, and Generalization.*

[Wittner, BS 88*] SEE: Denker, JS 88a. *Network Generality, Traning Required, and Precision Required.*

[Wittner, BS 88a] Wittner, Ben S.; Denker, John S. (1988). *Strategies for Teaching Layered Networks Classification Tasks.* In Anderson, Dana Z. (ed.). Neural Information Processing Systems. pp 850-859. New York: American Institute of Physics.

[Wobcke, WR 86*] SEE: Sharkey, NE 86a. *Mixing Binary and Continuous Connection Schemes for Knowledge Access.*

[Wolf, L 87a] Wolf, Luzian; Jaron, D.; Moore, T. (1987). *A Trainable Model for Near Periodic Signals.* Northeast Biomedical Engineering Conference, 13th, Philadelphia, PA, 1987: Proceedings.

[Wolf, L 87b] Wolf, Luzian. (1987). *Recurrent Nets for the Storage of Cyclic Sequences.* IEEE International Conference on Neural Networks (ICNN-88), San Diego, CA, July 24-27, 1988: Proceedings. Volume 1. pp 53-60. Perchtoldsdorf, Austria: Wolf-Software.

[Wolff, JR 86*] SEE: Dammasch, IE 86b. *Self-Stabilization of Neuronal Networks, I. The Compensation Algorithm for Synaptogenesis.*

[Wolff, JR 88*] SEE: Dammasch, IE 88a. *A Model of Synaptogenetic Consequences of Afferent Input in Neuronal Nets.*

[Wolff, JR 88*] SEE: Dammasch, IE 88c. *Self-Stabilization of Neuronal Networks, II. Stability Conditions for Synaptogenesis.*

[Wolsey, LA 88*] SEE: Nemhauser, GL 88a. Integer and Combinatorial Optimizations.

[Wong, AJ 88a] Wong, A. Jun-Wei. (1988). *Recognition of General Patterns Using Neural Networks.* Biological Cybernetics. Volume 58. pp 361-372.

[Wong, MK 86a] Wong, Maurice K.; Chun, Hon Wai. (1986). *Toward a Massively Parallel System for Word Recognition.* IEEE International Conference on Acoustics, Speech, and Signal Processing (ICASSP-86), Tokyo, April 7-11, 1986: Proceedings. pp 1967-1970. New York: Institute of Electrical and Electronics Engineers.

[Wong, WS 88*] SEE: Morris, RJT 88a. *A Decentralized Tunable Short Term Neural Network Memory and Application to Tracking.*

[Wong, Y 87a] Wong, Yiu-fai; Banik, Jashojiban; Bower, James M. (1987). *Neural Networks for Template Matching: Application to Real-Time Classification of the Action Potentials of Real Neurons.* Neural Information Processing Systems. New York: American Institute of Physics.

[Wong, Y 88*] SEE: Bower, JM 88b. *Neural Networks for Template Matching: Application to Real-Time Classification of the Action Potentials of Real Neurons.*

[Wood, CC 78a] Wood, Charles C. (1978). *Variations on a Theme by Lashley: Lesion Experiments on the Neural Model of Anderson, Silverstein, Ritz and Jones.* Psychological Review. Volume 85. pp 582-591.

[Wood, CT 71*] SEE: Juola, JF 71a. *Recognition Time for Information Stored in Long-Term Memory.*

[Woodcock, R 60*] SEE: Gabor, D 60a. *A Universal Non-Linear Filter, Predictor, and Simulator which Optimizes Itself by a Learning Process.*

[Woodruff, HB 75*] SEE: Ritter, GL 75a. *An Algorithm for a Selective Nearest Neighbor Decision Rule.*

[Woods, JW 72a] Woods, John W. (1972). *Two-Dimensional Discrete Markovian Fields.* IEEE Transactions on Information Theory. Volume IT-18. pp 232-240.

[Wright, MH 81*] SEE: Gill, PE 81a. Practical Optimization.

[Wright, MJ 88*] SEE: Dobson, VG 88b. *Differences in Performance of Binary Hetero-Associative Networks Using Link-Enabling vs. Link-Disabling Local Modification Rules.*

[Wright, WA 88*] SEE: White, HJ 88c. *Holographic Implementation of a Discrete Hopfield Neural Network Model.*

[Wright, WA 88*] SEE: White, HJ 88d. *Holographic Implementation of a Hopfield Model with Discrete Weightings.*

[Wright, WA 89a] Wright, W. A. (in press). *A Markov Random Field Approach to Data Fusion and Colour Segmentation.* Image and Vision Computing.

[Wright, WA 89b] Wright, W. A. (in press). *Contextual Road Finding with a Neural Network.* Neural Networks.

[Wright, WA 89c] Wright, W. A. (in press). *Road Finding Neural Network.* PAMI.

[Wright, WA 89d] Wright, W. A. (in press). *Towards a Generic Learning Algorithm.* Neural Networks.

[Wu, H 83*] SEE: Salton, G 83a. *Extended Boolean Information Retrieval.*

[Wu, J 88*] SEE: Waxman, AM 88a. *Convected Activation Profiles and the Measurement of Visual Motion.*

[Wu, J 88*] SEE: Waxman, AM 88b. *The Neural Analog Diffusion-Enhancement Layer (NADEL) and Early Visual Processing.*

[Wu, PT 84*] SEE: Aipperspach, AG 84a. *Content-Addressable memory Cell Having Only Six Transitions.*

[Wu, ZL 88a] Wu, Zong Liang. (1988). *Simulating the Adaptation Phenomena in the Response of Auditory Nerve Fibers: A Functional Model.* Signal Processing IV: Theories and Applications. pp 1313-1316. New York: Elsevier.

[Wyant, JC 77*] SEE: Tamura, PN 77a. *Matrix Multiplication Using Coherent Optical Techniques.*

[Wyatt Jr, JL 88*] SEE: Standley, DL 88a. *Stability Theorem for Lateral Inhibition Networks that is Robust in the Presence of Circuit Parasitics.*

[Wyatt Jr, JL 88a] Wyatt Jr, John L.; Standley, D. L. (1988). *A Method for the Design of Stable Lateral Inhibition Networks that is Robust in the Presence of Circuit Parasitics.* In Anderson, Dana Z. (ed.). Neural Information Processing Systems. pp 860-868. New York: American Institute of Physics.

[Wyatt Jr, JL 88b] Wyatt Jr, John L.; Standley, D. L. (1988). *Circuit Design Criteria for Stable Lateral Inhibition Neural Networks.* IEEE International Symposium on Circuits and Systems, Helsinki, June 7-9, 1988: Proceedings. Volume 2. pp 997-1000. New York: Institute of Electrical and Electronics Engineers.

[Wyman, H 68*] SEE: Segundo, JP 68a. *Input-Output Relations in Computer-Simulated Nerve Cells: Influence of Statistical Properties, Strength, etc..*

[Wyner, AD 84a] Wyner, Aaron D.; Landau, Henry J. (1984). *Optimum Waveform Signal Sets with Amplitude and Energy Constraints.* IEEE Transactions on Information Theory. Volume IT-30. pp 615-622.

[Xu, J 86a] Xu, Jing-hua; Li, Wei. (1986). *The Dynamics of Large Scale Neuron-Glia Network and Its Relation to the Brain Functions (I).* Communication in Theoretical Physics. Volume 5. pp 339-346.

[Xu, X 88*] SEE: Chen, S 88a. A Case Study of Solving Optimization Problems Using Neural Network.

[Xu, X 88a] Xu, X.; Tsai, W. T.; Huang, Nai-Kuan. (1988). A Generalized Neural Network Model. Minneapolis, MN: University of Minnesota, Computer Science and Electrical Engineering.

[Xu, X 88b] Xu, X.; Tsai, W. T. (1988). Constructing Associative Memories Using Neural Networks. Minneapolis, MN: University of Minnesota, Computer Science Department.

[Xu, X 88c] Xu, X.; Tsai, W. T. (1988). On "Information Capacity of the Hopfield Model". Minneapolis, MN: University of Minnesota, Computer Science Department.

[Yager, RR 87a] Yager, Ronald R. (ed.); Ovchinnikov, S. (ed.); Tong, R. M. (ed.); Nguyen, H. T. (ed.). (1987). Fuzzy Sets and Applications: Selected Papers by L A Zadeh. New York: Wiley.

[Yager, RR 87b] Yager, Ronald R. (1987). *On the Aggregation of Processing Units in Neural Networks.* IEEE International Conference on Neural Networks (ICNN-87), San Diego, CA, June 21-24, 1987: Proceedings. Volume 2. pp 327-334. New York: Institute of Electrical and Electronics Engineers.

[Yajima, A 72*] SEE: Morishita, I 72a. *Analysis and Simulation of Networks of Mutually Inhibiting Neurons.*

[Yajima, S 68a] Yajima, S.; Ibaraki, T.; Kawano, I. (1968). *On Autonomous Logic Nets of Threshold Elements.* IEEE Transactions on Computers. Volume C-17. pp 385-391.

[Yakimovsky, Y 74*] SEE: Feldman, JA 74a. *Decision Theory and Artificial Intelligence: I. A Semantics-Based Region Analyzer.*

[Yamafuji, K 81*] SEE: Toko, K 81a. *Application of Bifurcation Theory to Nerve Excitations.*

[Yamajo, Y 88*] SEE: Murakami, K 88a. *Construction of Associative Memory from Incomplete Memorized Information Using Space Reduction Method.*

[Yamamura, A 87*] SEE: Psaltis, D 87b. *Neural Controllers.*

[Yang, H 87a] Yang, Hedong; Guest, Clark C. (1987). *Performance of Backpropagation for Rotation Invariant Pattern Recognition.* IEEE International Conference on Neural Networks (ICNN), San Diego, CA, June 21-24, 1987: Proceedings. Volume 4. pp 365-370. New York: Institute of Electrical and Electronics Engineers.

[Yang, L 88*] SEE: Chua, LO 88b. *Cellular Neural Networks: Theory.*

[Yang, R 88*] SEE: Pati, YC 88a. *Neural Networks for Tactile Perception.*

[Yanowich, M 83*] SEE: Grissel, R 83a. Oscillations in Mathematical Biology.

[Yao, CT 88*] SEE: Pati, YC 88a. *Neural Networks for Tactile Perception.*

[Yao, Y 87a] Yao, Yong. (1987). *A Neural Network Model of CAAM and its Applications to Handprinted Chinese Character Recognition.* IEEE International Conference on Neural Networks (ICNN-87), San Diego, CA, June 21-24, 1987: Proceedings. Volume 3. pp 309-316. New York: Institute of Electrical and Electronics Engineers.

[Yao, Y 88*] SEE: Freeman, WJ 88a. *Central Pattern Generating and Recognizing in Olfactory Bulb: A Correlation Learning Rule.*

[Yarbus, AL 67a] Yarbus, A. L. (1967). Eye Movements and Vision. New York: Plenum.

[Yariv, A 86a] Yariv, A.; Kwong, S. K. (1986). *Associative Memories Based on Message-Bearing Optical Modes in Phase-Conjugate Resonator.* Optics Letters. Volume 11. p 186-188.

[Yariv, A 87*] SEE: Agranat, A 87a. *A New Architecture for a Microelectronic Implementation of Neural Network Models.*

[Yee, RD 88*] SEE: Baloh, RW 88a. *Eye Movements Induced by Linear Acceleration on a Parallel Swing.*

[Yegani, P 87*] SEE: Ersoy, OK 87a. *Information Processing and Storage Based on Transforms and Neural Networks.*

[Yeshurun, Y 87*] SEE: Schwartz, EL 87a. *Towards a Non-Network Approach to "Neural Modeling": Some Basic Issues of Measurement, Simulation, and Computational Significance of Brain Maps.*

[Yeshurun, Y 87a] Yeshurun, Yehezkel; Schwartz, Eric L. (1987). *An Ocular Dominance Column Map as a Data Structure for Stereo Segmentation.* IEEE International Conference on Neural Networks (ICNN-87), San Diego, CA, June 21-24, 1987: Proceedings. Volume 4. pp 371-380. New York: Institute of Electrical and Electronics Engineers.

[Yoshida, A 74*] SEE: Okuda, M 74a. *A Dynamical Behaviour of Active Regions in Randomly Connected Neural Networks.*

[Yoshida, K 77*] SEE: Amari, S 77c. *Mathematical Foundation for Statistical Neurodynamics.*

[Yoshida, S 79*] SEE: Hoshimiya, N 79a. *Two-Terminal Electronic Circuit Neuron Model with Excitable Membrane V-I-t Characteristics.*

[Yoshida, T 78a] Yoshida, Toshinobu. (1978). *Response Characteristics of a Mathematical Neuron Model.* Transactions of the Institute of Electronics and Communication Engineers of Japan. Volume E61. pp 276-281.

[Yoshihara, K 80a] Yoshihara, Kunio. (1980). *Equivalent Circuits Model for Transmitter Release at Neuromuscular Junction.* Transactions of the Institute of Electronics and Communication Engineers of Japan. Volume E63. pp 844-845.

[Yoshimi, T 87*] SEE: Kawamura, S 87b. *Technical Issues in Learning Control for Robot Motions.*

[Yoshizawa, S 62*] SEE: Nagumo, J 62a. *An Active Pulse Transmission Line Simulating Nerve Axon.*

[Yoshizawa, S 74a] Yoshizawa, Shuji. (1974). *Some Properties of Randomly Connected Networks of Neuron-Like Elements with Refractory.* Kybernetik. Volume 16. pp 173-182.

[Yoshizawa, S 81*] SEE: Osada, H 81a. *Theoretical Study on Cantor Function-Like Response Characteristics of a Neuron Model.*

[Yoshizawa, S 82a] Yoshizawa, Shuji; Osada, Hiroyasu; Nagumo, Jin-ichi. (1982). *Pulse Sequences Generated by a Degenerate Analog Neuron Model.* Biological Cybernetics. Volume 45. pp 23-33.

[Young, G 40*] SEE: Offner, F 40a. *Nerve Conduction Theory: Some Mathematical Consequences of Bernstein's Model.*

[Young, TY 74a] Young, Tzay Y.; Calvert, T. W. (1974). Classification, Estimation, and Pattern Recognition. New York: Elsevier.

[Young, TY 86a] Young, Tzay Y. (ed.); Fu, King-Sun (ed.). (1986). Handbook of Pattern Recognition and Image Processing. New York: Academic Press.

[Youssef, AM 88*] SEE: Hassoun, MH 88b. *High-Performance Recording Algorithm for Hopfield Model Associative Memories.*

[Yovitz, MC 62a] Yovitz, M. C.; Jacobi, George T.; Goldstein, Gordon D. (1962). Self-Organizing Systems. Washington, DC: Spartan.

[Yu, P 87*] SEE: Ersoy, OK 87a. *Information Processing and Storage Based on Transforms and Neural Networks.*

[Yuille, AL 85*] SEE: Koch, C 85a. Analog Neuronal Networks in Early Vision.

[Yuille, AL 86*] SEE: Koch, C 86a. *Analog Neuronal Networks in Early Vision.*

[Yuille, AL 86*] SEE: Grzywacx, NM 86a. *Motion Correspondence and Analog Networks.*

[Yuille, AL 87*] SEE: Brady, M 87a. *An Extremum Principle for Shape from Contour.*

[Yuille, AL 88*] SEE: Kammen, DM 88a. *Spontaneous Symmetry-Breaking Energy Functions and the Emergence of Orientation Selective Cortical Cells.*

[Yunck, TP 76a] Yunck, Thomas P. (1976). *A Technique to Identify Nearest Neighbors.* IEEE Transactions on Systems, Man and Cybernetics. Volume SMC-6. pp 678-683.

[Zaagman, WH 83a] Zaagman, Willem H.; Masterbroek, Henk A. K.; de Ruyter van Steveninck, Rob. (1983). *Adaptive Strategies in Fly Vision: On Their Image-Processing Qualities.* IEEE Transactions on Systems, Man and Cybernetics. Volume SMC-13. pp 900-906.

[Zaagman, WH 88*] SEE: Schuling, FH 88a. *An Adaptive and Heterodyne Filtering Procedure for the Imaging of Moving Objects.*

[Zaccaria, R 88*] SEE: Mussa-Ivaldi, FA 88b. *Kinematic Networks: A Distributed Model for Representing and Regularizing Motor Redundancy.*

[Zaccaria, R 88*] SEE: Morasso, P 88a. *A Parallel Distributed Architecture for Motor Control.*

[Zadeh, LA 65a] Zadeh, Lotfi A. (1965). *Fuzzy Sets.* Information and Control. Volume 8. pp 338-353.

[Zadeh, LA 71a] Zadeh, Lofti A. (1971). *Similarity Relations and Fuzzy Orderings.* Information Sciences. Volume 3. pp 177-200.

[Zadeh, LA 83a] Zadeh, Lotfi A. (1983). *A Computational Approach to Fuzzy Quantifiers in Natural Languages.* Computers and Mathematics with Applications. Volume 9. pp 149-184.

[Zadeh, LA 83b] Zadeh, Lotfi A. (1983). *The Role of Fuzzy Logic in the Management of Uncertainty in Expert Systems.* Fuzzy Sets and Systems. Volume 11. pp 199-227.

[Zadeh, LA 84a] Zadeh, Lotfi A. (August 1984). *Making Computers that Think Like People.* IEEE Spectrum. Volume 24. pp 26-32.

[Zadeh, LA 85a] Zadeh, Lotfi A. (1985). *Syllogistic Reasoning in Fuzzy Logic and its Application to Usuality and Reasoning with Dispositions.* IEEE Transactions on Systems, Man and Cybernetics. Volume SMC-15. pp 754-763.

[Zadeh, LA 87*] SEE: Bellman, RE 87a. *Decision-Making in a Fuzzy Environment.*

[Zadeh, LA 87b] Zadeh, Lofti A. (1987). *A Theory of Approximate Reasoning.* Fuzzy Sets and Applications: Selected Papers by L A Zadeh. pp 367-412. New York: Wiley.

[Zadeh, LA 87c] Zadeh, Lofti A. (1987). *Fuzzy Sets.* Fuzzy Sets and Applications: Selected Papers by L A Zadeh. pp 29-44. New York: Wiley.

[Zadeh, LA 87d] Zadeh, Lofti A. (1987). *Fuzzy Sets as a Basis for a Theory of Possibility.* Fuzzy Sets and Applications: Selected Papers by L A Zadeh. pp 193-218. New York: Wiley.

[Zadeh, LA 87e] Zadeh, Lofti A. (1987). *Outline of a New Approach to the Analysis of Complex Systems and Decision Process.* Fuzzy Sets and Applications: Selected Papers by L A Zadeh. pp 105-146. New York: Wiley.

[Zadeh, LA 87f] Zadeh, Lofti A. (1987). *The Role of Fuzzy Logic in the Management of Uncertainty in Expert Systems.* Fuzzy Sets and Applications: Selected Papers by L A Zadeh. pp 413-441. New York: Wiley.

[Zaghoul, M 87*] SEE: El-Leithy, N 87a. *A Basic MOS Neural-Type Junction: A Perspective on Neural-Type Microsystems.*

[Zapalac, JJ 68*] SEE: Mendel, JM 68a. *The Application of Techniques of Artificial Intelligence to Control System Design.*

[Zargham, MR 85a] Zargham, M. R.; Tyman, M. (1985). *Neural Petri Nets.* International Workshop on Timed Petri Nets. Catalog 2187-3. pp 72-73. New York: Institute of Electrical and Electronics Engineers.

[Zee, A 88*] SEE: Scalettar, R 88a. *Emergence of Grandmother Memory in Feed Forward Networks: Learning with Noise and Forgetfulness.*

[Zee, A 88*] SEE: Scalettar, R 88b. *Perception of Left and Right by a Feed Forward Net.*

[Zee, DS 85*] SEE: Keller, EL 85a. Asilomar Conference on Adaptive Processes in Visual and Oculomoter Systems, June 16-20, 1985: Proceedings.

[Zeeman, CC 77a] Zeeman, C. C. (1977). Catastrophe Theory: Selected Papers. Reading, MA: Addison-Wesley.

[Zeeman, EC 62a] Zeeman, E. C. (1962). *The Topology of the Brain and Visual Perception.* Topology of 3-Manifolds and Related Topics. pp 240-256. Englewood Cliffs, NJ: Prentice-Hall.

[Zeevi, YY 79*] SEE: Bruckstein, AM 79a. *Analysis of "Integrate-to-Threshold" Neural Coding Schemes.*

[Zeevi, YY 80a] Zeevi, Yehoshua Y.; Bruckstein, Alfred M. (1980). *Modeling of Two-Phase Adaptation in Neuronal Encoders.* International Conference on Cybernetics and Society, Cambridge, MA, October 8-10, 1980: Proceedings. pp 605-607. New York: Institute of Electrical and Electronics Engineers.

[Zeevi, YY 81*] SEE: Bruckstein, AM 81a. *Stochastic Pulse Frequency Modulation Models of Neural Coding Processes.*

[Zeevi, YY 81a] Zeevi, Yehoshua Y.; Bruckstein, Alfred M. (1981). *Adaptive Neural Encoder Model with Selfinhibition and Threshold Control.* Biological Cybernetics. Volume 40. pp 79-92.

[Zeevi, YY 83*] SEE: Sanderson, AC 83a. *Special Issue on Neural and Sensory Information Processing.*

[Zeevi, YY 85*] SEE: Bruckstein, AM 85a. *An Adaptive Stochastic Model for the Neural Coding Process.*

[Zeevi, YY 88a] Zeevi, Yehoshua Y.; Ginosar, Ran. (1988). *Neural Computers in Vision: Processing of High Dimensional Data.* In Eckmiller, Rolf (ed.); von der Malsburg, Christoph (ed.). Neural Computers. NATO ASI Series, Volume F41. pp 169-178. Berlin: Springer-Verlag.

[Zeitouni, O 87*] SEE: Bachmann, CM 87a. *A Relaxation Model for Memory with High Storage Density.*

[Zeitouni, O 88*] SEE: Dembo, A 88a. *General Potential Surfaces and Neural Networks.*

[Zeitouni, O 88*] SEE: Dembo, A 88b. *High Density Associative Memories.*

[Zeki, S 80a] Zeki, S. (1980). *The Representation of Colors in the Cerebral Cortex.* Nature. Volume 284. pp 412-418.

[Zemanek, H 60a] Zemanek, H.; Kretz, H.; Angyan, A. J. (1961). *A Model for Neurophysiological Functions.* Information Theory, London Symposium, 4th, Held at the Royal Institution, London, August 29-September 2, 1960. pp 270-284. Washington: Butterworths.

[Zendeh, F 63*] SEE: Steinbuch, K 63b. *Self-Correcting Translator Circuits.*

[Zenone, S 79*] SEE: Papadopoullos, MS 79a. *Memory Topology of Neural Nets.*

[Zenone, S 80*] SEE: Anninos, PA 80a. *A Neural Net Model for the a-Rhythm.*

[Zenone, S 82*] SEE: Anninos, PA 82a. *Encoding of Visual Information: Correlation of EEG Response to Intracellular Response.*

[Zenone, S 83*] SEE: Anninos, PA 83b. *Artificial Neural Nets: Dependence of the EEG Amplitude's Probability Distribution on Statistical Parameters.*

[Zhang, W 87a] Zhang, Wen-Ran; Bezdek, James C.; Pettus, Robert O. *NPN Relations and Cognitive Map Development.* International Symposium on Methodologies for Intelligent Systems, 2nd, Charlotte, NC, October 17, 1987: Proceedings.

[Zhang, W 88a] Zhang, Wen-Ran; Chen, Su-Shing; Bezdek, James C. (1988). *On NPN Logic.* International Symposium on Multiple-Valued Logic, 18th, Palma de Mallorca, May 24-26, 1988: Proceedings. pp 381-388. New York: Institute of Electrical and Electronics Engineers.

[Zhang, X 88*] SEE: Bookman, L 88a. *A Data Flow Analysis of Connectionist Networks.*

[Zhou, Y 87a] Zhou, Yi-Tong; Chellappa, Ramalingam; Jenkins, B. K. (1987). *A Novel Approach to Image Restoration Based on a Neural Network.* IEEE International Conference on Neural Networks (ICNN-87), San Diego, CA, June 21-24, 1987: Proceedings. Volume 4. pp 269-276. New York: Institute of Electrical and Electronics Engineers.

[Zhou, Y 88a] Zhou, Yi-Tong; Chellappa, Ramalingam. (1988). *Computation of Optical Flow Using a Neural Network.* IEEE International Conference on Neural Networks (ICNN-87), San Diego, CA, June 21-24, 1987: Proceedings. Volume 2. pp 71-78. New York: Institute of Electrical and Electronics Engineers.

[Zhou, Y 88b] Zhou, Yi-Tong; Chellappa, Ramalingam; Vaid, Aseem; Jenkins, B. Keith. (1988). *Image Restoration Using a Neural Network.* IEEE Transactions on Acoustics, Speech, and Signal Processing. Volume 36. pp 1141-1151. New York: Institute of Electrical and Electronics Engineers.

[Zhou, Y 88c] Zhou, Yi-Tong; Chellappa, Ramalingam. (1988). *Stereo Matching Using a Neural Network.* IEEE International Conference on Acoustics, Speech, and Signal Processing (ICASSP-88), New York, NY, April 11-14, 1988: Proceedings. pp 940-943. New York: Institute of Electrical and Electronics Engineers.

[Zielenziger, D 88a] Zielenziger, David. (1988). *Neurons in Silicon.* Electronic Engineering Times. Issue 482. pp 85-86.

[Zimmerman, GL 87*] SEE: Wechsler, H 87a. *Fault-Tolerant Recognition Using DAM's.*

[Zimmerman, GL 88*] SEE: Char, JM 88a. *Distributed and Fault-Tolerant Computation for Retrieval Tasks Using Distributed*

Associative Memories.

[Zimmerman, GL 88*] SEE: Wechsler, H 88a. *Invariant Object Recognition Using a Distributed Associative Memory.*

[Zink, WT 88*] SEE: Vogl, TP 88a. *Accelerating the Convergence of the Back-Propagation Method.*

[Zippelius, A 87*] SEE: Derrida, B 87a. *An Exactly Solvable Asymmetric Neural Network Model.*

[Zipser, D 83a] Zipser, David. (1983). The Representation of Location. ICS-8301. San Diego, CA: University of California, San Diego, Institute for Cognitive Science.

[Zipser, D 83b] Zipser, David. (1983). The Representation of Maps. ICS-8304. San Diego, CA: University of California, San Diego, Institute for Cognitive Science.

[Zipser, D 85*] SEE: Rumelhart, DE 85a. *Feature Discovery by Competitive Learning.*

[Zipser, D 85a] Zipser, David. (1985). *A Computational Model of Hippocampus Place-Fields.* Behavioral Neuroscience. Volume 99. pp 1006-1018.

[Zipser, D 86*] SEE: Rumelhart, DE 86a. *Feature Discovery by Competitive Learning.*

[Zipser, D 86a] Zipser, David. (1986). *A Model of Hippocampal Learning During Classical Conditioning.* Behavioral Neuroscience. Volume 100. pp 764-776.

[Zipser, D 86b] Zipser, David. (1986). *Biologically Plausible Models of Place Recognition and Goal Location.* In Rumelhart, David E. (ed.). Parallel Distributed Processing: Explorations in the Microstructure of Cognition. Volume II. pp 432-470. Cambridge, MA: MIT Press.

[Zipser, D 86c] Zipser, David; Rabin, D. E. (1986). *P3: A Parallel Network Simulating System.* In Rumelhart, David E. (ed.). Parallel Distributed Processing: Explorations in the Microstructure of Cognition. Volume I, Foundations. pp 488-506. Cambridge, MA: MIT Press.

[Zipser, D 86d] Zipser, David. (1986). Programming Neural Nets to Do Spatial Computations. ICS-8608. San Diego, CA: University of California, San Diego, Institute for Cognitive Science.

[Zipser, D 87*] SEE: Cottrell, GW 87a. Image Compression by Back Propagation: An Example of Extensional Programming.

[Zipser, D 87*] SEE: Cottrell, GW 87b. *Learning Internal Representations from Gray-Scale Images: An Example of Extensional Programming.*

[Zipser, D 87*] SEE: Elman, JL 87a. Learning the Hidden Structure of Speech.

[Zipser, D 88*] SEE: Rumelhart, DE 88a. *Feature Discovery by Competitive Learning.*

[Zopf, GW 62*] SEE: von Foerster, H 62a. Principles of Self-Organization.

[Zucker, SW 79a] Zucker, Steven W.; Hummel, Robert A. (1979). *Toward a Low-Level Description of Dot Clusters: Labeling Edge, Interior, and Noise Points.* Computer Graphics and Image Processing. Volume 9. pp 213-233.

[Zucker, SW 83*] SEE: Hummel, RA 83a. *On the Foundations of Relaxation Labeling Processes.*

[Zucker, SW 86*] SEE: Rosenfeld, A 86a. *Scene Labeling by Relaxation Operations.*

[Zucker, SW 88a] Zucker, Steven W.; Iverson, Lee. (1988). *Computational Networks in Early Vision: From Orientation Selection to Optical Flow.* In Eckmiller, Rolf (ed.); von der Malsburg, Christoph (ed.). Neural Computers. NATO

ASI Series, Volume F41. pp 179–188. Berlin: Springer-Verlag.

[Zuse, K 88a] Zuse, Konrad. (1988). *Faust, Mephistopheles and Computer*. In Eckmiller, Rolf (ed.); von der Malsburg, Christoph (ed.). Neural Computers. NATO ASI Series, Volume F41. pp 9-16. Berlin: Springer-Verlag.

Keyword Cross-Reference

Neural Dynamics of Decision Making Under Risk:	Affective Balance and Cognitive-Emotional Interactions	Grossberg, S 87k
Neural Dynamics of Decision Making Under Risk:	Affective Balance Theory	Grossberg, S 87l
A Model of Synaptogenetic Consequences of	Afferent Input in Neuronal Nets.	Dammasch, IE 88a
Visual Illusions in Neural Networks: Line Neutralization, Tilt	After Effect, and Angle Expansion.	Levine, DS 76a
...of Adult Rat Motor Cortex Somatic Representation Patterns	after Motor Nerve Injury	Sanes, JN 88a
Neural Systems Make Comeback	After Nearly 50-Year Gestation	Johnson, RC 87k
Using Artificial Neural Nets for Statistical Discovery: Observations	after Using Back-propagation, Expert Systems, and Multiple-Linear.	Irani, EA 88a
Adaptation,	After-Effect and Contrast in the Perception of Curved Lines.	Gibson, JJ 33a
Neural Theory of Circadian Rhythms: Split Rhythms,	After-Effects, and Motivational Interactions	Carpenter, GA 86m
Random Cells: An Idea Whose Time Has Come and Gone... And Come	Again?	Gallant, SI 87b
A Microcomputer Neural Net Benchmarked	Against Standard Classification Techniques	Draper, JS 87a
A Mathematical Model for the Decay of Short-Term Memory with	Age.	Anninos, PA 83a
Interacting and Communicating Connectionist	Agents	Allen, RB 88a
On the	Aggregation of Processing Units in Neural Networks	Yager, RR 87b
Forecasting Applications of GMDH in	Agricultural and Meteorological Time Series.	Lebow, WM 84a
Machine Vision Lurches	Ahead.	Brown, C 87f
Cognitive Computer: On Language, Learning and	AI	Schank, RC 84a
First Commercial Nerve Nets Challenge Traditional	AI	Brown, C 86c
Goal-Seeking Systems from Goal-Seeking Components: Implications for	AI	Klopf, AH 79a
Neural Network Architectures for	AI	Hinton, GE 87c
Theories of Neural Nets Expose the Limits of	AI	Brown, C 87l
Massively Parallel Architectures for	AI: Netl, Thistle, and Boltzmann Machines.	Fahlman, SE 83a
Darwinian Evolution as a Paradigm for	AI Research.	Fenanzo Jr, AJ 86a
Superresolution for Ultrasonic Imaging in	Air Using Neural Networks.	Winters, JH 88b
The	Airline Marketing Tactician (AMT): A Commercial Application of	Hutchison, WR 87b
Introduction to Linear	Algebra in Parallel Distributed Processing	Jordan, MI 86b
Correlational Strength and Computational	Algebra of Synaptic Connections Between Neurons.	Fetz, EE 88a
Acoustooptic	Algebraic Processing Architectures	Rhodes, WT 84a
An	Algebraic Projection Analysis for Back-Propagation Learning.	Kung, S 88a
Neural Networks, Orientations of the Hypercube and	Algebraic Threshold Functions.	Baldi, P 88f
A FORTRAN Program for the GMDH	Algorithm.	Farlow, SJ 84a
A High-Storage Content-Addressable Memory and Its Learning	Algorithm.	Verleysen, M 88a
An Improved Three-Layer, Back Propagation	Algorithm.	Stornetta, WS 87a
Analysis of a Cooperative Stereo	Algorithm.	Marr, D 78a
...Bulb with Periodic Attractors Stored by a Sequence Learning	Algorithm.	Baird, B 87a
...Use of the Hopfield Neural Network Model as a Nearest-Neighbor	Algorithm.	Montgomery, BL 86a
Faster Learning Through a Probabilistic Approximation	Algorithm.	Kolen, JF 88a
Incorporating Fuzzy Membership Functions into the Perceptron	Algorithm.	Keller, JM 85a
Langevin Equation as a Global Minimization	Algorithm.	Gidas, B 86a
Mobile Robot Control by Neural Networks and Their Associated Learning	Algorithm.	Watanabe, N 88a
...Content-Addressable Memory: VLSI Circuit and Learning	Algorithm.	Verleysen, M 88g
Optical Neurocomputer for Implementation of the Marr-Poggio Stereo	Algorithm.	Gmitro, AF 87a
...in Small Neuronal Circuits: Simulations Based on the Compensation	Algorithm.	Dammasch, IE 87b
Reduction of Visual Processes to a Common	Algorithm.	Carman, GJ 87b
Reduction of Visual Processes to a Common	Algorithm.	Carman, GJ 88a
Scaling the Neural TSP	Algorithm.	Cuykendall, R 89a
Sensor-Based Control of Robotic Manipulators Using a General Learning	Algorithm.	Miller, WT 87c
...Greater) in Back-Propagation Networks with the Heuristic Learning	Algorithm.	Cater, JP 87a
The "Brain-State-in-a-Box" Neural Model Is a Gradient Descent	Algorithm.	Golden, RM 86c
The Complex LMS	Algorithm.	Widrow, B 75b
The GMDH	Algorithm.	Farlow, SJ 84c
Towards a Generic Learning	Algorithm.	Wright, WA 89d
Variations on the Boltzmann Machine Learning	Algorithm.	Derthick, MA 84a
The Trinocular General Support	Algorithm: A Three Camera Stereo Algorithm for Overcoming Binocular.	Stewart, CV 88b
Computing Motion in the Presence of Discontinuities:	Algorithm and Analog Networks.	Koch, C 88a
The Sum-Line Extrapolative	Algorithm and its Application to Statistical Classification.	Talbert, LR 70a
The Least Mean Fourth (LMF) Adaptive	Algorithm and Its Family	Walach, E 84a
Fundamental Relations Between the LMS	Algorithm and the DFT.	Widrow, B 87b
	Algorithm Development for Neural Networks.	Page, EW 88a
An	Algorithm for a Selective Nearest Neighbor Decision Rule	Ritter, GL 75a
Neural Network	Algorithm for an NP-Complete Problem: Map and Graph Coloring	Dahl, ED 87b
An Adaptive Training	Algorithm for Back Propagation Networks.	Chan, L 87a
A Learning	Algorithm for Boltzmann Machines	Ackley, DH 88a
A Learning	Algorithm for Boltzmann Machines	Ackley, DH 85b
A Learning	Algorithm for Boltzmann Machines	Ackley, DH 85a
A Branch and Bound	Algorithm for Computing k-Nearest Neighbors.	Fukunaga, K 75a
An	Algorithm for Finding Best Matches in Logarithmic Expected Time.	Friedman, JH 77a
An	Algorithm for Finding Nearest Neighbors.	Friedman, JH 75a
High-Performance Recording	Algorithm for Hopfield Model Associative Memories.	Hassoun, MH 88b

A Stochastic	Algorithm for Learning Real-Valued Functions via Reinforcement . . .	Gullapalli, V 88a
An	Algorithm for Learning without External Supervision and its.	Nicolic, ZJ 66a
A Screening	Algorithm for Levenglick's Consensus Ranking Model	Flood, MM 87a
Electronic 'Neural' Net	Algorithm for Maximum Entropy Deconvolution.	Marrian, CRK 87a
A Connectionist	Algorithm for Minimum Cost Network Flows	Marcus, R 87a
A New	Algorithm for Monte Carlo Simulation of Ising Spin Systems	Bortz, AB 75a
A Learning	Algorithm for Multilayer Neuronlike Network.	Gawronski, RR 88a
A Preprocessing	Algorithm for Nearest Neighbor Decision Rules.	Fischer, FP 70a
The Trinocular General Support Algorithm: A Three Camera Stereo	Algorithm for Overcoming Binocular Matching Errors	Stewart, CV 88b
Neocognitron: A New	Algorithm for Pattern Recognition Tolerant of Deformation and. . . .	Fukushima, K 82a
An	Algorithm for Pattern Recognition with VLSI Neural Networks.	Sirletti, B 88a
A New LMS-Based	Algorithm for Rapid Adaptive Classification in Dynamic Environments.	Nguyen, DD 88a
Fast	Algorithm for Recognizing Nearest Neighbors.	Sethi, IK 81a
An Adaptive Attentive Learning	Algorithm for Single-Layer Neural Networks	Hassoun, MH 88a
A Practical	Algorithm for Solving Dynamic Membrane Equations	Rush, S 78a
Self-Stabilization of Neuronal Networks, I. The Compensation	Algorithm for Synaptogenesis	Dammasch, IE 86b
An Effective Heuristic	Algorithm for the Traveling Salesman Problem	Lin, S 73a
Learning a Color	Algorithm from Examples.	Poggio, TA 88a
Synthesizing a Color	Algorithm from Examples.	Hurlbert, AC 88a
A Synaptic Modification	Algorithm in Consideration of the Generation of Rhythmic	Tsutsumi, K 84a
Simulation of a Connectionist Stereo	Algorithm on a Shared-Memory Multiprocessor.	Stewart, CV 88a
Application of Genetic	Algorithm to a Neural Network for Finding Optimal Paths.	Smith, T 87a
An Application of the GMDH	Algorithm to Economic Modeling	Scott, DE 84a
Extending Kohonen's Self-Organizing Mapping	Algorithm to Learn Ballistic Movements	Ritter, H 88b
Application of a General Learning	Algorithm to the Control of Robotic Manipulators	Miller, WT 87b
Simulation of a Distributed	Algorithm Using the Connection Machine	Aburdene, MF 88a
On the Statistical Efficiency of the LMS	Algorithm with Nonstationary Inputs.	Widrow, B 84a
A Comparison of Optical Character Recognition	Algorithms	Burton, MJ 88a
A Comparison of Two Unsupervised Learning	Algorithms	Bobrowski, L 80a
An Efficient Vector/Scalar Processing Structure for Adaptive	Algorithms	Guerin, A 88a
...Investigation of a Class of Pattern Recognition Synthesis	Algorithms	Mattson, RL 63a
Automatic Classification	Algorithms	Dorofeyuk, AM 71a
Classification	Algorithms	James, M 85a
Color Separation Using General Purpose Computer Vision	Algorithms	Walters, DKW 87c
Discrete-Time Parameter Estimation with Genetic	Algorithms	Das, R 88a
Neural Network Learning	Algorithms	Sejnowski, TJ 88b
On the Speed of Convergence of Adaptive	Algorithms	Widrow, B 80a
Programming Parallel Computers Using Energy Minimization	Algorithms	Simmes, SD 87a
Scaling Techniques for Neural	Algorithms	Cuykendall, R 88a
Self-Organizing Methods in Modeling: GMDH Type	Algorithms	Farlow, SJ 84b
Visual	Algorithms	Poggio, TA 82a
Learning	Algorithms and Applications.	Lakshmivarahan, S 81a
Application of Neural Network	Algorithms and Architectures to Correlation / Tracking and	Gardner, S 86a
Image Reconstruction from Coded Data: I. Reconstruction	Algorithms and Experimental Results.	Smith, WE 85a
Learning	Algorithms and Probability Distributions in Feed-Forward and Feed. .	Hopfield, JJ 87a
	Algorithms and Techniques for Automated Visual Inspection.	Chin, RT 86a
A Comparison of Adaptive	Algorithms Based on the Methods of Steepest Descent and Random . . .	Widrow, B 76a
Optimal	Algorithms for Adaptive Networks: Second Order Back Propagation. . .	Parker, DB 87a
New "Neural"	Algorithms for Associative Memory.	Baum, EB 87a
Concurrent Neuromorphic	Algorithms for Autonomous Robots	Barhen, J 88b
Learning	Algorithms for Connectionist Networks: Applied Gradient Methods of .	Watrous, RL 87c
Bit-Vector Optimization	Algorithms for Control of Learning in Neurons with Second-Messenger.	Kirby, KG 88a
Adaptive	Algorithms for Discovering and Using General Patterns in Growing . .	Holland, JH 80a
Kalman Filter-Based	Algorithms for Estimating Depth from Image Sequences	Matthies, L 87a
Three Constructive	Algorithms for Network Learning.	Gallant, SI 86c
Comparison of	Algorithms for Neuron-Like Cells	Parker, DB 86a
Iterative Restoration	Algorithms for Nonlinear Constraint Computing.	Szu, HH 86e
Concurrent Neuromorphic	Algorithms for Optimization of the Computational Load of a	Barhen, J 87a
Analysis and Comparison of Different Learning	Algorithms for Pattern Association Problems.	Bernasconi, J 88a
A Class of Gradient-Estimating	Algorithms for Reinforcement Learning in Neural Networks	Williams, RJ 87a
Cooperative	Algorithms for Solving Random Dot Stereograms.	Szeliski, R 86a
A Comparison of Some Hierarchial Monothetic Divisive Clustering	Algorithms for Structure-Property Correlation.	Rubin, V 83a
Systolic	Algorithms for the CMU Warp Processor.	Kung, HT 84a
On Pattern Classification	Algorithms: Introduction and Survey.	Ho, Y 68a
e-Optimal Learning	Algorithms: Non-Absorbing Barrier Type	Lakshmivarahan, S 79a
On Revised	Algorithms of GMDH With Applications	Tamura, H 84a
...Networks: Development and Application in the United States of	Algorithms Related to GMDH	Barron, RL 84a
Applying Genetic	Algorithms to Neural Net Learning.	Whitley, D 88a
Efficient	Algorithms with Neural Network Behavior.	Omohundro, SM 87a

Recent Developments in a Neural Model of Real-Time Speech	Analysis and Synthesis	Cohen, MA 87c
Qualitative	Analysis and Synthesis of a Class of Neural Networks	Li, JH 88a
The	Analysis and Synthesis of Adaptive Systems Which Use Networks of . .	Mattson, RL 62a
An Algebraic Projection	Analysis for Back-Propagation Learning	Kung, S 88a
Non-Parametric Prediction	Analysis for Binary Data	Davis, BR 84a
Configuration Space	Analysis for Optimization Problems	Solla, SA 86a
Time Series	Analysis: Forecasting and Control.	Box, G 76a
EEG	Analysis Gives Models of Neuronal Template-Matching Mechanism for. .	Freeman, WJ 79a
Ill-Posed Problems and Regularization	Analysis in Early Vision	Poggio, TA 84b
Implementing Stages of Motion	Analysis in Neural Networks.	Sereno, ME 87a
Independent Components	Analysis in Neural Networks.	Herault, J 88b
Neural Networks and Principal Component	Analysis: Learning from Examples Without Local Minima.	Baldi, P 88e
	Analysis of "Integrate-to-Threshold" Neural Coding Schemes	Bruckstein, AM 79a
A Connectivity	Analysis of a Class of Simple Associative Neural Networks.	Hammerstrom, D 88a
	Analysis of a Competition-Based Connectionist Model.	Seidman, SB 87a
	Analysis of a Cooperative Stereo Algorithm	Marr, D 78a
	Analysis of a Distributed Associative Memory Model	Wiles, J 87a
	Analysis of a Four-Layer Series-Coupled Perceptron, II	Block, HD 62a
Bifurcation	Analysis of a Network Model of Rabit Olfactory Bulb with Periodic. .	Baird, B 87a
On Language and Connectionism:	Analysis of a Parallel Distributed Processing Model of Language. . .	Pinker, S 88b
Simulation and	Analysis of a Simple Cortical Network.	Lynch, G 88a
	Analysis of a Simple Self-Organizing Process	Kohonen, T 82b
	Analysis of a Verbal Conditioning Situation in Terms of Statistical.	Estes, WK 54a
	Analysis of an Adaptive Threshold Logic Unit	Morishita, I 70a
Input-Output	Analysis of an IPFM Neural Model: Effects of Spike Regularity and. .	Sanderson, AC 80a
Performance	Analysis of Associative Memories with Nonlinearities in the.	Marks II, RJ 88e
The Microscopic	Analysis of Behavior: Toward a Synthesis of Instrumental	Grossberg, S 84d
Exhaustive Thermodynamical	Analysis of Boolean Learning Networks.	Carnevali, P 87a
	Analysis of Competition-Based Spreading Activation in Connectionist.	Wang, PY 88a
Outline of a New Approach to the	Analysis of Complex Systems and Decision Process	Zadeh, LA 87e
A Data Flow	Analysis of Connectionist Networks	Bookman, L 88a
Response Mapping Functions: Classification and	Analysis of Connectionist Representations.	Walters, DKW 87f
A Continuous Information Theoretic Approach to the	Analysis of Cutaneous Receptor Neurons	Fuller, MS 83a
...Currents in a Central Neuron Match Predictions from Binomial	Analysis of Evoked Responses	Korn, H 87a
A Mathematical	Analysis of Excitable Membrane Phenomena	Carpenter, GA 78a
	Analysis of Hidden Units in a Layered network Trained to Classify. .	Gorman, RP 88a
Mean-field	Analysis of Hierarchical Associative Networks with Magnetisation . .	Krogh, A 88a
Sensitivity	Analysis of Hopfield Net	David, GW 87a
Spatial Interaction and the Statistical	Analysis of Lattice Systems.	Besag, J 74a
Mathematical	Analysis of Learning Behavior of Neuronal Models	Cheung, JY 88a
The	Analysis of Moving Visual Patterns	Movshon, JA 85a
Spatio-Temporal	Analysis of Multi-channel Alpha EEG Map Series	Fuchs, A 87a
Geometric	Analysis of Neural Network Capabilities.	Wieland, A 87a
Activity	Analysis of Neural Networks.	Oguztoreli, MN 79a
	Analysis of Neural Networks.	an der Heiden, U 80a
Qualitative	Analysis of Neural Networks.	Michel, AN 88a
Modeling and	Analysis of Neural Networks in the Visuomotor System of the Anuran .	Cervantes-Perez, F 85a
Bifurcation	Analysis of Oscillating Network Model of Pattern Recognition in the.	Baird, B 86a
	Analysis of Perceptrons.	Block, HD 61a
Harmonic	Analysis of Polynomial Threshold Functions	Bruck, J 88c
A Maximization Technique Occurring in the Statistical	Analysis of Probabilistic Functions of Markov Chains	Baum, LE 70a
A Connectivity	Analysis of Recursive, Auto-Associative Connection Networks.	Hammerstrom, D 86a
Preliminary	Analysis of Recursively Generated Networks	Mjolsness, E 86a
The Connectivity	Analysis of Simple Association	Hammerstrom, D 88b
Opposition Space as a Structuring Concept for the	Analysis of Skilled Hand Movements	Iberall, T 86a
Low Temperature	Analysis of the Axial Next-Nearest Neighbor Ising Model Near its . .	Fisher, ME 81a
	Analysis of the Delta Rule and Learning of Statistical Associations.	Stone, GO 86a
Nonlinear Systems	Analysis of the Hippocampal Perforant Path-Dentate Projection. I . .	Sclabassi, RJ 88b
Nonlinear Systems	Analysis of the Hippocampal Perforant Path-Dentate Projection. II. .	Berger, TW 88a
Nonlinear Systems	Analysis of the Hippocampal Perforant Path-Dentate Projection. III .	Berger, TW 88b
A Method of	Analysis of the Models of Neural Nets.	Burattini, E 72a
Consensus versus Competition in Neural Networks: A Comparative	Analysis of Three Models	Montalvo, FS 75a
The	Analysis of Visual Motion: From Computational Theory to Neuronal . .	Hildreth, E 86a
A Macroscopic	Analysis on Large Scale Networks on Neuronlike Elements with	Aihara, K 80a
Scene	Analysis Research: Optical Pattern Recognition and Artificial. . . .	Casasent, DP 86b
Introduction to Factor	Analysis: What It Is and How to Do It.	Kim, J 78a
Digital Spectral	Analysis with Applications	Marple, SL 87a
	Analytic Study of Assemblies of Neurons in Memory Storage.	Roney, KJ 80a
	Analytic Study of the Storage Capacity of a Neural Network	Little, WA 78a

	Application of Orthogonal Projection Principles to Simultaneous.	Jalanko, M 78a
Interpretation of Natural Language: A Potential	Application of Parallelism	Pollack, JB 86a
	Application of Stochastic Averaging to Learning Systems.	Geman, S 78a
The	Application of Techniques of Artificial Intelligence to Control.	Mendel, JM 68a
An	Application of the GMDH Algorithm to Economic Modeling	Scott, DE 84a
	Application of the Neural Network for the Trajectory Planning of a	Kitamura, S 88a
	Application of the Subspace Method to Speech Recognition	Jalanko, M 78b
An Introduction to the	Application of the Theory of Probabilistic Functions of a Markov	Levinson, SE 83a
STRIPS: A New Approach to the	Application of Theorem Proving to Problem Solving.	Fikes, RE 71a
The	Application of Wafer-Scale Technology to Neuromorphic Systems.	Raffel, JI 88b
Combinations of Neural Systems for Particular	Application Situations	Josin, GM 87a
General Principles of Operations in Neuron Nets with	Application to Acoustical Pattern Recognition.	Mueller, P 62a
A Mean Field Theory of Layer IV of Visual Cortex and Its	Application to Artificial Neural Networks.	Scofield, CL 88a
Self-Organizing Model of Learning and Its	Application to Eyesight System of Robot.	Suzuki, H 88b
Neural Network Models and Their	Application to Handwritten Digit Recognition	Pawlicki, TF 88a
An Algorithm for Learning without External Supervision and its	Application to Learning Control Systems.	Nicolic, ZJ 66a
Neural Net	Application to Optical Character Recognition	Mehr, D 87a
Linear Decision Functions, with	Application to Pattern Recognition	Highleyman, WH 62a
An Artificial Neural Network for Spatio-Temporal Bipolar Patterns:	Application to Phoneme Classification.	Homma, T 88a
An Artificial Neural Network for Spatio-Temporal Bipolar Patterns:	Application to Phoneme Classification.	Atlas, LE 88a
Machine for Neural Computation of Acoustical Patterns with	Application to Real Time Speech Recognition.	Mueller, P 86a
Neural Networks for Template Matching:	Application to Real-Time Classification of the Action Potentials of.	Bower, JM 88b
Neural Networks for Template Matching:	Application to Real-Time Classification of the Action Potentials of.	Wong, Y 87a
...Characteristics of the Peripheral Auditory System, Part IV,	Application to Response.	Lachs, G 83a
Self-Organizing Model for Pattern Learning and Its	Application to Robot Eyesight.	Suzuki, H 88a
A Hierarchical Model for Voluntary Movement and Its	Application to Robotics.	Kawato, M 87a
Hierarchical Neural Network Model for Voluntary Movement with	Application to Robotics.	Kawato, M 88a
Static and Dynamic Error Propagation Networks with	Application to Speech Coding	Robinson, AJR 88b
The Sum-Line Extrapolative Algorithm and its	Application to Statistical Classification Problems	Talbert, LR 70a
Theory of Neural-Analog Reinforcement Systems and its	Application to the Brain-Model Problem	Minsky, ML 54a
Modeling Networks of Morishita Neurons With	Application to the Cerebellum.	Dimopoulos, N 79a
Dynamically Expanding Context, with	Application to the Correction of Symbol Strings in the Recognition	Kohonen, T 86a
...Mechanism for the Establishment of Ordered Neural Mapping: Its	Application to the Retinotectal Connections.	Willshaw, DJ 79a
A Decentralized Tunable Short Term Neural Network Memory and	Application to Tracking.	Morris, RJT 88a
Syllogistic Reasoning in Fuzzy Logic and its	Application to Usuality and Reasoning with Dispositions.	Zadeh, LA 85a
Training with Noise:	Application to Word and Text Storage	Gardner, E 88a
	Application-Oriented Connection Systems for Pattern Acquisition and.	Horne, W 87a
...for Simulating a Network of Interacting Neurons, Part III:	Applications	Perkel, DH 76c
A Frequency-Domain Adaptive Pole-Zero Filter with	Applications	Shynk, JJ 85a
A Logarithmic Vector Processor for Neural Net	Applications	Richfield, S 87a
Adaptation, Learning, and Pattern Recognition Systems: Theory and	Applications	Mendel, JM 70a
Adaptive Noise Canceling: Principles and	Applications	Widrow, B 75a
Adaptive Signal Processing: Theory and	Applications	Alexander, ST 86a
Automatic Subclass Determination for Pattern-Recognition	Applications	Firschein, O 63a
Character Recognition by Computer and	Applications	Suen, CY 86a
Digital Spectral Analysis with	Applications	Marple, SL 87a
Fuzzy Mathematical Techniques with	Applications	Kandel, A 86a
Generalized Inverse of a Matrix and Its	Applications	Rao, CR 72a
...the Explorer / Odyssey for Image Processing and Understanding	Applications	Gove, RJ 87a
Learning Algorithms and	Applications	Lakshmivarahan, S 81a
Learning Matrices and Their	Applications	Steinbuch, K 63a
Neural Net Architectures for Robotic	Applications	Kung, S 89a
Neural Network Based Architectures for Aerospace	Applications	Ricart, R 87a
Neurocomputer	Applications	Hecht-Nielsen, R 88b
Neurocomputer	Applications	Hecht-Nielsen, R 87h
Neurocomputer Workstations and Processors: Approaches and	Applications	Kuczewski, RM 87e
Neurocomputer Workstations and Processors: Approaches and	Applications	Kuczewski, RM 87f
On Revised Algorithms of GMDH With	Applications	Tamura, H 84a
On Tropistic Processing and Its	Applications	Fernandez, MF 88a
Polynomial and Neural Networks: Analogies and Engineering	Applications	Shrier, S 87a
Self-Organization of Associative Database and Its	Applications	Suzuji, H 88a
Self-Tuning and Adaptive Control: Theory and	Applications	Harris, CJ 81a
The Fourier Transform and its	Applications	Bracewell, R 78a
...of the Parametric Avalanche for Advanced Electronic Warfare	Applications	Dawes, RL 88b
A Review of Artificial Neural Systems II: Paradigms,	Applications, and Implementations.	Simpson, PK 87a
Les	Applications des Modeles Connexionistes a la Reconnaissance de la.	Wellekens, CJ 88b
Neurocomputing	Applications Examples.	Hecht-Nielsen, R 87i
Practical	Applications for Adaptive Data-Processing Systems.	Widrow, B 63b
A Generalized Convergence Theorem for Neural Networks and Its	Applications in Combinatorial Optimization	Bruck, J 87a

	Architecture of Rod and Cone Circuits to the On-Beta Ganglion Cell	Sterling, P 88a
Computational Anatomy and Functional	Architecture of Striate Cortex: A Spatial Mapping Approach to. . . .	Schwartz, EL 80a
Microcircuitry of the Dark-Adapted Cat Retina: Functional	Architecture of the Rod-Cone Network	Smith, RG 86a
The OGC Cognitive	Architecture Project: Silicon Implementation of Connectionist. . . .	Hammerstrom, D 86e
Hecht-Nielsen Unveils First New Neural Net	Architecture Since '84	Schwartz, TJ 87b
A Factored	Architecture to Solve Illusory Conjunctions in Neural Networks . . .	Strong, GW 87a
Complex Cognitive Information-Processing: A Computational	Architecture with a Connectionist Implementation	Barnden, JA 86a
An Associative Co-Processor	Architecture with the UNIX Operating System for the Acceleration of.	Schulz, M 88a
Acoustooptic Algebraic Processing	Architectures.	Rhodes, WT 84a
Computation Power of Parallelism in Optical	Architectures.	Abu-Mostafa, YS 85a
...to a System for Implementing Neural Net Connections on SIMD	Architectures.	Tomboulian, S 88a
Memory Requirements for Balanced Computer	Architectures.	Kung, HT 85b
Neuromolecular Computing: Mechanisms and	Architectures.	Conrad, M 88a
Optical Adaptive Associative Computer	Architectures.	Fisher, AD 85b
...Developments in the Design of Real-Time Non-Linear Neural Network	Architectures.	Grossberg, S 87n
Silicon Brains: Whither Neuromimetic Computer	Architectures.	Vidal, JJ 87c
Symbol Structures in Connectionist Networks: Five Properties and Two	Architectures.	Touretzky, DS 87c
...to Binary Vector Transformation Suitable for Neural Network	Architectures.	Penz, PA 87a
The Case for Conceptual and Operational Separation of Network	Architectures and Learning Mechanisms.	White, H 88b
	Architectures and Methodologies for Self-Organization and.	Farhat, NH 87a
Neural Network	Architectures and Their Implications for Next Generation	Will, CA 87a
Attentive Associative	Architectures and Their Implications to Optical Computing.	Athale, RA 86a
New Computer	Architectures and Their Relationships to Physics, or Why Computer. .	Hillis, WD 82a
Learning System	Architectures Composed of Multiple Learning Modules.	Reilly, DL 87a
	Architectures for a Continuous Level Neural Network Based on	Marks II, RJ 88a
Compact	Architectures for Adaptive Neural Nets	Athale, RA 88a
Neural Network Based	Architectures for Aerospace Applications	Ricart, R 87a
Neural Network	Architectures for AI	Hinton, GE 87c
Massively Parallel	Architectures for AI: Netl, Thistle, and Boltzmann Machines.	Fahlman, SE 83a
Optical-Processor	Architectures for Alternating-Projection Neural Networks	Marks II, RJ 88d
Connectionist	Architectures for Artificial Intelligence.	Fahlman, SE 87a
Digital VLSI	Architectures for Artificial Neural Nets	Kung, S 88b
Parallel	Architectures for Image Processing, Computer Vision, and Pattern . .	Uhr, L 86a
VLSI	Architectures for Implementation of Neural Networks.	Sivilotti, MA 86a
Parallel	Architectures for Neural Computers	Recce, M 88a
Associative Processor	Architectures for Optical Computing.	Fisher, AD 84a
Neural Net	Architectures for Robotic Applications	Kung, S 89a
Evaluation of Network	Architectures on Test Learning Tasks	Soulie, FF 87a
Application of Neural Network Algorithms and	Architectures to Correlation / Tracking and Identification	Gardner, S 86a
...Description of Retinotopic Mapping in the Cat's Visual Cortex	Areas 17, 18, and 19	Mallot, HA 85a
Some	Arguments for a Theory of Cell Assemblies in the Cerebral Cortex . .	Braitenberg, V 89a
A Neural Net Model for the	a-Rhythm	Anninos, PA 80a
Network Model of Shape-From-Shading: Neural Function	Arises from Both Receptive and Projective Fields	Lehky, SR 88a
...and Stability of a Limit Cycle for a Third Order Dynamical System	Arising in Neuron Modelling.	Li, B 81a
PLATO /	ARISTOTLE: A Neural Net Knowledge Processor.	Voevodsky, J 87a
Asynchronous VLSI Neural Networks Using Pulse-Stream	Arithmetic	Murray, AF 88b
Asynchronous	Arithmetic for VLSI Neural Systems	Murray, AF 87a
	Arithmetical Operations Performed by Nerve Cells	Blomfield, S 74a
The Vite Model: A Neural Command Circuit for Generating	Arm and Articulator Trajectories	Bullock, D 88d
Human Movement Initiation: Specification of	Arm, Direction, and Extent	Rosenbaum, DA 80a
Using a Neural Network to Learn the Dynamics of the CMU Direct-Drive	Arm II	Goldberg, K 88a
...and Coordination of Multiple Effectors During Variable-Speed	Arm Movements.	Bullock, D 87a
Neural Dynamics of Planned	Arm Movements: Emergent Invariants and Speed-Accuracy Properties . .	Bullock, D 88c
Neural Dynamics of Planned	Arm Movements: Emergent Invariants and Speed-Accuracy Properties . .	Bullock, D 88b
Neural Nets to	Arm NASA Space Robots?	Johnson, RC 87h
Theoretical and Experimental Perspectives on	Arm Trajectory Formation: A Distributed Model of Motor Redundancy. .	Mussa-Ivaldi, FA 88c
...Processing Principles to Pattern Classification: Final Report, US	Army Research Office	Myers, MH 87a
Perceived Lightness Depends on Perceived Spatial	Arrangement.	Gilchrist, AL 77a
Spatial	Arrangement of the Vestibular and the Oculomotor System in the Rat .	Daunicht, WJ 87a
When Does Perceived Lightness Depend on Perceived Spatial	Arrangement?	Gilchrist, AL 80a
Simulated Annealing in Networks for Computing Possible	Arrangements for Red and Green Cones	Ahumada, AJ 87a
How Sensory Maps Could Enhance Resolution Through Ordered	Arrangements of Broadly Tuned Receivers.	Baldi, P 88c
...of 3D Motion and 3D Layout Information from the Flow of the Optic	Array.	Tsao, T 88a
...of Neural Networks for Speech Recognition on a Transputer	Array.	Chong, MWH 88a
Modelling the Retina as a Stochastic Receptor	Array.	Wardell, RW 82a
Rod Bipolar	Array in the Cat Retina: Pattern of Input from Rods and GABA	Freed, MA 87a
Finding Circles by an	Array of Accumulators.	Kimme, C 75a
Model of the Cerebellum as an	Array of Adjustable Pattern Generators	Houk, JC 87a
Parallel Asynchronous Focal Plane	Array Processing	Coon, DD 87b

Dynamic Control of an	Artificial Neural System: The Property Inheritance Network	Ryan, TW 87a
A Darwinian Approach to	Artificial Neural Systems.	Dress, WB 87a
Learning in	Artificial Neural Systems.	Matheus, CJ 87a
Nonlinear Dynamics of	Artificial Neural Systems.	Maxwell, T 86a
Pattern Recognition Using	Artificial Neural Systems.	Gutschow, T 87a
Sensor Processing Using	Artificial Neural Systems.	Hecht-Nielsen, R 86b
Survey on	Artificial Neural Systems.	Miller, RK 88a
A Survey of	Artificial Neural Systems I: Foundations	Simpson, PK 87b
A Review of	Artificial Neural Systems II: Paradigms, Applications, and	Simpson, PK 87a
Fast Learning in	Artificial Neural Systems: Multilayer Perceptron Training Using.	Shepanski, JF 88a
Teaching	Artificial Neural Systems to Drive: Manual Training Techniques for	Shepanski, JF 88b
	Artificial Neuron.	Hiltz, FF 63a
	Artificial Neuron.	Harmon, LD 59a
Entropy Driven	Artificial Neuronal Networks and Sensorial Representation: A	Van Hulle, MM 88c
Studies with	Artificial Neurons, I: Properties and Functions of an Artificial	Harmon, LD 61a
	Artificial Organisms: A Neural Modeling Approach	Hood, G 86a
Biological vs	Artificial Representations of Variables in Fine-Grain Parallel	Walters, DKW 87b
Adaptation in Natural and	Artificial Systems	Holland, JH 75a
Relational Models in Natural and	Artificial Vision.	Bienenstock, EL 88a
Processing of Moving Images in Natural and	Artificial Visual Systems.	Schuling, FH 88b
	Artina: A Flexible Computer Model of Retinal Processes for Moving.	Gawronski, RP 87a
A Neural Theory of Circadian Rhythms:	Aschoff's Rule in Diurnal and Nocturnal Mammals.	Carpenter, GA 84a
Neural Theory of Circadian Rhythms:	Aschoff's Rule in Diurnal and Nocturnal Mammals.	Carpenter, GA 861
Binary Synaptic Connections Based on Memory Switching in	a-Si:H for Artificial Neural Networks.	Thakoor, AP 87a
	Asilomar Conference on Adaptive Processes in Visual and Oculomoter	Keller, EL 85a
Let's	Ask GMDH What Effect the Environment Has on Fisheries.	Brooks, HA 84a
Progress on the Spin Chip	Asociative Processor	Goodwin, JM 88b
...Simulating a Network of Interacting Neurons, Part II: Programming	Aspects.	Perkel, DH 76b
Theoretical and Experimental	Aspects of a Cerebellar Model.	Albus, JS 72b
Topological and Dynamical	Aspects of a Neural Network Model for Generation of Pursuit Motor.	Eckmiller, R 86b
Temporal	Aspects of Credit Assignment in Reinforcement Learning: PhD Thesis	Sutton, RS 84a
Physico-Mathematical	Aspects of Excitation and Conduction in Nerves	Rashevsky, N 36a
Electronic Device	Aspects of Neural Network Memories	Lambe, J 85a
Mathematical	Aspects of Periodic Catatonic Schizophrenia.	Cronin, J 77a
Simple Changes in Reflex Threshold Cannot Explain All	Aspects of Rapid Voluntary Movements	Gielen, CCAM 86a
Certain	Aspects of the Anatomy and Physiology of the Cerebral Cortex	Crick, FHC 86a
Theoretical and Linguistic	Aspects of the Fuzzy Logic Controller.	Braae, M 79a
Morphological	Aspects of the Vestibular Nerve of the Squirrel Monkey	Honrubia, V 87a
Shifter Circuits: A Computational Strategy for Dyanmic	Aspects of Visual Processing	Anderson, CH 87a
An	Assembler for a Neural Network	Noos, FA 87a
Associative Networks and Cell	Assemblies	Palm, G 86a
Dynamics of Image Formation by Nerve Cell	Assemblies	Freeman, WJ 83a
Operational Significance of Receptive Field	Assemblies	Koenderink, JJ 88b
Towards a Theory of Cell	Assemblies	Palm, G 81a
Neural	Assemblies: An Alternative Approach to Artificial Intelligence	Palm, G 82a
Memory Storage Capacity and the Role of	Assemblies in Neural Networks.	Shaw, GL 77a
Some Arguments for a Theory of Cell	Assemblies in the Cerebral Cortex.	Braitenberg, V 89a
Cooperativity in Brain Function:	Assemblies of Approximately 30 Neurons	Shaw, GL 82a
Analytic Study of	Assemblies of Neurons in Memory Storage.	Roney, KJ 80a
...Collateral Inhibition Simulated in a Simple Neuronal Automata	Assembly	Axelrad, H 86a
...of Behavior: The First Stage of Perception: Growth of the	Assembly	Hebb, DO 49a
A Task-Level Robot Planner for	Assembly Operations.	Vijaykumar, R 86a
Tests on a Cell	Assembly Theory of the Action of the Brain, Using a Large Digital.	Rochester, N 56b
Tests on a Cell	Assembly Theory of the Action of the Brain, Using a Large Digital.	Rochester, N 56a
Goal Seeking Components for Adaptive Intelligence: An Initial	Assessment	Barto, AG 81b
Bayesian	Assessment of a Connectionist Model for Fault Detection.	Gallant, SI 88a
Mechanisms of Sentence Processing:	Assigning Roles to Constituents.	McClelland, JL 86f
Parallel Distributed Processing and Role	Assignment Constraints	McClelland, JL 86g
Temporal Credit	Assignment in Reinforcement Learning	Sutton, RS 84b
Temporal Aspects of Credit	Assignment in Reinforcement Learning: PhD Thesis	Sutton, RS 84a
A Neural-Network Solution to the Concentrator	Assignment Problem	Tagliarini, GA 88a
Learning from Solution Paths: An Approach to the Credit	Assignment Problem	Sleeman, D 82a
Learning in Connectionist Networks: Has the Credit	Assignment Problem Been Solved?.	Goel, A 89a
A Parallel Computation that	Assigns Canonical Object-Based Frames of Reference	Hinton, GE 81a
Neural Networks, Potential	Assistants to Knowledge Engineers.	Lendaris, GG 88b
	Associate Networks Underlying Problem-Solving: Representation of	Hirai, Y 87a
Representation of	Associated Data by Matrix Operators.	Kohonen, T 73b
Mobile Robot Control by Neural Networks and Their	Associated Learning Algorithm.	Watanabe, N 88a
Statistical Properties of Flip-Flop Processes	Associated to the Chaotic Behavior of Systems with Strange	Aicardi, F 87a

The Capacity of the Kanerva	Associative Memory is Exponential.	Chou, PA 88b
Analysis of a Distributed	Associative Memory Model	Wiles, J 87a
The Adaptive, Dynamic, and	Associative Memory Model: A Possible Future Tool for Vocal Human	Beroule, D 87b
An	Associative Memory Model Based on a Quantification Method.	Kato, K 88a
Optical	Associative Memory Model Based on Neural Networks Having Variable.	Macukow, B 87b
Optical Information Processing Based on an	Associative Memory Model of Neural Nets with Thresholding and.	Psaltis, D 85a
	Associative Memory Models.	Willshaw, DJ 70a
...for Patterns and Sequences in Kanerva's SDM as Compared to Other	Associative Memory Models.	Keeler, JD 88a
Neural Networks: Implementing	Associative Memory Models in Neurocomputers.	Miller, RK 87a
Stability and Attractivity in	Associative Memory Networks.	Cottrell, M 88a
Class of Continuous Level	Associative Memory Neural Nets	Marks II, RJ 87b
Construction of a Distributed	Associative Memory on the Basis of Bayes Discriminant Rule	Murakami, K 81a
An Adaptive	Associative Memory Principle	Kohonen, T 74a
The Correlation	Associative Memory Realizes Hebbian Learning	Karayiannis, NB 88a
Self-Organization and	Associative Memory, Second Edition	Kohonen, T 88c
	Associative Memory Storage Using a Variant of the Generalized Delta.	Samad, T 87b
Learning Control Structures with Neuron-Like	Associative Memory Systems	Ersu, E 87a
Storage and Processing of Information in Distributed	Associative Memory Systems	Kohonen, T 81c
An	Associative Memory Technique for the Recognition of Patterns	Pao, Y 78a
Design for a Massive All-Optical Bidirectional	Associative Memory: the Big BAM.	Kinser, JM 88a
Optical Holographic	Associative Memory Using a Phase Conjugate Resonator	Dunning, GJ 86a
Optical Implementation of	Associative Memory with Controlled Bit-Significance.	Lee, S 88a
Optical Implementation of	Associative Memory with Controlled Nonlinearity in the Correlation	Athale, RA 86c
An	Associative Memory with Heterogeneous Input Fields	Winter, CL 88b
Holographic	Associative Memory with Nonlinearities in the Correlation Domain	Owechko, Y 87a
Optical Implementation of Quadratic	Associative Memory with Outer-Product Storage.	Jang, J 88b
Video Rate Optical	Associative Memory with Shift Invariance	Owechko, Y 87c
All-Optical	Associative Memory with Shift Invariance and Multiple-Image Recall	Dunning, GJ 87a
A Very Fast	Associative Method for the Recognition and Correction of Misspelt.	Kohonen, T 78a
Investigations into the Pattern Processing Capabilities of	Associative Nets	Lansner, A 86a
Prototype Extraction and Matching in	Associative Nets	Lansner, A 85a
Reliability and Speed of Recall in an	Associative Network.	Lansner, A 85b
	Associative Network Applications to Robot Vision	Oyster, JM 86a
An	Associative Network Solving the "4-Bit ADDER Problem".	Lansner, A 87a
A Program Package Implementing a System of Interacting Adaptive	Associative Networks	Lansner, A 83a
	Associative Networks	Findler, NV 79a
	Associative Networks	Cruz, CA 82a
Decrementing	Associative Networks	Dobson, VG 88a
Hierarchical	Associative Networks	Cortes, C 87a
History of	Associative Networks	Hecht-Nielsen, R 87d
NEP: An Emulation-Assist Processor for Parallel	Associative Networks	Cruz, CA 85a
Flow-of-Activation Processing: Parallel	Associative Networks (PAN)	Cruz, CA 86a
On Applying	Associative Networks: An Approach for Representation and Goal.	Fisher, AD 87a
	Associative Networks and Cell Assemblies	Palm, G 86a
Modeling the Process of Problem-Solving by	Associative Networks Capable of Improving the Performance.	Hirai, Y 88a
	Associative Networks, II	Cruz, CA 83a
	Associative Networks: Representation and Use of Knowledge by	Findler, NV 79b
Superior Accuracy of Decrementing over Incrementing	Associative Networks when Operating in Initially Random.	Dobson, VG 87a
Mean-field Analysis of Hierarchical	Associative Networks with Magnetisation.	Krogh, A 88a
Optical Implementations of	Associative Networks with Versatile Adaptive Learning Capabilities	Fisher, AD 87b
A Connectivity Analysis of a Class of Simple	Associative Neural Networks.	Hammerstrom, D 88a
Spatial EEG Correlates of Non-Associative and	Associative Olfactory Learning in Rabbits.	Grajski, KA 88b
Implementations of Adaptive	Associative Optical Computing Elements	Fisher, AD 86a
On a Model of Human	Associative Processor.	Hirai, Y 86a
A Model of Human	Associative Processor (HASP)	Hirai, Y 83a
	Associative Processor Architectures for Optical Computing.	Fisher, AD 84a
An	Associative Processor as a Design Rule Check Accelerator	Strugala, M 87a
Concept of a Flag-Oriented Fully Parallel	Associative Processor Based on Flag-Algebra.	Tavangarian, D 88a
...Representation of the Process of Addition by a Model of an	Associative Processor, HASP.	Hirai, Y 87a
Representation of Control Structures by a Model of an	Associative Processor, HASP.	Hirai, Y 87b
...Model for Selective Attention in Visual Pattern Recognition and	Associative Recall	Fukushima, K 87c
...Model for Selective Attention in Visual Pattern Recognition	Associative Recall	Fukushima, K 87d
On Optimal Nonlinear	Associative Recall	Poggio, TA 75a
Theories of	Associative Recall	Longuet-Higgins, HC 70a
A Composite Holographic	Associative Recall Model	Eich, JM 82a
	Associative Recall of Images	Kohonen, T 76a
	Associative Recall Properties of the Trion Model of Cortical	Silverman, DJ 86a
On the Use of Backpropagation in	Associative Reinforcement Learning	Williams, RJ 88a
Landmark Learning: An Illustration of	Associative Search	Barto, AG 81c

Finite	Automata, Pattern Recognition and Perceptrons.	Keller, HB 61a
An	Automata-Theoretical Approach to Developing Learning Neural.	Veelenturf, LPJ 81a
Pseudo-Unsupervised Neural Network Based	Automated EMG Decomposition.	Hassoun, MH 88c
	Automated Fake Color Separation: Combining Computer Vision and . . .	Walters, DKW 87a
	Automated Generation of Connectionist Expert Systems for Problems. .	Gallant, SI 87a
	Automated Reasoning on Neural Networks: A Probabilistic Approach . .	Chen, S 87a
Using	Automated Techniques to Generate an Expert System for R&D Project. .	Gallant, SI 86d
Algorithms and Techniques for	Automated Visual Inspection.	Chin, RT 86a
Towards Integrating	Automatic and Controlled Problem Solving	Day, DS 87a
	Automatic Character Reading Using a Holographic Data Processing. . .	Nakajima, M 72a
	Automatic Character Recognition: A State-of-the-Art Report	Stevens, ME 61a
	Automatic Classification Algorithms.	Dorofeyuk, AM 71a
	Automatic Computation of Nerve Excitation.	Cole, KS 55a
	Automatic Computation of Nerve Excitation - Detailed Corrections . .	FitzHugh, R 59a
	Automatic Formation of Topological Maps of Patterns in a Self. . . .	Kohonen, T 81a
A Neural Network System with an	Automatic Generation Mechanism for Distorted Patterns.	Shimohara, K 88a
	Automatic Generation of Expert Systems from Examples	Gallant, SI 85a
	Automatic Generation of Time Efficient Digital Signal Processing . .	Morris, LR 77a
	Automatic Learning, Rule Extraction, and Generalization.	Denker, JS 87a
Predicted Squared Error: A Criterion for	Automatic Model Selection.	Barron, AR 84a
On the Control of	Automatic Processes: A Parallel Distributed Processing Model of the.	Cohen, JC 88a
Toward a Model of Attention and the Development of	Automatic Processing	Schneider, W 85a
The AUROS Project:	Automatic Recognition of Speakers by Computers	Bunge, E 77b
...of the Theory of Probabilistic Functions of a Markov Process to	Automatic Speech Recognition	Levinson, SE 83a
Multilayer Perceptrons and	Automatic Speech Recognition	Bourland, H 87a
Stochastic Modeling as a Means of	Automatic Speech Recognition	Baker, JK 75a
Structural Methods in	Automatic Speech Recognition	Levinson, SE 85a
The Modified Kanerva Model for	Automatic Speech Recognition	Prager, RW 88a
	Automatic Subclass Determination for Pattern-Recognition	Firschein, O 63a
Adaptation and Learning in	Automatic Systems.	Tsypkin, YZ 71a
A Neural Network Architecture for	Automatic Trajectory Formation and Coordination of Multiple.	Bullock, D 87a
Unitization,	Automaticity, Temporal Order, and Word Recognition	Grossberg, S 84e
Generalization Learning Techniques for	Automating the Learning of Heuristics.	Waterman, DA 70a
...Adaptive Systems: A Statistical/Numerical Approach to Factory	Automation and Brain Research.	Werbos, PJ 87a
An Intelligent	Automaton.	Rosen, CA 67a
The Perceptron: A Perceiving and Recognizing	Automaton, Project PARA.	Rosenblatt, F 57a
Teaching a Stochastic	Automaton to Skillfully Play Hand / Eye Games.	Jarvis, RA 77a
Dynamical Properties of an	Automaton With Memory.	Cosnard, M 86a
...Global Search in a Time Varying Environment Using a Probabilistic	Automaton with Pattern Recognition Supervision	Jarvis, RA 70a
Directional Entropies of Celluar	Automaton-Maps	Milnor, J 86a
Behaviour of	Autonomous, (Synchronous), Boolean Networks.	Martland, D 87b
A Neuron Model of a Memory System for	Autonomous Exploration of an Environment	Trenary, R 87a
On	Autonomous Logic Nets of Threshold Elements.	Yajima, S 68a
Visual Control of	Autonomous Mobile Robot Based on Self-Organizing Model for Pattern .	Suzuki, H 88c
Cognitive Control of	Autonomous Mobile Robots: Nested Hierarchical Intelligent Module . .	Meystel, A 86a
Neural Network Representation of Sensor Graphs in	Autonomous Robot Path Planning	Jorgensen, CC 87a
Concurrent Neuromorphic Algorithms for	Autonomous Robots.	Barhen, J 88b
Connectionist Networks for Learning Coordinated Motion in	Autonomous Systems	Pabon, JA 88a
...Neural Systems to Drive: Manual Training Techniques for	Autonomous Systems	Shepanski, JF 88b
Manual Training Technqiues of	Autonomous Systems Based on Artificial Neural Networks	Shepanski, JF 87a
An Artificial Neural System for	Autonomous Undersea Vehicles	Blackburn, MR 88a
Theory and Feasibility of the Parametric	Avalanche for Advanced Electronic Warfare Applications	Dawes, RL 88b
Exact Results for the	Average Dynamic Behavior of some Non-Linear Neural Networks.	Rossler, J 87a
Some	Averaging and Stability Results for Random Differential Equations. .	Geman, S 79a
A Signal	Averaging Model for Concept Formation.	Knapp, AG 84a
Application of Stochastic	Averaging to Learning Systems.	Geman, S 78a
There Exists a Neural Network That Does Not Make	Avoidable Mistakes	Gallant, AR 88a
A Neural Theory of Punishment and	Avoidance, Part I: Qualitative Theory.	Grossberg, S 72a
A Neural Theory of Punishment and	Avoidance, Part II: Quantitative Theory.	Grossberg, S 72b
Computers That Come	Awfully Close to Thinking.	Port, O 86a
Representational	Axes and Temporal Cooperative Processes.	Tsotsos, JK 87a
Low Temperature Analysis of the	Axial Next-Nearest Neighbor Ising Model Near its Multiphase Point. .	Fisher, ME 81a
Old Dogmas and New	Axioms in Brain Theory	Pellionisz, AJ 86a
An Active Pulse Transmission Line Simulating Nerve	Axon	Nagumo, J 62a
Possibilities for Signal Processing in	Axon Systems	Crane, HD 64a
Development of Ocularity Domains and Growth Behaviour of	Axon Terminals	von der Malsburg, C 79a
...Model for Conduction of Action Potentials Along Bifurcating	Axons.	Parnas, I 79a
An Improved Three-Layer,	Back Propagation Algorithm	Stornetta, WS 87a
Image Compression by	Back Propagation: An Example of Extensional Programming.	Cottrell, GW 87a

A Comparison of the Boltzmann Machine and the	Back Propagation Network as Recognizers of Static Speech patterns. .	Prager, RW 87a
An Adaptive Training Algorithm for	Back Propagation Networks.	Chan, L 87a
Optimal Algorithms for Adaptive Networks: Second Order	Back Propagation, Second Order Direct Propagaion and Second Order. .	Parker, DB 87a
Generalizing	Back Propagation to Computation.	Baum, EB 86a
Neural Network Startups	Backed by Venture Capital.	Brown, C 87i
An Associative	Backend Machine for Data Base Management	Hurson, A 81a
A 'Neural' Network that Learns to Play	Backgammon	Tesauro, G 88a
Learning Representations by	Back-Propagating Errors.	Rumelhart, DE 86e
Learning Representations by	Back-Propagating Errors.	Rumelhart, DE 86f
Experiments on	Back-Propagation	Plaut, DC 86a
Factors Influencing Learning by	Backpropagation.	Von Lehmen, A 88a
Forming Global Representations with Extended	Backpropagation.	Miikkulainen, R 88b
Self-Adapting	Back-Propagation	Devos, MR 88a
Several Studies on Natural Language and	Back-Propagation	Allen, RB 87a
Some Asymptotic Results for	Back-Propagation	White, H 87a
	Back-Propagation: A Generalized Delta Learning Rule.	Jones, WP 87a
Two Problems with	Back-Propagation and Other Steepest-Descent Learning Procedures for.	Sutton, RS 86a
...Neural Nets for Statistical Discovery: Observations after Using	Back-propagation, Expert Systems, and Multiple-Linear Regression . .	Irani, EA 88a
Performance of	Backpropagation for Rotation Invariant Pattern Recognition	Yang, H 87a
On the Use of	Backpropagation in Associative Reinforcement Learning.	Williams, RJ 88a
Gradient Following Without	Back-Propagation in Layered Networks	Barto, AG 87b
	Backpropagation in Nonfeedforward Networks	Almeida, LB 88a
	Backpropagation in Perceptrons with Feedback	Almeida, LB 88b
	Back-Propagation is Significantly Faster if the Expected Value of. .	Samad, T 88a
An Algebraic Projection Analysis for	Back-Propagation Learning.	Kung, S 88a
Scaling Relationships in	Back-Propagation Learning: Dependence on Predicate Order	Tesauro, G 88c
Scaling Relationships in	Back-Propagation Learning: Dependence on Training Set Size	Tesauro, G 87a
Minkowski-r	Back-Propagation: Learning in Connectionist Models with Non.	Hanson, SJ 88a
Refining and Redefining the	Back-Propagation Learning Rule for Connectionist Networks.	Samad, T 87c
Accelerating the Convergence of the	Back-Propagation Method.	Vogl, TP 88a
Tactile Information Processing: A	Back-Propagation Model of Pattern Recognition in the Human	Grajski, KA 88c
Teaching a Minimally Structured	Back-Propagation Network to Recognize Speech Sounds.	Landauer, TK 87a
An Empirical Study of Learning Speed in	Back-Propagation Networks.	Fahlman, SE 88a
	Back-Propagation Networks for Event-Driven Temporal Sequence	Shimohara, K 88b
Successfully Using Peak Learning Rates of 10 (and Greater) in	Back-Propagation Networks with the Heuristic Learning Algorithm. . .	Cater, JP 87a
Inverting a Connectionist Network Mapping by	Back-Propagation of Error.	Williams, RJ 86a
Cascaded	Back-Propagation on Dynamic Connectionist Networks	Pollack, JB 87a
	Backpropagation: Past and Future	Werbos, PJ 88a
A Dual	Back-Propagation Scheme for Scalar Reward Learning	Munro, PW 87a
Generalization of	Backpropagation to Recurrent and Higher Order Neural Networks. . . .	Pineda, FJ 88b
Generalization of	Back-Propagation to Recurrent Neural Networks.	Pineda, FJ 87a
In	Back-to-Basics Movement, VLSI Expert Calls for More Simplicity in. .	Brown, C 87e
	Backward Conditioning: A Neural Network Model which Exhibits Both. .	Ricart, R 88a
Exploration of	Backward Error Propagation as a Self-Organizational Structure. . . .	Kuczewski, RM 87a
Exploration of	Backward Error Propagation as a Self-Organized Structure	Kuczewski, RM 87b
Weight Discretization in	Backward Error Propagation Neural Networks	Fiesler, E 88a
Japan's 6th Generation Project: Striking A Fine	Balance.	Johnson, RC 87d
Neural Dynamics of Decision Making Under Risk: Affective	Balance and Cognitive-Emotional Interactions	Grossberg, S 87k
Neural Dynamics of Decision Making Under Risk: Affective	Balance Theory	Grossberg, S 87l
	Balanced and Unbalanced, Complete and Partial Transparency	Metelli, F 85a
Memory Requirements for	Balanced Computer Architectures.	Kung, HT 85b
A	Balanced Scalable Parallel Processor	Pilpel, S 87a
Neural Dynamics of Adaptive Sensory-Motor Control:	Ballistic Eye Movements.	Grossberg, S 86h
Extending Kohonen's Self-Organizing Mapping Algorithm to Learn	Ballistic Movements.	Ritter, H 88b
A	Ballpark Approach to Modelling Human Prehension.	Iberall, T 87a
...a Massive All-Optical Bidirectional Associative Memory: the Big	BAM.	Kinser, JM 88a
e-Optimal Learning Algorithms: Non-Absorbing	Barrier Type	Lakshmivarahan, S 79a
An Associative Backend Machine for Data	Base Management.	Hurson, A 81a
The T5	Base Modulator Hypothesis: A Dynamic Model of T5 Neuron Function in.	Betts, B 88a
Convergence Issues in Perceptron	Based Adaptive Neural Network Models	Shrivastava, Y 87a
A Feature Projection	Based Adaptive Pattern Recognition Network	Lee, JSJ 88b
Neural Network	Based Architectures for Aerospace Applications	Ricart, R 87a
Nonlinearity Requirements for Correlation	Based Associative Memories	Marks II, RJ 88c
Cross-talk Minimization for a Walsh-Hadamard	Based Associative Memory	Oh, S 84b
Pseudo-Unsupervised Neural Network	Based Automated EMG Decomposition.	Hassoun, MH 88c
Evaluation of a Hypercube	Based Compact Neural Network	Somani, AK 88a
TITANIC: A VLSI	Based Content Addressable Parallel Array Processor	Weems, C 82a
A Neural Network	Based Controller for Robotic Manipulators.	Kamangar, FA 88a
A Walsh-Hadamard	Based Distributed Storage Device for the Associative Search of . . .	Oh, S 84a

Basic Results for the	Behavior of Discrete Iterations.	Robert, F 86a
	Behavior of Homogeneous Statistical Ensemble of Finite Automata. . .	Rotenberg, AR 71a
Synchronous vs Asynchronous	Behavior of Hopfield's CAM Neural Net.	Cheung, KF 87c
A Study of the Asymptotic	Behavior of Neural Networks.	Dimopoulos, N 82a
Temporal	Behavior of Neural Networks.	Gelenbe, E 87a
Mathematical Analysis of Learning	Behavior of Neuronal Models.	Cheung, JY 88a
Exact Results for the Average Dynamic	Behavior of some Non-Linear Neural Networks.	Rossler, J 87a
...Properties of Flip-Flop Processes Associated to the Chaotic	Behavior of Systems with Strange Attractors.	Aicardi, F 87a
The Basis of Lawful	Behavior: Rules or Connections?.	McClelland, JL 87c
Relation of Olfactory EEG to	Behavior: Spatial Analysis	Freeman, WJ 87c
Adaptive Self-Organization of Serial Order in	Behavior: Speech, Language, and Motor Control.	Grossberg, S 86c
The Adaptive Self-Organization of Serial Order in	Behavior: Speech, Language, and Motor Control.	Grossberg, S 86r
The Organization of	Behavior: The First Stage of Perception: Growth of the Assembly. . .	Hebb, DO 49a
Relation of Olfactory EEG to	Behavior: Time Series Analysis	Freeman, WJ 86a
The Microscopic Analysis of	Behavior: Toward a Synthesis of Instrumental, Perceptual, and. . . .	Grossberg, S 84d
...Reduced Withdrawal from Shadows in Hermissenda: A Direct	Behavioral Analog of Photoreceptor Responses to Brief Light Steps. .	Lederhendler, II 87a
Dynamic Pattern Generation in	Behavioral and Neural Systems.	Schoner, G 88d
A Dynamic Pattern Theory of	Behavioral Change.	Schoner, G 88a
Psychophysiological Substrates of Schedule Interaction and	Behavioral Contrast.	Grossberg, S 81b
	Behavioral Contrast in Short Term Memory: Serial Binary Memory . . .	Grossberg, S 78c
Towards an Information-Flow Model of Human	Behaviour.	MacKay, DM 56a
Bifurcation and Chaotic	Behaviour in Simple Feedback Systems	Sparrow, C 80a
A Dynamical	Behaviour of Active Regions in Randomly Connected Neural Networks. .	Okuda, M 74a
	Behaviour of Autonomous, (Synchronous), Boolean Networks	Martland, D 87b
Development of Ocularity Domains and Growth	Behaviour of Axon Terminals.	von der Malsburg, C 79a
Dynamical	Behaviour of Neural Networks	Goles, E 85a
Propagating	Belief Functions with Local Computations	Shenoy, P 86a
Fusion, Propagation, and Structuring in	Belief Networks.	Pearl, J 86b
Fusion, Propagation, and Structuring in	Belief Networks.	Pearl, J 86a
Distributed Revision of Composite	Beliefs.	Pearl, J 87a
Fuzzy Logic's Soft Start	Belies a Cutting Edge.	Chester, M 86a
What	Bell Laboratories is Learning from Slugs	Newport, JP 85a
	Bell Labs Models Parallel Processor on Neural Networks	Victor, J 86b
	Bell Labs Models Parallel Processor on Neural Networks	Victor, J 86a
A Microcomputer Neural Net	Benchmarked Against Standard Classification Techniques	Draper, JS 87a
Depth and Shape from Differential Perspective in the Presence of	Bending Deformations	Koenderink, JJ 86a
Melding Computers and Biology for Commercial	Benefit.	Hunter, L 88b
Expectations Increase the	Benefit Derived from Parafoveal Visual Information in Reading Words.	McClelland, JL 81c
CONSIM (Conflict Simulator): Risk, Cost and	Benefit in Political Simulations	Clema, JK 71a
Nerve Conduction Theory: Some Mathematical Consequences of	Bernstein's Model.	Offner, F 40a
A Digital Implementation of a	Best Match Classifier.	Mackie, S 88a
Comparison of a Pipelined "	Best Match" Content Addressable Memory with Neural Networks.	Clark, LT 87a
An Algorithm for Finding	Best Matches in Logarithmic Expected Time.	Friedman, JH 77a
Microcircuitry of	Beta Ganglion Cells in Cat Retina.	McGuire, BA 86a
The Relationship	Between Image Restoration by the Maximum A Posteriori Method and a .	Trussell, HJ 80a
Relations	Between Initial Conditions and Periodic Firing Modes in Reciprocal .	Nagashino, H 81a
...of Randomly Connected McCulloch-Pitts Networks: Differences	Between Input-Constant and Input-Variant Networks.	Dammasch, IE 84a
Comparison	Between Kanerva's SDM and Hopfield-type Neural Networks.	Keeler, JD 88b
The Relation	Between Linguistic Structure and Associative Theories of Language. .	Lachter, J 88a
Relationships	Between LMS Compensators for Intersymbol Interference in Linear and.	Kaye, AR 73a
Links	Between Markov Models and Multilayer Perceptrons	Bourland, H 88a
Comparisons	Between Neural Net and Conventional Classifiers.	Huang, WY 87a
...Strength and Computational Algebra of Synaptic Connections	Between Neurons.	Fetz, EE 88a
Dale's Principle and Communication	Between Neurons.	Osborne, NN 82a
Fundamental Relations	Between the LMS Algorithm and the DFT.	Widrow, B 87b
On the Relationship	Between Two Models of NeuralEntrainment.	Torras i Genis, C 87a
DMORPH: An Entropy-Based Measure of the Structural Similarity	Between Two Random Systems	Dawes, RL 88a
	Beyond Neural Networks - The Discipline of Intelligent Systems . . .	Johnson, RC 87a
Learning Symmetry Groups with Hidden Units:	Beyond the Perceptron.	Sejnowski, TJ 86b
	Bias and Anticipation in Neural Networks	Sherwood, DL 87a
Storing Sequences of	Biased Patterns in Neural Networks with Stochastic Dynamics.	Buhmann, J 88a
Some Developmental and Attentional	Biases in the Contrast Enhancement and Short-Term Memory of.	Grossberg, S 75b
Associative Memories and Processors: An Overview and Selected	Bibliography	Parhami, B 73a
Connectionist Models in Speech Recognition: A Brief Summary and	Bibliography	Fallside, F 88b
Connectionist/Neural Network	Bibliography	Hammerstrom, D 86d
A Connectionist/Neural Network	Bibliography, Volume II.	Hammerstrom, D 87a
Adaptive	Bidirectional Associative Memories	Kosko, B 87a
Competitive Adaptive	Bidirectional Associative Memories	Kosko, B 87d
Designs and Devices for Optical	Bidirectional Associative Memories	Guest, CC 87a

Noise Modulation of Synaptic Weights in a	Biological Neural Network.	Gardner, D 89a
Exaptation, Preadaptation, Evolution and	Biological Neural Networks	Stork, DG 87b
Synaptic Diversity Characterizes	Biological Neural Networks	Gardner, D 87a
Disordered Systems and	Biological Organization.	Bienenstock, EL 86a
Methods of	Biological Signal Processing	Shiavi, RG 86a
The Limits of	Biological Simulation.	Conrad, M 74c
...Parallel Computation in Electronic, Optical, Molecular and	Biological Systems	MacLennan, BJ 88a
Multisensor Integration in	Biological Systems	Gelfand, JJ 88a
Recognition of Hierarchically Encoded Images by Technical and	Biological Systems	Hartmann, G 87a
Where Does the Fundamental Information Processing Ability of	Biological Systems Come From?.	Kohonen, T 75a
	Biological vs Artificial Representations of Variables in Fine-Grain.	Walters, DKW 87b
A	Biologically Constrained Learning Mechanism in Networks of Formal. .	Personnaz, L 86a
	Biologically Plausible Models of Place Recognition and Goal.	Zipser, D 86b
Les Modeles Mathematiques de l'Organization	Biologique de la Memoire	Anderson, JA 78a
Oscillations in Mathematical	Biology.	Grissel, R 83a
Melding Computers and	Biology for Commercial Benefit	Hunter, L 88b
	Biomasscomp: A Procedure for Mapping the Architecture of a Living. .	Dawes, RL 87a
	Biomedical Image Analysis.	Sklansky, J 86a
Self-Processing Networks and Their	Biomedical Implications.	Reggia, JA 88a
	Biomolecular Information Processing.	Conrad, M 87a
Doppler Tolerant Classification of Distributed Targets - a	Bionic Sonar	Altes, RA 75a
Conditioning-Specific	Biophysical Alterations in Rabbit Hippocampus.	Disterhoft, JF 88a
A	Biophysical Approach to the Spatial Function of Eye Movements. . . .	Daunicht, WJ 88a
	Biophysics of Computation: Neurons, Synapses and Membranes	Koch, C 84a
A	Bipartite Connectionist Model to Represent N-ary Boolean and Linear.	Post, SD 87a
Application of the Neural Network for the Trajectory Planning of a	Biped Locomotive Robot	Kitamura, S 88a
Rod	Bipolar Array in the Cat Retina: Pattern of Input from Rods and. . .	Freed, MA 87a
Microcircuitry of	Bipolar Cells in Cat Retina.	McGuire, BA 84a
An Artificial Neural Network for Spatio-Temporal	Bipolar Patterns: Application to Phoneme Classification.	Atlas, LE 88a
An Artificial Neural Network for Spatio-Temporal	Bipolar Patterns: Application to Phoneme Classification.	Homma, T 88a
	Birth, Life, and Death in Microelectronic Systems.	Widrow, B 61a
	Bisensory Stimulation: Inferring Decision-Related Processes from . .	Squires, NK 77a
Programmable	Bistable Switches and Resistors for Neural Networks.	Spencer, EG 86a
	Bit-Serial Neural Networks	Murray, AF 88c
VLSI	Bit-Serial Neural Networks	Butler, ZF 88a
Optical Implementation of Associative Memory with Controlled	Bit-Significance	Lee, S 88a
	Bit-Vector Optimization Algorithms for Control of Learning in. . . .	Kirby, KG 88a
Programmable	Blackboard Model of Reading.	McClelland, JL 86i
Engineer A	Black-Box Solution in the First "Real Cognizer" Contest.	Brown, C 87c
	Blob Detection by Relaxation	Danker, AJ 81a
	Board Creates Neural AT.	Schwartz, TJ 88a
...System of Adult Hydra II. Conversion of Ganglion Cells of the	Body Column into Epidermal Sensory Cells of the Hypostome.	Koizumi, O 88a
	BoltzCONS: Reconciling Connectionism with the Recursive Nature of. .	Touretzky, DS 86b
Hopfield,	Boltzman Machines, and Vision.	Sejnowski, TJ 87d
A Mathematical Model for the	Boltzmann Machine.	Mazaika, PK 87a
Combinatorial Optimization Using a Continuous State	Boltzmann Machine.	Gutzmann, KM 87a
Separating Figure from Ground with a	Boltzmann Machine.	Sejnowski, TJ 87g
A Comparison of the	Boltzmann Machine and the Back Propagation Network as Recognizers. .	Prager, RW 87a
Variations on the	Boltzmann Machine Learning Algorithm	Derthick, MA 84a
Characterization of the	Boltzmann Machine Learning Rate.	Parks, M 87a
A Learning Algorithm for	Boltzmann Machines	Ackley, DH 85b
A Learning Algorithm for	Boltzmann Machines	Ackley, DH 85a
A Learning Algorithm for	Boltzmann Machines	Ackley, DH 88a
Higher-Order	Boltzmann Machines	Sejnowski, TJ 86a
Learning and Relearning in	Boltzmann Machines	Hinton, GE 86c
Massively Parallel Architectures for AI: Netl, Thistle, and	Boltzmann Machines	Fahlman, SE 83a
Numerical Simulations of	Boltzmann Machines	Bounds, DG 86a
Adaptive Pattern Recognition Systems and	Boltzmann Machines: A Rapprochement.	Aleksander, I 87a
	Boltzmann Machines: Constraint Satisfaction Networks that Learn. . .	Hinton, GE 84a
	Boltzmann Machines for Speech Recognition.	Prager, RW 86a
	Boltzmann Machines in which Weights Converge to Global Optimal . . .	Uchiyama, T 88a
Relating	Boltzmann Machines to Conventional Models of Computation	Parberry, I 88c
...Recursive Objects in a Neural Network, or "Trees DO Grow on	Boltzmann Machines".	Touretzky, DS 86c
`Ensemble'	Boltzmann Units have Collective Computational Properties like those.	Derthick, MA 88a
	Bond Rating: A Non-Conservative Application of Neural Networks . . .	Dutta, S 88a
An Artificial Intelligence Approach to Predicting	Bond Ratings	Dutta, S 89a
Memory Networks with Asymmetric	Bonds.	Hertz, JA 86a
	Book Review.	Hecht-Nielsen, R 83a
	Book Review.	Hinton, GE 79a

...Equation for the EEG and the General Solution within the	Brain and in Space	Anninos, PA 75a
	Brain and Neurocomputing	Herault, J 88a
A Model of the Brain for Robot Control, Part 3: A Comparison of the	Brain and our Model.	Albus, JS 79c
The Topology of the	Brain and Visual Perception.	Zeeman, EC 62a
Scientists Identify "Gate" in	Brain as Crucial to Memory	Johnson, G 88a
How Does a	Brain Build a Cognitive Code?.	Grossberg, S 80c
How Does a	Brain Build a Cognitive Code?.	Grossberg, S 80d
Electrotonic Processing of Information by	Brain Cells.	Schmitt, FO 76a
The Brains of Men and Machines, Part 2: How the	Brain Controls Outputs	Kent, EW 78b
	Brain Drain Siphons Neural Net Pioneers into Industry.	Brown, C 87a
Hardware Simulation of	Brain Dynamics in Learning: The SPOCK.	Freeman, WJ 87a
Human	Brain EEG Fields: Micro-States and Their Functional Significance	Lehmann, D 87b
Spontaneous and Information-Triggered Segments of Series of Human	Brain Electric Field Maps.	Lehmann, D 88a
A Multiplexed Multi-Electrode Semiconductor	Brain Electrode Implant.	Kabrisky, M 87a
A Model of the	Brain for Robot Control, Part 1: Defining Notation	Albus, JS 79a
A Model of the	Brain for Robot Control, Part 2: A Neurological Model.	Albus, JS 79b
A Model of the	Brain for Robot Control, Part 3: A Comparison of the Brain and our	Albus, JS 79c
A Model of the	Brain for Robot Control, Part 4: Mechanisms of Choice.	Albus, JS 79d
The Correlation Theory of	Brain Function	von der Malsburg, C 81b
	Brain Function and Adaptive Systems: A Heterostatic Theory	Klopf, AH 72a
Cooperativity in	Brain Function: Assemblies of Approximately 30 Neurons	Shaw, GL 82a
Tensor Theory of	Brain Function: The Cerebellum as a Space-Time Metric.	Pellionisz, AJ 82b
The Perceptron: A Model for	Brain Functioning. I	Block, HD 62b
...Chaos in Neural Systems: An Approach to the Dynamics of Higher	Brain Functions.	Harth, EM 83a
Neural Chips Emulate	Brain Functions (ATT VLSI)	Brown, C 86e
...of Large Scale Neuron-Glia Network and Its Relation to the	Brain Functions (I).	Xu, J 86a
	Brain Functions and Neural Dynamics.	Harth, EM 70a
Learning-Logic: Casting the Cortex of the Human	Brain in Silicon	Parker, DB 85a
The Cognitive Revolution and Mind/	Brain Issues	Pribram, KH 86a
...of Measurement, Simulation, and Computational Significance of	Brain Maps	Schwartz, EL 87a
Principles of Neurodynamics: Perceptrons and the Theory of	Brain Mechanisms	Rosenblatt, F 61a
Principles of Neurodynamics: Perceptrons and the Theory of	Brain Mechanisms	Rosenblatt, F 62a
	Brain Metaphors, Theories, and Facts	Grossberg, S 86e
EEG Alpha Map Series:	Brain Micro-States by Space-Oriented Adaptive Segmentation	Lehmann, D 87a
Neural and	Brain Modeling	MacGregor, RJ 87a
	Brain Modeling by Tensor Network Theory and Computer Simulation.	Pellionisz, AJ 79a
Studies of Mind and	Brain: Neural Principles of Learning, Perception, Development.	Grossberg, S 82c
Synaptic Plasticity as Basis of	Brain Organization	von der Malsburg, C 87b
Vision by Man and Machine: How the	Brain Processes Visual Information May Be Suggested by Studies in.	Poggio, TA 84c
Modularity and Interaction of	Brain Regions Underlying Visuomotor Coordination	Arbib, MA 85e
"Ignorance-Based" Neural-Network Signal Processing in	Brain Research	Gevins, AS 87a
Applications of Neural-Network (NN) Signal Processing in	Brain Research	Gevins, AS 88a
...A Statistical/Numerical Approach to Factory Automation and	Brain Research	Werbos, PJ 87a
...Acts: Results of Latest Research Using Neural Networks to Analyze	Brain Signals.	Gevins, AS 87b
On the Scheme by Which the Human	Brain Stores Information	Legendy, CR 67a
Space-Time Representation in the	Brain: The Cerebellum as a Predictive Space-Time Metric Tensor	Pellionisz, AJ 82a
A Chain of Suprasegmental Neuroscillatory Circuits: A Human	Brain Theory	Deshmukh, VD 88a
Old Dogmas and New Axioms in	Brain Theory	Pellionisz, AJ 86a
Plain Talk About Neurodevelopmental Epistemology and	Brain Theory	Fisher, AD 77a
	Brain Theory and Cooperative Computation	Arbib, MA 85a
	Brain Theory: Connecting Neurobiology to Robotics.	Pellionisz, AJ 83a
Sensorimotor Operations: A Ground for the Co-Evolution of	Brain Theory with Neurobotics and Neurocomputers	Pellionisz, AJ 87a
How the Human	Brain Thinks and Acts: Results of Latest Research Using Neural	Gevins, AS 87b
Tests on a Cell Assembly Theory of the Action of the	Brain, Using a Large Digital Computer.	Rochester, N 56b
Tests on a Cell Assembly Theory of the Action of the	Brain, Using a Large Digital Computer.	Rochester, N 56a
Adaptive	Brain, Volumes I & II.	Grossberg, S 86a
On Comparing the	Brain with Machines.	MacKay, DM 54a
How the	Brain Works: The Next Great Scientific Revolution.	Hestenes, D 83a
How the	Brain Works: The Next Great Scientific Revolution.	Hestenes, D 87a
What Do Drug-Induced Visual Hallucinations Tell Us About the	Brain?	Cowan, JD 85b
	Brain-Emulating Circuits Need 'Sleep' and 'Dreams'	Johnson, RC 86a
On the Stability of	Brain-Like Structures.	Griffith, JS 63b
...of Neural-Analog Reinforcement Systems and its Application to the	Brain-Model Problem.	Minsky, ML 54a
Real	Brains and Artificial Intelligence	Reeke Jr, GN 88a
Tensor Geometry: A Language of	Brains and Neurocomputers. Generalized Coordinates in Neuroscience	Pellionisz, AJ 88a
	Brains, Behavior and Robotics.	Albus, JS 81a
Chips for	Brains: Computers Built with Neural Networks	Newquist, HP 86a
	Brains, Machines, and Mathematics.	Arbib, MA 87a
How	Brains Make Chaos in Order to Make Sense of the World.	Skarda, CA 87a

The Information-Storage System	Called "Human Memory".	Simon, HA 76a
A Self-Organizing System	Called `Mickey Mouse'.	Browning, I 64a
Neural Net Simulation of the Corpus	Callosum	Anninos, PA 88b
In Back-to-Basics Movement, VLSI Expert	Calls for More Simplicity in Computation	Brown, C 87e
The Recognition of Orca	Calls with a Neural Network.	Taber, WR 88d
Synchronous vs Asynchronous Behavior of Hopfield's	CAM Neural Net	Cheung, KF 87c
The Trinocular General Support Algorithm: A Three	Camera Stereo Algorithm for Overcoming Binocular Matching Errors	Stewart, CV 88b
Adaptive Noise	Canceling for Speech Signals	Sambur, MR 78a
Adaptive Noise	Canceling: Principles and Applications	Widrow, B 75a
	Cancellation in a Parallel Semantic Network.	Fahlman, SE 81a
Simple Changes in Reflex Threshold	Cannot Explain All Aspects of Rapid Voluntary Movements.	Gielen, CCAM 86a
Unifying the Tank and Hopfield Linear Programming Circuit and the	Canonical Nonlinear Programming Circuit of Chua and Lin.	Kennedy, MP 87b
A Parallel Computation that Assigns	Canonical Object-Based Frames of Reference	Hinton, GE 81a
Theoretical Study on	Cantor Function-Like Response Characteristics of a Neuron Model.	Osada, H 81a
Geometric Analysis of Neural Network	Capabilities	Wieland, A 87a
Kohonen's Self-Organizing Maps: Exploring their Computational	Capabilities	Ritter, H 88c
...of Associative Networks with Versatile Adaptive Learning	Capabilities	Fisher, AD 87b
Cognitive	Capabilities of a Parallel System.	Anderson, JA 86a
Investigations into the Pattern Processing	Capabilities of Associative Nets	Lansner, A 86a
On the	Capabilities of Multilayer Perceptrons	Baum, EB 88c
A Computer Simulation of Cerebral Neocortex: Computational	Capabilities of Nonlinear Neural Networks.	Singer, A 88a
A Computer Simulation of Cerebral Neocortex: Computational	Capabilities of Nonlinear Neural Networks.	Singer, A 87a
	Capabilities of Three-Layered Perceptrons.	Irie, B 88a
A Neuron Model with Learning	Capability and its Relation to Mechanisms of Association	Wigstrom, H 73a
A Network of Multistate Units	Capable of Associative Memory and Pattern Classification	Banzhaf, W 89a
Neural Net	Capable of Competitive and Cooperative Computation	Morse, A 78a
Modeling the Process of Problem-Solving by Associative Networks	Capable of Improving the Performance	Hirai, Y 88a
A Simple Network	Capable of Inductive Generalization.	Willshaw, DJ 76a
Some Nonlinear Networks	Capable of Learning a Spatial Pattern of Arbitrary Complexity.	Grossberg, S 68e
Neocognitron: Self-Organizing Network	Capable of Position-Invariant Recognition of Patterns.	Fukushima, K 80b
Neocognitron: Self-Organizing Network	Capable of Position-Invariant Recognition of Patterns.	Fukushima, K 80c
Selective Networks	Capable of Representative Transformations, Limited Generalizations	Edelman, GM 82a
Neocognitron: A Hierarchical Neural Network	Capable of Visual Pattern Recognition.	Fukushima, K 88c
	Capacities of Multiconnected Memory Models	Horn, D 88a
Neural Network Model with Rhythm-Assimilation	Capacity	Torras i Genis, C 86b
Space-Time Correlations of Neuronal Firing Related to Memory Storage	Capacity	Shaw, GL 78b
...Bioelectric Potentials with Microelectrodes and Neutralized Input	Capacity Amplifiers.	Amatniek, E 58a
On the Stability, Storage	Capacity, and Design of Nonlinear Continuous Neural Networks	Guez, A 88b
Memory Storage	Capacity and the Role of Assemblies in Neural Networks	Shaw, GL 77a
	Capacity for Patterns and Sequences in Kanerva's SDM as Compared to.	Keeler, JD 88a
Memory	Capacity in Neural Network Models: Rigorous Lower Bounds	Newman, CM 88a
Computer Networking	Capacity in Robotic Neural Systems	Nizkik, CA 84a
Four	Capacity Models for Coarse-Coded Symbol Memories	Rosenfeld, R 87a
Enhanced Memory	Capacity of a Hopfield Neural Network.	Little, MJ 86a
Analytic Study of the Storage	Capacity of a Neural Network	Little, WA 78a
Information	Capacity of Associative Memories	Kuh, A 88a
Comparison of Information	Capacity of Hebbian Neural Networks.	Keeler, JD 87a
Long Term Memory Storage	Capacity of Multiconnected Neural Networks	Peretto, P 86b
The	Capacity of Multilevel Threshold Functions	Olafsson, S 88a
Epsilon	Capacity of Neural Networks.	Venkatesh, SS 86a
On the Asymptotic Information Storage	Capacity of Neural Networks.	Palm, G 88a
Information	Capacity of Outer-Product Neural Networks.	Keeler, JD 87b
	Capacity of the Hopfield Associative Memory.	McEliece, RJ 86a
The	Capacity of the Hopfield Associative Memory.	McEliece, RJ 87a
Information	Capacity of the Hopfield Model	Abu-Mostafa, YS 85b
On "Information	Capacity of the Hopfield Model".	Xu, X 88c
The	Capacity of the Kanerva Associative Memory	Chou, PA 88a
The	Capacity of the Kanerva Associative Memory is Exponential.	Chou, PA 88b
Neural Network Startups Backed by Venture	Capital.	Brown, C 87i
Quantum Monte	Carlo.	Ceperley, D 86a
Monte	Carlo Calculation of Phase Separation in a Two-Dimensional Ising	Flinn, PA 74a
A New Algorithm for Monte	Carlo Simulation of Ising Spin Systems	Bortz, AB 75a
Retrieval Strategies for a	Carousel Conveyor.	Bartholdi, JJ 86a
...of EEG Spatial Patterns with a Tree-Structured Methodology:	CART	Grajski, KA 86a
	Carver Mead Joins Forces with Neural Net Startup	Johnson, RC 86b
...of Mental Processes: An Examination of Systems of Processes in	Cascade.	McClelland, JL 79a
	Cascaded Back-Propagation on Dynamic Connectionist Networks.	Pollack, JB 87a
Fast Connectionist Learning: Words and	Case	Sharkey, NE 89b
The	Case for Conceptual and Operational Separation of Network.	White, H 88b

A Digital-Computer Model of Nerve	Cell Functioning	Perkel, DH 64a
Content-Addressable memory	Cell Having Only Six Transitions	Aipperspach, AG 84a
Microcircuitry of the On-Beta Ganglion	Cell in Daylight, Twilight, and Starlight.	Sterling, P 87a
Physiological Engineering Model of the Retinal Horizontal	Cell Layer	Usui, S 88a
The On-Alpha Ganglion	Cell of the Cat Retina and Its Presynaptic Cell Types.	Freed, MA 88a
Nerve	Cell Soma Model with High Reliability and Low Power Consumption.	Fernandes, PML 80a
The On-Alpha Ganglion Cell of the Cat Retina and Its Presynaptic	Cell Types	Freed, MA 88a
A Neural Network Model for the Developmental Process of Hypercomplex	Cells.	Nagano, T 83a
A Self-Organizing Neural Network Model for the Development of Complex	Cells.	Nagano, T 81a
Arithmetical Operations Performed by Nerve	Cells.	Blomfield, S 74a
Comparison of Algorithms for Neuron-Like	Cells.	Parker, DB 86a
Electrotonic Processing of Information by Brain	Cells.	Schmitt, FO 76a
...to Neural Architecture: Emergence of Orientation-Selective	Cells.	Linsker, R 86b
...Principles to Neural Architecture: Emergence of Spatial-Opponent	Cells.	Linsker, R 86c
Mathematical Theory on Formation of Category Detecting Nerve	Cells.	Amari, S 78a
...Behavior of Cortical Neurons-Simulation of Hippocampal Pyramidal	Cells.	Borg-Graham, LJ 87a
Normal and Abnormal Signal Patterns in Nerve	Cells.	Carpenter, GA 81a
...Retina: Pattern of Input from Rods and GABA-Accumulating Amacrine	Cells.	Freed, MA 87a
...Functions and the Emergence of Orientation Selective Cortical	Cells.	Kammen, DM 88a
The Dynamics of Electrically Interacting	Cells.	Torre, V 83a
The Stochastic Properties of Spike Trains Recorded from Nerve	Cells.	Stein, RB 72a
Random	Cells: An Idea Whose Time Has Come and Gone... And Come Again?	Gallant, SI 87b
Remarks About the "Singularity" of Nerve	Cells and its Ontogenesis.	Changeux, J 85a
Microelectronics in Nerve	Cells: Dendritic Morphology and Information Processing	Poggio, TA 81a
Neural Expectation: Cerebellar and Retinal Analogs of	Cells Fired by Learnable or Unlearned Pattern Classes.	Grossberg, S 72c
Microcircuitry of Beta Ganglion	Cells in Cat Retina.	McGuire, BA 86a
Microcircuitry of Bipolar	Cells in Cat Retina.	McGuire, BA 84a
Self-organization of Orientation Sensitive	Cells in the Striate Cortex.	von der Malsburg, C 73a
Self-Organization of Orientation Sensitive	Cells in the Striate Cortex.	von der Malsburg, C 73b
A Theory for the Development of Feature Detecting	Cells in the Visual Cortex	Nass, MM 75a
Input-Output Relations in Computer-Simulated Nerve	Cells: Influence of Statistical Properties, Strength, etc.	Segundo, JP 68a
A Computational Model of How Cortical Complex	Cells Multiplex Information About Position, Contrast, Orientation.	Grossberg, S 87a
Interpretation of Responses of Visual	Cells of Limulus	Borsellino, AI 68a
...in the Nervous System of Adult Hydra II. Conversion of Ganglion	Cells of the Body Column into Epidermal Sensory Cells of the	Koizumi, O 88a
...of Ganglion Cells of the Body Column into Epidermal Sensory	Cells of the Hypostome	Koizumi, O 88a
...Model of Pain Mechanisms: Functional Properties of the Network	Cells Providing the Power Law.	Minamitani, H 82a
How to Label Nerve	Cells so that they can Interconnect in an Ordered Fashion.	von der Malsburg, C 77a
Inhomogeneous	Celluar Automata (INCA).	Hartman, H 86a
Directional Entropies of	Celluar Automaton-Maps	Milnor, J 86a
Fibroblasts, Morphogenesis and	Cellular Automata.	Bouligand, Y 86a
Motion Sensitive	Cellular Automata.	Lopez, LR 88a
The Connection Machine: A Computer Architecture Based on	Cellular Automata.	Hillis, WD 84a
	Cellular Automata Models of Disorder and Organization.	Vichniac, GY 86a
...and Release of Chemical Transmitters and Related Topics in	Cellular Control	Grossberg, S 69g
	Cellular Logic Arrays for Image Processing	Preston Jr, K 86a
Conservation of	Cellular Mechanisms for Models of Learning and Memory.	Alkon, DL 88a
An Evaluation of Cell Function in Relation to Reverberatory Modes in	Cellular Networks.	Fairhurst, MC 80a
	Cellular Neural Networks: Theory	Chua, LO 88b
B.U.	Center Researching Neural Nets	Brown, C 86a
A Model of the Vertebrate	Central Command System	Kilmer, WL 69a
A Statistical Approach to the Theory of the	Central Nervous System	Shimbel, A 48a
Mathematical Modeling of Mechanisms of the	Central Nervous System	Gelfand, IM 71a
The Tracing of Pathways in the	Central Nervous System	Heimer, L 68a
The	Central Nervous System as a Low and High Level Control System.	Albus, JS 88a
Molecular Information Processing in the	Central Nervous System, Part I	Conrad, M 74b
OCCAM: A Computer Model for a Content-Addressable Memory in the	Central Nervous Systems.	Spinelli, DN 70a
Distributed Memory in the	Central Nervous Systems: Possible Test of Assumptions in Visual.	Cooper, LN 81a
Spontaneous Quantal Currents in a	Central Neuron Match Predictions from Binomial Analysis of Evoked.	Korn, H 87a
The Mammalian	Central Neurone is a Complex Computing Device.	Bras, H 87c
	Central Pattern Generating and Recognizing in Olfactory Bulb: A.	Freeman, WJ 88a
Are	Central Pattern Generators Understandable?	Selverston, AI 80a
The	Centrality of Instantiations	Barnden, JA 87c
	Centric Models of the Orientation Map in Primary Visual Cortex	Baxter, W 88a
	Centrifugal Control Mechanisms	Whitfield, IC 68a
Structuro-Functional Considerations of the	Cerebella Neuron Network	Szentagothai, J 68a
Neural Expectation:	Cerebellar and Retinal Analogs of Cells Fired by Learnable or.	Grossberg, S 72c
A Theory of	Cerebellar Cortex.	Marr, D 69a
...the Information Transfer Properties of a Simplified Model of the	Cerebellar Cortex.	Axelrad, H 87a
A Theory of	Cerebellar Function.	Albus, JS 71a

Programmable Synaptic	Chip for Electronic Neural Networks	Moopenn, A 88a
Neural Optical	Chip Modeled	Johnson, RC 88j
Electronic Neural Network	Chips	Jackel, LD 87b
Neural	Chips Emulate Brain Functions (ATT VLSI)	Brown, C 86e
	Chips for Brains: Computers Built with Neural Networks	Newquist, HP 86a
IBM Neural Network	Chips Make Everything Obsolete	Gantz, J 88a
Are Special	Chips Necessary for Neural Computers?	Aleksander, I 88a
A	Chipset for High Speed Simulation of Neural Network Systems	Garth, SCJ 87a
A Model of the Brain for Robot Control, Part 4: Mechanisms of	Choice	Albus, JS 79d
The Brains of Men and Machines, Part 4: The Machinery of Emotion and	Choice	Kent, EW 78d
Estimation and	Choice of Neighbors in Spatial-Interaction Models of Images	Kashyap, RL 83a
On	Choosing Non-Random Input Exemplars to Form Stable States in a	Feild Jr, WB 88a
The Illusion of Transparency and	Chromatic Subjective Contours	Meyer, GE 83a
A Connectionist Model of	Chronic Pain	Katz, BF 88a
...Circuit and the Canonical Nonlinear Programming Circuit of	Chua and Lin	Kennedy, MP 87b
	Chunking and Consolidation: A Theoretical Synthesis of Semantic	Wickelgren, WA 79a
Gated Dipoles: Photoreceptors and	Circadian Rhythms	Carpenter, GA 84b
The Non-Parametric Influence of Light on Mammalian	Circadian Rhythms	Carpenter, GA 85d
A Neural Theory of	Circadian Rhythms: Aschoff's Rule in Diurnal and Nocturnal Mammals	Carpenter, GA 84a
Neural Theory of	Circadian Rhythms: Aschoff's Rule in Diurnal and Nocturnal Mammals	Carpenter, GA 86l
Neural Theory of	Circadian Rhythms: Split Rhythms, After-Effects, and Motivational	Carpenter, GA 86m
A Neural Theory of	Circadian Rhythms: The Gated Pacemaker	Carpenter, GA 83a
Neural Theory of	Circadian Rhythms: The Gated Pacemaker	Carpenter, GA 86n
Neural Dynamics of	Circadian Rhythms: The Mammalian Hypothalamic Pacemaker	Carpenter, GA 85c
Finding	Circles by an Array of Accumulators	Kimme, C 75a
...An A/D Converter, Signal Decision Circuit, and Linear Programming	Circuit	Tank, DW 86a
The Properties of the Cerebello-Pontine Reverberating	Circuit	Tsukahara, N 72a
A Neuromime System for Neural	Circuit Analysis	Mitchell, CE 81a
Neural Network or High-Storage Content-Addressable Memory: VLSI	Circuit and Learning Algorithm	Verleysen, M 88g
...Neural Optimization Networks: An A/D Converter, Signal Decision	Circuit, and Linear Programming Circuit	Tank, DW 86a
...Network Model. II. Purkinje Cell Activity within the Basic	Circuit and Modified by Inhibitory Sys.	Pellionisz, AJ 74a
Unifying the Tank and Hopfield Linear Programming	Circuit and the Canonical Nonlinear Programming Circuit of Chua and.	Kennedy, MP 87b
Artificial Neural Network Integrated	Circuit Based on NMOS/CCD Principles	Sage, JP 86a
Neural-Type Junctions: A New	Circuit Concept	DeClaris, N 76a
	Circuit Design Criteria for Stable Lateral Inhibition Neural	Wyatt Jr, JL 88b
A Plausible Neural	Circuit for Classical Conditioning Without Synpatic Plasticity	Tesauro, G 88b
The Vite Model: A Neural Command	Circuit for Generating Arm and Articulator Trajectories	Bullock, D 88d
A Modifiable Weight	Circuit for Use in Adaptive Neuromorphic Networks	Shoemaker, PA 88a
...of Mass of Neurons Forming Simple Feedback Loops (I): Lumped	Circuit Model	Ahn, SM 74a
...of Mass of Neurons Forming Simple Feedback Loops (II): Lumped	Circuit Model	Ahn, SM 74b
Four Neural	Circuit Models and Their Role in the Organization of Voluntary	Hemami, H 83a
Two-Terminal Electronic	Circuit Neuron Model with Excitable Membrane V-I-t Characteristics	Hoshimiya, N 79a
...Programming Circuit and the Canonical Nonlinear Programming	Circuit of Chua and Lin	Kennedy, MP 87b
...Lateral Inhibition Networks that is Robust in the Presence of	Circuit Parasitics	Wyatt Jr, JL 88a
...for Lateral Inhibition Networks that is Robust in the Presence of	Circuit Parasitics	Standley, DL 88a
Analog MOS	Circuit Techniques in the VLSI Implementation of Neural Networks	Satyanarayana, S 88a
	Circuit Theoretic Solutions for Neural Networks - An Old Approach.	Kennedy, MP 87a
...Nonvolatile MOS Analogue Memory (MAM) Cell for Neural-Type	Circuitry	Shimabukuro, RL 88a
From Understanding Computation to Understanding Neural	Circuitry	Marr, D 77a
Analog	Circuitry for the van der Pol and FitzHugh-Nagumo Equations	Keener, JP 83a
A Semistate Model for Neural-Type Junction	Circuits	El-Leithy, N 87b
Adaptive Switching	Circuits	Widrow, B 60c
Adaptive Switching	Circuits	Widrow, B 60b
Collective Computation in Neuronlike	Circuits	Tank, DW 87a
Complex Dynamics in Simple Neural	Circuits	Babcock, KL 86a
Effects of Connection Delays in Two State Model Neural	Circuits	Conwell, PR 87a
Error Correcting System Based on Neural	Circuits	Takefuji, Y 87a
Learning Phenomena in Networks of Adaptive Switching	Circuits	Hoff, ME 62a
Self-Correcting Translator	Circuits	Steinbuch, K 63b
Shifter	Circuits: A Computational Strategy for Dyanmic Aspects of Visual	Anderson, CH 87a
A Chain of Suprasegmental Neuroscillatory	Circuits: A Human Brain Theory	Deshmukh, VD 88a
Computing With Neural	Circuits: A Model	Hopfield, JJ 85b
Integration of Multi-Microelectrode and Interface	Circuits by Silicon Planar and Three-Dimensional Fabrication	Takahashi, K 84a
Analog MOS Integrated	Circuits - Certain New Ideas, Trends, and Obstacles	Tsividis, Y 87a
International Solid State	Circuits Conference, February, 1987	Sage, JP 87b
Integrate-and-Fire Coding in Hodgkin-Huxley	Circuits Employing Silicon Diodes	Coon, DD 88b
Neural	Circuits for Generating Rhythmic Movements	Friesen, WO 78a
Analogue	Circuits for Variable-Synapse Electronic Neural Networks	Tsividis, Y 87b
The Role of Adaptive and Associative	Circuits in Future Computer Designs	Kohonen, T 88e

A	CMOS Associative Memory Chip	Graf, HP 87a
A	CMOS Associative Memory Chip Based on Neural Networks.	Graf, HP 87b
A New Discipline for	CMOS Design: An Architecture for Sound Synthesis	Mead, CA 85a
A	CMOS Implementation of a Neural Network Model.	Graf, HP 87c
Real-Time Visual Computations Using Analog	CMOS Processing Arrays	Sivilotti, MA 87a
Real-time Visual Computations Using Analog	CMOS Processing Arrays	Sivilotti, MA 87b
Using a Neural Network to Learn the Dynamics of the	CMU Direct-Drive Arm II.	Goldberg, K 88a
Systolic Algorithms for the	CMU Warp Processor	Kung, HT 84a
...A Vector-Matrix Description of Transformations of Overcomplete	CNS Coordinates and Tensorial Solution Using the Moore-Penrose	Pellionisz, AJ 84a
A Systems Theoretic Approach to the Study of	CNS Function	Sclabassi, RJ 88a
Four Capacity Models for	Coarse-Coded Symbol Memories	Rosenfeld, R 87a
Scaling Properties of	Coarse-Coded Symbol Memories	Rosenfeld, R 87b
Scaling Properties of	Coarse-Coded Symbol Memories	Rosenfeld, R 88a
	Coarticulation in Recent Speech Production Models.	Kent, RD 77a
An Analog Electronic	Cochlea.	Lyon, RF 88a
Using Neural Networks to Improve	Cochlear Implant Speech Perception	Tenorio, MF 87g
Simulating Nerves to Improve	Cochlear Implants for the Hearing Impaired	Tenorio, MF 87e
	Cochlear Nerve Filter Responses: Distribution Along the Cochlear	Pfeiffer, RR 75a
...of Spontaneous Spike Discharge Patterns of Neurons in the	Cochlear Nucleus	Molnar, CE 68a
Cochlear Nerve Filter Responses: Distribution Along the	Cochlear Partition	Pfeiffer, RR 75a
A Neural	Cocktail-Party Processor	von der Malsburg, C 86a
The Closeness	Code: An Integer to Binary Vector Transformation Suitable for.	Penz, PA 87a
How Does a Brain Build a Cognitive	Code?.	Grossberg, S 80d
How Does a Brain Build a Cognitive	Code?.	Grossberg, S 80c
Image Reconstruction from	Coded Data: I. Reconstruction Algorithms and Experimental Results.	Smith, WE 85a
Optimizing Digital Speech	Coders by Exploiting Masking Properties of the Human Ear	Schroeder, MR 79a
...Order in Chaos: Stable Self-Organization of Neural Recognition	Codes.	Carpenter, GA 86h
Identifying the Speech	Codes.	Foss, DJ 80a
Neuromorphic Networks Based on Sparse Optical Orthogonal	Codes.	Vecchi, MP 88a
Neural Networks, Error-Correcting	Codes and Polynomials Over the Binary n-Cube	Bruck, J 88d
Absolutely Stable Learning of Recognition	Codes by a Self-Organizing Neural Network.	Carpenter, GA 86b
Building Permanent Memory	Coces: Codification and Repetition Effects in Word Identification.	Salasoo, A 85a
ART-2: Self-Organization of Stable Category Recognition	Codes for Analog Input Patterns.	Carpenter, GA 87c
Unitized Recognition	Codes for Parts and Wholes: The Unique Cue in Configural	Cohen, MA 86e
...Resonance Theory: Stable Self-organization of Neural Recognition	Codes in Response to Arbitrary Lists of Input Patterns	Carpenter, GA 86d
Stable Self-Organization of Sensory Recognition	Codes: Is Chaos Necessary?	Grossberg, S 87o
...Human Memory: Self-Organization and Performance of Sensory-Motor	Codes, Maps, and Plans	Grossberg, S 78a
Building Permanent Memory Codes:	Codification and Repetition Effects in Word Identification	Salasoo, A 85a
A Neural Network Approach to Transform Image	Coding	Chua, LO 88a
...of Striate Cortex: A Spatial Mapping Approach to Perceptual	Coding	Schwartz, EL 80a
...Design: Spontaneous Node Activity, Node Plasticity and Temporal	Coding	Torras i Genis, C 88a
...and Dynamic Error Propagation Networks with Application to Speech	Coding	Robinson, AJR 88b
Vector Quantization in Speech	Coding	Makhoul, J 85a
Interpolation	Coding: A Representation for Numbers in Neural Models.	Ballard, DH 87b
Statistical	Coding and Short-Term Synaptic Plasticity: A Scheme for Knowledge.	von der Malsburg, C 86b
Context-Sensitive	Coding, Associative Memory, and Serial Order in (speech) Behavior.	Wickelgren, WA 69a
Spectral Information	Coding by Infrared Photoreceptors.	Coon, DD 86a
Hierarchical Contour	Coding by the Visual Cortex.	Hartmann, G 85a
Neural Dynamics of Speech and Language	Coding: Developmental Programs, Perceptual Grouping, and	Cohen, MA 86c
Noise-Like	Coding in Associative Memories	Bottini, S 87a
Integrate-and-Fire	Coding in Hodgkin-Huxley Circuits Employing Silicon Diodes	Coon, DD 88b
Speech	Coding in the Auditory Nerve: II. Processing Schemes for Vowel-Like.	Delgutte, B 84a
Neural	Coding in the Bullfrog's Auditory System: A Teleological Approach.	Frishkopf, LS 68a
A Theory of Visual	Coding, Memory, and Development.	Grossberg, S 78b
...and Universal Recoding: I. Parallel Development and	Coding of Neural Feature Detectors	Grossberg, S 76a
...and Universal Recording, Part I: Parallel Development and	Coding of Neural Feature Detectors	Grossberg, S 76b
The Problem of Sensing and the Neural	Coding of Sensory Events	Mountcastle, VB 67a
Auditory and Phonetic	Coding of Speech	Sawusch, JR 86a
An Adaptive Stochastic Model for the Neural	Coding Process	Bruckstein, AM 85a
Stochastic Pulse Frequency Modulation Models of Neural	Coding Processes	Bruckstein, AM 81a
Analysis of "Integrate-to-Threshold" Neural	Coding Schemes	Bruckstein, AM 79a
A Technique for Determining and	Coding Subclasses in Pattern Recognition Problems.	Mattson, RL 65a
A Neural Network Classifier Based on	Coding Theory.	Chiueh, T 88a
Space / Time Trade-Offs in Hash	Coding with Allowable Errors	Bloom, BH 70a
Sensorimotor Operations: A Ground for the	Co-Evolution of Brain Theory with Neurobotics and Neurocomputers	Pellionisz, AJ 87a
Soar: A Unified Theory of	Cogition?.	Waldrop, MM 88a
Adaptive Resonance in Development, Perception, and	Cognition.	Grossberg, S 81a
Aesthetics, Psychobiology, and	Cognition.	Martindale, C 88a
An Architecture for Mathematical	Cognition.	Ammon, K 86a

Context	Keyword	Author
	Coherent Optical Feedback for the Analog Solution of Partial	Cederquist, J 80a
	Coherent Optical Pattern Recognition: A Review	Casasent, DP 85a
Matrix Multiplication Using	Coherent Optical Techniques.	Tamura, PN 77a
...in Nervous Systems I. Relevance of Signal Covariances and Signal	Coincidences	Toet, A 87a
Recurrent	Collateral Inhibition Simulated in a Simple Neuronal Automata.	Axelrad, H 86a
A Novel Associative Memory Implemented Using	Collective Computation	Sivilotti, MA 85a
	Collective Computation in Neuronlike Circuits.	Tank, DW 87a
	Collective Computation with Continuous Variables	Hopfield, JJ 85a
	Collective Computation with Continuous Variables	Hopfield, JJ 86a
Neural Networks and Physical Systems with Emergent	Collective Computational Abilities	Hopfield, JJ 82b
Neural Networks and Physical Systems with Emergent	Collective Computational Abilities	Hopfield, JJ 82a
`Ensemble' Boltzmann Units have	Collective Computational Properties like those of Hopfield and Tank.	Derthick, MA 88a
Neurons with Graded Response Have	Collective Computational Properties Like Those of Two-State Neurons.	Hopfield, JJ 84a
Neurons With Graded Response Have	Collective Computational Properties Like Those of Two-state Neurons.	Hopfield, JJ 84b
	Collective Computational Properties of Neural Networks: New.	Personnaz, L 86b
Unlearning Has a Stabilizing Effect in	Collective Memories.	Hopfield, JJ 83a
	Collective Properties of Neural Networks	Peretto, P 86a
	Collective Properties of Neural Networks: A Statistical Physics.	Peretto, P 84a
Expectation and the Perception of	Color.	Bruner, JS 51a
Learning a	Color Algorithm from Examples.	Poggio, TA 88a
Synthesizing a	Color Algorithm from Examples.	Hurlbert, AC 88a
Cortical Dynamics of Three-Dimensional Form,	Color, and Brightness Perception, I: Monocular Theory.	Grossberg, S 87h
Cortical Dynamics of Three-Dimensional Form,	Color, and Brightness Perception, II: Binocular Theory	Grossberg, S 87i
	Color Atlas Theory	Koenderink, JJ 87b
	Color Constancy and Early Vision: A Connectionist Model.	Heisey, I 87a
	Color Representation Using Statistical Pattern Recognition	Parkkinen, J 87a
Perceptual Significance Hierarchy: A Computer Vision Theory for	Color Separation	Walters, DKW 87d
Automated Fake	Color Separation: Combining Computer Vision and Computer Graphics.	Walters, DKW 87a
	Color Separation Using General Purpose Computer Vision Algorithms.	Walters, DKW 87c
Acousto-Optic	Color Spectrometer	Hallikainen, J 88a
...Form Perception: Boundary Completion, Illusory Figures, and Neon	Color Spreading.	Grossberg, S 86j
The Retinex Theory of	Color Vision	Land, EH 77a
Seeing the Light: Optics in Nature, Photography,	Color, Vision, and Holography.	Stork, DG 85a
Neural Network Algorithm for an NP-Complete Problem: Map and Graph	Coloring	Dahl, ED 87b
The Perception of Transparency with Achromatic	Colors	Beck, J 84a
The Representation of	Colors in the Cerebral Cortex.	Zeki, S 80a
A Markov Random Field Approach to Data Fusion and	Colour Segmentation.	Wright, WA 89a
...of Adult Hydra II. Conversion of Ganglion Cells of the Body	Column into Epidermal Sensory Cells of the Hypostome	Koizumi, O 88a
An Ocular Dominance	Column Map as a Data Structure for Stereo Segmentation	Yeshurun, Y 87a
Formation of Retinotopy and	Columnar Microstructures by Self-Organization: A Mathematical Model.	Amari, S 85a
Formation of Topographic Maps and	Columnar Microstructures in Nerve Fields	Takeuchi, A 79a
...Mechanism for the Origin and Development of Iso-Orientation	Columns.	Cowan, JD 85a
...Principles to Neural Architecture: Emergence of Orientation	Columns.	Linsker, R 86a
Fuzzy Knowledge	Combination.	Kosko, B 86e
Neural Correlates of the Aural	Combination Tone 2f(1)-f(2).	Goldstein, JL 68a
Stochastic	Combinational Networks	Windecker, RC 78a
	Combinations of Neural Systems for Particular Application.	Josin, GM 87a
A Learning Rule for Asynchronous Perceptrons with Feedback in a	Combinatorial Environment.	Almeida, LB 87a
	Combinatorial Hypercompression	Hecht-Nielsen, R 87a
...Convergence Theorem for Neural Networks and Its Applications in	Combinatorial Optimization	Bruck, J 87a
Statistical Mechanics: A General Approach to	Combinatorial Optimization	Bonomi, E 86a
Towards Practical 'Neural' Computation for	Combinatorial Optimization Problems.	Baum, EB 86b
	Combinatorial Optimization Using a Continuous State Boltzmann.	Gutzmann, KM 87a
Integer and	Combinatorial Optimizations.	Nemhauser, GL 88a
Heuristics Based on Space-Filling Curves for	Combinatorial Problems in the Plane.	Bartholdi, JJ 84a
...Paper "Threshold Regulation - A Key to the Understanding of the	Combined Dynamics of EEG and Event-Related Potentials" by Elbert	Palm, G 87b
Hidden Patterns in	Combined Knowledge Networks.	Kosko, B 87h
Automated Fake Color Separation:	Combining Computer Vision and Computer Graphics.	Walters, DKW 87a
Random Cells: An Idea Whose Time Has Come and Gone... And	Come Again?.	Gallant, SI 87b
Random Cells: An Idea Whose Time Has	Come and Gone... And Come Again?	Gallant, SI 87b
Computers That	Come Awfully Close to Thinking	Port, O 86a
...Fundamental Information Processing Ability of Biological Systems	Come From?	Kohonen, T 75a
Neural Systems Make	Comeback After Nearly 50-Year Gestation.	Johnson, RC 87k
The Vite Model: A Neural	Command Circuit for Generating Arm and Articulator Trajectories.	Bullock, D 88d
A Model of the Vertebrate Central	Command System	Kilmer, WL 69a
Infinitely Many	Commensurate Phases in a Simple Ising Model.	Fisher, ME 80a
Do All Neural Networks Really Look Alike? A	Comment on Anderson, Silverstein, Ritz and Jones	Grossberg, S 78g
A Question of Levels:	Comment on McClelland and Rumelhart.	Broadbent, D 85a
Self-Organizing Systems - A Review and	Commentary	Hawkins, JK 61a

Context	Keyword	Reference
The Analysis of Visual Motion: From	Computational Theory to Neuronal Mechanisms.	Hildreth, E 86a
From	Computational Theory to Psychology and Neurophysiology	Marr, D 76c
Neural Networks and	Computational Vision	Chen, S 87b
Neural Networks for	Computational Vision	Mingolla, E 87a
	Computational Vision and Regularization Theory	Poggio, TA 85a
Probabilistic Characterization of Neural Model	Computations	Golden, RM 88c
Programming Neural Nets to Do Spatial	Computations	Zipser, D 86d
Propagating Belief Functions with Local	Computations	Shenoy, P 86a
Networks,	Computations, Logic and Noise.	Hopfield, JJ 87b
	Computations the Hippocampus Might Perform	O'Keefe, J 89a
Real-Time Visual	Computations Using Analog CMOS Processing Arrays	Sivilotti, MA 87a
Real-time Visual	Computations Using Analog CMOS Processing Arrays	Sivilotti, MA 87b
L'Uso di Modelli	Computazionali in Psicofisiologia Ed Il Principio Gnoseologico di.	Basti, G 86b
That Does Not	Compute.	Berke, P 87a
Synapses that	Compute Motion	Poggio, TA 87a
	Computed Tomography for Optical Computing.	Szu, HH 86b
A Full Associative Memory Chip for Data-Flow	Computer	Nishida, K 84a
Faust, Mephistopheles and	Computer	Zuse, K 88a
Growing 'Brains' in a	Computer	Amato, I 87a
Is the Cerebellum Really a	Computer	Ito, M 79a
...Research for Visicn - Simulation on a Parallel Processing	Computer	Miyake, S 88a
On Design Principles for a Molecular	Computer	Conrad, M 85a
...Gaussian Elimination cn an Optically Interconnected Data Flow	Computer	Nelken, I 87a
Simulation of Self-Organizing Systems by Digital	Computer	Farley, BG 54a
Speaker Recognition by	Computer	Bunge, E 77a
Speech Recognition by	Computer	Levinson, SE 81a
Technology-Independent Design of Neurocomputers: The Universal Field	Computer	MacLennan, BJ 87a
...Assembly Theory of the Action of the Brain, Using a Large Digital	Computer	Rochester, N 56b
...Assembly Theory of the Action of the Brain, Using a Large Digital	Computer	Rochester, N 56a
The NYU Ultracomputer-Designing an MIMD Shared Memory Parallel	Computer	Gottlieb, A 83a
States:	Computer & Brain	Anderson, JA 79b
The Cerebral	Computer: An Introduction to the Computational Structure of the.	Baron, RJ 87a
Character Recognition by	Computer and Applications.	Suen, CY 86a
The	Computer and the Brain	von Neumann, J 58a
The Connection Machine: A	Computer Architecture Based on Cellular Automata	Hillis, WD 84a
Memory Requirements for Balanced	Computer Architectures	Kung, HT 85b
Optical Adaptive Associative	Computer Architectures	Fisher, AD 85b
Silicon Brains: Whither Neuromimetic	Computer Architectures	Vidal, JJ 87c
New	Computer Architectures and Their Relationships to Physics, or Why.	Hillis, WD 82a
Neural Nets Inspire	Computer Designer.	Johnson, RC 86d
The Role of Adaptive and Associative Circuits in Future	Computer Designs	Kohonen, T 88e
Experience with Bayes Theorem for	Computer Diagnosis of Congenital Heart Disease	Warner, HR 64a
	Computer Firm is Closing in on Dream Machine	Kraul, C 87a
The CHASM: A Macromodular	Computer for Analyzing Neuron Models	Molnar, CE 67a
A Dedicated	Computer for Simulation of Large Systems of Neural Nets.	Garth, S 88a
Automated Fake Color Separation: Combining Computer Vision and	Computer Graphics.	Walters, DKW 87a
ANNA: An Adaptive Neural Network Anticipator for Human/	Computer Interfacing	Jones, WP 88a
New	Computer Learns on the Job	Rowe, B 87a
Parallel Structures in Human and	Computer Memory.	Kanerva, P 86a
OCCAM: A	Computer Model for a Content-Addressable Memory in the Central	Spinelli, DN 70a
A	Computer Model for Learning Processes and the Role of the Cerebral	Anninos, PA 84a
Synthesis of Fragmented Data on Neuronal Systems: A	Computer Model of Cerebellum	Pellionisz, AJ 78a
Artina: A Flexible	Computer Model of Retinal Processes for Moving Images.	Gawronski, RP 87a
A	Computer Model of Sentence Comprehension	Small, SL 83a
Tensor Network Theory and its Application in	Computer Modeling of the Metaorganization of Sensorimotor.	Pellionisz, AJ 86b
	Computer Models of Bursting Neural Networks.	Hartline, DK 78a
	Computer Networking Capacity in Robotic Neural Systems	Nizkik, CA 84a
Cognitive	Computer: On Language, Learning and AI	Schank, RC 84a
A	Computer Package for the Evaluation of Neuron Models Involving	Lee, MH 82a
A	Computer Program for Simulating a Network of Interacting Neurons	Perkel, DH 76a
A	Computer Program for Simulating a Network of Interacting Neurons	Perkel, DH 76b
A	Computer Program for Simulating a Network of Interacting Neurons	Perkel, DH 76c
	Computer Recognition of Speech	de Mori, R 86a
Human and	Computer Rules and Representations are not Equivalent.	Grossberg, S 80e
Structured Neural Networks in Nature and in	Computer Science	Feldman, JA 88d
New Computer Architectures and Their Relationships to Physics, or Why	Computer Science is No Good.	Hillis, WD 82a
Brain Modeling by Tensor Network Theory and	Computer Simulation.	Pellionisz, AJ 79a
Plasticity in the Organization of Adult Cerebral Cortical Maps: A	Computer Simulation Based on Neuronal Group Selection.	Pearson, JC 87a
Analog	Computer Simulation of a Neural Element.	Hiltz, FF 62a

A	Computer Simulation of Cerebral Neocortex: Computational	Singer, A 87a
A	Computer Simulation of Cerebral Neocortex: Computational	Singer, A 88a
	Computer Simulation of Event Related Potentials During Voluntary	Deshmukh, VD 88b
	Computer Simulation of Neural Networks for Perceptual Psychology	Grossberg, S 86f
A	Computer Simulation of Olfactory Cortex with Functional.	Bower, JM 88a
A	Computer Simulation of Olfactory Cortex with Functional.	Wilson, MA 87a
	Computer Simulation of Rhythmic Oscillation in Neuron Pools.	MacGregor, RJ 74a
	Computer Simulation of Short-Term Memory: A Component Decay Model.	Laughery, KR 69a
	Computer Simulation of Somatotopic Map Plasticity.	Edelman, GM 87a
	Computer Simulations of Learning in Neural Systems	Salu, Y 83a
	Computer Simulations of Pattern Recognition as a Dynamical Process	Fuchs, A 88a
Parallel Optical	Computer Solves Multivariable Problems in Real-Time.	Brown, C 86h
Analytical Versus	Computer Studies in Modeling Neural Networks	Shaw, GL 78a
Programming A Digital	Computer to Learn.	Oettinger, AG 52a
A	Computer Use for Establishing and Proving Conjectures in Uniform	Shingai, R 78a
	Computer Vision.	Ballard, DH 82a
Multiple-Window Parallel Adaptive Boundary Finding in	Computer Vision.	Cooper, DB 83a
Properties of Cortical Systems - A Basis for	Computer Vision.	von Seelen, W 87b
UCLA PUNNS - A Neural Network Machine for	Computer Vision.	Gunger, D 87a
...Brain Processes Visual Information May Be Suggested by Studies in	Computer Vision (and Vice Versa)	Poggio, TA 84c
Color Separation Using General Purpose	Computer Vision Algorithms	Walters, DKW 87c
Automated Fake Color Separation: Combining	Computer Vision and Computer Graphics.	Walters, DKW 87a
Parallel Architectures for Image Processing,	Computer Vision, and Pattern Perception.	Uhr, L 86a
A	Computer Vision Architecture for Intelligence Control of Neural.	Lee, JSJ 88a
Neural Nets and	Computer Vision Systems.	Tao, KM 87a
Perceptual Significance Hierarchy: A	Computer Vision Theory for Color Separation.	Walters, DKW 87d
	Computer vs the Brain.	Hammerstrom, D 86c
	Computer-Aided Design of Neurobiological Experiments	Dammasch, IE 88b
Reconstruction from Projections: Applications in	Computerized Tomography.	Kak, AC 86a
...Method Adapted for the Postprocessing and Error-Correction of	Computer-Recognized Speech	Reuhkala, E 79a
Associative Networks: Representation and Use of Knowledge by	Computers.	Findler, NV 79b
Complex Systems-Operational Approaches in Neurobiology, Physics, and	Computers.	Stork, DG 88b
Conventional Fault-tolerance and Neural	Computers.	Moore, WR 88a
Experimental Demonstrations of Optical Neural	Computers.	Hsu, K 88a
Goal and Architecture of Neural	Computers.	von der Malsburg, C 88a
Implementing Neural Network Models on Parallel	Computers.	Forrest, BM 87a
Learning in Optical Neural	Computers.	Psaltis, D 87a
Optical Neural	Computers.	Abu-Mostafa, YS 87a
Parallel Architectures for Neural	Computers.	Recce, M 88a
Pattern Recognition and Modern	Computers.	Selfridge, OG 55a
Synergetics and	Computers.	Haken, H 88d
The AUROS Project: Automatic Recognition of Speakers by	Computers.	Bunge, E 77b
The Design of Conditional Probability	Computers.	Uttley, AM 59b
Talking to	Computers: An Empirical Investigation.	Hauptmann, AG 87a
Melding	Computers and Biology for Commercial Benefit	Hunter, L 88b
	Computers and Intractability: A Guide to the Theory of NP.	Garey, MR 79a
Chips for Brains:	Computers Built with Neural Networks	Newquist, HP 86a
Connecting with the Human Mind: A New Approach to Building	Computers Draws on the Research of Neuroscientists	Hammerstrom, D 87b
Neural	Computers in Vision: Processing of High Dimensional Data	Zeevi, YY 88a
Why	Computers May Never Think Like People.	Dreyfus, H 86a
Mimicking the Human Mind: Radical New	Computers Teach Themselves	Rogers, M 87a
	Computers That Come Awfully Close to Thinking.	Port, O 86a
An Approach to	Computers that Perceive, Learn, and Reason	Greene, PH 59a
Making	Computers that Think Like People	Zadeh, LA 84a
Designing	Computers That Think the Way We Do	Allman, WF 87a
Neural and Massively Parallel	Computers: The Sixth Generation.	Soucek, B 88a
Programming Parallel	Computers Using Energy Minimization Algorithms	Simmes, SD 87a
Are Special Chips Necessary for Neural	Computers?	Aleksander, I 88a
A	Computer-Simulated Model of a Second Order Sensory Neurone	Walloe, L 69a
Input-Output Relations in	Computer-Simulated Nerve Cells: Influence of Statistical Properties.	Segundo, JP 68a
Dynamic Activity and Memory Traces in	Computer-Simulated Recurrently-Connected Neural Networks	Gaudiano, P 87a
Adaptive, Associative, and Self-Organizing Functions in Neural	Computing.	Kohonen, T 87a
An Introduction to Neural	Computing.	Kohonen, T 88a
Artificial Neural Networks for	Computing.	Jackel, LD 86a
Associative Processor Architectures for Optical	Computing.	Fisher, AD 84a
Attentive Associative Architectures and Their Implications to Optical	Computing.	Athale, RA 86a
Computed Tomography for Optical	Computing.	Szu, HH 86b
European Seminar on Neural	Computing.	ESNC 88a
Holographic Coordinate Transformations and Optical	Computing.	Szu, HH 86d

Iterative Restoration Algorithms for Nonlinear Constraint	Computing.	Szu, HH 86e
Optical and Hybrid	Computing.	Szu, HH 86g
Parallel Stochastic	Computing.	Tenorio, MF 87c
Schemas and Neurons: Two Levels of Neural	Computing.	Arbib, MA 88a
State of the Art in Neural	Computing.	Kohonen, T 87g
Super-Resolution and Neural	Computing.	Abbiss, JB 88a
Neural Net Models and Optical	Computing: A Brief Overview.	Farhat, NH 86a
Reliable, Trainable Networks for	Computing and Control.	Widrow, B 62c
The Second European Seminar on Neural	Computing: Commercial Prospects.	ESNC 89a
The Mammalian Central Neurone is a Complex	Computing Device	Bras, H 87c
An Adaptive Optical	Computing Element.	Fisher, AD 85a
Implementations of Adaptive Associative Optical	Computing Elements	Fisher, AD 86a
Learning by Statistical Cooperation of Self-Interested Neuron-Like	Computing Elements	Barto, AG 85c
Learning by Statistical Cooperation of Self-Interested Neuron-Like	Computing Elements	Barto, AG 85b
...of a Simultaneous Functional Order in Nervous Systems II.	Computing Geometrical Structures	Toet, A 87b
Neural	Computing: Hype or Reality?.	Aleksander, I 88d
Neural	Computing: Ideas from the Brain.	Aleksander, I 88e
Conditional Probability	Computing in a Nervous System.	Uttley, AM 59a
Recent Advances in Optical	Computing in Japan	Ishihara, S 86a
A Branch and Bound Algorithm for	Computing k-Nearest Neighbors.	Fukunaga, K 75a
Neuromolecular	Computing: Mechanisms and Architectures.	Conrad, M 88a
	Computing Motion in the Presence of Discontinuities: Algorithm and .	Koch, C 88a
	Computing Motion Using Analog and Binary Resistive Networks.	Hutchinson, J 88a
	Computing Motion Using Resistive Networks.	Koch, C 88b
Simulated Annealing in Networks for	Computing Possible Arrangements for Red and Green Cones.	Ahumada, AJ 87a
	Computing Shape with Neural Networks: A Proposal	Oyster, JM 87a
Neural Networks for	Computing, Snowbird, Utah, 1986.	Denker, JS 86b
Adaptation and Self-Repair in Parallel	Computing Structures	Huberman, BA 84a
Dynamics of	Computing Structures	Huberman, BA 85a
A Connectionist Architecture for	Computing Textural Segmentation.	Mesrobian, E 87a
Globally Connected Network Models for	Computing Using Fine-Grained Processing Elements	Szu, HH 85a
	Computing with Connections	Sejnowski, TJ 87b
	Computing with Connections	Feldman, JA 81a
	Computing with Connections in Visual Recognition of Origami Objects.	Sabbah, D 85a
	Computing with Connections in Visual Recognition of Origami Objects.	Sabbah, D 88a
	Computing With Neural Circuits: A Model.	Hopfield, JJ 85b
An Introduction to	Computing with Neural Nets	Lippmann, RP 87b
	Computing with Neural Nets	Lippmann, RP 87c
	Computing with Neural Networks	Palm, G 87c
	Computing with Neural Networks	Kohonen, T 87b
	Computing with Neural Networks	Kinoshita, J 87a
Technical Comments:	Computing with Neural Networks	Kohonen, T 87h
Technical Comments:	Computing with Neural Networks	Palm, G 87d
Technical Comments:	Computing with Neural Networks	Hopfield, JJ 87c
Technical Comments:	Computing with Neural Networks	Carpenter, GA 87f
	Computing with Neural Networks: The Role of Symmetry	Carpenter, GA 86g
	Computing with Structured Neural Networks.	Feldman, JA 88a
Why "	Computing" Requires Symbols.	Pylyshyn, ZW 84a
A Survey of Interconnection Networks (Concurrent	Computing)	Tse-Yun, F 81a
Neural Networks for	Computing?	Abu-Mostafa, YS 86b
	Concentrating Information in Time: Analog Neural Networks with . . .	Tank, DW 87b
A Neural-Network Solution to the	Concentrator Assignment Problem.	Tagliarini, GA 88a
Modelling the Visual Cortex with "Modulo System"	Concept.	Fuhrmann, G 81a
Neural-Type Junctions: A New Circuit	Concept.	DeClaris, N 76a
Neuristor: A Novel Device and System	Concept.	Crane, HD 62a
Syneuristor: A Device and	Concept.	Waldron, MB 71a
Opposition Space as a Structuring	Concept for the Analysis of Skilled Hand Movements	Iberall, T 86a
A Signal Averaging Model for	Concept Formation.	Knapp, AG 84a
Neural Theory of Association and	Concept Formation.	Amari, S 77d
Neural Theory of Association and	Concept Formation.	Amari, S 77e
...Nerve Nets with Special Reference to Association and	Concept Formation.	Amari, S 79b
	Concept Formation of Neural Networks: Implications for Evolution of.	Anderson, JA 88a
	Concept of a Flag-Oriented Fully Parallel Associative Processor. . .	Tavangarian, D 88a
A Distributed Connectionist Representation for	Concept Structures	Touretzky, DS 87a
Connectionist Representation of	Concepts	Feldman, JA 88c
Engineering Applications of Spin Glass	Concepts	Guyon, I 86a
Learning Distributed Representations of	Concepts	Hinton, GE 86d
Learning Efficient Recognizers for Analytically Derived	Concepts	Utgoff, PE 85a
Models of	Concepts	Cohen, B 85a

Chaos, Symbols, and	Connectionism.	Barnden, JA 87a
Internal Representation of Universal Regularities: A Challenge for	Connectionism.	Shepard, RN 89a
On the Proper Treatment of	Connectionism.	Smolensky, P 87b
On Language and	Connectionism: Analysis of a Parallel Distributed Processing Model	Pinker, S 88b
Ecological	Connectionism and Animal-Environment Mutuality	Treffner, PJ 87a
	Connectionism and Cognitive Architecture: A Critical Analysis.	Fodor, JA 88a
	Connectionism and Explanation.	Charniak, E 86b
	Connectionism at the Philips Research Laboratory Brussels.	Wellekens, CJ 88a
Is Distributed	Connectionism Compatible with the Physical Symbol System Hypothesis.	Derthick, MA 86a
BoltzCONS: Reconciling	Connectionism with the Recursive Nature of Stacks and Trees.	Touretzky, DS 86b
Interacting and Communicating	Connectionist Agents	Allen, RB 88a
A	Connectionist Algorithm for Minimum Cost Network Flows	Marcus, R 87a
...Analogical Ordering of Consideration: One Method for Integrating	Connectionist and Symbolic Processing.	Becker, LA 87b
...Intentionality and Realism: The Philosophical Importance of the	Connectionist Approach	Basti, G 87a
Early Parallel Processing in Reading: A	Connectionist Approach	Mozer, MC 86a
Organization of Action Sequences in Motor Learning: A	Connectionist Approach	Miyata, Y 87a
Viewing Parsing as Word Sense Discrimination: A	Connectionist Approach	Cottrell, GW 84a
A	Connectionist Approach to Motor Control.	Addanki, S 83a
A Simple VLSI	Connectionist Architecture	Cleary, JG 87a
A	Connectionist Architecture for Computing Textural Segmentation	Mesrobian, E 87a
A	Connectionist Architecture for Matching 3-D Models to Moving Edge.	Casasent, DP 86a
A	Connectionist Architecture for Matching 3-D Models to Moving Edge.	Gunger, D 86a
A	Connectionist Architecture for Representing and Reasoning About.	Derthick, MA 87a
	Connectionist Architectures for Artificial Intelligence.	Fahlman, SE 87a
A Comparison of Speech Recognition Front Ends Using a	Connectionist Classifier	Renals, S 88a
Encoding Input/Output Representations in	Connectionist Cognitive Systems.	Miikkulainen, R 88a
A	Connectionist Context-Free Parser Which is not Context-Free, But	Charniak, E 87a
The Right of Free Association: Relative-Position Encoding for	Connectionist Data Structures.	Barnden, JA 88b
...Parser Which is not Context-Free, But Then it is not Really	Connectionist Either	Charniak, E 87a
A	Connectionist Encoding of Semantic Networks.	Shastri, L 87a
RAMBOT: A	Connectionist Expert System that Learns by Example	Mozer, MC 86b
RAMBOT: A	Connectionist Expert System that Learns by Example	Mozer, MC 87a
	Connectionist Expert Systems	Gallant, SI 88b
Example-Based Knowledge Engineering with	Connectionist Expert Systems	Gallant, SI 88c
Automated Generation of	Connectionist Expert Systems for Problems Involving Noise and.	Gallant, SI 87a
...Information-Processing: A Computational Architecture with a	Connectionist Implementation	Barnden, JA 86a
Symbols Among the Neurons: Details of a	Connectionist Inference Architecture	Touretzky, DS 85a
A	Connectionist Learning Model for 3-Dimensional Mental Rotation	Mel, BW 86a
...Theories of Language Learning: A Constructive Critique of Some	Connectionist Learning Models.	Lachter, J 88a
	Connectionist Learning Procedures.	Hinton, GE 87a
Complexity of	Connectionist Learning with Various Node Functions	Judd, JS 87a
Fast	Connectionist Learning: Words and Case	Sharkey, NE 89b
ZZENN (Zig Zag Epigenetic Neural Networks), A New Approach to	Connectionist Machine Learning	Cochet, Y 88b
Symbolic Schemata, Role Binding, and the Evolution of Structure in	Connectionist Memories	Dolan, CP 87a
	Connectionist Methods in Speech Recognition.	Rohwer, R 88a
Analysis of a Competition-Based	Connectionist Model.	Seidman, SB 87a
Color Constancy and Early Vision: A	Connectionist Model.	Heisey, I 87a
Bayesian Assessment of a	Connectionist Model for Fault Detection.	Gallant, SI 88a
A Dynamic	Connectionist Model for Phoneme Recognition: Preliminary Results	Robinson, AJR 88a
A	Connectionist Model for Simple Visual Functions.	Mel, BW 86b
A	Connectionist Model for Stereo Vision.	Stewart, CV 87a
A	Connectionist Model for Unsupervised Feature Discovery and Pattern	Franzini, MA 87a
A	Connectionist Model of Chronic Pain.	Katz, BF 88a
Generation of Simple Sentences in English Using the	Connectionist Model of Computation	Kalita, J 87a
A	Connectionist Model of Form Perception	Hrechanyk, LM 82a
MUSACT: A	Connectionist Model of Musical Harmony	Bharucha, JJ 87a
A Dynamic	Connectionist Model of Problem Solving	Lundell, J 87a
Local Constraint Integration in a	Connectionist Model of Stereo Vision	Stewart, CV 87b
	Connectionist Model of Visual Memory	Feldman, JA 81b
A Bipartite	Connectionist Model to Represent N-ary Boolean and Linear.	Post, SD 87a
	Connectionist Modeling: Neural Computation/Mental Connections.	Smolensky, P 89a
	Connectionist Modeling of Syntactic Constraints on Sentence.	Parisi, D 87a
Tools for	Connectionist Modeling: The Dynamical Systems Methodology.	Munro, PW 88a
A General-Purpose Simulation Environment for Developing	Connectionist Models	D'Autrechy, CL 88a
An Information-Theoretic Approach to Unsupervised	Connectionist Models	Plumbley, MD 88a
Analysis of Competition-Based Spreading Activation in	Connectionist Models	Wang, PY 88a
Concepts in	Connectionist Models	Anderson, JA 86b
Energy and the Behavior of	Connectionist Models	Feldman, JA 85b
...in Symmetry Detection: Parallel Studies of Human Subjects and	Connectionist Models	Cohen, NJ 86a

	Connectionist Models and Cognitive Science: Goals, Directions and. .	McClelland, JL 86c
	Connectionist Models and Geometric Reasoning	Fallside, F 88a
Special Issue on	Connectionist Models and Their Application	Feldman, JA 85d
	Connectionist Models and Their Implications.	Waltz, DL 88a
	Connectionist Models and Their Implications: Readings from	Waltz, DL 88b
	Connectionist Models and their Properties.	Feldman, JA 82b
	Connectionist Models and Their Properties.	Feldman, JA 82a
	Connectionist Models and Their Properties.	Feldman, JA 88b
Positive Feedback in Hierarchical	Connectionist Models: Applications in Language Production.	Dell, GS 88a
	Connectionist Models in Speech Recognition: A Brief Summary and. .	Fallside, F 88b
	Connectionist Models of Learning	Sejnowski, TJ 87c
Process and	Connectionist Models of Pattern Recognition.	Massaro, DW 87a
Minkowski-r Back-Propagation: Learning in	Connectionist Models with Non-Euclidian Error Signals.	Hanson, SJ 88a
Detection of Format Transitions by	Connectionist Network.	Smythe, EJ 87a
Learning to Understand Sentences in a	Connectionist Network.	Nolfi, S 88a
Evaluating a	Connectionist Network as a Simulation of Speech Development. . . .	Schwartz, RG 88a
Inverting a	Connectionist Network Mapping by Back-Propagation of Error	Williams, RJ 86a
PARSNIP: A	Connectionist Network that Learns Natural Language Grammar from. .	Hanson, SJ 87a
A	Connectionist Network that Learns to Process Some (Very) Simple. .	Samad, T 87a
Solving Puzzles with a	Connectionist Network Using a Hill-Climbing Heuristic.	Kawamoto, AH 86a
A Data Flow Analysis of	Connectionist Networks	Bookman, L 88a
Abstraction and Representation of Continuous Variables in	Connectionist Networks	Saund, E 86a
Cascaded Back-Propagation on Dynamic	Connectionist Networks	Pollack, JB 87a
Context-Free Parsing in	Connectionist Networks	Fanty, MA 85a
Context-Free Parsing with	Connectionist Networks	Fanty, MA 86b
ISCON: An Interactive Simulator for	Connectionist Networks	Small, SL 82b
Learned Phonetic Discrimination Using	Connectionist Networks	Watrous, RL 87a
Learning Acoustic Features from Speech Data Using	Connectionist Networks	Watrous, RL 87b
Learning of World Models by	Connectionist Networks	Sutton, RS 85a
On the Probabilistic Semantics of	Connectionist Networks	Geffner, H 87a
Refining and Redefining the Back-Propagation Learning Rule for	Connectionist Networks	Samad, T 87c
The Representation of Variables in	Connectionist Networks	Walters, DKW 87j
Reinforcement Learning in	Connectionist Networks: A Mathematical Analysis.	Williams, RJ 86c
Learning Phonetic Features Using	Connectionist Networks: An Experiment in Speech Recognition. . . .	Watrous, RL 87d
Learning Algorithms for	Connectionist Networks: Applied Gradient Methods of Nonlinear. . .	Watrous, RL 87c
Symbol Structures in	Connectionist Networks: Five Properties and Two Architectures. . .	Touretzky, DS 87c
Sequential	Connectionist Networks for Answering Simple Questions about a. . .	Allen, RB 88b
	Connectionist Networks for Learning Coordinated Motion in.	Pabon, JA 88a
Learning in	Connectionist Networks: Has the Credit Assignment Problem Been . .	Goel, A 89a
Unstable	Connectionist Networks in Speech Recognition	Rohwer, R 88c
Microelectronic Implementations of	Connectionist Neural Networks.	Mackie, S 88c
	Connectionist Parsing.	Cottrell, GW 85a
Toward	Connectionist Parsing.	Small, SL 82c
A Rule-Based	Connectionist Parsing System	Selman, B 85a
A Distributed	Connectionist Production System.	Touretzky, DS 86a
On the	Connectionist Reduction of Conscious Rule Interpretation	Smolensky, P 87a
A Distributed	Connectionist Representation for Concept Structures.	Touretzky, DS 87a
	Connectionist Representation of Concepts	Feldman, JA 88c
Response Mapping Functions: Classification and Analysis of	Connectionist Representations.	Walters, DKW 87f
On the Methodology of	Connectionist Research Programmes in Vision.	Gonzales, MEQ 87a
Towards	Connectionist Rule-Based Systems	Samad, T 88b
A	Connectionist Scheme for Modeling Word Sense Disambiguation. . . .	Cottrell, GW 83a
A	Connectionist Scheme for Modelling Context	Bookman, LA 88a
Attractor Dynamics and Parallelism in a	Connectionist Sequential Machine	Jordan, MI 86a
A	Connectionist Simulator for the BBN Butterfly Multi-Processor. . .	Fanty, MA 86a
	Connectionist Speech Recognition Using the Temporal Flow Model . .	Watrous, RL 88a
A Model of Schema Selection Using Marker Passing and	Connectionist Spreading Activation	Chun, HW 87b
Simulation of a	Connectionist Stereo Algorithm on a Shared-Memory Multiprocessor .	Stewart, CV 88a
A	Connectionist Structure for Phoneme Recognition.	Harrison, TD 88a
Rule-Based Processing in a	Connectionist System for Natural Language Understanding.	Selman, B 85b
A Unified Framework for	Connectionist Systems.	Golden, RM 88b
An Approach to Learning Control Surfaces by	Connectionist Systems.	Barto, AG 87a
Designing Appropriate Learning Rules for	Connectionist Systems.	Belew, RK 87a
Learning and Problem Solving with Multilayer	Connectionist Systems.	Anderson, CW 86a
Logical	Connectionist Systems.	Aleksander, I 88c
Representing Causal Schemata in	Connectionist Systems.	Golden, RM 86b
Structural Learning in	Connectionist Systems.	Barto, AG 85f
The Logic of	Connectionist Systems.	Aleksander, I 88f
Toward a Theory of Reinforcement-Learning	Connectionist Systems.	Williams, RJ 88b

Experiments with a	Connectionist Text Reader	Burr, DJ 87a
On Short-Term Information-Processing in	Connectionist Theories	Barnden, JA 84a
	Connectionist Value Units: Some Concerns	Barnden, JA 86b
Properties of	Connectionist Variable Representations	Walters, DKW 87e
	Connectionist/Neural Network Bibliography	Hammerstrom, D 86d
A	Connectionist/Neural Network Bibliography, Volume II	Hammerstrom, D 87a
The OGC Cognitive Architecture Project: Silicon Implementation of	Connectionist/Neural Networks	Hammerstrom, D 86e
	Connectionistic Learning in Real Time: Sutton-Barto Adaptive	Moore, JW 85a
...of Ordered Neural Mapping: Its Application to the Retinotectal	Connections	Willshaw, DJ 79a
Computing with	Connections	Feldman, JA 81a
Computing with	Connections	Sejnowski, TJ 87b
Connectionist Modeling: Neural Computation/Mental	Connections	Smolensky, P 89a
	Connections	Feldman, JA 85a
Detection of Favored Patterns in the Temporal Structure of Nerve Cell	Connections	Dayhoff, JE 87a
...on Pattern Separability in a Random Neural Net with Inhibitory	Connections	Torioka, T 80a
How to Make a Billion	Connections	Bailey, J 86a
Pattern Separability in a Random Neural Net with Inhibitory	Connections	Torioka, T 79a
...of a Tightly Coupled Random Nerve Net with Inhibitory	Connections	Torioka, T 79b
...of Two Layered Random Nerve Nets with Feedforward Inhibitory	Connections	Torioka, T 88a
...Separation in a Two-Layer Random Neural Net with Inhibitory	Connections	Torioka, T 80b
Cortical	Connections and Parallel Processing: Structure and Function	Ballard, DH 86a
Cortical	Connections and Parallel Processing: Structure and Function	Ballard, DH 87a
	Connections and Symbols	Pinker, S 88a
Binary Synaptic	Connections Based on Memory Switching in a SI:H	Thakoor, AP 86a
Binary Synaptic	Connections Based on Memory Switching in a-Si:H for Artificial	Thakoor, AP 87a
Correlational Strength and Computational Algebra of Synaptic	Connections Between Neurons	Fetz, EE 88a
How Patterned Neural	Connections Can be Set Up by Self-Organization	Willshaw, DJ 76b
How Patterned Neural	Connections Can be Set Up by Self-Organization	Willshaw, DJ 75a
Clustered Intrinsic	Connections in Cat Visual Cortex	Gilbert, CD 83a
Dynamic	Connections in Neural Networks	Feldman, JA 82c
Computing with	Connections in Visual Recognition of Origami Objects	Sabbah, D 85a
Computing with	Connections in Visual Recognition of Origami Objects	Sabbah, D 88a
Neural	Connections, Mental Computation	Nadel, L 89a
Cerebral Cortex: Architecture, Intracortical	Connections, Motor Projections	Lorente de No, R 49a
Introduction to a System for Implementing Neural Net	Connections on SIMD Architectures	Tomboulian, S 88a
Exploded	Connections: Unchunking Schematic Knowledge	Small, SL 82a
Neuronal	Connections Underlying Orientation Selectivity in Cat Visual Cortex	Ferster, D 87a
The Basis of Lawful Behavior: Rules or	Connections?	McClelland, JL 87c
...Associative Networks when Operating in Initially Random	Connectivities	Dobson, VG 87a
Optimization of Neural Network	Connectivity	Cooper, LN 87a
A	Connectivity Analysis of a Class of Simple Associative Neural	Hammerstrom, D 88a
A	Connectivity Analysis of Recursive, Auto-Associative Connection	Hammerstrom, D 86a
The	Connectivity Analysis of Simple Association	Hammerstrom, D 88b
Anisotropic	Connectivity and Cooperative Phenomena as a Basis for Orientation	Finette, S 78a
Lower Bound for	Connectivity in Local-Learning Neural Networks	Abu-Mostafa, YS 88b
Effect of	Connectivity on the Activity of Neural Net Models	Leake, B 76a
Local Synaptic Modification Can Lead to Organized	Connectivity Patterns in Associative Memory	Palm, G 84a
	Connectivity Versus Entropy	Abu-Mostafa, YS 88a
Les Applications des Modeles	Connexionistes a la Reconnaissance de la Parole	Wellekens, CJ 88b
The Minds of Robots, Sense Data, Memory Images, and Behavior in	Conscious Automata	Culbertson, JT 63a
On the Connectionist Reduction of	Conscious Rule Interpretation	Smolensky, P 87a
	Consciousness and the Goals of Artificial Intelligence	Franzini, MA 86a
Competition, Decision,	Consensus	Grossberg, S 78e
A Screening Algorithm for Levenglick's	Consensus Ranking Model	Flood, MM 87a
	Consensus versus Competition in Neural Networks: A Comparative	Montalvo, FS 75a
A Statistical	Consequence of the Logical Calculus of Nervous Nets	Landahl, HD 43a
Equidistance Tendency and Its	Consequences	Gogel, WC 65a
A Model of Synaptogenetic	Consequences of Afferent Input in Neuronal Nets	Dammasch, IE 88a
Nerve Conduction Theory: Some Mathematical	Consequences of Bernstein's Model	Offner, F 40a
Some Physiological and Biochemical	Consequences of Psychological Postulates	Grossberg, S 68f
	Conservation of Cellular Mechanisms for Models of Learning and	Alkon, DL 88a
Topology	Conserving Mappings for Learning Motor Tasks	Ritter, H 86b
Convergence Properties of Kohonen's Topology	Conserving Maps: Fluctuations, Stability and Dimension Selection	Ritter, H 88a
A Synaptic Modification Algorithm in	Consideration of the Generation of Rhythmic Oscillation in a Ring	Tsutsumi, K 84a
Further	Consideration on Pattern Separability in a Random Neural Net with	Torioka, T 80a
Using Activation Networks for Analogical Ordering of	Consideration: One Method for Integrating Connectionist and	Becker, LA 87b
Attention: Some Theoretical	Considerations	Deutsch, JA 63a
Hybrid Optical Inference Machines: Architectural	Considerations	Warde, C 86a
Neural Network Emulation Hardware Design	Considerations	Cruz, CA 87c

Engineer A Black-Box Solution in the First "Real Cognizer"	Contest.	Brown, C 87c
A Connectionist Scheme for Modelling	Context.	Bookman, LA 88a
Dynamically Expanding	Context.	Kohonen, T 87d
Self-Learning Rules by Dynamically Expanding	Context.	Kohonen, T 87f
...of Varying the Proportion of Items Presented in an Appropriate	Context.	Tweedy, JR 77a
	Context and the Allocation of Resources in Word Recognition.	Larochelle, S 80a
An Interactive Activation Model of	Context Effects in Letter Perception: Part 1, An Account of Basic.	McClelland, JL 81b
An Interactive Activation Model of	Context Effects in Letter Perception, Part 1: An Account of Basic.	McClelland, JL 81a
An Interactive Activation Model of	Context Effects in Letter Perception: Part 2	Rumelhart, DE 82a
Conceptual Graph Knowledge Systems as Problem	Context for Neural Networks.	Lendaris, GG 88a
The Role of Syntactic	Context in Word Recognition.	Goodman, GO 81a
A	Context Sensitive Line Finder for Recognition of Polyhedra	Shirai, Y 73a
Dynamically Expanding	Context, with Application to the Correction of Symbol Strings in	Kohonen, T 86a
	Context-Dependent Modeling for Acoustic-Phonetic Recognition of.	Schwartz, RM 85a
A	Context-Driven World Model Appropriate for Training and Testing of	Keller, R 87a
A Connectionist Context-Free Parser Which is not	Context-Free, But Then it is not Really Connectionist Either	Charniak, E 87a
A Connectionist	Context-Free Parser Which is not Context-Free, But Then it is not.	Charniak, E 87a
	Context-Free Parsing in Connectionist Networks	Fanty, MA 85a
	Context-Free Parsing with Connectionist Networks	Fanty, MA 86b
	Context-Sensitive Coding, Associative Memory, and Serial Order in.	Wickelgren, WA 69a
Learning and Applying	Contextual Constraints in Sentence Comprehension	St John, MF 88a
On Visual and	Contextual Factors in Reading: A Reply to Rayner and Slowiaczek.	McClelland, JL 81d
Passing Markers: A Theory of	Contextual Influence in Language Comprehension	Charniak, E 83a
Stimulus Information and	Contextual Information as Determinants of Tachistoscopic	Tulving, E 63a
	Contextual Road Finding with a Neural Network.	Wright, WA 89b
Generalization in Neural Networks: The	Contiguity Problem	Maxwell, T 87a
Voltage-Dependent Calcium and Potassium Ion Conductances: A	Contingency Mechanism for an Associative Learning Model.	Alkon, DL 79a
Theory of Learning with Constant, Variable, or	Contingent Probabilities of Reinforcement.	Estes, WK 57a
Mixing Binary and	Continuous Connection Schemes for Knowledge Access	Sharkey, NE 86a
A	Continuous Information Theoretic Approach to the Analysis of	Fuller, MS 83a
Real-Time Holographic Associative Memory for Identifying Words in a	Continuous Letter String	Paek, EG 88c
Class of	Continuous Level Associative Memory Neural Nets.	Marks II, RJ 87b
A Class of	Continuous Level Neural Nets	Marks II, RJ 87a
Architectures for a	Continuous Level Neural Network Based on Alternating Orthogonal.	Marks II, RJ 88a
Processing of	Continuous Lines and Edges by the Visual System.	Hartmann, G 83a
...in Short Term Memory: Serial Binary Memory Models or Parallel	Continuous Memory Models?.	Grossberg, S 78c
Towards	Continuous Models of Memory.	Banzhaf, W 87b
On the Activities in a	Continuous Neural Network.	Oguztoreli, MN 75a
On the Stability, Storage Capacity, and Design of Nonlinear	Continuous Neural Networks	Guez, A 88b
A Network for Recognition and Classification of	Continuous Patterns.	Banzhaf, W 88a
...Principles to Simultaneous Phonemic Segmentation and Labeling of	Continuous Speech.	Jalanko, M 78a
Context-Dependent Modeling for Acoustic-Phonetic Recognition of	Continuous Speech.	Schwartz, RM 85a
...to the Correction of Symbol Strings in the Recognition of	Continuous Speech.	Kohonen, T 86a
Processing Interactions and Lexical Access During Word Recognition in	Continuous Speech.	Marslen-Wilson, WD 78a
A Maximum Likelihood Approach to	Continuous Speech Recognition.	Bahl, LR 83a
Improved Hidden Markov Modeling of Phonemes for	Continuous Speech Recognition.	Schwartz, RM 84a
	Continuous Speech Recognition by Statistical Methods	Jelinek, F 76a
Large-Vocabulary Speaker-Independent	Continuous Speech Recognition: The SPHINX System	Lee, K 88a
Speaker-Independent	Continuous Speech Recognition Using Hidden Markow Models	Lee, K 87a
Combinatorial Optimization Using a	Continuous State Boltzmann Machine	Gutzmann, KM 87a
Collective Computation with	Continuous Variables	Hopfield, JJ 85a
Collective Computation with	Continuous Variables	Hopfield, JJ 86a
Abstraction and Representation of	Continuous Variables in Connectionist Networks	Saund, E 86a
A	Continuous-Time Optical Neural Network	Stoll, HM 87a
A	Continuous-Time Optical Neural Network Associative Memory.	Lee, LS 87a
Analog Memory for	Continuous-Voltage, Discrete-Time Implementations of Neural.	Brown, PB 87a
An Extremum Principle for Shape from	Contour.	Brady, M 87a
...Solid Shape Based on the Topological Properties of the Apparent	Contour.	Koenderink, JJ 87a
Hierarchical	Contour Coding by the Visual Cortex.	Hartmann, G 85a
	Contour Enhancement, Short-Term Memory, and Constancies in	Grossberg, S 73a
On the Nature of Inducing Forms Generating Perceptions of Illusory	Contours	Prazdny, K 85a
Subjective	Contours	Kanizsa, G 76a
The Illusion of Transparency and Chromatic Subjective	Contours	Meyer, GE 83a
Psychophysiological Substrates of Schedule Interaction and Behavioral	Contrast	Grossberg, S 81b
Pattern Formation,	Contrast Control, and Oscillations in the Short Term Memory of	Ellias, SA 75a
Local Determination of a Moving	Contrast Edge.	Kahn, P 85a
Some Developmental and Attentional Biases in the	Contrast Enhancement and Short-Term Memory of Recurrent Neural	Grossberg, S 75b
Behavioral	Contrast in Short Term Memory: Serial Binary Memory Models or.	Grossberg, S 78c
Adaptation, After-Effect and	Contrast in the Perception of Curved Lines	Gibson, JJ 33a

	Distinctive Features, Categorical Perception, and Probability.	Anderson, JA 77a
	Distinctive Features, Categorical Perception, and Probability.	Anderson, JA 77b
A Neural Network System with an Automatic Generation Mechanism for	Distorted Patterns	Shimohara, K 88a
Unrestricted English Text-to-Speech Conversion with a	Distributed Adaptive Network	Tenorio, MF 88b
Simulation of a	Distributed Algorithm Using the Connection Machine	Aburdene, MF 88a
	Distributed and Fault-Tolerant Computation for Retrieval Tasks	Char, JM 88a
Integration of	Distributed and Symbolic Knowledge Representations	Hutchison, WR 87a
A Parallel	Distributed Architecture for Motor Control	Morasso, P 88a
Distributed and Fault-Tolerant Computation for Retrieval Tasks Using	Distributed Associative Memories	Char, JM 88a
Invariant Object Recognition Using a	Distributed Associative Memory	Wechsler, H 88a
	Distributed Associative Memory for Patterns.	Pao, Y 75a
ADAM: A	Distributed Associative Memory for Scene Analysis.	Austin, J 87a
	Distributed Associative Memory for Use in Scene Analysis	Austin, J 87b
Analysis of a	Distributed Associative Memory Model	Wiles, J 87a
Construction of a	Distributed Associative Memory on the Basis of Bayes Discriminant.	Murakami, K 81a
Storage and Processing of Information in	Distributed Associative Memory Systems	Kohonen, T 81c
Inductive Information Retrieval Using Parallel	Distributed Computation.	Mozer, MC 84a
A Task-Level Model of	Distributed Computation for Sensory-Based Control of Complex Robot	Lyons, D 85a
Is	Distributed Connectionism Compatible with the Physical Symbol.	Derthick, MA 86a
A	Distributed Connectionist Production System.	Touretzky, DS 86a
A	Distributed Connectionist Representation for Concept Structures.	Touretzky, DS 87a
A Parallel Natural Language Processing with	Distributed Control.	Berg, G 87a
Graphical Folding and	Distributed Failures in Large Network Simulations.	Dayhoff, JE 87b
Amnesia and	Distributed Memory	McClelland, JL 86a
Application of an Associatively Addressed,	Distributed Memory	Simmons, GJ 64a
Sparse	Distributed Memory	Kanerva, P 88a
	Distributed Memory and the Representation of General and Specific.	McClelland, JL 85a
Learning and	Distributed Memory: Getting Close to Neurons	Aleksander, I 88b
	Distributed Memory in the Central Nervous Systems: Possible Test of.	Cooper, LN 81a
A	Distributed Memory Model for Serial-Order Information.	Murdock, BB 83a
Order Information and	Distributed Memory Models.	Ratcliff, R 87a
...in Self-Organizing Feature Maps, and Relation of these Maps to	Distributed Memory Networks.	Kohonen, T 86d
Theory of Categorization Based on	Distributed Memory Storage	Knapp, AG 84b
Theory of Categorization Based on	Distributed Memory Storage	Knapp, AG 84c
Kinematic Networks: A	Distributed Model for Representing and Regularizing Motor.	Mussa-Ivaldi, FA 88b
	Distributed Model of Human Learning and Memory	McClelland, JL 86d
...and Experimental Perspectives on Arm Trajectory Formation: A	Distributed Model of Motor Redundancy.	Mussa-Ivaldi, FA 88c
Perceptual Structures and	Distributed Motor Control.	Arbib, MA 79b
	Distributed Neural Information Processing in the Vestibulo-Ocular.	Lau, CGY 88a
Natural and Social System Metaphors for	Distributed Problem Solving: Introduction to the Issue	Chandrasekaran, B 81a
A Parallel Simulation of a	Distributed Problem Solving Network.	Durfee, EH 84a
An Approach to Cooperation: Planning and Communication in a	Distributed Problem Solving Network.	Durfee, EH 86a
Increasing Coherence in a	Distributed Problem Solving Network.	Durfee, EH 85b
Appeal of Parallel	Distributed Processing	McClelland, JL 86b
General Framework for Parallel	Distributed Processing	Rumelhart, DE 86b
Introduction to Linear Algebra in Parallel	Distributed Processing	Jordan, MI 86b
Reflections on Cognition and Parallel	Distributed Processing	Norman, DA 86a
Explorations in Parallel	Distributed Processing: A Handbook of Models, Programs, and.	McClelland, JL 88a
Parallel	Distributed Processing and Role Assignment Constraints	McClelland, JL 86g
Serial Order: A Parallel	Distributed Processing Approach.	Jordan, MI 86c
Parallel	Distributed Processing: Explorations in the Microstructure of.	Rumelhart, DE 86h
Parallel	Distributed Processing: Implications for Cognition and Development	McClelland, JL 88b
On Language and Connectionism: Analysis of a Parallel	Distributed Processing Model of Language Acquisition	Pinker, S 88b
On the Control of Automatic Processes: A Parallel	Distributed Processing Model of the Stroop Effect.	Cohen, JC 88a
A Network that Induces a	Distributed Representation Useful for Encoding Kinship Relations	Hinton, GE 86a
	Distributed Representations.	Hinton, GE 84b
	Distributed Representations.	Hinton, GE 86b
Learning	Distributed Representations of Concepts.	Hinton, GE 86d
	Distributed Revision of Composite Beliefs.	Pearl, J 87a
A Walsh-Hadamard Based	Distributed Storage Device for the Associative Search of	Oh, S 84a
ACTORS: A Model of Concurrent Computation in	Distributed Systems.	Agha, GA 86a
Concepts in	Distributed Systems.	Anderson, JA 86c
Functionally-Accurate, Cooperative	Distributed Systems.	Lesser, VR 81a
Doppler Tolerant Classification of	Distributed Targets - a Bionic Sonar	Altes, RA 75a
Word Expert Parsing: A Theory of	Distributed Word-Based Natural Language Understanding.	Small, SL 80a
Cochlear Nerve Filter Responses:	Distribution Along the Cochlear Partition.	Pfeiffer, RR 75a
Artificial Neural Nets: Dependence of the EEG Amplitude's Probability	Distribution on Statistical Parameters	Anninos, PA 83b
Bayesian Image Processing, Gibbs	Distributions, and Neural Networks	Vold, TG 87a
Stochastic Relaxation, Gibbs	Distributions, and the Bayesian Restoration of Images.	Geman, S 84b

...Arm Movements: Emergent Invariants and Speed-Accuracy Properties	During Trajectory Formation.	Bullock, D 88b
...Arm Movements: Emergent Invariants and Speed-Accuracy Properties	During Trajectory Formation.	Bullock, D 88c
Rules for Forming Receptive Fields of Formal Neurons	During Unsupervised Learning Processes	Bobrowski, L 82a
...Trajectory Formation and Coordination of Multiple Effectors	During Variable-Speed Arm Movements.	Bullock, D 87a
Computer Simulation of Event Related Potentials	During Voluntary Cognition	Deshmukh, VD 88b
Processing Interactions and Lexical Access	During Word Recognition in Continuous Speech	Marslen-Wilson, WD 78a
Masking of Alphabetic Character Recognition by Dynamic Visual Noise (	DVN)	Uttal, WR 69a
Shifter Circuits: A Computational Strategy for	Dyanmic Aspects of Visual Processing	Anderson, CH 87a
	Dynamic Activity and Memory Traces in Computer-Simulated	Gaudiano, P 87a
The Adaptive,	Dynamic, and Associative Memory Model: A Possible Future Tool for.	Beroule, D 87b
Exact Results for the Average	Dynamic Behavior of some Non-Linear Neural Networks.	Rossler, J 87a
A	Dynamic Connectionist Model for Phoneme Recognition: Preliminary	Robinson, AJR 88a
A	Dynamic Connectionist Model of Problem Solving	Lundell, J 87a
Cascaded Back-Propagation on	Dynamic Connectionist Networks	Pollack, JB 87a
	Dynamic Connections in Neural Networks	Feldman, JA 82c
	Dynamic Control of an Artificial Neural System: The Property	Ryan, TW 87a
...Architecture for Intelligence Control of Neural Recognition in	Dynamic Environments	Lee, JSJ 88a
A New LMS-Based Algorithm for Rapid Adaptive Classification in	Dynamic Environments: Theory and Preliminary Results	Nguyen, DD 88a
The Utility Driven	Dynamic Error Propagation Network.	Robinson, AJR 87a
Static and	Dynamic Error Propagation Networks with Application to Speech.	Robinson, AJR 88b
Determining the	Dynamic Landscape of Hopfield Networks	Weisbuch, G 86a
A Practical Algorithm for Solving	Dynamic Membrane Equations	Rush, S 78a
	Dynamic Model of Olfactory Discrimination.	Grajski, KA 86b
The T5 Base Modulator Hypothesis: A	Dynamic Model of T5 Neuron Function in Toads	Betts, B 88a
Simulation of Chaotic EEG Patterns with a	Dynamic Model of the Olfactory System.	Freeman, WJ 87d
...Static or Periodic Attractors and Their Bifurcations in	Dynamic Neural Networks.	Baird, B 88a
	Dynamic Optical Interconnects: Volume Holograms as Optical Two-Port.	Anderson, DZ 87b
	Dynamic Pattern Generation in Behavioral and Neural Systems.	Schoner, G 88d
	Dynamic Pattern Matching with Filling-In	Ogmen, H 87a
A	Dynamic Pattern Theory of Behavioral Change.	Schoner, G 88a
	Dynamic Patterns of Biological Coordination: Theoretical Strategy.	Schoner, G 88e
	Dynamic Processes in the Resolution of Lexical Ambiguity: PhD.	Kawamoto, AH 85b
Neural Networks Mediating Linearizable	Dynamic Redundant Sensori-Motor Reflexes Characterized by Minimum.	Daunicht, WJ 88b
The	Dynamic Re-Evaluation of Alternatives and the Emulation of Human	Clema, JK 72a
Nonlinear Optical Neural Networks:	Dynamic Ring Oscillations.	Anderson, DZ 88b
	Dynamic Shape.	Koenderink, JJ 86b
Theory of	Dynamic Similarity in Neuronal Systems	MacGregor, RJ 88a
	Dynamic Single Unit Simulation of a Realistic Cerebellar Network	Pellionisz, AJ 73a
	Dynamic Single Unit Simulation of a Realistic Cerebellar Network	Pellionisz, AJ 74a
Planning a	Dynamic Trajectory via Path Finding in Discretized Phase Space	Ritter, H 87a
Masking of Alphabetic Character Recognition by	Dynamic Visual Noise (DVN)	Uttal, WR 69a
A	Dynamical Approach to Temporal Pattern Processing.	Stornetta, WS 88a
Topological and	Dynamical Aspects of a Neural Network Model for Generation of.	Eckmiller, R 86b
Characterization of	Dynamical Behavior Associated with a Single Neural Equation.	Nagami, H 76a
A	Dynamical Behaviour of Active Regions in Randomly Connected Neural	Okuda, M 74a
	Dynamical Behaviour of Neural Networks	Goles, E 85a
Spatiotemporal Receptive Fields: A	Dynamical Model Derived from Cortical Architectonics	Krone, G 86a
Computer Simulations of Pattern Recognition as a	Dynamical Process of a Synergetic System	Fuchs, A 88a
	Dynamical Properties of a New Type of Neural Network	Cotterill, RMJ 88a
	Dynamical Properties of an Automaton With Memory	Cosnard, M 86a
On Some	Dynamical Properties of Monotone Networks.	Robert, Y 86a
Spontaneous Pattern Classification by a	Dynamical System	Banzhaf, W 87a
Uniqueness and Stability of a Limit Cycle for a Third Order	Dynamical System Arising in Neuron Modelling	Li, B 81a
An Introduction to Chaotic	Dynamical Systems.	Devaney, RL 87a
Roads to Turbulence in Dissipative	Dynamical Systems.	Eckmann, J 81a
Strange Attractors and Chaotic Motions of	Dynamical Systems.	Ott, E 81a
The	Dynamical Systems Approach to Differential Equations	Hirsch, MW 84a
Information Processing in	Dynamical Systems: Foundations of Harmony Theory	Smolensky, P 86a
Tools for Connectionist Modeling: The	Dynamical Systems Methodology.	Munro, PW 88a
Neural Networks with	Dynamical Thresholds	Horn, D 88b
	Dynamically Expanding Context.	Kohonen, T 87d
Self-Learning Rules by	Dynamically Expanding Context.	Kohonen, T 87f
	Dynamically Expanding Context, with Application to the Correction.	Kohonen, T 86a
...and Learned Patterns of Movement Coordination, I. Relative Phase	Dynamics	Schoner, G 88b
...Patterns of Movement Coordination, II. Component Oscillator	Dynamics	Schoner, G 88c
...for Control of Learning in Neurons with Second-Messenger	Dynamics	Kirby, KG 88a
Brain Functions and Neural	Dynamics	Harth, EM 70a
...Response of a Neural Model with Threshold Modulations and Neural	Dynamics	Sibain, P 86a
Rapprochement of Artificial Intelligence and	Dynamics	Conrad, M 87b

Towards a Primal Sketch of Real World Scenes in	Early Vision	Korn, AF 88a
Color Constancy and	Early Vision: A Connectionist Model	Heisey, I 87a
Computational Networks in	Early Vision: From Orientation Selection to Optical Flow	Zucker, SW 88a
A Cortical Network Model for	Early Vision Processing	Moller, P 88a
Scientists Focus on "	Early Vision" for Smart Sensors	Brown, C 87k
A Silicon Model of	Early Visual Processing	Mead, CA 88a
The Neural Analog Diffusion-Enhancement Layer (NADEL) and	Early Visual Processing	Waxman, AM 88b
Macular Bioaccelerometers on	Earth and in Space	Ross, MD 88a
An Interpretation of Cortical Maps in	Echolocating Bats	Altes, RA 89a
Sonar for Generalized Target Description and its Similarity to Animal	Echolocation Systems	Altes, RA 76a
The	Ecological Approach to Visual Perception	Gibson, JJ 79a
	Ecological Connectionism and Animal-Environment Mutuality	Treffner, PJ 87a
Self-Organization of Cooperative and Competitive Networks in	Ecological Paradigm	Matsuo, K 87a
The Behavior of Computational	Ecologies	Huberman, BA 88b
An Application of the GMDH Algorithm to	Economic Modeling	Scott, DE 84a
	Economic Prediction Using Neural Networks: The Case of IBM Daily	White, H 88a
A Model for the	Economical Encoding of the Visual Image in Cerebral Cortex	Sakitt, B 82a
Neural Nets for	Economists	Baum, EB 88b
Oscillatory Neural	Ectoderm (One): The "One" Hypothesis for Human Holoencephalon and	Deshmukh, VD 87a
L'Uso di Modelli Computazionali in Psicofisiologia	Ed Il Principio Gnoseologico di Rappresentazione	Basti, G 86b
Fuzzy Logic's Soft Start Belies a Cutting	Edge	Chester, M 86a
Local Determination of a Moving Contrast	Edge	Kahn, P 85a
A Connectionist Architecture for Matching 3-D Models to Moving	Edge Features	Casasent, DP 86a
A Connectionist Architecture for Matching 3-D Models to Moving	Edge Features	Gunger, D 86a
Toward a Low-Level Description of Dot Clusters: Labeling	Edge, Interior, and Noise Points	Zucker, SW 79a
Fast	Edge-Only Matching Techniques for Robot Pattern Recognition	Caelli, TM 87a
...A Neural Network for the Detection and Representation of Oriented	Edges	Walters, DKW 87h
...A Neural Network for the Detection and Representation of Oriented	Edges	Walters, DKW 87g
The Classification and Integration of	Edges as Critical to the Perception of Reflectance and Illumination	Gilchrist, AL 83a
Processing of Continuous Lines and	Edges by the Visual System	Hartmann, G 83a
Asymptotic Properties of Nearest Neighbor Rules Using	Edited Data	Wilson, DL 72a
Perceptrons: Expanded	Edition	Minsky, ML 88a
Self-Organization and Associative Memory, Second	Edition	Kohonen, T 88c
Parallel Processing of Signals in Neural Nets as Manifested in the	EEG	Freeman, WJ 75b
	EEG Alpha Map Series: Brain Micro-States by Space-Oriented Adaptive	Lehmann, D 87a
Artificial Neural Nets: Dependence of the	EEG Amplitude's Probability Distribution on Statistical Parameters	Anninos, PA 83b
	EEG Analysis Gives Models of Neuronal Template-Matching Mechanism	Freeman, WJ 79a
...A Key to the Understanding of the Combined Dynamics of	EEG and Event-Related Potentials" by Elbert and Rockstroh	Palm, G 87b
Derivation of a Mathematical Equation for the	EEG and the General Solution within the Brain and in Space	Anninos, PA 75a
Spatial	EEG Correlates of Non-Associative and Associative Olfactory	Grajski, KA 88b
Spatial Patterns of Visual Cortical Fast	EEG During Conditioned Reflex in a Rhesus Monkey	Freeman, WJ 87e
Human Brain	EEG Fields: Micro-States and Their Functional Significance	Lehmann, D 87b
Spatio-Temporal Analysis of Multi-channel Alpha	EEG Map Series	Fuchs, A 87a
Simulation of Chaotic	EEG Patterns with a Dynamic Model of the Olfactory System	Freeman, WJ 87d
Encoding of Visual Information: Correlation of	EEG Response to Intracellular Response	Anninos, PA 82a
Classification of	EEG Spatial Patterns with a Tree-Structured Methodology: CART	Grajski, KA 86a
Relation of Olfactory	EEG to Behavior: Factor Analysis	Freeman, WJ 87b
Relation of Olfactory	EEG to Behavior: Spatial Analysis	Freeman, WJ 87c
Relation of Olfactory	EEG to Behavior: Time Series Analysis	Freeman, WJ 86a
Extended Fuzzy Cognitive Maps (	EFCM) Engines for the Diagnosis of Electro-Mechanical and	Tenorio, MF 87a
...Processes: A Parallel Distributed Processing Model of the Stroop	Effect	Cohen, JC 88a
Visual Illusions in Neural Networks: Line Neutralization, Tilt After	Effect, and Angle Expansion	Levine, DS 76a
Unlearning Has a Stabilizing	Effect in Collective Memories	Hopfield, JJ 83a
	Effect of Clock Skew in Iterative Neural Network and Optical	Oh, S 88a
	Effect of Connectivity on the Activity of Neural Net Models	Leake, B 76a
	Effect of Duration on the Perception of Voicing	Denes, P 55a
A Neural Model Theory Leading to Kindling	Effect of Epilepsy	Anninos, PA 87b
The Transmission of Information and the	Effect of Local Feedback in Theoretical and Neural Networks	Uttley, AM 66a
On the	Effect of Noise on the Moore-Penrose Generalized Inverse	Stiles, GS 85a
	Effect of Orientation and of Shape Similarity on Perceptual	Beck, J 66a
The	Effect of Stochastic Interconnects in Artificial Neural Network	Marks II, RJ 88f
The Spacing	Effect on NETtalk, a Massively-Parallel Network	Rosenberg, CR 86a
Let's Ask GMDH What	Effect the Environment Has on Fisheries	Brooks, HA 84a
An	Effective Heuristic Algorithm for the Traveling Salesman Problem	Lin, S 73a
On the	Effectiveness of Receptors in Recognition Systems	Marill, T 63a
Two Attentional Models of Classical Conditioning: Variations in CS	Effectiveness Revisited	Schmajuk, NA 87a
...for Automatic Trajectory Formation and Coordination of Multiple	Effectors During Variable-Speed Arm Movements	Bullock, D 87a
...Interactions in Two-Word Displays: Familiarity and Similarity	Effects	McClelland, JL 86h
A Model for Variability	Effects in Hand-Printing, with Implications for the Design of On	Ward, JR 88a

An Associative Memory Based on an	Electronic Neural Network Architecture	Howard, RE 87a
	Electronic Neural Network Chips.	Jackel, LD 87b
Analogue Circuits for Variable-Synapse	Electronic Neural Networks	Tsividis, Y 87b
Basins of Attraction for	Electronic Neural Networks	Marcus, CM 88a
	Electronic Neural Networks	Hubbard, W 86a
Programmable Synaptic Chip for	Electronic Neural Networks	Moopenn, A 88a
Resistive Synaptic Interconnects for	Electronic Neural Networks	Lamb, JL 87a
Research Challenges and Opportunities in	Electronic Neural Systems.	Lau, CGY 87b
	Electronic Neurocomputers.	Thakoor, AP 88a
Performance Limits of Optical, Electro-Optical, and	Electronic Neurocomputers.	Hecht-Nielsen, R 86a
Field Computation: A Model of Massively Parallel Computation in	Electronic, Optical, Molecular and Biological Systems.	MacLennan, BJ 88a
A Sensitive	Electronic Photoreceptor	Mead, CA 85b
	Electronic Simulation of the Vertebrate Retina	Eckmiller, R 75a
The	Electronic Spreadsheet as a Tool in Modeling and Simulating Neural	Hewett, TT 86a
Theory and Feasibility of the Parametric Avalanche for Advanced	Electronic Warfare Applications.	Dawes, RL 88b
Overview of Neural-Type	Electronics.	El-Leithy, N 85a
Post-Digital	Electronics.	Pettit, FR 87a
Performance Limits of Optical,	Electro-Optical, and Electronic Neurocomputers	Hecht-Nielsen, R 86a
Speech Prosthesis System for Laryngectomees Using "	Electropalatograph" and a Small Keyboard	Muraki, S 86a
The	Electrophysics of a Nerve Fiber.	Scott, AC 75a
	Electrophysiological Observations Which Help Describe an	Levy, WB 87a
Neuron Doctrine and	Electrophysiology.	Bullock, TH 59a
	Electrotonic Processing of Information by Brain Cells.	Schmitt, FO 76a
An Adaptive Optical Computing	Element.	Fisher, AD 85a
Analog Computer Simulation of a Neural	Element.	Hiltz, FF 62a
...Responses in Classical Conditioning by a Neuron-Like Adaptive	Element.	Barto, AG 82b
...Nictitating Membrane Response by a Neuron-Like Adaptive	Element.	Blazis, DEJ 86a
Connectionistic Learning in Real Time: Sutton-Barto Adaptive	Element and Classical Conditioning of the Rabbit Nictitating	Moore, JW 85a
The Use of an Adaptive Threshold	Element to Design a Linear Optimal Pattern Classifier.	Koford, JS 66a
The Rules of	Elemental Synaptic Plasticity.	Levy, WB 85b
Activity in Networks of Neuron-Like	Elements	Farley, BG 60a
Characteristics of Random Nets of Analog Neuron-Like	Elements	Amari, S 72a
...on Convergence Time and Tolerance for Adaptive Threshold	Elements	Mays, CH 64a
...and Associative Memory in Recurrent Networks of Neuron-Like	Elements	Kohonen, T 76b
...Network Models for Computing Using Fine-Grained Processing	Elements	Szu, HH 85a
Homogeneous Nets of Neuron-Like	Elements	Amari, S 75a
Implementations of Adaptive Associative Optical Computing	Elements	Fisher, AD 86a
...Statistical Cooperation of Self-Interested Neuron-Like Computing	Elements	Barto, AG 85b
...Statistical Cooperation of Self-Interested Neuron-Like Computing	Elements	Barto, AG 85c
Learning in Networks of Nondeterministic Adaptive Logic	Elements	Windecker, RC 88a
...and Pattern Sequences by Self-Organizing Nets of Threshold	Elements	Amari, S 72b
Nets of Threshold	Elements	Krohn, K 65a
On Autonomous Logic Nets of Threshold	Elements	Yajima, S 68a
On Properties of Networks of Neuron-Like	Elements	Baldi, P 88h
Simulation Experiments with Goal-Seeking Adaptive	Elements	Barto, AG 84b
Textural Segmentation, Second-Order Statistics, and Textural	Elements	Beck, J 83b
...and Synthesis of Adaptive Systems Which Use Networks of Threshold	Elements	Mattson, RL 62a
...of Symmetric Switching Functions with Linear-Input Logical	Elements	Kautz, WH 61a
...Feature Extraction by a Multi-Layered Network of Analog Threshold	Elements	Fukushima, K 69a
Scanning the Issue: Special Issue on Studies of Neural	Elements and Systems	Weiss, TF 68a
The Simulation of Neural	Elements by Electrical Networks Based on Multi-Aperture Magnetic	Brain, AE 61a
The	Elements of Single Neurons: A Review	Lewis, ER 83a
Textrons, the	Elements of Texture Perception and Their Interactions.	Julesz, B 81a
...and Selection by a Layered Network of Reinforcement Learning	Elements: Some Initial Experiments	Anderson, CW 82a
Neuron-like	Elements That Can Solve Difficult Learning and Control Problems.	Barto, AG 83c
Adaptive Neurons-Like	Elements that Can Solve Difficult Learning Control Problems.	Barto, AG 82a
Neuron-like Adaptive	Elements that Can Solve Difficult Learning Control Problems.	Barto, AG 83b
A Macroscopic Analysis on Large Scale Networks on Neuronlike	Elements with Refractory	Aihara, K 80a
Some Properties of Randomly Connected Networks of Neuron-Like	Elements with Refractory	Yoshizawa, S 74a
On the Nature and	Elimination of Stimulus Artifact in Nerve Signals Evoked and	McGill, KC 82a
Parallel Gaussian	Elimination on an Optically Interconnected Data Flow Computer.	Nelken, I 87a
	Emergence of Grandmother Memory in Feed Forward Networks: Learning	Scalettar, R 88a
	Emergence of Hierarchical Data Structures in Parallel Computation.	Kerszberg, M 86a
On the Spontaneous	Emergence of Neuronal Schemata	Harth, EM 82b
From Basic Network Principles to Neural Architecture:	Emergence of Orientation Columns	Linsker, R 86a
Spontaneous Symmetry-Breaking Energy Functions and the	Emergence of Orientation Selective Cortical Cells.	Kammen, DM 88a
From Basic Network Principles to Neural Architecture:	Emergence of Orientation-Selective Cells	Linsker, R 86b
From Basic Network Principles to Neural Architecture:	Emergence of Spatial-Opponent Cells.	Linsker, R 86c
Neural Networks and Physical Systems with	Emergent Collective Computational Abilities.	Hopfield, JJ 82b

...Neural Systems: Multilayer Perceptron Training Using Optimal	Estimation	Shepanski, JF 88a
Non-Parametric Probability Density	Estimation	Tapia, RA 78a
	Estimation and Choice of Neighbors in Spatial-Interaction Models of.	Kashyap, RL 83a
Detection,	Estimation, and Classification with Spectrograms	Altes, RA 80a
Classification,	Estimation, and Pattern Recognition.	Young, TY 74a
On	Estimation of a Probability Density Function and Mode.	Parzen, E 62a
Neural Network Implementation for Maximum-Likelihood Sequence	Estimation of Binary Signals in Gaussian Noise	Provence, JD 87a
	Estimation of Expert Weights Using Fuzzy Cognitive Maps.	Taber, WR 87a
Adaptive Training of Multilayer Neural Networks Using a Least Squares	Estimation Technique	Kollias, S 88a
Discrete-Time Parameter	Estimation with Genetic Algorithms	Das, R 88a
Unconstrained Minimum Mean-Square Error Parameter	Estimation with Hopfield Networks.	Altes, RA 88a
The Structure of Least Mean Square Linear	Estimators for Synchronous M-ary Signals	Gardner, WA 73a
Nonlinear	Etalons in Competitive Optical Learning Networks	Wagner, K 87b
	European Seminar on Neural Computing	ESNC 88a
The Second	European Seminar on Neural Computing: Commercial Prospects	ESNC 89a
A Pattern Recognition Program That Generates,	Evaluates, and Adjusts to its Own Operators.	Uhr, L 61a
	Evaluating a Connectionist Network as a Simulation of Speech	Schwartz, RG 88a
	Evaluating the Interface of a Document Processor: A Comparison of.	Hammond, N 84a
Effects of Reinforcement on Knowledge Retrieval and	Evaluation	Leven, SJ 87a
A Pattern Classification Approach to	Evaluation Function Learning	Lee, K 86a
	Evaluation of a Hypercube Based Compact Neural Network	Somani, AK 88a
An	Evaluation of Cell Function in Relation to Reverberatory Modes in.	Fairhurst, MC 80a
Simulation and	Evaluation of Ill Defined Neuronal Networks with Random Structure.	Benito, S 80a
	Evaluation of Network Architectures on Test Learning Tasks	Soulie, FF 87a
A Computer Package for the	Evaluation of Neuron Models Involving Large Uniform Networks	Lee, MH 82a
	Evaluation of the Use of the Hopfield Neural Network Model as a.	Montgomery, BL 86a
Linguistic Descriptions of Visual	Event Perceptions.	Maddox, AB 87a
Computer Simulation of	Event Related Potentials During Voluntary Cognition.	Deshmukh, VD 88b
Back-Propagation Networks for	Event-Driven Temporal Sequence Processing.	Shimohara, K 88b
Probing Cognitive Processes Through the Structure of	Event-Related Potentials During Learning: An Experimental and.	Banquet, J 87a
Structure of	Event-Related Potentials During Learning: An Experimental and.	Banquet, J 86a
...A Key to the Understanding of the Combined Dynamics of EEG and	Event-Related Potentials" by Elbert and Rockstroh.	Palm, G 87b
The Problem of Sensing and the Neural Coding of Sensory	Events	Mountcastle, VB 67a
Processing of Expected and Unexpected	Events During Conditioning and Attention: A Psychophysiological.	Grossberg, S 86n
The Processing of Expected and Unexpected	Events During Conditioning and Attention: A Psychophysiological.	Grossberg, S 82d
Representation of	Events in Nerve Nets and Finite Automata	Kleene, SC 56a
General Model for the Molecular	Events in Synapses During Learning	Huttunen, MO 73a
	Event-Train Correlation and Real-Time Microcomputer Systems.	Soucek, B 83a
Expert Focus, Neural Networks, Part 2: What are They and Why Is	Everybody So Interested in Them Now?	Wasserman, PD 88a
IBM Neural Network Chips Make	Everything Obsolete.	Gantz, J 88a
...that Are Competitive or Cooperative II: Convergence Almost	Everywhere	Hirsch, MW 85a
Semantic Priming and Retrieval from Lexical Memory:	Evidence for Facilitatory and Inhibitory Processes	Neely, JH 76a
Morphological	Evidence for Parallel Processing of Information in Rat Macula.	Ross, MD 88b
	Evidence of Chaotic Dynamics of Brain Activity During the Sleep.	Babloyantz, A 85a
Bayesian Networks: A Model of Self-Activated Memory for	Evidential Reasoning	Pearl, J 85a
Real-Time	Evidential Reasoning and Network Based Processing.	Green, PE 87a
On	Evidential Reasoning in a Hierarchy of Hypotheses.	Pearl, J 86d
On	Evidential Reasoning in a Hierarchy of Hypotheses.	Pearl, J 86c
	Evidential Reasoning in Semantic Networks: A Formal Theory	Shastri, L 85a
	Evidential Reasoning in Semantic Networks: A Formal Theory and its	Shastri, L 85b
On the Nature and Elimination of Stimulus Artifact in Nerve Signals	Evoked and Recorded Using Surface Electrodes	McGill, KC 82a
...of Category Learning and Recognition: Structural Invariants,	Evoked Potentials, and Reinforcement	Carpenter, GA 86k
	Evoked Potentials in Artificial Neural Nets.	Anninos, PA 73a
The Neural Generators of the Vestibular	Evoked Response.	Elidan, J 87a
...in a Central Neuron Match Predictions from Binomial Analysis of	Evoked Responses	Korn, H 87a
...Investigation of Process Identification by Competitive	Evolution.	Kaufman, H 67a
Boolean Systems, Adaptive Automata,	Evolution.	Kaufman, SA 86a
How Learning Can Guide	Evolution.	Hinton, GE 87b
Exaptation, Preadaptation,	Evolution and Biological Neural Networks	Stork, DG 87b
Darwinian	Evolution as a Paradigm for AI Research.	Fenanzo Jr, AJ 86a
Self Organizing Mathematical Models: Non-Linear	Evolution Equations with a Convolution Term.	Schatzman, M 86a
The Eye-Movement Control System:	Evolution of a Model	Fender, DH 62a
Concept Formation of Neural Networks: Implications for	Evolution of Cognitive Functions	Anderson, JA 88a
	Evolution of Cognitive Structure and Processes	Hayes-Roth, B 77a
The	Evolution of Data Processing Abilities in Competing Automata	Bergman, A 86a
Symbolic Schemata, Role Binding, and the	Evolution of Structure in Connectionist Memories	Dolan, CP 87a
	Evolution of the Adaptive Landscape.	Conrad, M 78a
Intraneuronal Dynamics as a Substrate for	Evolutionary Learning.	Kirby, KG 86a
Neuronal Dynamics and	Evolutionary Learning.	Conrad, M 88b

Computational Modeling of	Evolutionary Learning Processes in the Brain	Kampfner, RR 83a
	Evolutionary Pattern Recognition in Incomplete Nonlinear	Mucciardi, AN 66a
An	Evolutionary Pattern Recognition Network	Klopf, AH 69a
Prolegomena to	Evolutionary Programming	Conrad, M 88c
	Exact Results for the Average Dynamic Behavior of some Non-Linear.	Rossler, J 87a
	Exact Solutions for Feedforward Network Parameters	Atkinson, RC 88b
An	Exactly Solvable Asymmetric Neural Network Model	Derrida, B 87a
An	Exactly Solvable Model for Competitive Networks.	Bennett, A 88a
On the Time Relations of Mental Processes: An	Examination of Systems of Processes in Cascade	McClelland, JL 79a
Empirical	Examination of the Threshold Model of Neuron Firing.	Brillinger, DR 79a
RAMBOT: A Connectionist Expert System that Learns by	Example.	Mozer, MC 87a
RAMBOT: A Connectionist Expert System that Learns by	Example.	Mozer, MC 86b
The Chess Machine: An	Example of Dealing with a Complex Task by Adaptation	Newell, A 55a
Image Compression by Back Propagation: An	Example of Extensional Programming	Cottrell, GW 87a
Learning Internal Representations from Gray-Scale Images: An	Example of Extensional Programming	Cottrell, GW 87b
	Example-Based Knowledge Engineering with Connectionist Expert.	Gallant, SI 88c
Building a Hierarchy with Neural Networks: An	Example-Image Vector Quantization.	Jackel, LD 87a
A Game-Playing Program that Learns by Analyzing	Examples	Minton, S 85a
Automatic Generation of Expert Systems from	Examples	Gallant, SI 85a
Learning a Color Algorithm from	Examples	Poggio, TA 88a
Learning Structural Descriptions from	Examples	Winston, PH 75a
Neurocomputing Applications	Examples	Hecht-Nielsen, R 87i
Synthesizing a Color Algorithm from	Examples	Hurlbert, AC 88a
Attentional Mechanisms in Neural Information Processing:	Examples From Pavlovian Conditioning	Grossberg, S 87f
Concurrent Neural Network Simulation: Two	Examples within a Single, Integrated Neural Network Hardware	Bahr, C 88c
Neural Networks and Principal Component Analysis: Learning from	Examples Without Local Minima.	Baldi, P 88e
	Exaptation, Preadaptation, Evolution and Biological Neural Networks.	Stork, DG 87b
Supervised Learning and Systems with	Excess Degrees of Freedom.	Jordan, MI 88b
A Mathematical Analysis of	Excitable Membrane Phenomena	Carpenter, GA 78a
Two-Terminal Electronic Circuit Neuron Model with	Excitable Membrane V-I-t Characteristics	Hoshimiya, N 79a
Automatic Computation of Nerve	Excitation	Cole, KS 55a
Physico-Mathematical Aspects of	Excitation and Conduction in Nerves.	Rashevsky, N 36a
Mathematical Models of	Excitation and Propagation in Nerve.	FitzHugh, R 69a
Automatic Computation of Nerve	Excitation - Detailed Corrections and Additions.	FitzHugh, R 59a
A Biological Model of the	Excitation of a Second Order Sensory Neurone	Enger, PS 69a
Application of Bifurcation Theory to Nerve	Excitations.	Toko, K 81a
Existence and Stability of Local	Excitations in Neural Fields	Kishimoto, K 79a
Backward Conditioning: A Neural Network Model which Exhibits Both	Excitatory and Inhibitory Conditioning	Ricart, R 88a
Coupled Van der Pol Oscillators - A Model of	Excitatory and Inhibitory Neural Interactions.	Kawahara, T 80a
...Patterns by Networks with Ordered Sensory and Motor Components:	Excitatory Components of the Cerebellum.	Grossberg, S 69e
...Interactive Neural Populations-II: Stability and Dynamics of	Excitatory Sets.	Brannan, JR 82a
Optically	Excited Synapse for Neural Networks.	Boyd, GD 87a
	EXCLUSIVE OR Processing (Binary Image Subtraction) Using Thick	Guest, CC 84a
On Parallel	Execution of the Traveling Salesman Problem on a Neural Network.	Bagherzadeh, N 87a
...Forgetting of Prototypes and Old Instances: Simulation by an	Exemplar-Based Classification Model.	Hintzman, DL 80a
Limitations of	Exemplar-Based Generalization and the Abstraction of Categorical	Homa, D 81a
On Choosing Non-Random Input	Exemplars to Form Stable States in a Hopfield Neural Net	Feild Jr, WB 88a
Explaining PDP: A Handbook of Models, Programs and	Exercises.	McClelland, JL 87a
...Distributed Processing: A Handbook of Models, Programs, and	Exercises.	McClelland, JL 88a
From Chemotaxis to Cooperativity: Abstract	Exercises in Neuronal Learning Strategies.	Barto, AG 88a
	Exhaustive Thermodynamical Analysis of Boolean Learning Networks	Carnevali, P 87a
Noisy Neural Nets	Exhibiting Epileptic Features.	Kokkinidis, M 85a
Noisy Neural Nets	Exhibiting Memory Domains.	Anninos, PA 84c
Caging and	Exhibiting Ultrametric Structures.	Baldi, P 86b
Backward Conditioning: A Neural Network Model which	Exhibits Both Excitatory and Inhibitory Conditioning	Ricart, R 88a
	Existence and Stability of Local Excitations in Neural Fields.	Kishimoto, K 79a
	Existence of a Limiting Pattern for a System of Non-Linear	Levine, DS 79a
Kolmogorov's Mapping Neural Network	Existence Theorem.	Hecht-Nielsen, R 87e
There	Exists a Neural Network That Does Not Make Avoidable Mistakes.	Gallant, AR 88a
Perceptrons:	Expanded Edition	Minsky, ML 88a
Dynamically	Expanding Context.	Kohonen, T 87d
Self-Learning Rules by Dynamically	Expanding Context.	Kohonen, T 87f
Dynamically	Expanding Context, with Application to the Correction of Symbol.	Kohonen, T 86a
...Networks: Line Neutralization, Tilt After Effect, and Angle	Expansion.	Levine, DS 76a
	Expectancy as a Determinant of Functional Units in Perceptual.	Aderman, D 71a
Toward A Modern Theory of Adaptive Networks:	Expectation and Prediction	Sutton, RS 81b
	Expectation and the Perception of Color.	Bruner, JS 51a
Neural	Expectation: Cerebellar and Retinal Analogs of Cells Fired by.	Grossberg, S 72c
...Classification and Universal Recording, Part II: Feedback,	Expectation, Olfaction, and Illusions.	Grossberg, S 76c

...as a Guide to the Identification of Solution Strategy from	Eye-Movement Recordings.	Deffner, G 87a
Derivation of a Model for the Human	Eye-Positioning Mechanism.	Cook, G 67a
...Model for Pattern Learning and Its Application to Robot	Eyesight	Suzuki, H 88a
Self-Organizing Model of Learning and Its Application to	Eyesight System of Robot	Suzuki, H 88b
...and Interface Circuits by Silicon Planar and Three-Dimensional	Fabrication Technology	Takahashi, K 84a
Hierarchical Scene Structure Representations to	Facilitate Image Understanding	Maren, AJ 88b
Semantic Priming and Retrieval from Lexical Memory: Evidence for	Facilitatory and Inhibitory Processes.	Neely, JH 76a
Modern	Factor Analysis.	Harman, HH 76a
Relation of Olfactory EEG to Behavior:	Factor Analysis.	Freeman, WJ 87b
Introduction to	Factor Analysis: What It Is and How to Do It	Kim, J 78a
How Neural Networks	Factor Problems of Sensory-Motor Control	Bullock, D 88a
A	Factored Architecture to Solve Illusory Conjunctions in Neural	Strong, GW 87a
Structural	Factors in Figure Perception	McClelland, JL 79b
On Visual and Contextual	Factors in Reading: A Reply to Rayner and Slowiaczek	McClelland, JL 81d
Visual	Factors in Word Perception	Johnston, JC 73a
	Factors Influencing Learning by Backpropagation.	Von Lehmen, A 88a
State-Dependent	Factors Influencing Neural Plasticity: A Partial Account of the.	Munro, PW 86b
...Adaptive Systems: A Statistical/Numerical Approach to	Factory Automation and Brain Research.	Werbos, PJ 87a
Brain Metaphors, Theories, and	Facts.	Grossberg, S 86e
	Facts on Optic Flow.	Koenderink, JJ 87c
Committing	Facts to Long Term Memory.	Becker, LA 88a
	Factual and Counterfactual Reasoning by Constructing Plausible	Derthick, MA 87c
Graphical Folding and Distributed	Failures in Large Network Simulations.	Dayhoff, JE 87b
Automated	Fake Color Separation: Combining Computer Vision and Computer.	Walters, DKW 87a
The Role of	Familiar Units in Perception of Words and Nonwords	McClelland, JL 77b
Perceptual Interactions in Two-Word Displays:	Familiarity and Similarity Effects	McClelland, JL 86h
	Familiarity of Letter Sequences and Tachistoscopic Identification.	Miller, GA 54a
The Least Mean Fourth (LMF) Adaptive Algorithm and Its	Family	Walach, E 84a
Group Actions and Learning for a	Family of Automata	Baldi, P 88b
	Farey-Fraction Frequency Modulation in the Neuronlike Output of.	Coon, DD 87a
...of Second-Order Vestibular Neurons in Correlation with the	Far-Field Recorded Vestibular-Evoked Response.	Elidan, J 89a
How to Label Nerve Cells so that they can Interconnect in an Ordered	Fashion.	von der Malsburg, C 77a
	Fast Adaptive Formation of Orthogonalizing and Associative Memory.	Kohonen, T 76b
	Fast Algorithm for Recognizing Nearest Neighbors	Sethi, IK 81a
	Fast and Compact Associative Memory Cell	Chang, H 80a
A Very	Fast Associative Method for the Recognition and Correction of.	Kohonen, T 78a
	Fast Component Threshold for Vestibular Nystagmus in the Rabbit.	Lau, CGY 87a
	Fast Connectionist Learning: Words and Case.	Sharkey, NE 89b
	Fast Edge-Only Matching Techniques for Robot Pattern Recognition	Caelli, TM 87a
Spatial Patterns of Visual Cortical	Fast EEG During Conditioned Reflex in a Rhesus Monkey.	Freeman, WJ 87e
A Neural Network Simulation Method Using the	Fast Fourier Transform	Barto, AG 76a
	Fast Learning in Artificial Neural Systems: Multilayer Perceptron.	Shepanski, JF 88a
	Fast Learning in Networks of Locally-Tuned Processing Units.	Moody, J 88a
Architecture for	Fast Learning of Decision Regions in Neural Networks	Hayduk, JW 88a
Artificial Intelligence Applications of	Fast Optical Memory Access	Henshaw, PD 86a
	Fast Simulated Annealing	Szu, HH 87a
	Fast Simulated Annealing	Szu, HH 86c
Nonconvex Optimization by	Fast Simulated Annealing	Szu, HH 87e
Nonconvex Optimization by	Fast Simulated Annealing	Szu, HH 87f
...Neural Nets: Architecture and Methodologies for Implementing	Fast Stochastic Learning by Simulated Annealing.	Farhat, NH 87b
Using	Fast Weights to Deblur Old Memories.	Hinton, GE 87d
Back-Propagation is Significantly	Faster if the Expected Value of the Source Unit is Used for Update	Samad, T 88a
	Faster Learning Through a Probabilistic Approximation Algorithm.	Kolen, JF 88a
	Fastest Multipliers Zip Along at 210ps	Brown, C 87d
The	Fastest Software-only Neural Net?	Johnson, RC 88m
	Fat-Trees: Universal Networks for Hardware Efficient Supercomputing.	Leiserson, CE 85a
Using the Hyperbase Intelligent Processor for	Fault Analysis and Detection	Tenorio, MF 87i
Bayesian Assessment of a Connectionist Model for	Fault Detection.	Gallant, SI 88a
Neural Network Models Applied to Real-Time	Fault Diagnosis.	Dietz, WE 89a
Qualitative and Temporal Reasoning in Engine Behavior Analysis and	Fault Diagnosis.	Dietz, WE 87a
Pattern-Based	Fault Diagnosis Using Neural Networks.	Dietz, WE 88a
	Fault Simulation of a Wafer-Scale Integrated Neural Network.	May, N 88a
Conventional	Fault-tolerance and Neural Computers	Moore, WR 88a
	Fault-tolerance in Imaging-oriented Systolic Arrays.	Negrini, R 88a
Distributed and	Fault-Tolerant Computation for Retrieval Tasks Using Distributed	Char, JM 88a
	Fault-Tolerant Recognition Using DAM's	Wechsler, H 87a
	Faust, Mephistopheles and Computer	Zuse, K 88a
Detection of	Favored Patterns in the Temporal Structure of Nerve Cell	Dayhoff, JE 87a
Theory and	Feasibility of the Parametric Avalanche for Advanced Electronic.	Dawes, RL 88b

Learning Algorithms and Probability Distributions in Feed-Forward and	Feed-Back Networks	Hopfield, JJ 87a
Positive	Feedback Processes Underlying Functional Differentiation	Gaines, BR 87a
Effect of Clock Skew in Iterative Neural Network and Optical	Feedback Processors.	Oh, S 88a
Bifurcation and Chaotic Behaviour in Simple	Feedback Systems	Sparrow, C 80a
Self-Organizing Neural Network with a Function of Associative Memory:	Feedback Type Cognitron.	Fukushima, K 78a
...Neural Network with a Function of Associative Memory -	Feedback Type Cognitron -.	Fukushima, K 77a
All-Optical Associative Holographic Memory with	Feedback Using Phase Conjugate Mirrors	Dunning, GJ 85a
Associative Holographic Memory with	Feedback Using Phase-Conjugate Mirrors	Soffer, BH 86a
	Feedback-Induced Sequentiality in Neural Networks.	Norrod, FE 87a
Learning Algorithms and Probability Distributions in	Feed-Forward and Feed-Back Networks.	Hopfield, JJ 87a
Neural Network Based	Feed-Forward High Density Associative Memory	Daud, T 87a
Pattern Separating Functioning of Two Layered Random Nerve Nets with	Feedforward Inhibitory Connections	Torioka, T 88a
Some Asymptotic Results for Learning in Single Hidden Layer	Feedforward Network Models	White, H 87b
Exact Solutions for	Feedforward Network Parameters	Atkinson, RC 88b
A Methodology to Select Appropriate Learning Rate Parameters in	Feed-Forward Networks.	Pabon, JA 87a
Experience vs. Architecture in	Feedforward Networks: A Case Study	Atkinson, RC 88c
Multi-Layer	Feedforward Networks Are Universal Approximators	Hornik, K 88a
A Comparison of	Feedforward Neural Networks: Fuzzy Operators and Acoustic Ship . . .	Taber, WR 88a
Design of a	Feeling-Thinking Machine	Scanlon, RD 88a
Experiences with a	Feeling-Thinking Machine	Johnson, M 87a
Noise Reduction in	Fetal Electrocardiography.	Hon, E 63a
The	FFT: Fundamentals and Concepts	Ramirez, RW 85a
The Electrophysics of a Nerve	Fiber.	Scott, AC 75a
Nonlinear Signal Processing Using	Fiber-Optics Neurograms.	Szu, HH 86f
Simulating the Adaptation Phenomena in the Response of Auditory Nerve	Fibers: A Functional Model	Wu, ZL 88a
Differential Equations for the Development of Topological Nerve	Fibre Projections.	von der Malsburg, C 81a
	Fibroblasts, Morphogenesis and Cellular Automata	Bouligand, Y 86a
A Markov Random	Field Approach to Data Fusion and Colour Segmentation.	Wright, WA 89a
A	Field Approach to Higher Order ANNs for Sensorial Representation . .	Van Hulle, MM 88a
Operational Significance of Receptive	Field Assemblies	Koenderink, JJ 88b
The Description of a Random	Field by Means of Conditional Probabilities and Conditions of Its. .	Dobruschin, PL 68a
	Field Computation: A Model of Massively Parallel Computation in. . .	MacLennan, BJ 88a
Technology-Independent Design of Neurocomputers: The Universal	Field Computer	MacLennan, BJ 87a
A Field Theory for Neural Nets, Part I: Derivation of the	Field Equations.	Griffith, JS 63a
A Field Theory for Neural Nets, Part II: Properties of the	Field Equations.	Griffith, JS 65a
...Information-Triggered Segments of Series of Human Brain Electric	Field Maps	Lehmann, D 88a
Value Unit Interpolation: A Receptive	Field Model.	Ballard, DH 86c
Image Segmentation Using Simple Markov	Field Models	Hansen, FR 82a
Markov Random	Field Texture Models	Cross, GR 83a
A	Field Theory for Neural Nets, Part I: Derivation of the Field. . . .	Griffith, JS 63a
A	Field Theory for Neural Nets, Part II: Properties of the Field . . .	Griffith, JS 65a
A Mean	Field Theory of Layer IV of Visual Cortex and Its Application to . .	Scofield, CL 88a
	Field Theory of Self-Organizing Neural Networks.	Amari, S 83a
ALOPEX: A Stochastic Method for Determining Visual Receptive	Fields	Harth, EM 74a
An Associative Memory with Heterogeneous Input	Fields	Winter, CL 88b
An Introduction to Spatial Point Processes and Markov Random	Fields	Isham, V 81a
Dynamics of Pattern Formation in Lateral-Inhibition Type Neural	Fields	Amari, S 77b
Existence and Stability of Local Excitations in Neural	Fields	Kishimoto, K 79a
Formation of Topographic Maps and Columnar Microstructures in Nerve	Fields	Takeuchi, A 79a
Integration of Stereo Vision and Optical Flow Using Markov Random	Fields	Clifford, SP 88a
Learning Receptive	Fields	Barrow, HG 87a
Learning with Localized Receptive	Fields	Moody, J 88b
...Neural Function Arises from Both Receptive and Projective	Fields	Lehky, SR 88a
On the Topological Representation of Signals in Self-Organizing Nerve	Fields	Amari, S 87b
Recursive Features of Circular Receptive	Fields	Hartmann, G 82a
Topographic Organization of Nerve	Fields	Amari, S 80a
Two-Dimensional Discrete Markovian	Fields	Woods, JW 72a
Spatiotemporal Receptive	Fields: A Dynamical Model Derived from Cortical Architectonics . . .	Krone, G 86a
Masking	Fields: A Massively Parallel Architecture for Learning, Recognizing.	Cohen, MA 87a
Masking	Fields: A Massively Parallel Neural Architecture for Learning. . . .	Cohen, MA 87b
Enabling	Fields: A Theory of Learning with Physiological Implications	Grossberg, S 69a
Markov Random	Fields and Gibbs Ensembles	Spitzer, F 71a
Random	Fields and Spatial Renewal Potentials.	Demongeot, J 86a
The Use of Markov Random	Fields as Models of Texture.	Hassner, M 80a
A Model for the Temporal Organization of X- and Y-Type Receptive	Fields in the Primate Retina	Richter, J 82a
Human Brain EEG	Fields: Micro-States and Their Functional Significance	Lehmann, D 87b
Rules for Forming Receptive	Fields of Formal Neurons During Unsupervised Learning Processes. . .	Bobrowski, L 82a
The Structure of Two-Dimensional Scalar	Fields with Applications to Vision	Koenderink, JJ 79a
Separating	Figure from Ground with a Boltzmann Machine.	Sejnowski, TJ 87g

	Four Frames Do Not Suffice	Grossberg, S 85b
	Four Frames Suffice.	Feldman, JA 82d
	Four Frames Suffice: A Provisional Model of Vision and Space	Feldman, JA 85c
A	Four Mechanism Model for Threshold Spatial Vision.	Wilson, HR 79a
	Four Neural Circuit Models and Their Role in the Organization of	Hemami, H 83a
A Novel View of	Fourier Analysis	Overman, KC 81a
EXCLUSIVE OR Processing (Binary Image Subtraction) Using Thick	Fourier Holograms.	Guest, CC 84a
The Response of the Limulus Retina to Moving Stimuli: A Prediction by	Fourier Synthesis.	Brodie, SE 78a
A Neural Network Simulation Method Using the Fast	Fourier Transform.	Barto, AG 76a
The	Fourier Transform and its Applications	Bracewell, R 78a
Pattern Recognition of	Fourier Transform Infared Spectra of Organic Compounds	Frankel, DS 84a
Signal Reconstruction from	Fourier Transform Sign Information	Curtis, SR 85a
A Neural Net Approach to Discrete	Fourier Transforms	Culhane, AD 88a
A	Fourier-Based Textual Feature Extraction Procedure	Stromberg, W 86a
	Fourier-Mellin Transform and Mammalian Hearing	Altes, RA 78a
Analysis of a	Four-Layer Series-Coupled Perceptron, II	Block, HD 62a
The Least Mean	Fourth (LMF) Adaptive Algorithm and Its Family	Walach, E 84a
Influence of Contrast on	Foveal and Peripheral Detection of Coherent Motion in Moving Random.	van de Grind, WA 87a
Neural Control of	Foveal Pursuit Versus Saccadic Eye Movements in Primates - Single.	Eckmiller, R 83a
Maximum Entropy Formalism,	Fractals, Scaling Phenomena, and 1/f Noise: A Tale of Tails.	Montroll, EW 83a
Synthesis of	Fragmented Data on Neuronal Systems: A Computer Model of Cerebellum.	Pellionisz, AJ 78a
...Approach Towards Solving the Problem of Procedural Attachment in	Frame Theories	Hewitt, C 75a
Four	Frames Do Not Suffice.	Grossberg, S 85b
Task	Frames in Robot Manipulation	Ballard, DH 84c
A Parallel Computation that Assigns Canonical Object-Based	Frames of Reference.	Hinton, GE 81a
Four	Frames Suffice	Feldman, JA 82d
Three	Frames Suffice	Hinton, GE 85c
Four	Frames Suffice: A Provisional Model of Vision and Space.	Feldman, JA 85c
A	Framework for Adaptive Routing	Ngai, JY 87a
Proteus: An Activation	Framework for Cognitive Process Models	Levin, JA 76a
A Unified	Framework for Connectionist Systems.	Golden, RM 88b
General	Framework for Parallel Distributed Processing.	Rumelhart, DE 86b
A	Framework for Representing Knowledge	Minsky, ML 75a
A	Framework for Speech Recognition Using Neural Networks	Lippmann, RP 87a
Harmony Theory: A Mathematical	Framework for Stochastic Parallel Processing	Smolensky, P 83a
The Right of	Free Association: Relative-Position Encoding for Connectionist Data.	Barnden, JA 88b
Supervised Learning and Systems with Excess Degrees of	Freedom.	Jordan, MI 88b
Nestor	Frees Neural Nets from Hardware Chains	Johnson, RC 88c
Detection of Grating Patterns Containing Two Spatial	Frequencies: A Comparison of Single-Channel and Multiple-Channels.	Graham, N 71a
...of Orthographic Structure Defined by Single-Letter Positional	Frequency.	Mason, M 75a
A Synaptic Model for Spatial	Frequency Adaptation	Wilson, HR 75a
Spatial	Frequency Adaptation Under a Slit-Vision Condition	Hirai, Y 86b
Uncertainty Relation for Resolution in Space, Spatial	Frequency, and Orientation Optimized by Two-Dimensional Visual	Daugman, JG 85a
Probability,	Frequency, and Reasonable Expectations	Cox, RT 46a
Orthographic Regularity, Positional	Frequency, and Visual Processing of Letter Strings	Massaro, DW 79a
...Information About Position, Contrast, Orientation, Spatial	Frequency, Disparity	Grossberg, S 87a
Adaptive Filtering in the	Frequency Domain	Dentino, M 78a
Farey-Fraction	Frequency Modulation in the Neuronlike Output of Silicon p-i-n	Coon, DD 87a
Stochastic Pulse	Frequency Modulation Models of Neural Coding Processes	Bruckstein, AM 81a
Gabor:	Frequency, Time and Memory	Korpel, A 82a
Using	Frequency-and Orientation-Tuned Channels to Determine Surface Slant.	Kube, P 86a
	Frequency-Coded Artificial Neural Networks: An Approach to Self.	Dress, WB 87d
A	Frequency-Domain Adaptive Pole-Zero Filter with Applications	Shynk, JJ 85a
Modeling Neural Mechanisms of Visual-Motor Coordination in the	Frog and Toad.	Arbib, MA 82a
A Neural Network Model of	Frog Retina: A Discrete Time-Space Approach.	Lee, Y 86a
What the Frog's Eye Tells the	Frog's Brain	Lettvin, JY 59a
What the	Frog's Eye Tells the Frog's Brain.	Lettvin, JY 59a
On-Line Recognition of Spoken Words	from a Large Vocabulary.	Kohonen, T 84a
The QBKG System: Generating Explanations	from a Non-Discrete Knowledge Representation	Berliner, HJ 82a
Stable Matching with Preferences Derived	From a Psychological Model	Bartholdi, JJ 86b
Experimental Results	from a Simple Neural Network	White, HJ 88b
Neuroelectric Potentials Derived	from an Extended Version of the Hodgkin-Huxley Model	Lewis, ER 66a
Cortical Encoding of Memory: Hypotheses Derived	from Analysis and Simulation of Physiological Learning Rules in.	Lynch, G 89a
	From Basic Network Principles to Neural Architecture: Emergence of	Linsker, R 86a
	From Basic Network Principles to Neural Architecture: Emergence of	Linsker, R 86b
	From Basic Network Principles to Neural Architecture: Emergence of	Linsker, R 86c
Spontaneous Quantal Currents in a Central Neuron Match Predictions	from Binomial Analysis of Evoked Responses	Korn, H 87a
Network Model of Shape-From-Shading: Neural Function Arises	from Both Receptive and Projective Fields.	Lehky, SR 88a
	From Chemotaxis to Cooperativity: Abstract Exercises in Neuronal	Barto, AG 88a

Neural Models: Networks for	Fun and Profit	Anderson, JA 86d
A Self-Organizing Multi-Layered Neural Network Realizing a	Funciton in the Human Recognition and Memory Mechanisms	Mizuno, H 80a
A Learning Machine with New Structure and	Function	Nagano, T 73a
A Neural Network Theory of Frontal Lobe	Function	Levine, DS 86a
A Systems Theoretic Approach to the Study of CNS	Function	Sclabassi, RJ 88a
A Theory of Cerebellar	Function	Albus, JS 71a
An Organizing Principle for Cerebral	Function	Mountcastle, VB 79a
Cortical Connections and Parallel Processing: Structure and	Function	Ballard, DH 86a
Cortical Connections and Parallel Processing: Structure and	Function	Ballard, DH 87a
Model for Cortical	Function	Colvin, ME 86a
Speech Processing by Splicing of Autocorrelation	Function	Suzuki, J 76a
Speech Processing System by the use of Short-Time Cross-Correlation	Function	Suzuki, J 77a
The Correlation Theory of Brain	Function	von der Malsburg, C 81b
Two New Theories of Retinal	Function	Johnson, K 87a
Reconciling Stable Asymmetry with Recovery of	Function: An Adaptive Systems Perspective on Functional Plasticity	Bullock, D 87b
Drive-Reinforcement Model of Single Neuron	Function: An Alternate to the Hebbian Neuronal Model	Klopf, AH 86a
Brain	Function and Adaptive Systems: A Heterostatic Theory	Klopf, AH 72a
On Estimation of a Probability Density	Function and Mode	Parzen, E 62a
A Theory of Memory that Explains the	Function and Structure of the Cerebellum	Gilbert, PFC 74a
Network Model of Shape-From-Shading: Neural	Function Arises from Both Receptive and Projective Fields	Lehky, SR 88a
Cooperativity in Brain	Function: Assemblies of Approximately 30 Neurons	Shaw, GL 82a
Lyapunov	Function for Parallel Neural Networks	Goles, E 86b
A Trainable Non-Linear	Function Generator	Smith, FW 66a
An Evaluation of Cell	Function in Relation to Reverberatory Modes in Cellular Networks	Fairhurst, MC 80a
A Self-Organizing Multi-Layered Neural Network Realizing a	Function in the Human Recognition and Memory Mechanisms	Mizuno, H 78a
The T5 Base Modulator Hypothesis: A Dynamic Model of T5 Neuron	Function in Toads	Betts, B 88a
A Pattern Classification Approach to Evaluation	Function Learning	Lee, K 86a
Influence of Noise on the	Function of a "Physiological" Neural Network	Buhmann, J 87a
Self-Organizing Neural Network with a	Function of Associative Memory: Feedback Type Cognitron	Fukushima, K 78a
Self-Organizing Neural Network with a	Function of Associative Memory - Feedback Type Cognitron -	Fukushima, K 77a
The	Function of Dream Sleep	Crick, F 83a
Problem Solving in a Natural Task as a	Function of Experience	Lancaster, JS 87a
A Biophysical Approach to the Spatial	Function of Eye Movements, Extraocular Proprioception and the	Daunicht, WJ 88a
Perceptual Recognition as a	Function of Meaningfulness of Stimulus Material	Reicher, GM 69a
Prototype Abstraction and Classification of New Instances as a	Function of the Number of Instances Defining the Prototype	Homa, D 73a
	Function of the Thalamic Reticular Complex: The Searchlight	Crick, F 84a
Neural Network Processing as a Tool for	Function Optimization	Jeffrey, W 86a
Some Notes on the Probabilistic Semantics of Logistic	Function Parameters in Neural Networks	Davies, TR 88a
Tensor Theory of Brain	Function: The Cerebellum as a Space-Time Metric	Pellionisz, AJ 82b
A Simplified Model of Retinal	Function with Parallel Architecture and Integrated Space-Time	Johnson, K 86a
Computational Anatomy and	Functional Architecture of Striate Cortex: A Spatial Mapping	Schwartz, EL 80a
Microcircuitry of the Dark-Adapted Cat Retina:	Functional Architecture of the Rod-Cone Network	Smith, RG 86a
Positive Feedback Processes Underlying	Functional Differentiation	Gaines, BR 87a
A Mathematical Theory of the	Functional Dynamics of Cortical and Thalamic Nervous Tissue	Wilson, HR 73a
A Computer Simulation of Olfactory Cortex with	Functional Implications for Storage and Retrieval of Olfactory	Wilson, MA 87a
A Computer Simulation of Olfactory Cortex with	Functional Implications for Storage and Retrieval of Olfactory	Bower, JM 88a
Statistical Analysis and	Functional Interpretation of Neuronal Spike Data	Moore, GP 66a
Characteristics of the	Functional Link Net: A Higher Order Delta Rule Net	Klassen, M 88a
...Adaptation Phenomena in the Response of Auditory Nerve Fibers: A	Functional Model	Wu, ZL 88a
A	Functional Model of Vision and Space	Feldman, JA 87a
...III. The Influence of Environmental Constraints on the Resulting	Functional Order	Toet, A 87c
The Construction of a Simultaneous	Functional Order in Nervous Systems I. Relevance of Signal	Toet, A 87a
The Construction of a Simultaneous	Functional Order in Nervous Systems II. Computing Geometrical	Toet, A 87b
The Construction of a Simultaneous	Functional Order in Nervous Systems III. The Influence of	Toet, A 87c
	Functional Organization in Random Networks	Beurle, RL 62a
...with Recovery of Function: An Adaptive Systems Perspective on	Functional Plasticity	Bullock, D 87b
	Functional Properties of Models for Direction Selectivity in the	Grzywacz, N 87a
A Neural Network Model of Pain Mechanisms:	Functional Properties of the Network Cells Providing the Power Law	Minamitani, H 82a
Human Brain EEG Fields: Micro-States and Their	Functional Significance	Lehmann, D 87b
anaLOG: A	Functional Simulator for VLSI Neural Systems	Lazzaro, JP 86a
Expectancy as a Determinant of	Functional Units in Perceptual Recognition	Aderman, D 71a
Global Ratio Limit Theorems for Some Nonlinear	Functional-Differential Equations, I	Grossberg, S 68c
A Prediction Theory for Some Nonlinear	Functional-Differential Equations I. Learning of Lists	Grossberg, S 68a
Global Ratio Limit Theorems for Some Nonlinear	Functional-Differential Equations, II	Grossberg, S 68d
A Prediction Theory for Some Nonlinear	Functional-Differential Equations. II. Learning of Patterns	Grossberg, S 68b
Pattern Learning by	Functional-Differential Neural Networks with Arbitrary Path Weights	Grossberg, S 72d
Learning and Energy-Entropy Dependence in Some Nonlinear	Functional-Differential Systems	Grossberg, S 69b
	Functionally-Accurate, Cooperative Distributed Systems	Lesser, VR 81a
A Digital-Computer Model of Nerve Cell	Functioning	Perkel, DH 64a

Context	Keyword	Reference
The Perceptron: A Model for Brain	Functioning. I	Block, HD 62b
Pattern Separating	Functioning of Two Layered Random Nerve Nets with Feedforward.	Torioka, T 88a
Theoretical Study on Cantor	Function-Like Response Characteristics of a Neuron Model	Osada, H 81a
A Closed Set of Normal Orthogonal	Functions.	Walsh, JL 23a
A Connectionist Model for Simple Visual	Functions.	Mel, BW 86b
A Memory Storage Model Utilizing Spatial Correlation	Functions.	Anderson, JA 68a
A Model for Neurophysiological	Functions.	Zemanek, H 60a
...of Brownian Motion, and a General Theory of Gaussian Random	Functions.	Levy, P 56a
Complexity of Connectionist Learning with Various Node	Functions.	Judd, JS 87a
...of Neural Networks: Implications for Evolution of Cognitive	Functions.	Anderson, JA 88a
Harmonic Analysis of Polynomial Threshold	Functions.	Bruck, J 88c
...a Common Mechanism in Experience-Dependent Maturation of Cortical	Functions.	Singer, W 85a
Logic of Activation	Functions.	Williams, RJ 86b
...Networks, Orientations of the Hypercube and Algebraic Threshold	Functions.	Baldi, P 88f
On the Size of Weights Required for Linear-Input Switching	Functions.	Myhill, J 61a
...in Neural Systems: An Approach to the Dynamics of Higher Brain	Functions.	Harth, EM 83a
Parallel Computation with Threshold	Functions.	Parberry, I 88b
The Capacity of Multilevel Threshold	Functions.	Olafsson, S 88a
Neural Chips Emulate Brain	Functions (ATT VLSI)	Brown, C 86e
...of Large Scale Neuron-Glia Network and Its Relation to the Brain	Functions (I).	Xu, J 86a
Nonlinear Discriminant	Functions and Associative Memories	Psaltis, D 86a
Brain	Functions and Neural Dynamics.	Harth, EM 70a
Statistical Recognition	Functions and the Design of Pattern Recognizers.	Marill, T 60a
Spontaneous Symmetry-Breaking Energy	Functions and the Emergence of Orientation Selective Cortical Cells.	Kammen, DM 88a
Lyapunov	Functions and Their Use in Automata Networks	Soulie, FF 86a
Response Mapping	Functions: Classification and Analysis of Connectionist.	Walters, DKW 87f
Serial Position	Functions for Letter Identification at Brief and Extended Exposure	Estes, WK 76a
Comparisons of Sigmoid	Functions for Neural Network Convergence	Graziano, MJ 88a
Generation of Polynomial Discriminant	Functions for Pattern Recognition.	Specht, DF 67a
Neural Networks and Radial Basis	Functions in Classifying Static Speech Patterns.	Niranjan, M 88a
Adaptive, Associative, and Self-Organizing	Functions in Neural Computing.	Kohonen, T 87a
Incorporating Fuzzy Membership	Functions into the Perceptron Algorithm.	Keller, JM 85a
An Introduction to the Application of the Theory of Probabilistic	Functions of a Markov Process to Automatic Speech Recognition.	Levinson, SE 83a
Studies with Artificial Neurons, I: Properties and	Functions of an Artificial	Harmon, LD 61a
...Technique Occurring in the Statistical Analysis of Probabilistic	Functions of Markov Chains	Baum, LE 70a
A Stochastic Algorithm for Learning Real-Valued	Functions via Reinforcement Feedback	Gullapalli, V 88a
Linear Decision	Functions, with Application to Pattern Recognition	Highleyman, WH 62a
The Realization of Symmetric Switching	Functions with Linear-Input Logical Elements	Kautz, WH 61a
Propagating Belief	Functions with Local Computations.	Shenoy, P 86a
Where Does the	Fundamental Information Processing Ability of Biological Systems	Kohonen, T 75a
	Fundamental Relations Between the LMS Algorithm and the DFT.	Widrow, B 87b
On Some	Fundamental Theorems of Perceptron Theory and Their Geometry	Charnes, A 63a
Adaptive Filters I:	Fundamentals	Widrow, B 66b
The FFT:	Fundamentals and Concepts.	Ramirez, RW 85a
	Further Consideration on Pattern Separability in a Random Neural	Torioka, T 80a
On a method of Storing Information: II. A	Further Study of Model Properties.	Roy, AE 62a
Neural Network Approach to Sensory	Fusion	Pearson, JC 88a
A Markov Random Field Approach to Data	Fusion and Colour Segmentation	Wright, WA 89a
	Fusion, Propagation, and Structuring in Belief Networks.	Pearl, J 86a
	Fusion, Propagation, and Structuring in Belief Networks.	Pearl, J 86b
Binocular	Fusion Using Simulated Annealing	Landa, J 87a
Backpropagation: Past and	Future	Werbos, PJ 88a
Precise Past, Fuzzy	Future	Gaines, BR 83a
The Role of Adaptive and Associative Circuits in	Future Computer Designs.	Kohonen, T 88e
Past, Present, and	Future of GMDH	Ivakhnenko, AG 84a
	Future Perspectives on Neural Network Research	Szu, HH 87b
The Adaptive, Dynamic, and Associative Memory Model: A Possible	Future Tool for Vocal Human-Computer Communication	Beroule, D 87b
	Fuzziness Versus Probability	Kosko, B 87f
	Fuzzy Associative Memories	Kosko, B 87g
Estimation of Expert Weights Using	Fuzzy Cognitive Maps	Taber, WR 87a
	Fuzzy Cognitive Maps	Kosko, B 86c
Risk Weighting in Knowledge Bases Using	Fuzzy Cognitive Maps	Taber, WR 88c
Extended	Fuzzy Cognitive Maps (EFCM) Engines for the Diagnosis of Electro	Tenorio, MF 87a
Applications of Neural Networks to Underwater Acoustic Signatures:	Fuzzy Decision Making and the Neuron Ring.	Deich, RO 88a
	Fuzzy Entropy and Conditioning	Kosko, B 86d
Decision-Making in a	Fuzzy Environment.	Bellman, RE 87a
	Fuzzy Expert Systems	Kandel, A 87a
Precise Past,	Fuzzy Future	Gaines, BR 83a
Fuzzy Power Sets and	Fuzzy Implication Operators.	Bandler, W 80a

Context	Keyword	Author
	Fuzzy Knowledge Combination	Kosko, B 86e
Adaptive Inference in	Fuzzy Knowledge Networks	Kosko, B 87b
On Designing	Fuzzy Learning Neural-Automata	Shiue, LC 87a
NASA Unites Neural Nets,	Fuzzy Logic	Johnson, RC 88b
Syllogistic Reasoning in	Fuzzy Logic and its Application to Usuality and Reasoning with	Zadeh, LA 85a
Theoretical and Linguistic Aspects of the	Fuzzy Logic Controller	Braae, M 79a
The Role of	Fuzzy Logic in the Management of Uncertainty in Expert Systems	Zadeh, LA 83b
The Role of	Fuzzy Logic in the Management of Uncertainty in Expert Systems	Zadeh, LA 87f
	Fuzzy Logic's Soft Start Belies a Cutting Edge	Chester, M 86a
	Fuzzy Mathematical Techniques with Applications	Kandel, A 86a
Incorporating	Fuzzy Membership Functions into the Perceptron Algorithm	Keller, JM 85a
Operations on Fuzzy Sets and	Fuzzy Numbers Related to Triangular Norms	Klement, EP 81a
A Comparison of Feedforward Neural Networks:	Fuzzy Operators and Acoustic Ship Signatures	Taber, WR 88a
Similarity Relations and	Fuzzy Orderings	Zadeh, LA 71a
	Fuzzy Power Sets and Fuzzy Implication Operators	Bandler, W 80a
A Computational Approach to	Fuzzy Quantifiers in Natural Languages	Zadeh, LA 83a
	Fuzzy Requests: An Approach to Weighted Boolean Searches	Bookstein, A 80a
Counting with	Fuzzy Sets	Kosko, B 86a
	Fuzzy Sets	Zadeh, LA 87c
	Fuzzy Sets	Zadeh, LA 65a
	Fuzzy Sets and Applications: Selected Papers by L A Zadeh	Yager, RR 87a
Operations on	Fuzzy Sets and Fuzzy Numbers Related to Triangular Norms	Klement, EP 81a
	Fuzzy Sets and Neural Networks	Taber, WR 88b
	Fuzzy Sets as a Basis for a Theory of Possibility	Zadeh, LA 87d
A Definition of a Nonprobabilistic Entropy in the Setting of	Fuzzy Sets Theory	De Luca, A 72a
Rod Bipolar Array in the Cat Retina: Pattern of Input from Rods and	GABA-Accumulating Amacrine Cells	Freed, MA 87a
	Gabor: Frequency, Time and Memory	Korpel, A 82a
Complete Discrete 2-D	Gabor Transforms by Neural Networks for Image Analysis and	Daugman, JG 88a
Neural Networking Research	Gains New Life	Kane, B 87a
An N-Player Sequential Stochastic	Game with Identical Payoffs	Narendra, KS 83a
A	Game-Playing Program that Learns by Analyzing Examples	Minton, S 85a
Teaching a Stochastic Automaton to Skillfully Play Hand / Eye	Games	Jarvis, RA 77a
	Game-Theoretic Cooperativity in Networks of Self-Interested Units	Barto, AG 86b
A Prototype Crossbar Switch for the	Gamma Network	Farley, ET 87a
The	Gamma Network - A Physical and Logical Data Structure	Hennessey, AK 87a
The	Gamma Network as an Interconnection Architecture for Neural	Hennessey, AK 87b
Self-Inhibition Alters Firing Patterns of Neurons in Aplysia Buccal	Ganglia	White, RL 81a
Architecture of Rod and Cone Circuits to the On-Beta	Ganglion Cell	Sterling, P 88a
Microcircuitry of the On-Beta	Ganglion Cell in Daylight, Twilight, and Starlight	Sterling, P 87a
The On-Alpha	Ganglion Cell of the Cat Retina and Its Presynaptic Cell Types	Freed, MA 88a
Microcircuitry of Beta	Ganglion Cells in Cat Retina	McGuire, BA 86a
Plasticity in the Nervous System of Adult Hydra II. Conversion of	Ganglion Cells of the Body Column into Epidermal Sensory Cells of.	Koizumi, O 88a
	Gas Tungsten Arc Weld Modeling Using a Mapping Network	Karsai, G 88a
Image Statistics,	Gases, and Visual Neural Primitives	Daugman, JG 87a
Neural Net Developer Nestor Sprints From Startup	Gate	Johnson, RC 87f
Scientists Identify "	Gate" in Brain as Crucial to Memory	Johnson, G 88a
Early Spatio-Temporal Integration by	Gated Competitive Networks	Ogmen, H 87b
	Gated Dipoles: Photoreceptors and Circadian Rhythms	Carpenter, GA 84b
A Neural Theory of Circadian Rhythms: The	Gated Pacemaker	Carpenter, GA 83a
Neural Theory of Circadian Rhythms: The	Gated Pacemaker	Carpenter, GA 86n
Adaptation and Transmitter	Gating in Vertebrate Photoreceptors	Carpenter, GA 86c
Parallel	Gaussian Elimination on an Optically Interconnected Data Flow	Nelken, I 87a
...for Maximum-Likelihood Sequence Estimation of Binary Signals in	Gaussian Noise	Provence, JD 87a
A Special Problem of Brownian Motion, and a General Theory of	Gaussian Random Functions	Levy, P 56a
...Modeling of the Metaorganization of Sensorimotor Hierarchies of	Gaze	Pellionisz, AJ 86b
Distributed Memory and the Representation of	General and Specific Information	McClelland, JL 85a
Statistical Mechanics: A	General Approach to Combinatorial Optimization	Bonomi, E 86a
	General Framework for Parallel Distributed Processing	Rumelhart, DE 86b
PDP Models and	General Issues in Cognitive Science	Rumelhart, DE 86i
A Methodology for the Development of	General Knowledge-Based Vision Systems	Riseman, EM 87a
Sensor-Based Control of Robotic Manipulators Using a	General Learning Algorithm	Miller, WT 87c
Application of a	General Learning Algorithm to the Control of Robotic Manipulators	Miller, WT 87b
Correlation Memory Models: A First Approximation in a	General Learning Scheme	Pfaffelhuber, E 75a
Synthesis of Boolean Nets and Time Behavior of a	General Mathematical Neuron	Caianiello, ER 75a
	General Model for the Molecular Events in Synapses During Learning	Huttunen, MO 73a
Learning on a	General Network	Atiya, AF 87a
Self Organizing	General Pattern Class Separator and Identifier	Cooper, LN 82a
An Approach to	General Pattern Recognition	Fischler, M 62a
Adaptive Algorithms for Discovering and Using	General Patterns in Growing Knowledge-Bases	Holland, JH 80a

Automated	Generation of Connectionist Expert Systems for Problems Involving. .	Gallant, SI 87a
Automatic	Generation of Expert Systems from Examples	Gallant, SI 85a
The Relevance of Mechanical Sensors for Neural	Generation of Movements in Space	Daunicht, WJ 88c
The Relevance of Mechanical Sensors for Neural	Generation of Movements in Space	Daunicht, WJ 88d
	Generation of Polynomial Discriminant Functions for Pattern.	Specht, DF 67a
Topological and Dynamical Aspects of a Neural Network Model for	Generation of Pursuit Motor Programs	Eckmiller, R 86b
A Synaptic Modification Algorithm in Consideration of the	Generation of Rhythmic Oscillation in a Ring Neural Network.	Tsutsumi, K 84a
An Image Processing Architecture for Real Time	Generation of Scale and Rotation Invariant Patterns.	Messner, RA 85a
	Generation of Simple Sentences in English Using the Connectionist. .	Kalita, J 87a
Models for the	Generation of the Motor Output Pattern in Flying Locusts	Wilson, DM 68a
Automatic	Generation of Time Efficient Digital Signal Processing Software. . .	Morris, LR 77a
Japan's 6th	Generation Project: Striking A Fine Balance.	Johnson, RC 87d
A Trainable Non-Linear Function	Generator.	Smith, FW 66a
Computational Model of the Motor Program	Generator for Pursuit.	Eckmiller, R 87a
Ring Neural Network qua a	Generator of Rhythmic Oscillation with Period Control Mechanism. . .	Tsutsumi, K 84b
Model of the Cerebellum as an Array of Adjustable Pattern	Generators	Houk, JC 87a
The Neural	Generators of the Vestibular Evoked Response	Elidan, J 87a
Are Central Pattern	Generators Understandable?	Selverston, AI 80a
A	Generic Architecture for Wafer-Scale Neuromorphic Systems.	Raffel, JI 87a
Towards a	Generic Learning Algorithm	Wright, WA 89d
On the	Genesis of Abstract Ideas.	Posner, MI 68a
Application of	Genetic Algorithm to a Neural Network for Finding Optimal Paths. . .	Smith, T 87a
Discrete-Time Parameter Estimation with	Genetic Algorithms	Das, R 88a
Applying	Genetic Algorithms to Neural Net Learning.	Whitley, D 88a
Scaling, Machine Learning, and	Genetic Neural Nets.	Mjolsness, E 88b
	Geometric Analysis of Neural Network Capabilities.	Wieland, A 87a
A	Geometric Approach to Singular Perturbation Problems with.	Carpenter, GA 77a
A	Geometric Convergence Theorem for the Perceptron	Konheim, AG 63a
Encoding	Geometric Invariances in Higher-Order Neural Networks.	Giles, CL 88a
Connectionist Models and	Geometric Reasoning.	Fallside, F 88a
	Geometrical and Statistical Properties of Systems of Linear.	Cover, TM 65a
Feature-Extracting Network for	Geometrical Patterns: A Design with Reference to Visual Neural . . .	Fukushima, K 71a
...a Simultaneous Functional Order in Nervous Systems II. Computing	Geometrical Structures	Toet, A 87b
Order and Defects in	Geometrically Frustrated Systems	Mosseri, R 86a
On Some Fundamental Theorems of Perceptron Theory and Their	Geometry	Charnes, A 63a
Perceptrons: An Introduction to Computational	Geometry	Minsky, ML 69a
Tensor	Geometry: A Language of Brains and Neurocomputers. Generalized . . .	Pellionisz, AJ 88a
Representation of Local	Geometry in the Visual System.	Koenderink, JJ 87d
Quantized	Geometry of Visual Space: The Coherent Computation of Depth, Form. .	Grossberg, S 86p
The Quantized	Geometry of Visual Space: The Coherent Computation of Depth, Form. .	Grossberg, S 83b
Neuristor Line Made of S-Type	Germanium Diodes	Batalina, MA 80a
Neural Systems Make Comeback After Nearly 50-Year	Gestation.	Johnson, RC 87k
Neural Networks	Get a Conference to Call Their Own	Brown, C 87j
Neural Networks Finally	Get Real	Johnson, RC 88i
Learning and Distributed Memory:	Getting Close to Neurons	Aleksander, I 88b
	Gibbs and Markov Random Systems with Constraints	Moussouris, J 74a
Bayesian Image Processing,	Gibbs Distributions, and Neural Networks	Vold, TG 87a
Stochastic Relaxation,	Gibbs Distributions, and the Bayesian Restoration of Images.	Geman, S 84a
Stochastic Relaxation,	Gibbs Distributions, and the Bayesian Restoration of Images.	Geman, S 84b
Markov Random Fields and	Gibbs Ensembles.	Spitzer, F 71a
Designing a Neural Network Satisfying a	Given Set of Constraints	Personnaz, L 86c
Low Rank Matrices with a	Given Sign Pattern	Delsarte, P 87a
EEG Analysis	Gives Models of Neuronal Template-Matching Mechanism for Sensory . .	Freeman, WJ 79a
Self-Organization of a Neural Network Which	Gives Position-Invariant Response.	Fukushima, K 79c
Simplest Spin	Glass.	Gross, DJ 84a
Spin	Glass and Pseudo Boolean Optimization.	Rosenberg, IG 86a
Engineering Applications of Spin	Glass Concepts	Guyon, I 86a
Replica Symmetry Breaking and the Nature of the Spin	Glass Phase.	Mezard, M 84b
A Design for an Associative Spin	Glass Processor.	Goodwin, JM 87a
On the Statistical Physics of Spin	Glasses.	Mezard, M 86a
Zero Temperature Parallel Dynamics for Infinite Range Spin	Glasses and Neural Networks.	Gardner, E 87a
Symbolic Computation Methods for Some Spin	Glasses Problems	Lacolle, B 86a
Models of Hierarchically Constrained Dynamics for	Glassy Relaxation.	Palmer, RG 84a
On the	Global Limits and Oscillations of a System of Nonlinear.	Grossberg, S 69f
Pattern Formation by the	Global Limits of a Nonlinear Competitive Interactin in n-Dimensions.	Grossberg, S 77b
Local Versus	Global Minima, Hysteresis, Multiple Meanings	Kergosien, YL 86a
Langevin Equation as a	Global Minimization Algorithm.	Gidas, B 86a
Boltzmann Machines in which Weights Converge to	Global Optimal Solutions	Uchiyama, T 88a
	Global Optimization with Stochastic Neural Networks.	Levy, BC 87a

Neural Net Pioneer	Grossberg Extends ART-2 Architecture into Analog Domain.	Brown, C 87g
Sensorimotor Operations: A	Ground for the Co-Evolution of Brain Theory with Neurobotics and . .	Pellionisz, AJ 87a
Separating Figure from	Ground with a Boltzmann Machine.	Sejnowski, TJ 87g
Separating Figure from	Ground with a Parallel Network	Kienker, PK 86a
	Group Actions and Learning for a Family of Automata.	Baldi, P 88b
Modeling Accuracy Improvements for the	Group Method of Data Handling.	Molnar, DO 87a
...Cerebral Cortical Maps: A Computer Simulation Based on Neuronal	Group Selection.	Pearson, JC 87a
Neuronal	Group Selection: A Basis for Categorization by the Nervous System. .	Finkel, LH 88b
Effect of Orientation and of Shape Similarity on Perceptual	Grouping	Beck, J 66a
...of Speech and Language Coding: Developmental Programs, Perceptual	Grouping, and Competition for Short-Term Memory.	Cohen, MA 86c
Similarity	Grouping and Peripheral Discriminability Under Uncertainty	Beck, J 72a
Perceptual	Grouping Produced by Changes in Orientation and Shape.	Beck, J 66b
Neural Dynamics of Perceptual	Grouping: Textures, Boundaries, and Emergent Segmentations	Grossberg, S 85c
Neural Dynamics of Perceptual	Grouping: Textures, Boundaries, and Emergent Segmentations	Grossberg, S 86k
...Architecture for Learning, Recognizing, and Predicting Multiple	Groupings of Patterned Data.	Cohen, MA 87a
...Architecture for Learning, Recognizing, and Predicting Multiple	Groupings of Patterned Data.	Cohen, MA 87b
Synchronization in Small	Groups of Neurons: A Study with Electronic Models.	Lewis, ER 68b
Learning Symmetry	Groups with Hidden Units: Beyond the Perceptron.	Sejnowski, TJ 86b
...Transforming Recursive Objects in a Neural Network, or "Trees DO	Grow on Boltzmann Machines".	Touretzky, DS 86c
	Growing 'Brains' in a Computer	Amato, I 87a
Adaptive Algorithms for Discovering and Using General Patterns in	Growing Knowledge-Bases.	Holland, JH 80a
Development of Ocularity Domains and	Growth Behaviour of Axon Terminals	von der Malsburg, C 79a
The Organization of Behavior: The First Stage of Perception:	Growth of the Assembly	Hebb, DO 49a
A Test of the Sophisticated	Guessing Theory of Word Perception	Johnston, JC 78a
How Learning Can	Guide Evolution.	Hinton, GE 87b
Simulation of Task Performance as a	Guide to the Identification of Solution Strategy from Eye-Movement .	Deffner, G 87a
Computers and Intractability: A	Guide to the Theory of NP-Completeness	Garey, MR 79a
Depth and Detours: An Essay on Visually	Guided Behavior.	Arbib, MA 87b
Depth and Detours: An Essay on Visually	Guided Behavior.	Arbib, MA 85c
	Guided Propagation Inside a Topographic Memory	Beroule, D 87a
Expert System and Neural Network Approaches for Classical	Guitar Fingering	Sayegh, S 88a
Image	Halftoning and Reconstruction Using a Neural Network	Kollias, S 89a
Digital Image	Halftoning Using Neural Networks	Anastassiou, D 88a
Neural Networks, Pattern Recognition, and Fingerprint	Hallucination.	Mjolsness, E 85a
A Mathematical Theory of Visual	Hallucination Patterns	Ermentrout, GB 79a
What Do Drug-Induced Visual	Hallucinations Tell Us About the Brain?.	Cowan, JD 85b
A Comparison of	Hamming and Hopfield Neural Nets for Pattern Classification.	Lippman, RP 87a
Coordinated Control Programs for Movements of the	Hand	Arbib, MA 85b
Teaching a Stochastic Automaton to Skillfully Play	Hand / Eye Games	Jarvis, RA 77a
Schemas That Integrate Vision and Touch for	Hand Control	Arbib, MA 87c
A Stochastic Theory of Phase Transitions in Human	Hand Movement.	Schoner, G 86a
Opposition Space as a Structuring Concept for the Analysis of Skilled	Hand Movements	Iberall, T 86a
The Hierarchical Control of Skilled	Hand Movements	Dakin, GA 86a
	Handbook of Digital Signal Processing.	Elliot, DF 87a
Explaining PDP: A	Handbook of Models, Programs and Exercises	McClelland, JL 87a
Explorations in Parallel Distributed Processing: A	Handbook of Models, Programs, and Exercises.	McClelland, JL 88a
	Handbook of Pattern Recognition and Image Processing	Young, TY 86a
Modeling Accuracy Improvements for the Group Method of Data	Handling	Molnar, DO 87a
Supplier-Consumer Nets for Material	Handling	Parunak, HVD 87a
A Neural Network Model of CAAM and its Applications to	Handprinted Chinese Character Recognition.	Yao, Y 87a
A Model for Variability Effects in	Hand-Printing, with Implications for the Design of On Line	Ward, JR 88a
The Nature of Human Prehension: Three Dextrous	Hands in One	Iberall, T 87c
Designing a	Handwriting Reader	Burr, DJ 83a
Feature Extraction and Recognition of	Handwritten Characters by Homogeneous Layers	Giebel, H 71a
Recognition of	Handwritten Characters with a System of Homogeneous Layers	Marko, H 70a
Experiments with Neural Network Models for	Handwritten Digit Recognition.	Lee, D 88a
Neural Network Models and Their Application to	Handwritten Digit Recognition.	Pawlicki, TF 88a
Learning in Networks is	Hard	Judd, JS 87b
On the Power of Neural Networks for Solving	Hard Problems.	Bruck, J 88e
Implementing Semantic Networks in Parallel	Hardware	Hinton, GE 81b
Nestor Frees Neural Nets from	Hardware Chains.	Johnson, RC 88c
Neural Network Emulation	Hardware Design Considerations	Cruz, CA 87c
...Two Examples within a Single, Integrated Neural Network	Hardware Development Environment	Bahr, C 88c
Fat-Trees: Universal Networks for	Hardware Efficient Supercomputing.	Leiserson, CE 85a
New	Hardware for Massive Neural Networks	Coon, DD 88d
Neural	Hardware for Vision.	Mead, CA 87a
Electronic	Hardware Implementations of Neural Networks.	Thakoor, AP 87b
A Preliminary Study of the Phoneme Recognition with	Hardware Neural Network.	Hoshimiya, N 82a
A	Hardware Neuron Model as a Component of the Neural Network	Koga, K 80a

A Comparison of Some	Hierarchial Monothetic Divisive Clustering Algorithms for Structure.	Rubin, V 83a
	Hierarchical Associative Networks.	Cortes, C 87a
Mean-field Analysis of	Hierarchical Associative Networks with Magnetisation	Krogh, A 88a
Positive Feedback in	Hierarchical Connectionist Models: Applications in Language.	Dell, GS 88a
	Hierarchical Contour Coding by the Visual Cortex	Hartmann, G 85a
Theory and Practice of	Hierarchical Control	Albus, JS 81b
	Hierarchical Control for Robots, and Teleoperators	Albus, JS 85a
	Hierarchical Control of Rapid Movement Sequences	Rosenbaum, DA 83a
	Hierarchical Control of Robots Using Microcomputers.	Barbera, AJ 79a
The	Hierarchical Control of Skilled Hand Movements	Dakin, GA 86a
Mapping Images to a	Hierarchical Data Structure - A Way to Knowledge-based Pattern	Hartmann, G 88a
Emergence of	Hierarchical Data Structures in Parallel Computation	Kerszberg, M 86a
Cognitive Control of Autonomous Mobile Robots: Nested	Hierarchical Intelligent Module.	Meystel, A 86a
Network Processing of	Hierarchical Knowledge for Classification and Diagnosis.	Becker, LA 87a
	Hierarchical Learning Control: An Approach with Neuron-Like.	Ersu, E 88a
A	Hierarchical Model for Voluntary Movement and Its Application to	Kawato, M 87a
	Hierarchical Model of Memory and Memory Loss	Sutton, JP 88a
Experimental Tests of a	Hierarchical Model of Word Identification.	Johnston, JC 80a
Counterpropagation Networks: Adaptive	Hierarchical Networks for 'Near Optimal' Mappings.	Stork, DG 87a
Neocognitron: A	Hierarchical Neural Network Capable of Visual Pattern Recognition.	Fukushima, K 88c
A	Hierarchical Neural Network Model for Associative Memory	Fukushima, K 83a
A	Hierarchical Neural Network Model for Associative Memory	Fukushima, K 84a
A	Hierarchical Neural Network Model for Selective Attention.	Fukushima, K 88a
	Hierarchical Neural Network Model for Voluntary Movement with.	Kawato, M 88a
The Control of Behavior in a	Hierarchical Neural Network: Modelling Attention Channels and.	Eilbert, JL 87a
Exponential Storage and Retrieval in	Hierarchical Neural Networks	Willcox, CR 88a
Time-Sequential Self-Organization of	Hierarchical Neural Networks	Silverman, RH 88a
A	Hierarchical Neural-Network Model for Control and Learning of.	Kawato, M 87b
	Hierarchical Scene Structure Representations to Facilitate Image	Maren, AJ 88b
An Associative	Hierarchical Self-Organizing System.	Davis, BR 85a
A Multilayer Cooperative/Competitive Method for Creating	Hierarchical Structures by Clustering Maximally-Related Nodes.	Maren, AJ 88a
A	Hierarchical System for Character Recognition with Stochastic.	Vlontzos, JA 88a
Statistical Dynamics of Ultradiffusion in	Hierarchical Systems	Gardner, S 87a
Ultradiffusion: The Relaxation of	Hierarchical Systems	Huberman, BA 85b
Models of	Hierarchically Constrained Dynamics for Glassy Relaxation.	Palmer, RG 84a
Neural Networks with	Hierarchically Correlated Patterns	Gutfreund, H 88a
Recognition of	Hierarchically Encoded Images by Technical and Biological Systems.	Hartmann, G 87a
...in Computer Modeling of the Metaorganization of Sensorimotor	Hierarchies of Gaze.	Pellionisz, AJ 86b
Perceptual Significance	Hierarchy: A Computer Vision Theory for Color Separation	Walters, DKW 87d
On Evidential Reasoning in a	Hierarchy of Hypotheses.	Pearl, J 86c
On Evidential Reasoning in a	Hierarchy of Hypotheses.	Pearl, J 86d
Building a	Hierarchy with Neural Networks: An Example-Image Vector.	Jackel, LD 87a
	High Density Associative Memories.	Dembo, A 88b
Neural Network Based Feed-Forward	High Density Associative Memory.	Daud, T 87a
Neural Computers in Vision: Processing of	High Dimensional Data.	Zeevi, YY 88a
The Central Nervous System as a Low and	High Level Control System.	Albus, JS 88a
	High Order Correlation Model for Associative Memory.	Chen, HH 86a
	High Order Neural Networks for Efficient Associative Memory Design	Dreyfus, G 88a
Nerve Cell Soma Model with	High Reliability and Low Power Consumption	Fernandes, PML 80a
	High Resolution Micro-Fabrication and Neural Networks.	Jackel, LD 86b
Real Time Experiments in Neural Network Based Learning Control During	High Speed Nonrepetitive Robotic Operations.	Miller, WT 88a
A Chipset for	High Speed Simulation of Neural Network Systems.	Garth, SCJ 87a
A Relaxation Model for Memory with	High Storage Density	Bachmann, CM 87a
	High-Capacity Exponential Associative Memories	Chiueh, T 88b
Order and Chaos in Neural Systems: An Approach to the Dynamics of	Higher Brain Functions	Harth, EM 83a
A Field Approach to	Higher Order ANNs for Sensorial Representation	Van Hulle, MM 88a
	Higher Order Associative Memories and Their Optical Implementations.	Psaltis, D 88b
Characteristics of the Functional Link Net: A	Higher Order Delta Rule Net.	Klassen, M 88a
Correlative Binding in	Higher Order Network for Sensorial Representation.	Van Hulle, MM 88b
Generalization of Backpropagation to Recurrent and	Higher Order Neural Networks	Pineda, FJ 88b
Number of Stable Points for Spin-Glasses and Neural Networks of	Higher Orders.	Baldi, P 87a
	Higher-Order Boltzmann Machines.	Sejnowski, TJ 86a
Encoding Geometric Invariances in	Higher-Order Neural Networks	Giles, CL 88a
Neural Networks with	Higher-Order Nonlinearity.	Tai, H 88a
	Highlights and the Perception of Glossiness.	Beck, J 81a
Design of a	Highly Parallel Visual Recognition System.	Sabbah, D 81a
Learning Invariance, and Generalization in	High-Order Neural Networks	Giles, CL 87a
Two-Dimensional Pattern Sequence Prediction Using	High-Order Neural Networks	Collins, DH 88a
	High-Order Neural Networks: Information Storage without Errors	Personnaz, L 87a

Neural Networks with a	Hopf Bifurcation: Slowly Modulated Waves	Cohen, MS 87a
...Units have Collective Computational Properties like those of	Hopfield and Tank Neurons.	Derthick, MA 88a
Capacity of the	Hopfield Associative Memory.	McEliece, RJ 86a
The Capacity of the	Hopfield Associative Memory.	McEliece, RJ 87a
	Hopfield, Boltzman Machines, and Vision.	Sejnowski, TJ 87d
Unifying the Tank and	Hopfield Linear Programming Circuit and the Canonical Nonlinear.	Kennedy, MP 87b
A New Approach to Optical Information Processing Based on the	Hopfield Model	Psaltis, D 84a
Dynamics and Statistical Mechanics of the	Hopfield Model	Bruce, AD 87a
Information Capacity of the	Hopfield Model	Abu-Mostafa, YS 85b
...Threshold Condition for a Content-Addressable Memory Based on the	Hopfield Model	Macukow, B 87a
Optical Implementation of the	Hopfield Model	Farhat, NH 85b
Optical Implementation of the	Hopfield Model	Farhat, NH 85a
Structure of Metastable States in the	Hopfield Model	Gardner, E 86a
Ultrametricity,	Hopfield Model and all that.	Virasoro, MA 86a
	Hopfield Model Applied to Vowel and Consonant Discrimination	Gold, B 86b
	Hopfield Model Applied to Vowel and Consonant Discrimination	Gold, B 86a
High-Performance Recording Algorithm for	Hopfield Model Associative Memories.	Hassoun, MH 88b
On the Use of the	Hopfield Model for Optical Pattern Recognition	Bayley, JS 87a
Optical Implementation of the	Hopfield Model for Two-Dimensional Associative Memory.	Jang, J 88c
	Hopfield Model with Complementary Binary Representations	Jong, T 88a
Holographic Implementation of a	Hopfield Model with Discrete Weightings.	White, HJ 88d
The	Hopfield Model with Multi-Level Neurons.	Fleisher, M 88a
On "Information Capacity of the	Hopfield Model".	Xu, X 88c
Sensitivity Analysis of	Hopfield Net	David, GW 87a
The State Space of Memory Clusters in the	Hopfield Network	Krowitz, A 87a
A Large VLSI	Hopfield Network for Pattern Recognition Problems.	Verleysen, M 88c
Determining the Dynamic Landscape of	Hopfield Networks.	Weisbuch, G 86a
Unconstrained Minimum Mean-Square Error Parameter Estimation with	Hopfield Networks.	Altes, RA 88a
On Choosing Non-Random Input Exemplars to Form Stable States in a	Hopfield Neural Net.	Feild Jr, WB 88a
A Comparison of Hamming and	Hopfield Neural Nets for Pattern Classification.	Lippman, RP 87a
Enhanced Memory Capacity of a	Hopfield Neural Network.	Little, MJ 86a
Transputer Implementation of	Hopfield Neural Network.	Feild Jr, WB 88b
Holographic Implementation of a Discrete	Hopfield Neural Network Model.	White, HJ 88c
Evaluation of the Use of the	Hopfield Neural Network Model as a Nearest-Neighbor Algorithm.	Montgomery, BL 86a
Synchronous vs Asynchronous Behavior of	Hopfield's CAM Neural Net.	Cheung, KF 87c
	Hopfield's Nerve Nets Realize Biocomputing	Brown, C 86d
A New VLSI Architecture for Large	Hopfield's Neural Networks	Verleysen, M 88e
Comparison Between Kanerva's SDM and	Hopfield-type Neural Networks.	Keeler, JD 88b
Physiological Engineering Model of the Retinal	Horizontal Cell Layer.	Usui, S 88a
...of Single Brain-Stem Motoneurons Intracellularly Labelled with	Horseradish Peroxidase in the Cat: An Ultrastructural Analysis	Bras, H 87a
...of Single Brain-Stem Motoneurons Intracellularly Labelled with	Horseradish Peroxidase in the Cat: Morphological and Electrical.	Bras, H 87b
Use of the	Hough Transform to Detect Lines and Curves in Pictures	Duda, RO 72a
Physicists Explore	Human and Artificial Intelligence.	Buhmann, J 86c
Parallel Structures in	Human and Computer Memory.	Kanerva, P 86a
	Human and Computer Rules and Representations are not Equivalent.	Grossberg, S 80e
	Human Associative Memory	Anderson, JR 73a
On a Model of	Human Associative Processor.	Hirai, Y 86a
A Model of	Human Associative Processor (HASP)	Hirai, Y 83a
Towards an Information-Flow Model of	Human Behaviour.	MacKay, DM 56a
Neurocomputing: Picking the	Human Brain.	Hecht-Nielsen, R 88c
Optics and Neural Nets: Trying to Model the	Human Brain.	Williams, T 87a
...Computer: An Introduction to the Computational Structure of the	Human Brain.	Baron, RJ 87a
	Human Brain EEG Fields: Micro-States and Their Functional.	Lehmann, D 87b
Spontaneous and Information-Triggered Segments of Series of	Human Brain Electric Field Maps.	Lehmann, D 88a
Learning-Logic: Casting the Cortex of the	Human Brain in Silicon	Parker, DB 85a
On the Scheme by Which the	Human Brain Stores Information	Legendy, CR 67a
A Chain of Suprasegmental Neuroscillatory Circuits: A	Human Brain Theory	Deshmukh, VD 88a
How the	Human Brain Thinks and Acts: Results of Latest Research Using.	Gevins, AS 87b
Knowledge-Rich Solutions to the Binding Problem: A Simulation of Some	Human Computational Mechanisms	Stenning, K 88a
Two-Dimensional Adaptive Model of a	Human Controller Using Pattern Recognition Techniques.	Gilstad, DW 71a
Strange Attractors in the	Human Cortex	Babloyantz, A 87b
The Dynamic Re-Evaluation of Alternatives and the Emulation of	Human Decision Making.	Clema, JK 72a
...Digital Speech Coders by Exploiting Masking Properties of the	Human Ear.	Schroeder, MR 79a
A Psychologist Views Human Processing:	Human Errors and Other Phenomena Suggest Processing Mechanisms	Norman, DA 81a
The	Human Eye as an Optical Filter	Campbell, FW 68a
Derivation of a Model for the	Human Eye-Positioning Mechanism.	Cook, G 67a
A Stochastic Theory of Phase Transitions in	Human Hand Movement.	Schoner, G 86a
Oscillatory Neural Ectoderm (One): The "One" Hypothesis for	Human Holoencephalon and Behavior.	Deshmukh, VD 87a
Machine Recognition of	Human Language: Part 2, Theoretical Models of Speech Perception and.	Lindgren, N 65a

Distributed Model of	Human Learning and Memory.	McClelland, JL 86d
...Mechanisms and Processing of Organized Information in	Human Memory	Ratcliff, R 81a
	Human Memory: A Proposed System and its Control Processes.	Atkinson, RC 68a
On the Applied Use of	Human Memory Models: The Memory Extender Personal Filing System.	Jones, WP 86a
A Theory of	Human Memory: Self-Organization and Performance of Sensory-Motor	Grossberg, S 78a
The Information-Storage System Called "	Human Memory".	Simon, HA 76a
Connecting with the	Human Mind: A New Approach to Building Computers Draws on the.	Hammerstrom, D 87b
Mimicking the	Human Mind: Radical New Computers Teach Themselves	Rogers, M 87a
	Human Movement Initiation: Specification of Arm, Direction, and.	Rosenbaum, DA 80a
A Ballpark Approach to Modelling	Human Prehension	Iberall, T 87a
Grasp Planning for	Human Prehension	Iberall, T 87b
The Nature of	Human Prehension: Three Dextrous Hands in One.	Iberall, T 87c
A Psychologist Views	Human Processing: Human Errors and Other Phenomena Suggest	Norman, DA 81a
...Multi-Layered Neural Network Realizing a Funciton in the	Human Recognition and Memory Mechanisms.	Mizuno, H 80a
...Multi-Layered Neural Network Realizing a Function in the	Human Recognition and Memory Mechanisms.	Mizuno, H 78a
	Human Software	Blackmore, WR 81a
...A Back-Propagation Model of Pattern Recognition in the	Human Somatosensory System	Grajski, KA 88c
A Computational Theory of	Human Stereo Vision.	Marr, D 79a
...and Repetition Priming in Symmetry Detection: Parallel Studies of	Human Subjects and Connectionist Models.	Cohen, NJ 86a
ANNA: An Adaptive Neural Network Anticipator for	Human/Computer Interfacing	Jones, WP 88a
...and Associative Memory Model: A Possible Future Tool for Vocal	Human-Computer Communication	Beroule, D 87b
...Show Neural Networks: "Intelligent" Systems Learn, Reason Like	Humans	Brownstein, M 87a
Visual Pattern Recognition in	Humans	Caelli, TM 87b
Pattern Recognition by	Humans and Machines.	Schwab, EC 86a
Pattern Recognition by	Humans and Machines.	Schwab, EC 86b
VLSI Implementation of a Neural Network Memory with Several	Hundreds of Neurons.	Graf, HP 86a
A Neurocomputer Based on an Analog-Digital	Hybrid Architecture.	Moopenn, A 87a
Optical and	Hybrid Computing	Szu, HH 86g
Simple Analog and	Hybrid Networks for Surface Interpolation.	Hutchinson, JM 86a
	Hybrid Optical Inference Machines: Architectural Considerations.	Warde, C 86a
The Restricted Language Architecture of a	Hybrid Representation System	Vilain, MB 85a
Develpment of the Two-Part Pattern During Regeneration of the Head in	Hydra.	Bode, PM 88a
Plasticity in the Nervous System of Adult	Hydra I. The Position-Dependent Expression of FMRFamide-Like	Koizumi, O 86a
Plasticity in the Nervous System of Adult	Hydra II. Conversion of Ganglion Cells of the Body Column into	Koizumi, O 88a
Neural Computing:	Hype or Reality?	Aleksander, I 88d
Using the	Hyperbase Intelligent Processor for Fault Analysis and Detection	Tenorio, MF 87i
A Mathematical Model of the Primary Visual Cortex and	Hypercolumn.	Okajima, K 86a
A Neural Network Model for the Developmental Process of	Hypercomplex Cells	Nagano, T 83a
Combinatorial	Hypercompression	Hecht-Nielsen, R 87a
Piriform (Olfactory) Cortex Model on the	Hypercube.	Bower, JM 87a
Neural Networks, Orientations of the	Hypercube and Algebraic Threshold Functions.	Baldi, P 88f
Neural Networks, Acyclic Orientations of the	Hypercube and Sets of Orthogonal Vectors	Baldi, P 88d
Evaluation of a	Hypercube Based Compact Neural Network	Somani, AK 88a
...Algorithms for Optimization of the Computational Load of a	Hypercube Supercomputer.	Barhen, J 87a
Optimization of the Computational Load of a	Hypercube Supercomputer Onboard a Mobile Robot	Barhen, J 87b
...Cells of the Body Column into Epidermal Sensory Cells of the	Hypostome.	Koizumi, O 88a
Neural Dynamics of Circadian Rhythms: The Mammalian	Hypothalamic Pacemaker	Carpenter, GA 85c
On Evidential Reasoning in a Hierarchy of	Hypotheses	Pearl, J 86c
On Evidential Reasoning in a Hierarchy of	Hypotheses	Pearl, J 86d
Recognizing Partially Visible Objects Using Feature Indexed	Hypotheses	Knoll, TF 86a
...Learning Minimum Risk Pattern Classification for Dependent	Hypotheses and Dependent Measurements.	Hilborn, CG 69a
Cortical Encoding of Memory:	Hypotheses Derived from Analysis and Simulation of Physiological	Lynch, G 89a
...in the Motor Cortex Encode Movement Direction? An Alternative	Hypothesis	Mussa-Ivaldi, FA 88a
Function of the Thalamic Reticular Complex: The Searchlight	Hypothesis	Crick, F 84a
...Connectionism Compatible with the Physical Symbol System	Hypothesis	Derthick, MA 86a
The Action of the Corticofugal Pathway on Sensory Thalamic Nuclei: A	Hypothesis	Koch, C 87b
The Biochemistry of Memory: A New and Specific	Hypothesis	Lynch, G 84a
The T5 Base Modulator	Hypothesis: A Dynamic Model of T5 Neuron Function in Toads	Betts, B 88a
Oscillatory Neural Ectoderm (One): The "One"	Hypothesis for Human Holoencephalon and Behavior	Deshmukh, VD 87a
Neurons with	Hysteresis Form a Network that Can Learn without Any Changes in.	Hoffman, GW 86a
Local Versus Global Minima,	Hysteresis, Multiple Meanings.	Kergosien, YL 86a
Economic Prediction Using Neural Networks: The Case of	IBM Daily Stock Returns.	White, H 88a
	IBM Neural Network Chips Make Everything Obsolete.	Gantz, J 88a
	IBM Research Yields Artificial Neural Net Workstation.	Schwartz, TJ 86a
Pattern Classification by Neural Network: An Experimental System for	Icon Recognition	Gullichsen, E 87a
...Maps of the Oropharynx in the Vagal Lobe of the Channel Catfish,	Ictalurus punctatus.	Kanwal, JS 88b
Random Cells: An	Idea Whose Time Has Come and Gone... And Come Again?	Gallant, SI 87b
On the Genesis of Abstract	Ideas.	Posner, MI 68a
Retention of Abstract	Ideas.	Posner, MI 70a

...Toward a Synthesis of Instrumental, Perceptual, and Cognitive	Ideas.	Grossberg, S 84d
Neural Computing:	Ideas from the Brain	Aleksander, I 88e
A Logical Calculus of the	Ideas Immanent in Nervous Activity	McCulloch, WS 43b
A Logical Calculus of the	Ideas Immanent in Nervous Activity	McCulloch, WS 43a
Analog MOS Integrated Circuits - Certain New	Ideas, Trends, and Obstacles	Tsividis, Y 87a
An N-Player Sequential Stochastic Game with	Identical Payoffs.	Narendra, KS 83a
A Learning Method for System	Identification	Nagumo, J 67a
...Algorithms and Architectures to Correlation / Tracking and	Identification	Gardner, S 86a
...Memory Codes: Codification and Repetition Effects in Word	Identification	Salasoo, A 85a
Experimental Tests of a Hierarchical Model of Word	Identification	Johnston, JC 80a
Familiarity of Letter Sequences and Tachistoscopic	Identification	Miller, GA 54a
Letter and Configuration Information in Word	Identification	McClelland, JL 77a
Neural Networks for Medical Image Processing: A Study of Feature	Identification	Dayhoff, RE 88a
Serial Position Functions for Letter	Identification at Brief and Extended Exposure Durations.	Estes, WK 76a
An Experimental Investigation of Process	Identification by Competitive Evolution.	Kaufman, H 67a
Preliminary Letter	Identification in the Perception of Words and Nonwords	McClelland, JL 75a
Simulation of Task Performance as a Guide to the	Identification of Solution Strategy from Eye-Movement Recordings	Deffner, G 87a
Self Organizing Neural Networks for the	Identification Problem	Tenorio, MF 88a
Solving of Optimization and	Identification Problems by the Committee Methods	Mazurov, VD 87a
Self Organizing General Pattern Class Separator and	Identifier	Cooper, LN 82a
Scientists	Identify "Gate" in Brain as Crucial to Memory.	Johnson, G 88a
A Technique to	Identify Nearest Neighbors	Yunck, TP 76a
	Identifying the Speech Codes	Foss, DJ 80a
Real-Time Holographic Associative Memory for	Identifying Words in a Continuous Letter String.	Paek, EG 88c
	IEEE Puts Neural Nets into Focus	Johnson, RC 87b
	Ignition Phenomena in Random Nets.	Rapoport, A 52a
"	Ignorance-Based" Neural-Network Signal Processing in Brain Research.	Gevins, AS 87a
L'Uso di Modelli Computazionali in Psicofisiologia Ed	Il Principio Gnoseologico di Rappresentazione.	Basti, G 86b
Simulation and Evaluation of	Ill Defined Neuronal Networks with Random Structure and Parameter.	Benito, S 80a
Adaptive Control of	Ill-Defined Systems.	Selfridge, OG 84a
	Ill-Posed Problems and Regularization Analysis in Early Vision	Poggio, TA 84b
An Analog Model of Computation for the	Ill-Posed Problems of Early Vision	Poggio, TA 84a
Neural Dynamics of Surface Perception: Boundary Webs,	Illuminants, and Shape-From-Shading.	Grossberg, S 87m
Surface Lightness and Cues for the	Illumination	Beck, J 71a
...of Edges as Critical to the Perception of Reflectance and	Illumination	Gilchrist, AL 83a
...for Brightness Perception Under Constant and Variable	Illumination Conditions.	Grossberg, S 87c
The	Illusion of Transparency and Chromatic Subjective Contours	Meyer, GE 83a
...Recording, Part II: Feedback, Expectation, Olfaction, and	Illusions.	Grossberg, S 76c
Visual	Illusions in Neural Networks: Line Neutralization, Tilt After.	Levine, DS 76a
Shape Recognition and	Illusory Conjunctions.	Hinton, GE 85b
A Factored Architecture to Solve	Illusory Conjunctions in Neural Networks	Strong, GW 87a
On the Nature of Inducing Forms Generating Perceptions of	Illusory Contours.	Prazdny, K 85a
Neural Dynamics of Form Perception: Boundary Completion,	Illusory Figures, and Neon Color Spreading	Grossberg, S 86j
Landmark Learning: An	Illustration of Associative Search	Barto, AG 81c
Bayesian	Image Analysis	Geman, D 86a
Biomedical	Image Analysis	Sklansky, J 86a
Iterative Methods in	Image Analysis	Rosenfeld, A 77a
Iterative Methods in	Image Analysis	Rosenfeld, A 78a
Complete Discrete 2-D Gabor Transforms by Neural Networks for	Image Analysis and Compression	Daugman, JG 88a
	Image Analysis and Mathematical Morphology	Serra, J 82a
	Image Analysis and Mathematical Morphology	Serra, J 88a
A Neural Network Approach to Transform	Image Coding	Chua, LO 88a
	Image Compression by Back Propagation: An Example of Extensional	Cottrell, GW 87a
Bayesian Recursive	Image Estimation	Nahi, NE 72a
Dynamics of	Image Formation by Nerve Cell Assemblies	Freeman, WJ 83a
	Image Halftoning and Reconstruction Using a Neural Network	Kollias, S 89a
Digital	Image Halftoning Using Neural Networks	Anastassiou, D 88a
A Model for the Economical Encoding of the Visual	Image in Cerebral Cortex	Sakitt, B 82a
Multi-Layer Perceptron Inversion Applied to	Image Neighbourhood Operations	Dodd, N 87a
	Image Pre-Processing with a Generalised Grey Scale N-Tuple Operator.	Austin, J 88b
Selection of	Image Primitives for General-Purpose Visual Processing	Walters, DKW 87i
Advances in Mathematical Models for	Image Processing	Jain, AK 81a
Cellular Logic Arrays for	Image Processing	Preston Jr, K 86a
Handbook of Pattern Recognition and	Image Processing	Young, TY 86a
Multi-Resolution	Image Processing & Analysis.	Rosenfeld, A 84a
Neural Networks for Medical	Image Processing: A Study of Feature Identification.	Dayhoff, RE 88a
	Image Processing and Understanding	Habibi, A 86a
...Digital Signal Processors with the Explorer / Odyssey for	Image Processing and Understanding Applications.	Gove, RJ 87a
An	Image Processing Architecture for Real Time Generation of Scale and.	Messner, RA 85a

Preceding	Keyword	Reference
Forgetting as a Way to	Improve Neural-Net Behavior.	Sasiela, RJ 86a
Neural Circuits Will	Improve Vision, Speech Systems	Brown, C 86f
An	Improved Elastic Net Method for the Traveling Salesman Problem	Burr, DJ 88a
	Improved Hidden Markov Modeling of Phonemes for Continuous Speech.	Schwartz, RM 84a
	Improved Holographic Model of Temporal Recall.	Gabor, D 68a
An	Improved Neural-Network Model for the Neural Integrator of the	Cannon, SC 85a
An	Improved Three-Layer, Back Propagation Algorithm	Stornetta, WS 87a
	Improvement in Pattern-Selectivity of a Cognitron.	Fukushima, K 79a
On the Asymptotic	Improvement in the Outcome of Supervised Learning Provided by.	Cooper, DB 70a
Modeling Accuracy	Improvements for the Group Method of Data Handling	Molnar, DO 87a
	Improving the Learning Rate of Neural Networks	Nelken, I 88a
...the Process of Problem-Solving by Associative Networks Capable of	Improving the Performance.	Hirai, Y 88a
...to Singular Perturbation Problems with Applications to Nerve	Impulse Equations.	Carpenter, GA 77a
Nerve	Impulse Equations.	Carpenter, GA 76a
Periodic Solutions of Nerve	Impulse Equations.	Carpenter, GA 77b
...Projection. III. Comparison of Random Train and Paired	Impulse Stimulation.	Berger, TW 88b
...Perforant Path-Dentate Projection. II. Effects of Random	Impulse Train Stimulation.	Berger, TW 88a
Inhomogeneous Celluar Automata (	INCA).	Hartman, H 86a
Independent Components Analysis (	INCA) Versus Principal Components Analysis	Jutten, C 88a
Segmenting Intersecting and	Incomplete Boundaries.	Krishnan, G 88a
Construction of Associative Memory from	Incomplete Memorized Information Using Space Reduction Method.	Murakami, K 88a
Evolutionary Pattern Recognition in	Incomplete Nonlinear Multithreshold Networks	Mucciardi, AN 66a
Localizing Sound Source by	Incorporating Biological Auditory Mechanism.	Sugie, N 87a
	Incorporating Fuzzy Membership Functions into the Perceptron	Keller, JM 85a
A Model of the Motor Servo:	Incorporating Nonlinear Spindle Receptor and Muscle Mechanical	Gielen, CCAM 87a
Expectations	Increase the Benefit Derived from Parafoveal Visual Information in	McClelland, JL 81c
	Increasing Coherence in a Distributed Problem Solving Network.	Durfee, EH 85b
	Incremental Network Generation in Template-Based Word Recognition.	Lee, K 85a
Superior Accuracy of Decrementing over	Incrementing Associative Networks when Operating in Initially.	Dobson, VG 87a
Levels	Indeed! A Response to Broadbent.	Rumelhart, DE 85c
	Independent Components Analysis (INCA) Versus Principal Components	Jutten, C 88a
	Independent Components Analysis in Neural Networks	Herault, J 88b
Software Package for Interactive Text	Independent, Speaker Verification.	Cohen, A 86a
	Index Term Weighting	Jones, KS 73a
Recognizing Partially Visible Objects Using Feature	Indexed Hypotheses	Knoll, TF 86a
Upper and Lower Probabilities	Induced by a Multi-Valued Mapping.	Dempster, AP 67a
Eye Movements	Induced by Linear Acceleration on a Parallel Swing	Baloh, RW 88a
A Test for Modulation of Synaptic Efficacy by	Induced Membrane Potential Variance.	Gardner, D 88a
A Network that	Induces a Distributed Representation Useful for Encoding Kinship	Hinton, GE 86a
On the Nature of	Inducing Forms Generating Perceptions of Illusory Contours	Prazdny, K 85a
A Marker	Induction Mechanism for the Establishment of Ordered Neural Mapping.	Willshaw, DJ 79a
A Simple Network Capable of	Inductive Generalization	Willshaw, DJ 76a
Holography, Associative Memory, and	Inductive Generalization	Willshaw, DJ 81a
	Inductive Inference by Neural Networks	Hestenes, D 88a
An	Inductive Inference Machine.	Solomonoff, RJ 57a
	Inductive Information Retrieval Using Parallel Distributed	Mozer, MC 84a
Novel	Inductorless Neuristor Line.	Wilamowski, BM 75a
Brain Drain Siphons Neural Net Pioneers into	Industry	Brown, C 87a
Neurocomputing - A New	Industry	Rosenfeld, E 87a
Geometrical and Statistical Properties of Systems of Linear	Inequalities with Applications in Pattern Recognition.	Cover, TM 65a
On the	Inequality of Cover and Hart in Nearest Neighbor Discrimination.	Devroye, L 81a
A Model of	Inexact Reasoning in Medicine.	Shortliffe, EH 75a
A Neural Model of the Mechanisms Underlying	Infant Perception of Voice-Onset Time.	Pont, MJ 88b
A Neural Model of	Infant Speech Perception	Pont, MJ 88a
Pattern Recognition of Fourier Transform	Infared Spectra of Organic Compounds	Frankel, DS 84a
A Generalization of Bayesian	Inference.	Dempster, AP 68a
Optimal Perceptual	Inference.	Hinton, GE 83b
Parallel Logical	Inference.	Ballard, DH 84a
Parallel Logical	Inference and Energy Minimization.	Ballard, DH 86b
Symbols Among the Neurons: Details of a Connectionist	Inference Architecture	Touretzky, DS 85a
Inductive	Inference by Neural Networks	Hestenes, D 88a
Adaptive	Inference in Fuzzy Knowledge Networks.	Kosko, B 87b
Schema Selection and Stochastic	Inference in Modular Environments.	Smolensky, P 83c
Schema Selection and Stochastic	Inference in Modular Environments.	Smolensky, P 83b
An Inductive	Inference Machine.	Solomonoff, RJ 57a
Hybrid Optical	Inference Machines: Architectural Considerations	Warde, C 86a
TIN: A Trainable	Inference Network.	Winter, CL 87a
TIN: A Trainable	Inference Network.	Winter, CL 87b
	Inference of Spatial Relations by Self-Organizing Networks	Williams, RJ 85b

An	Inference Technique for Integrating Knowledge from Disparate	Garvey, TD 81a
An	Inference upon the Neural Network Finding Binocular Correspondence	Hirai, Y 78a
Optical Learning (	Inference) Machines	Eichmann, G 85a
Bisensory Stimulation:	Inferring Decision-Related Processes from the P300 Component	Squires, NK 77a
	Inferring the Meaning of Direct Perception	Hinton, GE 80a
	Inferring Three-Dimensional Shapes from Two-Dimensional Silhouettes.	Richards, WA 87a
1-D Optical Cauchy Machine	Infinite Film Spectrum Search	Scheff, K 87a
Storing	Infinite Numbers of Patterns in a Spin-Glass Model of Neural	Amit, DJ 85b
Zero Temperature Parallel Dynamics for	Infinite Range Spin Glasses and Neural Networks	Gardner, E 87a
	Infinitely Many Commensurate Phases in a Simple Ising Model	Fisher, ME 80a
	Infinite-Ranged Models of Spin-Glasses	Kirkpatrick, S 78a
Passing Markers: A Theory of Contextual	Influence in Language Comprehension	Charniak, E 83a
	Influence of Contrast on Foveal and Peripheral Detection of	van de Grind, WA 87a
The	Influence of English Spelling Patterns on Pronunciation	Smith, PT 76a
...of a Simultaneous Functional Order in Nervous Systems III. The	Influence of Environmental Constraints on the Resulting Functional	Toet, A 87c
	Influence of Interconnection Weight Discretization and Noise in an	Von Lehmen, A 87a
The Non-Parametric	Influence of Light on Mammalian Circadian Rhythms	Carpenter, GA 85d
	Influence of Noise on the Behavior of an Autoassociative Neural	Buhmann, J 86b
	Influence of Noise on the Function of a "Physiological" Neural	Buhmann, J 87a
Input-Output Relations in Computer-Simulated Nerve Cells:	Influence of Statistical Properties, Strength, etc.	Segundo, JP 68a
Semantic-Context Effects on Word Recognition:	Influence of Varying the Proportion of Items Presented in an	Tweedy, JR 77a
Factors	Influencing Learning by Backpropagation	Von Lehmen, A 88a
State-Dependent Factors	Influencing Neural Plasticity: A Partial Account of the Critical	Munro, PW 86b
...Functional Implications for Storage and Retrieval of Olfactory	Information	Wilson, MA 87a
...Functional Implications for Storage and Retrieval of Olfactory	Information	Bower, JM 88a
A Distributed Memory Model for Serial-Order	Information	Murdock, BB 83a
A New Optical Method of Storing and Retrieving	Information	van Heerden, PJ 63a
A Theory for the Storage and Retrieval of Item and Associative	Information	Murdock Jr, BB 82a
...Based Distributed Storage Device for the Associative Search of	Information	Oh, S 84a
Distributed Memory and the Representation of General and Specific	Information	McClelland, JL 85a
...Exemplar-Based Generalization and the Abstraction of Categorical	Information	Homa, D 81a
Obtaining Shape From Shading	Information	Horn, B 75a
On a Method of Storing	Information	Roy, AE 60a
On the Scheme by Which the Human Brain Stores	Information	Legendy, CR 67a
Representing Visual	Information	Marr, D 77b
Signal Reconstruction from Fourier Transform Sign	Information	Curtis, SR 85a
The Non-Local Storage of Temporal	Information	Longuet-Higgins, HC 68b
A Computational Model of How Cortical Complex Cells Multiplex	Information About Position, Contrast, Orientation, Spatial	Grossberg, S 87a
Stimulus	Information and Contextual Information as Determinants of	Tulving, E 63a
Order	Information and Distributed Memory Models	Ratcliff, R 87a
	Information and Self-Organization: A Macroscopic Approach to	Haken, H 88a
The Transmission of	Information and the Effect of Local Feedback in Theoretical and	Uttley, AM 66a
Stimulus Information and Contextual	Information as Determinants of Tachistoscopic Recognition of Words	Tulving, E 63a
Electrotonic Processing of	Information by Brain Cells	Schmitt, FO 76a
	Information Capacity of Associative Memories	Kuh, A 88a
Comparison of	Information Capacity of Hebbian Neural Networks	Keeler, JD 87a
	Information Capacity of Outer-Product Neural Networks	Keeler, JD 87b
	Information Capacity of the Hopfield Model	Abu-Mostafa, YS 85b
On "	Information Capacity of the Hopfield Model"	Xu, X 88c
Spectral	Information Coding by Infrared Photoreceptors	Coon, DD 86a
Encoding of Visual	Information: Correlation of EEG Response to Intracellular Response	Anninos, PA 82a
Complexity of	Information Extraction	Abu-Mostafa, YS 86a
Retrieval of	Information from Long-Term Memory	Anderson, JR 83a
A Novel Architecture for Real Time Pick Up of 3D Motion and 3D Layout	Information from the Flow of the Optic Array	Tsao, T 88a
On a method of Storing	Information: II. A Further Study of Model Properties	Roy, AE 62a
Storage and Processing of	Information in Distributed Associative Memory Systems	Kohonen, T 81c
Parallel-Processing Mechanisms and Processing of Organized	Information in Human Memory	Ratcliff, R 81a
Principles Underlying the Representation and Storage of	Information in Neuronal Networks in the Primate Hippocampus and	Rolls, ET 89a
Storage and Manipulation of	Information in Random Networks	Beurle, RL 62b
Morphological Evidence for Parallel Processing of	Information in Rat Macula	Ross, MD 88b
Expectations Increase the Benefit Derived from Parafoveal Visual	Information in Reading Words Aloud	McClelland, JL 81c
Representation of Sensory	Information in Self-Organizing Feature Maps	Kohonen, T 86c
Representation of Sensory	Information in Self-Organizing Feature Maps, and Relation of these	Kohonen, T 86d
Neural Dynamics of Attention Switching and Temporal Order	Information in Short-Term Memory	Grossberg, S 86i
Interaction of Two Sources of	Information in Tachistoscopic Word Recognition	Tulving, E 64a
Processing of Visual	Information in the Distal Neurons of the Vertebrate Retina	Curlander, JC 83a
Concentrating	Information in Time: Analog Neural Networks with Applications to	Tank, DW 87b
Letter and Configuration	Information in Word Identification	McClelland, JL 77a
On Learning,	Information, Lateral Inhibition, and Transmitters	Grossberg, S 69d

Neurophysiological Predictions of a Two-Pathway	Informon Theory of Neural Conditioning	Uttley, AM 76b
Spectral Information Coding by	Infrared Photoreceptors.	Coon, DD 86a
	Infrared Transient Sensing	Coon, DD 88a
Dynamic Control of an Artificial Neural System: The Property	Inheritance Network.	Ryan, TW 87a
Emulating the Dynamics for a Class of Laterally	Inhibited Neural Networks.	Lemmon, M 88a
Analysis and Simulation of Double Layer Neural Networks with Mutually	Inhibiting Interconnections.	Tokura, T 77a
Analysis and Simulation of Networks of Mutually	Inhibiting Neurons	Morishita, I 72a
A Mathematical Theory of Lateral Sensory	Inhibition	Coleman, BD 71a
A Model of a Neural Network with Recurrent	Inhibition	Wigstrom, H 74a
...of Feature Extraction - A Self-Organizing Network with Feedback	Inhibition	Miyake, S 83a
Natural Language Processing Using Spreading Activation and Lateral	Inhibition	Pollack, JB 82a
On the Theory of Lateral	Inhibition	Hadeler, KP 74a
...Pavlovian Conditioning: Conditioned Reinforcement,	Inhibition, and Opponent Processing.	Grossberg, S 87b
...Modulated Pavlovian Conditioning: Conditioned Reinforcement,	Inhibition, and Opponent Processing.	Grossberg, S 87j
On Learning, Information, Lateral	Inhibition, and Transmitters	Grossberg, S 69d
Self-Organizing Process Based on Lateral	Inhibition and Weight Redistribution	Miikkulainen, R 87a
Virtual lateral	Inhibition in Parallel Activation Models of Associative Memory	Reggia, JA 85a
A Method for the Design of Stable Lateral	Inhibition Networks that is Robust in the Presence of Circuit.	Wyatt Jr, JL 88a
Stability Theorem for Lateral	Inhibition Networks that is Robust in the Presence of Circuit.	Standley, DL 88a
Circuit Design Criteria for Stable Lateral	Inhibition Neural Networks	Wyatt Jr, JL 88b
...Initial Conditions and Periodic Firing Modes in Reciprocal	Inhibition Neural Networks	Nagashino, H 81a
Recurrent Collateral	Inhibition Simulated in a Simple Neuronal Automata Assembly.	Axelrad, H 86a
Semantic Priming and Retrieval from Lexical Memory: Roles of	Inhibitionless Spreading Activation and Limited-Capacity Attention	Neely, JH 77a
...A Neural Network Model which Exhibits Both Excitatory and	Inhibitory Conditioning.	Ricart, R 88a
...Consideration on Pattern Separability in a Random Neural Net with	Inhibitory Connections	Torioka, T 80a
Pattern Separability in a Random Neural Net with	Inhibitory Connections	Torioka, T 79a
Pattern Separability of a Tightly Coupled Random Nerve Net with	Inhibitory Connections	Torioka, T 79b
...Functioning of Two Layered Random Nerve Nets with Feedforward	Inhibitory Connections	Torioka, T 88a
...of Pattern Separation in a Two-Layer Random Neural Net with	Inhibitory Connections	Torioka, T 80b
...System Noise on an Optical Implementation of an Additive Lateral	Inhibitory Network	Stirk, CW 87a
Coupled Van der Pol Oscillators - A Model of Excitatory and	Inhibitory Neural Interactions	Kawahara, T 80a
Variations in Amplitude and Time Course of	Inhibitory Postsynaptic Currents	Gardner, D 86a
...and Retrieval from Lexical Memory: Evidence for Facilitatory and	Inhibitory Processes	Neely, JH 76a
...Purkinje Cell Activity within the Basic Circuit and Modified by	Inhibitory Sys..	Pellionisz, AJ 74a
	Inhomogeneous Celluar Automata (INCA).	Hartman, H 86a
Goal Seeking Components for Adaptive Intelligence: An	Initial Assessment	Barto, AG 81b
Relations Between	Initial Conditions and Periodic Firing Modes in Reciprocal	Nagashino, H 81a
...by a Layered Network of Reinforcement Learning Elements: Some	Initial Experiments.	Anderson, CW 82a
...over Incrementing Associative Networks when Operating in	Initially Random Connectivities.	Dobson, VG 87a
Human Movement	Initiation: Specification of Arm, Direction, and Extent.	Rosenbaum, DA 80a
...Motor Cortex Somatic Representation Patterns after Motor Nerve	Injury	Sanes, JN 88a
	Innate and Learned Components in a Simple Visuo-Motor Reflex	Hoffmann, KP 88a
Integrate-And-Fire Models of Nerve Membrane Response to Oscillatory	Input.	Keener, JP 81a
The Brains of Men and Machines, Part 3: How the Brain Analyzes	Input.	Kent, EW 78c
The Response of a Model Neurone to a White Noise	Input.	Rubio, JE 75a
...of Bioelectric Potentials with Microelectrodes and Neutralized	Input Capacity Amplifiers.	Amatniek, E 58a
On Choosing Non-Random	Input Exemplars to Form Stable States in a Hopfield Neural Net	Feild Jr, WB 88a
An Associative Memory with Heterogeneous	Input Fields	Winter, CL 88b
An Invariant	Input for a Pattern Recognition Machine.	Ball, GH 62a
Rod Bipolar Array in the Cat Retina: Pattern of	Input from Rods and GABA-Accumulating Amacrine Cells	Freed, MA 87a
A Model of Synaptogenetic Consequences of Afferent	Input in Neuronal Nets	Dammasch, IE 88a
Addition and Multiplication Theorems for the	Input of Two Neurons Converging on a Third	Rapoport, A 51a
...of Neural Recognition Codes in Response to Arbitrary Lists of	Input Patterns	Carpenter, GA 86d
...Self-Organization of Stable Category Recognition Codes for Analog	Input Patterns	Carpenter, GA 87c
Encoding	Input/Output Representations in Connectionist Cognitive Systems.	Miikkulainen, R 88a
...Randomly Connected McCulloch-Pitts Networks: Differences Between	Input-Constant and Input-Variant Networks.	Dammasch, IE 84a
	Input-Output Analysis of an IPFM Neural Model: Effects of Spike.	Sanderson, AC 80a
	Input-Output Relations in Computer-Simulated Nerve Cells: Influence.	Segundo, JP 68a
Synaptic Potential Summation and Repetitive Firing Mechanisms:	Input-Output Theory for the Recruitment of Neurons into Epileptic.	Calvin, WH 72a
On the Statistical Efficiency of the LMS Algorithm with Nonstationary	Inputs	Widrow, B 84a
...McCulloch-Pitts Networks: Differences Between Input-Constant and	Input-Variant Networks	Dammasch, IE 84a
Why Neural Networks Don't Fly:	Inquiry into the Neurodynamics of Biological Intelligence.	Freeman, WJ 88b
Guided Propagation	Inside a Topographic Memory.	Beroule, D 87a
	Inside Intelligence.	Sternberg, RJ 86a
Phonotopic Maps -	Insightful Representation of Phonological Features for Speech.	Kohonen, T 84b
Algorithms and Techniques for Automated Visual	Inspection	Chin, RT 86a
Neural Nets	Inspire Computer Designer.	Johnson, RC 86d
	Instability of Pulse Activity in a Net with Threshold.	Ashby, WR 62a
Low-Dimensional Chaos in an	Instance of Epilepsy	Babloyantz, A 86a

	Intelligent Control of Robot Motion Based on Learning Method	Kawamura, S 87a
Behavior Analysis and	Intelligent Machines	Stephens, KR 88a
Cognitive Control of Autonomous Mobile Robots: Nested Hierarchical	Intelligent Module	Meystel, A 86a
Intelligent Recognition and Learning Control: Integrating	Intelligent Processing and Neural Networks for Adaptive Moving . . .	Lee, JSJ 87a
Using the Hyperbase	Intelligent Processor for Fault Analysis and Detection	Tenorio, MF 87i
A Mechanism for Stabilizing	Intelligent Properties of Neural Networks.	Dammasch, IE 87a
	Intelligent Recognition and Learning Control: Integrating.	Lee, JSJ 87a
Preliminary Design of an	Intelligent Robot.	Nilsson, NJ 67a
Electronic Life and Synthetic	Intelligent Systems.	Dress, WB 88a
Beyond Neural Networks - The Discipline of	Intelligent Systems Design Emerges	Johnson, RC 87a
Developers Show Neural Networks: "	Intelligent" Systems Learn, Reason Like Humans	Brownstein, M 87a
Design and Implementation of an Associative Memory for	Intensive Pattern Matching	Bou-Ghazale, S 88a
	Intentional Homunculi.	Hinton, GE 79c
Cognitive Neuroscience,	Intentionality and Realism: The Philosophical Importance of the. . .	Basti, G 87a
Pattern Formation by the Global Limits of a Nonlinear Competitive	Interactin in n-Dimensions	Grossberg, S 77b
A Program Package Implementing a System of	Interacting Adaptive Associative Networks.	Lansner, A 83a
	Interacting and Communicating Connectionist Agents	Allen, RB 88a
The Dynamics of Electrically	Interacting Cells.	Torre, V 83a
A Computer Program for Simulating a Network of	Interacting Neurons, Part I: Organization and Physiological.	Perkel, DH 76a
A Computer Program for Simulating a Network of	Interacting Neurons, Part II: Programming Aspects.	Perkel, DH 76b
A Computer Program for Simulating a Network of	Interacting Neurons, Part III: Applications.	Perkel, DH 76c
Two Models for Memory Organization Using	Interacting Traces	Anderson, JA 70a
Self-Organization of Neural Nets with Competitive and Cooperative	Interaction.	Sawada, R 82a
Psychophysiological Substrates of Schedule	Interaction and Behavioral Contrast.	Grossberg, S 81b
Spatial	Interaction and the Statistical Analysis of Lattice Systems.	Besag, J 74a
...of Neuron Selectivity: Orientation Specificity and Binocular	Interaction in Visual Cortex	Bienenstock, EL 82b
Digital Image Restoration Using Spatial	Interaction Models	Chellappa, R 82a
Modularity and	Interaction of Brain Regions Underlying Visuomoter Coordination. . .	Arbib, MA 85e
	Interaction of Synaptic Modification Rules within Populations of . .	Finkel, LH 85a
	Interaction of Two Sources of Information in Tachistoscopic Word . .	Tulving, E 64a
	Interaction of Visual and Cognitive Effects in Word Recognition. . .	Becker, CA 77a
Two-Level Neural Networks: Learning by	Interaction with Environment	Fukaya, M 87a
The Case for	Interactionism in Language Processing.	McClelland, JL 87d
...and Duration in Massively Parallel Networks: Exploiting Link	Interactions	Chun, HW 86a
...der Pol Oscillators - A Model of Excitatory and Inhibitory Neural	Interactions	Kawahara, T 80a
...Making Under Risk: Affective Balance and Cognitive-Emotional	Interactions	Grossberg, S 87k
...Circadian Rhythms: Split Rhythms, After-Effects, and Motivational	Interactions	Carpenter, GA 86m
Textrons, the Elements of Texture Perception and Their	Interactions	Julesz, B 81a
Using Electronic Circuits to Model Simple Neuroelectric	Interactions	Lewis, ER 68d
Processing	Interactions and Lexical Access During Word Recognition in	Marslen-Wilson, WD 78a
Nonlinear	Interactions in a Dendritic Tree: Localization, Timing, and Role in.	Koch, C 83a
...of Neuron Selectivity: Orientation Selectivity and Binocular	Interactions in the Visual Cortex.	Bienenstock, EL 82a
Perceptual	Interactions in Two-Word Displays: Familiarity and Similarity. . . .	McClelland, JL 86h
Levels of Modeling of Neural	Interactions Underlying Visuomoter Coordination.	Arbib, MA 85d
How the Catfish Tracks Its Prey: An	Interactive "Pipelined" Processing System May Direct Foraging Via. .	Kanwal, JS 88a
Speech Perception as a Cognitive Process: The	Interactive Activation Model	Elman, JL 83a
An	Interactive Activation Model of Context Effects in Letter.	McClelland, JL 81b
An	Interactive Activation Model of Context Effects in Letter.	McClelland, JL 81a
An	Interactive Activation Model of Context Effects in Letter.	Rumelhart, DE 82a
Competitive Learning: From	Interactive Activation to Adaptive Resonance	Grossberg, S 88a
Competitive Learning: From	Interactive Activation to Adaptive Resonance	Grossberg, S 87g
A Simple Neural Network Generating an	Interactive Memory	Anderson, JA 72a
A Simple Neural Network Generating an	Interactive Memory	Anderson, JA 72b
Models of	Interactive Memory	Anderson, JA 72c
Massively Parallel Parsing: A Strongly	Interactive Model of Natural Language Interpretation	Waltz, DL 88c
Massively Parallel Parsing: A Strongly	Interactive Model of Natural Language Interpretation	Waltz, DL 85a
Spatially Localized	Interactive Neural Populations-I, A Mathematical Model	Brannan, JR 81a
Spatially Localized	Interactive Neural Populations-II: Stability and Dynamics of	Brannan, JR 82a
Unbounded Oscillatory Solutions for a System of	Interactive Populations.	Cohen, MA 85a
	Interactive Processes in Speech Perception: The TRACE Model.	McClelland, JL 86e
ISCON: An	Interactive Simulator for Connectionist Networks	Small, SL 82b
Software Package for	Interactive Text Independent, Speaker Verification	Cohen, A 86a
How to Label Nerve Cells so that they can	Interconnect in an Ordered Fashion	von der Malsburg, C 77a
A VLSI	Interconnect Structure for Neural Networks	Bailey, J 88a
An	Interconnect Structure for Wafer Scale Neurocomputers.	Rudnick, M 88a
Parallel Gaussian Elimination on an Optically	Interconnected Data Flow Computer.	Nelken, I 87a
Optoelectronic Neural Networks Based on Holographically	Interconnected Image Processors.	Owechko, Y 88a
A Large VLSI Fully	Interconnected Neural Network.	Verleysen, M 88b
A Locally	Interconnected Neural Network for Deconvolution.	Petsche, T 87a

An	Introduction to Biological and Mechanical Pattern Recognition. . . .	Gose, EE 69a
An	Introduction to Chaotic Dynamical Systems.	Devaney, RL 87a
Perceptrons: An	Introduction to Computational Geometry	Minsky, ML 69a
An	Introduction to Computing with Neural Nets	Lippmann, RP 87b
	Introduction to Factor Analysis: What It Is and How to Do It	Kim, J 78a
	Introduction to Linear Algebra in Parallel Distributed Processing. .	Jordan, MI 86b
An	Introduction to Mathematical Learning Theory	Atkinson, RC 65a
	Introduction to Mathematical Techniques in Pattern Recognition . . .	Andrews, HC 72a
An	Introduction to Neural Computing	Kohonen, T 88a
Teaching a Topic in Cybernetics with APL: An	Introduction to Neural Net Modeling.	Peelle, HA 81a
An	Introduction to Relaxation Methods	Williams, G 87a
An	Introduction to Spatial Point Processes and Markov Random Fields . .	Isham, V 81a
Neural Networks:	Introduction to the 1 December 1987 Issue of Applied Optics.	Carpenter, GA 87e
An	Introduction to the Application of the Theory of Probabilistic . . .	Levinson, SE 83a
The Cerebral Computer: An	Introduction to the Computational Structure of the Human Brain . . .	Baron, RJ 87a
Natural and Social System Metaphors for Distributed Problem Solving:	Introduction to the Issue.	Chandrasekaran, B 81a
A Brief	Introduction to the Operation of an Adaptive Resonance Network . . .	Stork, DG 88a
Sight and Mind: An	Introduction to Visual Perception.	Kaufman, L 74a
A Threshold Logic Network for Shape	Invariance	Brousil, JK 67a
Video Rate Optical Associative Memory with Shift	Invariance	Owechko, Y 87c
Research Note - Cortical Mapping and Perceptual	Invariance: A Reply to Cavanagh.	Schwartz, EL 83a
Learning	Invariance, and Generalization in High-Order Neural Networks	Giles, CL 87a
All-Optical Associative Memory with Shift	Invariance and Multiple-Image Recall	Dunning, GJ 87a
Shift	Invariance in Optical Associative Memories	Psaltis, D 86b
Encoding Geometric	Invariances in Higher-Order Neural Networks.	Giles, CL 88a
	Invariant Cycles in the Random Mapping of n Integers into.	Coste, J 86a
An	Invariant Input for a Pattern Recognition Machine.	Ball, GH 62a
Adaptive	Invariant Novelty Filters.	Szu, HH 86a
	Invariant Object Recognition Using a Distributed Associative Memory.	Wechsler, H 88a
Position, Rotation, and Scale	Invariant Optical Correlation.	Casasent, DP 76a
A Neural Net for Size and Position	Invariant Pattern Recognition.	Willson, NJ 87a
A Neural Network for	Invariant Pattern Recognition.	Bienenstock, EL 87a
An Artificial Neural System Design for Rotation and Scale	Invariant Pattern Recognition.	Reber, WL 87a
Performance of Backpropagation for Rotation	Invariant Pattern Recognition.	Yang, H 87a
	Invariant Pattern Recognition and Recall by an Attentive Self. . . .	Carpenter, GA 87d
...Architecture for Real Time Generation of Scale and Rotation	Invariant Patterns	Messner, RA 85a
Neural Dynamics of Planned Arm Movements: Emergent	Invariants and Speed-Accuracy Properties During Trajectory	Bullock, D 88b
Neural Dynamics of Planned Arm Movements: Emergent	Invariants and Speed-Accuracy Properties During Trajectory	Bullock, D 88c
Neural Dynamics of Category Learning and Recognition: Structural	Invariants, Evoked Potentials, and Reinforcement	Carpenter, GA 86k
	Invariants in Automata Networks.	Goles, E 86a
	Invariants in Experimental Data on Linear Kinetics and the	Berman, M 56a
...and Tensorial Solution Using the Moore-Penrose Generalized	Inverse.	Pellionisz, AJ 84a
On the Effect of Noise on the Moore-Penrose Generalized	Inverse Associative Memory	Stiles, GS 85a
Generalized	Inverse of a Matrix and Its Applications	Rao, CR 72a
The	Inverse Problem for Linear Boolean Nets.	Caianiello, ER 88a
An	Inverse Problem in Neural Processing	Oguztoreli, NN 86a
Probabilistic Solution of	Inverse Problems	Marroquin, JL 85a
The Tendency to See Objects as Equidistant and Its	Inverse Relation to Lateral Separation	Gogel, WC 56a
An Optimization Network for Matrix	Inversion.	Jang, J 88a
Multi-Layer Perceptron	Inversion Applied to Image Neighbourhood Operations.	Dodd, N 87a
Engineering Design Neural Networks Using Split	Inversion Learning	McAulay, AD 87a
Phototransduction in	Invertebrates.	Ogmen, H 87d
	Inverting a Connectionist Network Mapping by Back-Propagation of . .	Williams, RJ 86a
Talking to Computers: An Empirical	Investigation.	Hauptmann, AG 87a
An Experimental	Investigation of a Class of Pattern Recognition Synthesis.	Mattson, RL 63a
An	Investigation of Knowledge Representation in a Neural Network. . . .	Healy, MJ 88a
An Experimental	Investigation of Process Identification by Competitive Evolution . .	Kaufman, H 67a
	Investigation of the Timing of Velar Movements during Speech	Moll, KL 71a
	Investigations into the Pattern Processing Capabilities of	Lansner, A 86a
A Computer Package for the Evaluation of Neuron Models	Involving Large Uniform Networks	Lee, MH 82a
Automated Generation of Connectionist Expert Systems for Problems	Involving Noise and Redundancy	Gallant, SI 87a
Architectural Relationships	Involving Symbolic Substitution.	Johnson, JL 88a
Voltage-Dependent Calcium and Potassium	Ion Conductances: A Contingency Mechanism for an Associative	Alkon, DL 79a
...Model of the Neuron Based on the Dynamics of Potassium and Sodium	Ion Fluxes	Lewis, ER 64a
Input-Output Analysis of an	IPFM Neural Model: Effects of Spike Regularity and Record Length . .	Sanderson, AC 80a
The	Iron Wire Model of the Neuron: A Review.	Lewis, ER 68c
	Iron-Wire Neuron Model	Suzuki, R 67a
Mathematical Analysis and Application of	ISCON: An Interactive Simulator for Connectionist Networks	Small, SL 82b
Infinitely Many Commensurate Phases in a Simple	Ising Model.	Fisher, ME 80a

Cell Automata:	Key to Science Spreadsheet?	Brown, C 87b
Comments on the Paper "Threshold Regulation - A	Key to the Understanding of the Combined Dynamics of EEG and Event	Palm, G 87b
Optimal Association with Partly Missing	Key Vectors	Murakami, K 82a
...System for Laryngectomees Using "Electropalatograph" and a Small	Keyboard	Muraki, S 86a
A Neural Model Theory Leading to	Kindling Effect of Epilepsy	Anninos, PA 87b
An Extended First Approximation Model for the Amygdaloid	Kindling Phenomenon	Lieblich, I 78a
A Critical Comparison of Two	Kinds of Adaptive Classification Networks	Steinbuch, K 65a
	Kinematic Networks: A Distributed Model for Representing and	Mussa-Ivaldi, FA 88b
Efficient Method for Estimating	Kinetic Constants from Single Channel Patch Clamp Recordings	Chay, TR 87b
Efficient Method for Estimating	Kinetic Constants from Single Channel Patch Clamp Recordings	Chay, TR 87a
Invariants in Experimental Data on Linear	Kinetics and the Formulation of Models	Berman, M 56a
...that Induces a Distributed Representation Useful for Encoding	Kinship Relations	Hinton, GE 86a
	K-Lines: A Theory of Memory	Minsky, ML 80a
A Branch and Bound Algorithm for Computing	k-Nearest Neighbors	Fukunaga, K 75a
Optimization of	k-Nearest-Neighbor Density Estimates	Fukunaga, K 73a
How We	Know Universals: The Perception of Auditory and Visual Forms	Pitts, WH 47a
How We	Know Universals: The Perception of Auditory and Visual Forms	Pitts, WH 47b
Do We	Know What We Want? Do We Know When We Want It?	Grossberg, S 77a
Do We Know What We Want? Do We	Know When We Want It?	Grossberg, S 77a
...Architecture for Representing and Reasoning About Structured	Knowledge	Derthick, MA 87a
A Framework for Representing	Knowledge	Minsky, ML 75a
Exploded Connections: Unchunking Schematic	Knowledge	Small, SL 82a
NETL: A System for Representing and Using Real-World	Knowledge	Fahlman, SE 79a
Neural Representation of Conceptual	Knowledge	Feldman, JA 89a
Neural Representation of Conceptual	Knowledge	Feldman, JA 36a
Representing Implicit	Knowledge	Fahlman, SE 81b
Mixing Binary and Continuous Connection Schemes for	Knowledge Access	Sharkey, NE 86a
Alternative	Knowledge Acquisition: Developing a Pulse-Coded Neural Network	Dress, WB 87b
	Knowledge Acquisition Tool for the Diagnosis of Electro-Mechanical	Tenorio, MF 87b
	Knowledge Bases and Machine Learning	Clema, JK 86a
Risk Weighting in	Knowledge Bases Using Fuzzy Cognitive Maps	Taber, WR 88c
Associative Networks: Representation and Use of	Knowledge by Computers	Findler, NV 79b
Fuzzy	Knowledge Combination	Kosko, B 86e
Example-Based	Knowledge Engineering with Connectionist Expert Systems	Gallant, SI 88c
Neural Networks, Potential Assistants to	Knowledge Engineers	Lendaris, GG 88b
Network Processing of Hierarchical	Knowledge for Classification and Diagnosis	Becker, LA 87a
An Inference Technique for Integrating	Knowledge from Disparate Sources	Garvey, TD 81a
Putting	Knowledge in its Place: A Scheme for Programming Parallel	McClelland, JL 88c
Putting	Knowledge in its Place: A Scheme for Programming Parallel	McClelland, JL 85b
Catching	Knowledge in Neural Nets	Jorgensen, C 86a
The Organization and Activation of Orthographic	Knowledge in Reading Aloud	Glushko, RJ 79a
Adaptive Inference in Fuzzy	Knowledge Networks	Kosko, B 87b
Hidden Patterns in Combined	Knowledge Networks	Kosko, B 87h
	Knowledge Processing Through Flow-of-Activation	Cruz, CA 87b
PLATO / ARISTOTLE: A Neural Net	Knowledge Processor	Voevodsky, J 87a
"I Lied About the Trees" Or, Defaults and Definitions in	Knowledge Representation	Brachman, RJ 85a
A Hierarchical System for Character Recognition with Stochastic	Knowledge Representation	Vlontzos, JA 88a
The QBKG System: Generating Explanations from a Non-Discrete	Knowledge Representation	Berliner, HJ 82a
An Investigation of	Knowledge Representation in a Neural Network	Healy, MJ 88a
Statistical Coding and Short-Term Synaptic Plasticity: A Scheme for	Knowledge Representation in the Brain	von der Malsburg, C 86b
Integration of Distributed and Symbolic	Knowledge Representations	Hutchison, WR 87a
Effects of Reinforcement on	Knowledge Retrieval and Evaluation	Leven, SJ 87a
Conceptual Graph	Knowledge Systems as Problem Context for Neural Networks	Lendaris, GG 88a
Mapping Images to a Hierarchical Data Structure - A Way to	Knowledge-based Pattern Recognition	Hartmann, G 88a
Integration of	Knowledge-Based System and Neural Network Techniques for Robotic	Handelman, DA 88a
A Methodology for the Development of General	Knowledge-Based Vision Systems	Riseman, EM 87a
...Algorithms for Discovering and Using General Patterns in Growing	Knowledge-Bases	Holland, JH 80a
	Knowledge-Rich Solutions to the Binding Problem: A Simulation of	Stenning, K 88a
Vector Quantization of Images Based Upon the	Kohonen Self-Organizing Feature Maps	Nasrabadi, NM 88a
Extending	Kohonen's Self-Organizing Mapping Algorithm to Learn Ballistic	Ritter, H 88b
	Kohonen's Self-Organizing Maps: Exploring their Computational	Ritter, H 88c
On the Stationary State of	Kohonen's Self-Organizing Sensory Mapping	Ritter, H 86a
Convergence Properties of	Kohonen's Topology Conserving Maps: Fluctuations, Stability and	Ritter, H 88a
	Kolmogorov's Mapping Neural Network Existence Theorem	Hecht-Nielsen, R 87e
Der Neuere	Konnektionismus, Ein Uberblick	Kemke, C 88a
A Simplified Version of	Kunihiko Fukushima's Neocognitron	Deutsch, S 81a
Competition and Cooperation in Neural Nets,	Kyoto, 1982: Proceedings	Amari, S 82c
Les Modeles Mathematiques de l'Organization Biologique de	la Memoire	Anderson, JA 78a
Les Applications des Modeles Connexionistes a la Reconnaissance de	la Parole	Wellekens, CJ 88b

A	Learning Algorithm for Boltzmann Machines.	Ackley, DH 85a
A	Learning Algorithm for Boltzmann Machines.	Ackley, DH 85b
A	Learning Algorithm for Boltzmann Machines.	Ackley, DH 88a
A	Learning Algorithm for Multilayer Neuronlike Network	Gawronski, RR 88a
An Adaptive Attentive	Learning Algorithm for Single-Layer Neural Networks.	Hassoun, MH 88a
Application of a General	Learning Algorithm to the Control of Robotic Manipulators.	Miller, WT 87b
A Comparison of Two Unsupervised	Learning Algorithms.	Bobrowski, L 80a
Neural Network	Learning Algorithms.	Sejnowski, TJ 88b
	Learning Algorithms and Applications	Lakshmivarahan, S 81a
	Learning Algorithms and Probability Distributions in Feed-Forward.	Hopfield, JJ 87a
	Learning Algorithms for Connectionist Networks: Applied Gradient	Watrous, RL 87c
Analysis and Comparison of Different	Learning Algorithms for Pattern Association Problems	Bernasconi, J 88a
e-Optimal	Learning Algorithms: Non-Absorbing Barrier Type.	Lakshmivarahan, S 79a
...Through the Structure of Event-Related Potentials During	Learning: An Experimental and Theoretical Analysis	Banquet, J 87a
Structure of Event-Related Potentials During	Learning: An Experimental and Theoretical Analysis	Banquet, J 86a
Landmark	Learning: An Illustration of Associative Search.	Barto, AG 81c
Neural Network Models of	Learning and Adaptation.	Denker, JS 87b
Category	Learning and Adaptive Pattern Recognition: A Neural Network Model.	Carpenter, GA 85a
Cognitive Computer: On Language,	Learning and AI.	Schank, RC 84a
	Learning and Applying Contextual Constraints in Sentence	St John, MF 88a
Neuron-like Elements That Can Solve Difficult	Learning and Control Problems.	Barto, AG 83c
Associative and Competitive Principles of	Learning and Development: The Temporal Unfolding and Stability of.	Grossberg, S 82b
Associative and Competitive Principles of	Learning and Development: The Temporal Unfolding and Stability of.	Grossberg, S 86d
	Learning and Distributed Memory: Getting Close to Neurons.	Aleksander, I 88b
On	Learning and Energy-Entropy Dependence in Recurrent and.	Grossberg, S 69c
	Learning and Energy-Entropy Dependence in Some Nonlinear Functional.	Grossberg, S 69b
Scaling, Machine	Learning, and Genetic Neural Nets.	Mjolsness, E 88b
The Hedonistic Neuron: A Theory of Memory	Learning and Intelligence.	Klopf, AH 82a
Self-Organizing Model of	Learning and Its Application to Eyesight System of Robot	Suzuki, H 88b
Self-Organizing Model for Pattern	Learning and Its Application to Robot Eyesight	Suzuki, H 88a
Conservation of Cellular Mechanisms for Models of	Learning and Memory.	Alkon, DL 88a
Distributed Model of Human	Learning and Memory.	McClelland, JL 86d
Neural Mechanisms of	Learning and Memory.	Rosenzweig, MR 76a
	Learning and Memory: Mechanisms of Information Storage in the.	Matthies, H 86a
	Learning and Memory Properties in Fully Connected Networks	Bruce, AD 86a
Associative-Contingency	Learning and Neuronal Networks	Levy, WB 79a
Adaptation,	Learning, and Pattern Recognition Systems: Theory and Applications	Mendel, JM 70a
	Learning and Problem Solving with Multilayer Connectionist Systems	Anderson, CW 86a
...of Feature Detectors in the Visual Cortex with Applications to	Learning and Reaction-Diffusion Systems.	Grossberg, S 76d
Understanding Biological Computation: Reliable	Learning and Recognition	Hogg, T 84a
Neural Dynamics of Category	Learning and Recognition: Attention, Memory Consolidation, and	Carpenter, GA 86i
Neural Dynamics of Category	Learning and Recognition: Attention, Memory Consolidation, and	Carpenter, GA 86j
Neural Dynamics of Category	Learning and Recognition: Structural Invariants, Evoked Potentials	Carpenter, GA 86k
The "Whirling Dervish", A Simulation Study in	Learning and Recognition Systems	Hoffman, A 62a
	Learning and Relearning in Boltzmann Machines.	Hinton, GE 86c
Skill	Learning and Repetition Priming in Symmetry Detection: Parallel.	Cohen, NJ 86a
Neural Dynamics of Word Recognition and Recall: Attentional Priming,	Learning, and Resonance.	Grossberg, S 86l
Neural Dynamics of Word Recognition and Recall: Attentional Priming,	Learning and Resonance	Grossberg, S 86m
Supervised	Learning and Systems with Excess Degrees of Freedom.	Jordan, MI 88b
A PDP	Learning Approach to Natural Language Understanding.	Sharkey, NE 89a
Associative Search Network: A Reinforcement	Learning Associative Memory.	Barto, AG 81a
Cooperativity in Networks of Pattern Recognizing Stochastic	Learning Automata.	Barto, AG 86a
Pattern Recognizing Stochastic	Learning Automata.	Barto, AG 85e
Pattern Recognizing Stochastic	Learning Automata.	Barto, AG 84a
	Learning Automata: A Critique.	Narendra, KS 77a
	Learning Automata: A Survey.	Narendra, KS 74a
Mathematical Analysis of	Learning Behavior of Neuronal Models	Cheung, JY 88a
Factors Influencing	Learning by Backpropagation.	Von Lehmen, A 88a
Pattern	Learning by Functional-Differential Neural Networks with Arbitrary	Grossberg, S 72d
Two-Level Neural Networks:	Learning by Interaction with Environment	Fukaya, M 87a
Classical and Instrumental	Learning by Neural Networks.	Grossberg, S 74a
Pavlovian Pattern	Learning by Nonlinear Neural Networks.	Grossberg, S 71b
...Architecture and Methodologies for Implementing Fast Stochastic	Learning by Simulated Annealing.	Farhat, NH 87b
	Learning by State Recurrence Detection	Rosen, BE 88a
	Learning by Statistical Cooperation of Self-Interested Neuron-Like	Barto, AG 85b
	Learning by Statistical Cooperation of Self-Interested Neuron-Like	Barto, AG 85c
Acceleration of	Learning by Variable Generalization for on-Line Self-Organizing.	Ersu, E 86a
How	Learning Can Guide Evolution	Hinton, GE 87b
...Implementations of Associative Networks with Versatile Adaptive	Learning Capabilities.	Fisher, AD 87b

"Bootstrap	Learning" in Threshold Logic Systems	Widrow, B 66a
	Learning-Logic: Casting the Cortex of the Human Brain in Silicon	Parker, DB 85a
A Layered Network Model for	Learning-to-Learn and Configuration in Classical Conditioning.	Kehoe, EJ 86a
A Game-Playing Program that	Learns by Analyzing Examples	Minton, S 85a
MURPHY: A Robot that	Learns by Doing.	Mel, BW 88a
RAMBOT: A Connectionist Expert System that	Learns by Example.	Mozer, MC 86b
RAMBOT: A Connectionist Expert System that	Learns by Example.	Mozer, MC 87a
PARSNIP: A Connectionist Network that	Learns Natural Language Grammar from Exposure to Natural Language.	Hanson, SJ 87a
New Computer	Learns on the Job.	Rowe, B 87a
A Novel Net that	Learns Sequential Decision Process	Sun, GZ 88a
A 'Neural' Network that	Learns to Play Backgammon.	Tesauro, G 88a
A Connectionist Network that	Learns to Process Some (Very) Simple Sentences	Samad, T 87a
NETtalk: A Parallel Network that	Learns to Read Aloud	Sejnowski, TJ 86d
NETtalk: A Parallel Network that	Learns to Read Aloud	Sejnowski, TJ 86c
The	Least Mean Fourth (LMF) Adaptive Algorithm and Its Family.	Walach, E 84a
The Structure of	Least Mean Square Linear Estimators for Synchronous M-ary Signals.	Gardner, WA 73a
Adaptive Training of Multilayer Neural Networks Using a	Least Squares Estimation Technique	Kollias, S 88a
Perception of	Left and Right by a Feed Forward Net	Scalettar, R 88b
...of an IPFM Neural Model: Effects of Spike Regularity and Record	Length	Sanderson, AC 80a
List	Length and the Time Course of Recognition in Immediate Memory.	Reed, AV 76a
Die	Lernmatrix	Steinbuch, K 61a
	Les Applications des Modeles Connexionistes a la Reconnaissance de	Wellekens, CJ 88b
	Les Modeles Mathematiques de l'Organization Biologique de la	Anderson, JA 78a
Variations on a Theme by Lashley:	Lesion Experiments on the Neural Model of Anderson, Silverstein.	Wood, CC 78a
	Let's Ask GMDH What Effect the Environment Has on Fisheries.	Brooks, HA 84a
	Letter and Configuration Information in Word Identification.	McClelland, JL 77a
Serial Position Functions for	Letter Identification at Brief and Extended Exposure Durations	Estes, WK 76a
Preliminary	Letter Identification in the Perception of Words and Nonwords.	McClelland, JL 75a
The Role of Orthographic and Phonotactic Rules in Perceiving	Letter Patterns.	Spoehr, KT 75a
An Interactive Activation Model of Context Effects in	Letter Perception, Part 1: An Account of Basic Findings.	McClelland, JL 81a
An Interactive Activation Model of Context Effects in	Letter Perception: Part 1, An Account of Basic Findings.	McClelland, JL 81b
An Interactive Activation Model of Context Effects in	Letter Perception: Part 2.	Rumelhart, DE 82a
Reading Ability and	Letter Search Time: Effects of Orthographic Structure Defined by	Mason, M 75a
Familiarity of	Letter Sequences and Tachistoscopic Identification	Miller, GA 54a
...Associative Memory for Identifying Words in a Continuous	Letter String.	Paek, EG 88c
...Regularity, Positional Frequency, and Visual Processing of	Letter Strings	Massaro, DW 79a
Perception of	Letters in Words: Seek Not and Ye Shall Find	Johnston, JC 74a
Class of Continuous	Level Associative Memory Neural Nets	Marks II, RJ 87b
The Central Nervous System as a Low and High	Level Control System	Albus, JS 88a
Low	Level Encoding of Motion by Asynchronous Shunting Networks	Ogmen, H 87c
A Class of Continuous	Level Neural Nets.	Marks II, RJ 87a
Architectures for a Continuous	Level Neural Network Based on Alternating Orthogonal Projections	Marks II, RJ 88a
...Parallel Architecture and Integrated Space-Time Filtering for Low	Level Visual Processing.	Johnson, K 86a
A Question of	Levels: Comment on McClelland and Rumelhart.	Broadbent, D 85a
	Levels Indeed! A Response to Broadbent	Rumelhart, DE 85c
	Levels of Modeling of Neural Interactions Underlying Visuomoter.	Arbib, MA 85d
Schemas and Neurons: Two	Levels of Neural Computing	Arbib, MA 88a
A Screening Algorithm for	Levenglick's Consensus Ranking Model	Flood, MM 87a
...Recognition: Strategies Affect Orthographic Computation But Not	Lexical Access	Carr, TH 78a
Processing Interactions and	Lexical Access During Word Recognition in Continuous Speech.	Marslen-Wilson, WD 78a
	Lexical Access Using a Neural Network.	Kawamoto, AH 84a
Dynamic Processes in the Resolution of	Lexical Ambiguity: PhD Thesis.	Kawamoto, AH 85b
Semantic Priming and Retrieval from	Lexical Memory: Evidence for Facilitatory and Inhibitory Processes	Neely, JH 76a
Semantic Priming and Retrieval from	Lexical Memory: Roles of Inhibitionless Spreading Activation and	Neely, JH 77a
The Role of the	Lexicon in Speech Perception	Samuel, AG 86a
"I	Lied About the Trees" Or, Defaults and Definitions in Knowledge.	Brachman, RJ 85a
Neural Networking Research Gains New	Life	Kane, B 87a
Birth,	Life, and Death in Microelectronic Systems	Widrow, B 61a
Electronic	Life and Synthetic Intelligent Systems	Dress, WB 88a
A	Lifelike Model for Association Relevance	Post, PB 69a
Operating Modes of the Microchannel Spatial	Light Modulator.	Warde, C 83a
...Processing Characteristics of the Microchannel Spatial	Light Modulator.	Warde, C 81a
The Non-Parametric Influence of	Light on Mammalian Circadian Rhythms	Carpenter, GA 85d
Seeing the	Light: Optics in Nature, Photography, Color, Vision, and Holography.	Stork, DG 85a
...A Direct Behavioral Analog of Photoreceptor Responses to Brief	Light Steps.	Lederhendler, II 87a
...of Visual Space: The Coherent Computation of Depth, Form and	Lightness.	Grossberg, S 86p
...of Visual Space: The Coherent Computation of Depth, Form, and	Lightness.	Grossberg, S 83b
Surface	Lightness and Cues for the Illumination.	Beck, J 71a
When Does Perceived	Lightness Depend on Perceived Spatial Arrangement?	Gilchrist, AL 80a

...of Binary Hetero-Associative Networks Using Link-Enabling vs.	Link-Disabling Local Modification Rules	Dobson, VG 88b
...in Performance of Binary Hetero-Associative Networks Using	Link-Enabling vs. Link-Disabling Local Modification Rules	Dobson, VG 88b
	Links Between Markov Models and Multilayer Perceptrons	Bourland, H 88a
	List Length and the Time Course of Recognition in Immediate Memory	Reed, AV 76a
	Listening for Mispronunciations: A Measure of What We Hear During	Cole, RA 73a
The Cerebellum: A Substrate for	List-Processing in the Brain	Albus, JS 72a
...Some Nonlinear Functional-Differential Equations I. Learning of	Lists	Grossberg, S 68a
On the Serial Learning of	Lists	Grossberg, S 69h
...of Neural Recognition Codes in Response to Arbitrary	Lists of Input Patterns	Carpenter, GA 86d
Biomasscomp: A Procedure for Mapping the Architecture of a	Living Neural Network into a Machine	Dawes, RL 87a
The Least Mean Fourth (	LMF) Adaptive Algorithm and Its Family	Walach, E 84a
Stationary and Nonstationary Learning Characteristics of the	LMS Adaptive Filter	Widrow, B 76b
The Complex	LMS Algorithm	Widrow, B 75b
Fundamental Relations Between the	LMS Algorithm and the DFT	Widrow, B 87b
On the Statistical Efficiency of the	LMS Algorithm with Nonstationary Inputs	Widrow, B 84a
Relationships Between	LMS Compensators for Intersymbol Interference in Linear and Non	Kaye, AR 73a
Comments on "An Adaptive Recursive	LMS Filter"	Widrow, B 77a
A New	LMS-Based Algorithm for Rapid Adaptive Classification in Dynamic	Nguyen, DD 88a
...Neuromorphic Algorithms for Optimization of the Computational	Load of a Hypercube Supercomputer	Barhen, J 87a
Optimization of the Computational	Load of a Hypercube Supercomputer Onboard a Mobile Robot	Barhen, J 87b
On the Complexity of	Loading Shallow Neural Networks	Judd, JS 88a
A Neural Network Theory of Frontal	Lobe Function	Levine, DS 86a
Overlapping Taste and Tactile Maps of the Oropharynx in the Vagal	Lobe of the Channel Catfish, Ictalurus punctatus	Kanwal, JS 88b
Propagating Belief Functions with	Local Computations	Shenoy, P 86a
	Local Constraint Integration in a Connectionist Model of Stereo	Stewart, CV 87b
	Local Determination of a Moving Contrast Edge	Kahn, P 85a
Existence and Stability of	Local Excitations in Neural Fields	Kishimoto, K 79a
The Transmission of Information and the Effect of	Local Feedback in Theoretical and Neural Networks	Uttley, AM 66a
Representation of	Local Geometry in the Visual System	Koenderink, JJ 87d
...and Principal Component Analysis: Learning from Examples Without	Local Minima	Baldi, P 88e
...Networks Using Link-Enabling vs. Link-Disabling	Local Modification Rules	Dobson, VG 88b
Relaxation and Constrained Optimization by	Local Processes	Ullman, S 79a
	Local Synaptic Modification Can Lead to Organized Connectivity	Palm, G 84a
	Local Versus Global Minima, Hysteresis, Multiple Meanings	Kergosien, YL 86a
Neural Processes for the Detection of Acoustic Patterns and for Sound	Localization	Evans, EF 74a
Nonlinear Interactions in a Dendritic Tree:	Localization, Timing, and Role in Information Processing	Koch, C 83a
Spatially	Localized Interactive Neural Populations-I, A Mathematical Model	Brannan, JR 81a
Spatially	Localized Interactive Neural Populations-II: Stability and Dynamics	Brannan, JR 82a
Learning with	Localized Receptive Fields	Moody, J 88b
	Localizing Sound Source by Incorporating Biological Auditory	Sugie, N 87a
Lower Bound for Connectivity in	Local-Learning Neural Networks	Abu-Mostafa, YS 88b
A	Locally Interconnected Neural Network for Deconvolution	Petsche, T 87a
Fast Learning in Networks of	Locally-Tuned Processing Units	Moody, J 88a
Biologically Plausible Models of Place Recognition and Goal	Location	Zipser, D 86b
The Representation of	Location	Zipser, D 83a
Learning to Represent and Understand	Locative Prepositional Phrases	Cosic, C 88a
Simulation of the Segmental Burst Generating Network for	Locomotion in Lamprey	Grillner, S 88a
...of the Neural Network for the Trajectory Planning of a Biped	Locomotive Robot	Kitamura, S 88a
Models for the Generation of the Motor Output Pattern in Flying	Locusts	Wilson, DM 68a
An Algorithm for Finding Best Matches in	Logarithmic Expected Time	Friedman, JH 77a
A	Logarithmic Vector Processor for Neural Net Applications	Richfield, S 87a
Adaptive Threshold	Logic	Mays, CH 63a
Implementing Neural Nets with Programmable	Logic	Vidal, JJ 88a
Implementing Neural Nets with Programmable	Logic	Vidal, JJ 87a
Learning Networks of Neurons with Boolean	Logic	Patarnello, S 87a
Learning Networks of Neurons with Boolean	Logic	Patarnello, S 88a
Linear-Input	Logic	Minnick, RC 61a
NASA Unites Neural Nets, Fuzzy	Logic	Johnson, RC 88b
On NPN	Logic	Zhang, W 88a
Real-Time Parallel Processing of Pixel Arrays Using Adaptive	Logic	Vidal, JJ 87b
Threshold	Logic	Sheng, CL 69a
Threshold	Logic	Lewis, PM 67a
Threshold	Logic	Hu, ST 65a
Syllogistic Reasoning in Fuzzy	Logic and its Application to Usuality and Reasoning with	Zadeh, LA 85a
Networks, Computations,	Logic and Noise	Hopfield, JJ 87b
On the Implementation of	Logic and Tables by Linear Networks	Kohonen, T 74b
Cellular	Logic Arrays for Image Processing	Preston Jr, K 86a
Theoretical and Linguistic Aspects of the Fuzzy	Logic Controller	Braae, M 79a
Learning in Networks of Nondeterministic Adaptive	Logic Elements	Windecker, RC 88a

A Mathematical Model for the Boltzmann	Machine.	Mazaika, PK 87a
A Pattern Recognition	Machine.	Palmieri, G 68a
An Implementation of Network Learning on the Connection	Machine.	Rosenberg, CR 88a
An Inductive Inference	Machine.	Solomonoff, RJ 57a
An Invariant Input for a Pattern Recognition	Machine.	Ball, GH 62a
Associative Memories on the Connection	Machine.	Simmes, SD 88a
Attractor Dynamics and Parallelism in a Connectionist Sequential	Machine.	Jordan, MI 86a
...for Mapping the Architecture of a Living Neural Network into a	Machine.	Dawes, RL 87a
Combinatorial Optimization Using a Continuous State Boltzmann	Machine.	Gutzmann, KM 87a
Computer Firm is Closing in on Dream	Machine.	Kraul, C 87a
Design of a Feeling-Thinking	Machine.	Scanlon, RD 88a
Design Sketch for a Million-Element NETL	Machine.	Fahlman, SE 80a
Discovering Structure from Motion in Monkey, Man and	Machine.	Siegel, RM 88a
Experiences with a Feeling-Thinking	Machine.	Johnson, M 87a
Mapping Image Processing Operations onto a Linear Systolic	Machine.	Kung, HT 85a
Neural Network Implementation Approaches for the Connection	Machine.	Brown, NH 88a
Notes on a Self-Organizing	Machine.	Geman, S 81a
Pattern Recognition and Reading by	Machine.	Bledsoe, WW 59a
Pattern Recognition by	Machine.	Selfridge, OG 60a
Separating Figure from Ground with a Boltzmann	Machine.	Sejnowski, TJ 87g
Simulation of a Distributed Algorithm Using the Connection	Machine.	Aburdene, MF 88a
The Cerebellum as a Neuronal	Machine.	Eccles, JC 67a
Empirical Explorations of the Logic Theory	Machine: A Case Study in Heuristic	Newell, A 57a
The Connection	Machine: A Computer Architecture Based on Cellular Automata.	Hillis, WD 84a
The Chess	Machine: An Example of Dealing with a Complex Task by Adaptation	Newell, A 55a
A Comparison of the Boltzmann	Machine and the Back Propagation Network as Recognizers of Static.	Prager, RW 87a
UCLA PUNNS - A Neural Network	Machine for Computer Vision.	Gunger, D 87a
An Associative Backend	Machine for Data Base Management	Hurson, A 81a
Warp as a	Machine for Low-Level Vision	Gross, T 85a
	Machine for Neural Computation of Acoustical Patterns with	Mueller, P 86a
Vision by Man and	Machine: How the Brain Processes Visual Information May Be	Poggio, TA 84c
1-D Optical Cauchy	Machine Infinite Film Spectrum Search.	Scheff, K 87a
A Neurobiological Approach to	Machine Intelligence	Davis, BR 82a
On	Machine Intelligence	Michie, D 86a
Brittleness and	Machine Learning	Gallant, SI 86a
Expert Systems and	Machine Learning	Clema, JK 85a
Knowledge Bases and	Machine Learning	Clema, JK 86a
...Zag Epigenetic Neural Networks), A New Approach to Connectionist	Machine Learning	Cochet, Y 88b
Variations on the Boltzmann	Machine Learning Algorithm	Derthick, MA 84a
Scaling,	Machine Learning, and Genetic Neural Nets.	Mjolsness, E 88b
Characterization of the Boltzmann	Machine Learning Rate.	Parks, M 87a
	Machine Operant Conditioning	Rosen, BE 88b
A Learning	Machine: Part 1.	Friedberg, RM 58a
	Machine Perception of Three-Dimensional Solids	Roberts, LG 65a
	Machine Recognition of Human Language: Part 2, Theoretical Models.	Lindgren, N 65a
Can a	Machine Think?	Turing, AM 56a
New Approaches to	Machine Translation.	Carbonell, JG 85a
	Machine Vision Lurches Ahead	Brown, C 87f
Multidimensional	Machine Vision Using Neural Networks	Kolodzy, PJ 87a
A Learning	Machine with New Structure and Function.	Nagano, T 73a
The Brains of Men and Machines, Part 4: The	Machinery of Emotion and Choice.	Kent, EW 78d
A Learning Algorithm for Boltzmann	Machines	Ackley, DH 85b
A Learning Algorithm for Boltzmann	Machines	Ackley, DH 85a
A Learning Algorithm for Boltzmann	Machines	Ackley, DH 88a
A Test for Linear Separability as Applied to Self-Organizing	Machines	Singleton, RC 62a
Behavior Analysis and Intelligent	Machines	Stephens, KR 88a
Higher-Order Boltzmann	Machines	Sejnowski, TJ 86a
Learning and Relearning in Boltzmann	Machines	Hinton, GE 86c
Massively Parallel Architectures for AI: Netl, Thistle, and Boltzmann	Machines	Fahlman, SE 83a
Multinomial Conjunctoid Statistical Learning	Machines	Takefuji, Y 88a
Neural Nets: The Dream	Machines	House, J 87a
Numerical Simulations of Boltzmann	Machines	Bounds, DG 86a
On Comparing the Brain with	Machines	MacKay, DM 54a
Optical Learning (Inference)	Machines	Eichmann, G 85a
Optimum Performance of Learning	Machines	Gamba, A 61a
Outline of a Theory of Thought Processes and Thinking	Machines	Caianiello, ER 61a
Pattern Classifiers and Trainable	Machines	Sklansky, J 81a
Pattern Recognition by Humans and	Machines	Schwab, EC 86a
Pattern Recognition by Humans and	Machines	Schwab, EC 86b

Spontaneous Quantal Currents in a Central Neuron	Match Predictions from Binomial Analysis of Evoked Responses	Korn, H 87a
Comparison of a Pipelined "Best	Match" Content Addressable Memory with Neural Networks	Clark, LT 87a
Nearest	Matched Filter Classification of Spatiotemporal Patterns	Hecht-Nielsen, R 87f
Nearest	Matched Filter Classification of Spatio-Temporal Patterns.	Hecht-Nielsen, R 87g
A	Matched Filter Detection System for Complicated Doppler Shifted. . .	Lerner, RM 60a
Composite	Matched Filter Output Partitioning	Marks II, RJ 87c
Comparison of Neural Network and	Matched Filter Processing for Detecting Lines in Images.	Grant, PM 86b
Composite	Matched Filtering with Error Correction.	Marks II, RJ 87d
Conventional and Composite	Matched Filters with Error Correction: A Comparison.	Cheung, KF 87a
A Comparison of Nonlinear Associative Memory and	Matched-Filter Processing for Detecting Lines in Optical Images. . .	Grant, PM 86a
	Matched-Filter Theory for High-Velocity, Accelerating Targets. . . .	Kelly, EJ 65a
An Algorithm for Finding Best	Matches in Logarithmic Expected Time	Friedman, JH 77a
...and Implementation of an Associative Memory for Intensive Pattern	Matching	Bou-Ghazale, S 88a
Parallel and Serial Methods of Pattern	Matching	Willshaw, DJ 72a
Pattern Recognition by Labeled Graph	Matching	von der Malsburg, C 88b
A Connectionist Architecture for	Matching 3-D Models to Moving Edge Features.	Gunger, D 86a
A Connectionist Architecture for	Matching 3-D Models to Moving Edge Features.	Casasent, DP 86a
Optimization in Model	Matching and Perceptual Organization: A First Look	Mjolsness, E 88a
Neural Networks for Template	Matching: Application to Real-Time Classification of the Action. . .	Wong, Y 87a
Neural Networks for Template	Matching: Application to Real-Time Classification of the Action. . .	Bower, JM 88b
...Pattern Recognition: Self-Organization of Templates and Template	Matching by a Disinhibitory Neural Network	Hirai, Y 80a
Neural Substrates of Binocular Form Perception: Filtering,	Matching, Diffusion, and Resonance	Grossberg, S 83a
...A Three Camera Stereo Algorithm for Overcoming Binocular	Matching Errors.	Stewart, CV 88b
Some Global Properties of Binocular Resonances: Disparity	Matching, Filling-In, and Figure-Ground Synthesis.	Grossberg, S 84b
Some Global Properties of Binocular Resonances: Disparity	Matching, Filling-in, and Figure-Ground Synthesis.	Cohen, MA 84b
Prototype Extraction and	Matching in Associative Nets	Lansner, A 85a
A Template	Matching Model for Pattern Recognition: Self-Organization of	Hirai, Y 80a
Fast Edge-Only	Matching Techniques for Robot Pattern Recognition.	Caelli, TM 87a
Stereo	Matching Using a Neural Network.	Zhou, Y 88c
Dynamic Pattern	Matching with Filling-In	Ogmen, H 87a
Stable	Matching with Preferences Derived From a Psychological Model	Bartholdi, JJ 86b
Perceptual Recognition as a Function of Meaningfulness of Stimulus	Material	Reicher, GM 69a
Supplier-Consumer Nets for	Material Handling.	Parunak, HVD 87a
Reinforcement Learning in Connectionist Networks: A	Mathematical Analysis.	Williams, RJ 86c
	Mathematical Analysis and Application of Iron-Wire Neuron Model. . .	Suzuki, R 67a
A	Mathematical Analysis of Excitable Membrane Phenomena.	Carpenter, GA 78a
	Mathematical Analysis of Learning Behavior of Neuronal Models. . . .	Cheung, JY 88a
A	Mathematical Approach to Neural Systems.	Amari, S 77a
	Mathematical Aspects of Periodic Catatonic Schizophrenia	Cronin, J 77a
Oscillations in	Mathematical Biology	Grissel, R 83a
An Architecture for	Mathematical Cognition	Ammon, K 86a
Nerve Conduction Theory: Some	Mathematical Consequences of Bernstein's Model	Offner, F 40a
Derivation of a	Mathematical Equation for the EEG and the General Solution within. .	Anninos, PA 75a
Modeling and	Mathematical Formalism for Robotic Cerebellar Neural Network	Niznik, CA 87a
A	Mathematical Formalism for the Finite Difference Equation in Neural.	Anninos, PA 87a
	Mathematical Foundation for Statistical Neurodynamics.	Amari, S 77c
Harmony Theory: A	Mathematical Framework for Stochastic Parallel Processing.	Smolensky, P 83a
An Introduction to	Mathematical Learning Theory	Atkinson, RC 65a
Studies in	Mathematical Learning Theory	Bush, RR 59a
On	Mathematical Methods in the Theory of Neural Networks.	Amari, S 87a
...Retinotopy and Columnar Microstructures by Self-Organization: A	Mathematical Model	Amari, S 85a
Spatially Localized Interactive Neural Populations-I, A	Mathematical Model	Brannan, JR 81a
A	Mathematical Model for Conduction of Action Potentials Along	Parnas, I 79a
A	Mathematical Model for Simple Learning	Bush, RR 51b
A	Mathematical Model for Simple Learning	Bush, RR 51a
A	Mathematical Model for the Boltzmann Machine	Mazaika, PK 87a
A	Mathematical Model for the Decay of Short-Term Memory with Age . . .	Anninos, PA 83a
A	Mathematical Model of the Primary Visual Cortex and Hypercolumn. . .	Okajima, K 86a
	Mathematical Modeling of Mechanisms of the Central Nervous System. .	Gelfand, IM 71a
Advances in	Mathematical Models for Image Processing	Jain, AK 81a
	Mathematical Models for the All-Or-None Activity of Some Neurons . .	Gerstein, GL 62a
Self Organizing	Mathematical Models: Non-Linear Evolution Equations with a	Schatzman, M 86a
	Mathematical Models of Binocular Vision.	Sperling, G 81a
	Mathematical Models of Excitation and Propagation in Nerve	FitzHugh, R 69a
Some	Mathematical Models of Learning.	Papert, SA 60a
	Mathematical Models of Memory Trace and Forgetfulness.	Anninos, PA 72b
Simple	Mathematical Models with Very Complicated Dynamics	May, RM 76a
Image Analysis and	Mathematical Morphology.	Serra, J 82a
Image Analysis and	Mathematical Morphology.	Serra, J 88a

	Measurement of Bioelectric Potentials with Microelectrodes and . . .	Amatniek, E 58a
Convected Activation Profiles and the	Measurement of Visual Motion	Waxman, AM 88a
...a Non-Network Approach to "Neural Modeling": Some Basic Issues of	Measurement, Simulation, and Computational Significance of Brain . .	Schwartz, EL 87a
...Pattern Classification for Dependent Hypotheses and Dependent	Measurements	Hilborn, CG 69a
...for Robotic Cerebellar Neural Network Pathway Information	Measures	Niznik, CA 87a
Performance	Measures for Associative Memories that Learn and Forget.	Kuh, A 88b
An Introduction to Biological and	Mechanical Pattern Recognition	Gose, EE 69a
...Motor Servo: Incorporating Nonlinear Spindle Receptor and Muscle	Mechanical Properties.	Gielen, CCAM 87a
The Relevance of	Mechanical Sensors for Neural Generation of Movements in Space . . .	Daunicht, WJ 88d
The Relevance of	Mechanical Sensors for Neural Generation of Movements in Space . . .	Daunicht, WJ 88c
On Presenting the Output of a	Mechanical Speech Recognizer	Fry, DB 57a
Statistical	Mechanics: A General Approach to Combinatorial Optimization.	Bonomi, E 86a
A Statistical	Mechanics of Nervous Activity.	Cowan, JD 70a
Dynamics and Statistical	Mechanics of the Hopfield Model.	Bruce, AD 87a
Statistical	Mechanics of Visual Processing: A Model for Layer IVc of the	Moller, P 87a
Derivation of a Model for the Human Eye-Positioning	Mechanism.	Cook, G 67a
Localizing Sound Source by Incorporating Biological Auditory	Mechanism.	Sugie, N 87a
...qua a Generator of Rhythmic Oscillation with Period Control	Mechanism.	Tsutsumi, K 84b
Image Segregation by Motion: Cortical	Mechanism and Implementation in Neural Networks.	Orban, GA 88a
...Calcium and Potassium Ion Conductances: A Contingency	Mechanism for an Associative Learning Model.	Alkon, DL 79a
A Learning	Mechanism for an Associative Storage System.	Ersu, E 81a
A Neural Network System with an Automatic Generation	Mechanism for Distorted Patterns	Shimohara, K 88a
A Neural	Mechanism for Generalization over Equivalent Stimuli in the.	Freeman, WJ 81a
A Physiological	Mechanism for Hebb's Postulate of Learning	Stent, GS 73a
A	Mechanism for Optic Nerve Conduction and Form Perception: I.	Culbertson, JT 48a
EEG Analysis Gives Models of Neuronal Template-Matching	Mechanism for Sensory Search with Olfactory Bulb	Freeman, WJ 79a
A	Mechanism for Stabilizing Intelligent Properties of Neural Networks.	Dammasch, IE 87a
A Marker Induction	Mechanism for the Establishment of Ordered Neural Mapping: Its . . .	Willshaw, DJ 79a
A Proposed	Mechanism for the Origin and Development of Iso-Orientation Columns.	Cowan, JD 85a
Drive-Reinforcement Learning: A Real-Time Learning	Mechanism for Unsupervised Learning.	Klopf, AH 87b
Hebbian Modification of Synaptic Transmission as a Common	Mechanism in Experience-Dependent Maturation of Cortical Functions .	Singer, W 85a
A Biologically Constrained Learning	Mechanism in Networks of Formal Neurons.	Personnaz, L 86a
Properties of a Competition-Based Activation	Mechanism in Neuromimetic Network Models	Reggia, JA 87a
A Four	Mechanism Model for Threshold Spatial Vision	Wilson, HR 79a
The Neuronal Basis of Behavior in Tritonia, III Neuronal	Mechanism of a Fixed Action Pattern.	Willows, AOD 73a
Neural	Mechanism of Conditioning.	Kotliar, BE 83a
A Neural Network Model for the	Mechanism of Feature Extraction.	Fukushima, K 84b
A Neural Network Model for the	Mechanism of Feature Extraction - A Self-Organizing Network with . .	Miyake, S 83a
Self-Organizing Neural Network with the	Mechanism of Feedback Information Processing	Miyake, S 88c
Neural Network Model for the	Mechanism of Pattern Information Processing.	Miyake, S 86a
Neural Network Model for a	Mechanism of Pattern Recogition Unaffected by Shift in Position. . .	Fukushima, K 79b
Neocognitron: A Self-Organizing Neural Network Model for a	Mechanism of Pattern Recognition Unaffected by Shift in Position . .	Fukushima, K 80a
Neocognitron: A Neural Network Model for a	Mechanism of Visual Pattern Recognition.	Fukushima, K 83c
Neocognitron: A Neural Network Model for a	Mechanism of Visual Pattern Recognition.	Fukushima, K 83b
Neocognitron: A Self-Organizing Neural Network Model for a	Mechanism of Visual Pattern Recognition.	Fukushima, K 82b
Possible Synaptic	Mechanism Subserving Learning.	Eccles, JC 72a
...Processing: Human Errors and Other Phenomena Suggest Processing	Mechanisms	Norman, DA 81a
...Network Realizing a Funciton in the Human Recognition and Memory	Mechanisms	Mizuno, H 80a
...Network Realizing a Function in the Human Recognition and Memory	Mechanisms	Mizuno, H 78a
Centrifugal Control	Mechanisms	Whitfield, IC 68a
Collective Computational Properties of Neural Networks: New Learning	Mechanisms	Personnaz, L 86b
...to the Binding Problem: A Simulation of Some Human Computational	Mechanisms	Stenning, K 88a
Principles of Neurodynamics: Perceptrons and the Theory of Brain	Mechanisms	Rosenblatt, F 62a
Principles of Neurodynamics: Perceptrons and the Theory of Brain	Mechanisms	Rosenblatt, F 61a
The Analysis of Visual Motion: From Computational Theory to Neuronal	Mechanisms	Hildreth, E 86a
...and Operational Separation of Network Architectures and Learning	Mechanisms	White, H 88b
Neuromolecular Computing:	Mechanisms and Architectures	Conrad, M 88a
Parallel-Processing	Mechanisms and Processing of Organized Information in Human Memory .	Ratcliff, R 81a
Neural Network	Mechanisms for Generation and Learning of Motor Programs	Eckmiller, R 87b
Conservation of Cellular	Mechanisms for Models of Learning and Memory	Alkon, DL 88a
Neuronal	Mechanisms for Spatial Computation and Information Storage	McNaughton, BL 89a
Neural	Mechanisms for the Production of Cyclic Motor Patterns	Selverston, AI 83a
A Neural Network Model of Pain	Mechanisms: Functional Properties of the Network Cells Providing . .	Minamitani, H 82a
Attentional	Mechanisms in Neural Information Processing: Examples From	Grossberg, S 87f
Associative and Cooperative / Competitive	Mechanisms in Neural Networks.	Grossberg, S 87e
...Processing Model for the Study of Possible Pattern Recognition	Mechanisms in the Brain.	Muller, FJ 76a
Conditional Control Transfer	Mechanisms in the Neocortex: 1. The Basic Model.	Scott, PD 79a
Synaptic Potential Summation and Repetitive Firing	Mechanisms: Input-Output Theory for the Recruitment of Neurons into.	Calvin, WH 72a
Intracellular	Mechanisms of Adaptation and Self-Regulation in Self-Organizing. . .	Grossberg, S 80f

A Neuron Model with Learning Capability and its Relation to	Mechanisms of Association.	Wigstrom, H 73a
A Model of the Brain for Robot Control, Part 4:	Mechanisms of Choice	Albus, JS 79d
Learning and Memory:	Mechanisms of Information Storage in the Nervous System.	Matthies, H 86a
Neural	Mechanisms of Learning and Memory.	Rosenzweig, MR 76a
	Mechanisms of Planning and Problem Solving in the Brain.	Albus, JS 79e
	Mechanisms of Sentence Processing: Assigning Roles to Constituents	McClelland, JL 86f
	Mechanisms of Slow Synaptic Potentials	Hartzell, HC 81a
Mathematical Modeling of	Mechanisms of the Central Nervous System	Gelfand, IM 71a
Modeling Neural	Mechanisms of Visual-Motor Coordination in the Frog and Toad	Arbib, MA 82a
A Neural Model of the	Mechanisms Underlying Infant Perception of Voice-Onset Time.	Pont, MJ 88b
Neural Networks	Mediating Linearizable Dynamic Redundant Sensori-Motor Reflexes.	Daunicht, WJ 88b
Neural Network Implementation of a	Medical Diagnosis Expert System.	Kudrycki, TP 88a
	Medical Diagnostic Expert System Based on PDP Model.	Saito, K 88a
Neural Networks for	Medical Image Processing: A Study of Feature Identification.	Dayhoff, RE 88a
A Model of Inexact Reasoning in	Medicine	Shortliffe, EH 75a
	Melding Computers and Biology for Commercial Benefit	Hunter, L 88b
The Power Spectrum of the	Mellin Transformation with Applications to Scaling of Physical	Moses, HE 74a
Incorporating Fuzzy	Membership Functions into the Perceptron Algorithm	Keller, JM 85a
Recognition of	Membership in Classes.	Sebestyen, GS 61a
Fluctuation Phenomena in Nerve	Membrane	Verveen, AA 68a
Persistent Conditioning-Specific Changes of Post-Synaptic	Membrane Currents in Heermissenda and Hippocampus.	Coulter, DA 87a
Conditioning-Specific Modification of Postsynaptic	Membrane Currents in Mollusc and Mammal.	Alkon, DL 87a
A Practical Algorithm for Solving Dynamic	Membrane Equations	Rush, S 78a
Specifications for Nerve	Membrane Models.	Moore, JW 68a
A Mathematical Analysis of Excitable	Membrane Phenomena	Carpenter, GA 78a
	Membrane Potential Transients and Membrane Time Constant of.	Rall, W 60a
A Test for Modulation of Synaptic Efficacy by Induced	Membrane Potential Variance.	Gardner, D 88a
...Element and Classical Conditioning of the Rabbit Nictitating	Membrane Response.	Moore, JW 85a
The Hippocampus and the Classically Conditioned Nictitating	Membrane Response: A Real-Time Attentinal-Associative Model.	Schmajuk, NA 88a
Simulation of the Classically Conditioned Nictitating	Membrane Response by a Neuron-Like Adaptive Element.	Blazis, DEJ 86a
Integrate-And-Fire Models of Nerve	Membrane Response to Oscillatory Input	Keener, JP 81a
Membrane Potential Transients and	Membrane Time Constant of Motoneurons.	Rall, W 60a
Two-Terminal Electronic Circuit Neuron Model with Excitable	Membrane V-I-t Characteristics	Hoshimiya, N 79a
Regulation of Hermissenda K+ Channels by Cytoplasmic and	Membrane-Associated C-Kinase	Alkon, DL 88b
Biophysics of Computation: Neurons, Synapses and	Membranes.	Koch, C 84a
Electrical Network Modeling of Active	Membranes of Nerves.	Marsocci, VA 82a
An Adaptive "Adaline" Neuron Using Chemical "	Memistors"	Widrow, B 60d
Les Modeles Mathematiques de l'Organization Biologique de la	Memoire.	Anderson, JA 78a
A Neuronal Model with Synaptic Long-Term	Memories	Niki, K 81a
A New VLSI Architecture for Neural Associative	Memories	Verleysen, M 88f
Adaptive Bidirectional Associative	Memories	Kosko, B 87a
Associative Holographic	Memories	Gabor, D 69a
Associative	Memories	Hayes, JP 78a
Competitive Adaptive Bidirectional Associative	Memories	Kosko, B 87d
Content-Addressable	Memories	Kohonen, T 87c
Correlation Matrix	Memories	Kohonen, T 72a
Correlation Matrix	Memories	Kohonen, T 72b
Demon-Based Associative	Memories	Hudak, MJ 86a
Designs and Devices for Optical Bidirectional Associative	Memories	Guest, CC 87a
Digital and Analogue Holographic Associative	Memories	White, HJ 88a
...Computation for Retrieval Tasks Using Distributed Associative	Memories	Char, JM 88a
Electronic Device Aspects of Neural Network	Memories	Lambe, J 85a
Experiments with Sequential Associative	Memories	Gallant, SI 88d
Four Capacity Models for Coarse-Coded Symbol	Memories	Rosenfeld, R 87a
Fuzzy Associative	Memories	Kosko, B 87g
...Learning Control: An Approach with Neuron-Like Associative	Memories	Ersu, E 88a
High Density Associative	Memories	Dembo, A 88b
High-Capacity Exponential Associative	Memories	Chiueh, T 88b
High-Performance Recording Algorithm for Hopfield Model Associative	Memories	Hassoun, MH 88b
Information Capacity of Associative	Memories	Kuh, A 88a
Neural Networks as Associative	Memories	Lemmon, M 87a
Noise-Like Coding in Associative	Memories	Bottini, S 87a
Nonlinear Discriminant Functions and Associative	Memories	Psaltis, D 86a
Nonlinearity Requirements for Correlation Based Associative	Memories	Marks II, RJ 88c
Optical Bi-directional Associative	Memories	Kosko, B 87i
Pattern Recognition Using Associative	Memories	Altman, JL 77a
Perspectives on Associative	Memories	Moody, J 87a
Reflexive Associative	Memories	Loos, H 88a
Scaling Properties of Coarse-Coded Symbol	Memories	Rosenfeld, R 88a

Intelligent Control of Robot Motion Based on Learning	Method	Kawamura, S 87a
Realization of Robot Motion Based on a Learning	Method	Kawamura, S 88a
...by the Maximum A Posteriori Method and a Maximum Entropy	Method	Trussell, HJ 80a
A Redundant Hash Addressing	Method Adapted for the Postprocessing and Error-Correction of.	Reuhkala, E 79a
...Between Image Restoration by the Maximum A Posteriori	Method and a Maximum Entropy Method.	Trussell, HJ 80a
A Thousand-Word Recognition System Based on the Learning Subspace	Method and Redundant Hash Addressing	Kohonen, T 80a
A Multilayer Cooperative/Competitive	Method for Creating Hierarchical Structures by Clustering Maximally.	Maren, AJ 88a
A	Method for Determining Optimal Interelectrode Spacing for Cerebral	Spitzer, AR 88a
ALOPEX: A Stochastic	Method for Determining Visual Receptive Fields	Harth, EM 74a
Efficient	Method for Estimating Kinetic Constants from Single Channel Patch.	Chay, TR 87b
Efficient	Method for Estimating Kinetic Constants from Single Channel Patch.	Chay, TR 87a
...Activation Networks for Analogical Ordering of Consideration: One	Method for Integrating Connectionist and Symbolic Processing	Becker, LA 87b
A	Method for Predicting Sales Amount by the Use of IWSM and GMDH	Nomura, J 84a
A Learning	Method for System Identification	Nagumo, J 67a
A Non-Stochastic	Method for the Correction of Sentences	Venta, O 86a
A	Method for the Design of Stable Lateral Inhibition Networks that is.	Wyatt Jr, JL 88a
A Content-Addressing Software	Method for the Emulation of Neural Networks.	Venta, O 88a
A Very Fast Associative	Method for the Recognition and Correction of Misspelt Words, Based	Kohonen, T 78a
A Quasi-Topological	Method for the Recognition of Line Patterns.	Sherman, H 60a
An Improved Elastic Net	Method for the Traveling Salesman Problem.	Burr, DJ 88a
A Novel Computational and Signalling	Method for VLSI Neural Networks.	Murray, AF 88a
A	Method of Analysis of the Models of Neural Nets.	Burattini, E 72a
Modeling Accuracy Improvements for the Group	Method of Data Handling.	Molnar, DO 87a
Bayesian-Based Iterative	Method of Image Restoration.	Richardson, WH 72a
Vectorcardiographic Diagnosis Using the Polynomial Discriminant	Method of Pattern Recognition.	Specht, DF 67b
A	Method of Statistical Neurodynamics.	Amari, S 74b
A New Optical	Method of Storing and Retrieving Information	van Heerden, PJ 63a
On a	Method of Storing Information.	Roy, AE 60a
On a	method of Storing Information: II. A Further Study of Model.	Roy, AE 62a
Application of the Subspace	Method to Speech Recognition	Jalanko, M 78b
A Neural Network Simulation	Method Using the Fast Fourier Transform.	Barto, AG 76a
...Analogs of Self-Programming Neural Nets: Architecture and	Methodologies for Implementing Fast Stochastic Learning by	Farhat, NH 87b
Architectures and	Methodologies for Self-Organization and Stochastic Learning in Opto.	Farhat, NH 87a
Tools for Connectionist Modeling: The Dynamical Systems	Methodology.	Munro, PW 88a
Classification of EEG Spatial Patterns with a Tree-Structured	Methodology: CART.	Grajski, KA 86a
A Design	Methodology for High-Performance, General-Purpose VLSI	Hammerstrom, D 86b
A	Methodology for the Development of General Knowledge-Based Vision.	Riseman, EM 87a
On the	Methodology of Connectionist Research Programmes in Vision	Gonzales, MEQ 87a
A	Methodology to Select Appropriate Learning Rate Parameters in Feed	Pabon, JA 87a
An Introduction to Relaxation	Methods.	Williams, G 87a
Continuous Speech Recognition by Statistical	Methods.	Jelinek, F 76a
...for River Flows and Typhoon Precipitation by Self-Organizing	Methods.	Ikeda, S 84a
...Systems: A Mapping Principle and a Comparison to Gradient Based	Methods.	Leong, HM 88a
Solving of Optimization and Identification Problems by the Committee	Methods.	Mazurov, VD 87a
Problem Solving	Methods for Pattern Recognition.	Kanal, LN 86a
Bifurcation Theory	Methods for Programming Static or Periodic Attractors and Their.	Baird, B 88a
Symbolic Computation	Methods for Some Spin Glasses Problems	Lacolle, B 86a
Structural	Methods in Automatic Speech Recognition.	Levinson, SE 85a
Iterative	Methods in Image Analysis.	Rosenfeld, A 78a
Iterative	Methods in Image Analysis.	Rosenfeld, A 77a
Self-Organizing	Methods in Modeling: GMDH Type Algorithms.	Farlow, SJ 84b
Bayesian	Methods in Nonlinear Digital Image Restoration	Hunt, BR 77a
Connectionist	Methods in Speech Recognition.	Rohwer, R 88a
On Mathematical	Methods in the Theory of Neural Networks	Amari, S 87a
	Methods of Biological Signal Processing.	Shiavi, RG 86a
Learning Algorithms for Connectionist Networks: Applied Gradient	Methods of Nonlinear Optimization.	Watrous, RL 87c
Practical	Methods of Optimization, Volume 1: Unconstrained Optimization.	Fletcher, R 80a
Parallel and Serial	Methods of Pattern Matching.	Willshaw, DJ 72a
A Comparison of Adaptive Algorithms Based on the	Methods of Steepest Descent and Random Search.	Widrow, B 76a
Tensor Theory of Brain Function: The Cerebellum as a Space-Time	Metric	Pellionisz, AJ 82b
...in Neural Nets: Concepts, Implications and a Proposed	Metric	Tillery, SI 87a
The Universal	Metric Properties of Nonlinear Transformations	Feigenbaum, MJ 79a
...in the Brain: The Cerebellum as a Predictive Space-Time	Metric Tensor.	Pellionisz, AJ 82a
A Self-Organizing System Called	Mickey Mouse'.	Browning, I 64a
Operating Modes of the	Microchannel Spatial Light Modulator	Warde, C 83a
Optical Information Processing Characteristics of the	Microchannel Spatial Light Modulator	Warde, C 81a
Performance of a Stochastic Learning	Microchip.	Alspector, J 88b
	Microcircuitry of Beta Ganglion Cells in Cat Retina.	McGuire, BA 86a
	Microcircuitry of Bipolar Cells in Cat Retina.	McGuire, BA 84a

	Mixing Binary and Continuous Connection Schemes for Knowledge	Sharkey, NE 86a
Enhancement of Noisy Speech by Adaptive	MMSE Filtering	Ahmed, MS 86a
...of the Computational Load of a Hypercube Supercomputer Onboard a	Mobile Robot	Barhen, J 87b
Visual Control of Autonomous	Mobile Robot Based on Self-Organizing Model for Pattern Learning	Suzuki, H 88c
	Mobile Robot Control by Neural Networks and Their Associated	Watanabe, N 88a
Cognitive Control of Autonomous	Mobile Robots: Nested Hierarchical Intelligent Module	Meystel, A 86a
On Estimation of a Probability Density Function and	Mode	Parzen, E 62a
Coupled	Mode Theory for Neural Networks	Cohen, MA 86a
A CMOS Implementation of a Neural Network	Model	Graf, HP 87c
A Composite Holographic Associative Recall	Model	Eich, JM 82a
A Generalized Neural Network	Model	Xu, X 88a
A Model of the Brain for Robot Control, Part 2: A Neurological	Model	Albus, JS 79b
...for Robot Control, Part 3: A Comparison of the Brain and our	Model	Albus, JS 79c
...Approach to Optical Information Processing Based on the Hopfield	Model	Psaltis, D 84a
A Screening Algorithm for Levenglick's Consensus Ranking	Model	Flood, MM 87a
An Exactly Solvable Asymmetric Neural Network	Model	Derrida, B 87a
Analysis of a Competition-Based Connectionist	Model	Seidman, SB 87a
Analysis of a Distributed Associative Memory	Model	Wiles, J 87a
Category Learning and Adaptive Pattern Recognition: A Neural Network	Model	Carpenter, GA 85a
Characterizing Synaptic Plasticity with an Activity Dependent	Model	Starmer, CF 87a
Classical Spin-Glass	Model	van Hemmen, JL 82a
Cognitron: A Self-Organizing Multi-Layered Neural Network	Model	Fukushima, K 81a
Color Constancy and Early Vision: A Connectionist	Model	Heisey, I 87a
Computer Simulation of Short-Term Memory: A Component Decay	Model	Laughery, KR 69a
Computing With Neural Circuits: A	Model	Hopfield, JJ 85b
...Control Transfer Mechanisms in the Neocortex: 1. The Basic	Model	Scott, PD 79a
Connectionist Speech Recognition Using the Temporal Flow	Model	Watrous, RL 88a
Cyclic Sequences Generated by Neuron	Model	Wada, M 82a
...and Old Instances: Simulation by an Exemplar-Based Classification	Model	Hintzman, DL 80a
...and Probability Learning: Some Applications of a Neural	Model	Anderson, JA 77b
...and Probability Learning: Some Applications of a Neural	Model	Anderson, JA 77a
...of Single Neuron Function: An Alternate to the Hebbian Neuronal	Model	Klopf, AH 86a
Dynamic Single Unit Simulation of a Realistic Cerebellar Network	Model	Pellionisz, AJ 73a
Dynamics and Statistical Mechanics of the Hopfield	Model	Bruce, AD 87a
...and Columnar Microstructures by Self-Organization: A Mathematical	Model	Amari, S 85a
Holographic Implementation of a Discrete Hopfield Neural Network	Model	White, HJ 88c
Infinitely Many Commensurate Phases in a Simple Ising	Model	Fisher, ME 80a
Information Capacity of the Hopfield	Model	Abu-Mostafa, YS 85b
Interactive Processes in Speech Perception: The TRACE	Model	McClelland, JL 86e
Mathematical Analysis and Application of Iron-Wire Neuron	Model	Suzuki, R 67a
Medical Diagnostic Expert System Based on PDP	Model	Saito, K 88a
...Condition for a Content-Addressable Memory Based on the Hopfield	Model	Macukow, B 87a
...Conduction Theory: Some Mathematical Consequences of Bernstein's	Model	Offner, F 40a
...Potentials Derived from an Extended Version of the Hodgkin-Huxley	Model	Lewis, ER 66a
...Execution of the Traveling Salesman Problem on a Neural Network	Model	Bagherzadeh, N 87a
Optical Implementation of the Hopfield	Model	Farhat, NH 85b
Optical Implementation of the Hopfield	Model	Farhat, NH 85a
Perception of Depth Surfaces in Random-Dot Stereograms: A Neural	Model	Dev, P 75a
Pulse Sequences Generated by a Degenerate Analog Neuron	Model	Yoshizawa, S 82a
Real Time Noisy Image Segmentation Using an Artificial Neural Network	Model	Tenorio, MF 87d
Response Characteristics of a Mathematical Neuron	Model	Yoshida, T 78a
...Phenomena in the Response of Auditory Nerve Fibers: A Functional	Model	Wu, ZL 88a
Simulation of an Array-Based Neural Net	Model	Barnden, JA 87b
Simulation Studies of a Temporal Sequence Memory	Model	Stanley, JC 76a
Spatially Localized Interactive Neural Populations-I, A Mathematical	Model	Brannan, JR 81a
Speech Perception as a Cognitive Process: The Interactive Activation	Model	Elman, JL 83a
Stable Matching with Preferences Derived From a Psychological	Model	Bartholdi, JJ 86b
...Mass of Neurons Forming Simple Feedback Loops (I): Lumped Circuit	Model	Ahn, SM 74a
...of Neurons Forming Simple Feedback Loops (II): Lumped Circuit	Model	Ahn, SM 74b
Stimulus-Response Characteristics of a Mathematical Neuron	Model	Wada, M 82b
Structure of Metastable States in the Hopfield	Model	Gardner, E 86a
The Eye-Movement Control System: Evolution of a	Model	Fender, DH 62a
...Nictitating Membrane Response: A Real-Time Attentinal-Associative	Model	Schmajuk, NA 88a
The Neural Path Probabilistic Delay	Model	Niznik, CA 83a
Theoretical and Experimental Aspects of a Cerebellar	Model	Albus, JS 72b
...on Cantor Function-Like Response Characteristics of a Neuron	Model	Osada, H 81a
Value Unit Interpolation: A Receptive Field	Model	Ballard, DH 86c
VLSI Implementation of a Neural Network	Model	Graf, HP 88a
...Conductances: A Contingency Mechanism for an Associative Learning	Model	Alkon, DL 79a

A Dynamic Connectionist	Model for Phoneme Recognition: Preliminary Results	Robinson, AJR 88a
A Neural Network	Model for Phoneme Recognition Using the Generalized Delta Rule for .	Watson, M 87a
A New	Model for Randomly Organized Associative Memory.	Kohonen, T 73a
Kinematic Networks: A Distributed	Model for Representing and Regularizing Motor Redundancy	Mussa-Ivaldi, FA 88b
A Hierarchical Neural Network	Model for Selective Attention.	Fukushima, K 88a
A Neural Network	Model for Selective Attention.	Fukushima, K 87b
A Neural Network	Model for Selective Attention.	Fukushima, K 87a
A Neural Network	Model for Selective Attention in Visual Pattern Recognition.	Fukushima, K 86a
Neural Network	Model for Selective Attention in Visual Pattern Recognition and. . .	Fukushima, K 87c
Neural Network	Model for Selective Attention in Visual Pattern Recognition.	Fukushima, K 87d
A	Model for Sensorimotor Control and Learning.	Raibert, MH 78a
A Distributed Memory	Model for Serial-Order Information	Murdock, BB 83a
A Mathematical	Model for Simple Learning.	Bush, RR 51a
A Mathematical	Model for Simple Learning.	Bush, RR 51b
A Connectionist	Model for Simple Visual Functions.	Mel, BW 86b
A Synaptic	Model for Spatial Frequency Adaptation	Wilson, HR 75a
A Simple	Model for Spike-Train-Output Neurons	Atkinson, RC 88a
Abilities and Limitations of a Neural Network	Model for Spoken Word Recognition.	Kurogi, S 88a
Coupled Rotators as a	Model for Spontaneous Activity in Neural Networks.	Kowalski, JM 88a
A Connectionist	Model for Stereo Vision.	Stewart, CV 87a
A	Model for Stimulus Generalization and Discrimination	Bush, RR 51c
A	Model for Stimulus Generalization and Discrimination	Bush, RR 51d
An Adaptive Computational	Model for Texture Segmentation	Caelli, TM 88a
An Extended First Approximation	Model for the Amygdaloid Kindling Phenomenon	Lieblich, I 78a
A Neural Net	Model for the a-Rhythm	Anninos, PA 80a
A Mathematical	Model for the Boltzmann Machine.	Mazaika, PK 87a
Large-Scale Activity in Neural-Nets II: A	Model for the Brainstem Respiratory Oscillator	Feldman, JL 75a
A Mathematical	Model for the Decay of Short-Term Memory with Age.	Anninos, PA 83a
A Self-Organizing Neural Network	Model for the Development of Complex Cells	Nagano, T 81a
A Neural Network	Model for the Developmental Process of Hypercomplex Cells.	Nagano, T 83a
A	Model for the Economical Encoding of the Visual Image in Cerebral. .	Sakitt, B 82a
Derivation of a	Model for the Human Eye-Positioning Mechanism.	Cook, G 67a
A Neural Network	Model for the Mechanism of Feature Extraction.	Fukushima, K 84b
A Neural Network	Model for the Mechanism of Feature Extraction - A Self-Organizing. .	Miyake, S 83a
Neural Network	Model for the Mechanism of Pattern Information Processing.	Miyake, S 86a
General	Model for the Molecular Events in Synapses During Learning	Huttunen, MO 73a
An Adaptive Stochastic	Model for the Neural Coding Process.	Bruckstein, AM 85a
An Improved Neural-Network	Model for the Neural Integrator of the Oculomotor System: More . . .	Cannon, SC 85a
A Neural Net	Model for the Release of Prolactin During Pseudopregnancy in the . .	Anninos, PA 87c
A Parallel Processing	Model for the Study of Possible Pattern Recognition Mechanisms in. .	Muller, FJ 76a
A	Model for the Temporal Organization of X- and Y-Type Receptive . . .	Richter, J 82a
A Four Mechanism	Model for Threshold Spatial Vision	Wilson, HR 79a
Equivalent Circuits	Model for Transmitter Release at Neuromuscular Junction.	Yoshihara, K 80a
Optical Implementation of the Hopfield	Model for Two-Dimensional Associative Memory	Jang, J 88c
A Connectionist	Model for Unsupervised Feature Discovery and Pattern Recognition . .	Franzini, MA 87a
A	Model for Variability Effects in Hand-Printing, with Implications. .	Ward, JR 88a
Towards a	Model for Visual Information Processing.	Okajima, K 88a
A Hierarchical	Model for Voluntary Movement and Its Application to Robotics	Kawato, M 87a
Hierarchical Neural Network	Model for Voluntary Movement with Application to Robotics.	Kawato, M 88a
Dynamic Single Unit Simulation of a Realistic Cerebellar Network	Model. II. Purkinje Cell Activity within the Basic Circuit and . . .	Pellionisz, AJ 74a
The "Brain-State-in-a-Box" Neural	Model Is a Gradient Descent Algorithm.	Golden, RM 86c
Optimization in	Model Matching and Perceptual Organization: A First Look	Mjolsness, E 88a
Low Temperature Analysis of the Axial Next-Nearest Neighbor Ising	Model Near its Multiphase Point.	Fisher, ME 81a
Morphogenesis and Properties of Neuronal	Model Networks	Dammasch, IE 86a
Effects of Connection Delays in Two State	Model Neural Circuits.	Conwell, PR 87a
The Response of a	Model Neurone to a White Noise Input	Rubio, JE 75a
Two-Dimensional Adaptive	Model of a Human Controller Using Pattern Recognition Techniques . .	Gilstad, DW 71a
A Neuron	Model of a Memory System for Autonomous Exploration of an.	Trenary, R 87a
A	Model of a Neural Network with Recurrent Inhibition.	Wigstrom, H 74a
	Model of a Plausible Learning Scheme for CA3 Hippocampus	Kilmer, W 74a
A Computer-Simulated	Model of a Second Order Sensory Neurone.	Walloe, L 69a
A Boolean Complete Neural	Model of Adaptive Behavior	Hampson, S 83a
...Problem-Solving: Representation of the Process of Addition by a	Model of an Associative Processor, HASP.	Hirai, Y 87a
Representation of Control Structures by a	Model of an Associative Processor, HASP.	Hirai, Y 87b
Variations on a Theme by Lashley: Lesion Experiments on the Neural	Model of Anderson, Silverstein, Ritz and Jones	Wood, CC 78a
Associatron: A	Model of Associative Memory.	Nakano, K 72a
A	Model of Associative Memory in the Brain	Fukushima, K 73a
Toward a	Model of Attention and the Development of Automatic Processing . . .	Schneider, W 85a
A Neural	Model of Attention, Reinforcement, and Discrimination Learning . . .	Grossberg, S 75a

The TRACE	Model of Speech Perception	McClelland, JL 86k
Local Constraint Integration in a Connectionist	Model of Stereo Vision	Stewart, CV 87b
A	Model of Synaptogenetic Consequences of Afferent Input in Neuronal	Dammasch, IE 88a
The T5 Base Modulator Hypothesis: A Dynamic	Model of T5 Neuron Function in Toads	Betts, B 88a
Holographic	Model of Temporal Recall	Chopping, PT 68a
Holographic	Model of Temporal Recall	Longuet-Higgins, HC 68a
Improved Holographic	Model of Temporal Recall	Gabor, D 68a
A Computational	Model of the Auditory System	Travis, BJ 87a
A	Model of the Brain for Robot Control, Part 1: Defining Notation.	Albus, JS 79a
A	Model of the Brain for Robot Control, Part 2: A Neurological Model	Albus, JS 79b
A	Model of the Brain for Robot Control, Part 3: A Comparison of the.	Albus, JS 79c
A	Model of the Brain for Robot Control, Part 4: Mechanisms of Choice	Albus, JS 79d
...to Analyze the Information Transfer Properties of a Simplified	Model of the Cerebellar Cortex	Axelrad, H 87a
Schema for Motor Control Utilizing a Network	Model of the Cerebellum.	Houk, JC 88b
	Model of the Cerebellum as an Array of Adjustable Pattern.	Houk, JC 87a
A	Model of the Complex Cell Based on Recent Neurophysiological	Nagano, T 80a
A Biological	Model of the Excitation of a Second Order Sensory Neurone.	Enger, PS 69a
A Neurobiologically Based Neural Net	Model of the Hippocampus	Krieger, D 88a
A Neural	Model of the Mechanisms Underlying Infant Perception of Voice-Onset.	Pont, MJ 88b
Computational	Model of the Motor Program Generator for Pursuit	Eckmiller, R 87a
A	Model of the Motor Servo: Incorporating Nonlinear Spindle Receptor	Gielen, CCAM 87a
The Iron Wire	Model of the Neuron: A Review.	Lewis, ER 68c
An Electronic	Model of the Neuron Based on the Dynamics of Potassium and Sodium.	Lewis, ER 64a
Simulation of Chaotic EEG Patterns with a Dynamic	Model of the Olfactory System.	Freeman, WJ 87d
A Mathematical	Model of the Primary Visual Cortex and Hypercolumn	Okajima, K 86a
Physiological Engineering	Model of the Retinal Horizontal Cell Layer	Usui, S 88a
...Control of Automatic Processes: A Parallel Distributed Processing	Model of the Stroop Effect	Cohen, JC 88a
A	Model of the Vertebrate Central Command System	Kilmer, WL 69a
Neural Nets with Complex Structure: A	Model of the Visual System	Gremillion, M 87a
A Functional	Model of Vision and Space.	Feldman, JA 87a
Four Frames Suffice: A Provisional	Model of Vision and Space.	Feldman, JA 85c
Connectionist	Model of Visual Memory	Feldman, JA 81b
A Developmental Neural	Model of Visual Word Perception.	Golden, RM 86a
A Developmental Neural	Model of Visual Word Perception.	Golden, RM 88a
Experimental Tests of a Hierarchical	Model of Word Identification	Johnston, JC 80a
Piriform (Olfactory) Cortex	Model on the Hypercube	Bower, JM 87a
Neural	Model: One Equation Does It All.	Johnson, RC 88e
On a method of Storing Information: II. A Further Study of	Model Properties	Roy, AE 62a
Predicted Squared Error: A Criterion for Automatic	Model Selection.	Barron, AR 84a
Using Electronic Circuits to	Model Simple Neuroelectric Interactions.	Lewis, ER 68d
Optics and Neural Nets: Trying to	Model the Human Brain.	Williams, T 87a
A Neural	Model Theory Leading to Kindling Effect of Epilepsy.	Anninos, PA 87b
The Use of an Artificial Neural Network	Model to Analyze the Information Transfer Properties of a.	Axelrad, H 87a
A Bipartite Connectionist	Model to Represent N-ary Boolean and Linear Constraints.	Post, SD 87a
A Memory Storage	Model Utilizing Spatial Correlation Functions.	Anderson, JA 68a
Backward Conditioning: A Neural Network	Model which Exhibits Both Excitatory and Inhibitory Conditioning	Ricart, R 88a
Retinal	Model with Adaptive Contrast Sensitivity and Resolution.	Brill, MH 87a
Hopfield	Model with Complementary Binary Representations.	Jong, T 88a
Holographic Implementation of a Hopfield	Model with Discrete Weightings	White, HJ 88d
Two-Terminal Electronic Circuit Neuron	Model with Excitable Membrane V-I-t Characteristics.	Hoshimiya, N 79a
Nerve Cell Soma	Model with High Reliability and Low Power Consumption.	Fernandes, PML 80a
A Neuron	Model with Learning Capability and its Relation to Mechanisms of	Wigstrom, H 73a
The Hopfield	Model with Multi-Level Neurons	Fleisher, M 88a
A Neural	Model with Multiple Memory Domains	Anninos, PA 88a
Neural Network	Model with Rhythm-Assimilation Capacity.	Torras i Genis, C 86b
Adaptive Neural Encoder	Model with Selfinhibition and Threshold Control.	Zeevi, YY 81a
A Neuronal	Model with Synaptic Long-Term Memories	Niki, K 81a
Firing Response of a Neural	Model with Threshold Modulations and Neural Dynamics	Sibain, P 86a
On "Information Capacity of the Hopfield	Model"	Xu, X 88c
Neural Optical Chip	Modeled.	Johnson, RC 88j
Les Applications des	Modeles Connexionistes a la Reconnaissance de la Parole.	Wellekens, CJ 88b
Les	Modeles Mathematiques de l'Organization Biologique de la Memoire	Anderson, JA 78a
An Application of the GMDH Algorithm to Economic	Modeling	Scott, DE 84a
Neural and Brain	Modeling	MacGregor, RJ 87a
Neural	Modeling	MacGregor, RJ 77a
Neural	Modeling	Harmon, LD 66a
...a Topic in Cybernetics with APL: An Introduction to Neural Net	Modeling	Peelle, HA 81a
The Law of Large Numbers in Neural	Modeling	Geman, S 81b
Issues in Cognitive	Modeling: A Reader	Aitkenhead, AM 85a

Factual and Counterfactual Reasoning by Constructing Plausible	Models	Derthick, MA 87c
Image Segmentation Using Simple Markov Field	Models	Hansen, FR 82a
...Sensors to Real Time Processors Based on Neural Network	Models	Bandara, KMSV 87a
Interpolation Coding: A Representation for Numbers in Neural	Models	Ballard, DH 87b
...in Experimental Data on Linear Kinetics and the Formulation of	Models	Berman, M 56a
Markov Mesh	Models	Kanal, LN 80a
Markov Random Field Texture	Models	Cross, GR 83a
Mathematical Analysis of Learning Behavior of Neuronal	Models	Cheung, JY 88a
Neural and Conceptual Interpretation of PDP	Models	Smolensky, P 86b
...Versus Saccadic Eye Movements in Primates - Single-Unit Data and	Models	Eckmiller, R 83a
Order Information and Distributed Memory	Models	Ratcliff, R 87a
...a Competition-Based Activation Mechanism in Neuromimetic Network	Models	Reggia, JA 87a
Proteus: An Activation Framework for Cognitive Process	Models	Levin, JA 76a
Pulse Processing by Neuron	Models	Jenik, F 64a
Robust Tracking Novelty Filters Based on Linear	Models	Kadar, I 87a
Schemata and Sequential Thought Processes in PDP	Models	Rumelhart, DE 86j
...Detection: Parallel Studies of Human Subjects and Connectionist	Models	Cohen, NJ 86a
...Results for Learning in Single Hidden Layer Feedforward Network	Models	White, H 87b
Space or Time Adaptive Signal Processing by Neural Network	Models	Herault, J 86a
Speaker-Independent Continuous Speech Recognition Using Hidden Markow	Models	Lee, K 87a
Specifications for Nerve Membrane	Models	Moore, JW 68a
Symmetries and Learning in Neural Network	Models	Baldi, P 87b
Synchronization in Small Groups of Neurons: A Study with Electronic	Models	Lewis, ER 68b
Temporal-Pattern Learning in Neural	Models	Torras i Genis, C 85a
The CHASM: A Macromodular Computer for Analyzing Neuron	Models	Molnar, CE 67a
...Learning: A Constructive Critique of Some Connectionist Learning	Models	Lachter, J 88a
Unit Activation Rules for Cognitive Network	Models	Williams, RJ 83a
Correlation Memory	Models: A First Approximation in a General Learning Scheme	Pfaffelhuber, E 75a
Cerebellum	Models: An Interpretation of Some Features	Chauvet, G 86a
Connectionist	Models and Cognitive Science: Goals, Directions and Implications	McClelland, JL 86c
PDP	Models and General Issues in Cognitive Science	Rumelhart, DE 86i
Connectionist	Models and Geometric Reasoning	Fallside, F 88a
Neural	Models and Memory	Arbib, MA 76a
Links Between Markov	Models and Multilayer Perceptrons	Bourland, H 88a
Neural Net	Models and Optical Computing: A Brief Overview	Farhat, NH 86a
Learning in Neural Networks,	Models and Simulations	Salu, Y 84a
Computational	Models and the Development of Topographic Projections	Sejnowski, TJ 87a
Special Issue on Connectionist	Models and Their Application	Feldman, JA 85d
Neural Network	Models and Their Application to Handwritten Digit Recognition	Pawlicki, TF 88a
Connectionist	Models and Their Implications	Waltz, DL 88a
Connectionist	Models and Their Implications: Readings from Cognitive Science	Waltz, DL 88b
Connectionist	Models and Their Properties	Feldman, JA 88b
Connectionist	Models and Their Properties	Feldman, JA 82a
Connectionist	Models and their Properties	Feldman, JA 82b
Four Neural Circuit	Models and Their Role in the Organization of Voluntary Movement	Hemami, H 83a
Harmony Theory: Problem Solving, Parallel Cognitive	Models, and Thermal Plastics	Smolensky, P 84a
Positive Feedback in Hierarchical Connectionist	Models: Applications in Language Production	Dell, GS 88a
Neural Network	Models Applied to Real-Time Fault Diagnosis	Dietz, WE 89a
Learning of World	Models by Connectionist Networks	Sutton, RS 85a
Representing Neural Networks	Models by Finite Automata	Kemke, C 88b
Regression Without	Models: Directions in the Search for Structure	Malone, JM 84a
A Derivation of Neural Network	Models for Associative Memory	Paek, EG 88a
Real-Time Attentional	Models for Classical Conditioning and the Hippocampus	Schmajuk, NA 85a
Four Capacity	Models for Coarse-Coded Symbol Memories	Rosenfeld, R 87a
Neuronal	Models for Cognitive Processes: Networks for Learning, Perception	Trehub, A 77a
Globally Connected Network	Models for Computing Using Fine-Grained Processing Elements	Szu, HH 85a
Functional Properties of	Models for Direction Selectivity in the Retina	Grzywacz, N 87a
Experiments with Neural Network	Models for Handwritten Digit Recognition	Lee, D 88a
Advances in Mathematical	Models for Image Processing	Jain, AK 81a
Stochastic	Models for Learning	Bush, RR 55a
Two	Models for Memory Organization Using Interacting Traces	Anderson, JA 70a
Stochastic Spin	Models for Pattern Recognition	Divko, R 86a
Nonlinear Prediction	Models for River Flows and Typhoon Precipitation by Self-Organizing	Ikeda, S 84a
The Brains of Men and Machines, Part 1: Biological	Models for Robotics	Kent, EW 78a
Stochastic	Models for Single Neuron Firing Trains: A Survey	Fienberg, SE 74a
Mathematical	Models for the All-Or-None Activity of Some Neurons	Gerstein, GL 62a
	Models for the Brain	van Heerden, PJ 70a
	Models for the Generation of the Motor Output Pattern in Flying	Wilson, DM 68a
Self-Organizing Neural Network	Models for Visual Pattern Recognition	Fukushima, K 87e

...of Adult Rat Motor Cortex Somatic Representation Patterns after	Motor Nerve Injury	Sanes, JN 88a
Models for the Generation of the	Motor Output Pattern in Flying Locusts	Wilson, DM 68a
Neural Mechanisms for the Production of Cyclic	Motor Patterns	Selverston, AI 83a
Neural Networks for	Motor Program Generation	Eckmiller, R 88a
Computational Model of the	Motor Program Generator for Pursuit.	Eckmiller, R 87a
Motion Control of Robotic Manipulator Based on	Motor Program Learning	Arimoto, S 88a
Neural Network Mechanisms for Generation and Learning of	Motor Programs	Eckmiller, R 87b
...Aspects of a Neural Network Model for Generation of Pursuit	Motor Programs	Eckmiller, R 86b
Cerebral Cortex: Architecture, Intracortical Connections,	Motor Projections.	Lorente de No, R 49a
...Networks: A Distributed Model for Representing and Regularizing	Motor Redundancy	Mussa-Ivaldi, FA 88b
...Perspectives on Arm Trajectory Formation: A Distributed Model of	Motor Redundancy	Mussa-Ivaldi, FA 88c
A Model of the	Motor Servo: Incorporating Nonlinear Spindle Receptor and Muscle	Gielen, CCAM 87a
Transformation of Cortical	Motor Signals in Spinal Cord	Kostyuk, PG 68a
Topology Conserving Mappings for Learning	Motor Tasks.	Ritter, H 86b
A Self-Organizing System Called `Mickey	Mouse'	Browning, I 64a
...Neural-Network Model for Control and Learning of Voluntary	Movement	Kawato, M 87b
A Stochastic Theory of Phase Transitions in Human Hand	Movement	Schoner, G 86a
...Circuit Models and Their Role in the Organization of Voluntary	Movement	Hemami, H 83a
A Hierarchical Model for Voluntary	Movement and Its Application to Robotics	Kawato, M 87a
...Theory of Environmentally-Specified and Learned Patterns of	Movement Coordination, I. Relative Phase Dynamics.	Schoner, G 88b
...Theory of Environmentally-Specified and Learned Patterns of	Movement Coordination, II. Component Oscillator Dynamics	Schoner, G 88c
Do Neurons in the Motor Cortex Encode	Movement Direction? An Alternative Hypothesis.	Mussa-Ivaldi, FA 88a
Human	Movement Initiation: Specification of Arm, Direction, and Extent	Rosenbaum, DA 80a
Hierarchical Control of Rapid	Movement Sequences	Rosenbaum, DA 83a
In Back-to-Basics	Movement, VLSI Expert Calls for More Simplicity in Computation	Brown, C 87e
Hierarchical Neural Network Model for Voluntary	Movement with Application to Robotics.	Kawato, M 88a
...and Coordination of Multiple Effectors During Variable-Speed Arm	Movements.	Bullock, D 87a
...Kohonen's Self-Organizing Mapping Algorithm to Learn Ballistic	Movements.	Ritter, H 88b
Neural Circuits for Generating Rhythmic	Movements.	Friesen, WO 78a
Neural Dynamics of Adaptive Sensory-Motor Control: Ballistic Eye	Movements.	Grossberg, S 86h
...Space as a Structuring Concept for the Analysis of Skilled Hand	Movements.	Iberall, T 86a
...in Reflex Threshold Cannot Explain All Aspects of Rapid Voluntary	Movements.	Gielen, CCAM 86a
The Hierarchical Control of Skilled Hand	Movements.	Dakin, GA 86a
Eye	Movements and Vision	Yarbus, AL 67a
Investigation of the Timing of Velar	Movements during Speech.	Moll, KL 71a
Neural Dynamics of Planned Arm	Movements: Emergent Invariants and Speed-Accuracy Properties During.	Bullock, D 88b
Neural Dynamics of Planned Arm	Movements: Emergent Invariants and Speed-Accuracy Properties During.	Bullock, D 88c
A Biophysical Approach to the Spatial Function of Eye	Movements, Extraocular Proprioception and the Vestibulo-Ocular	Daunicht, WJ 88a
Neural Control of Foveal Pursuit Versus Saccadic Eye	Movements in Primates - Single-Unit Data and Models.	Eckmiller, R 83a
The Relevance of Mechanical Sensors for Neural Generation of	Movements in Space	Daunicht, WJ 88c
The Relevance of Mechanical Sensors for Neural Generation of	Movements in Space	Daunicht, WJ 88d
Eye	Movements Induced by Linear Acceleration on a Parallel Swing	Baloh, RW 88a
Coordinated Control Programs for	Movements of the Hand.	Arbib, MA 85b
Local Determination of a	Moving Contrast Edge	Kahn, P 85a
A Connectionist Architecture for Matching 3-D Models to	Moving Edge Features	Casasent, DP 86a
A Connectionist Architecture for Matching 3-D Models to	Moving Edge Features	Gunger, D 86a
Artina: A Flexible Computer Model of Retinal Processes for	Moving Images.	Gawronski, RP 87a
Processing of	Moving Images in Natural and Artificial Visual Systems	Schuling, FH 88b
...Intelligent Processing and Neural Networks for Adaptive	Moving Object Tracking	Lee, JSJ 87a
An Adaptive and Heterodyne Filtering Procedure for the Imaging of	Moving Objects	Schuling, FH 88a
...Contrast on Foveal and Peripheral Detection of Coherent Motion in	Moving Random-Dot Patterns	van de Grind, WA 87a
Controlling the Motion of a	Moving Robot Using a Neuroid Network	Gol'tsev, AD 82a
The Response of the Limulus Retina to	Moving Stimuli: A Prediction by Fourier Synthesis.	Brodie, SE 78a
The Analysis of	Moving Visual Patterns	Movshon, JA 85a
The Simulation of Neural Elements by Electrical Networks Based on	Multi-Aperture Magnetic Cores.	Brain, AE 61a
	Multichannel Adaptive Filtering for Signal Enhancement	Ferrara, Jr, ER 81a
Spatio-Temporal Analysis of	Multi-channel Alpha EEG Map Series	Fuchs, A 87a
A	Multicomponent Theory of the Perception of Briefly Exposed Visual.	Rumelhart, DE 70a
Capacities of	Multiconnected Memory Models	Horn, D 88a
Long Term Memory Storage Capacity of	Multiconnected Neural Networks	Peretto, P 86b
Shift-Invariant	Multiconnected Neural-Networks	Horn, D 88c
	Multidimensional Machine Vision Using Neural Networks.	Kolodzy, PJ 87a
A Multiplexed	Multi-Electrode Semiconductor Brain Electrode Implant.	Kabrisky, M 87a
Visual Masking in	Multielement Displays.	Eriksen, CW 70a
Learning and Problem Solving with	Multilayer Connectionist Systems	Anderson, CW 86a
A	Multilayer Cooperative/Competitive Method for Creating Hierarchical.	Maren, AJ 88a
	Multi-Layer Feedforward Networks Are Universal Approximators	Hornik, K 88a
A Self-Organizing	Multilayer Neural Network.	Fukushima, K 75a
Adaptive Training of	Multilayer Neural Networks Using a Least Squares Estimation.	Kollias, S 88a

Neural Networks	Naive, Says Minsky	Johnson, RC 87j
A Bipartite Connectionist Model to Represent	N-ary Boolean and Linear Constraints	Post, SD 87a
Neural Nets to Arm	NASA Space Robots?	Johnson, RC 87h
	NASA Unites Neural Nets, Fuzzy Logic	Johnson, RC 88b
Computational Systems -	Natural and Artificial	Haken, H 87a
The Role of Physical Constraints in	Natural and Artificial Manipulation.	Morasso, P 86a
Adaptation in	Natural and Artificial Systems	Holland, JH 75a
Relational Models in	Natural and Artificial Vision.	Bienenstock, EL 88a
Processing of Moving Images in	Natural and Artificial Visual Systems.	Schuling, FH 88b
	Natural and Social System Metaphors for Distributed Problem Solving.	Chandrasekaran, B 81a
A Representation for	Natural Category Systems	Peters, SL 87a
Perceptual Organization and the Representation of	Natural Form	Pentland, AP 86a
Neural Networks and	Natural Intelligence	Grossberg, S 88d
Parallel Interpretation of	Natural Language	Pollack, JB 84a
Interpretation of	Natural Language: A Potential Application of Parallelism	Pollack, JB 86a
Several Studies on	Natural Language and Back-Propagation.	Allen, RB 87a
PARSNIP: A Connectionist Network that Learns	Natural Language Grammar from Exposure to Natural Language	Hanson, SJ 87a
Massively Parallel Parsing: A Strongly Interactive Model of	Natural Language Interpretation.	Waltz, DL 88c
Massively Parallel Parsing: A Strongly Interactive Model of	Natural Language Interpretation.	Waltz, DL 85a
Massively Parallel Network-Based	Natural Language Parsing System.	Li, T 87a
Symbolic NeuroEngineering for	Natural Language Processing: A Multilevel Research Approach.	Dyer, MG 88a
	Natural Language Processing Using Spreading Activation and Lateral	Pollack, JB 82a
A Parallel	Natural Language Processing with Distributed Control	Berg, G 87a
...Network that Learns Natural Language Grammar from Exposure to	Natural Language Sentences	Hanson, SJ 87a
A PDP Learning Approach to	Natural Language Understanding	Sharkey, NE 89a
Rule-Based Processing in a Connectionist System for	Natural Language Understanding	Selman, B 85b
Word Expert Parsing: A Theory of Distributed Word-Based	Natural Language Understanding	Small, SL 80a
A Computational Approach to Fuzzy Quantifiers in	Natural Languages.	Zadeh, LA 83a
Extracting and Labeling Boundary Segments in	Natural Scenes	Prager, JM 80a
Segmentation of	Natural Scenes	Hanson, AR 78a
Mathematics Applied to Deterministic Problems in the	Natural Sciences	Lin, CC 74a
Problem Solving in a	Natural Task as a Function of Experience	Lancaster, JS 87a
Discrimination of	Natural Textures: A Neural Network Architecture.	Mesrobian, E 87b
Self-Supervised Learning: A Scheme for Discovery of "	Natural" Categories by Single Units.	Munro, PW 86a
A	Naturalistic Approach to Mental Representation	Gonzales, MEQ 88a
On the	Nature and Elimination of Stimulus Artifact in Nerve Signals Evoked.	McGill, KC 82a
Structured Neural Networks in	Nature and in Computer Science	Feldman, JA 88d
The	Nature of Human Prehension: Three Dextrous Hands in One.	Iberall, T 87c
On the	Nature of Inducing Forms Generating Perceptions of Illusory.	Prazdny, K 85a
BoltzCONS: Reconciling Connectionism with the Recursive	Nature of Stacks and Trees	Touretzky, DS 86b
Replica Symmetry Breaking and the	Nature of the Spin Glass Phase	Mezard, M 84b
	Nature of the Spin-Glass Phase	Mezard, M 84a
The	Nature of Thought: Essays in Honor of D.O. Hebb.	Jusczyk, PW 80a
Seeing the Light: Optics in	Nature, Photography, Color, Vision, and Holography	Stork, DG 85a
...Networks, Error-Correcting Codes and Polynomials Over the Binary	n-Cube	Bruck, J 88d
...by the Global Limits of a Nonlinear Competitive Interactin in	n-Dimensions	Grossberg, S 77b
...Analysis of the Axial Next-Nearest Neighbor Ising Model	Near its Multiphase Point.	Fisher, ME 81a
Counterpropagation Networks: Adaptive Hierarchical Networks for	Near Optimal' Mappings	Stork, DG 87a
A Trainable Model for	Near Periodic Signals.	Wolf, L 87a
	Nearest Matched Filter Classification of Spatiotemporal Patterns	Hecht-Nielsen, R 87f
	Nearest Matched Filter Classification of Spatio-Temporal Patterns.	Hecht-Nielsen, R 87g
Finding Prototypes for	Nearest Neighbor Classifiers	Chang, C 74a
An Algorithm for a Selective	Nearest Neighbor Decision Rule	Ritter, GL 75a
Some Convergence Properties of a	Nearest Neighbor Decision Rule	Peterson, D 70a
A Preprocessing Algorithm for	Nearest Neighbor Decision Rules.	Fischer, FP 70a
On the Inequality of Cover and Hart in	Nearest Neighbor Discrimination.	Devroye, L 81a
	Nearest Neighbor Pattern Classification.	Cover, TM 67a
Condensed	Nearest Neighbor Rule.	Hart, PE 68a
New Error Bounds with the	Nearest Neighbor Rule.	Devijver, PA 79a
On the	Nearest Neighbor Rule.	Devijver, PA 80a
Asymptotic Properties of	Nearest Neighbor Rules Using Edited Data	Wilson, DL 72a
A Technique to Identify	Nearest Neighbors.	Yunck, TP 76a
An Algorithm for Finding	Nearest Neighbors.	Friedman, JH 75a
Fast Algorithm for Recognizing	Nearest Neighbors.	Sethi, IK 81a
Design of	Nearest Prototype Signal Classifiers	Gardner, WA 81a
Evaluation of the Use of the Hopfield Neural Network Model as a	Nearest-Neighbor Algorithm	Montgomery, BL 86a
	Nearest-Neighbour Systems and the Auto-Logistic Model for Binary	Besag, JE 72a
Neural Systems Make Comeback After	Nearly 50-Year Gestation	Johnson, RC 87k
A	Neat Theory of Marker Passing.	Charniak, E 86a

Are Special Chips	Necessary for Neural Computers?	Aleksander, I 88a
Stable Self-Organization of Sensory Recognition Codes: Is Chaos	Necessary?	Grossberg, S 87o
Brain-Emulating Circuits	Need 'Sleep' and 'Dreams'	Johnson, RC 86a
Finding Prototypes for Nearest	Neighbor Classifiers	Chang, C 74a
An Algorithm for a Selective Nearest	Neighbor Decision Rule	Ritter, GL 75a
Some Convergence Properties of a Nearest	Neighbor Decision Rule	Peterson, D 70a
A Preprocessing Algorithm for Nearest	Neighbor Decision Rules.	Fischer, FP 70a
On the Inequality of Cover and Hart in Nearest	Neighbor Discrimination.	Devroye, L 81a
Low Temperature Analysis of the Axial Next-Nearest	Neighbor Ising Model Near its Multiphase Point	Fisher, ME 81a
Nearest	Neighbor Pattern Classification.	Cover, TM 67a
Condensed Nearest	Neighbor Rule.	Hart, PE 68a
New Error Bounds with the Nearest	Neighbor Rule.	Devijver, PA 79a
On the Nearest	Neighbor Rule.	Devijver, PA 80a
Asymptotic Properties of Nearest	Neighbor Rules Using Edited Data	Wilson, DL 72a
A Branch and Bound Algorithm for Computing k-Nearest	Neighbors.	Fukunaga, K 75a
A Technique to Identify Nearest	Neighbors.	Yunck, TP 76a
An Algorithm for Finding Nearest	Neighbors.	Friedman, JH 75a
Fast Algorithm for Recognizing Nearest	Neighbors.	Sethi, IK 81a
Estimation and Choice of	Neighbors in Spatial-Interaction Models of Images.	Kashyap, RL 83a
Multi-Layer Perceptron Inversion Applied to Image	Neighbourhood Operations	Dodd, N 87a
L'Istanza Analogica	Nelle Scienze Del Cervello Considerazioni Filosofiche.	Basti, G 86a
A Simplified Version of Kunihiko Fukushima's	Neocognitron	Deutsch, S 81a
...Mechanism of Pattern Recogition Unaffected by Shift in Position -	Neocognitron -	Fukushima, K 79b
	Neocognitron: A Hierarchical Neural Network Capable of Visual.	Fukushima, K 88c
	Neocognitron: A Neural Network Model for a Mechanism of Visual	Fukushima, K 83c
	Neocognitron: A Neural Network Model for a Mechanism of Visual	Fukushima, K 83b
	Neocognitron: A New Algorithm for Pattern Recognition Tolerant of.	Fukushima, K 82a
	Neocognitron: A Self-Organizing Neural Network Model for a	Fukushima, K 80a
	Neocognitron: A Self-Organizing Neural Network Model for a	Fukushima, K 82b
	Neocognitron: Self-Organizing Network Capable of Position-Invariant.	Fukushima, K 80c
	Neocognitron: Self-Organizing Network Capable of Position-Invariant.	Fukushima, K 80b
A Theory of Cerebral	Neocortex.	Marr, D 70a
Conditional Control Transfer Mechanisms in the	Neocortex: 1. The Basic Model.	Scott, PD 79a
A Computer Simulation of Cerebral	Neocortex: Computational Capabilities of Nonlinear Neural Networks	Singer, A 87a
A Computer Simulation of Cerebral	Neocortex: Computational Capabilities of Nonlinear Neural Networks	Singer, A 88a
...of Form Perception: Boundary Completion, Illusory Figures, and	Neon Color Spreading	Grossberg, S 86j
	NEP: An Emulation-Assist Processor for Parallel Associative.	Cruz, CA 85a
Learning Modules Permit Scalable	Nerual Systems	Johnson, RC 87e
Mathematical Models of Excitation and Propagation in	Nerve.	FitzHugh, R 69a
An Active Pulse Transmission Line Simulating	Nerve Axon	Nagumo, J 62a
Dynamics of Image Formation by	Nerve Cell Assemblies.	Freeman, WJ 83a
Detection of Favored Patterns in the Temporal Structure of	Nerve Cell Connections	Dayhoff, JE 87a
A Digital-Computer Model of	Nerve Cell Functioning	Perkel, DH 64a
	Nerve Cell Soma Model with High Reliability and Low Power.	Fernandes, PML 80a
Arithmetical Operations Performed by	Nerve Cells.	Blomfield, S 74a
Mathematical Theory on Formation of Category Detecting	Nerve Cells.	Amari, S 78a
Normal and Abnormal Signal Patterns in	Nerve Cells.	Carpenter, GA 81a
The Stochastic Properties of Spike Trains Recorded from	Nerve Cells.	Stein, RB 72a
Remarks About the "Singularity" of	Nerve Cells and its Ontogenesis.	Changeux, J 85a
Microelectronics in	Nerve Cells: Dendritic Morphology and Information Processing	Poggio, TA 81a
Input-Output Relations in Computer-Simulated	Nerve Cells: Influence of Statistical Properties, Strength, etc.	Segundo, JP 68a
How to Label	Nerve Cells so that they can Interconnect in an Ordered Fashion.	von der Malsburg, C 77a
A Mechanism for Optic	Nerve Conduction and Form Perception: I.	Culbertson, JT 48a
	Nerve Conduction Theory: Some Mathematical Consequences of	Offner, F 40a
Automatic Computation of	Nerve Excitation	Cole, KS 55a
Automatic Computation of	Nerve Excitation - Detailed Corrections and Additions.	FitzHugh, R 59a
Application of Bifurcation Theory to	Nerve Excitations.	Toko, K 81a
Robots with a Lot of	Nerve: Eye-hand Coordination May Help a Robot Learn to See	Peterson, I 87a
The Electrophysics of a	Nerve Fiber.	Scott, AC 75a
Simulating the Adaptation Phenomena in the Response of Auditory	Nerve Fibers: A Functional Model	Wu, ZL 88a
Differential Equations for the Development of Topological	Nerve Fibre Projections.	von der Malsburg, C 81a
Formation of Topographic Maps and Columnar Microstructures in	Nerve Fields	Takeuchi, A 79a
On the Topological Representation of Signals in Self-Organizing	Nerve Fields	Amari, S 87b
Topographic Organization of	Nerve Fields	Amari, S 80a
Cochlear	Nerve Filter Responses: Distribution Along the Cochlear Partition.	Pfeiffer, RR 75a
Speech Coding in the Auditory	Nerve: II. Processing Schemes for Vowel-Like Sounds.	Delgutte, B 84a
...Approach to Singular Perturbation Problems with Applications to	Nerve Impulse Equations.	Carpenter, GA 77a
	Nerve Impulse Equations.	Carpenter, GA 76a
Periodic Solutions of	Nerve Impulse Equations.	Carpenter, GA 77b

...Rat Motor Cortex Somatic Representation Patterns after Motor	Nerve Injury	Sanes, JN 88a
Fluctuation Phenomena in	Nerve Membrane	Verveen, AA 68a
Specifications for	Nerve Membrane Models.	Moore, JW 68a
Integrate-And-Fire Models of	Nerve Membrane Response to Oscillatory Input	Keener, JP 81a
	Nerve Net Models of Plausible Size that Perform Many Simple.	Brindley, GS 69a
	Nerve Net with Inhibitory Connections.	Torioka, T 79b
Pattern Separability of a Tightly Coupled Random	Nerve Nets	Amari, S 74a
A Mathematical Theory of	Nerve Nets	Amari, S 82a
A Mathematical Theory of Self-Organizing	Nerve Nets	Amari, S 79a
Theory of Self-Organizing	Nerve Nets	Kleene, SC 56a
Representation of Events in	Nerve Nets and Finite Automata	Johnson, RC 86c
	Nerve Nets Built On Chip	Brown, C 86c
First Commercial	Nerve Nets Challenge Traditional AI.	Brown, C 86d
Hopfield's	Nerve Nets Realize Biocomputing.	Torioka, T 88a
Pattern Separating Functioning of Two Layered Random	Nerve Nets with Feedforward Inhibitory Connections	Amari, S 79b
Theory of Self-Organizing	Nerve Nets with Special Reference to Association and Concept	Honrubia, V 87a
Morphological Aspects of the Vestibular	Nerve of the Squirrel Monkey	McGill, KC 82a
On the Nature and Elimination of Stimulus Artifact in	Nerve Signals Evoked and Recorded Using Surface Electrodes	Amari, S 82b
A Mathematical Theory of Self-Organizing	Nerve Systems.	Marsocci, VA 82a
Electrical Network Modeling of Active Membranes of	Nerves	Rashevsky, N 36a
Physico-Mathematical Aspects of Excitation and Conduction in	Nerves	Tenorio, MF 87e
Simulating	Nerves to Improve Cochlear Implants for the Hearing Impaired	McCulloch, WS 43b
A Logical Calculus of the Ideas Immanent in	Nervous Activity	McCulloch, WS 43a
A Logical Calculus of the Ideas Immanent in	Nervous Activity	Cowan, JD 70a
A Statistical Mechanics of	Nervous Activity	Landahl, HD 43a
A Statistical Consequence of the Logical Calculus of	Nervous Nets	Shimbel, A 48a
A Statistical Approach to the Theory of the Central	Nervous System	Black, IB 87a
Biochemistry of Information Storage in the	Nervous System	Uttley, AM 59a
Conditional Probability Computing in a	Nervous System	Kristan, WB 80a
Information Processing in the	Nervous System	Uttley, AM 79a
Information Transmission in the	Nervous System	Matthies, H 86a
Learning and Memory: Mechanisms of Information Storage in the	Nervous System	Gelfand, IM 71a
Mathematical Modeling of Mechanisms of the Central	Nervous System	Finkel, LH 88b
Neuronal Group Selection: A Basis for Categorization by the	Nervous System	Levy, WB 84a
Synaptic Change in the	Nervous System	Heimer, L 68a
The Tracing of Pathways in the Central	Nervous System	Naito, S 88a
...Changes of Protein Phosphorylation in a Molluscan	Nervous System	Albus, JS 88a
The Central	Nervous System as a Low and High Level Control System.	Koizumi, O 86a
Plasticity in the	Nervous System of Adult Hydra I. The Position-Dependent Expression	Koizumi, O 88a
Plasticity in the	Nervous System of Adult Hydra II. Conversion of Ganglion Cells of.	Conrad, M 74b
Molecular Information Processing in the Central	Nervous System, Part I	Spinelli, DN 70a
...A Computer Model for a Content-Addressable Memory in the Central	Nervous Systems.	Toet, A 87a
The Construction of a Simultaneous Functional Order in	Nervous Systems I. Relevance of Signal Covariances and Signal.	Toet, A 87b
The Construction of a Simultaneous Functional Order in	Nervous Systems II. Computing Geometrical Structures	Toet, A 87c
The Construction of a Simultaneous Functional Order in	Nervous Systems III. The Influence of Environmental Constraints on	Cooper, LN 81a
Distributed Memory in the Central	Nervous Systems: Possible Test of Assumptions in Visual Cortex	Wilson, HR 73a
...Theory of the Functional Dynamics of Cortical and Thalamic	Nervous Tissue	Meystel, A 86a
Cognitive Control of Autonomous Mobile Robots:	Nested Hierarchical Intelligent Module	Johnson, RC 88c
	Nestor Frees Neural Nets from Hardware Chains.	Brown, C 86k
Startup	Nestor Readies Neural Network-Based Products	Johnson, RC 87f
Neural Net Developer	Nestor Sprints From Startup Gate	Klassen, M 88a
Characteristics of the Functional Link Net: A Higher Order Delta Rule	Net.	Feild Jr, WB 88a
...Input Exemplars to Form Stable States in a Hopfield Neural	Net.	Scalettar, R 88b
Perception of Left and Right by a Feed Forward	Net.	David, GW 87a
Sensitivity Analysis of Hopfield	Net.	Cheung, KF 87c
Synchronous vs Asynchronous Behavior of Hopfield's CAM Neural	Net.	Klassen, M 88a
Characteristics of the Functional Link	Net: A Higher Order Delta Rule Net	Marrian, CRK 87a
Electronic 'Neural'	Net Algorithm for Maximum Entropy Deconvolution.	Huang, WY 87a
Comparisons Between Neural	Net and Conventional Classifiers	Huang, WY 88a
Neural	Net and Traditional Classifiers.	Mehr, D 87a
Neural	Net Application to Optical Character Recognition	Richfield, S 87a
A Logarithmic Vector Processor for Neural	Net Applications	Culhane, AD 88a
A Neural	Net Approach to Discrete Fourier Transforms.	Troudet, T 88a
An Adaptive Neural	Net Approach to the Segmentation of Mixed Gray-Level and Binary.	Schwartz, TJ 87b
Hecht-Nielsen Unveils First New Neural	Net Architecture Since '84	Kung, S 89a
Neural	Net Architectures for Robotic Applications	Cheung, KF 87b
Neural	Net Associative Memories Based on Convex Set Projections	Draper, JS 87a
A Microcomputer Neural	Net Benchmarked Against Standard Classification Techniques	Morse, A 78a
Neural	Net Capable of Competitive and Cooperative Computation	Mann, J 88a
A Self-Organizing Neural	Net Chip	

Keyword context	Title	Author
Mathematical Theory of Self-Organization in Neural	Nets	Amari, S 88a
Memory Topology of Neural	Nets	Papadopoullos, MS 79a
Neural	Nets	Cowan, JD 87a
Neural Petri	Nets	Zargham, MR 85a
Parameter	Nets	Ballard, DH 84b
Perceptrons and Neural	Nets	Reece, P 87a
Prototype Extraction and Matching in Associative	Nets	Lansner, A 85a
Report Outlines Commercialization of Neural	Nets	Schwartz, TJ 87e
Resource Requirements of Standard and Programmable	Nets	McClelland, JL 86j
Scaling, Machine Learning, and Genetic Neural	Nets	Mjolsness, E 88b
Secondary Bifurcation in Neuronal	Nets	Ermentrout, GB 80b
Self-Control in Neural	Nets	Josin, GM 77a
Semantic Networks and Neural	Nets	Shastri, L 84a
Synchronization in Neural	Nets	Vidal, JJ 88b
The Construction of Homogeneous	Nets	Muir, A 81a
The Dynamics of Symmetric	Nets	Muir, A 83a
The Inverse Problem for Linear Boolean	Nets	Caianiello, ER 88a
Theory of Self-Organizing Nerve	Nets	Amari, S 79a
Tolerance	Nets	Warner, MW 81a
Neural	Nets and Artificial Intelligence	Cowan, JD 88a
Neural	Nets and Computer Vision Systems	Tao, KM 87a
Representation of Events in Nerve	Nets and Finite Automata	Kleene, SC 56a
Synthesis of Boolean	Nets and Time Behavior of a General Mathematical Neuron.	Caianiello, ER 75a
Optoelectronic Analogs of Self-Programming Neural	Nets: Architecture and Methodologies for Implementing Fast	Farhat, NH 87b
Parallel Processing of Signals in Neural	Nets as Manifested in the EEG.	Freeman, WJ 75b
Nerve	Nets Built On Chip	Johnson, RC 86c
First Commercial Nerve	Nets Challenge Traditional AI.	Brown, C 86c
The Problem of Internal Representations in Neural	Nets: Concepts, Implications and a Proposed Metric	Tillery, SI 87a
Artificial Neural	Nets: Dependence of the EEG Amplitude's Probability Distribution on.	Anninos, PA 83b
Noisy Neural	Nets Exhibiting Epileptic Features	Kokkinidis, M 85a
Noisy Neural	Nets Exhibiting Memory Domains	Anninos, PA 84c
Theories of Neural	Nets Expose the Limits of AI	Brown, C 87l
Neural	Nets for Adaptive Filtering and Adaptive Pattern Recognition	Widrow, B 88b
Neural	Nets for Economists.	Baum, EB 88b
Supplier-Consumer	Nets for Material Handling	Parunak, HVD 87a
A Comparison of Hamming and Hopfield Neural	Nets for Pattern Classification.	Lippman, RP 87a
Layered Neural	Nets for Pattern Recognition	Widrow, B 88a
Neural	Nets for Speech Recognition.	Huang, WY 88b
Using Artificial Neural	Nets for Statistical Discovery: Observations after Using Back.	Irani, EA 88a
Recurrent	Nets for the Storage of Cyclic Sequences	Wolf, L 87b
Nestor Frees Neural	Nets from Hardware Chains.	Johnson, RC 88c
NASA Unites Neural	Nets, Fuzzy Logic.	Johnson, RC 88b
Random Logical	Nets I	Rozonder, LI 69a
Random Logical	Nets II.	Rozonder, LI 69b
Random Logical	Nets III	Rozonder, LI 69c
Neural	Nets Inspire Computer Designer	Johnson, RC 86d
IEEE Puts Neural	Nets into Focus.	Johnson, RC 87b
Competition and Cooperation in Neural	Nets, Kyoto, 1982: Proceedings	Amari, S 82c
Neural	Nets Learn to Outsmart a Salesman.	Schindler, M 87a
Neural	Nets Learn to Outsmart a Salesman.	Schindler, M 87b
Optics and Neural	Nets: Marriage of Convenience.	Johnson, RC 88k
Characteristics of Random	Nets of Analog Neuron-Like Elements.	Amari, S 72a
Random	Nets of Analog Neurons	Averbukh, DYA 69a
Homogeneous	Nets of Neuron-Like Elements	Amari, S 75a
Learning Patterns and Pattern Sequences by Self-Organizing	Nets of Threshold Elements	Amari, S 72b
	Nets of Threshold Elements	Krohn, K 65a
On Autonomous Logic	Nets of Threshold Elements	Yajima, S 68a
A Field Theory for Neural	Nets, Part I: Derivation of the Field Equations.	Griffith, JS 63a
A Field Theory for Neural	Nets, Part II: Properties of the Field Equations	Griffith, JS 65a
Hopfield's Nerve	Nets Realize Biocomputing.	Brown, C 86d
Neural	Nets: The Dream Machines	House, J 87a
Neural	Nets to Arm NASA Space Robots?	Johnson, RC 87h
Programming Neural	Nets to Do Spatial Computations.	Zipser, D 86d
Optics and Neural	Nets: Trying to Model the Human Brain.	Williams, T 87a
General Principles of Operations in Neuron	Nets with Application to Acoustical Pattern Recognition.	Mueller, P 62a
A Mathematical Formalism for the Finite Difference Equation in Neural	Nets with Chemical Markers	Anninos, PA 87a
Self-Organization of Neural	Nets with Competitive and Cooperative Interaction.	Sawada, R 82a
Neural	Nets with Complex Structure: A Model of the Visual System.	Gremillion, M 87a
Pattern Separating Functioning of Two Layered Random Nerve	Nets with Feedforward Inhibitory Connections	Torioka, T 88a

		Grossberg, S 69f
...Differential Equations Describing a Flow on a Probabilistic	Network.	Granger, R 88b
Partitioning of Sensory Data by a Cortical	Network.	Roberts, LC 60a
Pattern Recognition with an Adaptive	Network.	Lansner, A 85b
Reliability and Speed of Recall in an Associative	Network.	Touretzky, DS 87b
Representing Conceptual Structures in a Neural	Network.	Andersen, P 66a
Rhythmic Activity in a Simulated Neuronal	Network.	Wright, WA 89c
Road Finding Neural	Network.	Linsker, R 88a
Self-Organization in a Perceptual	Network.	Kienker, PK 86a
Separating Figure from Ground with a Parallel	Network.	Lynch, G 88a
Simulation and Analysis of a Simple Cortical	Network.	Puleo P, F 81a
Simulation of a Neural	Network.	Uttley, AM 76c
Simulation Studies of Learning in an Informon	Network.	Allanson, JT 56a
Some Properties of a Randomly Connected Neural	Network.	Zhou, Y 88c
Stereo Matching Using a Neural	Network.	Willwacher, G 82a
Storage of a Temporal Pattern Sequence in a	Network.	Szentagothai, J 68a
Structuro-Functional Considerations of the Cerebella Neuron	Network.	Barto, AG 82c
...of Non-Linear Control Surfaces by a Layered Associative Search	Network.	Carpenter, GA 88a
The ART of Adaptive Pattern Recognition by a Self-Organizing Neural	Network.	Taber, WR 88d
The Recognition of Orca Calls with a Neural	Network.	Ryan, TW 88a
The Resonance Correlation	Network.	Rosenberg, CR 86a
The Spacing Effect on NETtalk, a Massively-Parallel	Network.	Krowitz, A 87a
The State Space of Memory Clusters in the Hopfield	Network.	Parga, N 86a
The Ultrametric Organization of Memories in a Neural	Network.	Robinson, AJR 87a
The Utility Driven Dynamic Error Propagation	Network.	Winter, CL 87a
TIN: A Trainable Inference	Network.	Winter, CL 87b
TIN: A Trainable Inference	Network.	Linsker, R 88b
Towards an Organizing Principle for a Layered Perceptual	Network.	Feild Jr, WB 88b
Transputer Implementation of Hopfield Neural	Network.	Tenorio, MF 88b
...English Text-to-Speech Conversion with a Distributed Adaptive	Network.	Ryan, TW 88b
Vector Quantization Training by a Self-Organizing Neural	Network.	Plaut, DC 84a
Visual Recognition of Simple Objects by a Connection	Network.	Hennessey, AK 87a
The Gamma	Network - A Physical and Logical Data Structure.	Barto, AG 81a
Associative Search	Network: A Reinforcement Learning Associative Memory	Dahl, ED 87b
Neural	Network Algorithm for an NP-Complete Problem: Map and Graph.	Gardner, S 86a
Application of Neural	Network Algorithms and Architectures to Correlation / Tracking and	Gullichsen, E 87a
Pattern Classification by Neural	Network: An Experimental System for Icon Recognition	Xu, J 86a
The Dynamics of Large Scale Neuron-Glia	Network and Its Relation to the Brain Functions (I).	Grant, PM 86b
Comparison of Neural	Network and Matched Filter Processing for Detecting Lines in Images.	Oh, S 88a
Effect of Clock Skew in Iterative Neural	Network and Optical Feedback Processors.	Jones, WP 88a
ANNA: An Adaptive Neural	Network Anticipator for Human/Computer Interfacing	Nguyen, DD 87a
Stochastic Processing in a Neural	Network Application.	Oyster, JM 86a
Associative	Network Applications to Robot Vision	Ramamoorthy, PA 88a
A Neural	Network Approach for Implementing Expert Systems	Pearson, JC 88a
Neural	Network Approach to Sensory Fusion	Chua, LO 88a
A Neural	Network Approach to Transform Image Coding	Sayegh, S 88a
Expert System and Neural	Network Approaches for Classical Guitar Fingering.	Kuczewski, RM 87d
Neural	Network Approaches to Multi-Target Tracking.	Kuczewski, RM 87c
Neural	Network Approaches to Multi-Target Tracking.	Howard, RE 87a
An Associative Memory Based on an Electronic Neural	Network Architecture	Mesrobian, E 87b
Discrimination of Natural Textures: A Neural	Network Architecture	Grossberg, S 87b
A Neural	Network Architecture for Attentionally-Modulated Pavlovian	Bullock, D 87a
A Neural	Network Architecture for Automatic Trajectory Formation and.	Grossberg, S 87c
A Neural	Network Architecture for Brightness Perception Under Constant and.	Guez, A 88a
Neural	Network Architecture for Control	Grossberg, S 87d
A Neural	Network Architecture for Preattentive Vision: Multiple Scale	Grossberg, S 87n
Recent Developments in the Design of Real-Time Non-Linear Neural	Network Architectures.	Penz, PA 87a
...An Integer to Binary Vector Transformation Suitable for Neural	Network Architectures.	White, H 88b
The Case for Conceptual and Operational Separation of	Network Architectures and Learning Mechanisms.	Will, CA 87a
Neural	Network Architectures and Their Implications for Next Generation	Hinton, GE 87c
Neural	Network Architectures for AI	Soulie, FF 87a
Evaluation of	Network Architectures on Test Learning Tasks	Schwartz, RG 88a
Evaluating a Connectionist	Network as a Simulation of Speech Development.	Buhmann, J 86a
A Physiological Neural	Network as an Autoassociative Memory	Hennessey, AK 87b
The Gamma	Network as an Interconnection Architecture for Neural Networks	Prager, RW 87a
A Comparison of the Boltzmann Machine and the Back Propagation	Network as Recognizers of Static Speech patterns	Lee, LS 87a
A Continuous-Time Optical Neural	Network Associative Memory	Ricart, R 87a
Neural	Network Based Architectures for Aerospace Applications	Hassoun, MH 88c
Pseudo-Unsupervised Neural	Network Based Automated EMG Decomposition.	Kamangar, FA 88a
A Neural	Network Based Controller for Robotic Manipulators.	

Incremental	Network Generation in Template-Based Word Recognition.	Lee, K 85a
...Simulation: Two Examples within a Single, Integrated Neural	Network Hardware Development Environment	Bahr, C 88c
Neural	Network Implementation Approaches for the Connection Machine	Brown, NH 88a
Neural	Network Implementation for Maximum-Likelihood Sequence Estimation. .	Provence, JD 87a
Neural	Network Implementation of a Medical Diagnosis Expert System.	Kudrycki, TP 88a
Analog Nonvolatile Memory for Neural	Network Implementations.	Sage, JP 88a
A Neural	Network in the Visual Cortex	Nylen, M 87a
Artificial Neural	Network Integrated Circuit Based on NMOS/CCD Principles.	Sage, JP 86a
...A Procedure for Mapping the Architecture of a Living Neural	Network into a Machine	Dawes, RL 87a
Three Constructive Algorithms for	Network Learning	Gallant, SI 86c
Neural	Network Learning Algorithms.	Sejnowski, TJ 88b
An Implementation of	Network Learning on the Connection Machine	Rosenberg, CR 88a
Neural	Network Learning Techniques.	Sharkey, NE 88b
UCLA PUNNS - A Neural	Network Machine for Computer Vision.	Gunger, D 87a
Inverting a Connectionist	Network Mapping by Back-Propagation of Error	Williams, RJ 86a
Neural	Network Mechanisms for Generation and Learning of Motor Programs . .	Eckmiller, R 87b
Electronic Device Aspects of Neural	Network Memories	Lambe, J 85a
A Decentralized Tunable Short Term Neural	Network Memory and Application to Tracking	Morris, RJT 88a
VLSI Implementation of a Neural	Network Memory with Several Hundreds of Neurons.	Graf, HP 86a
A CMOS Implementation of a Neural	Network Model.	Graf, HP 87c
A Generalized Neural	Network Model.	Xu, X 88a
An Exactly Solvable Asymmetric Neural	Network Model.	Derrida, B 87a
Category Learning and Adaptive Pattern Recognition: A Neural	Network Model.	Carpenter, GA 85a
Cognitron: A Self-Organizing Multi-Layered Neural	Network Model.	Fukushima, K 81a
Dynamic Single Unit Simulation of a Realistic Cerebellar	Network Model.	Pellionisz, AJ 73a
Holographic Implementation of a Discrete Hopfield Neural	Network Model.	White, HJ 88c
On Parallel Execution of the Traveling Salesman Problem on a Neural	Network Model.	Bagherzadeh, N 87a
Real Time Noisy Image Segmentation Using an Artificial Neural	Network Model.	Tenorio, MF 87d
VLSI Implementation of a Neural	Network Model.	Graf, HP 88a
Layered Neural	Network Model Applied to the Auditory System	Travis, BJ 86a
Evaluation of the Use of the Hopfield Neural	Network Model as a Nearest-Neighbor Algorithm.	Montgomery, BL 86a
Neural	Network Model for a Mechanism of Pattern Recogition Unaffected by. .	Fukushima, K 79b
Neocognitron: A Self-Organizing Neural	Network Model for a Mechanism of Pattern Recognition Unaffected by .	Fukushima, K 80a
Neocognitron: A Neural	Network Model for a Mechanism of Visual Pattern Recognition.	Fukushima, K 83b
Neocognitron: A Neural	Network Model for a Mechanism of Visual Pattern Recognition.	Fukushima, K 83c
Neocognitron: A Self-Organizing Neural	Network Model for a Mechanism of Visual Pattern Recognition.	Fukushima, K 82b
A Hierarchical Neural	Network Model for Associative Memory	Fukushima, K 83a
A Hierarchical Neural	Network Model for Associative Memory	Fukushima, K 84a
A Neural	Network Model for Cognitive Activity	Nelson, TJ 83a
A Cortical	Network Model for Early Vision Processing.	Moller, P 88a
Topological and Dynamical Aspects of a Neural	Network Model for Generation of Pursuit Motor Programs	Eckmiller, R 86b
A Layered	Network Model for Learning-to-Learn and Configuration in Classical .	Kehoe, EJ 86a
Cerebellar Neural	Network Model for Manipulator Control.	Hosogi, S 88a
Self-Organizing Process of a Neural	Network Model for Pattern Detection.	Nagashino, H 87a
A Neural	Network Model for Phoneme Recognition Using the Generalized Delta. .	Watson, M 87a
A Hierarchical Neural	Network Model for Selective Attention.	Fukushima, K 88a
A Neural	Network Model for Selective Attention.	Fukushima, K 87b
A Neural	Network Model for Selective Attention.	Fukushima, K 87a
A Neural	Network Model for Selective Attention in Visual Pattern Recognition.	Fukushima, K 86a
Neural	Network Model for Selective Attention in Visual Pattern Recognition.	Fukushima, K 87c
Neural	Network Model for Selective Attention in Visual Pattern Recognition.	Fukushima, K 87d
Abilities and Limitations of a Neural	Network Model for Spoken Word Recognition.	Kurogi, S 88a
A Self-Organizing Neural	Network Model for the Development of Complex Cells	Nagano, T 81a
A Neural	Network Model for the Developmental Process of Hypercomplex Cells. .	Nagano, T 83a
A Neural	Network Model for the Mechanism of Feature Extraction.	Fukushima, K 84b
A Neural	Network Model for the Mechanism of Feature Extraction - A Self . . .	Miyake, S 83a
Neural	Network Model for the Mechanism of Pattern Information Processing. .	Miyake, S 86a
Hierarchical Neural	Network Model for Voluntary Movement with Application to Robotics. .	Kawato, M 88a
Dynamic Single Unit Simulation of a Realistic Cerebellar	Network Model. II. Purkinje Cell Activity within the Basic Circuit .	Pellionisz, AJ 74a
A Neural	Network Model of CAAM and its Applications to Handprinted Chinese. .	Yao, Y 87a
A Neural	Network Model of Frog Retina: A Discrete Time-Space Approach	Lee, Y 86a
A Neural	Network Model of Multistable Perception.	Kawamoto, AH 85a
A Neural	Network Model of Pain Mechanisms: Functional Properties of the . . .	Minamitani, H 82a
Bifurcation Analysis of Oscillating	Network Model of Pattern Recognition in the Rabbit Olefactory Bulb .	Baird, B 86a
Bifurcation Analysis of a	Network Model of Rabit Olfactory Bulb with Periodic Attractors . . .	Baird, B 87a
	Network Model of Shape-From-Shading: Neural Function Arises from . .	Lehky, SR 88a
Schema for Motor Control Utilizing a	Network Model of the Cerebellum.	Houk, JC 88b
The Use of an Artificial Neural	Network Model to Analyze the Information Transfer Properties of a. .	Axelrad, H 87a
Backward Conditioning: A Neural	Network Model which Exhibits Both Excitatory and Inhibitory.	Ricart, R 88a

Neural Network Startups Proliferate Across the U.S. . . . Johnson, RC 88g
DARPA Neural Network Study. . . . DARPA 88a
A Neural Network System with an Automatic Generation Mechanism for Distorted. Shimohara, K 88a
A Chipset for High Speed Simulation of Neural Network Systems. . . . Garth, SCJ 87a
Characteristics of Randomly Connected Threshold-Element Networks and Network Systems. . . . Amari, S 71a
Optimization with Artificial Neural Network Systems: A Mapping Principle and a Comparison to Gradient. . Leong, HM 88a
...System for Alpha-Numeric Characters Based on Learning Network Techniques . . . Fairhurst, MC 76a
Integration of Knowledge-Based System and Neural Network Techniques for Robotic Control . . . Handelman, DA 88a
The Implementation of Neural Network Technology . . . Savely, RT 87a
Neurons with Hysteresis Form a Network that Can Learn without Any Changes in Synaptic Connection. . Hoffman, GW 86a
An Adaptive Network that Constructs and Uses an Internal Model of its World. . . Sutton, RS 81a
There Exists a Neural Network That Does Not Make Avoidable Mistakes. . . . Gallant, AR 88a
An Adaptive Network that Flees Pursuit . . . Winter, CL 88a
A Network that Induces a Distributed Representation Useful for . . . Hinton, GE 86a
PARSNIP: A Connectionist Network that Learns Natural Language Grammar from Exposure to. . . . Hanson, SJ 87a
A 'Neural' Network that Learns to Play Backgammon . . . Tesauro, G 88a
A Connectionist Network that Learns to Process Some (Very) Simple Sentences. . . . Samad, T 87a
NETtalk: A Parallel Network that Learns to Read Aloud. . . . Sejnowski, TJ 86d
NETtalk: A Parallel Network that Learns to Read Aloud. . . . Sejnowski, TJ 86c
An Adaptive Neural Network, The Cerebral Cortex . . . Burnod, Y 88a
Brain Modeling by Tensor Network Theory and Computer Simulation . . . Pellionisz, AJ 79a
Tensor Network Theory and its Application in Computer Modeling of the . . . Pellionisz, AJ 86b
A Neural Network Theory of Frontal Lobe Function. . . . Levine, DS 86a
Using a Neural Network to Learn the Dynamics of the CMU Direct-Drive Arm II . . . Goldberg, K 88a
Teaching a Minimally Structured Back-Propagation Network to Recognize Speech Sounds . . . Landauer, TK 87a
Analysis of Hidden Units in a Layered network Trained to Classify Sonar Targets. . . . Gorman, RP 88a
Solving Puzzles with a Connectionist Network Using a Hill-Climbing Heuristic. . . . Kawamoto, AH 86a
Self-Organization of a Neural Network Which Gives Position-Invariant Response. . . . Fukushima, K 79c
Self-Organizing Neural Network with a Function of Associative Memory: Feedback Type . . . Fukushima, K 78a
Self-Organizing Neural Network with a Function of Associative Memory - Feedback Type. . . . Fukushima, K 77a
A Heteroassociative Memory Network with Feedback Connection . . . Okajima, K 87a
...Model for the Mechanism of Feature Extraction - A Self-Organizing Network with Feedback Inhibition . . . Miyake, S 83a
Learning in a Layered Network with Many Fixed-Function Hidden Nodes. . . . Littlestone, N 87a
Electronic Analog Neural Network with Noise and Time Delays . . . Meidan, M 87a
Optical Competitive Neural Network with Optical Feedback. . . . Te Kolste, RD 87a
A Model of a Neural Network with Recurrent Inhibition. . . . Wigstrom, H 74a
Self-Organizing Neural Network with the Mechanism of Feedback Information Processing. . . . Miyake, S 88c
Speech Recognition with Network-Based Classifiers. . . . Hogg, T 88a
Massively Parallel Network-Based Natural Language Parsing System. . . . Li, T 87a
Startup Nestor Readies Neural Network-Based Products . . . Brown, C 86k
...Marketing Tactician (AMT): A Commercial Application of Adaptive Networking . . . Hutchison, WR 87b
Computer Networking Capacity in Robotic Neural Systems. . . . Nizkik, CA 84a
Neural Networking Research Gains New Life . . . Kane, B 87a
Networking the Brain . . . Johnson, RC 88d
...Algorithms for Reinforcement Learning in Neural Networks . . . Williams, RJ 87a
A CMOS Associative Memory Chip Based on Neural Networks . . . Graf, HP 87b
...for the Evaluation of Neuron Models Involving Large Uniform Networks . . . Lee, MH 82a
...Neocortex: Computational Capabilities of Nonlinear Neural Networks . . . Singer, A 87a
...Neocortex: Computational Capabilities of Nonlinear Neural Networks . . . Singer, A 88a
...Use for Establishing and Proving Conjectures in Uniform Neural Networks . . . Shingai, R 78a
A Connectionist Encoding of Semantic Networks . . . Shastri, L 87a
A Connectivity Analysis of a Class of Simple Associative Neural Networks . . . Hammerstrom, D 88a
A Connectivity Analysis of Recursive, Auto-Associative Connection Networks . . . Hammerstrom, D 86a
A Content-Addressing Software Method for the Emulation of Neural Networks . . . Venta, O 88a
...Appropriate for Training and Testing of Vision-Based Learning Networks . . . Keller, R 87a
A Critical Comparison of Two Kinds of Adaptive Classification Networks . . . Steinbuch, K 65a
A Data Flow Analysis of Connectionist Networks . . . Bookman, L 88a
A Dynamical Behaviour of Active Regions in Randomly Connected Neural Networks . . . Okuda, M 74a
A Factored Architecture to Solve Illusory Conjunctions in Neural Networks . . . Strong, GW 87a
A Framework for Speech Recognition Using Neural Networks . . . Lippmann, RP 87a
A Fully Digital Architecture for Multi-State Neural Networks . . . Potu, B 88a
A Generalized Convergence Theorem for Neural Networks . . . Bruck, J 88a
A Graphical Design System for Neural Networks . . . Hood, G 87a
A Magnetic Variable-Gain Component for Adaptive Networks . . . Crafts, HS 62a
...IV of Visual Cortex and Its Application to Artificial Neural Networks . . . Scofield, CL 88a
A Mechanism for Stabilizing Intelligent Properties of Neural Networks . . . Dammasch, IE 87a
...to Select Appropriate Learning Rate Parameters in Feed-Forward Networks . . . Pabon, JA 87a
A Modifiable Weight Circuit for Use in Adaptive Neuromorphic Networks . . . Shoemaker, PA 88a
A New CMOS Architecture for Neural Networks . . . Verleysen, M 88d

Constructing Associative Memories Using Neural Networks . . . Xu, X 88b
Context-Free Parsing in Connectionist Networks . . . Fanty, MA 85a
Context-Free Parsing with Connectionist Networks . . . Fanty, MA 86b
...Short-Term Memory, and Constancies in Reverberating Neural Networks . . . Grossberg, S 73a
Control of Attention in Neural Networks . . . Mjolsness, E 87a
Counterpropagation Networks . . . Hecht-Nielsen, R 87c
Counterpropagation Networks . . . Hecht-Nielsen, R 87b
Coupled Mode Theory for Neural Networks . . . Cohen, MA 86a
Coupled Rotators as a Model for Spontaneous Activity in Neural Networks . . . Kowalski, JM 88a
Cycles: A Simulation Tool for Studying Cyclic Neural Networks . . . Gatley, MT 88a
Decrementing Associative Networks . . . Dobson, VG 88a
Determining the Dynamic Landscape of Hopfield Networks . . . Weisbuch, G 86a
Digital Image Halftoning Using Neural Networks . . . Anastassiou, D 88a
Dress Rehearsal for Real-Time Artificial Neural Networks . . . Michael, C 87a
...Memory Traces in Computer-Simulated Recurrently-Connected Neural Networks . . . Gaudiano, P 87a
Dynamic Connections in Neural Networks . . . Feldman, JA 82c
Dynamical Behaviour of Neural Networks . . . Goles, E 85a
Dynamics of Random Boolean Networks . . . Pellegrin, D 86a
Early Spatio-Temporal Integration by Gated Competitive Networks . . . Ogmen, H 87b
Electronic Hardware Implementations of Neural Networks . . . Thakoor, AP 87b
Electronic Neural Networks . . . Hubbard, W 86a
Emulating the Dynamics for a Class of Laterally Inhibited Neural Networks . . . Lemmon, M 88a
Encoding Geometric Invariances in Higher-Order Neural Networks . . . Giles, CL 88a
Encoding Sequential Structure in Simple Recurrent Networks . . . Servan-Schreiber, D 88a
Epsilon Capacity of Neural Networks . . . Venkatesh, SS 86a
...Pattern Recognition in Incomplete Nonlinear Multithreshold Networks . . . Mucciardi, AN 66a
...for the Average Dynamic Behavior of some Non-Linear Neural Networks . . . Rossler, J 87a
Exaptation, Preadaptation, Evolution and Biological Neural Networks . . . Stork, DG 87b
Exhaustive Thermodynamical Analysis of Boolean Learning Networks . . . Carnevali, P 87a
Experiments in Artificial Neural Networks . . . Rietman, E 88a
Exponential Storage and Retrieval in Hierarchical Neural Networks . . . Willcox, CR 88a
Feedback-Induced Sequentiality in Neural Networks . . . Norrod, FE 87a
Field Theory of Self-Organizing Neural Networks . . . Amari, S 83a
...Analogue VLSI Devices for the Implementation of Neural Networks . . . Murray, AF 88d
Functional Organization in Random Networks . . . Beurle, RL 62a
Fusion, Propagation, and Structuring in Belief Networks . . . Pearl, J 86b
Fusion, Propagation, and Structuring in Belief Networks . . . Pearl, J 86a
Fuzzy Sets and Neural Networks . . . Taber, WR 88b
General Potential Surfaces and Neural Networks . . . Dembo, A 88a
Generalization in Layered Classification Neural Networks . . . Marks II, RJ 88b
...of Backpropagation to Recurrent and Higher Order Neural Networks . . . Pineda, FJ 88b
Generalization of Back-Propagation to Recurrent Neural Networks . . . Pineda, FJ 87a
Global Optimization with Stochastic Neural Networks . . . Levy, BC 87a
Gradient Following Without Back-Propagation in Layered Networks . . . Barto, AG 87b
Hidden Patterns in Combined Knowledge Networks . . . Kosko, B 87h
Hierarchical Associative Networks . . . Cortes, C 87a
High Resolution Micro-Fabrication and Neural Networks . . . Jackel, LD 86b
High-Performance Neural Networks . . . Dress, WB 87e
History of Associative Networks . . . Hecht-Nielsen, R 87d
...by Motion: Cortical Mechanism and Implementation in Neural Networks . . . Orban, GA 88a
Implementing Stages of Motion Analysis in Neural Networks . . . Sereno, ME 87a
Improving the Learning Rate of Neural Networks . . . Nelken, I 88a
Independent Components Analysis in Neural Networks . . . Herault, J 88b
Inductive Inference by Neural Networks . . . Hestenes, D 88a
Inference of Spatial Relations by Self-Organizing Networks . . . Williams, RJ 85b
Information Capacity of Outer-Product Neural Networks . . . Keeler, JD 87b
Information Processing and Storage Based on Transforms and Neural Networks . . . Ersoy, OK 87a
Information Storage and Retrieval in Spin-Glass Like Neural Networks . . . Personnaz, L 85a
Invariants in Automata Networks . . . Goles, E 86a
ISCON: An Interactive Simulator for Connectionist Networks . . . Small, SL 82b
Learned Phonetic Discrimination Using Connectionist Networks . . . Watrous, RL 87a
Learning Acoustic Features from Speech Data Using Connectionist Networks . . . Watrous, RL 87b
...and Probability Distributions in Feed-Forward and Feed-Back Networks . . . Hopfield, JJ 87a
Learning and Memory Properties in Fully Connected Networks . . . Bruce, AD 86a
Learning in Parallel Networks . . . Hinton, GE 85a
Learning Invariance, and Generalization in High-Order Neural Networks . . . Giles, CL 87a
Learning of World Models by Connectionist Networks . . . Sutton, RS 85a
Learning Phenomena in Layered Neural Networks . . . Widrow, B 87c
Learning Procedures in Multilayered Networks . . . Sejnowski, TJ 87e

Self-Organization of Neural Networks	Clark, JW 84a
Sequential Behavior and Stability Properties of Enzymatic Neuron Networks	Kampfner, RR 83b
Signal Modeling and Prediction Using Neural Networks	Ramamoorthy, PA 88b
Simple Selectionist Learning Rule for Neural Networks	Personnaz, L 86f
Solving Constraint Satisfaction Problems with Neural Networks	Tagliarini, GA 87a
...Contrast Enhancement and Short-Term Memory of Recurrent Neural Networks	Grossberg, S 75b
...Probabilistic Semantics of Logistic Function Parameters in Neural Networks	Davies, TR 88a
Spin-Glass Models of Neural Networks	Amit, DJ 85a
Stability and Attractivity in Associative Memory Networks	Cottrell, M 88a
Stability Properties of a Class of Large Scale Neural Networks	Dimopoulos, N 80a
Stability Results for Neural Networks	Michel, AN 88b
Stochastic Combinational Networks	Windecker, RC 78a
Stochastic Dynamics of Neural Networks	Peretto, P 86c
Storage and Manipulation of Information in Random Networks	Beurle, RL 62b
Storing Infinite Numbers of Patterns in a Spin-Glass Model of Neural Networks	Amit, DJ 85b
...Transformation and Storage of Patterns in Reverberating Neural Networks	Levine, DS 75a
Superresolution for Ultrasonic Imaging in Air Using Neural Networks	Winters, JH 88b
Supervised Learning of Probability Distributions by Neural Networks	Baum, EB 88d
Sustained Oscillation in Symmetric Cooperative-Competitive Neural Networks	Cohen, MA 87d
Switched-Capacitor Neural Networks	Tsividis, Y 87c
Synaptic Diversity Characterizes Biological Neural Networks	Gardner, D 87a
Synthesis of Reverberatory Neural Networks	Aiello, A 70a
Technical Comments: Computing with Neural Networks	Carpenter, GA 87f
Technical Comments: Computing with Neural Networks	Kohonen, T 87h
Technical Comments: Computing with Neural Networks	Palm, G 87d
Technical Comments: Computing with Neural Networks	Hopfield, JJ 87c
Temporal Behavior of Neural Networks	Gelenbe, E 87a
Temporal Patterns of Activity in Neural Networks	Gaudiano, P 88a
...Spreadsheet as a Tool in Modeling and Simulating Neural Networks	Hewett, TT 86a
The Gamma Network as an Interconnection Architecture for Neural Networks	Hennessey, AK 87b
...Project: Silicon Implementation of Connectionist/Neural Networks	Hammerstrom, D 86e
The Performance of Convex Set Projection Based Neural Networks	Marks II, RJ 88g
The Representation of Variables in Connectionist Networks	Walters, DKW 87j
The Simulation of Energy and Memory in Neural Networks	Fraser, MD 88a
...and the Effect of Local Feedback in Theoretical and Neural Networks	Uttley, AM 66a
Time-Sequential Self-Organization of Hierarchical Neural Networks	Silverman, RH 88a
Topics in Neural Networks	Petsche, T 88b
Training Recurrent Networks	Rohwer, R 88b
Training Time-Dependence in Neural Networks	Rohwer, R 87a
...and Other Steepest-Descent Learning Procedures for Networks	Sutton, RS 86a
Two-Dimensional Pattern Sequence Prediction Using High-Order Neural Networks	Collins, DH 88a
...Scale Space Transformation, and the Morphology of Neural Networks	Gardner, S 88c
Uncertainty as a Foundation of Computational Power in Neural Networks	Gaines, BR 87b
...Minimum Mean-Square Error Parameter Estimation with Hopfield Networks	Altes, RA 88a
Using A Visual Programming Language to Simulate Adaptive Networks	Tenorio, MF 87f
Visual Programming as a User Interfaace for Programming Neural Networks	Tenorio, MF 87j
Visual-Cognitive Neuronal Networks	Trehub, A 87a
VLSI Architectures for Implementation of Neural Networks	Sivilotti, MA 86a
VLSI Bit-Serial Neural Networks	Butler, ZF 88a
VLSI Neural Networks	Murray, AF 88e
Weight Discretization in Backward Error Propagation Neural Networks	Fiesler, E 88a
...Parallel Dynamics for Infinite Range Spin Glasses and Neural Networks	Gardner, E 87a
Developers Show Neural Networks: "Intelligent" Systems Learn, Reason Like Humans	Brownstein, M 87a
A Survey of Interconnection Networks (Concurrent Computing)	Tse-Yun, F 81a
Flow-of-Activation Processing: Parallel Associative Networks (PAN)	Cruz, CA 86a
A Primer on the Complexity Theory of Neural Networks (Preliminary Version)	Parberry, I 88a
Experience vs. Architecture in Feedforward Networks: A Case Study	Atkinson, RC 88c
Consensus versus Competition in Neural Networks: A Comparative Analysis of Three Models	Montalvo, FS 75a
Kinematic Networks: A Distributed Model for Representing and Regularizing	Mussa-Ivaldi, FA 88b
Evidential Reasoning in Semantic Networks: A Formal Theory	Shastri, L 85a
Evidential Reasoning in Semantic Networks: A Formal Theory and its Parallel Implementation: PhD	Shastri, L 85b
Reinforcement Learning in Connectionist Networks: A Mathematical Analysis	Williams, RJ 86c
Bayesian Networks: A Model of Self-Activated Memory for Evidential Reasoning	Pearl, J 85a
Understanding Neural Networks, a Primer	Cruz, CA 88a
Automated Reasoning on Neural Networks: A Probabilistic Approach	Chen, S 87a
Spatial Organization of Neural Networks: A Probabilistic Modeling Approach	Stafylopatis, A 88a
Computing Shape with Neural Networks: A Proposal	Oyster, JM 87a
Collective Properties of Neural Networks: A Statistical Physics Approach	Peretto, P 84a
Neural Networks, Acyclic Orientations of the Hypercube and Sets of	Baldi, P 88d

Neural Models:	Networks for Fun and Profit.	Anderson, JA 86d
Using Neural	Networks for Generalization Problems	Dutta, S 88b
Fat-Trees: Universal	Networks for Hardware Efficient Supercomputing	Leiserson, CE 85a
Complete Discrete 2-D Gabor Transforms by Neural	Networks for Image Analysis and Compression.	Daugman, JG 88a
Integer Linear Programming Neural	Networks for Job-Shop Scheduling	Foo, YS 88a
Adaptive Neural	Networks for Learning Control: Some Computational Experiments.	Barto, AG 85a
Connectionist	Networks for Learning Coordinated Motion in Autonomous Systems	Pabon, JA 88a
Neuronal Models for Cognitive Processes:	Networks for Learning, Perception, and Imagination	Trehub, A 77a
Neural	Networks for Medical Image Processing: A Study of Feature.	Dayhoff, RE 88a
Neural	Networks for Motor Program Generation.	Eckmiller, R 88a
Neural	Networks for Nonlinear Programming	Kennedy, MP 88a
Three Layers of Vector Outer Product Neural	Networks for Optical Pattern Recognition	Szu, HH 86h
An Efficient Model of Neural	Networks for Optimization.	Kamgar-Parsi, B 87a
Stability of Neural	Networks for Optimization.	Stork, DG 88d
Minimum Distance Automata in Parallel	Networks for Optimum Classification.	Winters, JH 89a
On Parallel	Networks for Optimum Classification.	Winters, JH 88a
Computer Simulation of Neural	Networks for Perceptual Psychology	Grossberg, S 86f
On the Power of Neural	Networks for Solving Hard Problems	Bruck, J 88e
Stochastic Neural	Networks for Solving Job-Shop Scheduling: Part 1. Problem.	Foo, YS 88b
Stochastic Neural	Networks for Solving Job-Shop Scheduling: Part 2. Architecture and	Foo, YS 88c
Neural	Networks for Speech Processing and Recognition	Shamma, SA 87a
Implementation of Neural	Networks for Speech Recognition on a Transputer Array.	Chong, MWH 88a
Simple Analog and Hybrid	Networks for Surface Interpolation	Hutchinson, JM 86a
Neural	Networks for Tactile Perception.	Pati, YC 88a
Neural	Networks for Template Matching: Application to Real-Time	Wong, Y 87a
Neural	Networks for Template Matching: Application to Real-Time	Bower, JM 88b
Self Organizing Neural	Networks for the Identification Problem.	Tenorio, MF 88a
Parallel Analog Neural	Networks for Tree Searching.	Saylor, J 86a
Layered	Networks for Unsupervised Learning	d'Humieres, D 86a
TRW Unit Uses Neural	Networks for Voice Recognition	Schwartz, TJ 86b
A Comparison of Feedforward Neural	Networks: Fuzzy Operators and Acoustic Ship Signatures	Taber, WR 88a
Neural	Networks Get a Conference to Call Their Own.	Brown, C 87j
Learning in Connectionist	Networks: Has the Credit Assignment Problem Been Solved?	Goel, A 89a
Optical Associative Memory Model Based on Neural	Networks Having Variable Interconnection Weights	Macukow, B 87b
Self-Stabilization of Neuronal	Networks, I. The Compensation Algorithm for Synaptogenesis	Dammasch, IE 86b
Associative	Networks, II	Cruz, CA 83a
Self-Stabilization of Neuronal	Networks, II. Stability Conditions for Synaptogenesis.	Dammasch, IE 88c
Neural	Networks: Implementing Associative Memory Models in Neurocomputers	Miller, RK 87a
Concept Formation of Neural	Networks: Implications for Evolution of Cognitive Functions.	Anderson, JA 88a
Analog Neuronal	Networks in Early Vision	Koch, C 86a
Analog Neuronal	Networks in Early Vision	Koch, C 85a
Computational	Networks in Early Vision: From Orientation Selection to Optical.	Zucker, SW 88a
Self-Organization of Cooperative and Competitive	Networks in Ecological Paradigm.	Matsuo, K 87a
Neural	Networks in Japan: Part 2.	Johnson, RC 87i
Structured Neural	Networks in Nature and in Computer Science	Feldman, JA 88d
Implementing Semantic	Networks in Parallel Hardware.	Hinton, GE 81b
Unstable Connectionist	Networks in Speech Recognition	Rohwer, R 88c
Morphology of Neural	Networks in the Mammalian Visual System.	Gardner, S 88b
...the Representation and Storage of Information in Neuronal	Networks in the Primate Hippocampus and Cerebral Cortex.	Rolls, ET 89a
Modeling and Analysis of Neural	Networks in the Visuomotor System of the Anuran Amphibian.	Cervantes-Perez, F 85a
High-Order Neural	Networks: Information Storage without Errors	Personnaz, L 87a
Neural	Networks: Introduction to the 1 December 1987 Issue of Applied	Carpenter, GA 87e
Learning in	Networks is Hard	Judd, JS 87b
Two-Level Neural	Networks: Learning by Interaction with Environment	Fukaya, M 87a
Emergence of Grandmother Memory in Feed Forward	Networks: Learning with Noise and Forgetfulness.	Scalettar, R 88a
Visual Illusions in Neural	Networks: Line Neutralization, Tilt After Effect, and Angle.	Levine, DS 76a
Neural	Networks Mediating Linearizable Dynamic Redundant Sensori-Motor.	Daunicht, WJ 88b
Learning in Neural	Networks, Models and Simulations	Salu, Y 84a
Representing Neural	Networks Models by Finite Automata	Kemke, C 88b
Neural	Networks Naive, Says Minsky.	Johnson, RC 87j
Collective Computational Properties of Neural	Networks: New Learning Mechanisms.	Personnaz, L 86b
Optical Implementation of Winner-Take-All	Networks: Noise Considerations	Rovnyak, SM 88a
Generalization and Information Storage in	Networks of Adaline "Neurons".	Widrow, B 62a
Learning Phenomena in	Networks of Adaptive Switching Circuits.	Hoff, ME 62a
Modelling Neural Networks by Means of	Networks of Finite Automata.	Kemke, C 87a
A Biologically Constrained Learning Mechanism in	Networks of Formal Neurons	Personnaz, L 86a
The Most Complicated	Networks of Formal Neurons	Labos, E 87a
Number of Stable Points for Spin-Glasses and Neural	Networks of Higher Orders.	Baldi, P 87a
Fast Learning in	Networks of Locally-Tuned Processing Units	Moody, J 88a

Storing Sequences of Biased Patterns in Neural	Networks with Stochastic Dynamics.	Buhmann, J 88a
...Using Peak Learning Rates of 10 (and Greater) in Back-Propagation	Networks with the Heuristic Learning Algorithm	Cater, JP 87a
Optical Implementations of Associative	Networks with Versatile Adaptive Learning Capabilities	Fisher, AD 87b
How Neural	Networks Work.	Lapedes, A 88a
ZZENN (Zig Zag Epigenetic Neural	Networks), A New Approach to Connectionist Machine Learning.	Cochet, Y 88b
Der	Neuere Konnektionismus, Ein Uberblick.	Kemke, C 88a
Parallel-Processors Bypass Von	Neumann Bottleneck	Brown, C 86j
Parallel Optics: Solution to Von	Neumann Bottleneck?.	Brown, C 86i
Scaling Techniques for	Neural Algorithms.	Cuykendall, R 88a
The	Neural Analog Diffusion-Enhancement Layer (NADEL) and Early Visual	Waxman, AM 88b
	Neural Analog Information Processing	Hecht-Nielsen, R 81a
	Neural Analog Processing	Hecht-Nielsen, R 82a
	Neural and Brain Modeling.	MacGregor, RJ 87a
	Neural and Conceptual Interpretation of PDP Models	Smolensky, P 86b
	Neural and Massively Parallel Computers: The Sixth Generation.	Soucek, B 88a
Special Issue on	Neural and Sensory Information Processing.	Sanderson, AC 83a
From Basic Network Principles to	Neural Architecture: Emergence of Orientation Columns.	Linsker, R 86a
From Basic Network Principles to	Neural Architecture: Emergence of Orientation-Selective Cells.	Linsker, R 86b
From Basic Network Principles to	Neural Architecture: Emergence of Spatial-Opponent Cells	Linsker, R 86c
Masking Fields: A Massively Parallel	Neural Architecture for Learning, Recognizing, and Predicting.	Cohen, MA 87b
	Neural Assemblies: An Alternative Approach to Artificial	Palm, G 82a
A New VLSI Architecture for	Neural Associative Memories.	Verleysen, M 88f
Board Creates	Neural AT.	Schwartz, TJ 88a
A Population Approach to the	Neural Basis of Perceptual Categorization.	Finkel, LH 89a
	Neural Basis of Rhythmic Behavior in Animals	Delcomyn, F 80a
	Neural Chips Emulate Brain Functions (ATT VLSI).	Brown, C 86e
	Neural Circuit Analysis.	Mitchell, CE 81a
A Neuromime System for	Neural Circuit for Classical Conditioning Without Synpatic	Tesauro, G 88b
A Plausible	Neural Circuit Models and Their Role in the Organization of.	Hemami, H 83a
Four	Neural Circuitry	Marr, D 77a
From Understanding Computation to Understanding	Neural Circuits.	Babcock, KL 86a
Complex Dynamics in Simple	Neural Circuits.	Conwell, PR 87a
Effects of Connection Delays in Two State Model	Neural Circuits.	Takefuji, Y 87a
Error Correcting System Based on	Neural Circuits: A Model	Hopfield, JJ 85b
Computing With	Neural Circuits for Generating Rhythmic Movements.	Friesen, WO 78a
	Neural Circuits Will Improve Vision, Speech Systems.	Brown, C 86f
A	Neural Cocktail-Party Processor.	von der Malsburg, C 86a
	Neural Coding in the Bullfrog's Auditory System: A Teleological.	Frishkopf, LS 68a
The Problem of Sensing and the	Neural Coding of Sensory Events.	Mountcastle, VB 67a
An Adaptive Stochastic Model for the	Neural Coding Process.	Bruckstein, AM 85a
Stochastic Pulse Frequency Modulation Models of	Neural Coding Processes.	Bruckstein, AM 81a
Analysis of "Integrate-to-Threshold"	Neural Coding Schemes.	Bruckstein, AM 79a
The Vite Model: A	Neural Command Circuit for Generating Arm and Articulator.	Bullock, D 88d
Dynamics and Architecture in	Neural Computation	Pineda, FJ 88a
Neural Representation and	Neural Computation	Churchland, PS 89a
Silicon Models of	Neural Computation	Mead, CA 87b
Sigmoid Nonlinearity in	Neural Computation: An Experimental Approach	Eeckman, FH 86a
	Neural Computation and Learning Strategy for Manipulator Position.	Tsutsumi, K 87a
	Neural Computation for Controlling the Configuration of 2.	Tsutsumi, K 88a
Why Corticies?	Neural Computation in the Vertebrate Visual System	Mallot, HA 88a
Machine for	Neural Computation of Acoustical Patterns with Application to Real	Mueller, P 86a
	Neural Computation of Decisions in Optimization Problems	Hopfield, JJ 85c
	Neural Computation of Visual Images.	Mueller, P 87a
	Neural Computation Research for Vision - Simulation on a Parallel.	Miyake, S 88a
Connectionist Modeling:	Neural Computation/Mental Connections.	Smolensky, P 89a
Conventional Fault-tolerance and	Neural Computers	Moore, WR 88a
Experimental Demonstrations of Optical	Neural Computers	Hsu, K 88a
Goal and Architecture of	Neural Computers	von der Malsburg, C 88a
Learning in Optical	Neural Computers	Psaltis, D 87a
Optical	Neural Computers	Abu-Mostafa, YS 87a
Parallel Architectures for	Neural Computers	Recce, M 88a
	Neural Computers in Vision: Processing of High Dimensional Data.	Zeevi, YY 88a
Are Special Chips Necessary for	Neural Computers?.	Aleksander, I 88a
Adaptive, Associative, and Self-Organizing Functions in	Neural Computing	Kohonen, T 87a
An Introduction to	Neural Computing	Kohonen, T 88a
European Seminar on	Neural Computing	ESNC 88a
Schemas and Neurons: Two Levels of	Neural Computing	Arbib, MA 88a
State of the Art in	Neural Computing	Kohonen, T 87g
Super-Resolution and	Neural Computing	Abbiss, JB 88a

Implementing	Neural Nets with Programmable Logic.	Vidal, JJ 88a
Implementing	Neural Nets with Programmable Logic.	Vidal, JJ 87a
...Information Processing Based on an Associative Memory Model of	Neural Nets with Thresholding and Feedback	Psaltis, D 85a
A Case Study of Solving Optimization Problems Using	Neural Network	Chen, S 88a
A Continuous-Time Optical	Neural Network	Stoll, HM 87a
A Hardware Neuron Model as a Component of the	Neural Network	Koga, K 80a
A Large VLSI Fully Interconnected	Neural Network	Verleysen, M 88b
A Novel Approach to Image Restoration Based on a	Neural Network	Zhou, Y 87a
A Preliminary Study of the Phoneme Recognition with Hardware	Neural Network	Hoshimiya, N 82a
A Self-Organizing Multilayer	Neural Network	Fukushima, K 75a
...Consideration of the Generation of Rhythmic Oscillation in a Ring	Neural Network	Tsutsumi, K 84a
...of Templates and Template Matching by a Disinhibitory	Neural Network	Hirai, Y 80a
A Trellis-Structured	Neural Network	Petsche, T 88a
Absolutely Stable Learning of Recognition Codes by a Self-Organizing	Neural Network	Carpenter, GA 86b
Alternative Knowledge Acquisition: Developing a Pulse-Coded	Neural Network	Dress, WB 87b
An Assembler for a	Neural Network	Noos, FA 87a
An Investigation of Knowledge Representation in a	Neural Network	Healy, MJ 88a
Analytic Study of the Storage Capacity of a	Neural Network	Little, WA 78a
Asymptotics Applied to a	Neural Network	Silverstein, JW 76a
Attention Based	Neural Network	Sornborger, AT 88a
Cognitron: A Self-Organizing Multilayered	Neural Network	Fukushima, K 75b
Compact	Neural Network	Somani, AK 87a
Computation of Optical Flow Using a	Neural Network	Zhou, Y 88a
Contextual Road Finding with a	Neural Network	Wright, WA 89b
Deconvolution Using a CMAC	Neural Network	Glanz, FH 88a
Dynamical Properties of a New Type of	Neural Network	Cotterill, RMJ 88a
Enhanced Memory Capacity of a Hopfield	Neural Network	Little, MJ 86a
Evaluation of a Hypercube Based Compact	Neural Network	Somani, AK 88a
Experimental Results from a Simple	Neural Network	White, HJ 88b
Fault Simulation of a Wafer-Scale Integrated	Neural Network	May, N 88a
Image Halftoning and Reconstruction Using a	Neural Network	Kollias, S 89a
Image Restoration Using a	Neural Network	Zhou, Y 88b
...Weight Discretization and Noise in an Optoelectronic	Neural Network	Von Lehmen, A 87a
Influence of Noise on the Behavior of an Autoassociative	Neural Network	Buhmann, J 86b
Influence of Noise on the Function of a "Physiological"	Neural Network	Buhmann, J 87a
Isolated Word Recognition with an Artificial	Neural Network	Rogers, CH 87a
Lexical Access Using a	Neural Network	Kawamoto, AH 84a
Noise Modulation of Synaptic Weights in a Biological	Neural Network	Gardner, D 89a
On the Activities in a Continuous	Neural Network	Oguztoreli, MN 75a
Representing Conceptual Structures in a	Neural Network	Touretzky, DS 87b
Road Finding	Neural Network	Wright, WA 89c
Simulation of a	Neural Network	Puleo P, F 81a
Some Properties of a Randomly Connected	Neural Network	Allanson, JT 56a
Stereo Matching Using a	Neural Network	Zhou, Y 88c
The ART of Adaptive Pattern Recognition by a Self-Organizing	Neural Network	Carpenter, GA 88a
The Recognition of Orca Calls with a	Neural Network	Taber, WR 88d
The Ultrametric Organization of Memories in a	Neural Network	Parga, N 86a
Transputer Implementation of Hopfield	Neural Network	Feild Jr, WB 88b
Vector Quantization Training by a Self-Organizing	Neural Network	Ryan, TW 88b
	Neural Network Algorithm for an NP-Complete Problem: Map and Graph	Dahl, ED 87b
Application of	Neural Network Algorithms and Architectures to Correlation	Gardner, S 86a
Pattern Classification by	Neural Network: An Experimental System for Icon Recognition.	Gullichsen, E 87a
Comparison of	Neural Network and Matched Filter Processing for Detecting Lines in.	Grant, PM 86b
Effect of Clock Skew in Iterative	Neural Network and Optical Feedback Processors	Oh, S 88a
ANNA: An Adaptive	Neural Network Anticipator for Human/Computer Interfacing.	Jones, WP 88a
Stochastic Processing in a	Neural Network Application	Nguyen, DD 87a
A	Neural Network Approach for Implementing Expert Systems.	Ramamoorthy, PA 88a
	Neural Network Approach to Sensory Fusion.	Pearson, JC 88a
A	Neural Network Approach to Transform Image Coding.	Chua, LO 88a
Expert System and	Neural Network Approaches for Classical Guitar Fingering	Sayegh, S 88a
	Neural Network Approaches to Multi-Target Tracking	Kuczewski, RM 87c
	Neural Network Approaches to Multi-Target Tracking	Kuczewski, RM 87d
An Associative Memory Based on an Electronic	Neural Network Architecture.	Howard, RE 87a
Discrimination of Natural Textures: A	Neural Network Architecture.	Mesrobian, E 87b
A	Neural Network Architecture for Attentionally-Modulated Pavlovian.	Grossberg, S 87b
A	Neural Network Architecture for Automatic Trajectory Formation and	Bullock, D 87a
A	Neural Network Architecture for Brightness Perception Under.	Grossberg, S 87c
	Neural Network Architecture for Control.	Guez, A 88a
A	Neural Network Architecture for Preattentive Vision: Multiple Scale.	Grossberg, S 87d

Recent Developments in the Design of Real-Time Non-Linear	Neural Network Architectures	Grossberg, S 87n
...Code: An Integer to Binary Vector Transformation Suitable for	Neural Network Architectures	Penz, PA 87a
	Neural Network Architectures and Their Implications for Next	Will, CA 87a
	Neural Network Architectures for AI.	Hinton, GE 87c
A Physiological	Neural Network as an Autoassociative Memory.	Buhmann, J 86a
A Continuous-Time Optical	Neural Network Associative Memory.	Lee, LS 87a
	Neural Network Based Architectures for Aerospace Applications. . . .	Ricart, R 87a
Pseudo-Unsupervised	Neural Network Based Automated EMG Decomposition	Hassoun, MH 88c
A	Neural Network Based Controller for Robotic Manipulators	Kamangar, FA 88a
	Neural Network Based Feed-Forward High Density Associative Memory. .	Daud, T 87a
Real Time Experiments in	Neural Network Based Learning Control During High Speed.	Miller, WT 88a
Architectures for a Continuous Level	Neural Network Based on Alternating Orthogonal Projections	Marks II, RJ 88a
A	Neural Network Based on Co-Occurrence Probabilities.	Collins, JW 87a
Efficient Algorithms with	Neural Network Behavior.	Omohundro, SM 87a
Connectionist/	Neural Network Bibliography.	Hammerstrom, D 86d
A Connectionist/	Neural Network Bibliography, Volume II	Hammerstrom, D 87a
Geometric Analysis of	Neural Network Capabilities.	Wieland, A 87a
Neocognitron: A Hierarchical	Neural Network Capable of Visual Pattern Recognition	Fukushima, K 88c
Electronic	Neural Network Chips	Jackel, LD 87b
IBM	Neural Network Chips Make Everything Obsolete.	Gantz, J 88a
The Effect of Stochastic Interconnects in Artificial	Neural Network Classification.	Marks II, RJ 88f
A	Neural Network Classifier Based on Coding Theory	Chiueh, T 88a
	Neural Network Classifiers for Speech Recognition.	Lippmann, RP 88c
A	Neural Network Computational Map Approach to Reflexive Motor	Lane, SH 88a
Optimization of	Neural Network Connectivity.	Cooper, LN 87a
Comparisons of Sigmoid Functions for	Neural Network Convergence	Graziano, MJ 88a
A Developmental Approach to	Neural Network Design.	Lehar, S 87a
	Neural Network Design for Fficient Information Retrieval.	Personnaz, L 86e
	Neural Network Design for Efficient Information Retrieval.	Personnaz, L 86d
A	Neural Network Digit Recognizer.	Burr, DJ 86a
Semiconductor Electronic Concepts for	Neural Network Emulation	Coon, DD 87c
	Neural Network Emulation Hardware Design Considerations.	Cruz, CA 87c
ANNE: Another	Neural Network Emulator.	Bahr, C 88a
Kolmogorov's Mapping	Neural Network Existence Theorem	Hecht-Nielsen, R 87e
An Artificial	Neural Network Experiment.	Brown, RJ 87a
A Model of	Neural Network Extracting Binocular Parallax	Hirai, Y 75a
An Inference upon the	Neural Network Finding Binocular Correspondence.	Hirai, Y 78a
Firing Probability of a	Neural Network: First-order Differential Equation.	Oonuki, M 76a
A Locally Interconnected	Neural Network for Deconvolution	Petsche, T 87a
Two-Level	Neural Network for Deterministic Logic Processing.	Hassoun, MH 88d
Application of Genetic Algorithm to a	Neural Network for Finding Optimal Paths	Smith, T 87a
A	Neural Network for Invariant Pattern Recognition	Bienenstock, EL 87a
A	Neural Network for Isolated-Word Recognition	Gold, B 88a
Training of a	Neural Network for Pattern Classification Based on an Entropy. . . .	Koutsougeras, C 88b
An Artificial	Neural Network for Spatio-Temporal Bipolar Patterns: Application to.	Atlas, LE 88a
An Artificial	Neural Network for Spatio-Temporal Bipolar Patterns: Application to.	Homma, T 88a
A Model of	Neural Network for Spatiotemporal Pattern Recognition.	Kurogi, S 87a
RHO-Space: A	Neural Network for the Detection and Representation of Oriented. . .	Walters, DKW 87g
RHO-Space: A	Neural Network for the Detection and Representation of Oriented. . .	Walters, DKW 87h
A Proposed	Neural Network for the Integrator of the Oculomotor System	Cannon, SC 83a
A	Neural Network for the Optimization of Communications Network. . . .	Vichniac, GY 88a
A	Neural Network for the Retrieval of Superimposed Connection.	von der Malsburg, C 87a
Application of the	Neural Network for the Trajectory Planning of a Biped Locomotive . .	Kitamura, S 88a
A	Neural Network for Visual Pattern Recognition.	Fukushima, K 88b
A Simple	Neural Network Generating an Interactive Memory.	Anderson, JA 72b
A Simple	Neural Network Generating an Interactive Memory.	Anderson, JA 72a
...Network Simulation: Two Examples within a Single, Integrated	Neural Network Hardware Development Environment.	Bahr, C 88c
	Neural Network Implementation Approaches for the Connection Machine.	Brown, NH 88a
	Neural Network Implementation for Maximum-Likelihood Sequence. . . .	Provence, JD 87a
	Neural Network Implementation of a Medical Diagnosis Expert System .	Kudrycki, TP 88a
Analog Nonvolatile Memory for	Neural Network Implementations	Sage, JP 88a
A	Neural Network in the Visual Cortex.	Nylen, M 87a
Artificial	Neural Network Integrated Circuit Based on NMOS/CCD Principles . . .	Sage, JP 86a
Biomasscomp: A Procedure for Mapping the Architecture of a Living	Neural Network into a Machine.	Dawes, RL 87a
	Neural Network Learning Algorithms	Sejnowski, TJ 88b
	Neural Network Learning Techniques	Sharkey, NE 88b
UCLA PUNNS - A	Neural Network Machine for Computer Vision	Gunger, D 87a
	Neural Network Mechanisms for Generation and Learning of Motor . . .	Eckmiller, R 87b
Electronic Device Aspects of	Neural Network Memories.	Lambe, J 85a

Preceding	Keyword	Following	Reference
	Neural Network	Processing as a Tool for Function Optimization.	Jeffrey, W 86a
Ring	Neural Network	qua a Generator of Rhythmic Oscillation with Period	Tsutsumi, K 84b
A Self-Organizing Multi-Layered	Neural Network	Realizing a Funciton in the Human Recognition and	Mizuno, H 80a
A Self-Organizing Multi-Layered	Neural Network	Realizing a Function in the Human Recognition and	Mizuno, H 78a
	Neural Network	Refinements and Extensions.	Denker, JS 86a
	Neural Network	Representation of Sensor Graphs in Autonomous Robot	Jorgensen, CC 87a
Future Perspectives on	Neural Network	Research.	Szu, HH 87b
Designing a	Neural Network	Satisfying a Given Set of Constraints	Personnaz, L 86c
A	Neural Network	Scheme for 3-Dimensional Pattern Recognition and.	Foo, YS 87a
Addendum: A Self-Organizing	Neural Network	Sharing Features of the Mammalian Visual System	Frohn, H 87a
UCLA SFINX - A	Neural Network	Simulation Environment.	Paik, E 87a
A	Neural Network	Simulation Language Based on Multi-Pascal	Lester, BP 87a
A	Neural Network	Simulation Method Using the Fast Fourier Transform.	Barto, AG 76a
Entrainment in a	Neural Network	Simulation of Spontaneously Active Units with	Blackburn, MR 87a
Concurrent	Neural Network	Simulation: Two Examples within a Single, Integrated.	Bahr, C 88c
Digital Signal Processor Accelerators for	Neural Network	Simulations	Penz, PA 86a
	Neural Network	Startups Backed by Venture Capital.	Brown, C 87i
	Neural Network	Startups Proliferate Across the U.S.	Johnson, RC 88g
DARPA	Neural Network	Study	DARPA 88a
A	Neural Network	System with an Automatic Generation Mechanism for	Shimohara, K 88a
A Chipset for High Speed Simulation of	Neural Network	Systems	Garth, SCJ 87a
Optimization with Artificial	Neural Network	Systems: A Mapping Principle and a Comparison to.	Leong, HM 88a
Integration of Knowledge-Based System and	Neural Network	Techniques for Robotic Control.	Handelman, DA 88a
The Implementation of	Neural Network	Technology.	Savely, RT 87a
There Exists a	Neural Network	That Does Not Make Avoidable Mistakes	Gallant, AR 88a
An Adaptive	Neural Network,	The Cerebral Cortex.	Burnod, Y 88a
A	Neural Network	Theory of Frontal Lobe Function	Levine, DS 86a
Using a	Neural Network	to Learn the Dynamics of the CMU Direct-Drive Arm II.	Goldberg, K 88a
Self-Organization of a	Neural Network	Which Gives Position-Invariant Response	Fukushima, K 79c
Self-Organizing	Neural Network	with a Function of Associative Memory: Feedback Type.	Fukushima, K 78a
Self-Organizing	Neural Network	with a Function of Associative Memory - Feedback.	Fukushima, K 77a
Electronic Analog	Neural Network	with Noise and Time Delays.	Meidan, M 87a
Optical Competitive	Neural Network	with Optical Feedback	Te Kolste, RD 87a
A Model of a	Neural Network	with Recurrent Inhibition	Wigstrom, H 74a
Self-Organizing	Neural Network	with the Mechanism of Feedback Information.	Miyake, S 88c
Startup Nestor Readies	Neural Network-Based	Products.	Brown, C 86k
	Neural Networking	Research Gains New Life.	Kane, B 87a
...of Gradient-Estimating Algorithms for Reinforcement Learning in	Neural Networks.		Williams, RJ 87a
A CMOS Associative Memory Chip Based on	Neural Networks.		Graf, HP 87b
...of Cerebral Neocortex: Computational Capabilities of Nonlinear	Neural Networks.		Singer, A 88a
...of Cerebral Neocortex: Computational Capabilities of Nonlinear	Neural Networks.		Singer, A 87a
A Computer Use for Establishing and Proving Conjectures in Uniform	Neural Networks.		Shingai, R 78a
A Connectivity Analysis of a Class of Simple Associative	Neural Networks.		Hammerstrom, D 88a
A Content-Addressing Software Method for the Emulation of	Neural Networks.		Venta, O 88a
A Dynamical Behaviour of Active Regions in Randomly Connected	Neural Networks.		Okuda, M 74a
A Factored Architecture to Solve Illusory Conjunctions in	Neural Networks.		Strong, GW 87a
A Framework for Speech Recognition Using	Neural Networks.		Lippmann, RP 87a
A Fully Digital Architecture for Multi-State	Neural Networks.		Potu, B 88a
A Generalized Convergence Theorem for	Neural Networks.		Bruck, J 88a
A Graphical Design System for	Neural Networks.		Hood, G 87a
...of Layer IV of Visual Cortex and Its Application to Artificial	Neural Networks.		Scofield, CL 88a
A Mechanism for Stabilizing Intelligent Properties of	Neural Networks.		Dammasch, IE 87a
A New CMOS Architecture for	Neural Networks.		Verleysen, M 88d
A New VLSI Architecture for Large Hopfield's	Neural Networks.		Verleysen, M 88e
A Novel Computational and Signalling Method for VLSI	Neural Networks.		Murray, AF 88a
A Study of the Asymptotic Behavior of	Neural Networks.		Dimopoulos, N 82a
A Study on	Neural Networks.		Bruck, J 88b
A VLSI Interconnect Structure for	Neural Networks.		Bailey, J 88a
...Pattern Formation and Parallel Memory Storage by Competitive	Neural Networks.		Cohen, MA 83b
...Pattern Formation and Parallel Memory Storage by Competitive	Neural Networks.		Cohen, MA 83a
Activity Analysis of	Neural Networks.		Oguztoreli, MN 79a
Algorithm Development for	Neural Networks.		Page, EW 88a
An Adaptive Attentive Learning Algorithm for Single-Layer	Neural Networks.		Hassoun, MH 88a
An Algorithm for Pattern Recognition with VLSI	Neural Networks.		Sirletti, B 88a
An Automata-Theoretical Approach to Developing Learning	Neural Networks.		Veelenturf, LPJ 81a
Analog Decoding Using	Neural Networks.		Platt, JC 86a
...Memory for Continuous-Voltage, Discrete-Time Implementations of	Neural Networks.		Brown, PB 87a
Analog MOS Circuit Techniques in the VLSI Implementation of	Neural Networks.		Satyanarayana, S 88a
Analogue Circuits for Variable-Synapse Electronic	Neural Networks.		Tsividis, Y 87b

Title	Keyword	Author
Information Processing and Storage Based on Transforms and	Neural Networks.	Ersoy, OK 87a
Information Storage and Retrieval in Spin-Glass Like	Neural Networks.	Personnaz, L 85a
Learning Invariance, and Generalization in High-Order	Neural Networks.	Giles, CL 87a
Learning Phenomena in Layered	Neural Networks.	Widrow, B 87c
Learning Stereopsis with	Neural Networks.	Sun, GZ 87a
Long Term Memory Storage Capacity of Multiconnected	Neural Networks.	Peretto, P 86b
Lower Bound for Connectivity in Local-Learning	Neural Networks.	Abu-Mostafa, YS 88b
Lyapunov Function for Parallel	Neural Networks.	Goles, E 86b
Manual Training Technqiues of Autonomous Systems Based on Artificial	Neural Networks.	Shepanski, JF 87a
Memory Storage Capacity and the Role of Assemblies in	Neural Networks.	Shaw, GL 77a
Microelectronic Implementations of Connectionist	Neural Networks.	Mackie, S 88c
Multidimensional Machine Vision Using	Neural Networks.	Kolodzy, PJ 87a
	Neural Networks.	Johnson, RC 88h
Neuroscience as a Possible Research Standard for	Neural Networks.	Morse, KG 87a
New Hardware for Massive	Neural Networks.	Coon, DD 88d
On Mathematical Methods in the Theory of	Neural Networks.	Amari, S 87a
On the Aggregation of Processing Units in	Neural Networks.	Yager, RR 87b
On the Asymptotic Information Storage Capacity of	Neural Networks.	Palm, G 88a
On the Complexity of Loading Shallow	Neural Networks.	Judd, JS 88a
On the Design of a Content-Addressable Memory via Binary	Neural Networks.	Kam, M 87a
...Stability, Storage Capacity, and Design of Nonlinear Continuous	Neural Networks.	Guez, A 88b
Optical Resonators and	Neural Networks.	Anderson, DZ 86b
Optically Excited Synapse for	Neural Networks.	Boyd, GD 87a
Optical-Processor Architectures for Alternating-Projection	Neural Networks.	Marks II, RJ 88d
Optoelectronic Resonator	Neural Networks.	Owechko, Y 87b
Parallel Computation with Simple	Neural Networks.	Anderson, JA 79a
Parallelism and Redundancy in	Neural Networks.	von Seelen, W 88a
Partially Connected Models of	Neural Networks.	Canning, A 88b
Pattern-Based Fault Diagnosis Using	Neural Networks.	Dietz, WE 88a
Pavlovian Pattern Learning by Nonlinear	Neural Networks.	Grossberg, S 71b
Percolation and Frustration in	Neural Networks.	Noest, AJ 86a
Phase Transitions in	Neural Networks.	Chover, J 88a
Phase Transitions in Quasirandom	Neural Networks.	Kurten, KE 87a
Phasor	Neural Networks.	Noest, AJ 88a
PLEXUS: An On-Line System for Modeling	Neural Networks.	Dill, JC 68a
Processing of Temporal Sequences in	Neural Networks.	Gutfreund, H 88b
Programmable Bistable Switches and Resistors for	Neural Networks.	Spencer, EG 86a
Programmable Synaptic Chip for Electronic	Neural Networks.	Moopenn, A 88a
Qualitative Analysis and Synthesis of a Class of	Neural Networks.	Li, JH 88a
Qualitative Analysis of	Neural Networks.	Michel, AN 88a
Recognition Automata Based on Selective	Neural Networks.	Reeke Jr, GN 86a
Recognition of General Patterns Using	Neural Networks.	Wong, AJ 88a
Recursively Generated	Neural Networks.	Mjolsness, E 87b
Reduced Interconnection	Neural Networks.	Somani, AK 88b
...Conditions and Periodic Firing Modes in Reciprocal Inhibition	Neural Networks.	Nagashino, H 81a
Resistive Synaptic Interconnects for Electronic	Neural Networks.	Lamb, JL 87a
Reverberations and Control of	Neural Networks.	Caianiello, ER 67a
Self-Organization of	Neural Networks.	Clark, JW 84a
Signal Modeling and Prediction Using	Neural Networks.	Ramamoorthy, PA 88b
Simple Selectionist Learning Rule for	Neural Networks.	Personnaz, L 86f
Solving Constraint Satisfaction Problems with	Neural Networks.	Tagliarini, GA 87a
...in the Contrast Enhancement and Short-Term Memory of Recurrent	Neural Networks.	Grossberg, S 75b
...on the Probabilistic Semantics of Logistic Function Parameters in	Neural Networks.	Davies, TR 88a
Spin-Glass Models of	Neural Networks.	Amit, DJ 85a
Stability Properties of a Class of Large Scale	Neural Networks.	Dimopoulos, N 80a
Stability Results for	Neural Networks.	Michel, AN 88b
Stochastic Dynamics of	Neural Networks.	Peretto, P 86c
Storing Infinite Numbers of Patterns in a Spin-Glass Model of	Neural Networks.	Amit, DJ 85b
...in the Transformation and Storage of Patterns in Reverberating	Neural Networks.	Levine, DS 75a
Superresolution for Ultrasonic Imaging in Air Using	Neural Networks.	Winters, JH 88b
Supervised Learning of Probability Distributions by	Neural Networks.	Baum, EB 88d
Sustained Oscillation in Symmetric Cooperative-Competitive	Neural Networks.	Cohen, MA 87d
Switched-Capacitor	Neural Networks.	Tsividis, Y 87c
Synaptic Diversity Characterizes Biological	Neural Networks.	Gardner, D 87a
Synthesis of Reverberatory	Neural Networks.	Aiello, A 70a
Technical Comments: Computing with	Neural Networks.	Carpenter, GA 87f
Technical Comments: Computing with	Neural Networks.	Kohonen, T 87h
Technical Comments: Computing with	Neural Networks.	Palm, G 87d
Technical Comments: Computing with	Neural Networks.	Hopfield, JJ 87c

Comparison of Three	Neuron-Based Computation Schemes	Cervantes, JH 87a
A Biological Model of the Excitation of a Second Order Sensory	Neurone.	Enger, PS 69a
A Computer-Simulated Model of a Second Order Sensory	Neurone.	Walloe, L 69a
The Mammalian Central	Neurone is a Complex Computing Device.	Bras, H 87c
The Response of a Model	Neurone to a White Noise Input	Rubio, JE 75a
The Dynamics of Large Scale	Neuron-Glia Network and Its Relation to the Brain Functions (I).	Xu, J 86a
Simulation of Anticipatory Responses in Classical Conditioning by a	Neuron-Like Adaptive Element	Barto, AG 82b
...of the Classically Conditioned Nictitating Membrane Response by a	Neuron-Like Adaptive Element	Blazis, DEJ 86a
	Neuron-like Adaptive Elements that Can Solve Difficult Learning.	Barto, AG 83b
Hierarchical Learning Control: An Approach with	Neuron-Like Associative Memories	Ersu, E 88a
Learning Control Structures with	Neuron-Like Associative Memory Systems	Ersu, E 87a
Comparison of Algorithms for	Neuron-Like Cells.	Parker, DB 86a
Collective Computation in	Neuronlike Circuits.	Tank, DW 87a
Learning by Statistical Cooperation of Self-Interested	Neuron-Like Computing Elements	Barto, AG 85c
Learning by Statistical Cooperation of Self-Interested	Neuron-Like Computing Elements	Barto, AG 85b
Activity in Networks of	Neuron-Like Elements	Farley, BG 60a
Characteristics of Random Nets of Analog	Neuron-Like Elements	Amari, S 72a
...Orthogonalizing and Associative Memory in Recurrent Networks of	Neuron-Like Elements	Kohonen, T 76b
Homogeneous Nets of	Neuron-Like Elements	Amari, S 75a
On Properties of Networks of	Neuron-Like Elements	Baldi, P 88h
	Neuron-like Elements That Can Solve Difficult Learning and Control	Barto, AG 83c
A Macroscopic Analysis on Large Scale Networks on	Neuronlike Elements with Refractory.	Aihara, K 80a
Some Properties of Randomly Connected Networks of	Neuron-Like Elements with Refractory	Yoshizawa, S 74a
A Learning Algorithm for Multilayer	Neuronlike Network	Gawronski, RR 88a
Modelling of an Adaptive	Neuronlike Network for the Construction of a Learning Control.	Gawronski, RR 87a
Farey-Fraction Frequency Modulation in the	Neuronlike Output of Silicon p-i-n Diodes at 4.2 K	Coon, DD 87a
	Neuron-Like Transient Phenomena in Silicon p-i-n Structures.	Coon, DD 88c
...Computational Properties like those of Hopfield and Tank	Neurons.	Derthick, MA 88a
A Biologically Constrained Learning Mechanism in Networks of Formal	Neurons.	Personnaz, L 86a
...Theoretic Approach to the Analysis of Cutaneous Receptor	Neurons.	Fuller, MS 83a
A Simple Model for Spike-Train-Output	Neurons.	Atkinson, RC 88a
Analysis and Simulation of Networks of Mutually Inhibiting	Neurons.	Morishita, I 72a
Categorization and Selective	Neurons.	Anderson, JA 81a
Convergence, Cycling or Strange Motion in the Adaptive Synthesis of	Neurons.	Labos, E 86a
Cooperativity in Brain Function: Assemblies of Approximately 30	Neurons.	Shaw, GL 82a
...and Computational Algebra of Synaptic Connections Between	Neurons.	Fetz, EE 88a
Dale's Principle and Communication Between	Neurons.	Osborne, NN 82a
...Processing System May Direct Foraging Via Reticulospinal	Neurons.	Kanwal, JS 88a
Interaction of Synaptic Modification Rules within Populations of	Neurons.	Finkel, LH 85a
Learning and Distributed Memory: Getting Close to	Neurons.	Aleksander, I 88b
Mathematical Models for the All-Or-None Activity of Some	Neurons.	Gerstein, GL 62a
...to Real-Time Classification of the Action Potentials of Real	Neurons.	Bower, JM 88b
...to Real-Time Classification of the Action Potentials of Real	Neurons.	Wong, Y 87a
...Have Collective Computational Properties Like Those of Two-state	Neurons.	Hopfield, JJ 84b
...Have Collective Computational Properties Like Those of Two-State	Neurons.	Hopfield, JJ 84a
Optimal Mapping of the Configuration Space of a Network of	Neurons.	Carman, GJ 87a
Random Nets of Analog	Neurons.	Averbukh, DYA 69a
...Processing Roles for the Diverse Currents in Hippocampal	Neurons.	Borg-Graham, LJ 88a
The Hopfield Model with Multi-Level	Neurons.	Fleisher, M 88a
The Most Complicated Networks of Formal	Neurons.	Labos, E 87a
...of a Neural Network Memory with Several Hundreds of	Neurons.	Graf, HP 86a
The Elements of Single	Neurons: A Review.	Lewis, ER 83a
Synchronization in Small Groups of	Neurons: A Study with Electronic Models.	Lewis, ER 68b
Feature Extraction by	Neurons and Behavior	Werner, G 75a
Addition and Multiplication Theorems for the Input of Two	Neurons Converging on a Third.	Rapoport, A 51a
Symbols Among the	Neurons: Details of a Connectionist Inference Architecture	Touretzky, DS 85a
Rules for Forming Receptive Fields of Formal	Neurons During Unsupervised Learning Processes	Bobrowski, L 82a
Enhancement of Synaptic Potentials in Rabbit CA1 Pyramidal	Neurons Following Classical Conditioning	LoTurco, JJ 88a
Steady-State and Limit Cycle Activity of Mass of	Neurons Forming Simple Feedback Loops (I): Lumped Circuit Model.	Ahn, SM 74a
Steady-State and Limit Cycle Activity of Mass of	Neurons Forming Simple Feedback Loops (II): Lumped Circuit Model	Ahn, SM 74b
Studies with Artificial	Neurons, I: Properties and Functions of an Artificial.	Harmon, LD 61a
Self-Inhibition Alters Firing Patterns of	Neurons in Aplysia Buccal Ganglia.	White, RL 81a
The Firing Properties of Second-Order Vestibular	Neurons in Correlation with the Far-Field Recorded Vestibular.	Elidan, J 89a
Analytic Study of Assemblies of	Neurons in Memory Storage.	Roney, KJ 80a
	Neurons in Silicon	Zielenziger, D 88a
Interpretation of Spontaneous Spike Discharge Patterns of	Neurons in the Cochlear Nucleus.	Molnar, CE 68a
Do	Neurons in the Motor Cortex Encode Movement Direction? An.	Mussa-Ivaldi, FA 88a
...Firing Mechanisms: Input-Output Theory for the Recruitment of	Neurons into Epileptic Bursting Firing Patterns.	Calvin, WH 72a
Processing of Visual Information in the Distal	Neurons of the Vertebrate Retina	Curlander, JC 83a

Computers and Intractability: A Guide to the Theory of	NP-Completeness.	Garey, MR 79a
An	N-Player Sequential Stochastic Game with Identical Payoffs	Narendra, KS 83a
On	NPN Logic.	Zhang, W 88a
	NPN Relations and Cognitive Map Development.	Zhang, W 87a
Image Pre-Processing with a Generalised Grey Scale	N-Tuple Operator	Austin, J 88b
The Action of the Corticofugal Pathway on Sensory Thalamic	Nuclei: A Hypothesis	Koch, C 87b
...Spontaneous Spike Discharge Patterns of Neurons in the Cochlear	Nucleus.	Molnar, CE 68a
Some Networks That Can Learn, Remember, and Reproduce Any	Number of Complicated Space-Time Patterns, I	Grossberg, S 69j
Some Networks That Can Learn, Remember, and Reproduce Any	Number of Complicated Space-Time Patterns, II.	Grossberg, S 70c
...and Classification of New Instances as a Function of the	Number of Instances Defining the Prototype	Homa, D 73a
	Number of Stable Points for Spin-Glasses and Neural Networks of.	Baldi, P 87a
Simulating Memory by	Numbers.	Maddox, J 87a
The Law of Large	Numbers in Neural Modeling	Geman, S 81b
Interpolation Coding: A Representation for	Numbers in Neural Models	Ballard, DH 87b
Storing Infinite	Numbers of Patterns in a Spin-Glass Model of Neural Networks	Amit, DJ 85b
	Numbers of Specific Types of Neuron in Layer IVab of Cat Striate	Solnick, B 84a
Operations on Fuzzy Sets and Fuzzy	Numbers Related to Triangular Norms.	Klement, EP 81a
Building and Understanding Adaptive Systems: A Statistical/	Numerical Approach to Factory Automation and Brain Research.	Werbos, PJ 87a
	Numerical Calculation of Stable and Unstable Periodic Solutions to	Rinzel, J 80a
Integrated-Optical Approaches to	Numerical Optical Processing	Verber, CM 84a
	Numerical Simulations of Boltzmann Machines.	Bounds, DG 86a
Fast Component Threshold for Vestibular	Nystagmus in the Rabbit.	Lau, CGY 87a
The	NYU Ultracomputer-Designing an MIMD Shared Memory Parallel Computer.	Gottlieb, A 83a
Real-Time 3-D	Object Classification Using a Learning System.	Rimey, RD 86a
On the Acquisition of	Object Concepts from Sensory Data.	Phillips, WA 88a
NORA: Neural-network	Object Recognition Architecture.	Pawlicki, TF 88b
Invariant	Object Recognition Using a Distributed Associative Memory.	Wechsler, H 88a
...Intelligent Processing and Neural Networks for Adaptive Moving	Object Tracking.	Lee, JSJ 87a
A Parallel Computation that Assigns Canonical	Object-Based Frames of Reference	Hinton, GE 81a
...and Heterodyne Filtering Procedure for the Imaging of Moving	Objects.	Schuling, FH 88a
Computing with Connections in Visual Recognition of Origami	Objects.	Sabbah, D 88a
Computing with Connections in Visual Recognition of Origami	Objects.	Sabbah, D 85a
The Tendency to See	Objects as Equidistant and Its Inverse Relation to Lateral	Gogel, WC 56a
Visual Recognition of Simple	Objects by a Connection Network.	Plaut, DC 84a
Recovering Three-Dimensional Structure and Motion of Jointed	Objects from Orthographically Projected Optical Flow	Shibata, T 86a
Representing and Transforming Recursive	Objects in a Neural Network, or "Trees DO Grow on Boltzmann.	Touretzky, DS 86c
Recognizing Partially Visible	Objects Using Feature Indexed Hypotheses	Knoll, TF 86a
A Model for	Oblique Saccade Generation and Adaptation.	Fujita, M 87a
...a Document Processor: A Comparison of Expert Judgement and User	Observation.	Hammond, N 84a
	Observation of Phase Transitions in Spreading Activation Networks.	Shrager, J 87a
Using Artificial Neural Nets for Statistical Discovery:	Observations after Using Back-propagation, Expert Systems, and	Irani, EA 88a
Electrophysiological	Observations Which Help Describe an Associative Synaptic	Levy, WB 87a
IBM Neural Network Chips Make Everything	Obsolete	Gantz, J 88a
Analog MOS Integrated Circuits - Certain New Ideas, Trends, and	Obstacles.	Tsividis, Y 87a
	Obtaining Shape From Shading Information	Horn, B 75a
	OCCAM: A Computer Model for a Content-Addressable Memory in the.	Spinelli, DN 70a
A Maximization Technique	Occurring in the Statistical Analysis of Probabilistic Functions of.	Baum, LE 70a
An	Ocular Dominance Column Map as a Data Structure for Stereo	Yeshurun, Y 87a
Development of	Ocularity Domains and Growth Behaviour of Axon Terminals	von der Malsburg, C 79a
Asilomar Conference on Adaptive Processes in Visual and	Oculomoter Systems, June 16-20, 1985: Proceedings.	Keller, EL 85a
The	Oculomotor Control System: A Review.	Robinson, DA 68b
Adaptive Compensation to Changes in the	Oculomotor Plant	Grossberg, S 86b
A Proposed Neural Network for the Integrator of the	Oculomotor System.	Cannon, SC 83a
Spatial Arrangement of the Vestibular and the	Oculomotor System in the Rat	Daunicht, WJ 87a
An Improved Neural-Network Model for the Neural Integrator of the	Oculomotor System: More Realistic Neuron Behavior.	Cannon, SC 85a
...and Multiple Digital Signal Processors with the Explorer /	Odyssey for Image Processing and Understanding Applications.	Gove, RJ 87a
...to Pattern Classification: Final Report, US Army Research	Office	Myers, MH 87a
...and Oscillations in the Short Term Memory of Shunting On-Center	Off-Surround Networks.	Ellias, SA 75a
The Toad Optic Tectum as a Recurrent On-Center	Off-Surround Neural Net with Quenching Threshold	Betts, B 88b
The	OGC Cognitive Architecture Project: Silicon Implementation of.	Hammerstrom, D 86e
Circuit Theoretic Solutions for Neural Networks - An	Old Approach to a New Problem.	Kennedy, MP 87a
	Old Dogmas and New Axioms in Brain Theory.	Pellionisz, AJ 86a
Differential Forgetting of Prototypes and	Old Instances: Simulation by an Exemplar-Based Classification Model.	Hintzman, DL 80a
Using Fast Weights to Deblur	Old Memories	Hinton, GE 87d
...of Oscillating Network Model of Pattern Recognition in the Rabbit	Olefactory Bulb.	Baird, B 86a
...and Universal Recording, Part II: Feedback, Expectation,	Olfaction, and Illusions	Grossberg, S 76c
...of Neuronal Template-Matching Mechanism for Sensory Search with	Olfactory Bulb	Freeman, WJ 79a
...of Pattern Formation and Pattern Recognition in the Rabbit	Olfactory Bulb	Baird, B 86b
Central Pattern Generating and Recognizing in	Olfactory Bulb: A Correlation Learning Rule.	Freeman, WJ 88a

Reading Ability and Letter Search Time: Effects of	Orthographic Structure Defined by Single-Letter Positional	Mason, M 75a
...Three-Dimensional Structure and Motion of Jointed Objects from	Orthographically Projected Optical Flow.	Shibata, T 86a
Bifurcation Analysis of	Oscillating Network Model of Pattern Recognition in the Rabbit	Baird, B 86a
	Oscillation and Chaos in Physiological Control Systems	Mackey, MC 77a
...Algorithm in Consideration of the Generation of Rhythmic	Oscillation in a Ring Neural Network	Tsutsumi, K 84a
Computer Simulation of Rhythmic	Oscillation in Neuron Pools.	MacGregor, RJ 74a
Sustained	Oscillation in Symmetric Cooperative-Competitive Neural Networks	Cohen, MA 87d
Ring Neural Network qua a Generator of Rhythmic	Oscillation with Period Control Mechanism.	Tsutsumi, K 84b
Biological Competition: Decision Rules, Pattern Formation and	Oscillations	Grossberg, S 80a
Nonlinear Optical Neural Networks: Dynamic Ring	Oscillations	Anderson, DZ 88b
Stability and	Oscillations in a Neurological Servomechanism.	Stark, L 59a
	Oscillations in Mathematical Biology	Grissel, R 83a
Decisions, Patterns, and	Oscillations in the Dynamics of Competitive Systems with	Grossberg, S 78f
Pattern Formation, Contrast Control, and	Oscillations in the Short Term Memory of Shunting On-Center Off.	Ellias, SA 75a
On the Global Limits and	Oscillations of a System of Nonlinear Differential Equations	Grossberg, S 69f
...Activity in Neural-Nets II: A Model for the Brainstem Respiratory	Oscillator	Feldman, JL 75a
...and Learned Patterns of Movement Coordination, II. Component	Oscillator Dynamics.	Schoner, G 88c
Coupled Van der Pol	Oscillators - A Model of Excitatory and Inhibitory Neural.	Kawahara, T 80a
Integrate-And-Fire Models of Nerve Membrane Response to	Oscillatory Input.	Keener, JP 81a
	Oscillatory Neural Ectoderm (One): The "One" Hypothesis for Human.	Deshmukh, VD 87a
Unbounded	Oscillatory Solutions for a System of Interactive Populations.	Cohen, MA 85a
...Brain for Robot Control, Part 3: A Comparison of the Brain and	our Model.	Albus, JS 79c
On the Asymptotic Improvement in the	Outcome of Supervised Learning Provided by Additional Non.	Cooper, DB 70a
Optical Matrix-Matrix Multiplication Based	Outer Product Decomposition.	Athale, RA 82a
Three Layers of Vector	Outer Product Neural Networks for Optical Pattern Recognition.	Szu, HH 86h
Optical Processing Using	Outer-Product Concepts	Athale, RA 84a
Information Capacity of	Outer-Product Neural Networks.	Keeler, JD 87b
Optical Implementation of Quadratic Associative Memory with	Outer-Product Storage.	Jang, J 88b
	Outline of a New Approach to the Analysis of Complex Systems and	Zadeh, LA 87e
	Outline of a Theory for the Ontogenesis of Iso-Orientation Domains	von der Malsburg, C 82a
	Outline of a Theory of Thought Processes and Thinking Machines	Caianiello, ER 61a
Report	Outlines Commercialization of Neural Nets.	Schwartz, TJ 87e
Learning Control Systems: A Review and	Outlook.	Fu, K 70a
On Presenting the	Output of a Mechanical Speech Recognizer	Fry, DB 57a
Farey-Fraction Frequency Modulation in the Neuronlike	Output of Silicon p-i-n Diodes at 4.2 K.	Coon, DD 87a
Composite Matched Filter	Output Partitioning.	Marks II, RJ 87c
Models for the Generation of the Motor	Output Pattern in Flying Locusts	Wilson, DM 68a
Encoding Input/	Output Representations in Connectionist Cognitive Systems.	Miikkulainen, R 88a
The Brains of Men and Machines, Part 2: How the Brain Controls	Outputs.	Kent, EW 78b
Neural Nets Learn to	Outsmart a Salesman.	Schindler, M 87b
Neural Nets Learn to	Outsmart a Salesman.	Schindler, M 87a
A Neural Mechanism for Generalization	over Equivalent Stimuli in the Olfactory System.	Freeman, WJ 81a
Superior Accuracy of Decrementing	over Incrementing Associative Networks when Operating in Initially	Dobson, VG 87a
Neural Networks, Error-Correcting Codes and Polynomials	Over the Binary n-Cube	Bruck, J 88d
An	Overall Description of Retinotopic Mapping in the Cat's Visual	Mallot, HA 85a
Spiking Threshold and	Overarousal Effects in Serial Learning	Grossberg, S 71c
...General Support Algorithm: A Three Camera Stereo Algorithm for	Overcoming Binocular Matching Errors	Stewart, CV 88b
Coordination:A Vector-Matrix Description of Transformations of	Overcomplete CNS Coordinates and Tensorial Solution Using the Moore.	Pellionisz, AJ 84a
	Overlapping Taste and Tactile Maps of the Oropharynx in the Vagal.	Kanwal, JS 88b
Neural Net Models and Optical Computing: A Brief	Overview	Farhat, NH 86a
Neural-Type Robotics: An	Overview	Chande, A 83a
Associative Memories and Processors: An	Overview and Selected Bibliography	Parhami, B 73a
	Overview of Neural-Type Electronics.	El-Leithy, N 85a
	P3: A Parallel Network Simulating System	Zipser, D 86c
Bisensory Stimulation: Inferring Decision-Related Processes from the	P300 Component	Squires, NK 77a
A Neural Theory of Circadian Rhythms: The Gated	Pacemaker.	Carpenter, GA 83a
Neural Dynamics of Circadian Rhythms: The Mammalian Hypothalamic	Pacemaker.	Carpenter, GA 85c
Neural Theory of Circadian Rhythms: The Gated	Pacemaker.	Carpenter, GA 86n
Entrainment in	Pacemakers Characterized by a V-Shaped PRC	Torras i Genis, C 86a
Software	Package for Interactive Text Independent, Speaker Verification	Cohen, A 86a
A Computer	Package for the Evaluation of Neuron Models Involving Large Uniform.	Lee, MH 82a
A Program	Package Implementing a System of Interacting Adaptive Associative.	Lansner, A 83a
A Connectionist Model of Chronic	Pain	Katz, BF 88a
A Neural Network Model of	Pain Mechanisms: Functional Properties of the Network Cells.	Minamitani, H 82a
...Path-Dentate Projection. III. Comparison of Random Train and	Paired Impulse Stimulation	Berger, TW 88b
...Learning Model for 3-Dimensional Mental Rotation, Zoom, and	Pan.	Mel, BW 86a
Flow-of-Activation Processing: Parallel Associative Networks (	PAN)	Cruz, CA 86a
	Pandemonium: A Paradigm for Learning	Selfridge, OG 58a
	Pandemonium: A Paradigm for Learning	Selfridge, OG 59a

	Parallel Logical Inference and Energy Minimization	Ballard, DH 86b
Absolute Stability of Global Pattern Formation and	Parallel Memory Storage by Competitive Neural Networks	Cohen, MA 83a
Absolute Stability of Global Pattern Formation and	Parallel Memory Storage by Competitive Neural Networks	Cohen, MA 83b
A	Parallel Model for Low-Level Vision.	Waltz, DL 78a
Schemata in a Massively	Parallel Model of Computation: A Non-Committing Approach to Schema	Chun, HW 85a
A Massively	Parallel Model of Schema Selection	Chun, HW 87a
	Parallel Models of Associative Memory.	Hinton, GE 81c
	Parallel N**4 Weighted Optical Interconnections.	Caulfield, HJ 87a
A	Parallel Natural Language Processing with Distributed Control.	Berg, G 87a
Learned Classification of Sonar Targets Using a Massively	Parallel Network	Gorman, RP 88b
Separating Figure from Ground with a	Parallel Network	Kienker, PK 86a
P3: A	Parallel Network Simulating System	Zipser, D 86c
NETtalk: A	Parallel Network that Learns to Read Aloud	Sejnowski, TJ 86d
NETtalk: A	Parallel Network that Learns to Read Aloud	Sejnowski, TJ 86c
Massively	Parallel Network-Based Natural Language Parsing System	Li, T 87a
Learning in	Parallel Networks.	Hinton, GE 85a
A Representation for Temporal Sequence and Duration in Massively	Parallel Networks: Exploiting Link Interactions.	Chun, HW 86a
Minimum Distance Automata in	Parallel Networks for Optimum Classification	Winters, JH 89a
On	Parallel Networks for Optimum Classification	Winters, JH 88a
	Parallel Networks that Learn to Pronounce English Text	Sejnowski, TJ 87f
Masking Fields: A Massively	Parallel Neural Architecture for Learning, Recognizing, and.	Cohen, MA 87b
Lyapunov Function for	Parallel Neural Networks	Goles, E 86b
	Parallel Optical Computer Solves Multivariable Problems in Real.	Brown, C 86h
	Parallel Optics: Solution to Von Neumann Bottleneck?	Brown, C 86i
Massively	Parallel Parsing: A Strongly Interactive Model of Natural Language	Waltz, DL 85a
Massively	Parallel Parsing: A Strongly Interactive Model of Natural Language	Waltz, DL 88c
A	Parallel PDP Network Simulator	Smith, AB 87a
Harmony Theory: A Mathematical Framework for Stochastic	Parallel Processing.	Smolensky, P 83a
Representing Conceptual Graphs for	Parallel Processing.	Lendaris, GG 88c
Neural Computation Research for Vision - Simulation on a	Parallel Processing Computer	Miyake, S 88a
	Parallel Processing for Associative and Neuronal Networks.	Palm, G 84b
Early	Parallel Processing in Reading: A Connectionist Approach	Mozer, MC 86a
A	Parallel Processing Model for the Study of Possible Pattern.	Muller, FJ 76a
Morphological Evidence for	Parallel Processing of Information in Rat Macula	Ross, MD 88b
Real-Time	Parallel Processing of Pixel Arrays Using Adaptive Logic	Vidal, JJ 87b
	Parallel Processing of Signals in Neural Nets as Manifested in the	Freeman, WJ 75b
Cortical Connections and	Parallel Processing: Structure and Function.	Ballard, DH 86a
Cortical Connections and	Parallel Processing: Structure and Function.	Ballard, DH 87a
Putting Knowledge in its Place: A Scheme for Programming	Parallel Processing Structures on the Fly.	McClelland, JL 88c
Putting Knowledge in its Place: A Scheme for Programming	Parallel Processing Structures on the Fly.	McClelland, JL 85b
A Balanced Scalable	Parallel Processor	Pilpel, S 87a
Bell Labs Models	Parallel Processor on Neural Networks.	Victor, J 86b
Bell Labs Models	Parallel Processor on Neural Networks.	Victor, J 86a
A Design Tool for Modeling Large Networks of Simple	Parallel Processors.	Bock, P 88a
Workshop on Performance Efficient	Parallel Programming	Segall, Z 86a
Cancellation in a	Parallel Semantic Network.	Fahlman, SE 81a
A	Parallel Signal Processor System	Gass, W 86a
A	Parallel Simulation of a Distributed Problem Solving Network	Durfee, EH 84a
	Parallel Stochastic Computing.	Tenorio, MF 87c
	Parallel Structures in Human and Computer Memory	Kanerva, P 86a
Skill Learning and Repetition Priming in Symmetry Detection:	Parallel Studies of Human Subjects and Connectionist Models.	Cohen, NJ 86a
Eye Movements Induced by Linear Acceleration on a	Parallel Swing	Baloh, RW 88a
Cognitive Capabilities of a	Parallel System.	Anderson, JA 86a
Toward a Massively	Parallel System for Word Recognition	Wong, MK 86a
Biological vs Artificial Representations of Variables in Fine-Grain	Parallel Systems	Walters, DKW 87b
Psychological Concepts in	Parallel Systems	Anderson, JA 86e
Shape Representation in	Parallel Systems	Hinton, GE 81e
Shape Representation in	Parallel Systems	Hinton, GE 81d
	Parallel Versus Serial Processing in Rapid Pattern Discrimination.	Bergen, JR 83a
	Parallel Visual Computation.	Ballard, DH 83a
Design of a Highly	Parallel Visual Recognition System	Sabbah, D 81a
Interpretation of Natural Language: A Potential Application of	Parallelism.	Pollack, JB 86a
Three Flavors of	Parallelism.	Fahlman, SE 82a
	Parallelism and Redundancy in Neural Networks.	von Seelen, W 88a
	Parallelism Breeds a New Class of Supercomputers	Mokhoff, N 87a
Attractor Dynamics and	Parallelism in a Connectionist Sequential Machine.	Jordan, MI 86a
Computation Power of	Parallelism in Optical Architectures	Abu-Mostafa, YS 85a
	Parallelism in Production Systems.	Gupta, A 86a
	Parallel-Processing Mechanisms and Processing of Organized	Ratcliff, R 81a

	Parallel-Processors Bypass Von Neumann Bottleneck	Brown, C 86j
Discrete-Time	Parameter Estimation with Genetic Algorithms	Das, R 88a
Unconstrained Minimum Mean-Square Error	Parameter Estimation with Hopfield Networks	Altes, RA 88a
	Parameter Nets	Ballard, DH 84b
	Parameter Networks	Ballard, DH 81b
	Parameter Networks: Towards a Theory of Low-Level Vision	Ballard, DH 81c
...of Ill Defined Neuronal Networks with Random Structure and	Parameter Setting	Benito, S 80a
...of the EEG Amplitude's Probability Distribution on Statistical	Parameters	Anninos, PA 83b
Exact Solutions for Feedforward Network	Parameters	Atkinson, RC 88b
Associative Learning of Scene	Parameters from Images	Kersten, D 87a
A Methodology to Select Appropriate Learning Rate	Parameters in Feed-Forward Networks	Pabon, JA 87a
Some Notes on the Probabilistic Semantics of Logistic Function	Parameters in Neural Networks	Davies, TR 88a
Effects of Adaptation	Parameters on Convergence Time and Tolerance for Adaptive Threshold	Mays, CH 64a
Theory and Feasibility of the	Parametric Avalanche for Advanced Electronic Warfare Applications	Dawes, RL 88b
...Inhibition Networks that is Robust in the Presence of Circuit	Parasitics	Wyatt Jr, JL 88a
...Inhibition Networks that is Robust in the Presence of Circuit	Parasitics	Standley, DL 88a
Les Applications des Modeles Connexionistes a la Reconnaissance de la	Parole	Wellekens, CJ 88b
A Connectionist Context-Free	Parser Which is not Context-Free, But Then it is not Really	Charniak, E 87a
Connectionist	Parsing	Cottrell, GW 85a
Toward Connectionist	Parsing	Small, SL 82c
Massively Parallel	Parsing: A Strongly Interactive Model of Natural Language	Waltz, DL 88c
Massively Parallel	Parsing: A Strongly Interactive Model of Natural Language	Waltz, DL 85a
Word Expert	Parsing: A Theory of Distributed Word-Based Natural Language	Small, SL 80a
Viewing	Parsing as Word Sense Discrimination: A Connectionist Approach	Cottrell, GW 84a
Context-Free	Parsing in Connectionist Networks	Fanty, MA 85a
	Parsing Scale-Space and Spatial Stability Analysis	Bischof, WF 88a
A Rule-Based Connectionist	Parsing System	Selman, B 85a
Massively Parallel Network-Based Natural Language	Parsing System	Li, T 87a
Context-Free	Parsing with Connectionist Networks	Fanty, MA 86b
	PARSNIP: A Connectionist Network that Learns Natural Language	Hanson, SJ 87a
State-Dependent Factors Influencing Neural Plasticity: A	Partial Account of the Critical Period	Munro, PW 86b
Coherent Optical Feedback for the Analog Solution of	Partial Differential Equations	Cederquist, J 80a
Balanced and Unbalanced, Complete and	Partial Transparency	Metelli, F 85a
	Partially Connected Models of Neural Networks	Canning, A 88b
Recognizing	Partially Visible Objects Using Feature Indexed Hypotheses	Knoll, TF 86a
Combinations of Neural Systems for	Particular Application Situations	Josin, GM 87a
Cochlear Nerve Filter Responses: Distribution Along the Cochlear	Partition	Pfeiffer, RR 75a
Composite Matched Filter Output	Partitioning	Marks II, RJ 87c
	Partitioning of Sensory Data by a Cortical Network	Granger, R 88b
Optimal Association with	Partly Missing Key Vectors	Murakami, K 82a
A Neat Theory of Marker	Passing	Charniak, E 86a
A Model of Schema Selection Using Marker	Passing and Connectionist Spreading Activation	Chun, HW 87b
	Passing Markers: A Theory of Contextual Influence in Language	Charniak, E 83a
Information Processing System Using Threshold	Passive Modification	Cooper, LN 81b
Connecting to the	Past	MacDonald, B 88a
Backpropagation:	Past and Future	Werbos, PJ 88a
Precise	Past, Fuzzy Future	Gaines, BR 83a
	Past, Present, and Future of GMDH	Ivakhnenko, AG 84a
On Learning the	Past Tenses of English Verbs	Rumelhart, DE 86g
Efficient Method for Estimating Kinetic Constants from Single Channel	Patch Clamp Recordings	Chay, TR 87a
Efficient Method for Estimating Kinetic Constants from Single Channel	Patch Clamp Recordings	Chay, TR 87b
Planning a Dynamic Trajectory via	Path Finding in Discretized Phase Space	Ritter, H 87a
Neural Network Representation of Sensor Graphs in Autonomous Robot	Path Planning	Jorgensen, CC 87a
The Neural	Path Probabilistic Delay Model	Niznik, CA 83a
...by Functional-Differential Neural Networks with Arbitrary	Path Weights	Grossberg, S 72d
Nonlinear Systems Analysis of the Hippocampal Perforant	Path-Dentate Projection. I. Theoretical and Interpretational	Sclabassi, RJ 88b
Nonlinear Systems Analysis of the Hippocampal Perforant	Path-Dentate Projection. II. Effects of Random Impulse Train	Berger, TW 88a
Nonlinear Systems Analysis of the Hippocampal Perforant	Path-Dentate Projection. III. Comparison of Random Train and Paired	Berger, TW 88b
...of Genetic Algorithm to a Neural Network for Finding Optimal	Paths	Smith, T 87a
Learning from Solution	Paths: An Approach to the Credit Assignment Problem	Sleeman, D 82a
...and Mathematical Formalism for Robotic Cerebellar Neural Network	Pathway Information Measures	Niznik, CA 87a
The Action of the Corticofugal	Pathway on Sensory Thalamic Nuclei: A Hypothesis	Koch, C 87b
The Tracing of	Pathways in the Central Nervous System	Heimer, L 68a
Low Rank Matrices with a Given Sign	Pattern	Delsarte, P 87a
...of Behavior in Tritonia, III Neuronal Mechanism of a Fixed Action	Pattern	Willows, AOD 73a
Application-Oriented Connection Systems for	Pattern Acquisition and Recognition	Horne, W 87a
A Consistency Technique for	Pattern Association	Ullmann, JR 62a
Analysis and Comparison of Different Learning Algorithms for	Pattern Association Problems	Bernasconi, J 88a
	Pattern Class Degeneracy in an Unrestricted Storage Density Memory	Scofield, CL 88b

A	Pattern Recognition Program That Generates, Evaluates, and Adjusts	Uhr, L 61a
A Template Matching Model for	Pattern Recognition: Self-Organization of Templates and Template	Hirai, Y 80a
Syntactic	Pattern Recognition: Stochastic Languages	Thomason, MG 86a
...a Time Varying Environment Using a Probabilistic Automaton with	Pattern Recognition Supervision	Jarvis, RA 70a
An Experimental Investigation of a Class of	Pattern Recognition Synthesis Algorithms	Mattson, RL 63a
Reinforcement-Learning Control and	Pattern Recognition Systems	Mendel, JM 70b
Adaptive	Pattern Recognition Systems and Boltzmann Machines: A Rapprochement	Aleksander, I 87a
Adaptation, Learning, and	Pattern Recognition Systems: Theory and Applications	Mendel, JM 70a
Two-Dimensional Adaptive Model of a Human Controller Using	Pattern Recognition Techniques	Gilstad, DW 71a
Neocognitron: A New Algorithm for	Pattern Recognition Tolerant of Deformation and Shifts in Position	Fukushima, K 82a
...A Self-Organizing Neural Network Model for a Mechanism of	Pattern Recognition Unaffected by Shift in Position	Fukushima, K 80a
	Pattern Recognition Using Artificial Neural Systems	Gutschow, T 87a
	Pattern Recognition Using Associative Memories	Altman, JL 77a
An Approach to	Pattern Recognition Using Linear Threshold Devices	Mattson, RL 60a
	Pattern Recognition with an Adaptive Network	Roberts, LC 60a
An Algorithm for	Pattern Recognition with VLSI Neural Networks	Sirletti, B 88a
Statistical Recognition Functions and the Design of	Pattern Recognizers	Marill, T 60a
Cooperativity in Networks of	Pattern Recognizing Stochastic Learning Automata	Barto, AG 86a
	Pattern Recognizing Stochastic Learning Automata	Barto, AG 85e
	Pattern Recognizing Stochastic Learning Automata	Barto, AG 84a
Generalization of	Pattern Recogntion in a Self-Organizing System	Clark, WA 55a
Further Consideration on	Pattern Separability in a Random Neural Net with Inhibitory	Torioka, T 80a
	Pattern Separability in a Random Neural Net with Inhibitory	Torioka, T 79a
	Pattern Separability of a Tightly Coupled Random Nerve Net with	Torioka, T 79b
	Pattern Separating Functioning of Two Layered Random Nerve Nets	Torioka, T 88a
Property of the Degree of	Pattern Separation in a Two-Layer Random Neural Net with Inhibitory	Torioka, T 80b
Storage of a Temporal	Pattern Sequence in a Network	Willwacher, G 82a
Two-Dimensional	Pattern Sequence Prediction Using High-Order Neural Networks	Collins, DH 88a
Learning Patterns and	Pattern Sequences by Self-Organizing Nets of Threshold Elements	Amari, S 72b
The "Rubber-Mask" Technique - I.	Pattern Storage and Recognition	Widrow, B 73b
The "Rubber-Mask" Technique - II.	Pattern Storage and Recognition	Widrow, B 73c
Auto-Associative	Pattern Storage Using Synchronous Boolean Networks	Martland, D 87a
A Dynamic	Pattern Theory of Behavioral Change	Schoner, G 88a
	Pattern-Based Fault Diagnosis Using Neural Networks	Dietz, WE 88a
Learning Machines: Foundations of Trainable	Pattern-Classifying Systems	Nilsson, NJ 65a
...for Learning, Recognizing, and Predicting Multiple Groupings of	Patterned Data	Cohen, MA 87a
...for Learning, Recognizing, and Predicting Multiple Groupings of	Patterned Data	Cohen, MA 87b
How	Patterned Neural Connections Can be Set Up by Self-Organization	Willshaw, DJ 76b
How	Patterned Neural Connections Can be Set Up by Self-Organization	Willshaw, DJ 75a
Boxes as a Model of	Pattern-Formation	Michie, D 68b
Automatic Subclass Determination for	Pattern-Recognition Applications	Firschein, O 63a
	Pattern-Recognizing Control Systems	Widrow, B 64b
	Pattern-Recognizing Control Systems	Widrow, B 64c
...and the Back Propagation Network as Recognizers of Static Speech	patterns	Prager, RW 87a
A Mathematical Theory of Visual Hallucination	Patterns	Ermentrout, GB 79a
A Network for Recognition and Classification of Continuous	Patterns	Banzhaf, W 88a
A Neural Network for the Retrieval of Superimposed Connection	Patterns	von der Malsburg, C 87a
...System with an Automatic Generation Mechanism for Distorted	Patterns	Shimohara, K 88a
...Some Nonlinear Functional-Differential Equations. II. Learning of	Patterns	Grossberg, S 68b
A Quasi-Topological Method for the Recognition of Line	Patterns	Sherman, H 60a
...Neural Recognition Codes in Response to Arbitrary Lists of Input	Patterns	Carpenter, GA 86d
An Associative Memory Technique for the Recognition of	Patterns	Pao, Y 78a
...for Real Time Generation of Scale and Rotation Invariant	Patterns	Messner, RA 85a
...of Stable Category Recognition Codes for Analog Input	Patterns	Carpenter, GA 87c
...Development: The Temporal Unfolding and Stability of STM and LTM	Patterns	Grossberg, S 86d
...Development: The Temporal Unfolding and Stability of STM and LTM	Patterns	Grossberg, S 82b
Classification of Binary Random	Patterns	Abend, K 65a
Clustering, Taxonomy, and Topological Maps of	Patterns	Kohonen, T 82c
Clustering, Taxonomy, and Topological Maps of	Patterns	Kohonen, T 82d
Determination and Detection of Features in	Patterns	Block, HD 64a
Distributed Associative Memory for	Patterns	Pao, Y 75a
...and Peripheral Detection of Coherent Motion in Moving Random-Dot	Patterns	van de Grind, WA 87a
Nearest Matched Filter Classification of Spatiotemporal	Patterns	Hecht-Nielsen, R 87f
Nearest Matched Filter Classification of Spatio-Temporal	Patterns	Hecht-Nielsen, R 87g
...Network Capable of Position-Invariant Recognition of	Patterns	Fukushima, K 80c
...Network Capable of Position-Invariant Recognition of	Patterns	Fukushima, K 80b
Neural Mechanisms for the Production of Cyclic Motor	Patterns	Selverston, AI 83a
...Networks and Radial Basis Functions in Classifying Static Speech	Patterns	Niranjan, M 88a
Neural Networks with Hierarchically Correlated	Patterns	Gutfreund, H 88a

Monte Carlo Calculation of	Phase Separation in a Two-Dimensional Ising System	Flinn, PA 74a
Planning a Dynamic Trajectory via Path Finding in Discretized	Phase Space	Ritter, H 87a
	Phase Transitions in Artificial Intelligence Systems	Huberman, BA 87a
A Stochastic Theory of	Phase Transitions in Human Hand Movement	Schoner, G 86a
	Phase Transitions in Neural Networks	Chover, J 88a
Nonequilibrium	Phase Transitions in Pattern Recognition and Associative Memory	Haken, H 88c
	Phase Transitions in Quasirandom Neural Networks	Kurten, KE 87a
Observation of	Phase Transitions in Spreading Activation Networks	Shrager, J 87a
Associative Holographic Memory with Feedback Using	Phase-Conjugate Mirrors	Soffer, BH 86a
Associative Memories Based on Message-Bearing Optical Modes in	Phase-Conjugate Resonator	Yariv, A 86a
Volume Holographic Information Processing in a	Phase-Conjugating Resonator	Cohen, MS 86a
Infinitely Many Commensurate	Phases in a Simple Ising Model	Fisher, ME 80a
A Model for	Phasic and Tonic Repetitively Firing Neuronal Encoders	Fohlmeister, JF 73a
	Phasor Neural Networks	Noest, AJ 88a
A Mathematical Analysis of Excitable Membrane	Phenomena	Carpenter, GA 78a
A Note on the Neural Unit Model for Contrast	Phenomena	Bergstrom, SS 73a
...Brightness Perception: A Unified Model of Classical and Recent	Phenomena	Grossberg, S 88c
...Brightness Perception: A Unified Model of Classical and Recent	Phenomena	Grossberg, S 88b
Maximum Entropy Formalism, Fractals, Scaling	Phenomena, and 1/f Noise: A Tale of Tails	Montroll, EW 83a
Anisotropic Connectivity and Cooperative	Phenomena as a Basis for Orientation Sensitivity in the Visual	Finette, S 78a
Adaptive Superresolution	Phenomena in Aberrant Arrays	Jablon, N 85a
Learning	Phenomena in Layered Neural Networks	Widrow, B 87c
Fluctuation	Phenomena in Nerve Membrane	Verveen, AA 68a
Learning	Phenomena in Networks of Adaptive Switching Circuits	Hoff, ME 62a
Ignition	Phenomena in Random Nets	Rapoport, A 52a
Neuron-Like Transient	Phenomena in Silicon p-i-n Structures	Coon, DD 88c
Simulating the Adaptation	Phenomena in the Response of Auditory Nerve Fibers: A Functional	Wu, ZL 88a
A Psychologist Views Human Processing: Human Errors and Other	Phenomena Suggest Processing Mechanisms	Norman, DA 81a
	Phenomenology of Perception	McClelland, JL 78c
An Extended First Approximation Model for the Amygdaloid Kindling	Phenomenon	Lieblich, I 78a
Connectionism at the	Philips Research Laboratory Brussels	Wellekens, CJ 88a
Cognitive Neuroscience, Intentionality and Realism: The	Philosophical Importance of the Connectionist Approach	Basti, G 87a
...Network for Spatio-Temporal Bipolar Patterns: Application to	Phoneme Classification	Atlas, LE 88a
...Network for Spatio-Temporal Bipolar Patterns: Application to	Phoneme Classification	Homma, T 88a
A Connectionist Structure for	Phoneme Recognition	Harrison, TD 88a
A Dynamic Connectionist Model for	Phoneme Recognition: Preliminary Results	Robinson, AJR 88a
A Neural Network Model for	Phoneme Recognition Using the Generalized Delta Rule for Connection	Watson, M 87a
A Preliminary Study of the	Phoneme Recognition with Hardware Neural Network	Hoshimiya, N 82a
Spectral Classification of	Phonemes by Learning Subspaces	Kohonen, T 79a
Improved Hidden Markov Modeling of	Phonemes for Continuous Speech Recognition	Schwartz, RM 84a
Speech Perception and	Phonemic Restorations	Warren, RM 71a
Application of Orthogonal Projection Principles to Simultaneous	Phonemic Segmentation and Labeling of Continuous Speech	Jalanko, M 78a
Auditory and	Phonetic Coding of Speech	Sawusch, JR 86a
Learned	Phonetic Discrimination Using Connectionist Networks	Watrous, RL 87a
Learning	Phonetic Features Using Connectionist Networks: An Experiment in	Watrous, RL 87d
The "Neural"	Phonetic Typewriter	Kohonen, T 88d
...Implementation of a Large Vocabulary Speech Recognizer and	Phonetic Typewriter for Finnish and Japanese	Kohonen, T 87e
	Phonetic Typewriter for Finnish and Japanese	Kohonen, T 88b
Phonotopic Maps - Insightful Representation of	Phonological Features for Speech Recognition	Kohonen, T 84b
The Role of Orthographic and	Phonotactic Rules in Perceiving Letter Patterns	Spoehr, KT 75a
	Phonotopic Maps - Insightful Representation of Phonological	Kohonen, T 84b
Transient and Persistent Depolarization-Induced Changes of Protein	Phosphorylation in a Molluscan Nervous System	Naito, S 88a
Seeing the Light: Optics in Nature,	Photography, Color, Vision, and Holography	Stork, DG 85a
A Sensitive Electronic	Photoreceptor	Mead, CA 85b
...from Shadows in Hermissenda: A Direct Behavioral Analog of	Photoreceptor Responses to Brief Light Steps	Lederhendler, II 87a
Adaptation and Transmitter Gating in Vertebrate	Photoreceptors	Carpenter, GA 86c
Electrical Response of Single	Photoreceptors	Tomita, T 68a
Spectral Information Coding by Infrared	Photoreceptors	Coon, DD 86a
Gated Dipoles:	Photoreceptors and Circadian Rhythms	Carpenter, GA 84b
Electrical Coupling of	Photoreceptors in Retinal Network Models	Marcelja, S 80a
Adaptive Optical Networks Using	Photorefractive Crystals	Psaltis, D 88a
Adaptable Interconnects for Optical Neuromorphs: Demonstration of a	Photorefractive Projection Operator	Anderson, DZ 87a
	Phototransduction in Invertebrates	Ogmen, H 87d
Learning to Represent and Understand Locative Prepositional	Phrases	Cosic, C 88a
The Gamma Network - A	Physical and Logical Data Structure	Hennessey, AK 87a
	Physical Broadcast Structure	Rudnick, M 88b
A	Physical Charge-Controlled Model for MOS Transistors	Maher, MA 87a
The Role of	Physical Constraints in Natural and Artificial Manipulation	Morasso, P 86a
...of the Mellin Transformation with Applications to Scaling of	Physical Quantities	Moses, HE 74a

A Parallel Processing Model for the Study of	Possible Pattern Recognition Mechanisms in the Brain	Muller, FJ 76a
Neuroscience as a	Possible Research Standard for Neural Networks	Morse, KG 87a
	Possible Synaptic Mechanism Subserving Learning.	Eccles, JC 72a
Distributed Memory in the Central Nervous Systems:	Possible Test of Assumptions in Visual Cortex.	Cooper, LN 81a
	Post-Digital Electronics	Pettit, FR 87a
The Relationship Between Image Restoration by the Maximum A	Posteriori Method and a Maximum Entropy Method	Trussell, HJ 80a
A Redundant Hash Addressing Method Adapted for the	Postprocessing and Error-Correction of Computer-Recognized Speech.	Reuhkala, E 79a
Variations in Amplitude and Time Course of Inhibitory	Postsynaptic Currents.	Gardner, D 86a
Persistent Conditioning-Specific Changes of	Post-Synaptic Membrane Currents in Heermissenda and Hippocampus.	Coulter, DA 87a
Conditioning-Specific Modification of	Postsynaptic Membrane Currents in Mollusc and Mammal	Alkon, DL 87a
A Physiological Mechanism for Hebb's	Postulate of Learning.	Stent, GS 73a
Some Physiological and Biochemical Consequences of Psychological	Postulates	Grossberg, S 68f
An Electronic Model of the Neuron Based on the Dynamics of	Potassium and Sodium Ion Fluxes.	Lewis, ER 64a
Single Channel	Potassium Currents of the Anomalous Rectifier.	Fukushima, Y 81a
Voltage-Dependent Calcium and	Potassium Ion Conductances: A Contingency Mechanism for an	Alkon, DL 79a
Interpretation of Natural Language: A	Potential Application of Parallelism	Pollack, JB 86a
Neural Networks,	Potential Assistants to Knowledge Engineers.	Lendaris, GG 88b
Synaptic	Potential Summation and Repetitive Firing Mechanisms: Input-Output	Calvin, WH 72a
General	Potential Surfaces and Neural Networks	Dembo, A 88a
Membrane	Potential Transients and Membrane Time Constant of Motoneurons	Rall, W 60a
A Test for Modulation of Synaptic Efficacy by Induced Membrane	Potential Variance	Gardner, D 88a
Mechanisms of Slow Synaptic	Potentials	Hartzell, HC 81a
Random Fields and Spatial Renewal	Potentials	Demongeot, J 86a
A Mathematical Model for Conduction of Action	Potentials Along Bifurcating Axons	Parnas, I 79a
...Category Learning and Recognition: Structural Invariants, Evoked	Potentials, and Reinforcement.	Carpenter, GA 86k
Neuroelectric	Potentials Derived from an Extended Version of the Hodgkin-Huxley.	Lewis, ER 66a
Probing Cognitive Processes Through the Structure of Event-Related	Potentials During Learning: An Experimental and Theoretical.	Banquet, J 87a
Structure of Event-Related	Potentials During Learning: An Experimental and Theoretical.	Banquet, J 86a
Computer Simulation of Event Related	Potentials During Voluntary Cognition.	Deshmukh, VD 88b
Evoked	Potentials in Artificial Neural Nets	Anninos, PA 73a
Enhancement of Synaptic	Potentials in Rabbit CA1 Pyramidal Neurons Following Classical	LoTurco, JJ 88a
...Matching: Application to Real-Time Classification of the Action	Potentials of Real Neurons	Bower, JM 88b
...Matching: Application to Real-Time Classification of the Action	Potentials of Real Neurons	Wong, Y 87a
Measurement of Bioelectric	Potentials with Microelectrodes and Neutralized Input Capacity	Amatniek, E 58a
...Understanding of the Combined Dynamics of EEG and Event-Related	Potentials" by Elbert and Rockstroh.	Palm, G 87b
Reseaux Neuronzux	Pour Bases D'Images.	Cochet, Y 88a
Nerve Cell Soma Model with High Reliability and Low	Power Consumption.	Fernandes, PML 80a
Uncertainty as a Foundation of Computational	Power in Neural Networks	Gaines, BR 87b
...Functional Properties of the Network Cells Providing the	Power Law.	Minamitani, H 82a
A Comparison of the Computational	Power of Neural Network Models	Hartley, RL 87a
On the	Power of Neural Networks for Solving Hard Problems	Bruck, J 88e
Computation	Power of Parallelism in Optical Architectures.	Abu-Mostafa, YS 85a
Fuzzy	Power Sets and Fuzzy Implication Operators	Bandler, W 80a
The	Power Spectrum of the Mellin Transformation with Applications to	Moses, HE 74a
Towards	Practical 'Neural' Computation for Combinatorial Optimization.	Baum, EB 86b
A	Practical 24 Channel Microelectrode for Neural Recording in Vivo	Kuperstein, M 81a
A	Practical Algorithm for Solving Dynamic Membrane Equations	Rush, S 78a
	Practical Applications for Adaptive Data-Processing Systems.	Widrow, B 63b
Predictive GMDH Models of Shrimp Catches: Some	Practical Considerations	Prager, MH 84b
	Practical Methods of Optimization, Volume 1: Unconstrained	Fletcher, R 80a
	Practical Optimization	Gill, PE 81a
Speech Recognition: Turning Theory into	Practice	Doddington, GR 81a
Theory and	Practice of Hierarchical Control	Albus, JS 81b
Entrainment in Pacemakers Characterized by a V-Shaped	PRC.	Torras i Genis, C 86a
Exaptation,	Preadaptation, Evolution and Biological Neural Networks.	Stork, DG 87b
A Neural Network Architecture for	Preattentive Vision: Multiple Scale Segmentation and Regularization.	Grossberg, S 87d
Nonlinear Prediction Models for River Flows and Typhoon	Precipitation by Self-Organizing Methods	Ikeda, S 84a
	Precise Past, Fuzzy Future	Gaines, BR 83a
Network Generality, Traning Required, and	Precision Required	Denker, JS 88a
Scaling Relationships in Back-Propagation Learning: Dependence on	Predicate Order.	Tesauro, G 88c
	Predicted Squared Error: A Criterion for Automatic Model Selection	Barron, AR 84a
An Artificial Intelligence Approach to	Predicting Bond Ratings.	Dutta, S 89a
...A Massively Parallel Architecture for Learning, Recognizing, and	Predicting Multiple Groupings of Patterned Data.	Cohen, MA 87a
...Parallel Neural Architecture for Learning, Recognizing, and	Predicting Multiple Groupings of Patterned Data.	Cohen, MA 87b
On	Predicting Perceptron Performance.	Joseph, RD 60a
A Method for	Predicting Sales Amount by the Use of IWSM and GMDH.	Nomura, J 84a
Toward A Modern Theory of Adaptive Networks: Expectation and	Prediction	Sutton, RS 81b
Non-Parametric	Prediction Analysis for Binary Data.	Davis, BR 84a
Nonlinear Difference-Differential Equations in	Prediction and Learning Theory	Grossberg, S 67a

Independent Components Analysis (INCA) Versus	Principal Components Analysis.	Jutten, C 88a
Speech Perception: Research, Theory, and the	Principal Issues	Pisoni, DB 86a
L'Uso di Modelli Computazionali in Psicofisiologia Ed Il	Principio Gnoseologico di Rappresentazione	Basti, G 86b
A New Microscopic	Principle.	Gabor, D 48a
An Adaptive Associative Memory	Principle.	Kohonen, T 74a
Optimization with Artificial Neural Network Systems: A Mapping	Principle and a Comparison to Gradient Based Methods	Leong, HM 88a
Dale's	Principle and Communication Between Neurons.	Osborne, NN 82a
Towards an Organizing	Principle for a Layered Perceptual Network	Linsker, R 88b
An Organizing	Principle for Cerebral Function.	Mountcastle, VB 79a
An Extremum	Principle for Shape from Contour	Brady, M 87a
Artificial Neural Network Integrated Circuit Based on NMOS/CCD	Principles	Sage, JP 86a
Adaptive Noise Canceling:	Principles and Applications.	Widrow, B 75a
Design	Principles for a Front-end Visual System	Koenderink, JJ 88a
On Design	Principles for a Molecular Computer.	Conrad, M 85a
Cybernetic	Principles in Perception and Motor Control	Harth, EM 82a
	Principles of Cortical Network Organization.	von der Malsburg, C 88c
Associative and Competitive	Principles of Learning and Development: The Temporal Unfolding and	Grossberg, S 86d
Associative and Competitive	Principles of Learning and Development: The Temporal Unfolding and	Grossberg, S 82b
Studies of Mind and Brain: Neural	Principles of Learning, Perception, Development, Cognition and	Grossberg, S 82c
	Principles of Neurodynamics: Perceptrons and the Theory of Brain	Rosenblatt, F 62a
	Principles of Neurodynamics: Perceptrons and the Theory of Brain	Rosenblatt, F 61a
General	Principles of Operations in Neuron Nets with Application to.	Mueller, P 62a
	Principles of Self-Organization.	von Foerster, H 62a
From Basic Network	Principles to Neural Architecture: Emergence of Orientation Columns.	Linsker, R 86a
From Basic Network	Principles to Neural Architecture: Emergence of Orientation.	Linsker, R 86b
From Basic Network	Principles to Neural Architecture: Emergence of Spatial-Opponent	Linsker, R 86c
Application of New Artificial Neural System Information Processing	Principles to Pattern Classification: Final Report, US Army.	Myers, MH 87a
Application of Orthogonal Projection	Principles to Simultaneous Phonemic Segmentation and Labeling of	Jalanko, M 78a
	Principles Underlying the Representation and Storage of Information.	Rolls, ET 89a
	Prior Probabilities.	Jaynes, ET 68a
Automated Reasoning on Neural Networks: A	Probabilistic Approach	Chen, S 87a
Faster Learning Through a	Probabilistic Approximation Algorithm.	Kolen, JF 88a
Adaptive Global Search in a Time Varying Environment Using a	Probabilistic Automaton with Pattern Recognition Supervision	Jarvis, RA 70a
	Probabilistic Characterization of Neural Model Computations.	Golden, RM 88c
The Neural Path	Probabilistic Delay Model.	Niznik, CA 83a
An Introduction to the Application of the Theory of	Probabilistic Functions of a Markov Process to Automatic Speech.	Levinson, SE 83a
A Maximization Technique Occurring in the Statistical Analysis of	Probabilistic Functions of Markov Chains	Baum, LE 70a
	Probabilistic Logics and the Synthesis of Reliable Organisms from.	von Neumann, J 56a
The Perceptron: A	Probabilistic Model for Information Storage and Organization in the.	Rosenblatt, F 58b
The Perceptron: A	Probabilistic Model for Information Storage and Organization in the.	Rosenblatt, F 58a
Spatial Organization of Neural Networks: A	Probabilistic Modeling Approach.	Stafylopatis, A 88a
...System of Nonlinear Differential Equations Describing a Flow on a	Probabilistic Network.	Grossberg, S 69f
	Probabilistic Neural Networks for Classification, Mapping, or.	Specht, DF 88a
A New	Probabilistic Relaxation Scheme.	Peleg, S 80a
On the	Probabilistic Semantics of Connectionist Networks.	Geffner, H 87a
Some Notes on the	Probabilistic Semantics of Logistic Function Parameters in Neural.	Davies, TR 88a
	Probabilistic Solution of Inverse Problems	Marroquin, JL 85a
Rapid Speaker Adaptation Using a	Probabilistic Spectral Mapping	Schwartz, RM 87a
A Neural Network Based on Co-Occurrence	Probabilities.	Collins, JW 87a
Prior	Probabilities.	Jaynes, ET 68a
Research and Theory on the Learning of	Probabilities.	Estes, WK 72a
The Description of a Random Field by Means of Conditional	Probabilities and Conditions of Its Regularity	Dobruschin, PL 68a
Upper and Lower	Probabilities Induced by a Multi-Valued Mapping.	Dempster, AP 67a
Theory of Learning with Constant, Variable, or Contingent	Probabilities of Reinforcement	Estes, WK 57a
How to Do with	Probabilities What People Say You Can't.	Pearl, J 85b
Fuzziness Versus	Probability.	Kosko, B 87f
In Defense of	Probability.	Cheeseman, P 85a
Maximum Entropy and Conditional	Probability.	van Campenhout, JM 81a
The	Probability Approach to the Treatment of Uncertainty in Artificial	Lindley, DV 87a
The Design of Conditional	Probability Computers.	Uttley, AM 59b
Conditional	Probability Computing in a Nervous System.	Uttley, AM 59a
Non-Parametric	Probability Density Estimation	Tapia, RA 78a
On Estimation of a	Probability Density Function and Mode.	Parzen, E 62a
Artificial Neural Nets: Dependence of the EEG Amplitude's	Probability Distribution on Statistical Parameters	Anninos, PA 83b
Supervised Learning of	Probability Distributions by Neural Networks	Baum, EB 88d
Learning Algorithms and	Probability Distributions in Feed-Forward and Feed-Back Networks	Hopfield, JJ 87a
	Probability, Frequency, and Reasonable Expectations.	Cox, RT 46a
Search Behavior: A Correction Procedure for Three-Choice	Probability Learning	Cole, M 65a
Distinctive Features, Categorical Perception, and	Probability Learning: Some Applications of a Neural Model.	Anderson, JA 77b

Multi-Resolution Image	Processing & Analysis.	Rosenfeld, A 84a
EXCLUSIVE OR	Processing (Binary Image Subtraction) Using Thick Fourier Holograms.	Guest, CC 84a
Tactile Information	Processing: A Back-Propagation Model of Pattern Recognition in the .	Grajski, KA 88c
Explorations in Parallel Distributed	Processing: A Handbook of Models, Programs, and Exercises.	McClelland, JL 88a
Statistical Mechanics of Visual	Processing: A Model for Layer IVc of the Primary Visual Cortex of.	Moller, P 87a
Symbolic NeuroEngineering for Natural Language	Processing: A Multilevel Research Approach	Dyer, MG 88a
Neural Networks for Medical Image	Processing: A Study of Feature Identification.	Dayhoff, RE 88a
The Evolution of Data	Processing Abilities in Competing Automata	Bergman, A 86a
Where Does the Fundamental Information	Processing Ability of Biological Systems Come From?.	Kohonen, T 75a
The Neurophysiology of Information	Processing and Cognition	John, ER 78a
On Tropistic	Processing and Its Applications.	Fernandez, MF 88a
Real Time Learned Sensor	Processing and Motor Control for a Robot with Vision	Miller, WT 88b
Intelligent Recognition and Learning Control: Integrating Intelligent	Processing and Neural Networks for Adaptive Moving Object Tracking .	Lee, JSJ 87a
Neural Networks for Speech	Processing and Recognition	Shamma, SA 87a
Parallel Distributed	Processing and Role Assignment Constraints	McClelland, JL 86g
Information	Processing and Storage Based on Transforms and Neural Networks	Ersoy, OK 87a
Image	Processing and Understanding	Habibi, A 86a
...Digital Signal Processors with the Explorer / Odyssey for Image	Processing and Understanding Applications.	Gove, RJ 87a
Serial Order: A Parallel Distributed	Processing Approach.	Jordan, MI 86c
An Image	Processing Architecture for Real Time Generation of Scale and.	Messner, RA 85a
Acoustooptic Algebraic	Processing Architectures	Rhodes, WT 84a
Real-Time Visual Computations Using Analog CMOS	Processing Arrays.	Sivilotti, MA 87a
Real-time Visual Computations Using Analog CMOS	Processing Arrays.	Sivilotti, MA 87b
Neural Network	Processing as a Tool for Function Optimization	Jeffrey, W 86a
Mechanisms of Sentence	Processing: Assigning Roles to Constituents.	McClelland, JL 86f
Optical Information	Processing Based on an Associative Memory Model of Neural Nets with.	Psaltis, D 85a
A New Approach to Optical Information	Processing Based on the Hopfield Model	Psaltis, D 84a
Space or Time Adaptive Signal	Processing by Neural Network Models.	Herault, J 86a
Pulse	Processing by Neuron Models.	Jenik, F 64a
Speech	Processing by Splicing of Autocorrelation Function	Suzuki, J 76a
Investigations into the Pattern	Processing Capabilities of Associative Nets.	Lansner, A 86a
Optical Information	Processing Characteristics of the Microchannel Spatial Light	Warde, C 81a
Neural Computation Research for Vision - Simulation on a Parallel	Processing Computer.	Miyake, S 88a
Parallel Architectures for Image	Processing, Computer Vision, and Pattern Perception.	Uhr, L 86a
	Processing Determinants of Reading Speed	Jackson, MD 79a
Globally Connected Network Models for Computing Using Fine-Grained	Processing Elements.	Szu, HH 85a
Attentional Mechanisms in Neural Information	Processing: Examples From Pavlovian Conditioning	Grossberg, S 87f
Parallel Distributed	Processing: Explorations in the Microstructure of Cognition.	Rumelhart, DE 86h
Parallel	Processing for Associative and Neuronal Networks	Palm, G 84b
Comparison of Neural Network and Matched Filter	Processing for Detecting Lines in Images	Grant, PM 86b
A Comparison of Nonlinear Associative Memory and Matched-Filter	Processing for Detecting Lines in Optical Images	Grant, PM 86a
Bayesian Image	Processing, Gibbs Distributions, and Neural Networks	Vold, TG 87a
A Psychologist Views Human	Processing: Human Errors and Other Phenomena Suggest Processing.	Norman, DA 81a
Parallel Distributed	Processing: Implications for Cognition and Development	McClelland, JL 88b
Rule-Based	Processing in a Connectionist System for Natural Language.	Selman, B 85b
Stochastic	Processing in a Neural Network Application	Nguyen, DD 87a
Volume Holographic Information	Processing in a Phase-Conjugating Resonator.	Cohen, MS 86a
Possibilities for Signal	Processing in Axon Systems	Crane, HD 64a
"Ignorance-Based" Neural-Network Signal	Processing in Brain Research	Gevins, AS 87a
Applications of Neural-Network (NN) Signal	Processing in Brain Research	Gevins, AS 88a
Information	Processing in Dynamical Systems: Foundations of Harmony Theory	Smolensky, P 86a
The Neuroanatomical Substrate for Information	Processing in Macular Endorgans.	Ross, MD 88c
Information	Processing in Molecular Systems.	Conrad, M 72a
Parallel Versus Serial	Processing in Rapid Pattern Discrimination	Bergen, JR 83a
Early Parallel	Processing in Reading: A Connectionist Approach.	Mozer, MC 86a
The Role of Attention and Active	Processing in Speech Perception.	Nusbaum, HC 86a
Models of Information	Processing in the Brain.	Anderson, JA 81b
Molecular Information	Processing in the Central Nervous System, Part I	Conrad, M 74b
Information	Processing in the Nervous System	Kristan, WB 80a
Distributed Neural Information	Processing in the Vestibulo-Ocular System.	Lau, CGY 88a
	Processing Interactions and Lexical Access During Word Recognition .	Marslen-Wilson, WD 78a
...Views Human Processing: Human Errors and Other Phenomena Suggest	Processing Mechanisms.	Norman, DA 81a
A Parallel	Processing Model for the Study of Possible Pattern Recognition	Muller, FJ 76a
On Language and Connectionism: Analysis of a Parallel Distributed	Processing Model of Language Acquisition	Pinker, S 88b
On the Control of Automatic Processes: A Parallel Distributed	Processing Model of the Stroop Effect.	Cohen, JC 88a
	Processing of Continuous Lines and Edges by the Visual System.	Hartmann, G 83a
Comparison of Internal	Processing of Different Network Paradigms.	Caudill, M 88a
	Processing of Expected and Unexpected Events During Conditioning	Grossberg, S 86n
The	Processing of Expected and Unexpected Events During Conditioning	Grossberg, S 82d

Associative Memories and	Processors: An Overview and Selected Bibliography	Parhami, B 73a
Neurocomputer Workstations and	Processors: Approaches and Applications	Kuczewski, RM 87e
Neurocomputer Workstations and	Processors: Approaches and Applications	Kuczewski, RM 87f
Interfacing Multispectral Sensors to Real Time	Processors Based on Neural Network Models	Bandara, KMSV 87a
Integration of Symbolic and Multiple Digital Signal	Processors with the Explorer / Odyssey for Image Processing and	Gove, RJ 87a
Perceptual Grouping	Produced by Changes in Orientation and Shape	Beck, J 66b
Optical Matrix-Matrix Multiplication Based Outer	Product Decomposition	Athale, RA 82a
Three Layers of Vector Outer	Product Neural Networks for Optical Pattern Recognition	Szu, HH 86h
...in Hierarchical Connectionist Models: Applications in Language	Production	Dell, GS 88a
On the	Production and Release of Chemical Transmitters and Related Topics	Grossberg, S 69g
Coarticulation in Recent Speech	Production Models	Kent, RD 77a
Neural Mechanisms for the	Production of Cyclic Motor Patterns	Selverston, AI 83a
A Distributed Connectionist	Production System	Touretzky, DS 86a
Parallelism in	Production Systems	Gupta, A 86a
Startup Nestor Readies Neural Network-Based	Products	Brown, C 86k
Convected Activation	Profiles and the Measurement of Visual Motion	Waxman, AM 88a
	Profiles: Marvin Minsky	Bernstein, J 81a
Neural Models: Networks for Fun and	Profit	Anderson, JA 86d
The Price of	Progammability	Conrad, M 88d
A Magnetic Integrator for the Perceptron	Program	Hawkins, JK 60a
An Endorsement-Based Plan Recognition	Program	Sullivan, M 85a
Speech Recognition: A Model and a	Program for Research	Halle, M 62a
An SAS	Program for Simplified GMDH Models	Prager, MH 84a
A Computer	Program for Simulating a Network of Interacting Neurons, Part I	Perkel, DH 76a
A Computer	Program for Simulating a Network of Interacting Neurons, Part II	Perkel, DH 76b
A Computer	Program for Simulating a Network of Interacting Neurons, Part III	Perkel, DH 76c
A FORTRAN	Program for the GMDH Algorithm	Farlow, SJ 84a
Neural Networks for Motor	Program Generation	Eckmiller, R 88a
Computational Model of the Motor	Program Generator for Pursuit	Eckmiller, R 87a
Motion Control of Robotic Manipulator Based on Motor	Program Learning	Arimoto, S 88a
A	Program Package Implementing a System of Interacting Adaptive	Lansner, A 83a
A Pattern Recognition	Program That Generates, Evaluates, and Adjusts to its Own Operators	Uhr, L 61a
A Game-Playing	Program that Learns by Analyzing Examples	Minton, S 85a
	Programmable Bistable Switches and Resistors for Neural Networks	Spencer, EG 86a
	Programmable Blackboard Model of Reading	McClelland, JL 86i
Implementing Neural Nets with	Programmable Logic	Vidal, JJ 88a
Implementing Neural Nets with	Programmable Logic	Vidal, JJ 87a
Resource Requirements of Standard and	Programmable Nets	McClelland, JL 86j
	Programmable Synaptic Chip for Electronic Neural Networks	Moopenn, A 88a
On the Methodology of Connectionist Research	Programmes in Vision	Gonzales, MEQ 87a
Image Compression by Back Propagation: An Example of Extensional	Programming	Cottrell, GW 87a
...Representations from Gray-Scale Images: An Example of Extensional	Programming	Cottrell, GW 87b
Neural Networks for Nonlinear	Programming	Kennedy, MP 88a
Prolegomena to Evolutionary	Programming	Conrad, M 88c
Workshop on Performance Efficient Parallel	Programming	Segall, Z 86a
	Programming A Digital Computer to Learn	Oettinger, AG 52a
	Programming a Massively Parallel, Computation Universal System	Lapedes, A 86a
Visual	Programming as a User Interfaace for Programming Neural Networks	Tenorio, MF 87j
...Program for Simulating a Network of Interacting Neurons, Part II:	Programming Aspects	Perkel, DH 76b
...Networks: An A/D Converter, Signal Decision Circuit, and Linear	Programming Circuit	Tank, DW 86a
Unifying the Tank and Hopfield Linear	Programming Circuit and the Canonical Nonlinear Programming Circuit	Kennedy, MP 87b
...Hopfield Linear Programming Circuit and the Canonical Nonlinear	Programming Circuit of Chua and Lin	Kennedy, MP 87b
Using A Visual	Programming Language to Simulate Adaptive Networks	Tenorio, MF 87f
	Programming Neural Nets to Do Spatial Computations	Zipser, D 86d
Visual Programming as a User Interfaace for	Programming Neural Networks	Tenorio, MF 87j
Integer Linear	Programming Neural Networks for Job-Shop Scheduling	Foo, YS 88a
	Programming Parallel Computers Using Energy Minimization Algorithms	Simmes, SD 87a
Putting Knowledge in its Place: A Scheme for	Programming Parallel Processing Structures on the Fly	McClelland, JL 85b
Putting Knowledge in its Place: A Scheme for	Programming Parallel Processing Structures on the Fly	McClelland, JL 88c
	Programming Pattern Recognition	Dinneen, GP 55a
Bifurcation Theory Methods for	Programming Static or Periodic Attractors and Their Bifurcations in	Baird, B 88a
Neural Network Mechanisms for Generation and Learning of Motor	Programs	Eckmiller, R 87b
...Aspects of a Neural Network Model for Generation of Pursuit Motor	Programs	Eckmiller, R 86b
Explaining PDP: A Handbook of Models,	Programs and Exercises	McClelland, JL 87a
...in Parallel Distributed Processing: A Handbook of Models,	Programs, and Exercises	McClelland, JL 88a
Coordinated Control	Programs for Movements of the Hand	Arbib, MA 85b
Neural Dynamics of Speech and Language Coding: Developmental	Programs, Perceptual Grouping, and Competition for Short-Term	Cohen, MA 86c
	Progress in Retinal Research	Osborne, NN 86a
	Progress in Robot Road-Following	Wallace, R 86a

Aesthetics,	Psychobiology, and Cognition	Martindale, C 88a
Cognitive and	Psychological Computation with Neural Models	Anderson, JA 83a
	Psychological Concepts in Parallel Systems	Anderson, JA 86e
Stable Matching with Preferences Derived From a	Psychological Model.	Bartholdi, JJ 86b
Some Physiological and Biochemical Consequences of	Psychological Postulates	Grossberg, S 68f
A	Psychologist Views Human Processing: Human Errors and Other.	Norman, DA 81a
Computer Simulation of Neural Networks for Perceptual	Psychology	Grossberg, S 86f
Neural Population Modeling and	Psychology: A Review	Levine, DS 83a
From Computational Theory to	Psychology and Neurophysiology	Marr, D 76c
Cognitive	Psychology: The Way We Were.	McClelland, JL 78a
Single Units and Sensation: A Neuron Doctrine for Perceptual	Psychology?.	Barlow, HB 72a
Some	Psychophysiological and Pharmacological Correlates of a.	Grossberg, S 86q
Some	Psychophysiological and Pharmacological Correlates of a.	Grossberg, S 84c
	Psychophysiological Substrates of Schedule Interaction and	Grossberg, S 81b
...and Unexpected Events During Conditioning and Attention: A	Psychophysiological Theory	Grossberg, S 86n
...and Unexpected Events During Conditioning and Attention: A	Psychophysiological Theory	Grossberg, S 82d
A	Psychophysiological Theory of Reinforcement, Drive, Motivation, and.	Grossberg, S 82a
	Psychophysiological Theory of Reinforcement, Drive, Motivation and	Grossberg, S 86o
Instability of	Pulse Activity in a Net with Threshold	Ashby, WR 62a
Stochastic	Pulse Frequency Modulation Models of Neural Coding Processes	Bruckstein, AM 81a
	Pulse Processing by Neuron Models.	Jenik, F 64a
	Pulse Sequences Generated by a Degenerate Analog Neuron Model.	Yoshizawa, S 82a
An Active	Pulse Transmission Line Simulating Nerve Axon.	Nagumo, J 62a
Alternative Knowledge Acquisition: Developing a	Pulse-Coded Neural Network	Dress, WB 87b
Asynchronous VLSI Neural Networks Using	Pulse-Stream Arithmetic.	Murray, AF 88b
...Oropharynx in the Vagal Lobe of the Channel Catfish, Ictalurus	punctatus.	Kanwal, JS 88b
	Punish / Reward: Learning With a Critic in Adaptive Threshold.	Widrow, B 73a
A Neural Theory of	Punishment and Avoidance, Part I: Qualitative Theory	Grossberg, S 72a
A Neural Theory of	Punishment and Avoidance, Part II: Quantitative Theory	Grossberg, S 72b
UCLA	PUNNS - A Neural Network Machine for Computer Vision	Gunger, D 87a
...Unit Simulation of a Realistic Cerebellar Network Model. II.	Purkinje Cell Activity within the Basic Circuit and Modified by.	Pellionisz, AJ 74a
Color Separation Using General	Purpose Computer Vision Algorithms	Walters, DKW 87c
An Adaptive Network that Flees	Pursuit.	Winter, CL 88a
Computational Model of the Motor Program Generator for	Pursuit.	Eckmiller, R 87a
...and Dynamical Aspects of a Neural Network Model for Generation of	Pursuit Motor Programs	Eckmiller, R 86b
Neural Control of Foveal	Pursuit Versus Saccadic Eye Movements in Primates - Single-Unit.	Eckmiller, R 83a
IEEE	Puts Neural Nets into Focus.	Johnson, RC 87b
	Putting Knowledge in its Place: A Scheme for Programming Parallel.	McClelland, JL 85b
	Putting Knowledge in its Place: A Scheme for Programming Parallel.	McClelland, JL 88c
Solving	Puzzles with a Connectionist Network Using a Hill-Climbing	Kawamoto, AH 86a
...Electrical Behavior of Cortical Neurons-Simulation of Hippocampal	Pyramidal Cells.	Borg-Graham, LJ 87a
Enhancement of Synaptic Potentials in Rabbit CA1	Pyramidal Neurons Following Classical Conditioning	LoTurco, JJ 88a
The	QBKG System: Generating Explanations from a Non-Discrete Knowledge	Berliner, HJ 82a
Ring Neural Network	qua a Generator of Rhythmic Oscillation with Period Control.	Tsutsumi, K 84b
Optical Implementation of	Quadratic Associative Memory with Outer-Product Storage.	Jang, J 88b
	Qualitative Analysis and Synthesis of a Class of Neural Networks	Li, JH 88a
	Qualitative Analysis of Neural Networks.	Michel, AN 88a
	Qualitative and Temporal Reasoning in Engine Behavior Analysis and	Dietz, WE 87a
A Neural Theory of Punishment and Avoidance, Part I:	Qualitative Theory	Grossberg, S 72a
Adaptive Strategies in Fly Vision: On Their Image-Processing	Qualities.	Zaagman, WH 83a
Spontaneous	Quantal Currents in a Central Neuron Match Predictions from.	Korn, H 87a
An Associative Memory Model Based on a	Quantification Method.	Kato, K 88a
A Computational Approach to Fuzzy	Quantifiers in Natural Languages	Zadeh, LA 83a
A Neural Theory of Punishment and Avoidance, Part II:	Quantitative Theory.	Grossberg, S 72b
	Quantitative Universality for a Class of Nonlinear Transformations	Feigenbaum, MJ 78a
...Mellin Transformation with Applications to Scaling of Physical	Quantities	Moses, HE 74a
Building a Hierarchy with Neural Networks: An Example-Image Vector	Quantization	Jackel, LD 87a
Speaker Adaptation Through Vector	Quantization	Shikano, K 86a
Learning Vector	Quantization for Pattern Recognition	Kohonen, T 86b
Vector	Quantization in Speech Coding.	Makhoul, J 85a
Vector	Quantization of Images Based Upon the Kohonen Self-Organizing.	Nasrabadi, NM 88a
Vector	Quantization Training by a Self-Organizing Neural Network.	Ryan, TW 88b
	Quantized Geometry of Visual Space: The Coherent Computation of.	Grossberg, S 86p
The	Quantized Geometry of Visual Space: The Coherent Computation of.	Grossberg, S 83b
	Quantum Monte Carlo.	Ceperley, D 86a
Design of	Quasi-Optimal Minimum-Time Controllers	Smith, FW 66b
Synthesis of	Quasi-Optimal Switching Surfaces by Means of Training Techniques	Mendel, JM 70c
Phase Transitions in	Quasirandom Neural Networks.	Kurten, KE 87a
A	Quasi-Topological Method for the Recognition of Line Patterns.	Sherman, H 60a

...Tectum as a Recurrent On-Center Off-Surround Neural Net with	Quenching Threshold.	Betts, B 88b
A Leading	Question	Wilks, RS 87a
A	Question of Levels: Comment on McClelland and Rumelhart.	Broadbent, D 85a
Sequential Connectionist Networks for Answering Simple	Questions about a Microworld	Allen, RB 88b
Open	Questions About Computation in Cerebral Cortex	Sejnowski, TJ 86e
Fast Component Threshold for Vestibular Nystagmus in the	Rabbit	Lau, CGY 87a
Enhancement of Synaptic Potentials in	Rabbit CA1 Pyramidal Neurons Following Classical Conditioning.	LoTurco, JJ 88a
Conditioning-Specific Biophysical Alterations in	Rabbit Hippocampus	Disterhoft, JF 88a
...Sutton-Barto Adaptive Element and Classical Conditioning of the	Rabbit Nictitating Membrane Response	Moore, JW 85a
...of Oscillating Network Model of Pattern Recognition in the	Rabbit Olefactory Bulb	Baird, B 86a
...Dynamics of Pattern Formation and Pattern Recognition in the	Rabbit Olfactory Bulb.	Baird, B 86b
...Attentional-Associative Network for Classical Conditioning of the	Rabbit's NMR	Schmajuk, NA 86a
...of Non-Associative and Associative Olfactory Learning in	Rabbits.	Grajski, KA 88b
Bifurcation Analysis of a Network Model of	Rabit Olfactory Bulb with Periodic Attractors Stored by a Sequence	Baird, B 87a
...Neural Networks and Their Application in Recognition of	Radar Targets.	Farhat, NH 86b
Neural Networks and	Radial Basis Functions in Classifying Static Speech Patterns	Niranjan, M 88a
Mimicking the Human Mind:	Radical New Computers Teach Themselves	Rogers, M 87a
	RAMBOT: A Connectionist Expert System that Learns by Example	Mozer, MC 86b
	RAMBOT: A Connectionist Expert System that Learns by Example	Mozer, MC 87a
	RAM-Unit Learning Networks	Milligan, DK 87a
Dynamics of	Random Boolean Networks.	Pellegrin, D 86a
	Random Cells: An Idea Whose Time Has Come and Gone... And Come	Gallant, SI 87b
...Incrementing Associative Networks when Operating in Initially	Random Connectivities.	Dobson, VG 87a
Some Averaging and Stability Results for	Random Differential Equations.	Geman, S 79a
Perception of	Random Dot Interference Patterns	Glass, L 73a
Cooperative Algorithms for Solving	Random Dot Stereograms	Szeliski, R 86a
Learning to Recognize Patterns in a	Random Environment	Abramson, N 62a
A Markov	Random Field Approach to Data Fusion and Colour Segmentation	Wright, WA 89a
The Description of a	Random Field by Means of Conditional Probabilities and Conditions.	Dobruschin, PL 68a
Markov	Random Field Texture Models.	Cross, GR 83a
An Introduction to Spatial Point Processes and Markov	Random Fields.	Isham, V 81a
Integration of Stereo Vision and Optical Flow Using Markov	Random Fields.	Clifford, SP 88a
Markov	Random Fields and Gibbs Ensembles.	Spitzer, F 71a
	Random Fields and Spatial Renewal Potentials	Demongeot, J 86a
The Use of Markov	Random Fields as Models of Texture	Hassner, M 80a
...Problem of Brownian Motion, and a General Theory of Gaussian	Random Functions	Levy, P 56a
...the Hippocampal Perforant Path-Dentate Projection. II. Effects of	Random Impulse Train Stimulation	Berger, TW 88a
	Random Logical Nets I.	Rozonder, LI 69a
	Random Logical Nets II	Rozonder, LI 69b
	Random Logical Nets III.	Rozonder, LI 69c
Invariant Cycles in the	Random Mapping of n Integers into Themselves. Comparison with.	Coste, J 86a
Pattern Separability of a Tightly Coupled	Random Nerve Net with Inhibitory Connections	Torioka, T 79b
Pattern Separating Functioning of Two Layered	Random Nerve Nets with Feedforward Inhibitory Connections.	Torioka, T 88a
Ignition Phenomena in	Random Nets.	Rapoport, A 52a
Learning in	Random Nets.	Minsky, ML 60a
Characteristics of	Random Nets of Analog Neuron-Like Elements	Amari, S 72a
	Random Nets of Analog Neurons.	Averbukh, DYA 69a
Functional Organization in	Random Networks.	Beurle, RL 62a
Storage and Manipulation of Information in	Random Networks.	Beurle, RL 62b
Further Consideration on Pattern Separability in a	Random Neural Net with Inhibitory Connections.	Torioka, T 80a
Pattern Separability in a	Random Neural Net with Inhibitory Connections.	Torioka, T 79a
Property of the Degree of Pattern Separation in a Two-Layer	Random Neural Net with Inhibitory Connections.	Torioka, T 80b
Classification of Binary	Random Patterns.	Abend, K 65a
	Random Problems.	Abu-Mostafa, YS 88c
...Adaptive Algorithms Based on the Methods of Steepest Descent and	Random Search.	Widrow, B 76a
Simulation and Evaluation of Ill Defined Neuronal Networks with	Random Structure and Parameter Setting	Benito, S 80a
...An Entropy-Based Measure of the Structural Similarity Between Two	Random Systems	Dawes, RL 88a
Gibbs and Markov	Random Systems with Constraints.	Moussouris, J 74a
...Hippocampal Perforant Path-Dentate Projection. III. Comparison of	Random Train and Paired Impulse Stimulation.	Berger, TW 88b
...on Foveal and Peripheral Detection of Coherent Motion in Moving	Random-Dot Patterns.	van de Grind, WA 87a
Perception of Depth Surfaces in	Random-Dot Stereograms: A Neural Model	Dev, P 75a
Solving	Random-Dot Stereograms Using the Heat Equation	Szeliski, R 85a
On the Properties of	Randomly Connected McCulloch-Pitts Networks: Differences Between	Dammasch, IE 84a
Some Properties of	Randomly Connected Networks of Neuron-Like Elements with Refractory.	Yoshizawa, S 74a
Some Properties of a	Randomly Connected Neural Network.	Allanson, JT 56a
A Dynamical Behaviour of Active Regions in	Randomly Connected Neural Networks	Okuda, M 74a
Characteristics of	Randomly Connected Threshold-Element Networks and Network Systems.	Amari, S 71a
A New Model for	Randomly Organized Associative Memory.	Kohonen, T 73a
Zero Temperature Parallel Dynamics for Infinite	Range Spin Glasses and Neural Networks	Gardner, E 87a

The	Recognition of Orca Calls with a Neural Network.	Taber, WR 88d
Computing with Connections in Visual	Recognition of Origami Objects	Sabbah, D 85a
Computing with Connections in Visual	Recognition of Origami Objects	Sabbah, D 88a
An Associative Memory Technique for the	Recognition of Patterns.	Pao, Y 78a
Neocognitron: Self-Organizing Network Capable of Position-Invariant	Recognition of Patterns.	Fukushima, K 80b
Neocognitron: Self-Organizing Network Capable of Position-Invariant	Recognition of Patterns.	Fukushima, K 80c
Nonsupervised Sequential Classification and	Recognition of Patterns.	Patrick, EA 66a
A Context Sensitive Line Finder for	Recognition of Polyhedra	Shirai, Y 73a
...of Two-Dimensional Neural Networks and Their Application in	Recognition of Radar Targets	Farhat, NH 86b
Visual	Recognition of Simple Objects by a Connection Network.	Plaut, DC 84a
The AUROS Project: Automatic	Recognition of Speakers by Computers	Bunge, E 77b
Computer	Recognition of Speech.	de Mori, R 86a
Experiments on Neural Net	Recognition of Spoken and Written Text	Burr, DJ 88b
On-Line	Recognition of Spoken Words from a Large Vocabulary.	Kohonen, T 84a
Representation and	Recognition of the Spatial Organization of Three-Dimensional Shapes.	Marr, D 78b
...and Contextual Information as Determinants of Tachistoscopic	Recognition of Words	Tulving, E 63a
Implementation of Neural Networks for Speech	Recognition on a Transputer Array.	Chong, MWH 88a
A Dynamic Connectionist Model for Phoneme	Recognition: Preliminary Results	Robinson, AJR 88a
Neural Dynamics of Adaptive Pattern	Recognition: Priming, Search, Attention, and Category Formation. . .	Carpenter, GA 85b
A Large VLSI Hopfield Network for Pattern	Recognition Problems	Verleysen, M 88c
A Review of Studies Applying Neural Net Models to Speech	Recognition Problems	Lippmann, RP 88a
A Technique for Determining and Coding Subclasses in Pattern	Recognition Problems	Mattson, RL 65a
...in Time: Analog Neural Networks with Applications to Speech	Recognition Problems	Tank, DW 87b
A Class of Non-Linear	Recognition Procedures	Chow, CK 66a
An Endorsement-Based Plan	Recognition Program.	Sullivan, M 85a
A Pattern	Recognition Program That Generates, Evaluates, and Adjusts to its. .	Uhr, L 61a
Some Neural Net	Recognition Results on Isolated Words.	Gold, B 87a
A Template Matching Model for Pattern	Recognition: Self-Organization of Templates and Template Matching. .	Hirai, Y 80a
Syntactic Pattern	Recognition: Stochastic Languages.	Thomason, MG 86a
Perceptual Flexibility in Word	Recognition: Strategies Affect Orthographic Computation But Not. . .	Carr, TH 78a
Neural Dynamics of Category Learning and	Recognition: Structural Invariants, Evoked Potentials, and	Carpenter, GA 86k
...Varying Environment Using a Probabilistic Automaton with Pattern	Recognition Supervision.	Jarvis, RA 70a
An Experimental Investigation of a Class of Pattern	Recognition Synthesis Algorithms	Mattson, RL 63a
Design of a Highly Parallel Visual	Recognition System	Sabbah, D 81a
Self-Corrective Character	Recognition System	Nagy, G 66a
A Thousand-Word	Recognition System Based on the Learning Subspace Method and	Kohonen, T 80a
	Recognition System Relies on Neurals	Sakurai, M 88b
...with Implications for the Design of On Line Character	Recognition Systems.	Ward, JR 88a
On the Effectiveness of Receptors in	Recognition Systems.	Marill, T 63a
Optimization of Reference Signals for Character	Recognition Systems.	Flores, I 60a
Reinforcement-Learning Control and Pattern	Recognition Systems.	Mendel, JM 70b
The "Whirling Dervish", A Simulation Study in Learning and	Recognition Systems.	Hoffman, A 62a
The Characteristic Selection Problem in	Recognition Systems.	Lewis, PM 62a
Adaptive Pattern	Recognition Systems and Boltzmann Machines: A Rapprochement.	Aleksander, I 87a
Creation of Efficient	Recognition Systems and the Problems of Microelectronics	Elinson, MI 81a
Adaptation, Learning, and Pattern	Recognition Systems: Theory and Applications	Mendel, JM 70a
Two-Dimensional Adaptive Model of a Human Controller Using Pattern	Recognition Techniques	Gilstad, DW 71a
Priming in Item	Recognition: The Organization of Propositions in Memory for Text . .	McKoon, G 80a
Large-Vocabulary Speaker-Independent Continuous Speech	Recognition: The SPHINX System	Lee, K 88a
	Recognition Time for Information Stored in Long-Term Memory.	Juola, JF 71a
Neocognitron: A New Algorithm for Pattern	Recognition Tolerant of Deformation and Shifts in Position	Fukushima, K 82a
Speech	Recognition: Turning Theory into Practice.	Doddington, GR 81a
...A Self-Organizing Neural Network Model for a Mechanism of Pattern	Recognition Unaffected by Shift in Position.	Fukushima, K 80a
Shape	Recognition Using a CMAC Based Learning.	Glanz, FH 87a
Invariant Object	Recognition Using a Distributed Associative Memory	Wechsler, H 88a
Pattern	Recognition Using Artificial Neural Systems.	Gutschow, T 87a
Pattern	Recognition Using Associative Memories	Altman, JL 77a
Fault-Tolerant	Recognition Using DAM's.	Wechsler, H 87a
Speaker-Independent Continuous Speech	Recognition Using Hidden Markow Models	Lee, K 87a
An Approach to Pattern	Recognition Using Linear Threshold Devices	Mattson, RL 60a
A Framework for Speech	Recognition Using Neural Networks.	Lippmann, RP 87a
A Neural Network Model for Phoneme	Recognition Using the Generalized Delta Rule for Connection.	Watson, M 87a
Connectionist Speech	Recognition Using the Temporal Flow Model.	Watrous, RL 88a
Pattern	Recognition with an Adaptive Network	Roberts, LC 60a
Isolated Word	Recognition with an Artificial Neural Network.	Rogers, CH 87a
A Preliminary Study of the Phoneme	Recognition with Hardware Neural Network	Hoshimiya, N 82a
Speech	Recognition with Network-Based Classifiers	Hogg, T 88a
A Hierarchical System for Character	Recognition with Stochastic Knowledge Representation	Vlontzos, JA 88a
An Algorithm for Pattern	Recognition with VLSI Neural Networks.	Sirletti, B 88a

Context	Keyword	Author
...Annealing in Networks for Computing Possible Arrangements for	Red and Green Cones	Ahumada, AJ 87a
Refining and	Redefining the Back-Propagation Learning Rule for Connectionist	Samad, T 87c
Self-Organizing Process Based on Lateral Inhibition and Weight	Redistribution	Miikkulainen, R 87a
	Reduced Interconnection Neural Networks	Somani, AK 88b
Associatively	Reduced Withdrawal from Shadows in Hermissenda: A Direct Behavioral	Lederhendler, II 87a
Noise	Reduction in Fetal Electrocardiography	Hon, E 63a
...Memory from Incomplete Memorized Information Using Space	Reduction Method	Murakami, K 88a
On the Connectionist	Reduction of Conscious Rule Interpretation	Smolensky, P 87a
	Reduction of Visual Processes to a Common Algorithm	Carman, GJ 88a
	Reduction of Visual Processes to a Common Algorithm	Carman, GJ 87b
...of Connectionist Expert Systems for Problems Involving Noise and	Redundancy	Gallant, SI 87a
...A Distributed Model for Representing and Regularizing Motor	Redundancy	Mussa-Ivaldi, FA 88b
...on Arm Trajectory Formation: A Distributed Model of Motor	Redundancy	Mussa-Ivaldi, FA 88c
Parallelism and	Redundancy in Neural Networks	von Seelen, W 88a
...Recognition System Based on the Learning Subspace Method and	Redundant Hash Addressing	Kohonen, T 80a
...for the Recognition and Correction of Misspelt Words, Based on	Redundant Hash Addressing	Kohonen, T 78a
A	Redundant Hash Addressing Method Adapted for the Postprocessing and	Reuhkala, E 79a
Neural Networks Mediating Linearizable Dynamic	Redundant Sensori-Motor Reflexes Characterized by Minimum of	Daunicht, WJ 88b
The Dynamic	Re-Evaluation of Alternatives and the Emulation of Human Decision	Clema, JK 72a
A Parallel Computation that Assigns Canonical Object-Based Frames of	Reference	Hinton, GE 81a
Consistency and Variation in Spatial	Reference	Douglas, SA 87a
An Adaptive	Reference Monitor for Information System Security	Kellum, C 87a
Optimization of	Reference Signals for Character Recognition Systems	Flores, I 60a
Theory of Self-Organizing Nerve Nets with Special	Reference to Association and Concept Formation	Amari, S 79b
Feature-Extracting Network for Geometrical Patterns: A Design with	Reference to Visual Neural Systems	Fukushima, K 71a
Neural Network	Refinements and Extensions	Denker, JS 86a
	Refining and Redefining the Back-Propagation Learning Rule for	Samad, T 87c
...and Integration of Edges as Critical to the Perception of	Reflectance and Illumination	Gilchrist, AL 83a
Hill-Shading and the	Reflectance Map	Horn, BKP 81a
	Reflections on Cognition and Parallel Distributed Processing	Norman, DA 86a
...Movements, Extraocular Proprioception and the Vestibulo-Ocular	Reflex	Daunicht, WJ 88a
Innate and Learned Components in a Simple Visuo-Motor	Reflex	Hoffmann, KP 88a
Spatial Patterns of Visual Cortical Fast EEG During Conditioned	Reflex in a Rhesus Monkey	Freeman, WJ 87e
Simple Changes in	Reflex Threshold Cannot Explain All Aspects of Rapid Voluntary	Gielen, CCAM 86a
Conditional Probability Machines and Conditioned	Reflexes	Uttley, AM 56a
...Networks Mediating Linearizable Dynamic Redundant Sensori-Motor	Reflexes Characterized by Minimum of Hermitian Norm	Daunicht, WJ 88b
	Reflexive Associative Memories	Loos, H 88a
A Neural Network Computational Map Approach to	Reflexive Motor Control	Lane, SH 88a
...Analysis on Large Scale Networks on Neuronlike Elements with	Refractory	Aihara, K 80a
...of Randomly Connected Networks of Neuron-Like Elements with	Refractory	Yoshizawa, S 74a
Develpment of the Two-Part Pattern During	Regeneration of the Head in Hydra	Bode, PM 88a
Diffraction	Regimes of Phase Gratings	Jaaskelainen, T 87a
Decision Theory and Artificial Intelligence: I. A Semantics-Based	Region Analyzer	Feldman, JA 74a
Network	Regions: Alternatives to the Winner-Take-All Structure	Chun, HW 87c
Architecture for Fast Learning of Decision	Regions in Neural Networks	Hayduk, JW 88a
A Dynamical Behaviour of Active	Regions in Randomly Connected Neural Networks	Okuda, M 74a
Modularity and Interaction of Brain	Regions Underlying Visuomoter Coordination	Arbib, MA 85e
...after Using Back-propagation, Expert Systems, and Multiple-Linear	Regression	Irani, EA 88a
	Regression and The Moore-Penrose Pseudoinverse	Albert, A 72a
	Regression Without Models: Directions in the Search for Structure	Malone, JM 84a
G-Maximization: An Unsupervised Learning Procedure for Discovering	Regularities	Pearlmutter, BA 86a
Internal Representation of Universal	Regularities: A Challenge for Connectionism	Shepard, RN 89a
...Field by Means of Conditional Probabilities and Conditions of Its	Regularity	Dobruschin, PL 68a
Input-Output Analysis of an IPFM Neural Model: Effects of Spike	Regularity and Record Length	Sanderson, AC 80a
Orthographic	Regularity, Positional Frequency, and Visual Processing of Letter	Massaro, DW 79a
...for Preattentive Vision: Multiple Scale Segmentation and	Regularization	Grossberg, S 87d
Ill-Posed Problems and	Regularization Analysis in Early Vision	Poggio, TA 84b
Computational Vision and	Regularization Theory	Poggio, TA 85a
Kinematic Networks: A Distributed Model for Representing and	Regularizing Motor Redundancy	Mussa-Ivaldi, FA 88b
Comments on the Paper "Threshold	Regulation - A Key to the Understanding of the Combined Dynamics of	Palm, G 87b
	Regulation of Hermissenda K+ Channels by Cytoplasmic and Membrane	Alkon, DL 88b
Dress	Rehearsal for Real-Time Artificial Neural Networks	Michael, C 87a
...and Recognition: Structural Invariants, Evoked Potentials, and	Reinforcement	Carpenter, GA 86k
...Learning with Constant, Variable, or Contingent Probabilities of	Reinforcement	Estes, WK 57a
A Neural Model of Attention,	Reinforcement, and Discrimination Learning	Grossberg, S 75a
A Psychophysiological Theory of	Reinforcement, Drive, Motivation, and Attention	Grossberg, S 82a
Psychophysiological Theory of	Reinforcement, Drive, Motivation and Attention	Grossberg, S 86o
A Stochastic Algorithm for Learning Real-Valued Functions via	Reinforcement Feedback	Gullapalli, V 88a
...for Attentionally-Modulated Pavlovian Conditioning: Conditioned	Reinforcement, Inhibition, and Opponent Processing	Grossberg, S 87b

A Neural Net Model for the	Release of Prolactin During Pseudopregnancy in the Rat	Anninos, PA 87c
A Lifelike Model for Association	Relevance.	Post, PB 69a
The	Relevance of Mechanical Sensors for Neural Generation of Movements .	Daunicht, WJ 88c
The	Relevance of Mechanical Sensors for Neural Generation of Movements .	Daunicht, WJ 88d
...of a Simultaneous Functional Order in Nervous Systems I.	Relevance of Signal Covariances and Signal Coincidences.	Toet, A 87a
Nerve Cell Soma Model with High	Reliability and Low Power Consumption.	Fernandes, PML 80a
	Reliability and Speed of Recall in an Associative Network.	Lansner, A 85b
The Role of Chaos in	Reliable Information Processing.	Nicolis, JS 84a
Understanding Biological Computation:	Reliable Learning and Recognition.	Hogg, T 84a
Probabilistic Logics and the Synthesis of	Reliable Organisms from Unreliable Components.	von Neumann, J 56a
	Reliable, Trainable Networks for Computing and Control	Widrow, B 62c
Recognition System	Relies on Neurals.	Sakurai, M 88b
Time Decay of the	Remanent Magnetization in Spin-Glasses	Chamberlin, RV 84a
	Remarks About the "Singularity" of Nerve Cells and its Ontogenesis .	Changeux, J 85a
Some Networks That Can Learn,	Remember, and Reproduce Any Number of Complicated Space-Time	Grossberg, S 69j
Some Networks That Can Learn,	Remember, and Reproduce Any Number of Complicated Space-Time	Grossberg, S 70c
The Neurophysiology of	Remembering.	Pribram, KH 69a
	Remote Sensing	Swain, PH 86a
Random Fields and Spatial	Renewal Potentials	Demongeot, J 86a
Rapid	Reorganization of Adult Rat Motor Cortex Somatic Representation. . .	Sanes, JN 88a
Building Permanent Memory Codes: Codification and	Repetition Effects in Word Identification.	Salasoo, A 85a
Skill Learning and	Repetition Priming in Symmetry Detection: Parallel Studies of Human.	Cohen, NJ 86a
Synaptic Potential Summation and	Repetitive Firing Mechanisms: Input-Output Theory for the.	Calvin, WH 72a
A Model for Phasic and Tonic	Repetitively Firing Neuronal Encoders.	Fohlmeister, JF 73a
	Replica Symmetry Breaking and the Nature of the Spin Glass Phase . .	Mezard, M 84b
Research Note - Cortical Mapping and Perceptual Invariance: A	Reply to Cavanagh.	Schwartz, EL 83a
On Visual and Contextual Factors in Reading: A	Reply to Rayner and Slowiaczek	McClelland, JL 81d
Automatic Character Recognition: A State-of-the-Art	Report	Stevens, ME 61a
	Report Outlines Commercialization of Neural Nets	Schwartz, TJ 87e
...Processing Principles to Pattern Classification: Final	Report, US Army Research Office.	Myers, MH 87a
Learning to	Represent and Understand Locative Prepositional Phrases.	Cosic, C 88a
A Bipartite Connectionist Model to	Represent N-ary Boolean and Linear Constraints	Post, SD 87a
"I Lied About the Trees" Or, Defaults and Definitions in Knowledge	Representation	Brachman, RJ 85a
A Field Approach to Higher Order ANNs for Sensorial	Representation	Van Hulle, MM 88a
...System for Character Recognition with Stochastic Knowledge	Representation	Vlontzos, JA 88a
A Naturalistic Approach to Mental	Representation	Gonzales, MEQ 88a
Correlative Binding in Higher Order Network for Sensorial	Representation	Van Hulle, MM 88b
...Neural Networks for Solving Job-Shop Scheduling: Part 1. Problem	Representation	Foo, YS 88b
...System: Generating Explanations from a Non-Discrete Knowledge	Representation	Berliner, HJ 82a
Entropy Driven Artificial Neuronal Networks and Sensorial	Representation: A Proposal	Van Hulle, MM 88c
On Applying Associative Networks: An Approach for	Representation and Goal-Directed Learning.	Fisher, AD 87a
Neural	Representation and Neural Computation.	Churchland, PS 89a
	Representation and Recognition of the Spatial Organization of Three.	Marr, D 78b
Principles Underlying the	Representation and Storage of Information in Neuronal Networks in. .	Rolls, ET 89a
Associative Networks:	Representation and Use of Knowledge by Computers	Findler, NV 79b
A Distributed Connectionist	Representation for Concept Structures.	Touretzky, DS 87a
A	Representation for Natural Category Systems.	Peters, SL 87a
Interpolation Coding: A	Representation for Numbers in Neural Models.	Ballard, DH 87b
Adaptive	Representation for Pattern Recognition	Sears, RW 65a
An International	Representation for Solid Shape Based on the Topological Properties .	Koenderink, JJ 87a
A	Representation for Temporal Sequence and Duration in Massively . . .	Chun, HW 86a
An Investigation of Knowledge	Representation in a Neural Network	Healy, MJ 88a
Shape	Representation in Parallel Systems	Hinton, GE 81d
Shape	Representation in Parallel Systems	Hinton, GE 81e
...Coding and Short-Term Synaptic Plasticity: A Scheme for Knowledge	Representation in the Brain.	von der Malsburg, C 86b
Space-Time	Representation in the Brain: The Cerebellum as a Predictive Space. .	Pellionisz, AJ 82a
	Representation of Associated Data by Matrix Operators.	Kohonen, T 73b
The	Representation of Colors in the Cerebral Cortex.	Zeki, S 80a
Connectionist	Representation of Concepts	Feldman, JA 88c
Neural	Representation of Conceptual Knowledge	Feldman, JA 89a
Neural	Representation of Conceptual Knowledge	Feldman, JA 86a
Abstraction and	Representation of Continuous Variables in Connectionist Networks . .	Saund, E 86a
	Representation of Control Structures by a Model of an Associative. .	Hirai, Y 87b
	Representation of Events in Nerve Nets and Finite Automata	Kleene, SC 56a
Distributed Memory and the	Representation of General and Specific Information	McClelland, JL 85a
	Representation of Local Geometry in the Visual System.	Koenderink, JJ 87d
The	Representation of Location	Zipser, D 83a
The	Representation of Maps	Zipser, D 83b
Perceptual Organization and the	Representation of Natural Form	Pentland, AP 86a

RHO-Space: A Neural Network for the Detection and	Representation of Oriented Edges	Walters, DKW 87g
RHO-Space: A Neural Network for the Detection and	Representation of Oriented Edges	Walters, DKW 87h
Phonotopic Maps - Insightful	Representation of Phonological Features for Speech Recognition	Kohonen, T 84b
Neural Network	Representation of Sensor Graphs in Autonomous Robot Path Planning.	Jorgensen, CC 87a
	Representation of Sensory Information in Self-Organizing Feature	Kohonen, T 86c
	Representation of Sensory Information in Self-Organizing Feature	Kohonen, T 86d
On the Topological	Representation of Signals in Self-Organizing Nerve Fields.	Amari, S 87b
Associate Networks Underlying Problem-Solving:	Representation of the Process of Addition by a Model of an	Hirai, Y 87a
Internal	Representation of Universal Regularities: A Challenge for.	Shepard, RN 89a
The	Representation of Variables in Connectionist Networks.	Walters, DKW 87j
Rapid Reorganization of Adult Rat Motor Cortex Somatic	Representation Patterns after Motor Nerve Injury	Sanes, JN 88a
The Restricted Language Architecture of a Hybrid	Representation System.	Vilain, MB 85a
A Network that Induces a Distributed	Representation Useful for Encoding Kinship Relations	Hinton, GE 86a
Color	Representation Using Statistical Pattern Recognition	Parkkinen, J 87a
	Representational Axes and Temporal Cooperative Processes	Tsotsos, JK 87a
Distributed	Representations.	Hinton, GE 86b
Distributed	Representations.	Hinton, GE 84b
Hopfield Model with Complementary Binary	Representations.	Jong, T 88a
Integration of Distributed and Symbolic Knowledge	Representations.	Hutchison, WR 87a
Properties of Connectionist Variable	Representations.	Walters, DKW 87e
...Mapping Functions: Classification and Analysis of Connectionist	Representations.	Walters, DKW 87f
Revealing the Structure of NETtalk's Internal	Representations.	Rosenberg, CR 87a
Human and Computer Rules and	Representations are not Equivalent	Grossberg, S 80e
Learning	Representations by Back-Propagating Errors	Rumelhart, DE 86f
Learning	Representations by Back-Propagating Errors	Rumelhart, DE 86e
Learning Internal	Representations by Error Propagation	Rumelhart, DE 86c
Learning Internal	Representations by Error Propagation	Rumelhart, DE 86d
Learning Internal	Representations by Error Propagation	Rumelhart, DE 85b
Learning	Representations by Recirculation	Hinton, GE 88a
Internal	Representations for Associative Memory	Baum, EB 88a
Learning Internal	Representations from Gray-Scale Images: An Example of Extensional.	Cottrell, GW 87b
Encoding Input/Output	Representations in Connectionist Cognitive Systems	Miikkulainen, R 88a
The Problem of Internal	Representations in Neural Nets: Concepts, Implications and a	Tillery, SI 87a
In Defense of Segmental	Representations in Speech Perception	Pisoni, DB 81a
Learning Distributed	Representations of Concepts.	Hinton, GE 86d
Two-Dimensional Markov	Representations of Sampled Images.	Stuller, JA 76a
Cognitive	Representations of Semantic Categories	Rosch, E 75a
Biological vs Artificial	Representations of Variables in Fine-Grain Parallel Systems.	Walters, DKW 87b
Hierarchical Scene Structure	Representations to Facilitate Image Understanding.	Maren, AJ 88b
Forming Global	Representations with Extended Backpropagation.	Miikkulainen, R 88b
Selective Networks Capable of	Representative Transformations, Limited Generalizations, and	Edelman, GM 82a
A Connectionist Architecture for	Representing and Reasoning About Structured Knowledge.	Derthick, MA 87a
Kinematic Networks: A Distributed Model for	Representing and Regularizing Motor Redundancy	Mussa-Ivaldi, FA 88b
	Representing and Transforming Recursive Objects in a Neural Network.	Touretzky, DS 86c
NETL: A System for	Representing and Using Real-World Knowledge.	Fahlman, SE 79a
	Representing Causal Schemata in Connectionist Systems.	Golden, RM 86b
	Representing Conceptual Graphs for Parallel Processing	Lendaris, GG 88c
	Representing Conceptual Structures in a Neural Network	Touretzky, DS 87b
	Representing Implicit Knowledge.	Fahlman, SE 81b
A Framework for	Representing Knowledge	Minsky, ML 75a
	Representing Magnitude by Memory Resonance	den Uyl, MJ 86a
	Representing Neural Networks Models by Finite Automata	Kemke, C 88b
	Representing Visual Information.	Marr, D 77b
Some Networks That Can Learn, Remember, and	Reproduce Any Number of Complicated Space-Time Patterns, I	Grossberg, S 69j
Some Networks That Can Learn, Remember, and	Reproduce Any Number of Complicated Space-Time Patterns, II.	Grossberg, S 70c
Fuzzy	Requests: An Approach to Weighted Boolean Searches	Bookstein, A 80a
Network Generality, Traning Required, and Precision	Required	Denker, JS 88a
Network Generality, Traning	Required, and Precision Required	Denker, JS 88a
On the Size of Weights	Required for Linear-Input Switching Functions.	Myhill, J 61a
Memory	Requirements for Balanced Computer Architectures	Kung, HT 85b
Nonlinearity	Requirements for Correlation Based Associative Memories.	Marks II, RJ 88c
Resource	Requirements of Standard and Programmable Nets	McClelland, JL 86j
Why "Computing"	Requires Symbols	Pylyshyn, ZW 84a
An Estimate of the Complexity	Requisite in a Universal Decision Network.	Cameron, SH 60a
"Ignorance-Based" Neural-Network Signal Processing in Brain	Research	Gevins, AS 87a
Applications of Neural-Network (NN) Signal Processing in Brain	Research	Gevins, AS 88a
...A Statistical/Numerical Approach to Factory Automation and Brain	Research	Werbos, PJ 87a
Darwinian Evolution as a Paradigm for AI	Research	Fenanzo Jr, AJ 86a
Future Perspectives on Neural Network	Research	Szu, HH 87b

Progress in Retinal	Research	Osborne, NN 86a
Speech Recognition: A Model and a Program for	Research	Halle, M 62a
	Research and Theory on the Learning of Probabilities	Estes, WK 72a
...NeuroEngineering for Natural Language Processing: A Multilevel	Research Approach.	Dyer, MG 88a
Optical Systems	Research at Texas Tech University.	Krile, TF 86a
	Research Challenges and Opportunities in Electronic Neural Systems	Lau, CGY 87b
Neural Computation	Research for Vision - Simulation on a Parallel Processing Computer	Miyake, S 88a
Neural Networking	Research Gains New Life.	Kane, B 87a
Connectionism at the Philips	Research Laboratory Brussels	Wellekens, CJ 88a
	Research Note - Cortical Mapping and Perceptual Invariance: A Reply.	Schwartz, EL 83a
...the Human Mind: A New Approach to Building Computers Draws on the	Research of Neuroscientists.	Hammerstrom, D 87b
...Principles to Pattern Classification: Final Report, US Army	Research Office.	Myers, MH 87a
Some Results of	Research on Speech Perception.	Liberman, AM 57a
Scene Analysis	Research: Optical Pattern Recognition and Artificial Intelligence.	Casasent, DP 86b
On the Methodology of Connectionist	Research Programmes in Vision.	Gonzales, MEQ 87a
Neuroscience as a Possible	Research Standard for Neural Networks.	Morse, KG 87a
Speech Perception:	Research, Theory, and the Principal Issues	Pisoni, DB 86a
Neural Systems for Use as a	Research Tool.	Josin, GM 87b
How the Human Brain Thinks and Acts: Results of Latest	Research Using Neural Networks to Analyze Brain Signals.	Gevins, AS 87b
IBM	Research Yields Artificial Neural Net Workstation.	Schwartz, TJ 86a
B.U. Center	Researching Neural Nets.	Brown, C 86a
	Reseaux Neuronzux Pour Bases D'Images.	Cochet, Y 88a
Computing Motion Using Analog and Binary	Resistive Networks	Hutchinson, J 88a
Computing Motion Using	Resistive Networks	Koch, C 88b
	Resistive Synaptic Interconnects for Electronic Neural Networks.	Lamb, JL 87a
Programmable Bistable Switches and	Resistors for Neural Networks.	Spencer, EG 86a
Retinal Model with Adaptive Contrast Sensitivity and	Resolution	Brill, MH 87a
Uncertainty Relation for	Resolution in Space, Spatial Frequency, and Orientation Optimized.	Daugman, JG 85a
High	Resolution Micro-Fabrication and Neural Networks	Jackel, LD 86b
Dynamic Processes in the	Resolution of Lexical Ambiguity: PhD Thesis.	Kawamoto, AH 85b
How Sensory Maps Could Enhance	Resolution Through Ordered Arrangements of Broadly Tuned Receivers	Baldi, P 88c
A Learning Network	Resolving Multiple Match in Associative Memory	Hirai, Y 82a
Competitive Learning: From Interactive Activation to Adaptive	Resonance.	Grossberg, S 88a
Competitive Learning: From Interactive Activation to Adaptive	Resonance.	Grossberg, S 87g
...of Brightness Perception: Features, Boundaries, Diffusion and	Resonance.	Cohen, MA 86b
...of Brightness Perception: Features, Boundaries, Diffusion, and	Resonance.	Cohen, MA 84a
...Word Recognition and Recall: Attentional Priming, Learning, and	Resonance.	Grossberg, S 86l
...of Word Recognition and Recall: Attentional Priming, Learning and	Resonance.	Grossberg, S 86m
...of Binocular Form Perception: Filtering, Matching, Diffusion, and	Resonance.	Grossberg, S 83a
Representing Magnitude by Memory	Resonance.	den Uyl, MJ 86a
Variations on Adaptive	Resonance.	Ryan, TW 87b
The	Resonance Correlation Network.	Ryan, TW 88a
Adaptive	Resonance in Development, Perception, and Cognition.	Grossberg, S 81a
A Brief Introduction to the Operation of an Adaptive	Resonance Network.	Stork, DG 88a
Adpative	Resonance Networks	Carpenter, GA 87b
Adaptive	Resonance Theory: Stable Self-organization of Neural Recognition	Carpenter, GA 86d
Direct Perception or Adaptive	Resonance?	Grossberg, S 80b
Some Global Properties of Binocular	Resonances: Disparity Matching, Filling-In, and Figure-Ground.	Grossberg, S 84b
Some Global Properties of Binocular	Resonances: Disparity Matching, Filling-in, and Figure-Ground.	Cohen, MA 84b
...Based on Message-Bearing Optical Modes in Phase-Conjugate	Resonator.	Yariv, A 86a
Optical Holographic Associative Memory Using a Phase Conjugate	Resonator.	Dunning, GJ 86a
...Association, and Categorization in a Phase Conjugating	Resonator.	Cohen, MA 86d
Volume Holographic Information Processing in a Phase-Conjugating	Resonator.	Cohen, MS 86a
Associative Memory in a Phase Conjugate	Resonator Cavity Utilizing a Hologram.	Owechko, Y 86b
Associative Memory in a Phase Conjugate	Resonator Cavity Utilizing a Hologram.	Owechko, Y 86a
Optoelectronic	Resonator Neural Networks.	Owechko, Y 87b
Optical	Resonators and Neural Networks	Anderson, DZ 86b
	Resource Requirements of Standard and Programmable Nets.	McClelland, JL 86j
Context and the Allocation of	Resources in Word Recognition.	Larochelle, S 80a
Large-Scale Activity in Neural-Nets II: A Model for the Brainstem	Respiratory Oscillator	Feldman, JL 75a
...of the Peripheral Auditory System, Part IV, Application to	Response	Lachs, G 83a
Adaptive Timing in Neural Networks: The Conditioned	Response	Desmond, JE 88a
...and Classical Conditioning of the Rabbit Nictitating Membrane	Response	Moore, JW 85a
Control of the Immune	Response	Weisbuch, G 88a
...Visual Information: Correlation of EEG Response to Intracellular	Response	Anninos, PA 82a
Neural Network Models of an Escape	Response	Wheeler, BC 88a
Self-Organization of a Neural Network Which Gives Position-Invariant	Response	Fukushima, K 79c
...in Correlation with the Far-Field Recorded Vestibular-Evoked	Response	Elidan, J 89a
The Neural Generators of the Vestibular Evoked	Response	Elidan, J 87a

A Simplified Model of	Retinal Function with Parallel Architecture and Integrated Space	Johnson, K 86a
Physiological Engineering Model of the	Retinal Horizontal Cell Layer	Usui, S 88a
	Retinal Model with Adaptive Contrast Sensitivity and Resolution	Brill, MH 87a
Electrical Coupling of Photoreceptors in	Retinal Network Models	Marcelja, S 80a
Artina: A Flexible Computer Model of	Retinal Processes for Moving Images	Gawronski, RP 87a
Progress in	Retinal Research	Osborne, NN 86a
Neural Models of Brightness Perception and	Retinal Rivalry in Binocular Vision	Sugie, N 82a
	Retinal-Type Neuristor Sections	Ajmera, RC 80a
Concurrency Assurance in Vertebrate	Retinas	Loebner, EE 87a
The	Retinex Theory of Color Vision	Land, EH 77a
Seeing-More-Than-is-There: A Probe of	Retinoid Networks	Trehub, A 86a
...Establishment of Ordered Neural Mapping: Its Application to the	Retinotectal Connections	Willshaw, DJ 79a
An Overall Description of	Retinotopic Mapping in the Cat's Visual Cortex Areas 17, 18, and 19	Mallot, HA 85a
Formation of	Retinotopy and Columnar Microstructures by Self-Organization: A	Amari, S 85a
Descriptions: An Intermediate Stage in Memory	Retrieval	Norman, DA 79a
Extended Boolean Information	Retrieval	Salton, G 83a
Neural Network Design for Efficient Information	Retrieval	Personnaz, L 86e
Neural Network Design for Efficient Information	Retrieval	Personnaz, L 86d
Effects of Reinforcement on Knowledge	Retrieval and Evaluation	Leven, SJ 87a
Semantic Priming and	Retrieval from Lexical Memory: Evidence for Facilitatory and	Neely, JH 76a
Semantic Priming and	Retrieval from Lexical Memory: Roles of Inhibitionless Spreading	Neely, JH 77a
Exponential Storage and	Retrieval in Hierarchical Neural Networks	Willcox, CR 88a
Information Storage and	Retrieval in Spin-Glass Like Neural Networks	Personnaz, L 85a
	Retrieval of Information from Long-Term Memory	Anderson, JR 83a
A Theory for the Storage and	Retrieval of Item and Associative Information	Murdock Jr, BB 82a
...of Olfactory Cortex with Functional Implications for Storage and	Retrieval of Olfactory Information	Bower, JM 88a
...of Olfactory Cortex with Functional Implications for Storage and	Retrieval of Olfactory Information	Wilson, MA 87a
A Neural Network for the	Retrieval of Superimposed Connection Patterns	von der Malsburg, C 87a
	Retrieval Strategies for a Carousel Conveyor	Bartholdi, JJ 86a
A Model for a Weighted	Retrieval System	Buell, DA 81a
Distributed and Fault-Tolerant Computation for	Retrieval Tasks Using Distributed Associative Memories	Char, JM 88a
	Retrieval Time from Semantic Memory	Collins, AM 69a
Inductive Information	Retrieval Using Parallel Distributed Computation	Mozer, MC 84a
A New Optical Method of Storing and	Retrieving Information	van Heerden, PJ 63a
...Prediction Using Neural Networks: The Case of IBM Daily Stock	Returns	White, H 88a
Neural Populations	Revealed	Sejnowski, TJ 88c
Memory-Scanning: Mental Processes	Revealed by Reaction-Time Experiments	Sternberg, S 69a
	Revealing the Structure of NETtalk's Internal Representations	Rosenberg, CR 87a
The Properties of the Cerebello-Pontine	Reverberating Circuit	Tsukahara, N 72a
Contour Enhancement, Short-Term Memory, and Constancies in	Reverberating Neural Networks	Grossberg, S 73a
Studies in the Transformation and Storage of Patterns in	Reverberating Neural Networks	Levine, DS 75a
Decision Equation and	Reverberations	Caianiello, ER 66a
	Reverberations and Control of Neural Networks	Caianiello, ER 67a
An Evaluation of Cell Function in Relation to	Reverberatory Modes in Cellular Networks	Fairhurst, MC 80a
Synthesis of	Reverberatory Neural Networks	Aiello, A 70a
	Reverse Engineering the Brain	Stevens, JK 85a
A Neuro-Biological	Review	Sperling, G 87a
Book	Review	Carpenter, GA 86f
Book	Review	Hinton, GE 79a
Book	Review	Hecht-Nielsen, R 83a
Book	Review	Grajski, KA 88a
Coherent Optical Pattern Recognition: A	Review	Casasent, DP 85a
Neural Population Modeling and Psychology: A	Review	Levine, DS 83a
	Review	Anderson, JA 80a
The Elements of Single Neurons: A	Review	Lewis, ER 83a
The Iron Wire Model of the Neuron: A	Review	Lewis, ER 68c
The Oculomotor Control System: A	Review	Robinson, DA 68b
Self-Organizing Systems - A	Review and Commentary	Hawkins, JK 61a
Learning Control Systems: A	Review and Outlook	Fu, K 70a
A	Review of Adaptive Antennas	Widrow, B 79a
A	Review of Artificial Neural Systems II: Paradigms, Applications	Simpson, PK 87a
A	Review of Studies Applying Neural Net Models to Speech Recognition	Lippmann, RP 88a
On	Revised Algorithms of GMDH With Applications	Tamura, H 84a
Distributed	Revision of Composite Beliefs	Pearl, J 87a
...Models of Classical Conditioning: Variations in CS Effectiveness	Revisited	Schmajuk, NA 87a
How the Brain Works: The Next Great Scientific	Revolution	Hestenes, D 87a
How the Brain Works: The Next Great Scientific	Revolution	Hestenes, D 83a
The Cognitive	Revolution and Mind/Brain Issues	Pribram, KH 86a
A Dual Back-Propagation Scheme for Scalar	Reward Learning	Munro, PW 87a

Real Time Learned Sensor Processing and Motor Control for a	Robot with Vision.	Miller, WT 88b
Neural Net Architectures for	Robotic Applications	Kung, S 89a
Modeling and Mathematical Formalism for	Robotic Cerebellar Neural Network Pathway Information Measures	Niznik, CA 87a
...of Knowledge-Based System and Neural Network Techniques for	Robotic Control.	Handelman, DA 88a
Motion Control of	Robotic Manipulator Based on Motor Program Learning.	Arimoto, S 88a
A Neural Network Based Controller for	Robotic Manipulators	Kamangar, FA 88a
A Nonlinear Learning Controller for	Robotic Manipulators	Miller, WT 86a
Application of a General Learning Algorithm to the Control of	Robotic Manipulators	Miller, WT 87b
Sensor-Based Control of	Robotic Manipulators Using a General Learning Algorithm.	Miller, WT 87c
Computer Networking Capacity in	Robotic Neural Systems	Nizkik, CA 84a
A Learning Controller for Nonrepetitive	Robotic Operations	Miller, WT 87a
...Network Based Learning Control During High Speed Nonrepetitive	Robotic Operations	Miller, WT 88a
A Hierarchical Model for Voluntary Movement and Its Application to	Robotics	Kawato, M 87a
Brain Theory: Connecting Neurobiology to	Robotics	Pellionisz, AJ 83a
Brains, Behavior and	Robotics	Albus, JS 81a
...Neural Network Model for Voluntary Movement with Application to	Robotics	Kawato, M 88a
...and Neurocomputers. Generalized Coordinates in Neuroscience &	Robotics	Pellionisz, AJ 88a
The Brains of Men and Machines, Part 1: Biological Models for	Robotics	Kent, EW 78a
Neural-Type	Robotics: An Overview.	Chande, A 83a
A Measurement and Control Model for Adaptive	Robots	Albus, JS 80a
Concurrent Neuromorphic Algorithms for Autonomous	Robots	Barhen, J 88b
Hierarchical Control for	Robots, and Teleoperators.	Albus, JS 85a
Cognitive Control of Autonomous Mobile	Robots: Nested Hierarchical Intelligent Module	Meystel, A 86a
The Minds of	Robots, Sense Data, Memory Images, and Behavior in Conscious	Culbertson, JT 63a
Hierarchical Control of	Robots Using Microcomputers.	Barbera, AJ 79a
	Robots with a Lot of Nerve: Eye-hand Coordination May Help a Robot	Peterson, I 87a
Neural Nets to Arm NASA Space	Robots?.	Johnson, RC 87h
A Method for the Design of Stable Lateral Inhibition Networks that is	Robust in the Presence of Circuit Parasitics	Wyatt Jr, JL 88a
Stability Theorem for Lateral Inhibition Networks that is	Robust in the Presence of Circuit Parasitics	Standley, DL 88a
	Robust Tracking Novelty Filters Based on Linear Models	Kadar, I 87a
...Dynamics of EEG and Event-Related Potentials" by Elbert and	Rockstroh.	Palm, G 87b
Architecture of	Rod and Cone Circuits to the On-Beta Ganglion Cell	Sterling, P 88a
	Rod Bipolar Array in the Cat Retina: Pattern of Input from Rods and.	Freed, MA 87a
...of the Dark-Adapted Cat Retina: Functional Architecture of the	Rod-Cone Network	Smith, RG 86a
Rod Bipolar Array in the Cat Retina: Pattern of Input from	Rods and GABA-Accumulating Amacrine Cells.	Freed, MA 87a
Neural Net Startup Takes Teaching	Role	Brown, C 87h
Parallel Distributed Processing and	Role Assignment Constraints.	McClelland, JL 86g
Symbolic Schemata,	Role Binding, and the Evolution of Structure in Connectionist.	Dolan, CP 87a
Nonlinear Interactions in a Dendritic Tree: Localization, Timing, and	Role in Information Processing	Koch, C 83a
Four Neural Circuit Models and Their	Role in the Organization of Voluntary Movement	Hemami, H 83a
Relaxation and its	Role in Vision	Hinton, GE 80b
The	Role of Adaptive and Associative Circuits in Future Computer	Kohonen, T 88e
Memory Storage Capacity and the	Role of Assemblies in Neural Networks.	Shaw, GL 77a
The	Role of Attention and Active Processing in Speech Perception	Nusbaum, HC 86a
The	Role of Chaos in Reliable Information Processing	Nicolis, JS 84a
...Adaptation and Self-Regulation in Self-Organizing Networks: The	Role of Chemical Transducers	Grossberg, S 80f
The	Role of Familiar Units in Perception of Words and Nonwords	McClelland, JL 77b
	Role of Feedback in Visual Perception.	Unnikrishnan, KP 87a
The	Role of Fuzzy Logic in the Management of Uncertainty in Expert	Zadeh, LA 87f
The	Role of Fuzzy Logic in the Management of Uncertainty in Expert	Zadeh, LA 83b
The	Role of Learning in Sensory-Motor Control.	Grossberg, S 85e
The	Role of Learning in Sensory-Motor Control.	Grossberg, S 85d
The	Role of Orthographic and Phonotactic Rules in Perceiving Letter.	Spoehr, KT 75a
The	Role of Physical Constraints in Natural and Artificial Manipulation.	Morasso, P 86a
The Mathematical	Role of Self-Consistency in Parallel Computation	Smolensky, P 84b
Computing with Neural Networks: The	Role of Symmetry	Carpenter, GA 86g
The	Role of Syntactic Context in Word Recognition.	Goodman, GO 81a
A Computer Model for Learning Processes and the	Role of the Cerebral Commissures	Anninos, PA 84a
The	Role of the Critic in Learning Systems	Dietterich, TD 81a
The	Role of the Lexicon in Speech Perception	Samuel, AG 86a
Simulations Suggest Information Processing	Roles for the Diverse Currents in Hippocampal Neurons.	Borg-Graham, LJ 88a
Semantic Priming and Retrieval from Lexical Memory:	Roles of Inhibitionless Spreading Activation and Limited-Capacity.	Neely, JH 77a
Mechanisms of Sentence Processing: Assigning	Roles to Constituents.	McClelland, JL 86f
Position,	Rotation, and Scale Invariant Optical Correlation.	Casasent, DP 76a
An Artificial Neural System Design for	Rotation and Scale Invariant Pattern Recognition	Reber, WL 87a
Performance of Backpropagation for	Rotation Invariant Pattern Recognition	Yang, H 87a
...Processing Architecture for Real Time Generation of Scale and	Rotation Invariant Patterns.	Messner, RA 85a
A Connectionist Learning Model for 3-Dimensional Mental	Rotation, Zoom, and Pan.	Mel, BW 86a
Coupled	Rotators as a Model for Spontaneous Activity in Neural Networks.	Kowalski, JM 88a

	Seeing the Light: Optics in Nature, Photography, Color, Vision, and.	Stork, DG 85a
	Seeing-More-Than-is-There: A Probe of Retinoid Networks.	Trehub, A 86a
Perception of Letters in Words:	Seek Not and Ye Shall Find	Johnston, JC 74a
Goal	Seeking Components for Adaptive Intelligence: An Initial Assessment.	Barto, AG 81b
Simulation of the	Segmental Burst Generating Network for Locomotion in Lamprey	Grillner, S 88a
In Defense of	Segmental Representations in Speech Perception	Pisoni, DB 81a
A Connectionist Architecture for Computing Textural	Segmentation	Mesrobian, E 87a
A Markov Random Field Approach to Data Fusion and Colour	Segmentation	Wright, WA 89a
A Theory of Textural	Segmentation	Beck, J 83a
An Adaptive Computational Model for Texture	Segmentation	Caelli, TM 88a
An Ocular Dominance Column Map as a Data Structure for Stereo	Segmentation	Yeshurun, Y 87a
EEG Alpha Map Series: Brain Micro-States by Space-Oriented Adaptive	Segmentation	Lehmann, D 87a
Integration of Detector Responses for Texture	Segmentation	Lively, R 88a
...of Orthogonal Projection Principles to Simultaneous Phonemic	Segmentation and Labeling of Continuous Speech	Jalanko, M 78a
A Neural Network Architecture for Preattentive Vision: Multiple Scale	Segmentation and Regularization.	Grossberg, S 87d
An Adaptive Neural Net Approach to the	Segmentation of Mixed Gray-Level and Binary Pictures	Troudet, T 88a
	Segmentation of Natural Scenes	Hanson, AR 78a
Application of Novelty Filter to	Segmentation of Speech	Haltsonen, S 78a
Textural	Segmentation, Second-Order Statistics, and Textural Elements	Beck, J 83b
Real Time Noisy Image	Segmentation Using an Artificial Neural Network Model.	Tenorio, MF 87d
Image	Segmentation Using Simple Markov Field Models.	Hansen, FR 82a
Image	Segmentation with Neurocomputers	Bilbro, GL 88a
...of Perceptual Grouping: Textures, Boundaries, and Emergent	Segmentations.	Grossberg, S 86k
...of Perceptual Grouping: Textures, Boundaries, and Emergent	Segmentations.	Grossberg, S 85c
	Segmenting Intersecting and Incomplete Boundaries.	Krishnan, G 88a
Extracting and Labeling Boundary	Segments in Natural Scenes	Prager, JM 80a
Spontaneous and Information-Triggered	Segments of Series of Human Brain Electric Field Maps.	Lehmann, D 88a
Stereopsis Without Image	Segregation.	Kaufman, L 73a
Image	Segregation by Motion: Cortical Mechanism and Implementation in. . .	Orban, GA 88a
	Seismic and Underwater Acoustic Waveform Analysis.	Chen, CH 86a
A Methodology to	Select Appropriate Learning Rate Parameters in Feed-Forward.	Pabon, JA 87a
Associative Memories and Processors: An Overview and	Selected Bibliography.	Parhami, B 73a
Catastrophe Theory:	Selected Papers.	Zeeman, CC 77a
Fuzzy Sets and Applications:	Selected Papers by L A Zadeh	Yager, RR 87a
A Massively Parallel Model of Schema	Selection.	Chun, HW 87a
...s Topology Conserving Maps: Fluctuations, Stability and Dimension	Selection.	Ritter, H 88a
On an Extended Fisher Criterion for Feature	Selection.	Malina, W 81a
...Cortical Maps: A Computer Simulation Based on Neuronal Group	Selection.	Pearson, JC 87a
Predicted Squared Error: A Criterion for Automatic Model	Selection.	Barron, AR 84a
...Model of Computation: A Non-Committing Approach to Schema	Selection.	Chun, HW 85a
Self-Organization Through	Selection.	Steels, L 87a
Neuronal Group	Selection: A Basis for Categorization by the Nervous System.	Finkel, LH 88b
Creativity,	Selection, and Categorization.	Finkel, LH 88a
Feature	Selection and Extraction	Kittler, J 86a
Schema	Selection and Stochastic Inference in Modular Environments	Smolensky, P 83b
Schema	Selection and Stochastic Inference in Modular Environments	Smolensky, P 83c
Feature Generation and	Selection by a Layered Network of Reinforcement Learning Elements. .	Anderson, CW 82a
	Selection of Image Primitives for General-Purpose Visual Processing.	Walters, DKW 87i
The Characteristic	Selection Problem in Recognition Systems	Lewis, PM 62a
Computational Networks in Early Vision: From Orientation	Selection to Optical Flow.	Zucker, SW 88a
A Model of Schema	Selection Using Marker Passing and Connectionist Spreading	Chun, HW 87b
Simple	Selectionist Learning Rule for Neural Networks	Personnaz, L 86f
	Selective Adaptation of Linguistic Feature Detectors	Eimas, PD 73a
A Hierarchical Neural Network Model for	Selective Attention.	Fukushima, K 88a
A Neural Network Model for	Selective Attention.	Fukushima, K 87b
A Neural Network Model for	Selective Attention.	Fukushima, K 87a
A Neural Network Model for	Selective Attention in Visual Pattern Recognition.	Fukushima, K 86a
Neural Network Model for	Selective Attention in Visual Pattern Recognition and Associative. .	Fukushima, K 87c
Neural Network Model for	Selective Attention in Visual Pattern Recognition Associative. . . .	Fukushima, K 87d
...Energy Functions and the Emergence of Orientation	Selective Cortical Cells	Kammen, DM 88a
An Algorithm for a	Selective Nearest Neighbor Decision Rule	Ritter, GL 75a
	Selective Networks and Recognition Automata.	Reeke Jr, GN 84a
	Selective Networks Capable of Representative Transformations	Edelman, GM 82a
Recognition Automata Based on	Selective Neural Networks.	Reeke Jr, GN 86a
	Selective Neural Networks and Their Implications for Recognition . .	Reeke Jr, GN 87a
Categorization and	Selective Neurons.	Anderson, JA 81a
A Theory for the Development of Neuron Selectivity: Orientation	Selectivity and Binocular Interactions in the Visual Cortex.	Bienenstock, EL 82a
Neuronal Connections Underlying Orientation	Selectivity in Cat Visual Cortex	Ferster, D 87a
Functional Properties of Models for Direction	Selectivity in the Retina.	Grzywacz, N 87a

A Theory for the Development of Neuron	Selectivity: Orientation Selectivity and Binocular Interactions in	Bienenstock, EL 82a
Theory for the Development of Neuron	Selectivity: Orientation Specificity and Binocular Interaction in.	Bienenstock, EL 82b
Neuron	Selectivity: Single Neuron and Neuron Networks	Cooper, LN 85a
	Self Organization and Associative Memory	Kohonen, T 84c
	Self Organization, Association, and Categorization in a Phase.	Cohen, MA 86d
Dynamics and	Self Organization in One-Dimensional Arrays.	Tchuente, M 86a
	Self Organizing General Pattern Class Separator and Identifier	Cooper, LN 82a
	Self Organizing Mathematical Models: Non-Linear Evolution Equations.	Schatzman, M 86a
	Self Organizing Neural Networks for the Identification Problem	Tenorio, MF 88a
Bayesian Networks: A Model of	Self-Activated Memory for Evidential Reasoning	Pearl, J 85a
	Self-Adapting Back-Propagation	Devos, MR 88a
The Mathematical Role of	Self-Consistency in Parallel Computation	Smolensky, P 84b
	Self-Control in Neural Nets.	Josin, GM 77a
	Self-Correcting Translator Circuits.	Steinbuch, K 63b
	Self-Corrective Character Recognition System	Nagy, G 66a
	Self-Inhibition Alters Firing Patterns of Neurons in Aplysia Buccal.	White, RL 81a
Adaptive Neural Encoder Model with	Selfinhibition and Threshold Control	Zeevi, YY 81a
Learning by Statistical Cooperation of	Self-Interested Neuron-Like Computing Elements	Barto, AG 85c
Learning by Statistical Cooperation of	Self-Interested Neuron-Like Computing Elements	Barto, AG 85b
Game-Theoretic Cooperativity in Networks of	Self-Interested Units.	Barto, AG 86b
	Self-Learning Rules by Dynamically Expanding Context	Kohonen, T 87f
	Self-Learning - What Is It?.	Tsypkin, YZ 68a
How Patterned Neural Connections Can be Set Up by	Self-Organization.	Willshaw, DJ 76b
How Patterned Neural Connections Can be Set Up by	Self-Organization.	Willshaw, DJ 75a
Principles of	Self-Organization.	von Foerster, H 62a
Information and	Self-Organization: A Macroscopic Approach to Complex Systems	Haken, H 88a
Formation of Retinotopy and Columnar Microstructures by	Self-Organization: A Mathematical Model.	Amari, S 85a
	Self-Organization and Associative Memory, Second Edition	Kohonen, T 88c
Cognitive	Self-Organization and Neural Modularity.	Grossberg, S 85a
A Theory of Human Memory:	Self-Organization and Performance of Sensory-Motor Codes, Maps, and.	Grossberg, S 78a
Architectures and Methodologies for	Self-Organization and Stochastic Learning in Opto-Electronic	Farhat, NH 87a
	Self-Organization in a Perceptual Network.	Linsker, R 88a
Mathematical Theory of	Self-Organization in Neural Nets	Amari, S 88a
	Self-Organization of a Neural Network Which Gives Position	Fukushima, K 79c
	Self-Organization of Associative Database and Its Applications	Suzuji, H 88a
	Self-Organization of Cooperative and Competitive Networks in	Matsuo, K 87a
Theory of	Self-Organization of Cortical Maps	Tanaka, S 88a
Time-Sequential	Self-Organization of Hierarchical Neural Networks.	Silverman, RH 88a
Cooperative	Self-Organization of Multiple Neural Systems During Adaptive	Grossberg, S 86g
	Self-Organization of Neural Nets with Competitive and Cooperative.	Sawada, R 82a
	Self-Organization of Neural Networks	Clark, JW 84a
Discovering Order in Chaos: Stable	Self-Organization of Neural Recognition Codes.	Carpenter, GA 86h
Adaptive Resonance Theory: Stable	Self-organization of Neural Recognition Codes in Response to	Carpenter, GA 86d
	Self-organization of Orientation Sensitive Cells in the Striate.	von der Malsburg, C 73a
	Self-Organization of Orientation Sensitive Cells in the Striate.	von der Malsburg, C 73b
Stable	Self-Organization of Sensory Recognition Codes: Is Chaos Necessary?.	Grossberg, S 87o
Adaptive	Self-Organization of Serial Order in Behavior: Speech, Language.	Grossberg, S 86c
The Adaptive	Self-Organization of Serial Order in Behavior: Speech, Language.	Grossberg, S 86r
ART-2:	Self-Organization of Stable Category Recognition Codes for Analog.	Carpenter, GA 87c
A Template Matching Model for Pattern Recognition:	Self-Organization of Templates and Template Matching by a.	Hirai, Y 80a
	Self-Organization Through Selection.	Steels, L 87a
Exploration of Backward Error Propagation as a	Self-Organizational Structure.	Kuczewski, RM 87a
	Self-Organized Formation of Feature Maps	Kohonen, T 81b
A Simple Paradigm for the	Self-Organized Formation of Structured Feature Maps.	Kohonen, T 82a
	Self-organized Formation of Topologically Correct Feature Maps	Kohonen, T 82f
	Self-Organized Formation of Topologically Correct Feature Maps	Kohonen, T 82e
Exploration of Backward Error Propagation as a	Self-Organized Structure	Kuczewski, RM 87b
Invariant Pattern Recognition and Recall by an Attentive	Self-Organizing ART Architecture in a Nonstationary World.	Carpenter, GA 87d
Acceleration of Learning by Variable Generalization for on-Line	Self-Organizing Control.	Ersu, E 86a
	Self-Organizing Control of Stochastic Systems.	Saridis, GN 77a
Representation of Sensory Information in	Self-Organizing Feature Maps	Kohonen, T 86c
Vector Quantization of Images Based Upon the Kohonen	Self-Organizing Feature Maps	Nasrabadi, NM 88a
	Self-Organizing Feature Maps and Abstractions.	Kohonen, T 84d
Representation of Sensory Information in	Self-Organizing Feature Maps, and Relation of these Maps to.	Kohonen, T 86d
	Self-Organizing Feature Maps and the Travelling Salesman Problem	Angeniol, B 88a
Adaptive and	Self-Organizing Flight Control Systems	Ostgaard, MA 62a
Adaptive, Associative, and	Self-Organizing Functions in Neural Computing.	Kohonen, T 87a
A	Self-Organizing Logical System	Mattson, RL 59a
Notes on a	Self-Organizing Machine.	Geman, S 81a

Neural Patterns as Detection Units in	Sensory Systems.	Kalay, A 87a
The Action of the Corticofugal Pathway on	Sensory Thalamic Nuclei: A Hypothesis.	Koch, C 87b
A Task-Level Model of Distributed Computation for	Sensory-Based Control of Complex Robot Systems	Lyons, D 85a
A Theory of Human Memory: Self-Organization and Performance of	Sensory-Motor Codes, Maps, and Plans	Grossberg, S 78a
...Self-Organization of Multiple Neural Systems During Adaptive	Sensory-Motor Control.	Grossberg, S 86g
How Neural Networks Factor Problems of	Sensory-Motor Control.	Bullock, D 88a
Skin & Muscles: The Basic Wet-Ware for	Sensory-Motor Control.	De Rossi, D 88a
The Role of Learning in	Sensory-Motor Control.	Grossberg, S 85e
The Role of Learning in	Sensory-Motor Control.	Grossberg, S 85d
Neural Dynamics of Adaptive	Sensory-Motor Control: Ballistic Eye Movements	Grossberg, S 86h
A Computer Model of	Sentence Comprehension	Small, SL 83a
Learning and Applying Contextual Constraints in	Sentence Comprehension	St John, MF 88a
Connectionist Modeling of Syntactic Constraints on	Sentence Processing.	Parisi, D 87a
Mechanisms of	Sentence Processing: Assigning Roles to Constituents	McClelland, JL 86f
A Connectionist Network that Learns to Process Some (Very) Simple	Sentences.	Samad, T 87a
A Non-Stochastic Method for the Correction of	Sentences.	Venta, O 86a
...Learns Natural Language Grammar from Exposure to Natural Language	Sentences.	Hanson, SJ 87a
Reconstructive Memory for	Sentences: A PDP Approach.	McClelland, JL 87b
Learning to Understand	Sentences in a Connectionist Network	Nolfi, S 88a
N-Gram Driven Search for	Sentences in a Syntactic Network	Venta, O 86b
Generation of Simple	Sentences in English Using the Connectionist Model of Computation. .	Kalita, J 87a
A Test for Linear	Separability as Applied to Self-Organizing Machines.	Singleton, RC 62a
Further Consideration on Pattern	Separability in a Random Neural Net with Inhibitory Connections. . .	Torioka, T 80a
Pattern	Separability in a Random Neural Net with Inhibitory Connections. . .	Torioka, T 79a
Pattern	Separability of a Tightly Coupled Random Nerve Net with Inhibitory .	Torioka, T 79b
	Separating Figure from Ground with a Boltzmann Machine	Sejnowski, TJ 87g
	Separating Figure from Ground with a Parallel Network.	Kienker, PK 86a
Pattern	Separating Functioning of Two Layered Random Nerve Nets with	Torioka, T 88a
Perceptual Significance Hierarchy: A Computer Vision Theory for Color	Separation	Walters, DKW 87d
...to See Objects as Equidistant and Its Inverse Relation to Lateral	Separation	Gogel, WC 56a
Automated Fake Color	Separation: Combining Computer Vision and Computer Graphics.	Walters, DKW 87a
Monte Carlo Calculation of Phase	Separation in a Two-Dimensional Ising System	Flinn, PA 74a
Property of the Degree of Pattern	Separation in a Two-Layer Random Neural Net with Inhibitory.	Torioka, T 80b
The Case for Conceptual and Operational	Separation of Network Architectures and Learning Mechanisms.	White, H 88b
Color	Separation Using General Purpose Computer Vision Algorithms.	Walters, DKW 87c
Self Organizing General Pattern Class	Separator and Identifier	Cooper, LN 82a
The Ising Model and The Rudin Shapiro	Sequence	Allouche, JP 86a
A Representation for Temporal	Sequence and Duration in Massively Parallel Networks: Exploiting . .	Chun, HW 86a
Neural Network Implementation for Maximum-Likelihood	Sequence Estimation of Binary Signals in Gaussian Noise.	Provence, JD 87a
Storage of a Temporal Pattern	Sequence in a Network.	Willwacher, G 82a
...of Rabit Olfactory Bulb with Periodic Attractors Stored by a	Sequence Learning Algorithm.	Baird, B 87a
Simulation Studies of a Temporal	Sequence Memory Model.	Stanley, JC 76a
Two-Dimensional Pattern	Sequence Prediction Using High-Order Neural Networks	Collins, DH 88a
Back-Propagation Networks for Event-Driven Temporal	Sequence Processing.	Shimohara, K 88b
Hierarchical Control of Rapid Movement	Sequences.	Rosenbaum, DA 83a
Kalman Filter-Based Algorithms for Estimating Depth from Image	Sequences.	Matthies, L 87a
Recurrent Nets for the Storage of Cyclic	Sequences.	Wolf, L 87b
Familiarity of Letter	Sequences and Tachistoscopic Identification.	Miller, GA 54a
Learning Patterns and Pattern	Sequences by Self-Organizing Nets of Threshold Elements.	Amari, S 72b
Pulse	Sequences Generated by a Degenerate Analog Neuron Model.	Yoshizawa, S 82a
Cyclic	Sequences Generated by Neuron Model.	Wada, M 82a
Capacity for Patterns and	Sequences in Kanerva's SDM as Compared to Other Associative Memory .	Keeler, JD 88a
Organization of Action	Sequences in Motor Learning: A Connectionist Approach.	Miyata, Y 87a
Processing of Temporal	Sequences in Neural Networks	Gutfreund, H 88b
Storing	Sequences of Biased Patterns in Neural Networks with Stochastic. . .	Buhmann, J 88a
Experiments with	Sequential Associative Memories.	Gallant, SI 88d
	Sequential Associative Memories.	Gallant, SI 87c
	Sequential Behavior and Stability Properties of Enzymatic Neuron . .	Kampfner, RR 83b
Nonsupervised	Sequential Classification and Recognition of Patterns.	Patrick, EA 66a
	Sequential Connectionist Networks for Answering Simple Questions . .	Allen, RB 88b
A Novel Net that Learns	Sequential Decision Process.	Sun, GZ 88a
Attractor Dynamics and Parallelism in a Connectionist	Sequential Machine	Jordan, MI 86a
	Sequential Processing Using Attractor Transitions.	Bell, T 88a
An N-Player	Sequential Stochastic Game with Identical Payoffs.	Narendra, KS 83a
Encoding	Sequential Structure in Simple Recurrent Networks.	Servan-Schreiber, D 88a
Schemata and	Sequential Thought Processes in PDP Models	Rumelhart, DE 86j
Feedback-Induced	Sequentiality in Neural Networks	Norrod, FE 87a
Behavioral Contrast in Short Term Memory:	Serial Binary Memory Models or Parallel Continuous Memory Models?. .	Grossberg, S 78c
Spiking Threshold and Overarousal Effects in	Serial Learning.	Grossberg, S 71c

Sequential Connectionist Networks for Answering	Simple Questions about a Microworld.	Allen, RB 88b
Encoding Sequential Structure in	Simple Recurrent Networks.	Servan-Schreiber, D 88a
	Simple Selectionist Learning Rule for Neural Networks.	Personnaz, L 86f
Analysis of a	Simple Self-Organizing Process	Kohonen, T 82b
A Connectionist Network that Learns to Process Some (Very)	Simple Sentences	Samad, T 87a
Generation of	Simple Sentences in English Using the Connectionist Model of	Kalita, J 87a
A Connectionist Model for	Simple Visual Functions.	Mel, BW 86b
Innate and Learned Components in a	Simple Visuo-Motor Reflex.	Hoffmann, KP 88a
A	Simple VLSI Architecture for Neurocomputing.	Pacheco, M 88a
A	Simple VLSI Connectionist Architecture	Cleary, JG 87a
	Simplest Spin Glass.	Gross, DJ 84a
In Back-to-Basics Movement, VLSI Expert Calls for More	Simplicity in Computation.	Brown, C 87e
An SAS Program for	Simplified GMDH Models	Prager, MH 84a
A	Simplified Model of Retinal Function with Parallel Architecture and.	Johnson, K 86a
...Network Model to Analyze the Information Transfer Properties of a	Simplified Model of the Cerebellar Cortex.	Axelrad, H 87a
A	Simplified Version of Kunihiko Fukushima's Neocognitron.	Deutsch, S 81a
Using A Visual Programming Language to	Simulate Adaptive Networks	Tenorio, MF 87f
Binocular Fusion Using	Simulated Annealing.	Landa, J 87a
Fast	Simulated Annealing.	Szu, HH 87a
Fast	Simulated Annealing.	Szu, HH 86c
Nonconvex Optimization by Fast	Simulated Annealing.	Szu, HH 87e
Nonconvex Optimization by Fast	Simulated Annealing.	Szu, HH 87f
Optimization by	Simulated Annealing.	Kirkpatrick, S 83a
Optimization by	Simulated Annealing.	Kirkpatrick, S 83b
...and Methodologies for Implementing Fast Stochastic Learning by	Simulated Annealing.	Farhat, NH 87b
	Simulated Annealing.	McBeth, JH 87a
Training of HMM Recognizers by	Simulated Annealing.	Douglas, P 85a
	Simulated Annealing in Networks for Computing Possible Arrangements.	Ahumada, AJ 87a
Recurrent Collateral Inhibition	Simulated in a Simple Neuronal Automata Assembly	Axelrad, H 86a
Rhythmic Activity in a	Simulated Neuronal Network	Andersen, P 66a
A Computer Program for	Simulating a Network of Interacting Neurons, Part I: Organization. .	Perkel, DH 76a
A Computer Program for	Simulating a Network of Interacting Neurons, Part II: Programming. .	Perkel, DH 76b
A Computer Program for	Simulating a Network of Interacting Neurons, Part III: Applications.	Perkel, DH 76c
	Simulating a Skilled Typist: A Study of Skilled Cognitive-Motor. . .	Rumelhart, DE 82b
	Simulating Memory by Numbers	Maddox, J 87a
An Active Pulse Transmission Line	Simulating Nerve Axon.	Nagumo, J 62a
	Simulating Nerves to Improve Cochlear Implants for the Hearing . . .	Tenorio, MF 87e
The Electronic Spreadsheet as a Tool in Modeling and	Simulating Neural Networks	Hewett, TT 86a
P3: A Parallel Network	Simulating System.	Zipser, D 86c
	Simulating the Adaptation Phenomena in the Response of Auditory. . .	Wu, ZL 88a
Brain Modeling by Tensor Network Theory and Computer	Simulation	Pellionisz, AJ 79a
The Limits of Biological	Simulation	Conrad, M 74c
	Simulation and Analysis of a Simple Cortical Network	Lynch, G 88a
...Approach to "Neural Modeling": Some Basic Issues of Measurement,	Simulation, and Computational Significance of Brain Maps	Schwartz, EL 87a
	Simulation and Evaluation of Ill Defined Neuronal Networks with. . .	Benito, S 80a
...in the Organization of Adult Cerebral Cortical Maps: A Computer	Simulation Based on Neuronal Group Selection	Pearson, JC 87a
Differential Forgetting of Prototypes and Old Instances:	Simulation by an Exemplar-Based Classification Model	Hintzman, DL 80a
UCLA SFINX - A Neural Network	Simulation Environment	Paik, E 87a
A General-Purpose	Simulation Environment for Developing Connectionist Models	D'Autrechy, CL 88a
Perceptron	Simulation Experiments	Rosenblatt, F 60a
	Simulation Experiments with Goal-Seeking Adaptive Elements	Barto, AG 84b
A Neural Network	Simulation Language Based on Multi-Pascal.	Lester, BP 87a
A Neural Network	Simulation Method Using the Fast Fourier Transform	Barto, AG 76a
	Simulation of a Connectionist Stereo Algorithm on a Shared-Memory. .	Stewart, CV 88a
	Simulation of a Distributed Algorithm Using the Connection Machine .	Aburdene, MF 88a
A Parallel	Simulation of a Distributed Problem Solving Network.	Durfee, EH 84a
Analog Computer	Simulation of a Neural Element	Hiltz, FF 62a
	Simulation of a Neural Network	Puleo P, F 81a
Dynamic Single Unit	Simulation of a Realistic Cerebellar Network Model	Pellionisz, AJ 73a
Dynamic Single Unit	Simulation of a Realistic Cerebellar Network Model. II. Purkinje . .	Pellionisz, AJ 74a
Fault	Simulation of a Wafer-Scale Integrated Neural Network.	May, N 88a
	Simulation of an Array-Based Neural Net Model.	Barnden, JA 87b
	Simulation of Anticipatory Responses in Classical Conditioning by a.	Barto, AG 82b
Hardware	Simulation of Brain Dynamics in Learning: The SPOCK.	Freeman, WJ 87a
A Computer	Simulation of Cerebral Neocortex: Computational Capabilities of. . .	Singer, A 87a
A Computer	Simulation of Cerebral Neocortex: Computational Capabilities of. . .	Singer, A 88a
	Simulation of Chaotic EEG Patterns with a Dynamic Model of the . . .	Freeman, WJ 87d
A Neural	Simulation of Classical Conditioning in Aplysia.	Henry, HN 86a
Analysis and	Simulation of Double Layer Neural Networks with Mutually Inhibiting.	Tokura, T 77a

The	Simulation of Energy and Memory in Neural Networks	Fraser, MD 88a
Computer	Simulation of Event Related Potentials During Voluntary Cognition. .	Deshmukh, VD 88b
A New Algorithm for Monte Carlo	Simulation of Ising Spin Systems	Bortz, AB 75a
A Dedicated Computer for	Simulation of Large Systems of Neural Nets	Garth, S 88a
	Simulation of Learning Processes in Neuronal Networks of the	Melkonian, DS 82a
A Rashevsky-Landahl Neural Net:	Simulation of Metacontrast	Weisstein, N 68a
Analysis and	Simulation of Networks of Mutually Inhibiting Neurons.	Morishita, I 72a
The	Simulation of Neural Elements by Electrical Networks Based on Multi.	Brain, AE 61a
A Chipset for High Speed	Simulation of Neural Network Systems	Garth, SCJ 87a
Computer	Simulation of Neural Networks for Perceptual Psychology.	Grossberg, S 86f
A Computer	Simulation of Olfactory Cortex with Functional Implications for. . .	Bower, JM 88a
A Computer	Simulation of Olfactory Cortex with Functional Implications for. . .	Wilson, MA 87a
Cortical Encoding of Memory: Hypotheses Derived from Analysis and	Simulation of Physiological Learning Rules in Anatomical Structures.	Lynch, G 89a
Computer	Simulation of Rhythmic Oscillation in Neuron Pools	MacGregor, RJ 74a
	Simulation of Self-Organizing Systems by Digital Computer.	Farley, BG 54a
Computer	Simulation of Short-Term Memory: A Component Decay Model	Laughery, KR 69a
Computer	Simulation of Somatotopic Map Plasticity	Edelman, GM 87a
Knowledge-Rich Solutions to the Binding Problem: A	Simulation of Some Human Computational Mechanisms.	Stenning, K 88a
Evaluating a Connectionist Network as a	Simulation of Speech Development	Schwartz, RG 88a
Entrainment in a Neural Network	Simulation of Spontaneously Active Units with Accommodation.	Blackburn, MR 87a
	Simulation of Task Performance as a Guide to the Identification of .	Deffner, G 87a
	Simulation of the Classically Conditioned Nictitating Membrane . . .	Blazis, DEJ 86a
Neural Net	Simulation of the Corpus Callosum.	Anninos, PA 88b
	Simulation of the Segmental Burst Generating Network for Locomotion.	Grillner, S 88a
Electronic	Simulation of the Vertebrate Retina.	Eckmiller, R 75a
Modeling and	Simulation of Vertebrate Primary Visual System: Basic Network. . . .	Oguztoreli, NN 83a
Neural Computation Research for Vision -	Simulation on a Parallel Processing Computer	Miyake, S 88a
	Simulation Results of Multiple Layer Difference Learning	Scheff, K 87b
	Simulation Studies of a Temporal Sequence Memory Model	Stanley, JC 76a
	Simulation Studies of Learning in an Informon Network.	Uttley, AM 76c
The "Whirling Dervish", A	Simulation Study in Learning and Recognition Systems	Hoffman, A 62a
Cycles: A	Simulation Tool for Studying Cyclic Neural Networks.	Gatley, MT 88a
Concurrent Neural Network	Simulation: Two Examples within a Single, Integrated Neural Network.	Bahr, C 88c
CONSIM (Conflict Simulator): Risk, Cost and Benefit in Political	Simulations.	Clema, JK 71a
Digital Signal Processor Accelerators for Neural Network	Simulations.	Penz, PA 86a
Graphical Folding and Distributed Failures in Large Network	Simulations.	Dayhoff, JE 87b
Learning in Neural Networks, Models and	Simulations.	Salu, Y 84a
...for Solving Job-Shop Scheduling: Part 2. Architecture and	Simulations.	Foo, YS 88c
Reactive Synaptogenesis in Small Neuronal Circuits:	Simulations Based on the Compensation Algorithm.	Dammasch, IE 87b
Numerical	Simulations of Boltzmann Machines.	Bounds, DG 86a
	Simulations of Composit, A Supra-Connectionist Architecture for. . .	Barnden, JA 88a
Computer	Simulations of Learning in Neural Systems.	Salu, Y 83a
Computer	Simulations of Pattern Recognition as a Dynamical Process of a . . .	Fuchs, A 88a
	Simulations Suggest Information Processing Roles for the Diverse . .	Borg-Graham, LJ 88a
A Parallel PDP Network	Simulator.	Smith, AB 87a
ISCON: An Interactive	Simulator for Connectionist Networks	Small, SL 82b
A Connectionist	Simulator for the BBN Butterfly Multi-Processor.	Fanty, MA 86a
anaLOG: A Functional	Simulator for VLSI Neural Systems.	Lazzaro, JP 86a
A Universal Non-Linear Filter, Predictor, and	Simulator which Optimizes Itself by a Learning Process	Gabor, D 60a
CONSIM (Conflict	Simulator): Risk, Cost and Benefit in Political Simulations.	Clema, JK 71a
The Construction of a	Simultaneous Functional Order in Nervous Systems I. Relevance of . .	Toet, A 87a
The Construction of a	Simultaneous Functional Order in Nervous Systems II. Computing . . .	Toet, A 87b
The Construction of a	Simultaneous Functional Order in Nervous Systems III. The Influence.	Toet, A 87c
Application of Orthogonal Projection Principles to	Simultaneous Phonemic Segmentation and Labeling of Continuous. . . .	Jalanko, M 78a
Hecht-Nielsen Unveils First New Neural Net Architecture	Since '84.	Schwartz, TJ 87b
The Dendrites of	Single Brain-Stem Motoneurons Intracellularly Labelled with.	Bras, H 87a
The Dendrites of	Single Brain-Stem Motoneurons Intracellularly Labelled with.	Bras, H 87b
Efficient Method for Estimating Kinetic Constants from	Single Channel Patch Clamp Recordings.	Chay, TR 87a
Efficient Method for Estimating Kinetic Constants from	Single Channel Patch Clamp Recordings.	Chay, TR 87b
	Single Channel Potassium Currents of the Anomalous Rectifier	Fukushima, Y 81a
	Single Channel Theory: A Neuronal Theory of Learning	Sutton, RS 78a
Some Asymptotic Results for Learning in	Single Hidden Layer Feedforward Network Models	White, H 87b
Concurrent Neural Network Simulation: Two Examples within a	Single, Integrated Neural Network Hardware Development Environment .	Bahr, C 88c
Characterization of Dynamical Behavior Associated with a	Single Neural Equation	Nagami, H 76a
Neuron Selectivity:	Single Neuron and Neuron Networks.	Cooper, LN 85a
Stochastic Models for	Single Neuron Firing Trains: A Survey.	Fienberg, SE 74a
Drive-Reinforcement Model of	Single Neuron Function: An Alternate to the Hebbian Neuronal Model .	Klopf, AH 86a
The Elements of	Single Neurons: A Review	Lewis, ER 83a
Electrical Response of	Single Photoreceptors.	Tomita, T 68a

On the Process of	Speech Perception.	Licklider, JCR 52a
Some Results of Research on	Speech Perception.	Liberman, AM 57a
	Speech Perception.	Studdert-Kennedy, M 80a
The Role of Attention and Active Processing in	Speech Perception.	Nusbaum, HC 86a
The Role of the Lexicon in	Speech Perception.	Samuel, AG 86a
The TRACE Model of	Speech Perception.	McClelland, JL 86k
Using Neural Networks to Improve Cochlear Implant	Speech Perception.	Tenorio, MF 87g
Machine Recognition of Human Language: Part 2, Theoretical Models of	Speech Perception and Language	Lindgren, N 65a
	Speech Perception and Phonemic Restorations.	Warren, RM 71a
	Speech Perception as a Cognitive Process: The Interactive.	Elman, JL 83a
	Speech Perception: Research, Theory, and the Principal Issues.	Pisoni, DB 86a
	Speech Perception: The TRACE Model	McClelland, JL 86e
Interactive Processes in	Speech Processing and Recognition.	Shamma, SA 87a
Neural Networks for	Speech Processing by Splicing of Autocorrelation Function.	Suzuki, J 76a
	Speech Processing System by the use of Short-Time Cross-Correlation.	Suzuki, J 77a
	Speech Production Models	Kent, RD 77a
Coarticulation in Recent	Speech Prosthesis System for Laryngectomees Using.	Muraki, S 86a
A Maximum Likelihood Approach to Continuous	Speech Recognition	Bahl, LR 83a
...of Probabilistic Functions of a Markov Process to Automatic	Speech Recognition	Levinson, SE 83a
Application of the Subspace Method to	Speech Recognition	Jalanko, M 78b
Bayesian Adaptation in	Speech Recognition	Brown, PF 83a
Boltzmann Machines for	Speech Recognition	Prager, RW 86a
Connectionist Methods in	Speech Recognition	Rohwer, R 88a
Improved Hidden Markov Modeling of Phonemes for Continuous	Speech Recognition	Schwartz, RM 84a
...Phonetic Features Using Connectionist Networks: An Experiment in	Speech Recognition	Watrous, RL 87d
...Computation of Acoustical Patterns with Application to Real Time	Speech Recognition	Mueller, P 86a
Multilayer Perceptrons and Automatic	Speech Recognition	Bourland, H 87a
Neural Nets for	Speech Recognition	Huang, WY 88b
Neural Network Classifiers for	Speech Recognition	Lippmann, RP 88c
Neural-Net Classifiers Useful for	Speech Recognition	Lippmann, RP 87d
...Maps - Insightful Representation of Phonological Features for	Speech Recognition	Kohonen, T 84b
Stochastic Modeling as a Means of Automatic	Speech Recognition	Baker, JK 75a
Structural Methods in Automatic	Speech Recognition	Levinson, SE 85a
The Modified Kanerva Model for Automatic	Speech Recognition	Prager, RW 88a
Unstable Connectionist Networks in	Speech Recognition	Rohwer, R 88c
Connectionist Models in	Speech Recognition: A Brief Summary and Bibliography	Fallside, F 88b
	Speech Recognition: A Model and a Program for Research	Halle, M 62a
	Speech Recognition by Computer	Levinson, SE 81a
Continuous	Speech Recognition by Statistical Methods.	Jelinek, F 76a
	Speech Recognition Experiments with Perceptrons.	Burr, DJ 88c
A Comparison of	Speech Recognition Front Ends Using a Connectionist Classifier	Renals, S 88a
Implementation of Neural Networks for	Speech Recognition on a Transputer Array	Chong, MWH 88a
A Review of Studies Applying Neural Net Models to	Speech Recognition Problems.	Lippmann, RP 88a
...Information in Time: Analog Neural Networks with Applications to	Speech Recognition Problems.	Tank, DW 87b
Large-Vocabulary Speaker-Independent Continuous	Speech Recognition: The SPHINX System.	Lee, K 88a
	Speech Recognition: Turning Theory into Practice	Doddington, GR 81a
Speaker-Independent Continuous	Speech Recognition Using Hidden Markow Models.	Lee, K 87a
A Framework for	Speech Recognition Using Neural Networks	Lippmann, RP 87a
Connectionist	Speech Recognition Using the Temporal Flow Model	Watrous, RL 88a
	Speech Recognition with Network-Based Classifiers.	Hogg, T 88a
On Presenting the Output of a Mechanical	Speech Recognizer.	Fry, DB 57a
Microprocessor Implementation of a Large Vocabulary	Speech Recognizer and Phonetic Typewriter for Finnish and Japanese	Kohonen, T 87e
Adaptive Noise Canceling for	Speech Signals	Sambur, MR 78a
Perceptual Restoration of Missing	Speech Sounds.	Warren, RM 70a
Teaching a Minimally Structured Back-Propagation Network to Recognize	Speech Sounds.	Landauer, TK 87a
Neural Net	Speech System In Works	Johnson, RC 87g
Neural Circuits Will Improve Vision,	Speech Systems	Brown, C 86f
Exploiting Lawful Variability in the	Speech Wave.	Elman, JL 86a
Context-Sensitive Coding, Associative Memory, and Serial Order in (	speech) Behavior	Wickelgren, WA 69a
A Real-Time Adaptive	Speech-Recognition System.	Talbert, LR 63a
Processing Determinants of Reading	Speed.	Jackson, MD 79a
Sensory and Cognitive Determinants of Reading	Speed.	Jackson, MD 75a
An Empirical Study of Learning	Speed in Back-Propagation Networks	Fahlman, SE 88a
...Experiments in Neural Network Based Learning Control During High	Speed Nonrepetitive Robotic Operations	Miller, WT 88a
On the	Speed of Convergence of Adaptive Algorithms.	Widrow, B 80a
Reliability and	Speed of Recall in an Associative Network.	Lansner, A 85b
A Chipset for High	Speed Simulation of Neural Network Systems	Garth, SCJ 87a
Neural Dynamics of Planned Arm Movements: Emergent Invariants and	Speed-Accuracy Properties During Trajectory Formation.	Bullock, D 88c
Neural Dynamics of Planned Arm Movements: Emergent Invariants and	Speed-Accuracy Properties During Trajectory Formation.	Bullock, D 88b

...Properties of Kohonen's Topology Conserving Maps: Fluctuations,	Stability and Dimension Selection.	Ritter, H 88a
Spatially Localized Interactive Neural Populations-II:	Stability and Dynamics of Excitatory Sets.	Brannan, JR 82a
	Stability and Oscillations in a Neurological Servomechanism.	Stark, L 59a
Self-Stabilization of Neuronal Networks, II.	Stability Conditions for Synaptogenesis.	Dammasch, IE 88c
Uniqueness and	Stability of a Limit Cycle for a Third Order Dynamical System. . . .	Li, B 81a
	Stability of Analog Neural Networks with Delay	Marcus, CM 88b
On the	Stability of Brain-Like Structures	Griffith, JS 63b
The	Stability of Generalized Volterra Equations.	Takeuchi, Y 78a
Absolute	Stability of Global Pattern Formation and Parallel Memory Storage. .	Cohen, MA 83b
Absolute	Stability of Global Pattern Formation and Parallel Memory Storage. .	Cohen, MA 83a
Existence and	Stability of Local Excitations in Neural Fields.	Kishimoto, K 79a
	Stability of Neural Networks for Optimization.	Stork, DG 88d
...of Learning and Development: The Temporal Unfolding and	Stability of STM and LTM Patterns.	Grossberg, S 82b
...of Learning and Development: The Temporal Unfolding and	Stability of STM and LTM Patterns.	Grossberg, S 86d
	Stability Properties of a Class of Large Scale Neural Networks . . .	Dimopoulos, N 80a
Sequential Behavior and	Stability Properties of Enzymatic Neuron Networks.	Kampfner, RR 83b
	Stability Results for Neural Networks.	Michel, AN 88b
Some Averaging and	Stability Results for Random Differential Equations.	Geman, S 79a
On the	Stability, Storage Capacity, and Design of Nonlinear Continuous. . .	Guez, A 88b
	Stability Theorem for Lateral Inhibition Networks that is Robust in.	Standley, DL 88a
Unlearning Has a	Stabilizing Effect in Collective Memories.	Hopfield, JJ 83a
A Mechanism for	Stabilizing Intelligent Properties of Neural Networks.	Dammasch, IE 87a
Numerical Calculation of	Stable and Unstable Periodic Solutions to the Hodgkin-Huxley	Rinzel, J 80a
Reconciling	Stable Asymmetry with Recovery of Function: An Adaptive Systems. . .	Bullock, D 87b
ART-2: Self-Organization of	Stable Category Recognition Codes for Analog Input Patterns.	Carpenter, GA 87c
A Method for the Design of	Stable Lateral Inhibition Networks that is Robust in the Presence. .	Wyatt Jr, JL 88a
Circuit Design Criteria for	Stable Lateral Inhibition Neural Networks.	Wyatt Jr, JL 88b
Absolutely	Stable Learning of Recognition Codes by a Self-Organizing Neural . .	Carpenter, GA 86b
	Stable Matching with Preferences Derived From a Psychological Model.	Bartholdi, JJ 86b
Number of	Stable Points for Spin-Glasses and Neural Networks of Higher Orders.	Baldi, P 87a
Discovering Order in Chaos:	Stable Self-Organization of Neural Recognition Codes	Carpenter, GA 86h
Adaptive Resonance Theory:	Stable Self-organization of Neural Recognition Codes in Response to.	Carpenter, GA 86d
	Stable Self-Organization of Sensory Recognition Codes: Is Chaos. . .	Grossberg, S 87o
On Choosing Non-Random Input Exemplars to Form	Stable States in a Hopfield Neural Net	Feild Jr, WB 88a
BoltzCONS: Reconciling Connectionism with the Recursive Nature of	Stacks and Trees	Touretzky, DS 86b
Descriptions: An Intermediate	Stage in Memory Retrieval.	Norman, DA 79a
The Organization of Behavior: The First	Stage of Perception: Growth of the Assembly.	Hebb, DO 49a
Implementing	Stages of Motion Analysis in Neural Networks	Sereno, ME 87a
Where Do We	Stand on Maximum Entropy?.	Jaynes, ET 79a
Resource Requirements of	Standard and Programmable Nets	McClelland, JL 86j
A Microcomputer Neural Net Benchmarked Against	Standard Classification Techniques	Draper, JS 87a
Neuroscience as a Possible Research	Standard for Neural Networks	Morse, KG 87a
...of the On-Beta Ganglion Cell in Daylight, Twilight, and	Starlight.	Sterling, P 87a
Fuzzy Logic's Soft	Start Belies a Cutting Edge.	Chester, M 86a
Carver Mead Joins Forces with Neural Net	Startup.	Johnson, RC 86b
Neural Net Developer Nestor Sprints From	Startup Gate	Johnson, RC 87f
	Startup Nestor Readies Neural Network-Based Products	Brown, C 86k
Neural Net	Startup Takes Teaching Role.	Brown, C 87h
Neural Network	Startups Backed by Venture Capital	Brown, C 87i
Neural Network	Startups Proliferate Across the U.S.	Johnson, RC 88g
Combinatorial Optimization Using a Continuous	State Boltzmann Machine.	Gutzmann, KM 87a
International Solid	State Circuits Conference, February, 1987.	Sage, JP 87b
Finite	State Markovian Decision Processes	Derman, C 70a
Effects of Connection Delays in Two	State Model Neural Circuits.	Conwell, PR 87a
On the Stationary	State of Kohonen's Self-Organizing Sensory Mapping	Ritter, H 86a
	State of the Art in Neural Computing	Kohonen, T 87g
The	State of the Art in Pattern Recognition.	Nagy, G 68a
	State Recurrence Detection	Rosen, BE 88a
Learning by	State Recurrence Learning.	Rosen, BE 88c
The	State Space of Memory Clusters in the Hopfield Network	Krowitz, A 87a
	State-Dependent Factors Influencing Neural Plasticity: A Partial . .	Munro, PW 86b
Automatic Character Recognition: A	State-of-the-Art Report.	Stevens, ME 61a
	States: Computer & Brain	Anderson, JA 79b
On Choosing Non-Random Input Exemplars to Form Stable	States in a Hopfield Neural Net.	Feild Jr, WB 88a
Structure of Metastable	States in the Hopfield Model	Gardner, E 86a
GMDH Forecasting of United	States Interest Rates.	Ohashi, K 84a
Adaptive Learning Networks: Development and Application in the United	States of Algorithms Related to GMDH	Barron, RL 84a
	Static and Dynamic Error Propagation Networks with Application to. .	Robinson, AJR 88b
Programming a Massively Parallel, Computation Universal System:	Static Behavior.	Lapedes, A 86a

Intraneuronal Dynamics as a	Substrate for Evolutionary Learning	Kirby, KG 86a
The Neuroanatomical	Substrate for Information Processing in Macular Endorgans	Ross, MD 88c
The Cerebellum: A	Substrate for List-Processing in the Brain	Albus, JS 72a
Neural	Substrates of Binocular Form Perception: Filtering, Matching	Grossberg, S 83a
Psychophysiological	Substrates of Schedule Interaction and Behavioral Contrast	Grossberg, S 81b
EXCLUSIVE OR Processing (Binary Image	Subtraction) Using Thick Fourier Holograms	Guest, CC 84a
	Successfully Using Peak Learning Rates of 10 (and Greater) in Back	Cater, JP 87a
Four Frames Do Not	Suffice	Grossberg, S 85b
Four Frames	Suffice	Feldman, JA 82d
Three Frames	Suffice	Hinton, GE 85c
Four Frames	Suffice: A Provisional Model of Vision and Space	Feldman, JA 85c
Simulations	Suggest Information Processing Roles for the Diverse Currents in	Borg-Graham, LJ 88a
...Views Human Processing: Human Errors and Other Phenomena	Suggest Processing Mechanisms	Norman, DA 81a
...and Machine: How the Brain Processes Visual Information May Be	Suggested by Studies in Computer Vision (and Vice Versa)	Poggio, TA 84c
A Feature Extractor for Curvilinear Patterns: A Design	Suggested by the Mammalian Visual System	Fukushima, K 70a
The Closeness Code: An Integer to Binary Vector Transformation	Suitable for Neural Network Architectures	Penz, PA 87a
The	Sum-Line Extrapolative Algorithm and its Application to Statistical	Talbert, LR 70a
Connectionist Models in Speech Recognition: A Brief	Summary and Bibliography	Fallside, F 88b
Synaptic Potential	Summation and Repetitive Firing Mechanisms: Input-Output Theory for	Calvin, WH 72a
...for Optimization of the Computational Load of a Hypercube	Supercomputer	Barhen, J 87a
Optimization of the Computational Load of a Hypercube	Supercomputer Onboard a Mobile Robot	Barhen, J 87b
Parallelism Breeds a New Class of	Supercomputers	Mokhoff, N 87a
Fat-Trees: Universal Networks for Hardware Efficient	Supercomputing	Leiserson, CE 85a
A Neural Network for the Retrieval of	Superimposed Connection Patterns	von der Malsburg, C 87a
	Superior Accuracy of Decrementing over Incrementing Associative	Dobson, VG 87a
	Super-Resolution and Neural Computing	Abbiss, JB 88a
	Superresolution for Ultrasonic Imaging in Air Using Neural Networks	Winters, JH 88b
Adaptive	Superresolution Phenomena in Aberrant Arrays	Jablon, N 85a
	Supervised Learning and Systems with Excess Degrees of Freedom	Jordan, MI 88b
	Supervised Learning of Probability Distributions by Neural Networks	Baum, EB 88d
On the Asymptotic Improvement in the Outcome of	Supervised Learning Provided by Additional Non-Supervised Learning	Cooper, DB 70a
...Using a Probabilistic Automaton with Pattern Recognition	Supervision	Jarvis, RA 70a
An Algorithm for Learning without External	Supervision and its Application to Learning Control Systems	Nicolic, ZJ 66a
	Supplier-Consumer Nets for Material Handling	Parunak, HVD 87a
The Trinocular General	Support Algorithm: A Three Camera Stereo Algorithm for Overcoming	Stewart, CV 88b
Simulations of Composit, A	Supra-Connectionist Architecture for Commonsense Reasoning	Barnden, JA 88a
A Chain of	Suprasegmental Neuroscillatory Circuits: A Human Brain Theory	Deshmukh, VD 88a
	Suprasegmentals in Very Large Vocabulary Word Recognition	Waibel, A 86a
...of Stimulus Artifact in Nerve Signals Evoked and Recorded Using	Surface Electrodes	McGill, KC 82a
Simple Analog and Hybrid Networks for	Surface Interpolation	Hutchinson, JM 86a
	Surface Lightness and Cues for the Illumination	Beck, J 71a
Neural Dynamics of	Surface Perception: Boundary Webs, Illuminants, and Shape-From	Grossberg, S 87m
Using Frequency-and Orientation-Tuned Channels to Determine	Surface Slant	Kube, P 86a
General Potential	Surfaces and Neural Networks	Dembo, A 88a
Synthesis of Non-Linear Control	Surfaces by a Layered Associative Search Network	Barto, AG 82c
An Approach to Learning Control	Surfaces by Connectionist Systems	Barto, AG 87a
Synthesis of Quasi-Optimal Switching	Surfaces by Means of Training Techniques	Mendel, JM 70c
Perception of Depth	Surfaces in Random-Dot Stereograms: A Neural Model	Dev, P 75a
Cooperating Processes for Low-Level Vision: A	Survey	Davis, LS 81a
Learning Automata: A	Survey	Narendra, KS 74a
On Pattern Classification Algorithms: Introduction and	Survey	Ho, Y 68a
Speaker Recognition: A	Survey	Corsi, P 82a
Stochastic Models for Single Neuron Firing Trains: A	Survey	Fienberg, SE 74a
A	Survey of Artificial Neural Systems I: Foundations	Simpson, PK 87b
A	Survey of Interconnection Networks (Concurrent Computing)	Tse-Yun, F 81a
A	Survey of Neural Network Models	Lippmann, RP 88b
	Survey on Artificial Neural Systems	Miller, RK 88a
	Sustained Oscillation in Symmetric Cooperative-Competitive Neural	Cohen, MA 87d
Connectionistic Learning in Real Time:	Sutton-Barto Adaptive Element and Classical Conditioning of the	Moore, JW 85a
Eye Movements Induced by Linear Acceleration on a Parallel	Swing	Baloh, RW 88a
A Prototype Crossbar	Switch for the Gamma Network	Farley, ET 87a
	Switched-Capacitor Neural Networks	Tsividis, Y 87c
Programmable Bistable	Switches and Resistors for Neural Networks	Spencer, EG 86a
Neural Dynamics of Attention	Switching and Temporal Order Information in Short-Term Memory	Grossberg, S 86i
Adaptive	Switching Circuits	Widrow, B 60c
Adaptive	Switching Circuits	Widrow, B 60b
Learning Phenomena in Networks of Adaptive	Switching Circuits	Hoff, ME 62a
On the Size of Weights Required for Linear-Input	Switching Functions	Myhill, J 61a
The Realization of Symmetric	Switching Functions with Linear-Input Logical Elements	Kautz, WH 61a

...of Neuronal Networks, I. The Compensation Algorithm for	Synaptogenesis	Dammasch, IE 86b
Self-Stabilization of Neuronal Networks, II. Stability Conditions for	Synaptogenesis	Dammasch, IE 88c
Reactive	Synaptogenesis in Small Neuronal Circuits: Simulations Based on the.	Dammasch, IE 87b
A Model of	Synaptogenetic Consequences of Afferent Input in Neuronal Nets	Dammasch, IE 88a
	Synchronization in Neural Nets	Vidal, JJ 88b
	Synchronization in Small Groups of Neurons: A Study with Electronic.	Lewis, ER 68b
Auto-Associative Pattern Storage Using	Synchronous Boolean Networks	Martland, D 87a
The Structure of Least Mean Square Linear Estimators for	Synchronous M-ary Signals.	Gardner, WA 73a
	Synchronous vs Asynchronous Behavior of Hopfield's CAM Neural Net.	Cheung, KF 87c
Behaviour of Autonomous, (	Synchronous), Boolean Networks	Martland, D 87b
...Simulations of Pattern Recognition as a Dynamical Process of a	Synergetic System.	Fuchs, A 88a
Learning in	Synergetic Systems for Pattern Recognition and Associative Action.	Haken, H 88b
A	Synergetic Theory of Environmentally-Specified and Learned Patterns.	Schoner, G 88b
A	Synergetic Theory of Environmentally-Specified and Learned Patterns.	Schoner, G 88c
	Synergetics and Computers.	Haken, H 88d
	Synergetics in Pattern Recognition and Associative Action.	Haken, H 88e
	Syneuristor: A Device and Concept.	Waldron, MB 71a
A Plausible Neural Circuit for Classical Conditioning Without	Synpatic Plasticity.	Tesauro, G 88b
Connectionist Modeling of	Syntactic Constraints on Sentence Processing	Parisi, D 87a
The Role of	Syntactic Context in Word Recognition.	Goodman, GO 81a
N-Gram Driven Search for Sentences in a	Syntactic Network.	Venta, O 86b
	Syntactic Pattern Recognition.	Fu, K 86a
	Syntactic Pattern Recognition: Stochastic Languages.	Thomason, MG 86a
A New Discipline for CMOS Design: An Architecture for Sound	Synthesis.	Mead, CA 85a
...Developments in a Neural Model of Real-Time Speech Analysis and	Synthesis.	Cohen, MA 87c
...Resonances: Disparity Matching, Filling-in, and Figure-Ground	Synthesis.	Cohen, MA 84b
...Resonances: Disparity Matching, Filling-In, and Figure-Ground	Synthesis.	Grossberg, S 84b
...of the Limulus Retina to Moving Stimuli: A Prediction by Fourier	Synthesis.	Brodie, SE 78a
An Experimental Investigation of a Class of Pattern Recognition	Synthesis Algorithms	Mattson, RL 63a
Qualitative Analysis and	Synthesis of a Class of Neural Networks.	Li, JH 88a
The Analysis and	Synthesis of Adaptive Systems Which Use Networks of Threshold.	Mattson, RL 62a
	Synthesis of Boolean Nets and Time Behavior of a General	Caianiello, ER 75a
	Synthesis of Fragmented Data on Neuronal Systems: A Computer Model	Pellionisz, AJ 78a
The Microscopic Analysis of Behavior: Toward a	Synthesis of Instrumental, Perceptual, and Cognitive Ideas	Grossberg, S 84d
Convergence, Cycling or Strange Motion in the Adaptive	Synthesis of Neurons	Labos, E 86a
	Synthesis of Non-Linear Control Surfaces by a Layered Associative.	Barto, AG 82c
	Synthesis of Quasi-Optimal Switching Surfaces by Means of Training	Mendel, JM 70c
Probabilistic Logics and the	Synthesis of Reliable Organisms from Unreliable Components	von Neumann, J 56a
	Synthesis of Reverberatory Neural Networks	Aiello, A 70a
Chunking and Consolidation: A Theoretical	Synthesis of Semantic Networks, Configuring in Conditioning, S-R	Wickelgren, WA 79a
	Synthesizing a Color Algorithm from Examples	Hurlbert, AC 88a
Electronic Life and	Synthetic Intelligent Systems.	Dress, WB 88a
Darwinian Optimization of	Synthetic Neural Systems	Dress, WB 87c
Limited Interconnectivity in	Synthetic Neural Systems	Akers, LA 88a
...Cell Activity within the Basic Circuit and Modified by Inhibitory	Sys.	Pellionisz, AJ 74a
A Computational Model of the Auditory	System	Travis, BJ 87a
A Distributed Connectionist Production	System	Touretzky, DS 86a
...Curvilinear Patterns: A Design Suggested by the Mammalian Visual	System	Fukushima, K 70a
A Learning Mechanism for an Associative Storage	System	Ersu, E 81a
A Model for a Weighted Retrieval	System	Buell, DA 81a
A Model of the Vertebrate Central Command	System	Kilmer, WL 69a
...for Generalization over Equivalent Stimuli in the Olfactory	System	Freeman, WJ 81a
A Neuromorphic VLSI Learning	System	Alspector, J 87a
A Parallel Signal Processor	System	Gass, W 86a
A Proposed Neural Network for the Integrator of the Oculomotor	System	Cannon, SC 83a
A Real-Time Adaptive Speech-Recognition	System	Talbert, LR 63a
A Rule-Based Connectionist Parsing	System	Selman, B 85a
A Self-Organizing Logical	System	Mattson, RL 59a
A Statistical Approach to the Theory of the Central Nervous	System	Shimbel, A 48a
A Trainable Weather-Forecasting	System	Hu, MJC 63a
...Neural Network Sharing Features of the Mammalian Visual	System	Frohn, H 87a
An Associative Hierarchical Self-Organizing	System	Davis, BR 85a
...Formation of Topological Maps of Patterns in a Self-Organizing	System	Kohonen, T 81a
Biochemistry of Information Storage in the Nervous	System	Black, IB 87a
Cognitive Capabilities of a Parallel	System	Anderson, JA 86a
...of Pattern Recognition as a Dynamical Process of a Synergetic	System	Fuchs, A 88a
Conditional Probability Computing in a Nervous	System	Uttley, AM 59a
Design of a Highly Parallel Visual Recognition	System	Sabbah, D 81a
Design Principles for a Front-end Visual	System	Koenderink, JJ 88a

Distributed Neural Information Processing in the Vestibulo-Ocular	System	Lau, CGY 88a
Generalization of Pattern Recogntion in a Self-Organizing	System	Clark, WA 55a
Information Processing in the Nervous	System	Kristan, WB 80a
Information Processing	System	Cooper, LN 76a
Information Transmission in the Nervous	System	Uttley, AM 79a
...Speaker-Independent Continuous Speech Recognition: The SPHINX	System	Lee, K 88a
Layered Neural Network Model Applied to the Auditory	System	Travis, BJ 86a
Learning and Memory: Mechanisms of Information Storage in the Nervous	System	Matthies, H 86a
Massively Parallel Network-Based Natural Language Parsing	System	Li, T 87a
Mathematical Modeling of Mechanisms of the Central Nervous	System	Gelfand, IM 71a
...Neuronlike Network for the Construction of a Learning Control	System	Gawronski, RR 87a
...Carlo Calculation of Phase Separation in a Two-Dimensional Ising	System	Flinn, PA 74a
Morphology of Neural Networks in the Mammalian Visual	System	Gardner, S 88b
Neural Nets with Complex Structure: A Model of the Visual	System	Gremillion, M 87a
Neural Network Implementation of a Medical Diagnosis Expert	System	Kudrycki, TP 88a
Neuronal Group Selection: A Basis for Categorization by the Nervous	System	Finkel, LH 88b
...Use of Human Memory Models: The Memory Extender Personal Filing	System	Jones, WP 86a
P3: A Parallel Network Simulating	System	Zipser, D 86c
Processing of Continuous Lines and Edges by the Visual	System	Hartmann, G 83a
Real-Time 3-D Object Classification Using a Learning	System	Rimey, RD 86a
Representation of Local Geometry in the Visual	System	Koenderink, JJ 87d
Self-Corrective Character Recognition	System	Nagy, G 66a
...of Chaotic EEG Patterns with a Dynamic Model of the Olfactory	System	Freeman, WJ 87d
Spontaneous Pattern Classification by a Dynamical	System	Banzhaf, W 87a
Synaptic Change in the Nervous	System	Levy, WB 84a
...Model of Pattern Recognition in the Human Somatosensory	System	Grajski, KA 88c
The Central Nervous System as a Low and High Level Control	System	Albus, JS 88a
The Restricted Language Architecture of a Hybrid Representation	System	Vilain, MB 85a
The Tracing of Pathways in the Central Nervous	System	Heimer, L 68a
...Changes of Protein Phosphorylation in a Molluscan Nervous	System	Naito, S 88a
Why Corticies? Neural Computation in the Vertebrate Visual	System	Mallot, HA 88a
The Oculomotor Control	System: A Review	Robinson, DA 68b
Neural Coding in the Bullfrog's Auditory	System: A Teleological Approach	Frishkopf, LS 68a
Human Memory: A Proposed	System and its Control Processes	Atkinson, RC 68a
Expert	System and Neural Network Approaches for Classical Guitar Fingering	Sayegh, S 88a
Integration of Knowledge-Based	System and Neural Network Techniques for Robotic Control	Handelman, DA 88a
Learning	System Architectures Composed of Multiple Learning Modules	Reilly, DL 87a
Uniqueness and Stability of a Limit Cycle for a Third Order Dynamical	System Arising in Neuron Modelling	Li, B 81a
The Central Nervous	System as a Low and High Level Control System	Albus, JS 88a
Error Correcting	System Based on Neural Circuits	Takefuji, Y 87a
Medical Diagnostic Expert	System Based on PDP Model	Saito, K 88a
A Thousand-Word Recognition	System Based on the Learning Subspace Method and Redundant Hash	Kohonen, T 80a
Modeling and Simulation of Vertebrate Primary Visual	System: Basic Network	Oguztoreli, NN 83a
Speech Processing	System by the use of Short-Time Cross-Correlation Function	Suzuki, J 77a
The Information-Storage	System Called "Human Memory"	Simon, HA 76a
A Self-Organizing	System Called `Mickey Mouse'	Browning, I 64a
NeuralWare Expert	System Classified Stock Patterns	Anthes, GH 88a
Neuristor: A Novel Device and	System Concept	Crane, HD 62a
The Application of Techniques of Artificial Intelligence to Control	System Design	Mendel, JM 68a
An Artificial Neural	System Design for Rotation and Scale Invariant Pattern Recognition	Reber, WL 87a
The Eye-Movement Control	System: Evolution of a Model	Fender, DH 62a
A Classification	System for Alpha-Numeric Characters Based on Learning Network	Fairhurst, MC 76a
MOSAIC: A Macro-Connectionist Organization	System for Artificial Intelligence Computation	Pham, KM 88a
A Neuron Model of a Memory	System for Autonomous Exploration of an Environment	Trenary, R 87a
An Artificial Neural	System for Autonomous Undersea Vehicles	Blackburn, MR 88a
A Hierarchical	System for Character Recognition with Stochastic Knowledge	Vlontzos, JA 88a
A Matched Filter Detection	System for Complicated Doppler Shifted Signals	Lerner, RM 60a
A PDP	System for Goal-Plan Decisions	Sharkey, NE 88a
Pattern Classification by Neural Network: An Experimental	System for Icon Recognition	Gullichsen, E 87a
Introduction to a	System for Implementing Neural Net Connections on SIMD	Tomboulian, S 88a
Speech Prosthesis	System for Laryngectomees Using "Electropalatograph" and a Small	Muraki, S 86a
A Minimal Technology Routing	System for Meals on Wheels	Bartholdi, JJ 83a
PLEXUS: An On-Line	System for Modeling Neural Networks	Dill, JC 68a
Rule-Based Processing in a Connectionist	System for Natural Language Understanding	Selman, B 85b
A Neuromime	System for Neural Circuit Analysis	Mitchell, CE 81a
A Graphical Design	System for Neural Networks	Hood, G 87a
Using Automated Techniques to Generate an Expert	System for R&D Project Monitoring	Gallant, SI 86d
NETL: A	System for Representing and Using Real-World Knowledge	Fahlman, SE 79a
An Associative Co-Processor Architecture with the UNIX Operating	System for the Acceleration of CAD Tools	Schulz, M 88a

Robot Systems. . . . . Albus, JS 76a
Scanning the Issue: Special Issue on Studies of Neural Elements and Systems. . . . . Weiss, TF 68a
Self-Organizing Control of Stochastic Systems. . . . . Saridis, GN 77a
Self-Organizing Systems. . . . . Yovitz, MC 62a
Sensor Processing Using Artificial Neural Systems. . . . . Hecht-Nielsen, R 86b
Shape Representation in Parallel Systems. . . . . Hinton, GE 81d
Shape Representation in Parallel Systems. . . . . Hinton, GE 81e
...Target Description and its Similarity to Animal Echolocation Systems. . . . . Altes, RA 76a
Spatial Interaction and the Statistical Analysis of Lattice Systems. . . . . Besag, J 74a
Statistical Dynamics of Ultradiffusion in Hierarchical Systems. . . . . Gardner, S 87a
...and Processing of Information in Distributed Associative Memory Systems. . . . . Kohonen, T 81c
Strange Attractors and Chaotic Motions of Dynamical Systems. . . . . Ott, E 81a
Structural Learning in Connectionist Systems. . . . . Barto, AG 85f
Survey on Artificial Neural Systems. . . . . Miller, RK 88a
...Systems to Drive: Manual Training Techniques for Autonomous Systems. . . . . Shepanski, JF 88b
...Whirling Dervish", A Simulation Study in Learning and Recognition Systems. . . . . Hoffman, A 62a
The Application of Wafer-Scale Technology to Neuromorphic Systems. . . . . Raffel, JI 88b
The Characteristic Selection Problem in Recognition Systems. . . . . Lewis, PM 62a
The Logic of Connectionist Systems. . . . . Aleksander, I 88f
...Treatment of Uncertainty in Artificial Intelligence and Expert Systems. . . . . Lindley, DV 87a
The Role of Fuzzy Logic in the Management of Uncertainty in Expert Systems. . . . . Zadeh, LA 83b
The Role of Fuzzy Logic in the Management of Uncertainty in Expert Systems. . . . . Zadeh, LA 87f
The Role of the Critic in Learning Systems. . . . . Dietterich, TD 81a
The Senses Considered as Perceptual Systems. . . . . Gibson, JJ 66a
Theory of Dynamic Similarity in Neuronal Systems. . . . . MacGregor, RJ 88a
Toward a Theory of Reinforcement-Learning Connectionist Systems. . . . . Williams, RJ 88b
Towards Connectionist Rule-Based Systems. . . . . Samad, T 88b
Ultradiffusion: The Relaxation of Hierarchical Systems. . . . . Huberman, BA 85b
...Zig Zag Epigenetic Neural Networks and Their Use in Information Systems. . . . . Cochet, Y 88c
Properties of Cortical Systems - A Basis for Computer Vision. . . . . von Seelen, W 87b
Synthesis of Fragmented Data on Neuronal Systems: A Computer Model of Cerebellum. . . . . Pellionisz, AJ 78a
Brain Function and Adaptive Systems: A Heterostatic Theory . . . . . Klopf, AH 72a
Optimization with Artificial Neural Network Systems: A Mapping Principle and a Comparison to Gradient Based. . . . Leong, HM 88a
Self-Organizing Systems - A Review and Commentary. . . . . Hawkins, JK 61a
Learning Control Systems: A Review and Outlook. . . . . Fu, K 70a
Building and Understanding Adaptive Systems: A Statistical/Numerical Approach to Factory Automation and. Werbos, PJ 87a
Boolean Systems, Adaptive Automata, Evolution. . . . . Kaufman, SA 86a
Order and Chaos in Neural Systems: An Approach to the Dynamics of Higher Brain Functions . . . Harth, EM 83a
Biological Control Systems Analysis . . . . . Milsum, J 66a
Nonlinear Systems Analysis of the Hippocampal Perforant Path-Dentate . . . . . Sclabassi, RJ 88b
Nonlinear Systems Analysis of the Hippocampal Perforant Path-Dentate . . . . . Berger, TW 88a
Nonlinear Systems Analysis of the Hippocampal Perforant Path-Dentate . . . . . Berger, TW 88b
Disordered Systems and Biological Organization. . . . . Bienenstock, EL 86a
Adaptive Pattern Recognition Systems and Boltzmann Machines: A Rapprochement. . . . . Aleksander, I 87a
Outline of a New Approach to the Analysis of Complex Systems and Decision Process . . . . . Zadeh, LA 87e
Theory of Neural-Analog Reinforcement Systems and its Application to the Brain-Model Problem . . . . . Minsky, ML 54a
Expert Systems and Machine Learning . . . . . Clema, JK 85a
...Discovery: Observations after Using Back-propagation, Expert Systems, and Multiple-Linear Regression. . . . . Irani, EA 88a
Nearest-Neighbour Systems and the Auto-Logistic Model for Binary Data. . . . . Besag, JE 72a
Creation of Efficient Recognition Systems and the Problems of Microelectronics . . . . . Elinson, MI 81a
Systems Approach in Vision: Proceedings of a Workshop, Amsterdam . . Regan, D 84a
The Dynamical Systems Approach to Differential Equations . . . . . Hirsch, MW 84a
How Many Memory Systems are there? . . . . . Tulving, E 85a
Conceptual Graph Knowledge Systems as Problem Context for Neural Networks . . . . . Lendaris, GG 88a
Manual Training Technqiues of Autonomous Systems Based on Artificial Neural Networks. . . . . Shepanski, JF 87a
Simulation of Self-Organizing Systems by Digital Computer. . . . . Farley, BG 54a
...Does the Fundamental Information Processing Ability of Biological Systems Come From? . . . . . Kohonen, T 75a
Beyond Neural Networks - The Discipline of Intelligent Systems Design Emerges . . . . . Johnson, RC 87a
Cooperative Self-Organization of Multiple Neural Systems During Adaptive Sensory-Motor Control. . . . . Grossberg, S 86g
Combinations of Neural Systems for Particular Application Situations. . . . . Josin, GM 87a
Application-Oriented Connection Systems for Pattern Acquisition and Recognition. . . . . Horne, W 87a
Learning in Synergetic Systems for Pattern Recognition and Associative Action . . . . . Haken, H 88b
Automated Generation of Connectionist Expert Systems for Problems Involving Noise and Redundancy. . . . . Gallant, SI 87a
Neural Systems for Use as a Research Tool . . . . . Josin, GM 87b
Information Processing in Dynamical Systems: Foundations of Harmony Theory . . . . . Smolensky, P 86a
Automatic Generation of Expert Systems from Examples. . . . . Gallant, SI 85a
Goal-Seeking Systems from Goal-Seeking Components: Implications for AI. . . . . Klopf, AH 79a
A Survey of Artificial Neural Systems I: Foundations . . . . . Simpson, PK 87b
The Construction of a Simultaneous Functional Order in Nervous Systems I. Relevance of Signal Covariances and Signal Coincidences . Toet, A 87a

Nerve Net Models of Plausible Size that Perform Many Simple Learning	Tasks.	Brindley, GS 69a
Strategies for Teaching Layered Networks Classification	Tasks.	Wittner, BS 88a
Topology Conserving Mappings for Learning Motor	Tasks.	Ritter, H 86b
Distributed and Fault-Tolerant Computation for Retrieval	Tasks Using Distributed Associative Memories	Char, JM 88a
Overlapping	Taste and Tactile Maps of the Oropharynx in the Vagal Lobe of the.	Kanwal, JS 88b
Clustering,	Taxonomy, and Topological Maps of Patterns	Kohonen, T 82c
Clustering,	Taxonomy, and Topological Maps of Patterns	Kohonen, T 82d
Mimicking the Human Mind: Radical New Computers	Teach Themselves	Rogers, M 87a
Learning to Recognize Patterns Without a	Teacher.	Fralick, SC 67a
Learning Without a	Teacher.	Spragins, J 66a
	Teaching a Minimally Structured Back-Propagation Network to.	Landauer, TK 87a
	Teaching a Stochastic Automaton to Skillfully Play Hand / Eye Games.	Jarvis, RA 77a
	Teaching a Topic in Cybernetics with APL: An Introduction to Neural.	Peelle, HA 81a
	Teaching Artificial Neural Systems to Drive: Manual Training	Shepanski, JF 88b
Strategies for	Teaching Layered Networks Classification Tasks	Wittner, BS 88a
Neural Net Startup Takes	Teaching Role.	Brown, C 87h
CCDs, NMOS	Team to Build Neural Net Models On Chip.	Brown, C 86b
Optical Systems Research at Texas	Tech University.	Krile, TF 86a
Recognition of Hierarchically Encoded Images by	Technical and Biological Systems	Hartmann, G 87a
	Technical Comments: Computing with Neural Networks	Hopfield, JJ 87c
	Technical Comments: Computing with Neural Networks	Kohonen, T 87h
	Technical Comments: Computing with Neural Networks	Carpenter, GA 87f
	Technical Comments: Computing with Neural Networks	Palm, G 87d
	Technical Issues in Learning Control for Robot Motions	Kawamura, S 87b
A Neural Unsupervised Learning	Technique.	Atiya, AF 88a
...of Multilayer Neural Networks Using a Least Squares Estimation	Technique.	Kollias, S 88a
Automatic Character Reading Using a Holographic Data Processing	Technique.	Nakajima, M 72a
A	Technique for Determining and Coding Subclasses in Pattern	Mattson, RL 65a
An Inference	Technique for Integrating Knowledge from Disparate Sources	Garvey, TD 81a
A Model and Proof	Technique for Message-Based Systems.	Feldman, JA 80a
A Consistency	Technique for Pattern Association.	Ullmann, JR 62a
An Associative Memory	Technique for the Recognition of Patterns.	Pao, Y 78a
The "Rubber-Mask"	Technique - I. Pattern Storage and Recognition	Widrow, B 73b
The "Rubber-Mask"	Technique - II. Pattern Storage and Recognition.	Widrow, B 73c
A Maximization	Technique Occurring in the Statistical Analysis of Probabilistic	Baum, LE 70a
A	Technique to Identify Nearest Neighbors.	Yunck, TP 76a
...System for Alpha-Numeric Characters Based on Learning Network	Techniques	Fairhurst, MC 76a
...Neural Net Benchmarked Against Standard Classification	Techniques	Draper, JS 87a
Matrix Multiplication Using Coherent Optical	Techniques	Tamura, PN 77a
Micropower	Techniques	Vittoz, EA 85a
Neural Network Learning	Techniques	Sharkey, NE 88b
Synthesis of Quasi-Optimal Switching Surfaces by Means of Training	Techniques	Mendel, JM 70c
...Adaptive Model of a Human Controller Using Pattern Recognition	Techniques	Gilstad, DW 71a
Algorithms and	Techniques for Automated Visual Inspection	Chin, RT 86a
Generalization Learning	Techniques for Automating the Learning of Heuristics	Waterman, DA 70a
Teaching Artificial Neural Systems to Drive: Manual Training	Techniques for Autonomous Systems.	Shepanski, JF 88b
Scaling	Techniques for Neural Algorithms	Cuykendall, R 88a
Fast Edge-Only Matching	Techniques for Robot Pattern Recognition	Caelli, TM 87a
Integration of Knowledge-Based System and Neural Network	Techniques for Robotic Control	Handelman, DA 88a
Introduction to Mathematical	Techniques in Pattern Recognition.	Andrews, HC 72a
Analog MOS Circuit	Techniques in the VLSI Implementation of Neural Networks	Satyanarayana, S 88a
The Application of	Techniques of Artificial Intelligence to Control System Design	Mendel, JM 68a
Using Automated	Techniques to Generate an Expert System for R&D Project Monitoring	Gallant, SI 86d
Fuzzy Mathematical	Techniques with Applications	Kandel, A 86a
...Circuits by Silicon Planar and Three-Dimensional Fabrication	Technology	Takahashi, K 84a
Some Speculations on Artificial Neural System	Technology	Myers, MH 86a
The Implementation of Neural Network	Technology	Savely, RT 87a
A Minimal	Technology Routing System for Meals on Wheels.	Bartholdi, JJ 83a
The Application of Wafer-Scale	Technology to Neuromorphic Systems	Raffel, JI 88b
	Technology-Independent Design of Neurocomputers: The Universal	MacLennan, BJ 87a
Manual Training	Technqiues of Autonomous Systems Based on Artificial Neural.	Shepanski, JF 87a
The Toad Optic	Tectum as a Recurrent On-Center Off-Surround Neural Net with	Betts, B 88b
Neural Coding in the Bullfrog's Auditory System: A	Teleological Approach.	Frishkopf, LS 68a
Hierarchical Control for Robots, and	Teleoperators.	Albus, JS 85a
What Do Drug-Induced Visual Hallucinations	Tell Us About the Brain?	Cowan, JD 85b
What the Frog's Eye	Tells the Frog's Brain	Lettvin, JY 59a
Low	Temperature Analysis of the Axial Next-Nearest Neighbor Ising Model.	Fisher, ME 81a
Zero	Temperature Parallel Dynamics for Infinite Range Spin Glasses and.	Gardner, E 87a
Neural Networks for	Template Matching: Application to Real-Time Classification of the.	Bower, JM 88b

...of Associational Span, Bowing, and Primacy vs. Recency on Spiking	Threshold.	Grossberg, S 70b
...as a Recurrent On-Center Off-Surround Neural Net with Quenching	Threshold.	Betts, B 88b
Spiking	Threshold and Overarousal Effects in Serial Learning	Grossberg, S 71c
Simple Changes in Reflex	Threshold Cannot Explain All Aspects of Rapid Voluntary Movements.	Gielen, CCAM 86a
Modification of the	Threshold Condition for a Content-Addressable Memory Based on the.	Macukow, B 87a
Adaptive Neural Encoder Model with Selfinhibition and	Threshold Control.	Zeevi, YY 81a
Associative Memory and	Threshold Control in Neural Networks	Palm, G 87a
An Approach to Pattern Recognition Using Linear	Threshold Devices.	Mattson, RL 60a
The Use of an Adaptive	Threshold Element to Design a Linear Optimal Pattern Classifier.	Koford, JS 66a
...Parameters on Convergence Time and Tolerance for Adaptive	Threshold Elements	Mays, CH 64a
Learning Patterns and Pattern Sequences by Self-Organizing Nets of	Threshold Elements	Amari, S 72b
Nets of	Threshold Elements	Krohn, K 65a
On Autonomous Logic Nets of	Threshold Elements	Yajima, S 68a
The Analysis and Synthesis of Adaptive Systems Which Use Networks of	Threshold Elements	Mattson, RL 62a
Visual Feature Extraction by a Multi-Layered Network of Analog	Threshold Elements	Fukushima, K 69a
Fast Component	Threshold for Vestibular Nystagmus in the Rabbit	Lau, CGY 87a
Harmonic Analysis of Polynomial	Threshold Functions.	Bruck, J 88c
Neural Networks, Orientations of the Hypercube and Algebraic	Threshold Functions.	Baldi, P 88f
Parallel Computation with	Threshold Functions.	Parberry, I 88b
The Capacity of Multilevel	Threshold Functions.	Olafsson, S 88a
Adaptive	Threshold Logic.	Mays, CH 63a
	Threshold Logic.	Lewis, PM 67a
	Threshold Logic.	Hu, ST 65a
	Threshold Logic.	Sheng, CL 69a
A	Threshold Logic Network for Shape Invariance	Brousil, JK 67a
"Bootstrap Learning" in	Threshold Logic Systems.	Widrow, B 66a
Analysis of an Adaptive	Threshold Logic Unit	Morishita, I 70a
Empirical Examination of the	Threshold Model of Neuron Firing	Brillinger, DR 79a
Firing Response of a Neural Model with	Threshold Modulations and Neural Dynamics.	Sibain, P 86a
Learning Processes in Multi-Layer	Threshold Nets	Bobrowski, L 78a
Learning Process in an Asymmetric	Threshold Network.	Le Cun, Y 86a
Information Processing System Using	Threshold Passive Modification	Cooper, LN 81b
Comments on the Paper "	Threshold Regulation - A Key to the Understanding of the Combined.	Palm, G 87b
A Four Mechanism Model for	Threshold Spatial Vision	Wilson, HR 79a
Punish / Reward: Learning With a Critic in Adaptive	Threshold Systems.	Widrow, B 73a
Characteristics of Randomly Connected	Threshold-Element Networks and Network Systems	Amari, S 71a
...Based on an Associative Memory Model of Neural Nets with	Thresholding and Feedback.	Psaltis, D 85a
Neural Networks with Dynamical	Thresholds	Horn, D 88b
Faster Learning	Through a Probabilistic Approximation Algorithm.	Kolen, JF 88a
Feature Discovery	Through Error-Correction Learning.	Williams, RJ 85a
Knowledge Processing	Through Flow-of-Activation	Cruz, CA 87b
How Sensory Maps Could Enhance Resolution	Through Ordered Arrangements of Broadly Tuned Receivers.	Baldi, P 88c
Self-Organization	Through Selection.	Steels, L 87a
Probing Cognitive Processes	Through the Structure of Event-Related Potentials During Learning.	Banquet, J 87a
Speaker Adaptation	Through Vector Quantization.	Shikano, K 86a
Pattern Separability of a	Tightly Coupled Random Nerve Net with Inhibitory Connections	Torioka, T 79b
Visual Illusions in Neural Networks: Line Neutralization,	Tilt After Effect, and Angle Expansion	Levine, DS 76a
...of the Mechanisms Underlying Infant Perception of Voice-Onset	Time	Pont, MJ 88b
An Algorithm for Finding Best Matches in Logarithmic Expected	Time	Friedman, JH 77a
Virtual	Time	Jefferson, DR 85a
Space or	Time Adaptive Signal Processing by Neural Network Models	Herault, J 86a
Concentrating Information in	Time: Analog Neural Networks with Applications to Speech	Tank, DW 87b
Gabor: Frequency,	Time and Memory.	Korpel, A 82a
Effects of Adaptation Parameters on Convergence	Time and Tolerance for Adaptive Threshold Elements	Mays, CH 64a
Synthesis of Boolean Nets and	Time Behavior of a General Mathematical Neuron	Caianiello, ER 75a
Membrane Potential Transients and Membrane	Time Constant of Motoneurons	Rall, W 60a
Variations in Amplitude and	Time Course of Inhibitory Postsynaptic Currents.	Gardner, D 86a
List Length and the	Time Course of Recognition in Immediate Memory	Reed, AV 76a
	Time Decay of the Remanent Magnetization in Spin-Glasses	Chamberlin, RV 84a
Electronic Analog Neural Network with Noise and	Time Delays.	Meidan, M 87a
Reading Ability and Letter Search	Time: Effects of Orthographic Structure Defined by Single-Letter	Mason, M 75a
Automatic Generation of	Time Efficient Digital Signal Processing Software.	Morris, LR 77a
Real	Time Experiments in Neural Network Based Learning Control During	Miller, WT 88a
Recognition	Time for Information Stored in Long-Term Memory.	Juola, JF 71a
Retrieval	Time from Semantic Memory.	Collins, AM 69a
An Image Processing Architecture for Real	Time Generation of Scale and Rotation Invariant Patterns	Messner, RA 85a
Random Cells: An Idea Whose	Time Has Come and Gone... And Come Again?.	Gallant, SI 87b
Real	Time Learned Sensor Processing and Motor Control for a Robot with.	Miller, WT 88b
Real	Time Noisy Image Segmentation Using an Artificial Neural Network	Tenorio, MF 87d

	Toward a Theory of Reinforcement-Learning Connectionist Systems. . .	Williams, RJ 88b
	Toward a Unified Theory of Cognition	Waldrop, MM 88b
Steps	Toward Artificial Intelligence	Minsky, ML 61b
Steps	Toward Artificial Intelligence	Minsky, ML 61a
Steps	Toward Common Sense.	Steels, L 88a
	Toward Connectionist Parsing	Small, SL 82c
	Towards a Generic Learning Algorithm	Wright, WA 89d
	Towards a Model for Visual Information Processing.	Okajima, K 88a
	Towards a Non-Network Approach to "Neural Modeling": Some Basic. . .	Schwartz, EL 87a
	Towards a Primal Sketch of Real World Scenes in Early Vision	Korn, AF 88a
	Towards a Theory of Cell Assemblies.	Palm, G 81a
Parameter Networks:	Towards a Theory of Low-Level Vision	Ballard, DH 81c
	Towards an Information-Flow Model of Human Behaviour	MacKay, DM 56a
	Towards an Organizing Principle for a Layered Perceptual Network . .	Linsker, R 88b
	Towards Connectionist Rule-Based Systems	Samad, T 88b
	Towards Continuous Models of Memory.	Banzhaf, W 87b
	Towards Integrating Automatic and Controlled Problem Solving	Day, DS 87a
	Towards Practical 'Neural' Computation for Combinatorial	Baum, EB 86b
Stereotypes as an Actor Approach	Towards Solving the Problem of Procedural Attachment in Frame. . . .	Hewitt, C 75a
Mathematical Models of Memory	Trace and Forgetfulness.	Anninos, PA 72b
Interactive Processes in Speech Perception: The	TRACE Model.	McClelland, JL 86e
The	TRACE Model of Speech Perception	McClelland, JL 86k
Two Models for Memory Organization Using Interacting	Traces	Anderson, JA 70a
Dynamic Activity and Memory	Traces in Computer-Simulated Recurrently-Connected Neural Networks .	Gaudiano, P 87a
The	Tracing of Pathways in the Central Nervous System.	Heimer, L 68a
...Tunable Short Term Neural Network Memory and Application to	Tracking	Morris, RJT 88a
...Processing and Neural Networks for Adaptive Moving Object	Tracking	Lee, JSJ 87a
Neural Network Approaches to Multi-Target	Tracking	Kuczewski, RM 87c
Neural Network Approaches to Multi-Target	Tracking	Kuczewski, RM 87d
...of Neural Network Algorithms and Architectures to Correlation /	Tracking and Identification.	Gardner, S 86a
Robust	Tracking Novelty Filters Based on Linear Models.	Kadar, I 87a
How the Catfish	Tracks Its Prey: An Interactive "Pipelined" Processing System May. .	Kanwal, JS 88a
Space / Time	Trade-Offs in Hash Coding with Allowable Errors.	Bloom, BH 70a
First Commercial Nerve Nets Challenge	Traditional AI	Brown, C 86c
Neural Net and	Traditional Classifiers.	Huang, WY 88a
Relating Neural Networks to	Traditional Engineering Approaches	Golden, RM 88d
...Perforant Path-Dentate Projection. III. Comparison of Random	Train and Paired Impulse Stimulation	Berger, TW 88b
...Perforant Path-Dentate Projection. II. Effects of Random Impulse	Train Stimulation.	Berger, TW 88a
TIN: A	Trainable Inference Network.	Winter, CL 87a
TIN: A	Trainable Inference Network.	Winter, CL 87b
Pattern Classifiers and	Trainable Machines	Sklansky, J 81a
A	Trainable Model for Near Periodic Signals.	Wolf, L 87a
Reliable,	Trainable Networks for Computing and Control	Widrow, B 62c
A	Trainable Non-Linear Function Generator.	Smith, FW 66a
Learning Machines: Foundations of	Trainable Pattern-Classifying Systems.	Nilsson, NJ 65a
A	Trainable Weather-Forecasting System	Hu, MJC 63a
Analysis of Hidden Units in a Layered network	Trained to Classify Sonar Targets.	Gorman, RP 88a
An Adaptive	Training Algorithm for Back Propagation Networks	Chan, L 87a
A Context-Driven World Model Appropriate for	Training and Testing of Vision-Based Learning Networks	Keller, R 87a
Vector Quantization	Training by a Self-Organizing Neural Network	Ryan, TW 88b
	Training of a Neural Network for Pattern Classification Based on an.	Koutsougeras, C 88b
	Training of HMM Recognizers by Simulated Annealing	Douglas, P 85a
Adaptive	Training of Multilayer Neural Networks Using a Least Squares	Kollias, S 88a
	Training Recurrent Networks.	Rohwer, R 88b
Scaling Relationships in Back-Propagation Learning: Dependence on	Training Set Size.	Tesauro, G 87a
Synthesis of Quasi-Optimal Switching Surfaces by Means of	Training Techniques.	Mendel, JM 70c
Teaching Artificial Neural Systems to Drive: Manual	Training Techniques for Autonomous Systems	Shepanski, JF 88b
Manual	Training Technqiues of Autonomous Systems Based on Artificial. . . .	Shepanski, JF 87a
	Training Time-Dependence in Neural Networks.	Rohwer, R 87a
Fast Learning in Artificial Neural Systems: Multilayer Perceptron	Training Using Optimal Estimation.	Shepanski, JF 88a
	Training with Noise: Application to Word and Text Storage.	Gardner, E 88a
Stochastic Models for Single Neuron Firing	Trains: A Survey	Fienberg, SE 74a
The Stochastic Properties of Spike	Trains Recorded from Nerve Cells	Stein, RB 72a
...A Neural Command Circuit for Generating Arm and Articulator	Trajectories	Bullock, D 88d
...Emergent Invariants and Speed-Accuracy Properties During	Trajectory Formation	Bullock, D 88c
...Emergent Invariants and Speed-Accuracy Properties During	Trajectory Formation	Bullock, D 88b
Theoretical and Experimental Perspectives on Arm	Trajectory Formation: A Distributed Model of Motor Redundancy. . . .	Mussa-Ivaldi, FA 88c
A Neural Network Architecture for Automatic	Trajectory Formation and Coordination of Multiple Effectors During .	Bullock, D 87a
Application of the Neural Network for the	Trajectory Planning of a Biped Locomotive Robot.	Kitamura, S 88a

Value	Unit Interpolation: A Receptive Field Model	Ballard, DH 86c
...is Significantly Faster if the Expected Value of the Source	Unit is Used for Update	Samad, T 88a
A Note on the Neural	Unit Model for Contrast Phenomena	Bergstrom, SS 73a
Dynamic Single	Unit Simulation of a Realistic Cerebellar Network Model	Pellionisz, AJ 73a
Dynamic Single	Unit Simulation of a Realistic Cerebellar Network Model. II	Pellionisz, AJ 74a
TRW	Unit Uses Neural Networks for Voice Recognition	Schwartz, TJ 86b
GMDH Forecasting of	United States Interest Rates	Ohashi, K 84a
Adaptive Learning Networks: Development and Application in the	United States of Algorithms Related to GMDH	Barron, RL 84a
NASA	Unites Neural Nets, Fuzzy Logic	Johnson, RC 88b
	Unitization, Automaticity, Temporal Order, and Word Recognition	Grossberg, S 84e
	Unitized Recognition Codes for Parts and Wholes: The Unique Cue in	Cohen, MA 86e
Fast Learning in Networks of Locally-Tuned Processing	Units	Moody, J 88a
Game-Theoretic Cooperativity in Networks of Self-Interested	Units	Barto, AG 86b
...A Scheme for Discovery of "Natural" Categories by Single	Units	Munro, PW 86a
Single	Units and Sensation: A Neuron Doctrine for Perceptual Psychology?	Barlow, HB 72a
Learning Symmetry Groups with Hidden	Units: Beyond the Perceptron	Sejnowski, TJ 86b
A Network of Multistate	Units Capable of Associative Memory and Pattern Classification	Banzhaf, W 89a
`Ensemble' Boltzmann	Units have Collective Computational Properties like those of	Derthick, MA 88a
Analysis of Hidden	Units in a Layered network Trained to Classify Sonar Targets	Gorman, RP 88a
On the Aggregation of Processing	Units in Neural Networks	Yager, RR 87b
The Role of Familiar	Units in Perception of Words and Nonwords	McClelland, JL 77b
Expectancy as a Determinant of Functional	Units in Perceptual Recognition	Aderman, D 71a
Neural Patterns as Detection	Units in Sensory Systems	Kalay, A 87a
Connectionist Value	Units: Some Concerns	Barnden, JA 86b
Entrainment in a Neural Network Simulation of Spontaneously Active	Units with Accommodation	Blackburn, MR 87a
Multi-Layer Feedforward Networks Are	Universal Approximators	Hornik, K 88a
An Estimate of the Complexity Requisite in a	Universal Decision Network	Cameron, SH 60a
Technology-Independent Design of Neurocomputers: The	Universal Field Computer	MacLennan, BJ 87a
The	Universal Metric Properties of Nonlinear Transformations	Feigenbaum, MJ 79a
Fat-Trees:	Universal Networks for Hardware Efficient Supercomputing	Leiserson, CE 85a
A	Universal Non-Linear Filter, Predictor, and Simulator which	Gabor, D 60a
Adaptive Pattern Classification and	Universal Recoding: I. Parallel Development and Coding of Neural	Grossberg, S 76a
Adaptive Pattern Classification and	Universal Recording, Part I: Parallel Development and Coding of	Grossberg, S 76b
Adaptive Pattern Classification and	Universal Recording, Part II: Feedback, Expectation, Olfaction, and	Grossberg, S 76c
Internal Representation of	Universal Regularities: A Challenge for Connectionism	Shepard, RN 89a
	Universal Sub-goaling	Laird, JE 84a
Programming a Massively Parallel, Computation	Universal System: Static Behavior	Lapedes, A 86a
Quantitative	Universality for a Class of Nonlinear Transformations	Feigenbaum, MJ 78a
How We Know	Universals: The Perception of Auditory and Visual Forms	Pitts, WH 47a
How We Know	Universals: The Perception of Auditory and Visual Forms	Pitts, WH 47b
Optical Systems Research at Texas Tech	University	Krile, TF 86a
An Associative Co-Processor Architecture with the	UNIX Operating System for the Acceleration of CAD Tools	Schulz, M 88a
...Cerebellar and Retinal Analogs of Cells Fired by Learnable or	Unlearned Pattern Classes	Grossberg, S 72c
	Unlearning Has a Stabilizing Effect in Collective Memories	Hopfield, JJ 83a
Probabilistic Logics and the Synthesis of Reliable Organisms from	Unreliable Components	von Neumann, J 56a
	Unrestricted English Text-to-Speech Conversion with a Distributed	Tenorio, MF 88b
Pattern Class Degeneracy in an	Unrestricted Storage Density Memory	Scofield, CL 88b
	Unstable Connectionist Networks in Speech Recognition	Rohwer, R 88c
Numerical Calculation of Stable and	Unstable Periodic Solutions to the Hodgkin-Huxley Equations	Rinzel, J 80a
An Information-Theoretic Approach to	Unsupervised Connectionist Models	Plumbley, MD 88a
A Connectionist Model for	Unsupervised Feature Discovery and Pattern Recognition	Franzini, MA 87a
A Modified Mean Square Error Criterion for Use in	Unsupervised Learning	Miller, WC 67a
Drive-Reinforcement Learning: A Real-Time Learning Mechanism for	Unsupervised Learning	Klopf, AH 87b
Layered Networks for	Unsupervised Learning	d'Humieres, D 86a
A Comparison of Two	Unsupervised Learning Algorithms	Bobrowski, L 80a
	Unsupervised Learning Minimum Risk Pattern Classification for	Hilborn, CG 69a
G-Maximization: An	Unsupervised Learning Procedure for Discovering Regularities	Pearlmutter, BA 86a
Rules for Forming Receptive Fields of Formal Neurons During	Unsupervised Learning Processes	Bobrowski, L 82a
A Neural	Unsupervised Learning Technique	Atiya, AF 88a
Hecht-Nielsen	Unveils First New Neural Net Architecture Since '84	Schwartz, TJ 87b
...Faster if the Expected Value of the Source Unit is Used for	Update	Samad, T 88a
Vector Quantization of Images Based	Upon the Kohonen Self-Organizing Feature Maps	Nasrabadi, NM 88a
An Inference	upon the Neural Network Finding Binocular Correspondence	Hirai, Y 78a
	Upper and Lower Probabilities Induced by a Multi-Valued Mapping	Dempster, AP 67a
...Significantly Faster if the Expected Value of the Source Unit is	Used for Update	Samad, T 88a
Visual Programming as a	User Interfaace for Programming Neural Networks	Tenorio, MF 87j
Anne	User Manual	Bahr, C 88b
...of a Document Processor: A Comparison of Expert Judgement and	User Observation	Hammond, N 84a
Speech Prosthesis System for Laryngectomees	Using "Electropalatograph" and a Small Keyboard	Muraki, S 86a

Shape Recognition	Using a CMAC Based Learning	Glanz, FH 87a
Deconvolution	Using a CMAC Neural Network	Glanz, FH 88a
A Comparison of Speech Recognition Front Ends	Using a Connectionist Classifier	Renals, S 88a
Combinatorial Optimization	Using a Continuous State Boltzmann Machine	Gutzmann, KM 87a
Invariant Object Recognition	Using a Distributed Associative Memory	Wechsler, H 88a
Sensor-Based Control of Robotic Manipulators	Using a General Learning Algorithm	Miller, WT 87c
Solving Puzzles with a Connectionist Network	Using a Hill-Climbing Heuristic	Kawamoto, AH 86a
Automatic Character Reading	Using a Holographic Data Processing Technique	Nakajima, M 72a
Tests on a Cell Assembly Theory of the Action of the Brain,	Using a Large Digital Computer	Rochester, N 56a
Tests on a Cell Assembly Theory of the Action of the Brain,	Using a Large Digital Computer	Rochester, N 56b
Real-Time 3-D Object Classification	Using a Learning System	Rimey, RD 86a
Adaptive Training of Multilayer Neural Networks	Using a Least Squares Estimation Technique	Kollias, S 88a
Gas Tungsten Arc Weld Modeling	Using a Mapping Network	Karsai, G 88a
Learned Classification of Sonar Targets	Using a Massively Parallel Network	Gorman, RP 88b
Computation of Optical Flow	Using a Neural Network	Zhou, Y 88a
Image Halftoning and Reconstruction	Using a Neural Network	Kollias, S 89a
Image Restoration	Using a Neural Network	Zhou, Y 88b
Lexical Access	Using a Neural Network	Kawamoto, AH 84a
Stereo Matching	Using a Neural Network	Zhou, Y 88c
	Using a Neural Network to Learn the Dynamics of the CMU Direct	Goldberg, K 88a
Controlling the Motion of a Moving Robot	Using a Neuroid Network	Gol'tsev, AD 82a
Optical Holographic Associative Memory	Using a Phase Conjugate Resonator	Dunning, GJ 86a
Adaptive Global Search in a Time Varying Environment	Using a Probabilistic Automaton with Pattern Recognition	Jarvis, RA 70a
Rapid Speaker Adaptation	Using a Probabilistic Spectral Mapping	Schwartz, RM 87a
Associative Memory Storage	Using a Variant of the Generalized Delta Rule	Samad, T 87b
	Using A Visual Programming Language to Simulate Adaptive Networks	Tenorio, MF 87f
	Using Activation Networks for Analogical Ordering of Consideration	Becker, LA 87b
Real-Time Parallel Processing of Pixel Arrays	Using Adaptive Logic	Vidal, JJ 87b
Real Time Noisy Image Segmentation	Using an Artificial Neural Network Model	Tenorio, MF 87d
An Analogue Approach to the Traveling Salesman Problem	Using an Elastic Net Method	Durbin, R 87a
Computing Motion	Using Analog and Binary Resistive Networks	Hutchinson, J 88a
Real-time Visual Computations	Using Analog CMOS Processing Arrays	Sivilotti, MA 87b
Real-Time Visual Computations	Using Analog CMOS Processing Arrays	Sivilotti, MA 87a
Neural Networks	Using Analog Multipliers	Paulos, JJ 88a
	Using Artificial Neural Nets for Statistical Discovery	Irani, EA 88a
Pattern Recognition	Using Artificial Neural Systems	Gutschow, T 87a
Sensor Processing	Using Artificial Neural Systems	Hecht-Nielsen, R 86b
Pattern Recognition	Using Associative Memories	Altman, JL 77a
Sequential Processing	Using Attractor Transitions	Bell, T 88a
	Using Automated Techniques to Generate an Expert System for R&D	Gallant, SI 86d
...Neural Nets for Statistical Discovery: Observations after	Using Back-propagation, Expert Systems, and Multiple-Linear	Irani, EA 88a
An Adaptive "Adaline" Neuron	Using Chemical "Memistors"	Widrow, B 60d
Matrix Multiplication	Using Coherent Optical Techniques	Tamura, PN 77a
A Novel Associative Memory Implemented	Using Collective Computation	Sivilotti, MA 85a
Learned Phonetic Discrimination	Using Connectionist Networks	Watrous, RL 87a
Learning Acoustic Features from Speech Data	Using Connectionist Networks	Watrous, RL 87b
Learning Phonetic Features	Using Connectionist Networks: An Experiment in Speech Recognition	Watrous, RL 87d
Fault-Tolerant Recognition	Using DAM's	Wechsler, H 87a
Optical Neurocomputers Implementation	Using Deformable Mirror Arrays	Collins, DR 87a
Distributed and Fault-Tolerant Computation for Retrieval Tasks	Using Distributed Associative Memories	Char, JM 88a
Asymptotic Properties of Nearest Neighbor Rules	Using Edited Data	Wilson, DL 72a
	Using Electronic Circuits to Model Simple Neuroelectric	Lewis, ER 68d
Programming Parallel Computers	Using Energy Minimization Algorithms	Simmes, SD 87a
	Using Fast Weights to Deblur Old Memories	Hinton, GE 87d
Recognizing Partially Visible Objects	Using Feature Indexed Hypotheses	Knoll, TF 86a
Nonlinear Signal Processing	Using Fiber-Optics Neurograms	Szu, HH 86f
Globally Connected Network Models for Computing	Using Fine-Grained Processing Elements	Szu, HH 85a
	Using Frequency-and Orientation-Tuned Channels to Determine Surface	Kube, P 86a
Estimation of Expert Weights	Using Fuzzy Cognitive Maps	Taber, WR 87a
Risk Weighting in Knowledge Bases	Using Fuzzy Cognitive Maps	Taber, WR 88c
Adaptive Algorithms for Discovering and	Using General Patterns in Growing Knowledge-Bases	Holland, JH 80a
Color Separation	Using General Purpose Computer Vision Algorithms	Walters, DKW 87c
Speaker-Independent Continuous Speech Recognition	Using Hidden Markow Models	Lee, K 87a
Two-Dimensional Pattern Sequence Prediction	Using High-Order Neural Networks	Collins, DH 88a
Two Models for Memory Organization	Using Interacting Traces	Anderson, JA 70a
An Approach to Pattern Recognition	Using Linear Threshold Devices	Mattson, RL 60a
Differences in Performance of Binary Hetero-Associative Networks	Using Link-Enabling vs. Link-Disabling Local Modification Rules	Dobson, VG 88b
A Model of Schema Selection	Using Marker Passing and Connectionist Spreading Activation	Chun, HW 87b
Integration of Stereo Vision and Optical Flow	Using Markov Random Fields	Clifford, SP 88a

Hierarchical Control of Robots	Using Microcomputers	Barbera, AJ 79a
A Case Study of Solving Optimization Problems	Using Neural Network	Chen, S 88a
A Framework for Speech Recognition	Using Neural Networks.	Lippmann, RP 87a
Analog Decoding	Using Neural Networks.	Platt, JC 86a
Constructing Associative Memories	Using Neural Networks.	Xu, X 88b
Digital Image Halftoning	Using Neural Networks.	Anastassiou, D 88a
Multidimensional Machine Vision	Using Neural Networks.	Kolodzy, PJ 87a
Pattern-Based Fault Diagnosis	Using Neural Networks.	Dietz, WE 88a
Recognition of General Patterns	Using Neural Networks.	Wong, AJ 88a
Signal Modeling and Prediction	Using Neural Networks.	Ramamoorthy, PA 88b
Superresolution for Ultrasonic Imaging in Air	Using Neural Networks.	Winters, JH 88b
	Using Neural Networks for Generalization Problems.	Dutta, S 88b
Nonlinear Signal Processing	Using Neural Networks: Prediction and System Modelling	Lapedes, A 87a
Economic Prediction	Using Neural Networks: The Case of IBM Daily Stock Returns	White, H 88a
How the Human Brain Thinks and Acts: Results of Latest Research	Using Neural Networks to Analyze Brain Signals	Gevins, AS 87b
	Using Neural Networks to Improve Cochlear Implant Speech Perception.	Tenorio, MF 87g
...in Artificial Neural Systems: Multilayer Perceptron Training	Using Optimal Estimation	Shepanski, JF 88a
Optical Processing	Using Outer-Product Concepts	Athale, RA 84a
Inductive Information Retrieval	Using Parallel Distributed Computation	Mozer, MC 84a
Two-Dimensional Adaptive Model of a Human Controller	Using Pattern Recognition Techniques	Gilstad, DW 71a
Successfully	Using Peak Learning Rates of 10 (and Greater) in Back-Propagation.	Cater, JP 87a
All-Optical Associative Holographic Memory with Feedback	Using Phase Conjugate Mirrors.	Dunning, GJ 85a
Associative Holographic Memory with Feedback	Using Phase-Conjugate Mirrors.	Soffer, BH 86a
Adaptive Optical Networks	Using Photorefractive Crystals	Psaltis, D 88a
Asynchronous VLSI Neural Networks	Using Pulse-Stream Arithmetic.	Murray, AF 88b
NETL: A System for Representing and	Using Real-World Knowledge	Fahlman, SE 79a
Computing Motion	Using Resistive Networks	Koch, C 88b
	Using SARCENet for Optimal Solutions of Scheduling Problems.	Tenorio, MF 87h
Image Segmentation	Using Simple Markov Field Models	Hansen, FR 82a
Binocular Fusion	Using Simulated Annealing.	Landa, J 87a
...of Associative Memory from Incomplete Memorized Information	Using Space Reduction Method	Murakami, K 88a
Digital Image Restoration	Using Spatial Interaction Models	Chellappa, R 82a
Engineering Design Neural Networks	Using Split Inversion Learning	McAulay, AD 87a
Natural Language Processing	Using Spreading Activation and Lateral Inhibition.	Pollack, JB 82a
Color Representation	Using Statistical Pattern Recognition.	Parkkinen, J 87a
...of Stimulus Artifact in Nerve Signals Evoked and Recorded	Using Surface Electrodes	McGill, KC 82a
Auto-Associative Pattern Storage	Using Synchronous Boolean Networks	Martland, D 87a
Simulation of a Distributed Algorithm	Using the Connection Machine	Aburdene, MF 88a
Generation of Simple Sentences in English	Using the Connectionist Model of Computation	Kalita, J 87a
A Neural Network Simulation Method	Using the Fast Fourier Transform	Barto, AG 76a
Accelerated Learning	Using the Generalized Delta Rule	Dahl, ED 87a
A Neural Network Model for Phoneme Recognition	Using the Generalized Delta Rule for Connection Strength	Watson, M 87a
Solving Random-Dot Stereograms	Using the Heat Equation.	Szeliski, R 85a
	Using the Hyperbase Intelligent Processor for Fault Analysis and	Tenorio, MF 87i
...of Overcomplete CNS Coordinates and Tensorial Solution	Using the Moore-Penrose Generalized Inverse.	Pellionisz, AJ 84a
Vectorcardiographic Diagnosis	Using the Polynomial Discriminant Method of Pattern Recognition.	Specht, DF 67b
Connectionist Speech Recognition	Using the Temporal Flow Model.	Watrous, RL 88a
EXCLUSIVE OR Processing (Binary Image Subtraction)	Using Thick Fourier Holograms.	Guest, CC 84a
Information Processing System	Using Threshold Passive Modification	Cooper, LN 81b
Semi-Distributed Neuristor Line	Using Unijunction Transistors.	Ambroziak, A 64a
L'	Uso di Modelli Computazionali in Psicofisiologia Ed Il Principio	Basti, G 86b
Syllogistic Reasoning in Fuzzy Logic and its Application to	Usuality and Reasoning with Dispositions	Zadeh, LA 85a
Neural Networks for Computing, Snowbird,	Utah, 1986	Denker, JS 86b
The	Utility Driven Dynamic Error Propagation Network	Robinson, AJR 87a
Associative Memory in a Phase Conjugate Resonator Cavity	Utilizing a Hologram	Owechko, Y 86a
Associative Memory in a Phase Conjugate Resonator Cavity	Utilizing a Hologram	Owechko, Y 86b
Schema for Motor Control	Utilizing a Network Model of the Cerebellum.	Houk, JC 88b
A Memory Storage Model	Utilizing Spatial Correlation Functions.	Anderson, JA 68a
Overlapping Taste and Tactile Maps of the Oropharynx in the	Vagal Lobe of the Channel Catfish, Ictalurus punctatus	Kanwal, JS 88b
Back-Propagation is Significantly Faster if the Expected	Value of the Source Unit is Used for Update.	Samad, T 88a
	Value Unit Interpolation: A Receptive Field Model.	Ballard, DH 86c
Connectionist	Value Units: Some Concerns	Barnden, JA 86b
Lattice	Valued Relations and Automata.	Muir, A 84a
Analog Circuitry for the	van der Pol and FitzHugh-Nagumo Equations.	Keener, JP 83a
Coupled	Van der Pol Oscillators - A Model of Excitatory and Inhibitory	Kawahara, T 80a
A Model for	Variability Effects in Hand-Printing, with Implications for the.	Ward, JR 88a
Exploiting Lawful	Variability in the Speech Wave	Elman, JL 86a
Acceleration of Learning by	Variable Generalization for on-Line Self-Organizing Control.	Ersu, E 86a
...Network Architecture for Brightness Perception Under Constant and	Variable Illumination Conditions	Grossberg, S 87c

Processing of	Visual Information in the Distal Neurons of the Vertebrate Retina.	Curlander, JC 83a
Vision by Man and Machine: How the Brain Processes	Visual Information May Be Suggested by Studies in Computer Vision.	Poggio, TA 84c
Towards a Model for	Visual Information Processing.	Okajima, K 88a
Algorithms and Techniques for Automated	Visual Inspection.	Chin, RT 86a
	Visual Masking in Multielement Displays.	Eriksen, CW 70a
Connectionist Model of	Visual Memory.	Feldman, JA 81b
Convected Activation Profiles and the Measurement of	Visual Motion.	Waxman, AM 88a
The Analysis of	Visual Motion: From Computational Theory to Neuronal Mechanisms.	Hildreth, E 86a
Image Statistics, Gases, and	Visual Neural Primitives	Daugman, JG 87a
...Network for Geometrical Patterns: A Design with Reference to	Visual Neural Systems.	Fukushima, K 71a
Masking of Alphabetic Character Recognition by Dynamic	Visual Noise (DVN)	Uttal, WR 69a
A Neural Network for	Visual Pattern Recognition	Fukushima, K 88b
A Neural Network Model for Selective Attention in	Visual Pattern Recognition	Fukushima, K 86a
Neocognitron: A Hierarchical Neural Network Capable of	Visual Pattern Recognition	Fukushima, K 88c
Neocognitron: A Neural Network Model for a Mechanism of	Visual Pattern Recognition	Fukushima, K 83c
Neocognitron: A Neural Network Model for a Mechanism of	Visual Pattern Recognition	Fukushima, K 83b
...A Self-Organizing Neural Network Model for a Mechanism of	Visual Pattern Recognition	Fukushima, K 82b
Self-Organizing Neural Network Models for	Visual Pattern Recognition	Fukushima, K 87e
Neural Network Model for Selective Attention in	Visual Pattern Recognition and Associative Recall.	Fukushima, K 87c
Neural Network Model for Selective Attention in	Visual Pattern Recognition Associative Recall.	Fukushima, K 87d
	Visual Pattern Recognition in Humans	Caelli, TM 87b
The Analysis of Moving	Visual Patterns.	Movshon, JA 85a
Role of Feedback in	Visual Perception.	Unnikrishnan, KP 87a
Sight and Mind: An Introduction to	Visual Perception.	Kaufman, L 74a
The Ecological Approach to	Visual Perception.	Gibson, JJ 79a
The Topology of the Brain and	Visual Perception.	Zeeman, EC 62a
Reduction of	Visual Processes to a Common Algorithm	Carman, GJ 87b
Reduction of	Visual Processes to a Common Algorithm	Carman, GJ 88a
A Silicon Model of Early	Visual Processing.	Mead, CA 88a
...Architecture and Integrated Space-Time Filtering for Low Level	Visual Processing.	Johnson, K 86a
Selection of Image Primitives for General-Purpose	Visual Processing.	Walters, DKW 87i
Shifter Circuits: A Computational Strategy for Dyanmic Aspects of	Visual Processing.	Anderson, CH 87a
The Neural Analog Diffusion-Enhancement Layer (NADEL) and Early	Visual Processing.	Waxman, AM 88b
Statistical Mechanics of	Visual Processing: A Model for Layer IVc of the Primary Visual	Moller, P 87a
Orthographic Regularity, Positional Frequency, and	Visual Processing of Letter Strings.	Massaro, DW 79a
	Visual Programming as a User Interfaace for Programming Neural	Tenorio, MF 87j
Using A	Visual Programming Language to Simulate Adaptive Networks.	Tenorio, MF 87f
ALOPEX: A Stochastic Method for Determining	Visual Receptive Fields.	Harth, EM 74a
Computing with Connections in	Visual Recognition of Origami Objects.	Sabbah, D 85a
Computing with Connections in	Visual Recognition of Origami Objects.	Sabbah, D 88a
	Visual Recognition of Simple Objects by a Connection Network	Plaut, DC 84a
Design of a Highly Parallel	Visual Recognition System.	Sabbah, D 81a
	Visual Search with Decrementing Network Competitive Windows.	Dobson, VG 88c
Quantized Geometry of	Visual Space: The Coherent Computation of Depth, Form and Lightness.	Grossberg, S 86p
The Quantized Geometry of	Visual Space: The Coherent Computation of Depth, Form, and	Grossberg, S 83b
...for Curvilinear Patterns: A Design Suggested by the Mammalian	Visual System.	Fukushima, K 70a
...Self-Organizing Neural Network Sharing Features of the Mammalian	Visual System.	Frohn, H 87a
Design Principles for a Front-end	Visual System.	Koenderink, JJ 88a
Morphology of Neural Networks in the Mammalian	Visual System.	Gardner, S 88b
Neural Nets with Complex Structure: A Model of the	Visual System.	Gremillion, M 87a
Processing of Continuous Lines and Edges by the	Visual System.	Hartmann, G 83a
Representation of Local Geometry in the	Visual System.	Koenderink, JJ 87d
Why Corticies? Neural Computation in the Vertebrate	Visual System.	Mallot, HA 88a
Modeling and Simulation of Vertebrate Primary	Visual System: Basic Network	Oguztoreli, NN 83a
Processing of Moving Images in Natural and Artificial	Visual Systems	Schuling, FH 88b
A Developmental Neural Model of	Visual Word Perception	Golden, RM 88a
A Developmental Neural Model of	Visual Word Perception	Golden, RM 86a
Perception of the	Visual World	Gibson, JJ 50a
	Visual-Cognitive Neuronal Networks	Trehub, A 87a
Depth and Detours: An Essay on	Visually Guided Behavior	Arbib, MA 87b
Depth and Detours: An Essay on	Visually Guided Behavior	Arbib, MA 85c
Modeling Neural Mechanisms of	Visual-Motor Coordination in the Frog and Toad	Arbib, MA 82a
Levels of Modeling of Neural Interactions Underlying	Visuomotor Coordination.	Arbib, MA 85d
Modularity and Interaction of Brain Regions Underlying	Visuomoter Coordination.	Arbib, MA 85e
Spatial and Temporal Transformations in	Visuo-Motor Coordination	Droulez, J 88a
Innate and Learned Components in a Simple	Visuo-Motor Reflex	Hoffmann, KP 88a
Modeling and Analysis of Neural Networks in the	Visuomotor System of the Anuran Amphibian.	Cervantes-Perez, F 85a
Two-Terminal Electronic Circuit Neuron Model with Excitable Membrane	V-I-t Characteristics.	Hoshimiya, N 79a
The	Vite Model: A Neural Command Circuit for Generating Arm and.	Bullock, D 88d

...Feasibility of the Parametric Avalanche for Advanced Electronic	Warfare Applications	Dawes, RL 88b
	Warp Architecture and Implementation	Annaratone, M 86a
	Warp as a Machine for Low-Level Vision	Gross, T 85a
Systolic Algorithms for the CMU	Warp Processor	Kung, HT 84a
Exploiting Lawful Variability in the Speech	Wave	Elman, JL 86a
Microscopy by Reconstructed	Wave Fronts: II.	Gabor, D 51a
Seismic and Underwater Acoustic	Waveform Analysis.	Chen, CH 86a
Optimum	Waveform Signal Sets with Amplitude and Energy Constraints	Wyner, AD 84a
Reconstructed	Wavefronts and Communication Theory.	Leith, EN 62a
Neural Networks with a Hopf Bifurcation: Slowly Modulated	Waves.	Cohen, MS 87a
Forgetting as a	Way to Improve Neural-Net Behavior	Sasiela, RJ 86a
Mapping Images to a Hierarchical Data Structure - A	Way to Knowledge-based Pattern Recognition	Hartmann, G 88a
Designing Computers That Think the	Way We Do.	Allman, WF 87a
Cognitive Psychology: The	Way We Were.	McClelland, JL 78a
Designing Computers That Think the Way	We Do.	Allman, WF 87a
Listening for Mispronunciations: A Measure of What	We Hear During Speech.	Cole, RA 73a
How	We Know Universals: The Perception of Auditory and Visual Forms.	Pitts, WH 47a
How	We Know Universals: The Perception of Auditory and Visual Forms.	Pitts, WH 47b
Do	We Know What We Want? Do We Know When We Want It?.	Grossberg, S 77a
Do We Know What We Want? Do	We Know When We Want It?	Grossberg, S 77a
Where Do	We Stand on Maximum Entropy?	Jaynes, ET 79a
Do We Know What We Want? Do We Know When	We Want It?.	Grossberg, S 77a
Do We Know What	We Want? Do We Know When We Want It?	Grossberg, S 77a
Cognitive Psychology: The Way	We Were.	McClelland, JL 78a
A Trainable	Weather-Forecasting System	Hu, MJC 63a
Neural Dynamics of Surface Perception: Boundary	Webs, Illuminants, and Shape-From-Shading.	Grossberg, S 87m
A Modifiable	Weight Circuit for Use in Adaptive Neuromorphic Networks	Shoemaker, PA 88a
Influence of Interconnection	Weight Discretization and Noise in an Optoelectronic Neural Network.	Von Lehmen, A 87a
	Weight Discretization in Backward Error Propagation Neural Networks.	Fiesler, E 88a
Self-Organizing Process Based on Lateral Inhibition and	Weight Redistribution.	Miikkulainen, R 87a
Fuzzy Requests: An Approach to	Weighted Boolean Searches.	Bookstein, A 80a
Parallel N**4	Weighted Optical Interconnections.	Caulfield, HJ 87a
A Model for a	Weighted Retrieval System.	Buell, DA 81a
Index Term	Weighting.	Jones, KS 73a
Risk	Weighting in Knowledge Bases Using Fuzzy Cognitive Maps.	Taber, WR 88c
Holographic Implementation of a Hopfield Model with Discrete	Weightings	White, HJ 88d
...Model Based on Neural Networks Having Variable Interconnection	Weights.	Macukow, B 87b
...by Functional-Differential Neural Networks with Arbitrary Path	Weights.	Grossberg, S 72d
Boltzmann Machines in which	Weights Converge to Global Optimal Solutions	Uchiyama, T 88a
Noise Modulation of Synaptic	Weights in a Biological Neural Network	Gardner, D 89a
On the Size of	Weights Required for Linear-Input Switching Functions.	Myhill, J 61a
Using Fast	Weights to Deblur Old Memories	Hinton, GE 87d
Estimation of Expert	Weights Using Fuzzy Cognitive Maps	Taber, WR 87a
Gas Tungsten Arc	Weld Modeling Using a Mapping Network.	Karsai, G 88a
Skin & Muscles: The Basic	Wet-Ware for Sensory-Motor Control	De Rossi, D 88a
Expert Focus, Neural Networks, Part 2:	What are They and Why Is Everybody So Interested in Them Now?.	Wasserman, PD 88a
	What Bell Laboratories is Learning from Slugs.	Newport, JP 85a
Neural Networks:	What Can They Do?.	Cromie, WJ 88a
	What Do Drug-Induced Visual Hallucinations Tell Us About the Brain?.	Cowan, JD 85b
Let's Ask GMDH	What Effect the Environment Has on Fisheries	Brooks, HA 84a
	What Hebb Synapses Build	Anderson, JA 85a
	What Is a Distinctive Feature?	Anderson, JA 74a
Self-Learning -	What Is It?.	Tsypkin, YZ 68a
Introduction to Factor Analysis:	What It Is and How to Do It.	Kim, J 78a
How to Do with Probabilities	What People Say You Can't.	Pearl, J 85b
	What the Frog's Eye Tells the Frog's Brain	Lettvin, JY 59a
Listening for Mispronunciations: A Measure of	What We Hear During Speech	Cole, RA 73a
Do We Know	What We Want? Do We Know When We Want It?.	Grossberg, S 77a
A Minimal Technology Routing System for Meals on	Wheels	Bartholdi, JJ 83a
	When Does Perceived Lightness Depend on Perceived Spatial.	Gilchrist, AL 80a
...Accuracy of Decrementing over Incrementing Associative Networks	when Operating in Initially Random Connectivities.	Dobson, VG 87a
Do We Know What We Want? Do We Know	When We Want It?	Grossberg, S 77a
	Where Do We Stand on Maximum Entropy?.	Jaynes, ET 79a
	Where Does the Fundamental Information Processing Ability of	Kohonen, T 75a
Backward Conditioning: A Neural Network Model	which Exhibits Both Excitatory and Inhibitory Conditioning	Ricart, R 88a
Self-Organization of a Neural Network	Which Gives Position-Invariant Response.	Fukushima, K 79c
Electrophysiological Observations	Which Help Describe an Associative Synaptic Modification Rule.	Levy, WB 87a
A Connectionist Context-Free Parser	Which is not Context-Free, But Then it is not Really Connectionist	Charniak, E 87a
A Universal Non-Linear Filter, Predictor, and Simulator	which Optimizes Itself by a Learning Process	Gabor, D 60a